IMPORTANT.

HERE IS YOUR REGISTRATION CODE TO ACCESS YOUR PREMIUM McGRAW-HILL ONLINE RESOURCES.

For key premium online resources you need THIS CODE to gain access. Once the code is entered, you will be able to use the Web resources for the length of your course.

If your course is using **WebCT** or **Blackboard**, you'll be able to use this code to access the McGraw-Hill content within your instructor's online course.

Access is provided if you have purchased a new book. If the registration code is missing from this book, the registration screen on our Website, and within your WebCT or Blackboard course, will tell you how to obtain your new code.

Registering for McGraw-Hill Online Resources

To gain access to your McGraw-Hill web resources simply follow the steps below:

1. USE YOUR WEB BROWSER TO GO TO: **http://www.mhhe.com/streeter/**
2. CLICK ON **FIRST TIME USER**.
3. ENTER THE REGISTRATION CODE* PRINTED ON THE TEAR-OFF BOOKMARK ON THE RIGHT.
4. AFTER YOU HAVE ENTERED YOUR REGISTRATION CODE, CLICK **REGISTER**.
5. FOLLOW THE INSTRUCTIONS TO SET-UP YOUR PERSONAL UserID AND PASSWORD.
6. WRITE YOUR UserID AND PASSWORD DOWN FOR FUTURE REFERENCE. KEEP IT IN A SAFE PLACE.

TO GAIN ACCESS to the McGraw-Hill content in your instructor's WebCT or Blackboard course simply log in to the course with the UserID and Password provided by your instructor. Enter the registration code exactly as it appears in the box to the right when prompted by the system. You will only need to use the code the first time you click on McGraw-Hill content.

Thank you, and welcome to your McGraw-Hill online resources!

* YOUR REGISTRATION CODE CAN BE USED ONLY ONCE TO ESTABLISH ACCESS. IT IS NOT TRANSFERABLE.

0-07-031772-0 STREETER: PREALGEBRA: AN INTEGRATED EQUATIONS APPROACH, 1/E

ONLINE RESOURCES

REGISTRATION CODE

compliance-33370666

Mc Graw Hill Higher Education

PREALGEBRA
An Integrated Equations Approach

James Streeter
Late Professor of Mathematics
Clackamas Community College

Donald Hutchison
Clackamas Community College

Barry Bergman
Clackamas Community College

Boston Burr Ridge, IL Dubuque, IA Madison, WI New York
San Francisco St. Louis Bangkok Bogotá Caracas Kuala Lumpur
Lisbon London Madrid Mexico City Milan Montreal New Delhi
Santiago Seoul Singapore Sydney Taipei Toronto

McGraw-Hill Higher Education

A Division of The ***McGraw-Hill*** *Companies*

PREALGEBRA: AN INTEGRATED EQUATIONS APPROACH

Published by McGraw-Hill, a business unit of The McGraw-Hill Companies, Inc., 1221 Avenue of the Americas, New York, NY 10020.

This book is printed on acid-free paper.

1 2 3 4 5 6 7 8 9 0 VNH/VNH 0 9 8 7 6 5 4 3 2

ISBN 0–07–031772–0
ISBN 0–07–254686–7 (AIE)

Publisher: *William K. Barter*
Senior sponsoring editor: *David Dietz*
Developmental editor: *Peter Galuardi*
Senior marketing manager: *Mary K. Kittell*
Senior project manager: *Susan J. Brusch*
Senior production supervisor: *Laura Fuller*
Media project manager: *Sandra M. Schnee*
Media technology producer: *Jeff Huettman*
Designer: *K. Wayne Harms*
Interior designer: *Sheilah Barrett*
Cover designer: *Jamie E. O'Neal*
Cover image: *Corbis*
Lead photo research coordinator: *Carrie K. Burger*
Supplement producer: *Brenda A. Ernzen*
Compositor: *Interactive Composition Corporation*
Typeface: *10/12 NewTimes Roman*
Printer: *Von Hoffmann Press, Inc.*

Photo Credits

Chapter Openers: 1: © Michael Newman/Photo Edit; 2: © R. Lord/The Image Works; 3: © Michael Newman/Photo Edit; 4: © Alexander Walther/Getty Images; 5: © David Frazier/Stone/Getty; 6: © Blair Seitz/Photo Researchers, Inc.; 7: © Brownie Harris/The Stock Market/Corbis; 8: © Paul Conklin; 9: © Stanley Rowin/Index Stock Imager; 10: © Small Business/Corbis CD

Interior: Page 64: © Environmental Concerns/PhotoDisc; p. 218: © Education/PhotoDisc, Inc.; p. 347: © Lifestyles Today/PhotoDisc, Inc.; p. 446: © Business & Agriculture/Corbis CD; p. 449: © Homes & Gardens/PhotoDisc, Inc.; p. 469: © Lifestyles Today/PhotoDisc, Inc.; p. 506: © Business Hands/Corbis CD; p. 575: © Eat, Drink, Dine/PhotoDisc, Inc.; p. 598: © Business & Industry/Corbis CD; p. 629: © Eat, Drink, Dine/PhotoDisc, Inc.; p. 665: © Recreational Sports/Corbis CD; p. 732: © Business on the Go/PhotoDisc, Inc.

www.mhhe.com

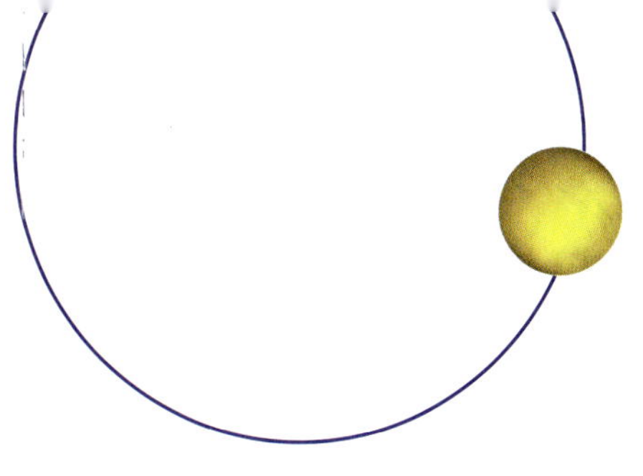

About the Authors

Don Hutchison spent his first ten years of teaching working with disadvantaged students. He taught in an intercity elementary school and an intercity high school. Don also worked with physically and mentally challenged children in state agencies in New York and Oregon.

In 1982, Don completed his graduate work in mathematics. He was then hired by Jim Streeter to teach at Clackamas Community College. Through Jim's tutelage, Don developed a fascination with the relationship between teaching and writing mathematics. He has come to believe that his best writing is a result of his classroom experience, and his best teaching is a result of the thinking involved in manuscript preparation.

Don is also active in several professional organizations. He was a member of the ACM committee that undertook the writing of computer curriculum for the two-year college. From 1989 to 1994 he was chair of the AMATYC Technology in Mathematics Education committee. He was President of ORMATYC from 1996 to 1998.

Barry Bergman has enjoyed teaching mathematics to a wide variety of students over the years. He began in the field of adult basic education, and moved into the teaching of high school mathematics in 1977. He taught at that level for 11 years, at which point he served for a year as a K–12 mathematics specialist for his county. This work allowed him the opportunity to help promote the emerging NCTM Standards in his region.

In 1990 Barry began the present portion of his career, having been hired to teach at Clackamas Community College. He maintains a strong interest in the appropriate use of technology in the learning of mathematics.

Throughout the past 25 years, Barry has played an active role in professional organizations. As a member of OCTM, he contributed several articles and activities to the group's journal. Recently, he has served as an officer of ORMATYC for 4 years, and has participated on an AMATYC committee to provide feedback and reactions to NCTM's new revision of the Standards.

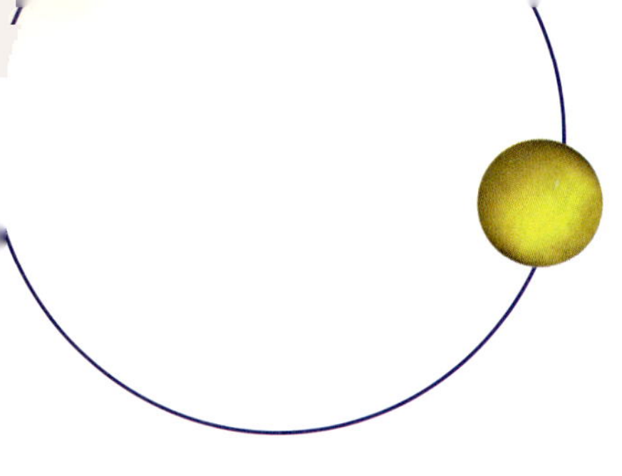

Dedication

This text is dedicated to the many fine teachers with whom we have worked. We are regularly inspired by their compassion, their energy, and their ingenuity. They have also provided help in the preparation of this text in a multitude of ways. Some of them have read manuscript and made valuable suggestions. Others have helped us to see new ways of introducing topics. Still others have provided us with excellent examples of applications of elementary mathematics. All of them have our respect and our thanks.

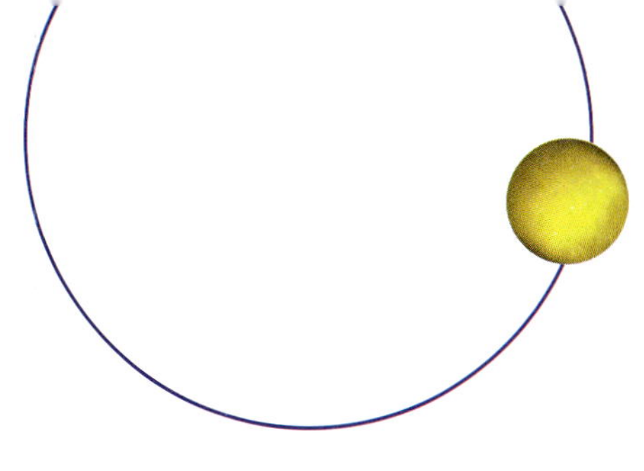

To the Student

You are about to begin a course in mathematics. We made every attempt to provide a text that will help you understand what mathematics is about and how to effectively use it. We made no assumptions about your previous experience with mathematics. Your progress through the course will depend on the amount of time and effort you devote to the course and your previous background. There are some specific features in this book that will aid you in your studies. Here are some suggestions about how to use this book. (Keep in mind that a review of *all* the chapter and summary material will further enhance your ability to grasp later topics and to move more effectively through the text.)

1. If you are in a lecture class, make sure that you take the time to read the appropriate text section *before* your instructor's lecture on the subject. Then take careful notes on the examples that your instructor presents during class.
2. After class, work through similar examples in the text, making sure that you understand each of the steps shown. Examples are followed in the text by *Check Yourself* exercises. Mathematics is best learned by being involved in the process, and that is the purpose of these exercises. Always have a pencil and paper at hand, and work out the problems presented and check your results immediately. If you have difficulty, go back and carefully review the previous exercises. Make sure you understand what you are doing and why. The best test of whether you understand a concept lies in your ability to explain that concept to one of your classmates. Try working together.
3. At the end of each chapter section you will find a set of exercises. Work these carefully to check your progress on the section you have just finished. You will find the solutions for the odd-numbered exercises following the exercise set. If you have difficulties with any of the exercises, review the appropriate parts of the chapter section. If your questions are not completely cleared up, by all means do not become discouraged. Ask your instructor or an available tutor for further assistance. A word of caution: Work the exercises on a regular (preferably daily) basis. Again, learning mathematics requires becoming involved. As is the case with learning any skill, the main ingredient is practice.
4. When you complete a chapter, review by using the *Summary.* You will find all the important terms and definitions in this section, along with examples illustrating all the techniques developed in the chapter. Following the summary are *Summary and Review Exercises* for further practice. The exercises are keyed to chapter sections, so you will know where to turn if you are still having problems.
5. When you finish with the *Summary and Review Exercises,* try the *Chapter Test* that appears at the end of each chapter. This test will give you an actual practice test to work as you review for in-class testing. Again, answers are provided.
6. Finally, an important element of success in studying mathematics is the process of regular review. We provide a series of *Cumulative Tests* throughout the textbook, beginning at the end of Chapter 2. These tests will help you review not only the concepts of the chapter that you have just completed but those of previous chapters. Use these tests in preparation for any midterm or final exams. If it appears that you have forgotten some concepts that are being tested, don't worry. Go back and review the sections where the idea was initially explained, or the appropriate chapter summary. That is the purpose of the cumulative tests.

We hope that you will find our suggestions helpful as you work through this material, and we wish you the best of luck in the course.

Donald Hutchison
Barry Bergman

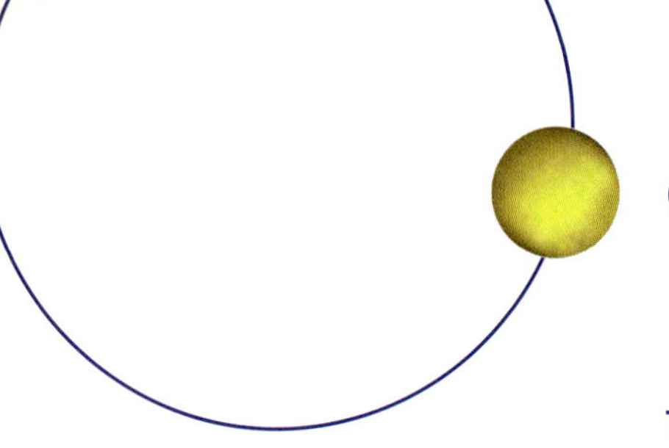

Contents

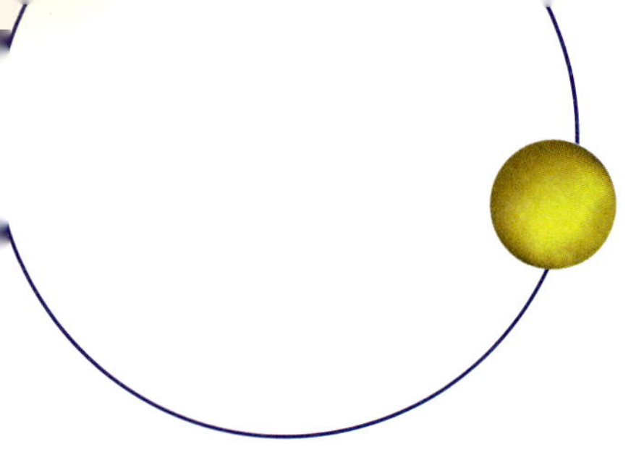

Preface

OUR PHILOSOPHY

We believe the key to learning mathematics, at any level, is active participation. When students are active participants in the learning process, they have the opportunity to construct their own mathematical ideas and make connections to previously studied material. Such participation leads to understanding, success, and confidence. We developed this text with that philosophy in mind and integrated many features throughout the book to reflect that philosophy. The *Check Yourself* exercises are designed to keep the students active and involved with every page of exposition. The optional calculator references involve students actively in the development of mathematical ideas. Almost every exercise set has application problems, challenging exercises, writing exercises, and/or collaborative exercises. Answer blanks for these exercises appear in the margin allowing the student to actively use the text, making it more than just a reference tool. These exercises are designed to awaken interest and insight within students. Not all of the exercises will be appropriate for every student, but each one provides another opportunity for both the instructor and the student. Our hope is that every student who uses this text will be a better mathematical thinker as a result.

FEATURES OF THIS TEXT

When preparing to write this text, the authors first solicited answers to the question, "How can we better prepare students to be successful in their algebra classes?" Two suggestions dominated the replies. Students need to be more comfortable with signed numbers and with word problems upon entering their first algebra class. We made these pursuits themes of this new text.

Signed numbers are not just introduced early in the text, they are thoroughly integrated throughout it. From the second chapter on, almost half of the examples contain at least one negative number.

The consistent use of dimensional analysis, again starting early in the text, gives the student experience in translating applications into mathematical statements. The "Units Analysis" boxes ensure that students see the importance of developing this skill.

The concepts of equations and expressions are gradually developed throughout the text. Equations are first presented at the end of Chapter 1, and at least one section per chapter thereafter is devoted to the topic. A number of features have contributed to the success of the Streeter series. All of these features are included in this text. Every one of those features and every supplement to this text was thoroughly discussed by the authors and reviewer panels. More than ever, we are confident that the entire learning package is of value to your students. We will describe each of the features of that package.

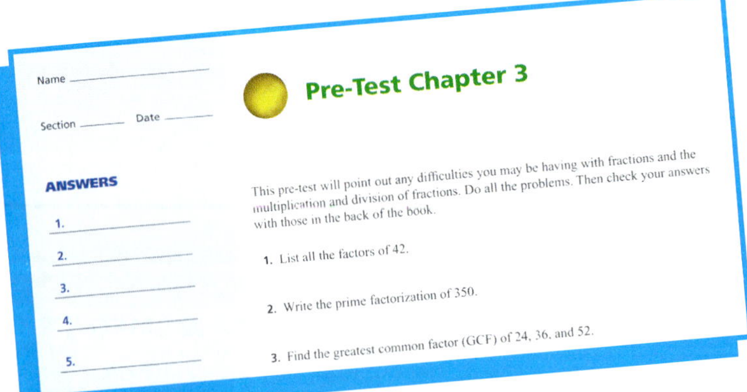

Name ______

Section ______ Date ______

ANSWERS

1. ______
2. ______
3. ______
4. ______
5. ______

Pre-Test Chapter 3

This pre-test will point out any difficulties you may be having with fractions and the multiplication and division of fractions. Do all the problems. Then check your answers with those in the back of the book.

1. List all the factors of 42.
2. Write the prime factorization of 350.
3. Find the greatest common factor (GCF) of 24, 36, and 52.

1. Every chapter begins with a Pre-Test covering material in the upcoming chapters.

Overcoming Math Anxiety

Working Together

How many of your classmates do you know? Whether you are by nature gregarious or shy, you have much to gain by getting to know your classmates.

1. It is important to have someone to call when you have missed class or if you are unclear on an assignment.
2. Working with another person is almost always beneficial to both people. If you don't understand something, it helps to have someone to ask about it. If you do understand something, nothing will cement that understanding more than explaining the idea to another person.
3. Sometimes we need to commiserate. If an assignment is particularly frustrating, it is reassuring to find that it is also frustrating for other students.
4. Have you ever thought you had the right answer, but it doesn't match the answer in the text? Frequently the answers are equivalent, but that's not always easy to see. A different perspective can help you see that. Occasionally there is an error in a textbook (here, we are talking about *other* textbooks). In such cases it is wonderfully reassuring to find that someone else has the same answer as you do.

2. Integers have been introduced early in the text and are used throughout the examples, exercises, and tests.
3. Units Analysis is consistently used throughout the text, especially in worked examples.
4. A Cumulative Test appears at the end of every chapter beginning with Chapter 2.
5. Geometric topics are integrated throughout the text in examples, exercises, and tests. Chapter 8 is devoted entirely to geometry and contains sections covering area, volume, lines and angles.
6. Supplementary Exercises are integrated into the exercise sets. As a result, answers to the even-numbered exercises may only be found in the *Instructor's Solutions Manual*.
7. Suggestions on "Overcoming Math Anxiety" can be found throughout the first three chapters. These suggestions are designed to be timely and useful. They are the same suggestions most of us make in class, but sometimes those words are given extra weight when students see them in print.

PEDAGOGICAL FEATURES

Chapter Openers

Each chapter opens with a real-world vignette that showcases an example of how mathematics is used in a job or profession. Exercise sets for each section then feature one or more modeling/word exercises that relate to the chapter-opening vignette. The application areas and the chapter each area appears in are:

Application Area	Chapter
Retail	Chapter 1
Anthropology	Chapter 2
Land Development	Chapter 3
Weights/Measurement	Chapter 4
Transportation/Navigation	Chapter 5
Bookkeeping/Accounting	Chapter 6
Sales	Chapter 7
Journalism	Chapter 8
Graphic Design	Chapter 9
Postal Work	Chapter 10

8.1 OBJECTIVES

1. Identify lines, line segments, and angles
2. Determine when lines are parallel or perpendicular
3. Determine whether an angle is right, acute, obtuse, or straight
4. Use a protractor to measure an angle
5. Determine when pairs of angles are complementary or supplementary
6. Identify a transversal, corresponding angles, vertical angles, and adjacent angles
7. Find the measure of the third angle of a triangle

Section Objectives

Objectives for each section are clearly identified. These numbered objectives are used to organize lessons on the accompanying tutorial CD-ROM and website.

Marginal Notes and Caution Icons

Marginal notes provided throughout are designed to help students focus on important topics and techniques. Caution icons point out potential trouble spots.

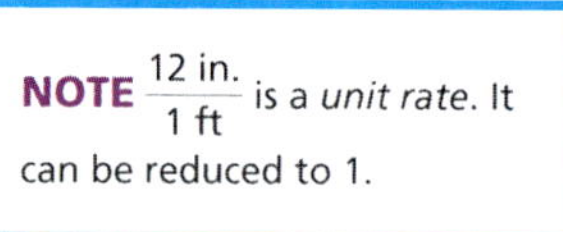

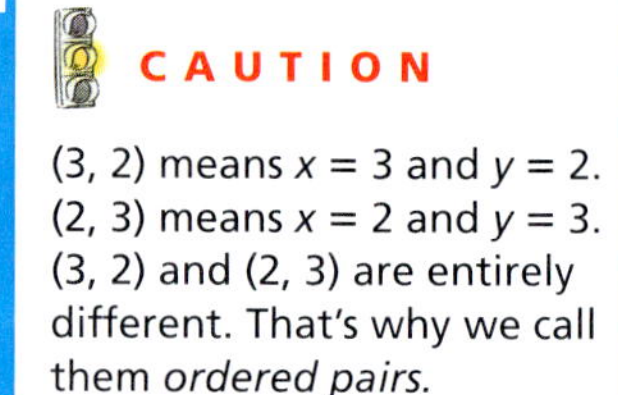

Check Yourself Exercises

These exercises have been the hallmark of the Streeter series; they are designed to actively involve students throughout the learning process. Each example is followed by an exercise that encourages students to solve a problem similar to the one just presented. Answers are provided at the end of the section for immediate feedback.

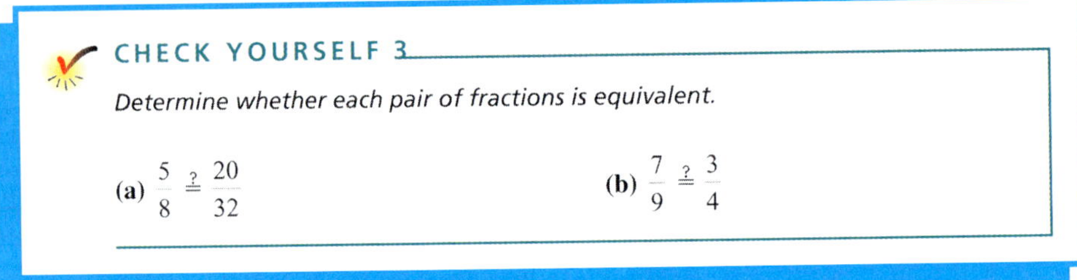

Comprehensive Exercise Sets and Challenge Exercises

Complete exercise sets are provided at the end of each section as well as after the summary at the end of each chapter. These exercises were designed to reinforce basic skills and develop critical thinking and communication abilities. Exercise sets include writing and word problems, collaborative and group exercises, and challenge exercises, all denoted by distinctive icons. The calculator icon points out examples and exercises that illustrate when the calculator can best be used for further understanding of the concept at hand.

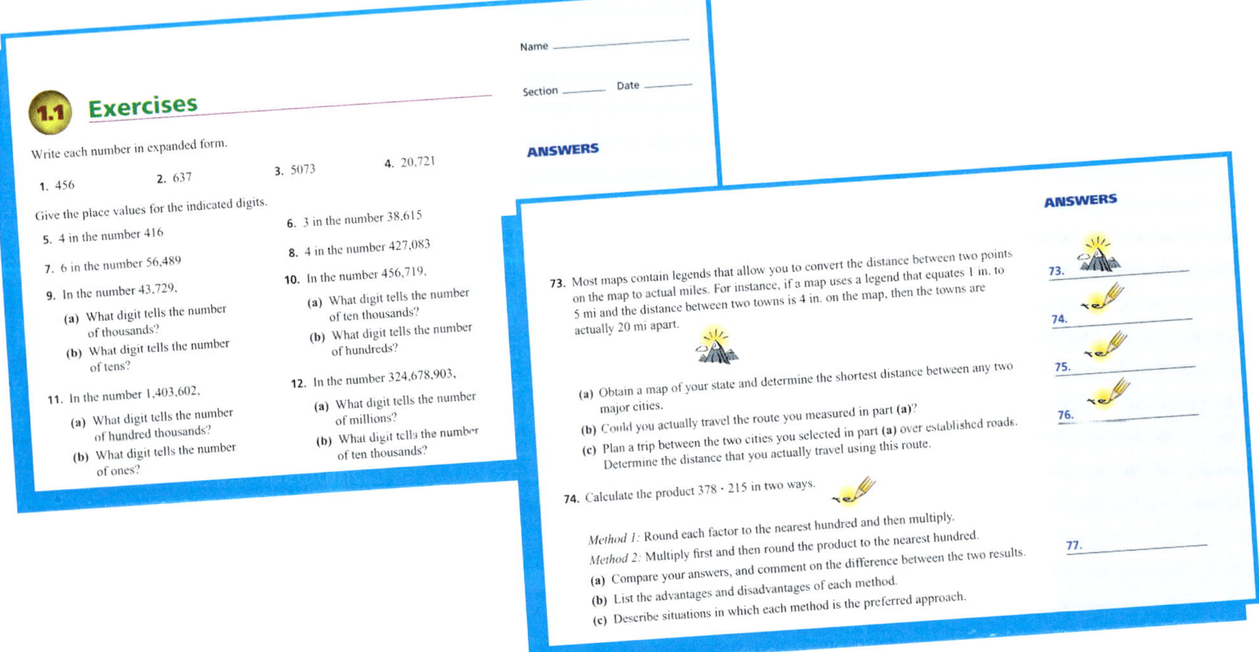

Name ______

Section ______ Date ______

1.1 Exercises

Write each number in expanded form.

1. 456 2. 637 3. 5073 4. 20,721

Give the place values for the indicated digits.

5. 4 in the number 416
6. 3 in the number 38,615
7. 6 in the number 56,489
8. 4 in the number 427,083
9. In the number 43,729,
   (a) What digit tells the number of thousands?
   (b) What digit tells the number of tens?
10. In the number 456,719,
   (a) What digit tells the number of ten thousands?
   (b) What digit tells the number of hundreds?
11. In the number 1,403,602,
   (a) What digit tells the number of hundred thousands?
   (b) What digit tells the number of ones?
12. In the number 324,678,903,
   (a) What digit tells the number of millions?
   (b) What digit tells the number of ten thousands?

ANSWERS

73. Most maps contain legends that allow you to convert the distance between two points on the map to actual miles. For instance, if a map uses a legend that equates 1 in. to 5 mi and the distance between two towns is 4 in. on the map, then the towns are actually 20 mi apart.
   (a) Obtain a map of your state and determine the shortest distance between any two major cities.
   (b) Could you actually travel the route you measured in part (a)?
   (c) Plan a trip between the two cities you selected in part (a) over established roads. Determine the distance that you actually travel using this route.
74. Calculate the product $378 \cdot 215$ in two ways.
   Method 1: Round each factor to the nearest hundred and then multiply.
   Method 2: Multiply first and then round the product to the nearest hundred.
   (a) Compare your answers, and comment on the difference between the two results.
   (b) List the advantages and disadvantages of each method.
   (c) Describe situations in which each method is the preferred approach.

ANSWERS

73. ______
74. ______
75. ______
76. ______
77. ______

Summary and Summary and Review Exercises

The comprehensive summaries at the end of each chapter enable students to review important concepts. The Summary and Review Exercises provide an opportunity for the student to practice these important concepts. The answers to these exercises are only provided in the *Instructor's Solutions Manual* and *Annotated Instructor's Edition*.

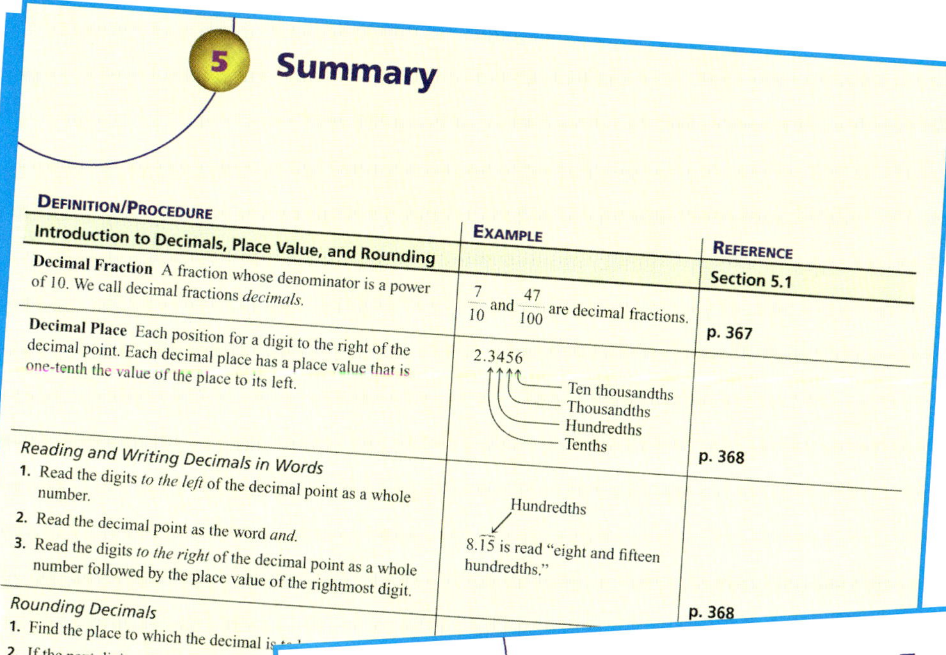

5 **Summary**

DEFINITION/PROCEDURE	EXAMPLE	REFERENCE
Introduction to Decimals, Place Value, and Rounding		**Section 5.1**
Decimal Fraction A fraction whose denominator is a power of 10. We call decimal fractions *decimals*.	$\frac{7}{10}$ and $\frac{47}{100}$ are decimal fractions.	p. 367
Decimal Place Each position for a digit to the right of the decimal point. Each decimal place has a place value that is one-tenth the value of the place to its left.	2.3456 (Ten thousandths, Thousandths, Hundredths, Tenths)	p. 368
Reading and Writing Decimals in Words 1. Read the digits *to the left* of the decimal point as a whole number. 2. Read the decimal point as the word *and.* 3. Read the digits *to the right* of the decimal point as a whole number followed by the place value of the rightmost digit.	Hundredths 8.15 is read "eight and fifteen hundredths."	p. 368
Rounding Decimals 1. Find the place to which the decimal is to b… 2. If the next digit to the right is 5 or m… the place you are rounding to by 1. D… digits to the right. 3. If the next digit to the right is less tha… digit and any remaining digits to the r…		

Summary and Review Exercises

You should now be reviewing the material in Chapter 5. These exercises will help in that process. Work all the exercises carefully. References are provided to the section for each exercise. If you made an error, go back and review the related material.

[5.1] In exercises 1 and 2, find the indicated place values.

1. 7 in 3.5742

2. 3 in 0.5273

In exercises 3 and 4, write the fractions in decimal form.

3. $\frac{37}{100}$

4. $\frac{307}{10000}$

In exercises 5 and 6, write the decimals in words.

5. 0.071

6. 12.39

In exercises 7 and 8, write the fractions in decimal form.

7. Four and five tenths

8. Four hundred and thirty-seven thousandths

In exercises 9 to 12, complete each statement using the symbol <, =, or >.

9. 0.79 ______ 0.785

10. 1.25 ______ 1.250

11. 12.8 ______ 13

12. 0.832 ______ 0.83

In exercises 13 to 15, round to the indicated place.

13. 5.837 hundredths

14. 9.5723 thousandths

Chapter Tests

Each chapter includes a chapter test to give students confidence and guidance in preparing for in-class tests. Answers are at the back of the book.

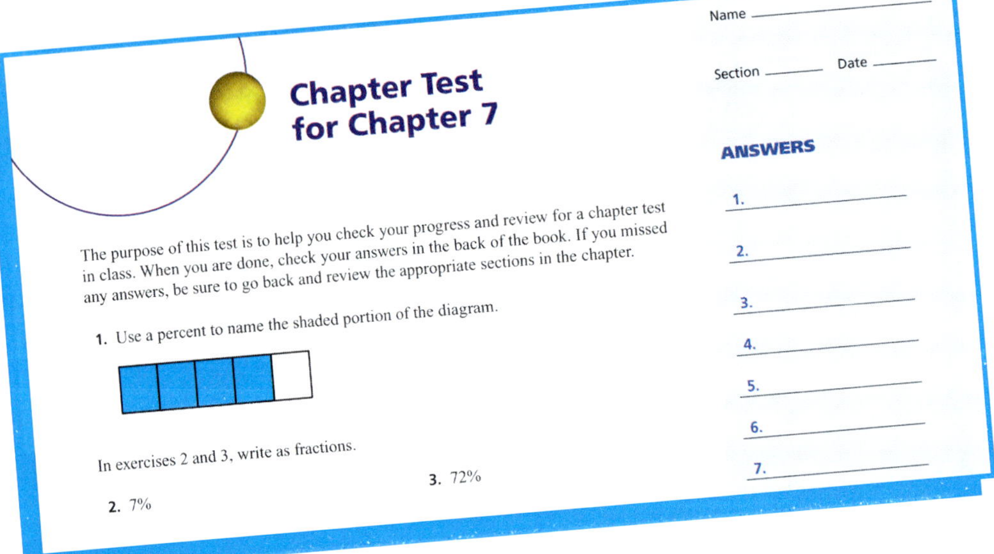

Cumulative Tests

These tests help students build on what was previously covered and give them more opportunity to reinforce skills necessary in preparing for midterm and final exams. Answers are at the back of the book.

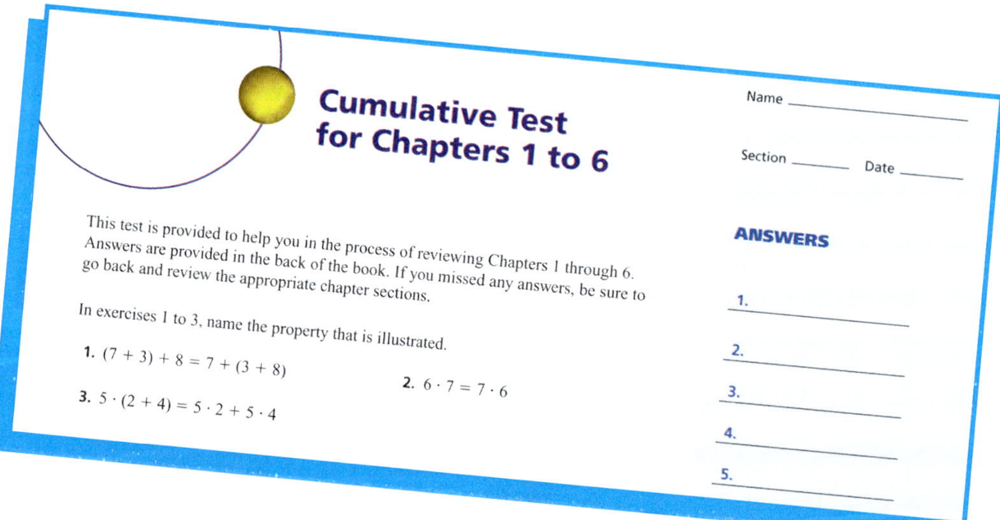

SUPPLEMENTS

A comprehensive set of ancillary materials for both the student and the instructor is available with this text.

Annotated Instructor's Edition

This ancillary includes answers to all exercises and tests. These answers are printed in a second color for ease of use by the instructor and are located on the appropriate pages throughout the text.

Instructor's Solutions Manual

The manual provides worked-out solutions to all the section exercises, Summary and Review Exercises, the Pre-Tests, Chapter Tests, and Cumulative Tests.

Print and Computerized Testing

The testing materials provide an array of formats that enable the instructor to create tests using both algorithmically generated and static test questions. This testing system enables the instructor to choose questions either manually or randomly by section, question type, difficulty level, and other criteria. Additionally, an instructor may administer Internet-based tests through a secure website provided by the testing software. Testing is available for Windows and Macintosh computers.

Student's Solutions Manual

The *Student's Solutions Manual* provides worked-out solutions to the odd-numbered section exercises in the text.

Video Series

The video series gives students additional reinforcement of the topics presented in the book. The videos were developed especially for this textbook, and features are tied directly to the main text's individual chapters and section objectives. The videos feature an effective combination of teaching techniques, including personal instruction, state-of-the-art graphics, and real-world applications. Students are encouraged to work examples on their own and check their results with those provided.

Streeter SMART Tutorial CD-ROM

This interactive CD-ROM is a self-paced tutorial specifically linked to the text and reinforces topics through unlimited opportunities to review concepts and practice problem solving. The CD-ROM contains chapter-specific and section-specific tutorials, multiple-choice questions with feedback, and algorithmically generated questions. It requires virtually no computer training on the part of students and supports Windows and Macintosh computers.

In addition, a number of other technology and Web-based ancillaries are under development; they will support the ever-changing technology needs in developmental mathematics. For further information about these or any supplements, please contact your local McGraw-Hill sales representative.

Online Learning Center

Web-based, interactive learning is available on the *Online Learning Center*, located at www.mhhe.com/streeter. Student resources are located in the OLC's **Student Center** and include interactive applications, algorithmically generated practice exams, online quizzing, audio/visual tutorials, all chapters of the text in PDF format, and web links. Instructor resources are loaded in the **Instructor Center** and include PowerPoint® slides, an AMATYC standards correlation, and links to PageOut and author-recommended sites. The Course Integration Guide is also located on the OLC. The password for the **Instructor Center** is located in the preface to the *Instructor's Solutions Manual*.

Course Integration Guide

This supplement, located on the OLC, integrates the multimedia and print supplements that accompany the main text into a useful and well-organized guide. The *Course Integrations Guide* includes helpful information about resources found in the text, video, CD-ROM, and Online Learning Center. The guide also contains a section-level correlation with the ELM, CLAST, and TASP standards.

NetTutor

NetTutor is a revolutionary system that enables students to interact with a live tutor over the World Wide Web. Students can receive instructions from qualified tutors using NetTutor's web-based, graphical chat capabilities. They can also submit questions and receive answers, browse previously answered questions, and view previous live chat sessions.

ALEKS®

ALEKS® (**A**ssessment and **LE**arning in **K**nowledge **S**paces) is an artificial intelligence-based system for individualized math learning, available over the World Wide Web. ALEKS® delivers precise, qualitative diagnostic assessments of students' math knowledge, guides them in the selection of appropriate new study material, and records their progress toward mastery of curricular goals in a robust classroom management system. It interacts with the student much as a skilled human tutor would, moving between explanation and practice as needed, correcting and analyzing errors, defining terms, and changing topics on request. By sophisticated modeling of a student's "knowledge state" for a given subject matter, ALEKS® can focus clearly on what the student is most ready to learn next, building a learning momentum that fuels success.

To learn more about ALEKS®, including purchasing information, visit the ALEKS® website at www.highed.aleks.com.

PageOut

Need a course website? More than 50,000 professors have chosen PageOut to create a custom course website. Their feedback is used to continually enhance PageOut.

New features based on customer feedback:

- Timed tests and the ability to author original questions
- Ability to insert diacritical marks and html codes with the click of a button
- Ability to choose a pre-built page design, or create a custom design

Short on time? Let us do the work.

Our McGraw-Hill service team is ready to build your PageOut website and provide any necessary training.

Learn more about PageOut and other McGraw-Hill digital solutions at www.mhhe.com/solutions.

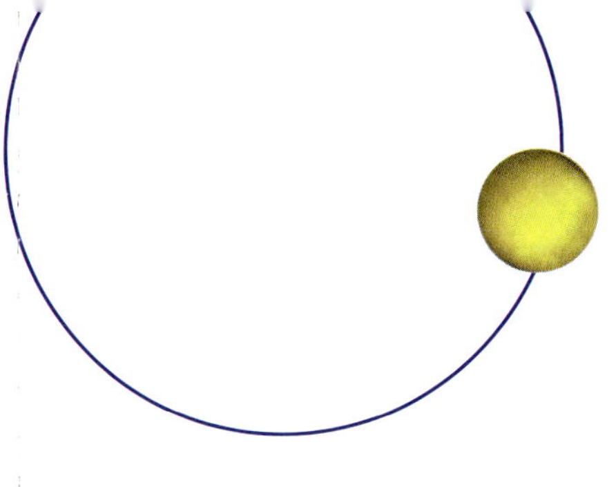

Acknowledgments

Those familiar with the publishing process will attest that change is inevitable. The same is true at MHHE. The amazing thing is that change invariably seems to lead to something positive. Throughout the development of this text, we have been most fortunate to work with Bill Barter, David Dietz, Bea Wikander and Peter Galuardi. All of them manage to be both demanding managers and supportive coworkers. We have appreciated their talents, energy, and time. As always, we encourage prospective authors to talk with the staff at MHHE. It will be a valuable use of your time.

In this age of the high-tech mathematics lab, the supplements and the supplement authors have become almost as important as the text itself. We are proud that our names appear on these supplements together with the following authors:

Brownstone Research Group
John Hunt
Laurel Tech, L.L.C.
Link Systems International, Inc.
VPG Integrated Media, Inc.

We would like to thank the many reviewers who took the time to review and help improve this text:

Debra Bryant, Tennessee Technological Univ. (TN)
Camille Cochrane, Shelton State Community College (AL)
Pat Cook, Weatherford College (TX)
Nancy Desilet, Carroll Community College (MD)
Carol Flakus, Lower Columbia College (WA)
Alberto Guerra, St. Philip's College (TX)
Celeste Hernandez, Richland College (TX)
Tammy Higson, Hillsborough Community College (FL)
Lori Holdren, Manatee Community College (FL)
Paul Hrabovsky, Indiana University of Pennsylvania (PA)
Mathew Hudock, St. Phillip's College (TX)
Nicholas Huerta, Fullerton College (CA)
John D. Jarvis, Utah Valley State College (UT)
Dr. Nancy Johnson, Manatee Community College (FL)
Judy Kasabian, El Camino College (CA)
Paul Wayne Lee, St. Philip's College (TX)
Nancy Lehmann, Austin Community College (TX)
Jean-Marie Magnier, Springfield Tech. Community College (MA)
Robert Maynard, Tidewater Community College (VA)
Philip Meurer, Palo Alto College (CA)
Mary Romans, Kent State University (OH)
Liz Russell, Glendale Community College (CA)
Sally Sestini, Cerritos College (CA)
Renée Starr, Beaver College (PA)
Daryl Stephens, East Tennessee State University (TN)
Patricia Taylor, Thomas Nelson Community College (VA)
Sharon Testone, Onondaga Community College (NY)
Dr. Joseph Tripp, Ferris State University (MI)
Marjorie Whitmore, Northwest Arkansas Community College (AK)
Alma Wlazlinski, McLennan Community College (TX)
Kevin Yokoyama, College of the Redwoods (CA)

We would also like to thank those who contributed to the development of the fifth editions of *Basic Mathematical Skills with Geometry* and *Beginning Algebra* from which this text is derived:

Basic Mathematical Skills with Geometry, Fifth Edition

Mary Kay Abbey, Montgomery College Takoma Park Campus (MD)
Della Bell, Texas Southern University (TX)
Palma Benko, Passaic County Community College (NJ)
R. Benn, Prince George Community College (MD)
Alberto Beron, Moorpark College (CA)
Robert L. Caldwell, El Camino College (CA)
Ethel Cameron, Shepherd College (WV)
P. Carlson, Delta College (MI)
Natalia Casper, College of Lake County (IL)
Laurence Chernoff, Miami-Dade Community College—Kendall Campus (FL)
Emily Cosby, Napa Valley College (CA)
Katheryne Earl, Eastern Wyoming College (WY)
Grace Foster, Beaufort Community College (NC)
Jacqueline B. Giles, Houston Community College—Central (TX)
Myriam Haddad, Riverside Community College—Norco (CA)
Lynn Hargrove, Sierra Community College (CA)
Lissa Hawkins, Chicago City College (IL)
Celeste Hernandez, Richland College (TX)
Doris Holland, Tarrant County College—South (TX)
Linda Horner, Broward Community College—North (FL)
Tracey Hoy, College of Lake County (IL)
Sally Jackman, Richland College (TX)
Karen Jensen, Southeastern Community College (IA)
Joseph Kazimir, East Los Angeles College (CA)
Surinder Kumar Khurana, Fullerton College (CA)
Lynda Laningham, Rend Lake College (IL)
Ann Lyndon, Northwest Shoals Community College—Muscle Shoals (AL)
Steven Mahoney, Shasta College (CA)
Bob Martin, Tarrant County College (TX)
Victor Mastrovincenzo, Hudson County Community College (NJ)
C. McDonald, St. Louis Community College—Florissant Valley (MO)
Carol McKillip, Southwestern Oregon Community College (OR)
Michael Montano, Riverside Community College (CA)
Brenda Moore, Indian Hills Community College—Ottumwa (IA)
M. Najafi, Kent State—Ashtabula (OH)
Kamilia Nemri, Spokane Community College (WA)
Jerry Chris Neve, Front Range Community College—Larimer Campus (CO)
Sharon Ostdiek, Central College—Hastings Campus (NE)
Maria Parker, Oxnard College (CA)
Michael Polley, Southeastern Community College (IA)
Jody Porter, Treasure Valley Community College (OR)
Scott Rippy, San Bernadino Valley College (CA)
C. Don Rogers, State University of New York—Erie Campus (NY)
Glenn R. Sandifer, San Jacinto College—Central Campus (TX)
Phyllis R. Schott, Catonsville Community College (MD)
Richard D. Semmler, Northern Virginia Community College (VA)
Bruce Sisko, Belleville Area College (IL)
Barbara Jane Sparks, Camden County College (NJ)
David D. Staten, Galveston College (TX)

Michael Tran, Antelope Valley College (CA)
Kathy Wellborn, Fort Lewis College (CO)
Judith Wolff, Cuyahoga Community College—Metro Campus (OH)
Paul Wozniak, El Camino College (CA)

Beginning Algebra, Fifth Edition

Sharon Abramson, Nassau Community College (NY)
Patricia Allaire, Queensborough Community College (NY)
Sharon Berrian, Northwest Shoals Community College (AL)
Matthews Chakkanakuzh, Palomar Community College (CA)
Alan Chutsky, Queensborough Community College (NY)
John Davidson, Southern State Community College (OH)
Katherine D'Orazio, Cumberland County Community College (NJ)
Bill Dunn, Las Positas College (CA)
Ellen Freedman, Camden County Community College (NJ)
Kelly Jackson, Camden County Community College (NJ)
Karen Jensen, Southeastern Community College (IA)
Ginny Licata, Camden County Community College (NJ)
S. Maheshwari, William Paterson University (NJ)
Laurie McManus, St. Louis Community College-Meramec (MO)
Diane Metzger, Rend Lake College (IL)
Wayne Miller, Lee College (TX)
Jeff Mock, Diablo Valley College (CA)
Ellen Musen, Brookdale Community College (NJ)
Larry Newman, Holyoke Community College (MA)
Lilia Orlova, Nassau Community College (NY)
Betty Pate, St. Petersburg Junior College (FL)
Kathryn Pletsch, Antelope Valley College (CA)
Larry Pontaski, Pueblo Community College (CO)
Donna Russo, Quincy College (MA)
Bruce Sisko, Belleville Area College (IL)
Barbara Jane Sparks, Camden County Community College (NJ)
Peter Speier, Prince George's Community College (MD)
Sharon Testone, Onandaga Community College (NY)
Patricia Wake, San Jacinto College (TX)

We thank all of the students whom we have taught, talked to, questioned, and tested. This text was created for them. We also thank our community college compatriots. Professionals such as Stefan Baratto, Medy Saqueton, Denise Conklin, and Wes Bruning are constantly providing us with both intentional and inadvertent guidance in our writing projects.

What Is ALEKS®?

ALEKS is...

- A comprehensive course management system. It tells you exactly what your students know and don't know.
- Artificial intelligence. It totally individualizes assessment and learning.
- Customizable. Click on or off each course topic.
- Web based. Use a standard browser for easy Internet access.
- Inexpensive. There are no setup fees or site license fees.

ALEKS (**A**ssessment and **LE**arning in **K**nowledge **S**paces) is an artificial intelligence-based system for individualized math learning available via the World Wide Web.

http://www.highed.aleks.com

ALEKS delivers precise, qualitative diagnostic assessments of students' math knowledge, guides them in the selection of appropriate new study material, and records their progress toward mastery of curricular goals in a robust course management system.

ALEKS interacts with the student much as a skilled human tutor would, moving between explanation and practice as needed, correcting and analyzing errors, defining terms and changing topics on request. By sophisticated modeling of a student's "knowledge state" for a given subject matter, ALEKS can focus clearly on what the student is most ready to learn next, thereby building a learning momentum that fuels success.

The ALEKS system was developed with a multimillion-dollar grant from the National Science Foundation. It has little to do with what is commonly thought of as educational software. The theory behind ALEKS is a specialized field of mathematical cognitive science called "Knowledge Spaces."

Knowledge Space theory, which concerns itself with the mathematical dynamics of knowledge acquisition, has been under development by researchers in cognitive science since the early 1980s. The Chairman and founder of ALEKS Corporation, Jean-Claude Falmagne, is an internationally recognized leader in scientific work in this field.

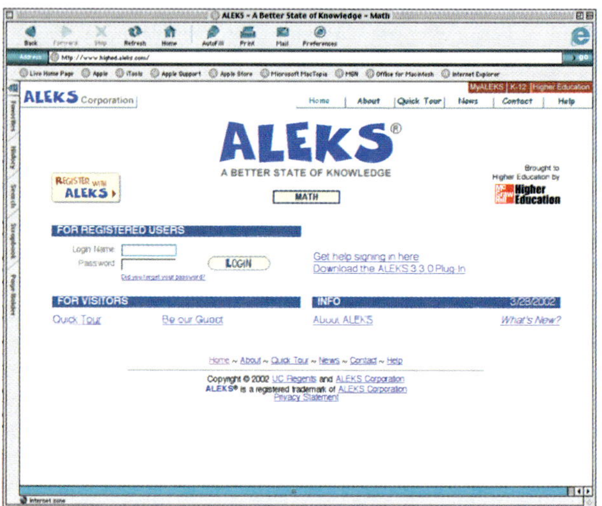

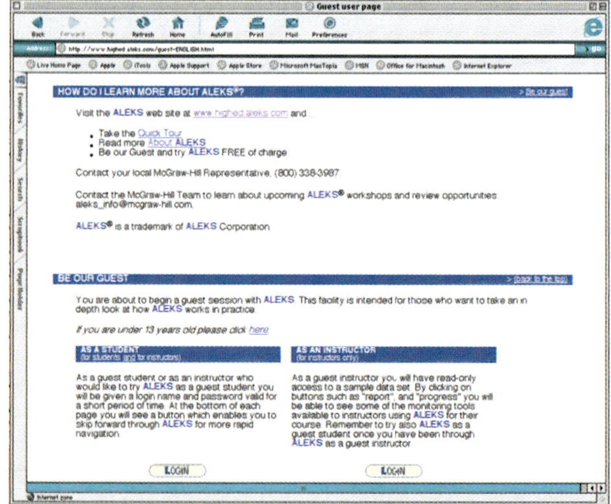

Please visit http://www.highed.aleks.com for a trial run.

1 WHOLE NUMBERS

INTRODUCTION

Saturday is always chaotic at Home Base Building Supply. Not only is the store bustling with activity, all five phone lines are constantly in use. Most of the calls are questions from people in the midst of some do-it-yourself project.

Nadia started working at Home Base right after she received her A.S. degree. At first she was concerned that she did not have enough experience in all the trades that are represented at the store. Since she started, Nadia has learned to answer most questions about plumbing, carpentry, and electricity. She is surprised at the great number of questions that really require only a math background. While working full time, Nadia has been taking night classes. Next year she will be getting her B.A. in business administration. She has enjoyed working the floor, but she is ready to move into administration.

Name ______________________

Section ________ Date ________

ANSWERS

1. ______________________
2. ______________________
3. ______________________
4. ______________________
5. ______________________
6. ______________________
7. ______________________
8. ______________________
9. ______________________
10. ______________________
11. ______________________
12. ______________________
13. ______________________
14. ______________________
15. ______________________

Pre-Test Chapter 1

This pre-test will point out any difficulties you may be having in performing operations with whole numbers. Do all the exercises without using a calculator, and then check your answers with those in the back of the book.

1. Write 107,945 in words.

2. The statement $2 + (3 + 5) = (2 + 3) + 5$ illustrates which property of addition?

3. The statement $5 \times 7 = 7 \times 5$ is an illustration of which property of multiplication?

4. $35{,}147 + 2873 - 7305 - 3101 = ?$

5. $392 \times 51 = ?$

6. $187 \times 300 = ?$

7. $3234 \div 7 = ?$

8. **(a)** $5 + 3 \times 7 = ?$ **(b)** $7 \times (3 + 5) = ?$

Divide by long division.

9. $7\overline{)8431}$

10. $267\overline{)21{,}758}$

11. **Test scores.** Suppose that you need a total of 360 points on four tests during the semester to receive an A for the course. Your scores on the first three tests were 84, 91, and 92. What is the lowest score you can get on the fourth test and still receive the A?

12. **Installment buying.** A refrigerator is advertised as follows: "Pay \$50 down and \$30 a month for 24 months." If the cash price of the refrigerator is \$619, how much extra will you pay if you buy on the installment plan?

13. Evaluate $2^3 \div 2 \times 3 - (5 - 2 + 3)$.

14. Find the perimeter (P) and area (A) of the figure.

5 yd

2 yd

15. Estimate the following sum by first rounding each value to the nearest hundred. Also, find the actual sum.

$$\begin{array}{r} 921 \\ 377 \\ 855 \\ +\ 649 \\ \hline \end{array}$$

1.1 Introduction to Whole Numbers, Place Value

1.1 OBJECTIVES

1. Write numbers in expanded form
2. Determine the place value of a digit
3. Write a number in words
4. Write a number, given its word name

Overcoming Math Anxiety

Throughout this text, we will present you with a series of class-tested techniques that are designed to improve your performance in this math class.

Become familiar with your textbook.

Perform each of the following tasks.

1. Use the Table of Contents to find the title of Section 5.1.
2. Use the Index to find the earliest reference to the term *mean*. (By the way, this term has nothing to do with the personality of either your instructor or the textbook authors!)
3. Find the answer to the first Check Yourself exercise in Section 1.1.
4. Find the answers to the pre-test for Chapter 1.
5. Find the answers to the odd-numbered exercises in Section 1.1.
6. In the margin notes for Section 1.1, find the origin for the term *digit*.

Now you know where some of the most important features of the text are. When you have a moment of confusion, think about using one of these features to help you clear up that confusion.

Number systems have been developed throughout human history. Starting with simple tally systems used to count and keep track of possessions, more and more complex systems were developed. The Egyptians used a set of picturelike symbols called **hieroglyphics** to represent numbers. The Romans and Greeks had their own systems of numeration. We see the Roman system today in the form of Roman numerals. Some examples of these systems are shown in the table.

Numerals	Egyptian	Greek	Roman
1	I	I	I
10	∩	Δ	X
100	⚲	H	C

NOTE The prefix *deci* means 10. Our word *digit* comes from the Latin word *digitus,* which means finger.

Any number system provides a way of naming numbers. The system we use is described as a **decimal place-value system.** This system is based on the number 10 and uses symbols called **digits.** (Other numbers have also been used as bases. The Mayans used 20, and the Babylonians used 60.)

The basic symbols of our system are the digits 0, 1, 2, 3, 4, 5, 6, 7, 8, 9. These basic symbols, or digits, were first used in India and then adopted by the Arabs. For this reason, our system is called the Hindu-Arabic numeration system.

NOTE Any number, no matter how large, can be represented using the 10 digits of our system.

Numbers may consist of one or more *digits*. 3, 45, 567, and 2359 are examples of the **standard form** for numbers. We say that 45 is a two-digit number, 567 is a three-digit number, and so on.

As we said, our decimal system uses a *place-value* concept based on the number 10. Understanding how this system works will help you see the reasons for the rules and methods of arithmetic that we will be introducing.

Example 1

Writing a Number in Expanded Form

NOTE Each digit in a number has its own place value.

Look at the number 438. We call 8 the *ones digit.* Moving to the left, the digit 3 is the *tens digit.* Again moving to the left, 4 is the *hundreds digit.*

NOTE Here the parentheses are used for emphasis. (4 × 100) means 4 is multiplied by 100. (3 × 10) means 3 is multiplied by 10. (8 × 1) means 8 is multiplied by 1.

If we rewrite a number such that each digit is written with its units, we have used the **expanded form** for the number. First think $400 + 30 + 8$; then we write 438 in expanded form as

$(4 \times 100) + (3 \times 10) + (8 \times 1)$

CHECK YOURSELF 1

Write 593 in expanded form.

The place-value diagram shows the place value of digits as we write larger numbers. For the number 3,156,024,798, we have

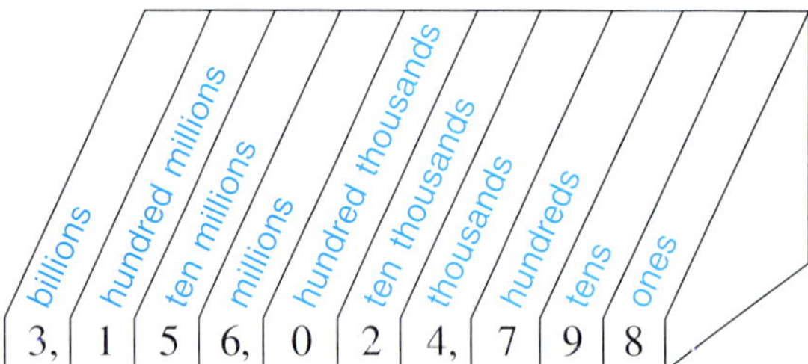

Of course, the naming of place values continues for larger and larger numbers beyond the chart.

For the number 3,156,024,798, the place value of 4 is thousands. As we move to the left, each place value is 10 times the value of the previous place. The place value of 2 is ten thousands, the place value of 0 is hundred thousands, and so on.

Example 2

Identifying Place Value

Identify the place value of each digit in the number 418,295.

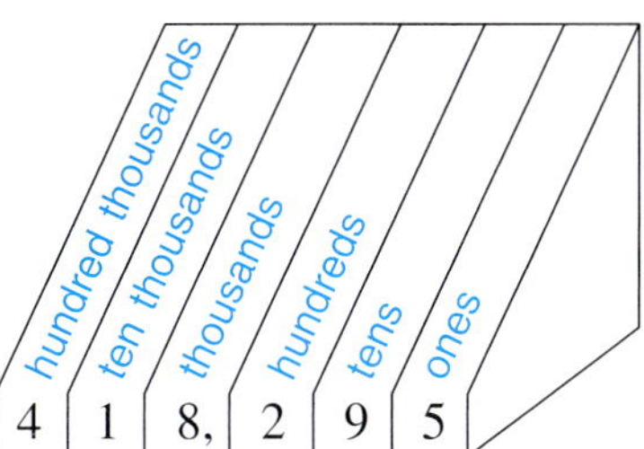

CHECK YOURSELF 2

Use a place value diagram to answer the following questions for the number 6,831,425,097.

(a) What is the place value of 2?
(b) What is the place value of 4?
(c) What is the place value of 3?
(d) What is the place value of 6?

Understanding place value will help you read or write numbers in word form. Look at the number

7 2, 3 5 8, 6 9 4
Millions Thousands Ones

NOTE A four-digit number, such as 3,456, can be written with or without a comma. We have chosen to include the comma in these materials.

Commas are used to set off groups of three digits in the number. The name of each group—for example, millions, thousands, or ones—is then used as we write the number in words. To write a word name for a number, we work from left to right, writing the numbers in each group, followed by the group name. The chart summarizes the group names.

Billions Group			Millions Group			Thousands Group			Ones Group		
Hundreds	Tens	Ones	Hundreds	Tens	Ones	Hundreds	Tens	Ones	Hundreds	Tens	Ones

Example 3

Writing Numbers in Words

NOTE Notice that the commas in the word statements are in the same place as the commas in the number.

27,345 is written in words as twenty-seven *thousand,* three hundred forty-five.
2,305,273 is two *million,* three hundred five *thousand,* two hundred seventy-three.

Note: We do *not* write the name of the ones group. Also, "and" is not used when a whole number is written in words. It will have a special meaning later.

CHECK YOURSELF 3

Write each of the following numbers in words.

(a) 658,942 **(b)** 2305

We reverse the process to write the standard form for numbers given in word form. Consider Example 4.

Example 4

Translating Words into Numbers

Forty-eight thousand, five hundred seventy-nine in standard form is

48,579

Five hundred three thousand, two hundred thirty-eight in standard form is

503,238

Note the use of 0 as a placeholder in writing the number.

CHECK YOURSELF 4

Write twenty-three thousand, seven hundred nine in standard form.

CHECK YOURSELF ANSWERS

1. $(5 \times 100) + (9 \times 10) + (3 \times 1)$ **2.** **(a)** Ten thousands; **(b)** hundred thousands; **(c)** ten millions; **(d)** billions **3.** **(a)** Six hundred fifty-eight thousand, nine hundred forty-two; **(b)** two thousand three hundred five **4.** 23,709

Name ______________

Section ________ Date ________

1.1 Exercises

Write each number in expanded form.

1. 456 **2.** 637 **3.** 5073 **4.** 20,721

Give the place values for the indicated digits.

5. 4 in the number 416

6. 3 in the number 38,615

7. 6 in the number 56,489

8. 4 in the number 427,083

9. In the number 43,729,

(a) What digit tells the number of thousands?

(b) What digit tells the number of tens?

10. In the number 456,719,

(a) What digit tells the number of ten thousands?

(b) What digit tells the number of hundreds?

11. In the number 1,403,602,

(a) What digit tells the number of hundred thousands?

(b) What digit tells the number of ones?

12. In the number 324,678,903,

(a) What digit tells the number of millions?

(b) What digit tells the number of ten thousands?

Write each number in words.

13. 5618 **14.** 21,812 **15.** 200,304 **16.** 103,900

Give the standard (numerical) form for each of the written numbers.

17. Two hundred fifty-three thousand, four hundred eighty-three

18. Three hundred fifty thousand, three hundred fifty-nine

19. Five hundred two million, seventy-eight thousand

20. Four billion, two hundred thirty million

ANSWERS

1. ______________
2. ______________
3. ______________
4. ______________
5. ______________
6. ______________
7. ______________
8. ______________
9. ______________
10. ______________
11. ______________
12. ______________
13. ______________
14. ______________
15. ______________
16. ______________
17. ______________
18. ______________
19. ______________
20. ______________

ANSWERS

21. ______________________

22. ______________________

23. ______________________

24. ______________________

Write the whole number in each sentence in standard form.

21. The first place finisher in the 2002 U.S. Senior Open won four hundred fifteen thousand dollars.

22. Some scientists speculate that the universe originated in the explosion of a primordial fireball approximately twenty billion years ago.

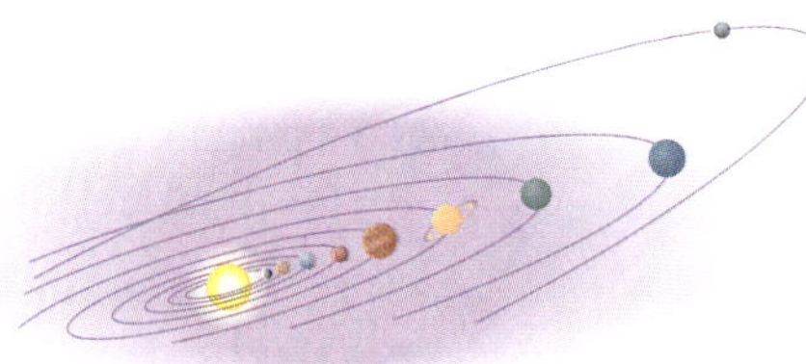

23. The population of Kansas City, Missouri, in 2000 was approximately four hundred forty-five thousand, six hundred.

24. The Nile river in Egypt is about four thousand, one hundred forty-five miles long.

Sometimes numbers found in charts and tables are abbreviated. The given table represents the population of the 10 largest cities in the 2000 U.S. census. Note that the numbers represent thousands. Thus Detroit had a population of 952 thousand or 952,000.

City Name	Rank	Population (thousands)
New York, NY	1	8008
Los Angeles, CA	2	3695
Chicago, IL	3	2896
Houston, TX	4	1954
Philadelphia, PA	5	1518
Phoenix, AZ	6	1321
San Diego, CA	7	1223
Dallas, TX	8	1119
San Antonio, TX	9	1145
Detroit, MI	10	952

In exercises 25 to 28, write your answers in standard form using the given table.

25. What was the population of Phoenix in 2000?

26. What was the population of Chicago in 2000?

27. What was the population of Philadelphia in 2000?

28. What was the population of Dallas in 2000?

Assume that you have alphabetized the word names for every number from one to one thousand.

29. Which number would appear first in the list?

30. Which number would appear last?

Determine the number represented by the scrambled place values.

31. 4 thousands
1 ten
3 ten thousands
5 ones
2 hundreds

32. 7 hundreds
4 ten thousands
9 ones
8 tens
6 thousands

33. Inci has to write a check for $2565. There is a space on the check to write out the amount of the check in words. What should she write in this space?

ANSWERS

25. ______________

26. ______________

27. ______________

28. ______________

29. ______________

30. ______________

31. ______________

32. ______________

33. ______________

34. ______

35. ______

36. ______

37. ______

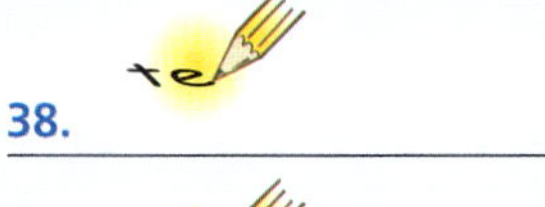

38. ______

39. ______

40. ______

41. ______

34. In addition to personal checks, name two other places where writing amounts in words is necessary.

35. In a rental agreement, the amount of the initial deposit required is two thousand, five hundred forty-five dollars. Write this amount as a number.

36. How many zeros are in the number for one billion?

37. Write the largest five-digit number that can be made using the digits 6, 3, and 9 if each digit is to be used at least once.

38. Several early numeration systems did not use place values. Do some research, and determine at least two of these systems. Describe the system that they used. What were the disadvantages?

39. What are the advantages of a place-value system of numeration?

40. The number 0 was not used initially by the Hindus in our number system (about 250 B.C.). Go to your library (or "surf the net"), and determine when a symbol for zero was introduced. What do you think is the importance of the role of 0 in a numeration system?

41. A *googol* is a very large number. Do some research to find out how big it is. Also try to find out where the name of this number comes from.

Answers

We will provide the answers (with some worked out in detail) for the odd-numbered exercises at the end of each exercise set. The answers for the even-numbered exercises are provided in the instructor's resource manual.

1. $(4 \times 100) + (5 \times 10) + (6 \times 1)$ **3.** $(5 \times 1000) + (7 \times 10) + (3 \times 1)$
5. Hundreds **7.** Thousands **9.** **(a)** 3; **(b)** 2 **11.** **(a)** 4; **(b)** 2
13. Five thousand, six hundred eighteen
15. Two hundred thousand, three hundred four **17.** 253,483 **19.** 502,078,000
21. \$415,000 **23.** 445,600 **25.** 1,321,000 **27.** 1,518,000
29. Eight **31.** 34,215 **33.** Two thousand, five hundred sixty-five
35. \$2545 **37.** 99,963 **39.** **41.**

Addition of Whole Numbers

1.2 OBJECTIVES

1. Use the language of addition
2. Add single-digit numbers
3. Identify the properties of addition
4. Add lists of numbers with no carrying
5. Add any lists of numbers
6. Find a perimeter
7. Solve applications that involve perimeter
8. Solve other applications

Overcoming Math Anxiety

Become familiar with your syllabus.

In the first class meeting, your instructor probably handed out a class syllabus. If you haven't done so already, you need to incorporate important information into your calendar and address book.

1. Write all important dates in your calendar. This includes homework due dates, quiz dates, test dates, and the date and time of the final exam. Never allow yourself to be surprised by any deadline!
2. Write your instructor's name, contact number, and office number in your address book. Also include the office hours. Make it a point to see your instructor early in the term. Although this is not the only person who can help clear up your confusion, he or she is the most important person.
3. Make note of other resources that are made available to you. These include CDs, videotapes, web pages, and tutoring.

Given all of these resources, it is important that you never let confusion or frustration mount. If you can't "get it" from the text, try another resource. All of the resources are there specifically for you, so take advantage of them!

The *natural* or *counting numbers* are the numbers we use to count objects.

NOTE The three dots (. . .) are called an **ellipsis;** they mean that the set continues the indicated pattern.

The natural numbers are 1, 2, 3, . . .

When we include the number 0, we then have the set of *whole numbers.*

The whole numbers are 0, 1, 2, 3, . . .

We will now look at the operation of *addition* on the whole numbers.

Definitions: Addition

Addition is the combining of two or more groups of the same kind of objects.

This concept is extremely important, as we will see in our later work with fractions. We can only combine or add numbers that represent the same kind of objects.

From your first encounter with arithmetic, you were taught to add "3 apples plus 2 apples."

On the other hand, you have probably encountered a phrase such as, "like combining apples and oranges." That is to say, what do you get when you add 3 apples and 2 oranges?

You could answer "5 fruits," or "5 objects," but you can't combine the apples and the oranges.

What if you walked 3 miles and then walked 2 more miles? Clearly, you have now walked 3 miles + 2 miles = 5 miles. The addition was possible because you add groups of the same kind.

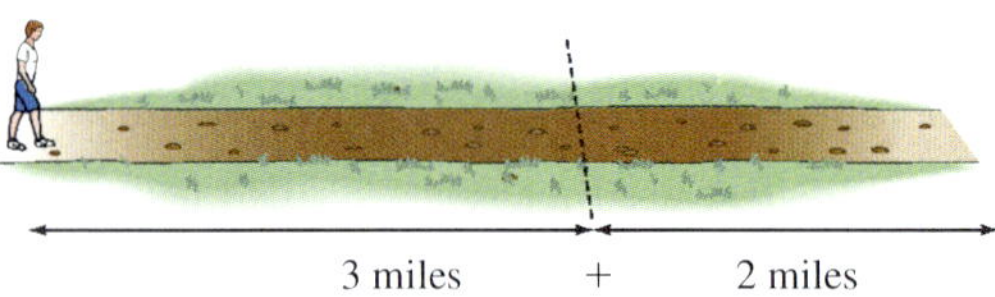

NOTE The first printed use of the symbol + dates back to 1500.

Each operation of arithmetic has its own special terms and symbols. The addition symbol + is read **plus.** When we write 3 + 4, 3 and 4 are called the **addends.**

We can use a number line to illustrate the addition process. To construct a number line, we pick a point on the line and label it 0. We then mark off evenly spaced units to the right, naming each point marked off with a successively larger whole number.

NOTE The point labeled 0 is called the **origin** of the number line.

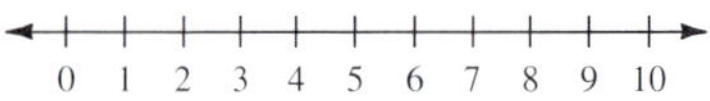

We use an arrowhead to show the number line continues.

Example 1

Representing Addition on a Number Line

Represent 3 + 4 on the number line.

To represent an addition, such as 3 + 4, on the number line, start by moving 3 spaces to the right of the origin. Then move 4 more spaces to the right to arrive at 7. The number 7 is called the *sum* of the addends.

NOTE Again, addition corresponds to combining groups of the same kind of objects.

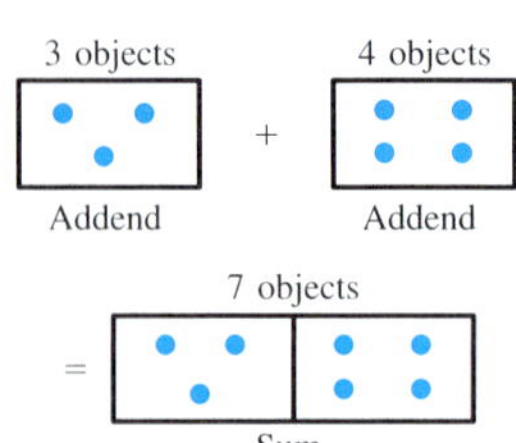

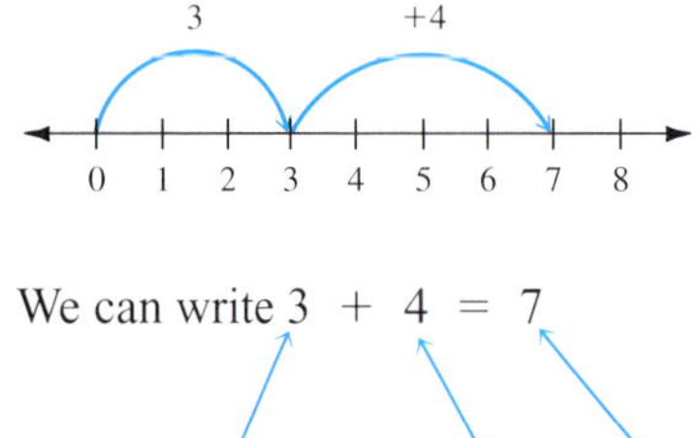

We can write 3 + 4 = 7

Addend Addend Sum

CHECK YOURSELF 1

Represent 5 + 6 *on the number line.*

A statement such as 3 + 4 = 7 is one of the **basic addition facts.** These facts include the sum of every possible pair of digits. Before you can add larger numbers correctly and quickly, you must memorize these basic facts.

Basic Addition Facts

+	0	1	2	3	4	5	6	7	8	9
0	0	1	2	3	4	5	6	7	8	9
1	1	2	3	4	5	6	7	8	9	10
2	2	3	4	5	6	7	8	9	10	11
3	3	4	5	6	7	8	9	10	11	12
4	4	5	6	7	8	9	10	11	12	13
5	5	6	7	8	9	10	11	12	13	14
6	6	7	8	9	10	11	12	13	14	15
7	7	8	9	10	11	12	13	14	15	16
8	8	9	10	11	12	13	14	15	16	17
9	9	10	11	12	13	14	15	16	17	18

NOTE To find the sum 5 + 8, start with the row labeled 5. Move along that row to the column headed 8 to find the sum, 13.

Examining the basic addition facts leads us to several important properties of addition on the whole numbers. For instance, we know that the sum 3 + 4 is 7. What about the sum 4 + 3? It is also 7. This is an illustration of the fact that addition is a **commutative** operation.

NOTE *Commute* means to move back and forth, as to school or work.

Rules and Properties: The Commutative Property of Addition

The order of two numbers around an addition sign *does not* affect the sum.

Example 2

Using the Commutative Property

NOTE The *order* does not affect the sum.

8 + 5 = 5 + 8 = 13

6 + 9 = 9 + 6 = 15

CHECK YOURSELF 2

Show that the sum on the left equals the sum on the right.

7 + 8 = 8 + 7

If we wish to add *more* than two numbers, we can group them and then add. In mathematics this grouping is indicated by a set of parentheses ( ). This symbol tells us to perform the operation inside the parentheses first.

Example 3

Using Grouping Symbols

NOTE We add 3 and 4 as the first step and then add 5.

$(3 + 4) + 5 = 7 + 5 = 12$ Here, the 4 is "associated" with the 3.

We also have

NOTE Here we add 4 and 5 as the first step and then add 3. Again the final sum is 12.

$3 + (4 + 5) = 3 + 9 = 12$ The 4 is "associated" with the 5.

This example suggests the following property of whole numbers.

Rules and Properties: The Associative Property of Addition

NOTE The two equations of Example 3 demonstrate that the 4 can be "associated" with the 3 or the 5.

The way in which several whole numbers are grouped *does not* affect the final sum when they are added.

CHECK YOURSELF 3

Show that the two expressions have the same value.

$(4 + 8) + 3$ and $4 + (8 + 3)$

The number 0 has a special property in addition.

Rules and Properties: The Additive Identity Property

The sum of 0 and any whole number is just that whole number.

Because of this property, we call 0 the **identity** for the addition operation.

Example 4

Adding Zero

Find the sum of **(a)** $3 + 0$ and **(b)** $0 + 8$.

(a) $3 + 0 = 3$

(b) $0 + 8 = 8$

CHECK YOURSELF 4

Find the sum.

(a) $4 + 0 =$ **(b)** $0 + 7 =$

When we are adding larger numbers, we will apply the following rule.

Rules and Properties: Adding Digits of the Same Place Value

We can add the digits of the same place value because they represent the same quantities.

Adding two numbers, such as $25 + 34$, can be done in expanded form. Here we write out the place value for each digit.

NOTE Remember that 25 means 2 tens and 5 ones; 34 means 3 tens and 4 ones.

$$\begin{array}{rl} 25 &= 2 \text{ tens} + 5 \text{ ones} \\ +\ 34 &= 3 \text{ tens} + 4 \text{ ones} \\ \hline &= 5 \text{ tens} + 9 \text{ ones} \\ &= 59 \end{array}$$

↓ Add down.

In actual practice, we use a more convenient short form and our basic addition facts to perform the addition.

Example 5

Adding Two Numbers

Add $352 + 546$.

NOTE In using the short form, be very careful to line up the numbers correctly so that each column contains digits of the same place value.

Step 1 Add in the ones column.

$$\begin{array}{r} 352 \\ +\ 546 \\ \hline 8 \end{array}$$

Step 2 Add in the tens column.

$$\begin{array}{r} 352 \\ +\ 546 \\ \hline 98 \end{array}$$

Step 3 Add in the hundreds column.

$$\begin{array}{r} 352 \\ +\ 546 \\ \hline 898 \end{array}$$

CHECK YOURSELF 5

Add.

$$\begin{array}{r} 245 \\ +\ 632 \\ \hline \end{array}$$

You have already seen that the word *sum* indicates addition. There are other words that also tell you to use the addition operation.

The *total* of 12 and 5 is written as

12 + 5 or 17

8 *more than* 10 is written as

10 + 8 or 18

12 *increased by* 3 is written as

12 + 3 or 15

Example 6

Translating Words That Indicate Addition

Find each number.

(a) 36 increased by 12.

36 increased by 12 is written as $36 + 12 = 48$.

(b) The total of 18 and 31.

The total of 18 and 31 is written as $18 + 31 = 49$.

CHECK YOURSELF 6

Find each number.

(a) 43 increased by 25 **(b)** The total of 22 and 73

In the examples and exercises of this section, the digits in each column added to 9 or less. What about the situation in which a column has a two-digit sum? This will involve the process of **carrying.** In Example 7, we will look at the process in expanded form.

Example 7

Adding in Expanded Form When Carrying Is Needed

NOTE Carrying in addition is also called **regrouping,** or **renaming.** Of course, the name makes no difference as long as you understand the process.

$$\begin{array}{rl} 67 = & 60 + 7 \\ +\ 28 = & 20 + 8 \\ \hline & 80 + 15 \end{array}$$

We have written 15 ones as 1 ten and 5 ones.

or $\underbrace{80 + 10}+ 5$

The 1 ten is then combined with the 8 tens.

or $90 + 5$

or 95

NOTE Of course, this is true for any size number. The place value thousands is 10 times the place value hundreds, and so on.

The more convenient short form carries the excess units from one column to the next column left. Recall that the place value of the next column left is 10 times the value of the original column. It is this property of our decimal place-value system that makes carrying work. Look at the problem again, this time done in the short, or "carrying," form.

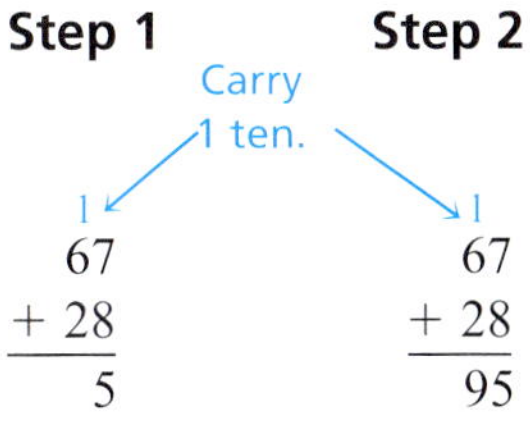

Step 1: The sum of the digits in the ones column is 15, so write 5 and carry 1 to the tens column. **Step 2:** Now add in the tens column, being sure to include the carried 1.

CHECK YOURSELF 7

Add.

$$\begin{array}{r} 58 \\ +\ 36 \\ \hline \end{array}$$

The addition process often requires more than one carrying step, as is shown in Example 8.

Example 8

Adding in Short Form When Carrying Is Needed

Add 285 and 378.

$$\begin{array}{r} {\scriptstyle 1} \\ 285 \\ +\ 378 \\ \hline 3 \end{array}$$

Carry 1 ten.

The sum of the digits in the ones column is 13, so write 3 and carry 1 to the tens column.

Carry 1 hundred.

$$\begin{array}{r} {\scriptstyle 1\,1} \\ 285 \\ +\ 378 \\ \hline 63 \end{array}$$

Now add in the tens column, being sure to include the carry. We have 16 tens, so write 6 in the tens place and carry 1 to the hundreds column.

$$\begin{array}{r} {\scriptstyle 1\,1} \\ 285 \\ +\ 378 \\ \hline 663 \end{array}$$

Finally, add in the hundreds column.

CHECK YOURSELF 8

Add.

$$\begin{array}{r} 479 \\ +\ 287 \\ \hline \end{array}$$

The carrying process is the same if we want to add more than two numbers.

Example 9

Adding in Short Form with Multiple Carrying Steps

Add 53, 2678, 587, and 27,009.

```
 1 1 2 2  ←—— Carries
      53
    2678
     587
+ 27,009
--------
  30,327
```

Add in the ones column: $3 + 8 + 7 + 9 = 27$.
Write 7 in the sum and carry 2 to the tens column.

Now add in the tens column, being sure to include the carry. The sum is 22. Write 2 tens and carry 2 to the hundreds column. Complete the addition by adding in the hundreds column, the thousands column, and the ten thousands column.

CHECK YOURSELF 9

Add 46, 365, 7254, *and* 24,006.

One application of addition is in finding the *perimeter* of a figure.

Definitions: Perimeter

The **perimeter** is the distance around a closed figure.

If the figure has straight sides, the perimeter is the sum of the lengths of its sides.

Example 10

Finding the Perimeter

We wish to fence in the field shown in the figure. How much fencing, in feet (ft), will be needed?

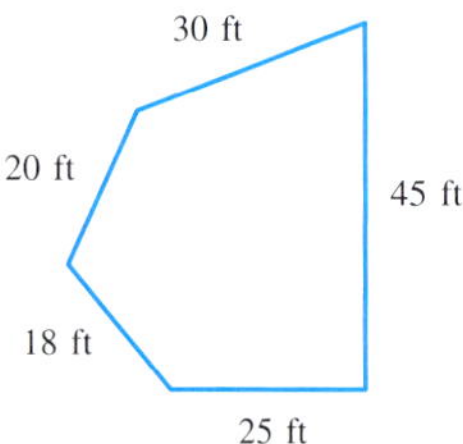

The fencing needed is the perimeter of (or the distance around) the field. We must add the lengths of the five sides.

NOTE Make sure to include the unit with each number.

20 ft + 30 ft + 45 ft + 25 ft + 18 ft = 138 ft

So the perimeter is 138 ft.

CHECK YOURSELF 10

What is the perimeter in inches (in.) of the region shown?

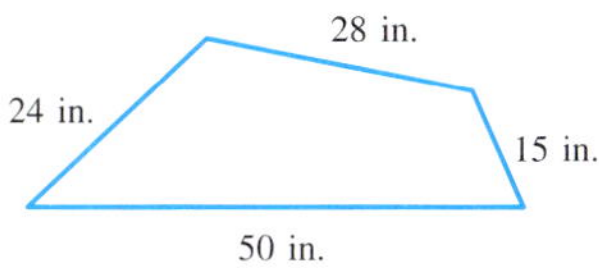

A **rectangle** is a figure, like a sheet of paper, with four sides and four equal corners. The perimeter of a rectangle is found by adding the lengths of the four sides.

Example 11

Finding the Perimeter of a Rectangle

Find the perimeter in inches of the rectangle pictured.

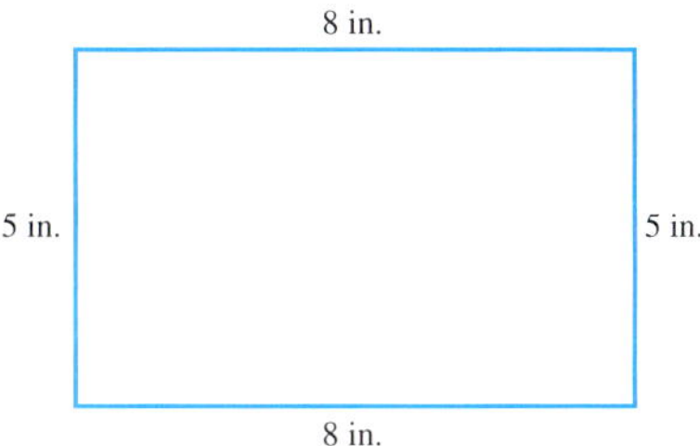

The perimeter is the sum of the lengths 8 in., 5 in., 8 in., and 5 in.

8 in. + 5 in. + 8 in. + 5 in. = 26 in.

The perimeter of the rectangle is 26 in.

CHECK YOURSELF 11

Find the perimeter of the rectangle pictured.

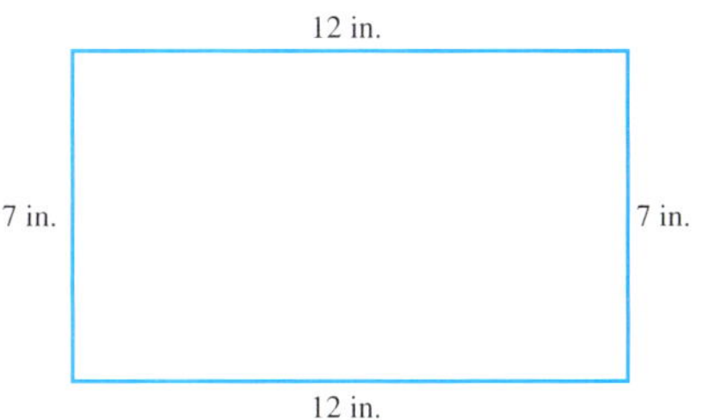

In general, we can find the perimeter of a rectangle by using a formula. A **formula** is a set of symbols that describe a general solution to a problem.

Look at a picture of a rectangle.

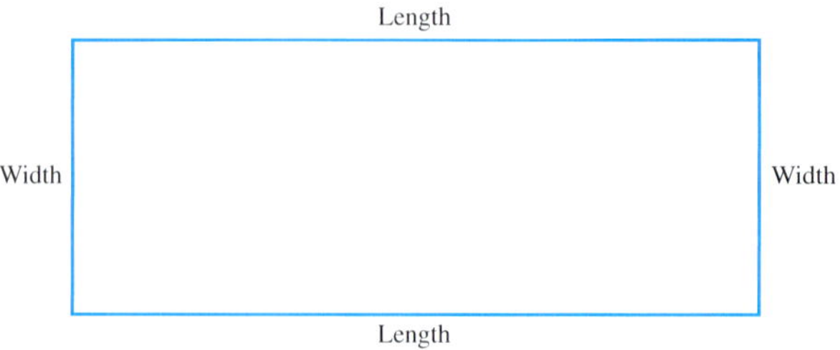

The perimeter can be found by adding the distances, so

Perimeter = length + width + length + width

To make this formula a little more readable, we abbreviate each of the words, using just the first letter.

Rules and Properties: Formula for the Perimeter of a Rectangle

$$P = L + W + L + W$$

Example 12

Finding the Perimeter of a Rectangle

A rectangle has length 11 in. and width 8 in. What is its perimeter?

Start by drawing a picture of the problem.

NOTE We say the rectangle is 8 in. by 11 in.

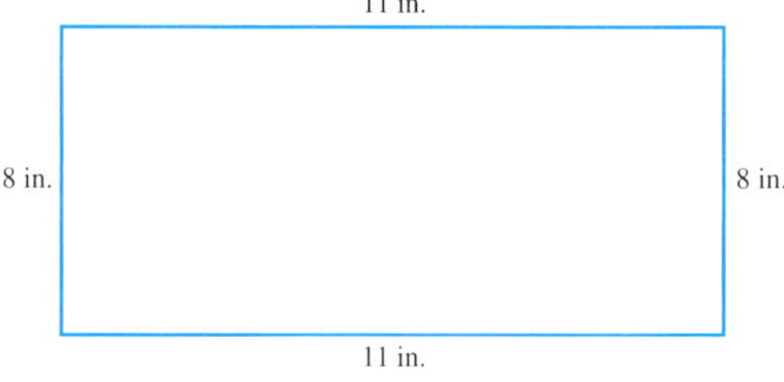

Use the formula $P = L + W + L + W$ with $L = 11$ in. and $W = 8$ in.

$$P = 11 \text{ in.} + 8 \text{ in.} + 11 \text{ in.} + 8 \text{ in.}$$
$$= 38 \text{ in.}$$

The perimeter is 38 in.

CHECK YOURSELF 12

A bedroom is 9 *ft by* 12 *ft. What is its perimeter?*

NOTE You may very well be able to do some of these problems in your head. Get into the habit of writing down *all* your work, rather than just an answer.

Now we consider other applications that will use the operation of addition. An organized approach is the key to successful problem solving; we would suggest this strategy.

Step by Step: Solving Addition Applications

Step 1 Read the problem carefully to determine the given information and what you are asked to find.

Step 2 Decide upon the operation (in this case, addition) to be used.

Step 3 Write down the complete statement necessary to solve the problem and do the calculations.

Step 4 Write your answer as a complete sentence. Check to make sure you have answered the question of the problem and that your answer seems reasonable.

We will use these steps in Example 13.

Example 13

Setting Up a Word Problem

Four sections of algebra were offered in the fall quarter, with enrollments of 33, 24, 20, and 22 students. What was the total number of students taking algebra?

Step 1 The given information is the number of students in each section. We want the total number.

Step 2 Since we wish a total, we use addition.

NOTE Remember to attach the proper unit (here "students") to your answer.

Step 3 Write $33 + 24 + 20 + 22 = 99$ students.

Step 4 There were 99 students taking algebra.

CHECK YOURSELF 13

Elva Ramos won an election for city council with 3110 *votes. Her two opponents had* 1022 *and* 1211 *votes. How many votes were cast in that election?*

Using a Scientific Calculator to Add

NOTE If you have a graphing calculator, follow the same steps, replacing [=] with [ENTER].

Although part of this book is intended to help you review the basic skills of arithmetic, many of you will want to be able to use a handheld calculator for some of the problems that are presented. Ideally you should learn to do the basic operations *by hand.* So in each section of this book, start by learning to do the work *without* your calculator. We will then provide these special calculator sections to show how you can use the calculator.

Example 14

Using a Scientific Calculator to Add

NOTE If you press the **clear** key first ([C]), the calculator is ready to evaluate a new expression.

Add 23 + 3456 + 7 + 985.

Enter:

23 [+] 3456 [+] 7 [+] 985 [=]

Enter each of the first three numbers followed by the plus key. Then enter the final number and press the equals key. The sum will be in the display.

Display 4471

If you have a calculator, try the addition of this example now. If you have difficulty getting the answer, check the operating manual or ask your instructor for assistance.

CHECK YOURSELF 14

Add 295 + 3 + 4162 + 84.

CHECK YOURSELF ANSWERS

1. 5 + 6 = 11

2. 7 + 8 = 15 and 8 + 7 = 15
3. (4 + 8) + 3 = 12 + 3 = 15; 4 + (8 + 3) = 4 + 11 = 15
4. **(a)** 4; **(b)** 7
5. 877
6. **(a)** 68; **(b)** 95
7. 94
8. 766
9. 31,671
10. 117 in.
11. 38 in.
12. 42 ft
13. 5343 votes
14. 295 [+] 3 [+] 4162 [+] 84 [=] 4544

Name ______________________

Section ________ Date ________

1.2 Exercises

1. In the statement $5 + 4 = 9$

5 is called the
4 is called the
9 is called the

2. In the statement $7 + 8 = 15$

7 is called the
8 is called the
15 is called the

Name the property of addition that is illustrated. Explain your choice of property.

3. $5 + 8 = 8 + 5$

4. $(4 + 5) + 8 = 4 + (5 + 8)$

5. $4 + (7 + 6) = 4 + (6 + 7)$

6. $5 + (2 + 3) = (2 + 3) + 5$

Perform the indicated addition.

7. $\begin{array}{r} 2792 \\ +\ 205 \\ \hline \end{array}$

8. $\begin{array}{r} 2345 \\ +\ 6053 \\ \hline \end{array}$

9. $\begin{array}{r} 2531 \\ +\ 5354 \\ \hline \end{array}$

10. $\begin{array}{r} 21{,}314 \\ +\ 43{,}042 \\ \hline \end{array}$

11. $\begin{array}{r} 3490 \\ 548 \\ +\ 25 \\ \hline \end{array}$

12. $\begin{array}{r} 2289 \\ 38 \\ 578 \\ +\ 3489 \\ \hline \end{array}$

13. $\begin{array}{r} 23{,}458 \\ +\ 32{,}623 \\ \hline \end{array}$

14. $\begin{array}{r} 26{,}735 \\ 259 \\ 3056 \\ +\ 35{,}489 \\ \hline \end{array}$

15. Find the number that is 356 more than 1213.

16. Find the number that is 567 more than 2322.

17. Find the sum of 3295, 9, 427, and 56.

18. Add 5637, 78, 690, 28, and 35,589.

19. Find the total of 124 and 2351.

20. Find the total of the three numbers 112, 24, and 532.

Find the perimeter of each figure.

21.

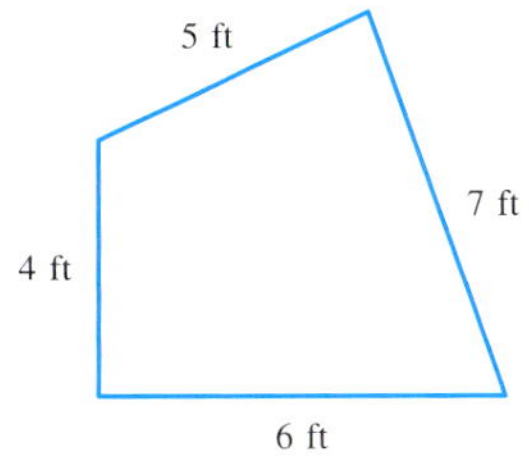

22.

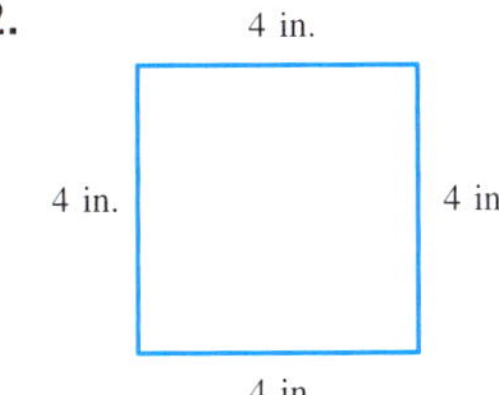

ANSWERS

1. ______________
2. ______________
3. ______________
4. ______________
5. ______________
6. ______________
7. ______________
8. ______________
9. ______________
10. ______________
11. ______________
12. ______________
13. ______________
14. ______________
15. ______________
16. ______________
17. ______________
18. ______________
19. ______________
20. ______________
21. ______________
22. ______________

ANSWERS

23. ______

24. ______

25. ______

26. ______

27. ______

28. ______

29. ______

30. ______

31. ______

23.

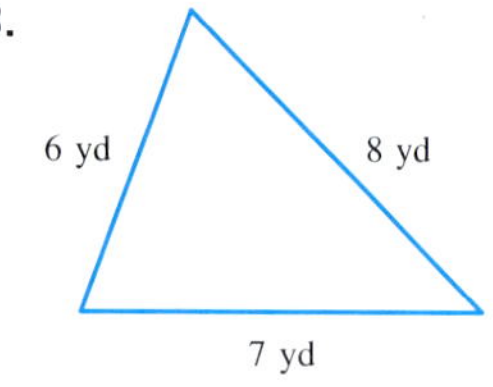

24.

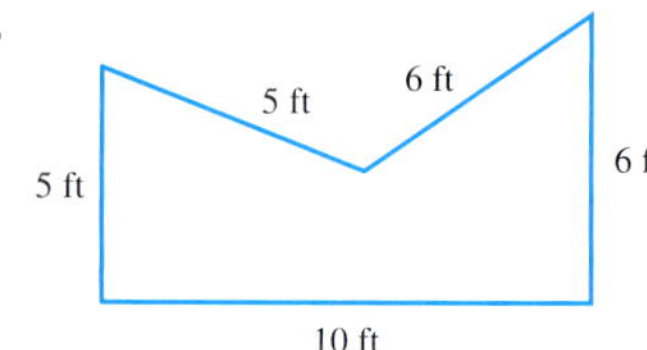

25.

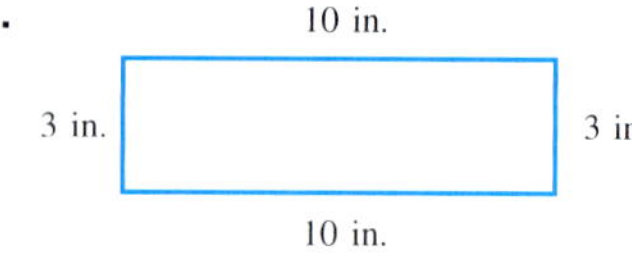

26.

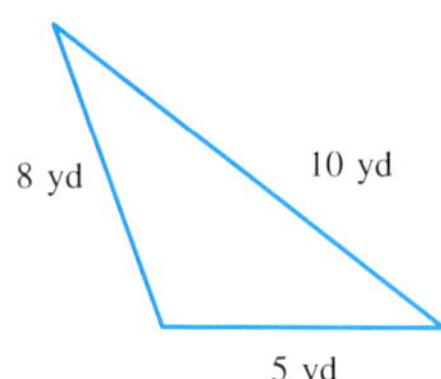

27.

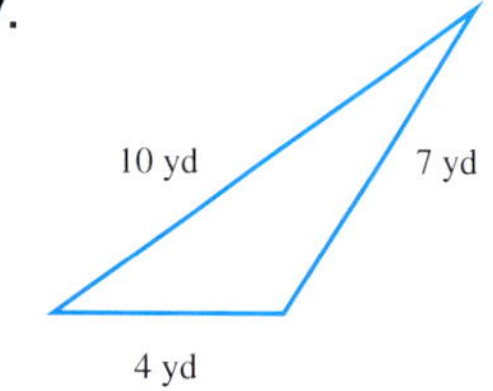

28.

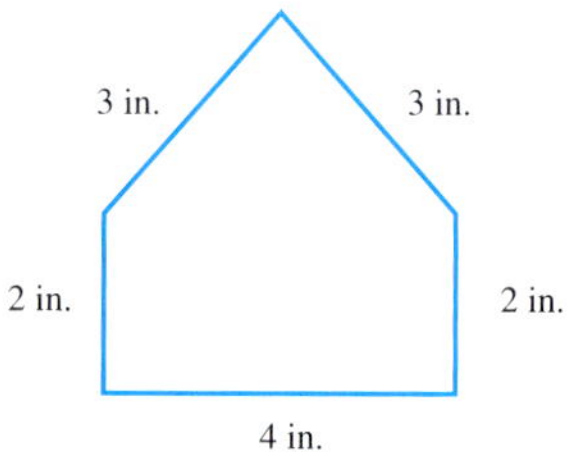

Solve each of the addition applications.

29. **Window size.** A rectangular picture window is 4 ft by 5 ft. Meg wants to put a trim molding around the window. How many feet of molding should she buy?

30. **Fencing material.** You are fencing in a backyard that measures 30 ft by 20 ft. How much fencing should you buy?

31. **Golf.** A golfer shot a score of 42 on the first nine holes and a score of 46 on the second nine holes. What was her total score for the round?

ANSWERS

32. ____________

33. ____________

34. ____________

35. ____________

36. ____________

32. Bowling. A bowler scored 201, 153, and 215 in three games. What was the total score for those games?

33. Shipping. Angelo's vineyard shipped 4200 pounds (lb) of grapes in August, 5970 lb in September, and 4850 lb in October. How many pounds were shipped?

34. Total distance. A salesperson drove 68 miles (mi) on Tuesday, 114 mi on Thursday, and 79 mi on Friday. What was the mileage for those 3 days?

35. Video rentals. The following chart shows Family Video's monthly rentals for the first three months of 2001 by category of film. Complete the totals.

Category of Film	Jan.	Feb.	Mar.	Category Totals
Comedy	4568	3269	2189	______
Drama	5612	4129	3879	______
Action/Adventure	2654	3178	1984	______
Musical	897	623	528	______
Monthly Totals	______	______	______	______

36. Business expenses. The chart shows Regina's Dress Shop's expenses by department for the last 3 months of the year. Complete the totals.

Department	Oct.	Nov.	Dec.	Department Totals
Office	\$31,714	\$32,512	\$30,826	______
Production	85,146	87,479	81,234	______
Sales	34,568	37,612	33,455	______
Warehouse	16,588	11,368	13,567	______
Monthly Totals	______	______	______	______

For exercises 37 to 40 use the given table, which ranks the top 10 areas for women-owned firms in the United States.

Metro Area	Number of Firms	Employment	Sales (in millions)
Los Angeles-Long Beach, Calif.	360,300	1,056,600	\$181,455,900
New York	282,000	1,077,900	193,572,200
Chicago	260,200	1,108,800	161,200,900
Washington, D.C.	193,600	440,000	56,644,000
Philadelphia	144,600	695,900	90,231,000
Atlanta	138,700	331,800	50,206,800
Houston	136,400	560,100	78,180,300
Dallas	123,900	431,900	63,114,900
Detroit	123,600	371,400	50,060,700
Minneapolis-St. Paul, Minn.	119,600	337,400	51,063,400

ANSWERS

37. ______

38. ______

39. ______

40. ______

41. ______

42. ______

43. ______

37. How many firms in total are located in Washington, Philadelphia, and New York?

38. What is the total number of employees in all 10 of the areas listed?

39. What is the total sales for firms in Houston and Dallas?

40. How many firms in total are located in Chicago and Detroit?

41. A magic square is a square in which the sum along any row, column, or diagonal is the same. For example

35	10	15
0	20	40
25	30	5

Use the numbers 1 to 9 to form a magic square.

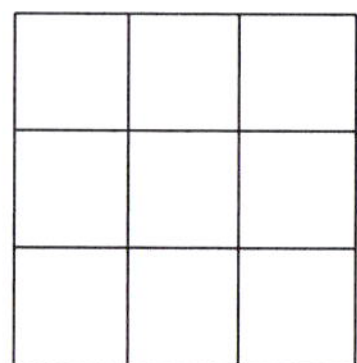

42. This puzzle will give you a chance to practice some of your addition skills.

Across

1. 23 + 22
3. 103 + 42
6. 29 + 58 + 19
8. 3 + 3 + 4
9. 1480 + 1624
11. 568 + 730
13. 25 + 25
14. 131 + 132
16. The total of 121, 146, 119, and 132
17. The perimeter of a 4 ft by 6 ft rug

Down

1. The sum of 224,000, 155, and 186,000
2. 20 + 30
4. 210 + 200
5. 500,000 + 4730
7. 130 + 509
10. 90 + 92
12. 100 + 101
15. The perimeter of a 15 ft by 16 ft room

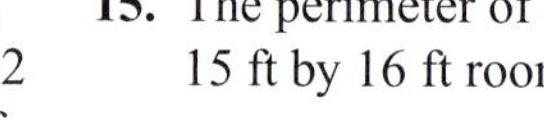

43. Adding of whole numbers is commutative. (The order in which you add does not affect the sum.) Can you think of two actions in your daily routine that are commutative? Explain. List two actions that are *not* commutative in your daily routine and explain.

ANSWERS

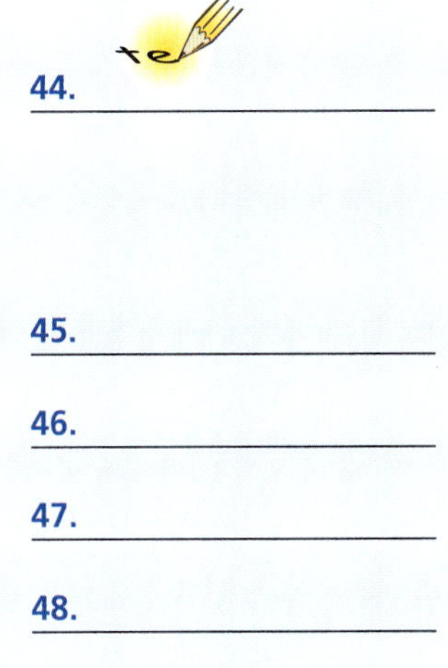

44. ____________

45. ____________

46. ____________

47. ____________

48. ____________

49. ____________

50. ____________

44. Adding of whole numbers is associative. (The way you group whole numbers does not affect the final sum.) If you are following a recipe that lists 10 ingredients that need to be combined, do you think that adding these ingredients is associative? *Be daring!* Find a recipe and combine the ingredients in different orders. Tell the class what happens in each case. (Better yet, bring in the completed product for all to sample.)

45. Complete each equation, using the given property.

(a) Associative property of addition: $5 + (8 + 0) =$ ________

(b) Commutative property of addition: $5 + (8 + 0) =$ ________

(c) Additive identity property: $5 + (8 + 0) =$ ________

Calculator Exercises

46.
$$\begin{array}{r} 458 \\ 273 \\ +\ 568 \\ \hline \end{array}$$

47.
$$\begin{array}{r} 2743 \\ 258 \\ 35 \\ +\ 5823 \\ \hline \end{array}$$

48.
$$\begin{array}{r} 3{,}295{,}153 \\ 573{,}128 \\ 21{,}257 \\ 2{,}586{,}241 \\ +\quad 5291 \\ \hline \end{array}$$

49. $23 + 5638 + 385 + 27{,}345$

50. Poker hands. The following table lists the number of possible types of poker hands. What is the total number possible?

Royal flush	4
Straight flush	36
Four of a kind	624
Full house	3744
Flush	5108
Straight	10,200
Three of a kind	54,912
Two pairs	123,552
One pair	1,098,240
Nothing	1,302,540

Answers

1. 5 is the addend, 4 is the addend, 9 is the sum **3.** Commutative property of addition **5.** Commutative property of addition **7.** 2997 **9.** 7885 **11.** 4063 **13.** 56,081 **15.** 1569 **17.** 3787 **19.** 2475 **21.** 22 ft **23.** 21 yd **25.** 26 in. **27.** 21 yd **29.** 18 ft **31.** 88 **33.** 15,020 lb

35.

Category of Film	Jan.	Feb.	Mar.	Category Totals
Comedy	4568	3269	2189	**10,026**
Drama	5612	4129	3879	**13,620**
Action/Adventure	2654	3178	1984	**7816**
Musical	897	623	528	**2048**
Monthly Totals	**13,731**	**11,199**	**8580**	**33,510**

37. 620,200 **39.** $141,295,200

41.

8	3	4
1	5	9
6	7	2

43.  

45. **(a)** $(5 + 8) + 0$; **(b)** $(8 + 0) + 5$ or $5 + (0 + 8)$; **(c)** $5 + 8$ **47.** 8859 **49.** 33,391

Subtraction of Whole Numbers

OBJECTIVES

1. Use the language of subtraction
2. Subtract whole numbers without borrowing
3. Use borrowing in subtracting whole numbers
4. Solve applications

Overcoming Math Anxiety

Don't procrastinate!

1. Do your math homework while you're still fresh. If you wait until too late at night, your tired mind will have all that much more difficulty understanding the concepts.
2. Do your homework the day it is assigned. The more recent the explanation is, the easier it is to recall.
3. When you've finished your homework, try reading the next section through once. This will give you a sense of direction when you next hear the material. This works whether you are in a lecture or lab setting.

Remember that, in a typical math class, you are expected to do two or three hours of homework for each weekly class hour. This means two or three hours per night. Schedule the time and stay to your schedule.

NOTE By *opposite* we mean that subtracting a number "undoes" an addition of that same number. Start with 1. Add 5 and then subtract 5. Where are you?

We are now ready to consider a second operation of arithmetic—subtraction. In Section 1.2, we described addition as the process of combining two or more groups of the same kind of objects. Subtraction can be thought of as the *opposite operation* to addition. Every arithmetic operation has its own notation. The symbol for subtraction, $-$, is called a **minus sign.**

When we write $8 - 5$, we wish to subtract 5 from 8. We call 5 the **subtrahend.** This is the number being subtracted. And 8 is the **minuend.** This is the number we are subtracting from. The **difference** is the result of the subtraction.

To find the *difference* of two numbers, we will assume that we wish to subtract the smaller number from the larger. Then we look for a number which, when added to the smaller number, will give us the larger number. For example,

$$8 - 5 = 3 \qquad \text{because} \qquad 3 + 5 = 8$$

This special relationship between addition and subtraction provides a method of checking subtraction.

Rules and Properties: Relationship Between Addition and Subtraction

The sum of the difference and the subtrahend must be equal to the minuend.

Example 1

Subtracting a Single-Digit Number

$12 - 5 = 7$

Check:

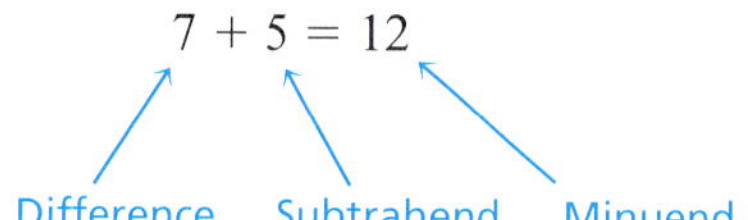

Our check works because 12 − 5 asks for the number that must be added to 5 to get 12.

CHECK YOURSELF 1

Subtract, and check your work.

$13 - 9 =$

The procedure for subtracting larger whole numbers is similar to the procedure for addition. We subtract digits of the same place value.

Example 2

Subtracting a Larger Number

Step 1	Step 2	Step 3
789	789	789
− 246	− 246	− 246
3	43	543

We subtract in the ones column, then in the tens column, and finally in the hundreds column.

To check:

$$\begin{array}{r} 789 \\ -\ 246 \\ \hline 543 \end{array}$$

Add 543 + 246 = 789

The sum of the difference and the subtrahend must be the minuend.

CHECK YOURSELF 2

Subtract, and check your work.

$$\begin{array}{r} 3468 \\ -\ 2248 \\ \hline \end{array}$$

You know that the word *difference* indicates subtraction. There are other words that also tell you to use the subtraction operation. For instance, 5 *less than* 12 is written as

12 − 5 or 7

20 *decreased* by 8 is written as

20 − 8 or 12

Example 3

Translating Words That Indicate Subtraction

Find each number.

(a) 4 less than 11

4 less than 11 is written 11 − 4, and 11 − 4 = 7.

(b) 27 decreased by 6

27 decreased by 6 is written 27 − 6, and 27 − 6 = 21.

CHECK YOURSELF 3

Find each number.

(a) 6 less than 19 **(b)** 18 decreased by 3

Difficulties can arise in subtraction if one or more of the digits of the subtrahend are larger than the corresponding digits in the minuend. We will solve this problem by using a process called **borrowing.**

First, we'll look at an example in expanded form.

Example 4

Subtracting When Borrowing Is Needed

$$\begin{array}{r} 52 = 50 + 2 \\ -\ 27 = 20 + 7 \\ \hline \end{array}$$

Do you see that we cannot subtract in the ones column?

Regrouping, we borrow 1 ten in the minuend and write that ten as 10 ones:

$$\begin{array}{ll} & \overbrace{\quad 50 \quad}^{} + 2 \\ \text{becomes} & 40 + \underbrace{10 + 2}_{} \\ \text{or} & 40 + \quad 12 \end{array}$$

We now have

$$\begin{array}{r} 52 = 40 + 12 \\ -\ 27 = 20 + \ \ 7 \\ \hline 20 + \ \ 5 \end{array}$$

or 25

We can now subtract as before.

In practice, we will use a more convenient short form for the subtraction.

$$\begin{array}{r} 52 \\ -\ 27 \\ \hline \end{array} \qquad \begin{array}{r} \overset{4}{\not{5}}\overset{1}{2} \\ -\ 27 \\ \hline 25 \end{array}$$

We indicate the fact that we have borrowed 1 ten by putting a slash through the 5 and then writing 4 tens. Add 10 ones to the original 2 ones to get 12 ones. We can then subtract.

Check: 25 + 27 = 52

CHECK YOURSELF 4

Subtract, and check your work.

$$\begin{array}{r} 64 \\ -\ 38 \\ \hline \end{array}$$

We will work through another subtraction example that will require a number of borrowing steps. Here, zero appears as a digit in the minuend.

Example 5

Subtracting When Borrowing Is Needed

Step 1

$$\begin{array}{r} 4\,\overset{4}{\cancel{5}}\,{}^{1}3 \\ 40\cancel{5}3 \\ -\ 2365 \\ \hline 8 \end{array}$$

In this first step we borrow 1 ten. This is written as 10 ones and combined with the original 3 ones. We can then subtract in the ones column.

NOTE Here we borrow 1 thousand; this is written as 10 hundreds.

Step 2

$$\begin{array}{r} {}^{3}\,{}^{10}\,{}^{4}\,{}^{1} \\ \cancel{4}\cancel{0}\cancel{5}3 \\ -\ 2365 \\ \hline 8 \end{array}$$

We must borrow again to subtract in the tens column. There are no hundreds, and so we move to the thousands column.

NOTE We now borrow 1 hundred; this is written as 10 tens and combined with the remaining 4 tens.

Step 3

$$\begin{array}{r} {}^{3}\,{}^{9}\,{}^{1}{}^{4}\,{}^{1} \\ \cancel{4}\cancel{0}\cancel{5}3 \\ -\ 2365 \\ \hline 8 \end{array}$$

The minuend is now renamed as 3 thousands, 9 hundreds, 14 tens, and 13 ones.

Step 4

$$\begin{array}{r} {}^{9}\,{}^{14} \\ {}^{3}\cancel{10}\cancel{4}\,{}^{1} \\ \cancel{4}\cancel{0}\cancel{5}3 \\ -\ 2365 \\ \hline 1688 \end{array}$$

The subtraction can now be completed.

To check our subtraction: 1688 + 2365 = 4053

CHECK YOURSELF 5

Subtract, and check your work.

$$\begin{array}{r} 5024 \\ -\ 1656 \\ \hline \end{array}$$

Units Analysis

This is the first in a series of essays that are designed to help you solve applications of mathematics. Questions in the exercise sets will require the skills that you build by reading these essays.

A number with a unit attached (like 7 **ft** or 26 **miles per gallon**) is called a **denominate** number. Any genuine application of mathematics will involve denominate numbers.

When adding or subtracting denominate numbers, the units must be identical for both numbers. The sum or difference will have those same units.

Examples:

\$4 + \$9 = \$13 — Notice that, although we write the dollar sign first, we read it after the quantity, as in "four dollars."

7 ft + 9 ft = 16 ft

39 degrees − 12 degrees = 27 degrees

7 ft + 12 degrees yields no meaningful answer!

3 ft + 9 in. yields a meaningful result only if the 3 ft is converted into 36 in. We will discuss conversion of units in later essays.

Now we consider subtraction word problems.

Step by Step: Solving Subtraction Applications

Step 1 Read the problem carefully to determine the given information and what you are asked to find.

Step 2 Decide upon the operation (in this case, subtraction) to be used.

Step 3 Write down the complete statement necessary to solve the problem and do the calculations.

Step 4 Check to make sure you have answered the question of the problem and that your answer seems reasonable.

You will need to use both addition and subtraction to solve some problems.

Example 6

Solving a Subtraction Application

Bernard wants to buy a new piece of stereo equipment. He has \$142 and can trade in his old amplifier for \$135. How much more does he need if the new equipment costs \$449?

First we must add to find out how much money Bernard has available. Then we subtract to find out how much more money he needs.

\$142 + \$135 = \$277 — The money available to Bernard

\$449 − \$277 = \$172 — The money Bernard still needs

CHECK YOURSELF 6

Martina spent \$239 in airfare, \$174 for lodging, and \$108 for food on a business trip. Her company allowed her \$375 for the expenses. How much of these expenses will she have to pay herself?

Using a Scientific Calculator to Subtract

Now that you have reviewed the process of subtracting by hand, we can look at the use of the calculator in performing that operation. Remember, the point is to brush up on your arithmetic skills by hand, *along with* learning to use the calculator in a variety of situations.

The calculator can be very helpful in a problem that involves both addition and subtraction operations.

Example 7

Using a Scientific Calculator to Subtract

NOTE As with addition, the same steps are used on a graphing calculator, except we press [ENTER] rather than [=].

Find

23 − 13 + 56 − 29

Enter the numbers and the operation signs exactly as they appear in the expression.

23 [−] 13 [+] 56 [−] 29 [=]

Display 37

An alternative approach would be to add 23 and 56 first, then subtract 13, and finally subtract 29. The result is the same in either case.

CHECK YOURSELF 7

Find.

58 − 12 + 93 − 67

CHECK YOURSELF ANSWERS

1. 13 − 9 = 4 Check: 4 + 9 = 13 **2.** 1220 **3.** **(a)** 13; **(b)** 15

4.
```
   5 1
   6 4
 − 3 8
 -----
   2 6
```
To check: 26 + 38 = 64

5. 3368 Check: 3368 + 1656 = 5024

6.
```
   $239
    174
 +  108
 ------
   $521   ← Total expenses
```
```
   $521   ← Total expenses
 −  375   ← Amount allowed
 ------
   $146
```

7. 58 [−] 12 [+] 93 [−] 67 [=] 72

1.3 Exercises

Name ______ Section ______ Date ______

1. In the statement $9 - 6 = 3$
   9 is called the
   6 is called the
   3 is called the
   Write the related addition statement.

2. In the statement $7 - 5 = 2$
   5 is called the
   2 is called the
   7 is called the
   Write the related addition statement.

In exercises 3 to 26, do the indicated subtraction, and check your results by addition.

3. $\begin{array}{r} 347 \\ -\ 201 \\ \hline \end{array}$

4. $\begin{array}{r} 575 \\ -\ 302 \\ \hline \end{array}$

5. $\begin{array}{r} 689 \\ -\ 245 \\ \hline \end{array}$

6. $\begin{array}{r} 598 \\ -\ 278 \\ \hline \end{array}$

7. $\begin{array}{r} 3446 \\ -\ 2326 \\ \hline \end{array}$

8. $\begin{array}{r} 5896 \\ -\ 3862 \\ \hline \end{array}$

9. $\begin{array}{r} 64 \\ -\ 27 \\ \hline \end{array}$

10. $\begin{array}{r} 73 \\ -\ 36 \\ \hline \end{array}$

11. $\begin{array}{r} 627 \\ -\ 358 \\ \hline \end{array}$

12. $\begin{array}{r} 642 \\ -\ 367 \\ \hline \end{array}$

13. $\begin{array}{r} 6423 \\ -\ 3678 \\ \hline \end{array}$

14. $\begin{array}{r} 5352 \\ -\ 2577 \\ \hline \end{array}$

15. $\begin{array}{r} 6034 \\ -\ 2569 \\ \hline \end{array}$

16. $\begin{array}{r} 5206 \\ -\ 1748 \\ \hline \end{array}$

17. $\begin{array}{r} 4000 \\ -\ 2345 \\ \hline \end{array}$

18. $\begin{array}{r} 6000 \\ -\ 4349 \\ \hline \end{array}$

19. $\begin{array}{r} 33{,}486 \\ -\ 14{,}047 \\ \hline \end{array}$

20. $\begin{array}{r} 53{,}487 \\ -\ 25{,}649 \\ \hline \end{array}$

21. $\begin{array}{r} 29{,}400 \\ -\ 17{,}900 \\ \hline \end{array}$

22. $\begin{array}{r} 53{,}500 \\ -\ 28{,}700 \\ \hline \end{array}$

23. $\begin{array}{r} 59{,}000 \\ -\ 23{,}458 \\ \hline \end{array}$

24. $\begin{array}{r} 41{,}000 \\ -\ 27{,}645 \\ \hline \end{array}$

25. $\begin{array}{r} 3537 \\ -\ 2675 \\ \hline \end{array}$

26. $\begin{array}{r} 4693 \\ -\ 2736 \\ \hline \end{array}$

27. Find the number that is 25 less than 76.

28. Find the number that results when 58 is decreased by 23.

29. Find the number that is the difference between 97 and 43.

30. Find the number that is 125 less than 265.

31. Find the number that results when 298 is decreased by 47.

32. Find the number that is the difference between 167 and 57.

ANSWERS

1. ______
2. ______
3. ______
4. ______
5. ______
6. ______
7. ______
8. ______
9. ______
10. ______
11. ______
12. ______
13. ______
14. ______ 15. ______
16. ______ 17. ______
18. ______ 19. ______
20. ______ 21. ______
22. ______ 23. ______
24. ______ 25. ______
26. ______ 27. ______
28. ______ 29. ______
30. ______ 31. ______
32. ______

ANSWERS

33. ______

34. ______

35. ______

36. ______

37. ______

38. ______

39. ______

40. ______

41. ______

42. ______

43. ______

44. ______

45. ______

46. ______

47. ______

48. ______

49. ______

Based on units, determine whether each of the operations in exercises 33 to 38 produces a meaningful result. (°F = degrees Fahrenheit; °C = degrees Celsius; yd = yard; mi/h = miles per hour; ft/s = feet per second).

33. 8 mi − 4 mi

34. \$560 + \$314

35. 7 ft + 11 in.

36. 18°F − 6°C

37. 17 yd − 10 yd

38. 4 mi/h + 6 ft/s

In exercises 39 to 42, for various treks by a hiker in a mountainous region, the starting elevations and various changes are given. Determine the final elevation of the hiker in each case.

39. Starting elevation 1053 ft, increase of 123 ft, decrease of 98 ft, increase of 63 ft.

40. Starting elevation 1231 ft, increase of 213 ft, decrease of 112 ft, increase of 78 ft.

41. Starting elevation 7302 ft, decrease of 623 ft, decrease of 123 ft, increase of 307 ft.

42. Starting elevation 6907 ft, decrease of 511 ft, decrease of 203 ft, increase of 419 ft.

Solve the applications in exercises 43 to 59.

43. Test scores. Shaka's score on a math test was 87 and Tony's score was 23 points less than Shaka's. What was Tony's score on the test?

44. New pay. Duardo's monthly pay of \$879 was decreased by \$175 for withholding. What amount of pay did he receive?

45. Number problem. The difference between two numbers is 134. If the larger number is 655, what is the smaller number?

46. Family budget. In Jason's monthly budget, he set aside \$375 for housing, and \$165 less than that for food. How much did he budget for food?

47. Consumer purchases. Inez has \$228 in cash and wants to buy a television set that costs \$449. How much more money does she need?

48. Construction. The Sears Tower in Chicago is 1454 ft tall. The Empire State Building is 1250 ft tall. How much taller is the Sears Tower than the Empire State Building?

49. Education. A college's enrollment was 2479 students in the fall of 2001 and 2653 students in the fall of 2002. What was the increase in enrollment?

ANSWERS

50. __________
51. __________
52. __________
53. __________
54. __________
55. __________
56. __________
57. __________

50. Net pay. In one week, Margaret earned $278 in regular pay and $53 for overtime work, and $49 was deducted from her paycheck for income taxes and $18 for social security. What was her take-home pay?

51. Savings. Rafael opened a checking account and made deposits of $85 and $272. He wrote checks during the month for $35, $27, $89, and $178. What was his balance at the end of the month?

52. Dieting. Dalila is trying to limit herself to 1500 calories per day (cal/day). Her breakfast was 270 cal, her lunch was 450 cal, and her dinner was 820 cal. By how much was she *under* or *over* her diet?

53. Recreation. A professional basketball team scored 98, 136, and 113 points in three games. If its opponents scored 102, 109, and 93 points, by how much did the team outscore its opponents?

54. Checking account balance. To keep track of a checking account, you must subtract the amount of each check from the current balance. Complete the given statement.

Beginning balance	$351
Check #1	29
Balance	
Check #2	139
Balance	
Check #3	75
Ending balance	

55. Expense accounts. Complete the given record of a monthly expense account.

Monthly income	$1620
House payment	343
Balance	
Car payment	183
Balance	
Food	312
Balance	
Clothing	89
Amount remaining	

56. Education. A course outline states that you must have 540 points on five tests during the term to receive an A for the course. Your scores on the first four tests have been 95, 84, 82, and 89. How many points must you score on the 200-point final to receive an A?

57. Travel. Carmen's frequent-flyer program requires 30,000 mi for a free flight. During 2001 she accumulated 13,850 mi. In 2002 she took three more flights of 2800, 1475, and 4280 mi. How much further must she fly for her free trip?

58. ______

59. ______

60. ______

61. ______

62. ______

63. ______

64. ______

58. Budget. Peter, Paul, and Mary all submitted advertising budgets for a student government dance.

Ad Medium	Peter	Paul	Mary
Radio ads	\$500	\$600	\$300
Newspaper ads	\$150	\$200	\$150
Posters	\$225	\$250	\$275
Handbills	\$175	\$150	\$250

If \$900 is available for advertising, how much over budget would each student be?

59. Farming. The value of all crops in the Salinas Valley in 2001 was about \$2 billion. The top four crops are listed below. **(a)** How much greater is the combined value of both types of lettuce than broccoli? **(b)** How much greater is the value of the lettuce and broccoli combined than the strawberries?

Crop	Crop value, in millions
Head lettuce	\$360
Broccoli	\$246
Leaf lettuce	\$210
Strawberries	\$198

Complete the magic squares.

60.

	7	2
	5	
8		

61.

4	3	
	5	
		6

62.

16	3		13
	10	11	
9	6	7	
			1

63.

7			14
2	13	8	11
16			
	6	15	

64. Efrain has lost track of his checking account transactions. He knows he started with \$50 and has deposited \$120, \$85, and \$120. He also knows he has withdrawn \$200 and \$55. He just can't remember the order in which he did all this.

(a) What is Efrain's balance after all these transactions?

(b) Does the order of the transactions make any difference from the math point of view?

(c) Does the order of transactions make any difference from the banking point of view?

Explain your answers.

65. Using the World Wide Web, determine the population of Arizona, California, Oregon, and Pennsylvania in each of the last three censuses.

(a) Find the total change in each state's population over this period.

(b) Which state shows the most change over the past three censuses?

(c) Write a brief essay describing the changes and any trends you see in this data. List any implications that they might have for future planning.

66. Describe in words each of the given equations. (Make sure you use a complete sentence.) Then exchange your sentence with other students and see if their interpretations result in the same equation you used.

(a) $69 - 23 = 46$ **(b)** $17 + 13 = 30$

67. Evaluate the given expressions:

(a) $8 - (4 - 2)$ **(b)** $(8 - 4) - 2$

Do you obtain the same answer? What conclusion can you draw about subtraction and an associative property?

68. Think of any whole number.

Add 5.
Subtract 3.
Subtract two less than the original number.
What number do you end up with?
Check with other people. Does everyone have the same answer? Can you explain the results?

Calculator Exercises

Do the indicated operations.

69. $\begin{array}{r} 89 \\ -\ 48 \\ \hline \end{array}$

70. $\begin{array}{r} 576 \\ -\ 389 \\ \hline \end{array}$

71. $\begin{array}{r} 5830 \\ -\ 3987 \\ \hline \end{array}$

72. $\begin{array}{r} 15{,}280 \\ -\ \ 7595 \\ \hline \end{array}$

73. $193{,}243 - 49{,}285$

74. $257{,}500 - 78{,}750$

75. Subtract 235 from the sum of 534 and 678.

76. Subtract 476 from the sum of 306 and 572.

ANSWERS

65. ______

66. ______

67. ______

68. ______

69. ______

70. ______

71. ______

72. ______

73. ______

74. ______

75. ______

76. ______

77. ______________

78. ______________

79. ______________

For exercises 77 to 79, solve the applications.

77. Gas sales. Readings from the Fast Service Station's storage tanks were taken at the beginning and the end of a month. How much of each type of gas was sold? What was the total sold?

	Regular	Unleaded	Super Unleaded	Total
Beginning reading	73,255	82,349	81,258	
End reading	28,387	19,653	8654	
Gallons used	______	______	______	______

The land areas, in square miles (mi^2), of three Pacific coast states are California, 158,693 mi^2; Oregon, 96,981 mi^2; Washington, 68,192 mi^2.

78. How much larger is California than Oregon?

79. How much larger is California than Washington?

Answers

1. 9 is the minuend, 6 is the subtrahend, and 3 is the difference. 3 + 6 = 9 **3.** 146 **5.** 444 **7.** 1120 **9.** 37 **11.** 269 **13.** 2745 **15.** 3465 **17.** 1655 **19.** 19,439 **21.** 11,500 **23.** 35,542 **25.** 862 **27.** 51 **29.** 54 **31.** 251 **33.** Yes **35.** No **37.** Yes **39.** 1141 ft **41.** 6863 ft **43.** 64 **45.** 521 **47.** \$221 **49.** 174 students **51.** \$28 **53.** 43 points **55.** Balance: 1277; balance: 1094; balance: 782; amount remaining: 693 **57.** 7595 mi **59.** **(a)** \$324,000,000; **(b)** \$618,000,000

61.

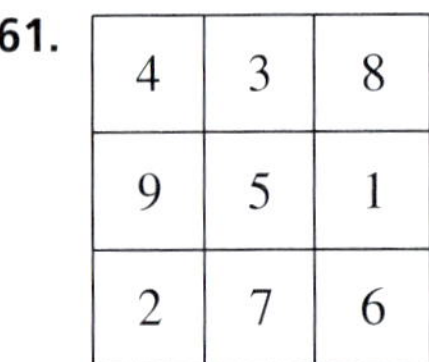

4	3	8
9	5	1
2	7	6

63.

7	12	1	14
2	13	8	11
16	3	10	5
9	6	15	4

65. 

67. 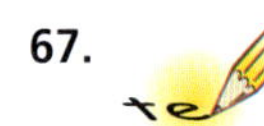 **69.** 41 **71.** 1843 **73.** 143,958 **75.** 977

77. Regular: 44,868 gal; unleaded: 62,696 gal; super unleaded: 72,604 gal; total: 180,168 gal
79. 90,501 mi^2

Rounding, Estimation, and Ordering of Whole Numbers

OBJECTIVES

1. Round a whole number at any place value
2. Estimate sums by rounding
3. Use the symbols < and >

It is a common practice to express numbers to the nearest hundred, thousand, and so on. For instance, the distance from Los Angeles to New York along one route is 2833 mi. We might say that the distance is 2800 mi. This is called **rounding,** because we have rounded the distance to the nearest hundred miles.

One way to picture this rounding process is with the use of a number line.

Example 1

Rounding to the Nearest Hundred

To round 2833 to the nearest hundred:

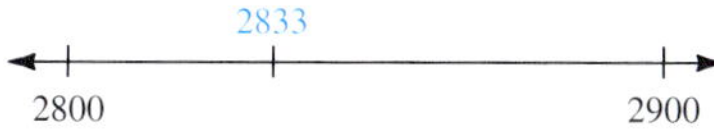

Because 2833 is closer to 2800, we round *down* to 2800.

CHECK YOURSELF 1

Round 587 *to the nearest hundred.*

Example 2

Rounding to the Nearest Thousand

To round 28,734 to the nearest thousand:

Because 28,734 is closer to 29,000, we round *up* to 29,000.

CHECK YOURSELF 2

Locate 1375 *and round to the nearest hundred.*

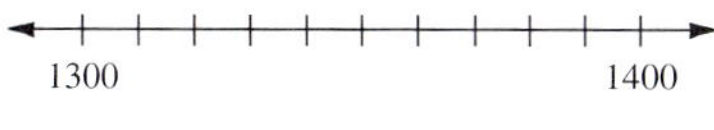

Instead of using a number line, we can apply a rule.

NOTE By a certain *place,* we mean tens, hundreds, thousands, and so on.

NOTE This is called **rounding up.**

NOTE This is called **rounding down.**

Step by Step: Rounding Whole Numbers

Step 1 Identify the place of the digit to be rounded.

Step 2 Look at the digit to the right of that place.

Step 3 **a.** If that digit is 5 or more, that digit and all digits to the right become 0. The digit in the place you are rounding to is increased by 1.

b. If that digit is less than 5, that digit and all digits to the right become 0. The digit in the place you are rounding to remains the same.

Example 3

Rounding to the Nearest Ten

Round 587 to the nearest ten:

Tens
↓
5 8 7
↑
The digit to the right of the tens place
↓
5 8 7 is rounded to 590

580 590
587

We identify the tens digit. The digit to the right of the tens place, 7, is 5 or more. So round up.

NOTE 587 is between 580 and 590. It is closer to 590, so it makes sense to round up.

CHECK YOURSELF 3

Round 847 to the nearest ten.

Example 4

Rounding to the Nearest Hundred

Round 2638 to the nearest hundred:

↓ ↓
2 6 3 8 is rounded to 2600

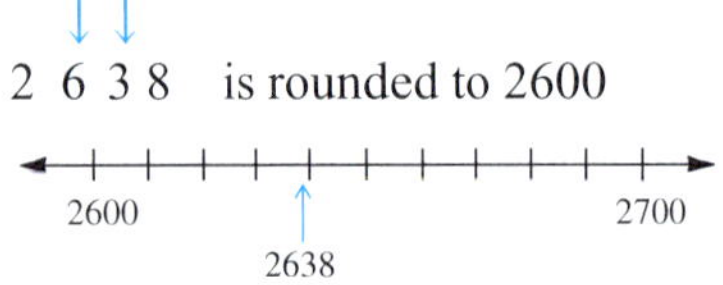

We identify the hundreds digit, 6. The digit to the right, 3, is less than 5. So round down.

NOTE 2638 is closer to 2600 than to 2700. So it makes sense to round down.

CHECK YOURSELF 4

Round 3482 to the nearest hundred.

We will look at some further examples of using the rounding rule.

Example 5

Rounding Whole Numbers

(a) Round 2378 to the nearest hundred:

↓ ↓
2 3 78 is rounded to 2400

We have identified the hundreds digit, 3. The digit to the right is 7. Because this is 5 or more, the 7 and all digits to the right become 0. The hundreds digit is increased by 1.

(b) Round 53,258 to the nearest thousand:

↓ ↓
5 3 ,258 is rounded to 53,000

We have identified the thousands digit, 3. Because the digit to the right is less than 5, it and all digits to the right become 0, and the thousands digit remains the same.

(c) Round 685 to the nearest ten:

↓ ↓
6 8 5 is rounded to 690

The digit to the right of the tens place is 5 or more. Round up by our rule.

(d) Round 52,813,212 to the nearest million:

↓ ↓
5 2 ,813,212 is rounded to 53,000,000

CHECK YOURSELF 5

(a) Round 568 to the nearest ten.
(b) Round 5446 to the nearest hundred.

In Example 6, we look at a case in which we round up a 9.

Example 6

Rounding to the Nearest Ten

Suppose we want to round 397 to the nearest ten. We identify the tens digit, which is 9, and look at the next digit to the right.

NOTE Which number is 397 closer to?

390 397 400

3 9 7

The digit to the right is 5 or more.
If this digit is 9, and it must be increased by 1, replace the 9 with 0 and increase the next digit to the *left* by 1.

So 397 is rounded to 400.

CHECK YOURSELF 6

Round 4961 to the nearest hundred.

NOTE An estimate is basically a good guess. If your answer is close to your estimate, then your answer is reasonable.

Whether you are doing an addition problem by hand or using a calculator, rounding numbers gives you a handy way of deciding if the answer seems reasonable. The process is called **estimating.** We will illustrate with an example.

Example 7

Estimating a Sum

NOTE Placing an arrow above the column to be rounded can be helpful.

Begin by rounding to the nearest hundred

456	500
235	200
976	1000
+ 344	+ 300
2011	2000 ← Estimate

By rounding to the nearest hundred and adding quickly, we get an estimate or guess of 2000. Because this is close to the actual sum calculated, 2011, our estimate seems reasonable.

CHECK YOURSELF 7

Round each addend to the nearest hundred and estimate the sum. Then find the actual sum.

287 + 526 + 311 + 378

Estimation is a wonderful tool to use while you're shopping. Every time you go to the store, you should try to estimate the total bill by rounding the price of each item. If you do this regularly, both your addition skills and your rounding skills will improve. The same holds true when you eat in a restaurant. It is always a good idea to know approximately how much you are spending.

Example 8

Estimating a Sum in a Word Problem

Samantha has taken the family out to dinner, and she's now ready to pay the bill. The dinner check has no total, only the individual entries, as below:

Soup	$2.95
Soup	$2.95
Salad	1.95
Salad	1.95
Salad	1.95
Lasagna	7.25
Spaghetti	4.95
Ravioli	5.95

What is the approximate cost of the dinner?

Rounding each entry to the nearest whole dollar, we can estimate the total by finding the sum

3 + 3 + 2 + 2 + 2 + 7 + 5 + 6 = $30

CHECK YOURSELF 8

Jason is doing the weekly food shopping at FoodWay. So far his basket has items that cost $3.99, $7.98, $2.95, $1.15, $2.99, *and* $1.95. *Approximate the total cost of these items.*

In the first four examples of this section, we used the number line to illustrate the idea of rounding numbers. The number line also gives us an excellent way to picture the concept of **order** for whole numbers, which means that numbers become larger as we move from left to right on the line.

For instance, we know that 3 is less than 5. On the number line

NOTE 3 is less than or smaller than 5.

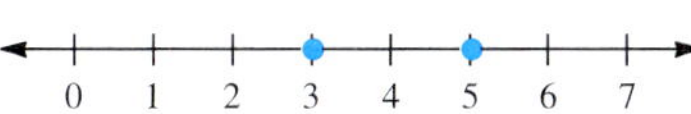

we see that 3 lies *to the left* of 5.

We also know that 4 is greater than 2. On the number line

NOTE 4 is greater than or larger than 2.

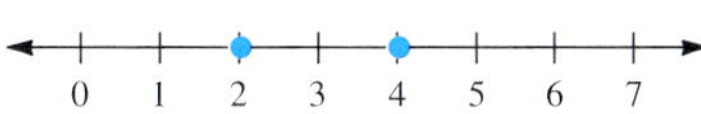

we see that 4 lies *to the right* of 2.

Two symbols, $<$ for "less than" and $>$ for "greater than," are used to indicate these relationships.

Definitions: Inequalities

For whole numbers, we can write

NOTE The inequality always "points at" the smaller number.

1. $2 < 5$ (read "2 is less than 5") because 2 is *to the left* of 5 on the number line.
2. $8 > 3$ (read "8 is greater than 3") because 8 is *to the right* of 3 on the number line.

Example 9 illustrates the use of this notation.

Example 9

Indicating Order with $<$ or $>$

Use the symbols $<$ or $>$ to complete each statement.

(a) 7 _____ 10
(b) 25 _____ 20
(c) 200 _____ 300
(d) 8 _____ 0

(a) $7 < 10$ 7 lies to the left of 10 on the number line.
(b) $25 > 20$ 25 lies to the right of 20 on the number line.
(c) $200 < 300$
(d) $8 > 0$

CHECK YOURSELF 9

Use one of the symbols $<$ or $>$ to complete each of the statements.

(a) 35___25

(b) 0 ___ 4

(c) 12___18

(d) 1000 ___ 100

CHECK YOURSELF ANSWERS

1. 600 **2.** 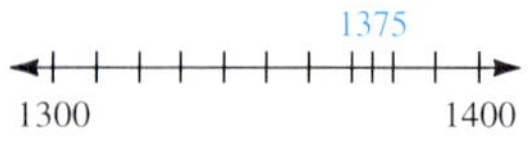

Round 1375 *up* to 1400

3. 850 **4.** 3500 **5.** **(a)** 570; **(b)** 5400 **6.** 5000 **7.** 1500; 1502

8. \$21 **9.** **(a)** $35 > 25$; **(b)** $0 < 4$; **(c)** $12 < 18$; **(d)** $1000 > 100$

Name ______________________

Section ________ Date ________

Exercises

In exercises 1 to 18, round each of the numbers to the indicated place.

1. 38, the nearest ten

2. 72, the nearest ten

3. 253, the nearest ten

4. 578, the nearest ten

5. 696, the nearest ten

6. 683, the nearest hundred

7. 3482, the nearest hundred

8. 6741, the nearest hundred

9. 5962, the nearest hundred

10. 4352, the nearest thousand

11. 4927, the nearest thousand

12. 39,621, the nearest thousand

13. 23,429, the nearest thousand

14. 38,589, the nearest thousand

15. 787,000, the nearest ten thousand

16. 582,000, the nearest hundred thousand

17. 21,800,000, the nearest million

18. 931,000, the nearest ten thousand

In exercises 19 to 40, estimate each of the sums or differences by rounding to the indicated place. Then do the addition or subtraction and use your estimate to see if your actual sum or difference seems reasonable.

Round to the nearest ten.

19.
$$\begin{array}{r} 58 \\ 27 \\ +\ 33 \\ \hline \end{array}$$

20.
$$\begin{array}{r} 92 \\ 37 \\ 85 \\ +\ 64 \\ \hline \end{array}$$

21.
$$\begin{array}{r} 87 \\ 53 \\ 41 \\ 93 \\ +\ 62 \\ \hline \end{array}$$

22.
$$\begin{array}{r} 78 \\ 67 \\ 53 \\ 42 \\ +\ 86 \\ \hline \end{array}$$

ANSWERS

1. ______
2. ______
3. ______
4. ______
5. ______
6. ______
7. ______
8. ______
9. ______
10. ______
11. ______
12. ______
13. ______
14. ______
15. ______
16. ______
17. ______
18. ______
19. ______
20. ______
21. ______
22. ______

ANSWERS

23. ______
24. ______
25. ______
26. ______
27. ______
28. ______
29. ______
30. ______
31. ______
32. ______
33. ______
34. ______
35. ______
36. ______
37. ______
38. ______
39. ______
40. ______

23. $\begin{array}{r} 83 \\ -\ 27 \\ \hline \end{array}$

24. $\begin{array}{r} 97 \\ -\ 31 \\ \hline \end{array}$

25. $\begin{array}{r} 33 \\ -\ 21 \\ \hline \end{array}$

26. $\begin{array}{r} 47 \\ -\ 36 \\ \hline \end{array}$

Round to the nearest hundred.

27. $\begin{array}{r} 379 \\ 1215 \\ +\ 528 \\ \hline \end{array}$

28. $\begin{array}{r} 967 \\ 2365 \\ 544 \\ +\ 738 \\ \hline \end{array}$

29. $\begin{array}{r} 1378 \\ 519 \\ 792 \\ +\ 2041 \\ \hline \end{array}$

30. $\begin{array}{r} 3145 \\ 889 \\ 259 \\ 692 \\ +\ 2518 \\ \hline \end{array}$

31. $\begin{array}{r} 679 \\ -\ 231 \\ \hline \end{array}$

32. $\begin{array}{r} 824 \\ -\ 358 \\ \hline \end{array}$

33. $\begin{array}{r} 915 \\ -\ 411 \\ \hline \end{array}$

34. $\begin{array}{r} 697 \\ -\ 539 \\ \hline \end{array}$

Round to the nearest thousand.

35. $\begin{array}{r} 2238 \\ 3925 \\ +\ 5217 \\ \hline \end{array}$

36. $\begin{array}{r} 3678 \\ 4215 \\ +\ 2032 \\ \hline \end{array}$

37. $\begin{array}{r} 9137 \\ 2315 \\ 7643 \\ +\ 3092 \\ \hline \end{array}$

38. $\begin{array}{r} 11{,}548 \\ 3874 \\ 14{,}435 \\ +\ 5398 \\ \hline \end{array}$

39. $\begin{array}{r} 4822 \\ -\ 2134 \\ \hline \end{array}$

40. $\begin{array}{r} 6120 \\ -\ 4890 \\ \hline \end{array}$

ANSWERS

41. ______
42. ______
43. ______
44. ______
45. ______
46. ______
47. ______
48. ______
49. ______
50. ______
51. ______
52. ______
53. ______
54. ______

Use the symbol $<$ or $>$ to complete each statement.

41. 4 _____ 8

42. 0 _____ 5

43. 500 _____ 400

44. 20 _____ 15

45. 100 _____ 1000

46. 3000 _____ 2000

In exercises 47 to 56, solve the applications.

47. Lunch bills. Ed and Sharon go to lunch. The lunch check has no total but only lists individual items:

Soup $1.95	Soup $1.95
Salad $1.80	Salad $1.80
Salmon $8.95	Flounder $6.95
Pecan pie $3.25	Vanilla ice cream $2.25

Estimate the total amount of the lunch check.

48. Consumer spending. Olivia will purchase several items at the stationery store. Thus far, the items she has collected cost $2.99, $6.97, $3.90, $2.15, $9.95, and $1.10. Approximate the total cost of these items.

49. Test scores. Oscar scored 78, 91, 79, 67, and 100 on his arithmetic tests. Round each score to the nearest ten to estimate his total score.

50. Pizza production. Luigi's pizza parlor makes 293 pizzas on an average day. Estimate (to the nearest hundred) how many pizzas were made on a 3-day holiday weekend.

51. Numeration. A whole number rounded to the nearest ten is 60. **(a)** What is the smallest possible number? **(b)** What is the largest possible number?

52. Numeration. A whole number rounded to the nearest hundred is 7700. **(a)** What is the smallest possible number? **(b)** What is the largest possible number?

53. Numeration. A whole number rounded to the nearest thousand is 5000. **(a)** What is the smallest possible number? **(b)** What is the largest possible number?

54. Consumer spending. Amir bought several items at the hardware store: hammer, $8.95; screwdriver, $3.15; pliers, $6.90; wire cutters, $4.25; and sandpaper; $1.89. Estimate the total cost of Amir's bill.

ANSWERS

55. __________

56. __________

57. __________

58. __________

55. **Clothes shopping.** Mrs. Gonzalez went shopping for clothes. She bought a sweater for $32.95, a scarf for $9.99, boots for $68.29, a coat for $125.90, and socks for $18.15. Estimate the total amount of Mrs. Gonzalez's purchases.

56. **Food shopping.** Maritza went to the local supermarket and purchased the following items: milk, $2.89; butter, $1.75; bread, $1.10; orange juice, $1.25; cereal, $3.95; and coffee, $3.80. Approximate the total cost of these items.

57. A bag contains 60 marbles. The number of blue marbles, rounded to the nearest 10, is 40, and the number of green marbles in the bag, rounded to the nearest 10, is 20. How many blue marbles are in the bag? (List all answers that satisfy the conditions of the problem.)

58. Describe some situations in which estimating and rounding would not produce a result that would be suitable or acceptable. Review the instructions for filing your federal income tax. What rounding rules are used in the preparation of your tax returns? Do the same rules apply to the filing of your state tax returns? If not, what are these rules?

Answers

1. 40 **3.** 250 **5.** 700 **7.** 3500 **9.** 6000 **11.** 5000
13. 23,000 **15.** 790,000 **17.** 22,000,000
19. Estimate: 120, actual sum: 118 **21.** Estimate: 330, actual sum: 336
23. Estimate: 50, actual difference: 56 **25.** Estimate: 10, actual difference: 12
27. Estimate: 2100, actual sum: 2122 **29.** Estimate: 4700, actual sum: 4730
31. Estimate: 500, actual difference: 448 **33.** Estimate: 500; actual difference: 504
35. Estimate: 11,000, actual sum: 11,380 **37.** Estimate: 22,000, actual sum: 22,187
39. Estimate: 3000, actual difference: 2688 **41.** $<$ **43.** $>$ **45.** $<$
47. $29 **49.** 420 **51.** **(a)** 55; **(b)** 64 **53.** **(a)** 4500; **(b)** 5499
55. $255 **57.** 36, 37, 38, 39, 40, 41, 42, 43, 44

1.5 Multiplication of Whole Numbers

1.5 OBJECTIVES

1. Use the language of multiplication
2. Multiply whole numbers
3. Estimate products
4. Identify the properties of multiplication
5. Solve applications involving multiplication
6. Find the area of a rectangular figure
7. Apply area formulas

NOTE The use of the symbol × dates back to the 1600s.

Our work in this section deals with multiplication, another of the basic operations of arithmetic. Multiplication is closely related to addition. In fact, we can think of multiplication as a shorthand method for repeated addition. The symbol × is used to indicate multiplication.

3×4 can be interpreted as 3 rows of 4 objects. By counting we see that $3 \times 4 = 12$. Similarly, 4 rows of 3 means $4 \times 3 = 12$.

	4			
3	★	★	★	★
	★	★	★	★
	★	★	★	★

	3		
4	★	★	★
	★	★	★
	★	★	★
	★	★	★

The fact that $3 \times 4 = 4 \times 3$ is an example of the **commutative property of multiplication.**

Rules and Properties: The Commutative Property of Multiplication

Given any two numbers, we can multiply them in either order and we get the same result.

In symbols, we say $a \cdot b = b \cdot a$

NOTE The centered dot is the same as the times sign (×). We use the centered dot so the times sign will not be confused with the letter *x*.

Example 1

Multiplying Single-Digit Numbers

$3 \cdot 5$ means 5 multiplied by 3. It is read 3 *times* 5. To find $3 \cdot 5$, we can add 5 three times.

$3 \cdot 5 = 5 + 5 + 5 = 15$

In a multiplication problem such as $3 \cdot 5 = 15$, we call 3 and 5 the **factors.** The answer, 15, is the **product** of the factors, 3 and 5.

$3 \cdot 5 = 15$

Factor (3), Factor (5), Product (15)

CHECK YOURSELF 1

Name the factors and the product in the statement.

$2 \cdot 9 = 18$

Statements such as $3 \cdot 4 = 12$ and $3 \cdot 5 = 15$ are called the **basic multiplication facts.** If you have difficulty with multiplication, it may be that you do not know some of these facts. This table will help you review before you go on. Notice that, because of the commutative property, you need memorize only half of these facts!

NOTE To use the table to find the product of $7 \cdot 6$: Find the row labeled 7, then move to the right in this row until you are in the column labeled 6 at the top. We see that $7 \cdot 6$ is 42.

Basic Multiplication Facts Table

·	0	1	2	3	4	5	6	7	8	9
0	0	0	0	0	0	0	0	0	0	0
1	0	1	2	3	4	5	6	7	8	9
2	0	2	4	6	8	10	12	14	16	18
3	0	3	6	9	12	15	18	21	24	27
4	0	4	8	12	16	20	24	28	32	36
5	0	5	10	15	20	25	30	35	40	45
6	0	6	12	18	24	30	36	42	48	54
7	0	7	14	21	28	35	42	49	56	63
8	0	8	16	24	32	40	48	56	64	72
9	0	9	18	27	36	45	54	63	72	81

Armed with these facts, you can become a better, and faster, problem solver. Take a look at Example 2.

Example 2

Multiplying Instead of Counting

Find the total number of squares on the following checkerboard.

NOTE This checkerboard is an example of a rectangular array, a series of rows or columns that form a rectangle. When you see such an arrangement, be prepared to multiply to find the total number of units.

You could find the number of squares by counting them. If you counted one per second, it would take you just over a minute. You could make the job a little easier by simply counting the squares in one row (8), then adding $8 + 8 + 8 + 8 + 8 + 8 + 8 + 8$. Multiplication, which is simply repeated addition, allows you to find the total number of squares by multiplying $8 \cdot 8$. How long that takes depends on how well you know the basic multiplication facts! By now, you know that there are 64 squares on the checkerboard.

CHECK YOURSELF 2

Find the number of windows on the displayed side of the building.

The next property involves *both* multiplication and addition.

Example 3

Using the Distributive Property

$2 \cdot (3 + 4) = 2 \cdot 7 = 14$ We have added 3 + 4 and then multiplied.

Also,

$2 \cdot (3 + 4) = (2 \cdot 3) + (2 \cdot 4)$ We have multiplied $2 \cdot 3$ and $2 \cdot 4$ as the first step.

$= 6 + 8$

$= 14$ The result is the same.

We see that $2 \cdot (3 + 4) = (2 \cdot 3) + (2 \cdot 4)$. This is an example of the **distributive property of multiplication over addition** because we distributed the multiplication (in this case by 2) over the "plus" sign.

Rules and Properties: The Distributive Property of Multiplication over Addition

To distribute a factor over a sum of numbers, multiply the factor by each number inside the parentheses. Then add the products. (The result will be the same if we find the sum and then multiply.)

CHECK YOURSELF 3

Show that

$3 \cdot (5 + 2) = (3 \cdot 5) + (3 \cdot 2)$

Carrying must often be used to multiply larger numbers. We will see how carrying works in multiplication by looking at an example in the expanded form.

Example 4

Multiplying by a Single-Digit Number

$3 \cdot 25 = 3 \cdot (20 + 5)$ We use the distributive property again.

$= 3 \cdot 20 + 3 \cdot 5$

$= 60 \quad + 15$ Write the 15 as 10 + 5.

$= 60 + 10 + 5$ Carry 10 ones or 1 ten to the tens place.

$= \quad 70 \quad + 5$

$= 75$

Here is the same multiplication problem using the short form.

Step 1

$$\begin{array}{r} {\scriptstyle 1} \\ 25 \\ \times\ 3 \\ \hline 5 \end{array}$$

1 ← Carry

Multiplying $3 \cdot 5$ gives us 15 ones. Write 5 ones and carry 1 ten.

Step 2

$$\begin{array}{r} \scriptstyle 1 \\ 25 \\ \times\ 3 \\ \hline 75 \end{array}$$

Now multiply $3 \cdot 2$ tens and add the carry to get 7, the tens digit of the product.

CHECK YOURSELF 4

Multiply.

$$\begin{array}{r} 34 \\ \times\ 6 \\ \hline \end{array}$$

Two numbers, 0 and 1, have special properties in multiplication.

Rules and Properties: Multiplicative Identity Property

The product of 1 and any number is that number. In symbols,

$a \cdot 1 = 1 \cdot a = a$

NOTE The number 1 is called the **multiplicative identity** for this reason.

Rules and Properties: Multiplicative Property of Zero

The product of 0 and any number is 0. In symbols,

$a \cdot 0 = 0 \cdot a = 0$

Example 5

Multiplying by 1 or 0

Find each product. Note that, in algebra, parentheses followed by another set of parentheses indicate multiplication.

(a) $(7)(1) = 7$ **(b)** $(1)(5) = 5$ **(c)** $(0)(7) = 0$ **(d)** $(3)(9)(0) = 0$

CHECK YOURSELF 5

Find each product.

(a) $(19)(1)$ **(b)** $(1)(9)$ **(c)** $(0)(8)$ **(d)** $(5)(6)(0)$

Units Analysis

When multiplying a denominate number, like 6 ft, by an abstract number, like 5, the result has the same units as the denominate number. Some examples are

$5 \cdot 6$ ft = 30 ft
$3 \cdot \$7 = \$21$
$9 \cdot 4$ A's = 36 A's

When multiplying two different denominate numbers, the units must also be multiplied. We will discuss this when we look at the area of a geometric figure.

NOTE Remember that it is best to write down the complete statement necessary for the solution of any application.

We will now review our discussion of applications, or word problems.

As you will see, the process of solving applications is the same no matter which operation is required for the solution. In fact, the four-step procedure we suggested in Section 1.2 can be effectively applied here.

Step by Step: Solving Applications

Step 1 Read the problem carefully to determine the given information and what you are asked to find.
Step 2 Decide upon the operation or operations to be used.
Step 3 Write down the complete statement necessary to solve the problem, and do the calculations.
Step 4 Check to make sure you have answered the question of the problem and that your answer seems reasonable.

Example 6

Solving an Application Involving Multiplication

A car rental agency orders a fleet of 7 new subcompact cars at a cost of $9258 per automobile. What will the company pay for the entire order?

Step 1 We know the number of cars and the price per car. We want to find the total cost.
Step 2 Multiplication is the best approach to the solution.
Step 3 Write

7 · \$9258 = \$64,806 — We could, of course, *add* \$9258, the cost, 7 times, but multiplication is certainly more efficient.

Step 4 The total cost of the order is $64,806.

CHECK YOURSELF 6

Tires sell for $47 apiece. What is the total cost for five tires?

There are some shortcuts that will let you simplify your work when you are multiplying by a number that ends in 0. We will see what we can discover by looking at some examples.

Example 7

Multiplying by Ten

First we'll multiply by 10.

$$\begin{array}{r} 67 \\ \times\ 10 \\ \hline 670 \end{array}$$

$10 \cdot 67 = 670$

Next we'll multiply by 100.

$$\begin{array}{r} 537 \\ \times\ 100 \\ \hline 53{,}700 \end{array}$$

$100 \cdot 537 = 53{,}700$

Finally, we'll multiply by 1000.

$$\begin{array}{r} 489 \\ \times\ 1000 \\ \hline 489{,}000 \end{array}$$

$1000 \cdot 489 = 489{,}000$

CHECK YOURSELF 7

Multiply.

$$\begin{array}{r} 257 \\ \times\ 100 \\ \hline \end{array}$$

NOTE We'll talk about powers of 10 in more detail in Section 1.7.

Do you see a pattern? Rather than writing out the multiplication, there is an easier way! We call the numbers 10, 100, 1000, and so on, **powers of 10.**

Rules and Properties: Multiplying by Powers of 10

When a whole number is multiplied by a power of 10, the product is just that number followed by as many zeros as there are in the power of 10.

Example 8

Multiplying by Numbers That End in Zero

Multiply 400 · 678.

Write

$$\begin{array}{r} 678 \\ \times\ \ 400 \\ \hline \end{array}$$

Shift 400 so that the two zeros are *to the right* of the digits above.

$$\begin{array}{r} {\scriptstyle 33}\ \ \ \\ 678 \\ \times\ \ 400 \\ \hline 271{,}200 \end{array}$$

Bring down the two zeros, then multiply 4 · 678 to find the product.

There is no mystery about why this works. We know that 400 is 4 · 100. In this method, we are multiplying 678 by 4 and then by 100, adding two zeros to the product by our earlier rule.

CHECK YOURSELF 8

Multiply.

300 · 574

Your work in this section, together with our earlier rounding techniques, provides a convenient means of using estimation to check the reasonableness of our results in multiplication, as Example 9 illustrates.

Example 9

Estimating a Product by Rounding

Estimate the product below by rounding each factor to the nearest hundred.

$$\begin{array}{rcr} & & \text{Rounded} \\ 512 & \longrightarrow & 500 \\ \times\ 289 & \longrightarrow & \times\ 300 \\ \hline & & 150{,}000 \end{array}$$

You might want to now find the *actual* product and use our estimate to see if your result seems reasonable.

CHECK YOURSELF 9

Estimate the product by rounding each factor to the nearest hundred.

$$\begin{array}{r} 689 \\ \times\ 425 \\ \hline \end{array}$$

Rounding the factors can be a very useful way of estimating the solution to an application problem.

Example 10

Estimating the Solution to a Multiplication Application

Bart is thinking of running an ad in the local newspaper for an entire year. The ad costs \$19.95 per week. Approximate the annual cost of the ad.

Rounding the charge to \$20 and rounding the number of weeks in a year to 50, we get

$50 \cdot 20 = 1000$

The ad would cost approximately \$1000.

CHECK YOURSELF 10

Phyllis is debating whether to join the health club for \$450 per year or just pay \$7 per visit. If she goes about once a week, approximately how much would she spend at \$7 per visit?

To multiply by numbers with more than one digit, we must multiply each digit of the first factor by each digit of the second. To do this, we form a series of partial products and then add them to arrive at the final product.

Example 11

Multiplying by a Two-Digit Number

Multiply $56 \cdot 47$.

Step 1

$$\begin{array}{r} {\scriptstyle 4} \\ 56 \\ \times\ 47 \\ \hline 392 \end{array}$$

The first partial product is $7 \cdot 56$, or 392. Note that we had to carry 4 to the tens column.

Step 2

$$\begin{array}{r} {\scriptstyle 2} \\ 56 \\ \times\ 47 \\ \hline 392 \\ 2240 \\ \hline \end{array}$$

The second partial product is $40 \cdot 56$, or 2240. We must carry 2 during the process.

Step 3

$$\begin{array}{r} {\scriptstyle 2} \\ 56 \\ \times\ 47 \\ \hline 392 \\ 2240 \\ \hline 2632 \end{array}$$

We add the partial products for our final result.

CHECK YOURSELF 11

Multiply.

$$\begin{array}{r} 38 \\ \times\ 76 \\ \hline \end{array}$$

NOTE The three partial products are formed when we multiply by the ones, tens, and then the hundreds digits.

If multiplication involves two three-digit numbers, another step is necessary. In this case we form three partial products. This will ensure that each digit of the first factor is multiplied by each digit of the second.

Example 12

Multiplying Two Three-Digit Numbers

Multiply.

$$\begin{array}{r} 2\,2 \\ 3\,3 \\ 2\,2 \\ 278 \\ \times\ 343 \\ \hline 834 \\ 11120 \\ 83400 \\ \hline 95354 \end{array}$$

In forming the third partial product, we must multiply by 300. To indicate this, we shift that product *two* places left.

CHECK YOURSELF 12

Multiply.

$$\begin{array}{r} 352 \\ \times\ 249 \\ \hline \end{array}$$

We will now look at an example of multiplying by a number involving 0 as a digit. There are several ways to arrange the work, as Example 13 shows.

Example 13

Multiplying Larger Numbers

Multiply 573 · 205.

Method 1

$$\begin{array}{r} 1 \\ 3\,1 \\ 573 \\ \times\ 205 \\ \hline 2865 \\ 0000 \\ 114600 \\ \hline 117465 \end{array}$$

We can write the second partial product as 0000 to indicate the multiplication by 0 in the tens place.

Next, we look at a second approach to the problem.

Method 2

```
      1
      3 1
      573
   × 205
     2865
   114600  ←
   117465
```

We can write a double 0 as our second step. If we place the third partial product on the same line, that product will be shifted *two* places left, indicating that we are multiplying by 200.

Because this second method is more compact, it is usually used.

CHECK YOURSELF 13

Multiply.

$$\begin{array}{r} 489 \\ \times\ 304 \\ \hline \end{array}$$

Example 14 will lead us to another property of multiplication.

Example 14

Using the Associative Property

$(2 \cdot 3) \cdot 4 = 6 \cdot 4 = 24$

We do the multiplication in the parentheses first, $2 \cdot 3 = 6$. Then multiply $6 \cdot 4$.

or,

$2 \cdot (3 \cdot 4) = 2 \cdot 12 = 24$

Here we multiply $3 \cdot 4$ as the first step. Then multiply $2 \cdot 12$.

We see that

$(2 \cdot 3) \cdot 4 = 2 \cdot (3 \cdot 4)$

The product is the same no matter which way we *group* the factors. This is called the **associative property** of multiplication.

Rules and Properties: The Associative Property of Multiplication

Multiplication is an *associative* operation. The way in which you group numbers in multiplication does not affect the final product.

CHECK YOURSELF 14

Find the products.

(a) $(5 \cdot 3) \cdot 6$ **(b)** $5 \cdot (3 \cdot 6)$

Units Analysis

What happens when we multiply two denominate numbers? The units of the result turn out to be the product of the units. This makes sense when we look at an example from geometry.

We write the area, A, of a square with sides of length s as

NOTE s^2 is read "s squared."

$A = s \cdot s = s^2$.

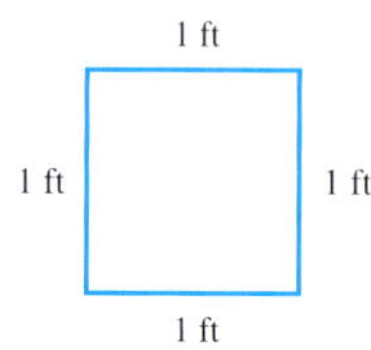

This tile is 1 ft by 1 ft.

$A = s^2 = (1 \text{ ft})^2 = 1 \text{ ft} \cdot 1 \text{ ft} = 1 \text{ (ft)} \cdot \text{(ft)} = 1 \text{ ft}^2$

In other words, its area is one square foot.

If we want to find the area of a room, we are actually finding how many of these square feet can be placed in the room.

We can look now at the idea of **area.** Area is a measure that we give to a surface. It is measured in terms of **square units.** The area is the number of square units that are needed to cover the surface.

One standard unit of area measure is the **square inch,** written in.^{2}. This is the measure of the surface contained in a square with sides of 1 in.

1 in.

1 in. 1 in.

1 in.

One square inch

NOTE The unit inch (in.) can be treated as though it were a number. So in. · in. can be written in.^{2}. It is read "square inches."

Other units of area measure are the square foot (ft^2), the square yard (yd^2), the square centimeter (cm^2), and the square meter (m^2).

Finding the area of a figure means finding the number of square units it contains. One simple case is a rectangle.

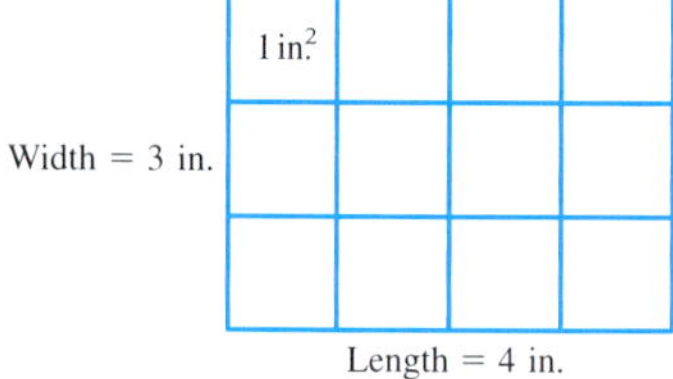

NOTE The length and width must be in terms of the same unit.

The length of the rectangle is 4 in., and the width is 3 in. The area of the rectangle is measured in terms of square inches. We can simply count to find the area, 12 in.^{2}. However, because each of the four vertical strips is 3 in. long, we can multiply:

Area = 4 in. · 3 in. = 12 in.^{2}

Rules and Properties: Formula for the Area of a Rectangle

In general, we can write the formula for the **area of a rectangle:** If the length of a rectangle is L units and the width is W units, then the formula for the area, A, of the rectangle can be written as

$A = L \cdot W$ (square units)

Example 15

Find the Area of a Rectangle

A room has dimensions 12 ft by 15 ft. Find its area.

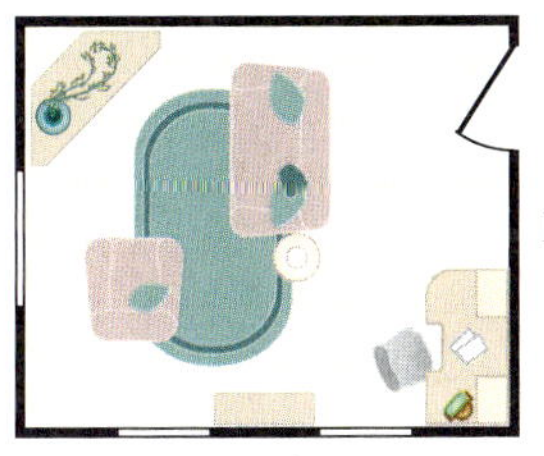

Use the formula $A = L \cdot W$, with $L = 15$ ft and $W = 12$ ft.

$$A = L \cdot W$$
$$= 15 \text{ ft} \cdot 12 \text{ ft} = 180 \text{ ft}^2$$

The area of the room is 180 ft^2.

CHECK YOURSELF 15

A desktop has dimensions 50 *in. by* 25 *in. What is the area of its surface?*

We can also write a convenient formula for the area of a square. If the sides of the square have length S, we can write

Rules and Properties: Formula for the Area of a Square

$A = S \cdot S = S^2$

Example 16

Finding Area

You wish to cover a square table with a plastic laminate that costs 60¢ a square foot. If each side of the table measures 3 ft, what will it cost to cover the table?

We first must find the area of the table. Use the formula $A = S^2$, with $S = 3$ ft.

$$A = S^2$$
$$= (3 \text{ ft})^2 = 3 \text{ ft} \cdot 3 \text{ ft} = 9 \text{ ft}^2$$

Now, multiply by the cost per square foot.

Cost = 9 · 60¢ = 540¢ = \$5.40

CHECK YOURSELF 16

You wish to carpet a room that is a square, 4 yd by 4 yd, with carpet that costs \$12 per square yard. What will be the total cost of the carpeting?

Sometimes the total area of an oddly shaped figure, frequently called a **composite figure,** is found by adding the smaller areas. Example 17 shows how this is done.

Example 17

Finding the Area of an Oddly Shaped Figure

Find the area of the figure.

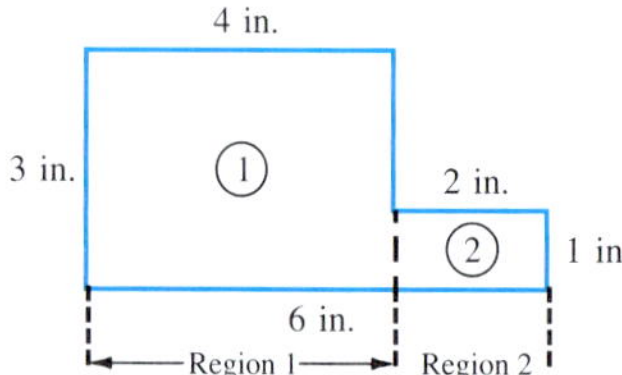

The area of the figure is found by adding the areas of regions 1 and 2. Region 1 is a 4 in. by 3 in. rectangle; the area of region 1 is 4 in. · 3 in. = 12 in.^{2} Region 2 is a 2 in. by 1 in. rectangle; the area of region 2 is 2 in. · 1 in. = 2 in.^{2}

The total area is the sum of the two areas:

Total area = 12 in.^{2} + 2 in.^{2} = 14 in.^{2}

CHECK YOURSELF 17

Find the area of the figure.

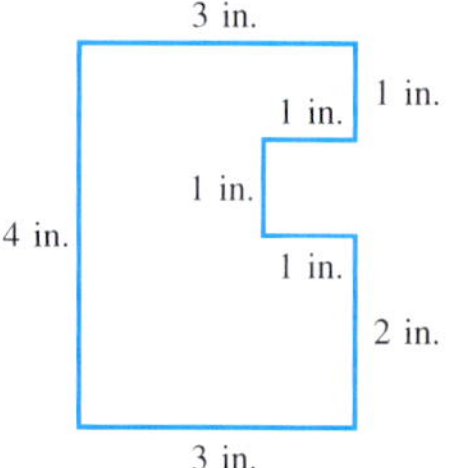

Hint: You can find the area by adding the areas of three rectangles, or by subtracting the area of the "missing" rectangle from the area of the "completed" larger rectangle.

CHECK YOURSELF ANSWERS

1. Factors 2, 9; product 18 **2.** 24 **3.** 3 · 7 = 21 and 15 + 6 = 21 **4.** 204 **5.** **(a)** 19; **(b)** 9; **(c)** 0; **(d)** 0 **6.** \$235 **7.** 25,700 **8.** 172,200 **9.** 280,000 **10.** \$350 **11.** 2888 **12.** 87,648 **13.** 148,656 **14.** **(a)** 90; **(b)** 90 **15.** 1250 in.^{2} **16.** \$192 **17.** 11 in.^{2}

Name ______________________

Section ________ Date ________

1.5 Exercises

ANSWERS

1. ______
2. ______
3. ______
4. ______
5. ______
6. ______
7. ______
8. ______
9. ______
10. ______
11. ______
12. ______
13. ______
14. ______
15. ______
16. ______
17. ______
18. ______
19. ______
20. ______
21. ______
22. ______
23. ______
24. ______

1. Find $3 \cdot 7$ and $7 \cdot 3$ by repeated addition.

2. Find $4 \cdot 5$ and $5 \cdot 4$ by repeated addition.

3. If $6 \cdot 7 = 42$, we call 6 and 7 ______ of 42. And 42 is the ______ of 6 and 7.

4. If $5 \cdot 8 = 40$, we call 5 and 8 ______ of 40. And 40 is the ______ of 5 and 8.

5. Find the number of 1-ft^2 tiles in a floor that is 9 ft long and 12 ft wide.

6. Find the number of squares in a quilt that has four squares in each of four rows.

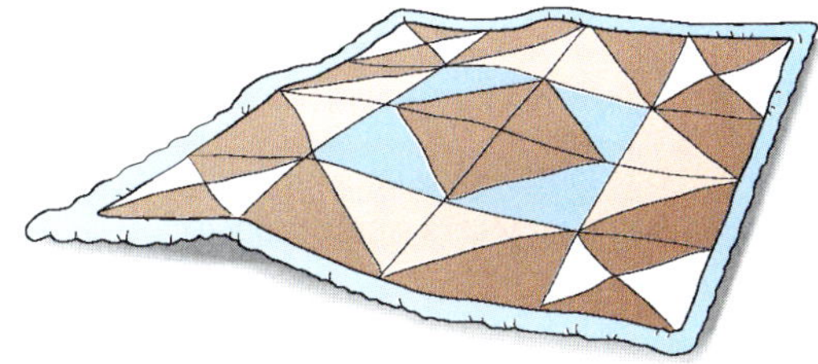

Multiply.

7. $8 \cdot 3$ yd

8. $9 \cdot \$15$

9. $6 \cdot 4°C$

10. $8 \cdot 11$ mi

11. $\begin{array}{r} 2 \\ \times 9 \\ \hline \end{array}$

12. $\begin{array}{r} 1 \\ \times 7 \\ \hline \end{array}$

13. $\begin{array}{r} 23 \\ \times \ \ 2 \\ \hline \end{array}$

14. $\begin{array}{r} 32 \\ \times \ \ 3 \\ \hline \end{array}$

15. $\begin{array}{r} 523 \\ \times \ \ 8 \\ \hline \end{array}$

16. $\begin{array}{r} 2035 \\ \times \ \ 9 \\ \hline \end{array}$

17. $\begin{array}{r} 327 \\ \times \ 59 \\ \hline \end{array}$

18. $\begin{array}{r} 2364 \\ \times \ 67 \\ \hline \end{array}$

19. $\begin{array}{r} 4075 \\ \times \ 84 \\ \hline \end{array}$

20. $\begin{array}{r} 315 \\ \times 243 \\ \hline \end{array}$

21. $\begin{array}{r} 58 \\ \times 40 \\ \hline \end{array}$

22. $\begin{array}{r} 562 \\ \times 400 \\ \hline \end{array}$

23. $\begin{array}{r} 907 \\ \times 900 \\ \hline \end{array}$

24. $\begin{array}{r} 345 \\ \times 230 \\ \hline \end{array}$

ANSWERS

25. ______
26. ______
27. ______
28. ______
29. ______
30. ______
31. ______
32. ______
33. ______
34. ______
35. ______
36. ______
37. ______
38. ______
39. ______
40. ______

25. Find the product of 304 and 7.

26. Find the product of 409 and 4.

27. What is the product of 21 and 551?

28. What is the product of 112 and 168?

Name the property of addition and/or multiplication that is illustrated.

29. $5 \cdot 8 = 8 \cdot 5$

30. $2 \cdot (3 \cdot 5) = (2 \cdot 3) \cdot 5$

31. $3 \cdot (2 + 8) = (3 \cdot 2) + (3 \cdot 8)$

32. $4 \cdot 1 = 4$

In exercises 33 and 34, complete the statement using the given property.

33. $3 \cdot (2 + 7) =$ Distributive property

34. $9 \cdot (8 \cdot 5) =$ Associative property of multiplication

In exercises 35 to 44, solve the applications.

35. Transportation. A convoy company can transport 8 new cars on one of its trucks. If 34 truck shipments were made in 1 week, how many cars were shipped?

36. Printing. A computer printer can print 40 mailing labels per minute. How many labels can be printed in 1 h?

37. Copying. A ream of paper is 500 sheets. If 29 reams of paper were used in a copy machine during 1 week, how many copies were made?

38. Science. If sound waves travel at a rate of 1088 feet per second (ft/s) and you hear thunder 23 s after seeing a lightning flash, how far away did the lightning flash?

39. Manufacturing. The manufacturer of woodburning stoves can make 15 stoves in 1 day. How many stoves can be made in 28 days?

40. Quantity of papers. Each bundle of newspapers contains 25 papers. If 43 bundles are delivered to Jose's house, how many papers are delivered?

ANSWERS

41. ______

42. ______

43. ______

44. ______

41. Total cost. Erin agrees to buy a boat by paying \$2500 down and \$139 a month for 36 months. What is the total cost of the boat?

42. Farming. Celeste harvested 34 bushels of corn per acre from 32 acres in June and 43 bushels of corn per acre from 36 acres in July. How many bushels of corn did Celeste harvest?

43. Construction. The table shows the hourly wages of four different types of jobs at a home remodeling company.

Job	Hourly Wage
Electrician	\$27
Plumber	\$22
Clerk	\$12
Accountant	\$18

Based on the architectural plans for an addition, it is estimated that it will require four electricians each working 50 h and two plumbers each working 21 h. In addition, 4 h of clerical work and 6 h of accounting are needed. What is the total cost of the job?

44. Sales. The monthly sales at the Magic Carpet used car dealership are given as

Car Model	Average Price per Sale	No. of Cars Sold
Honda Accord	\$29,000	23
Volkswagon Jetta	\$26,900	38
Pontiac Grand AM	\$31,700	18

What are the gross receipts for the month at the dealership?

ANSWERS

45. ______

46. ______

47. ______

48. ______

49. ______

50. ______

51. ______

52. ______

53. ______

54. ______

55. ______

56. ______

57. ______

58. ______

59. ______

60. ______

Estimate each of the products by rounding each factor to the nearest hundred.

45. $\begin{array}{r} 212 \\ \times\ 278 \\ \hline \end{array}$

46. $\begin{array}{r} 179 \\ \times\ 431 \\ \hline \end{array}$

47. $\begin{array}{r} 391 \\ \times\ 531 \\ \hline \end{array}$

48. $\begin{array}{r} 729 \\ \times\ 481 \\ \hline \end{array}$

Multiply. Be sure to use the proper units in your answer.

49. 3 ft · 2 ft

50. 5 mi · 13 mi

51. 17 in. · 11 in.

52. 143 yd · 26 yd

Label the statements true or false.

53. $(10 \text{ ft})^2 = 100 \text{ ft}$

54. $(5 \text{ mi})^2 = 25 \text{ mi}^2$

55. $(8 \text{ yd})^2 = 512 \text{ yd}^2$

56. $(9 \text{ in.})^2 = 9 \text{ in.}^2$

Find the area of each figure.

57.

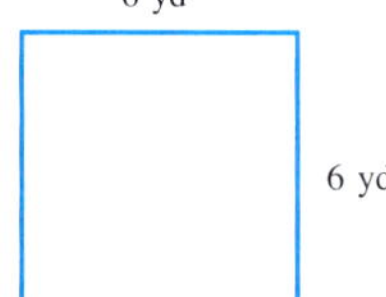

58.

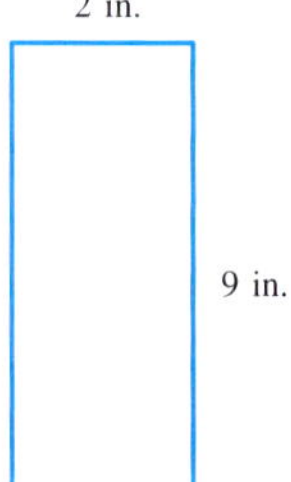

59.

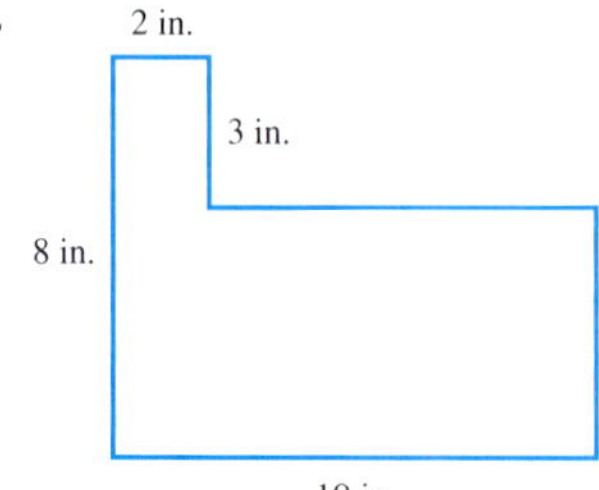

60.

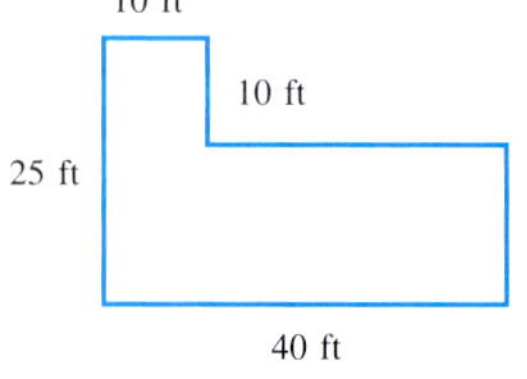

61.

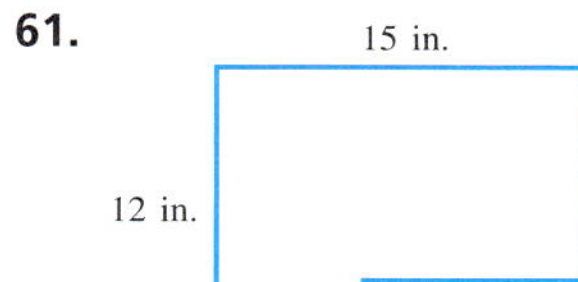

62.

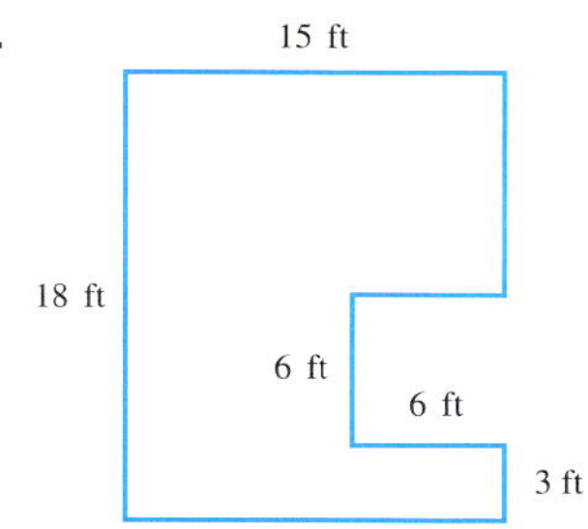

In exercises 63 to 68, solve the applications.

63. Manufacturing. A company can manufacture 45 sleds per day. Approximately how many can this company make in 128 days?

64. Recreation. The attendance at a basketball game was 2345. The cost of admission was $12 per person. Estimate the total gate receipts for the game.

65. House repairs. A plate glass window measures 5 ft by 7 ft. If glass costs $8 per square foot, how much will it cost to replace the window?

66. Paint costs. In a hallway, Bill is painting two walls that are 10 ft high by 22 ft long. The instructions on the paint can say that it will cover 400 ft^2 per gallon (gal). Will 1 gal be enough for the job?

67. Tile costs. Tile for a kitchen counter will cost $7 per square foot to install. If the counter measures 12 ft by 3 ft, what will the tile cost?

68. Carpet costs. You wish to cover a floor 4 yd by 5 yd with a carpet costing $13 per square yard. What will the carpeting cost?

ANSWERS

61. ____________
62. ____________
63. ____________
64. ____________
65. ____________
66. ____________
67. ____________
68. ____________

69.

70.

71.

72.

69. Suppose you wish to build a small, rectangular pen, and you have enough fencing for the pen's perimeter to be 36 ft. Assuming that the length and width are to be whole numbers, answer each question.

(a) What are the possible dimensions that the pen could have? (Note: a square is a type of rectangle.)

(b) For each set of dimensions (length and width), what is the area that the pen would enclose?

(c) Which dimensions give the greatest area?

(d) What is the greatest area?

70. Suppose you wish to build a rectangular kennel that encloses 100 ft^2. Assuming that the length and width are to be whole numbers, answer each question.

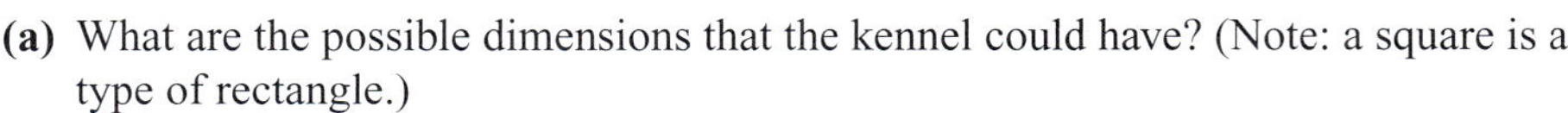

(a) What are the possible dimensions that the kennel could have? (Note: a square is a type of rectangle.)

(b) For each set of dimensions (length and width), what is the perimeter that would surround the kennel?

(c) Which dimensions give the least perimeter?

(d) What is the least perimeter?

71. We have seen that addition and multiplication are commutative operations. Decide which of the listed activities are commutative.

(a) Taking a shower and eating breakfast

(b) Getting dressed and taking a shower

(c) Putting on your shoes and your socks

(d) Brushing your teeth and combing your hair

(e) Putting your key in the ignition and starting your car

72. The associative properties of addition and multiplication indicate that the result of the operation is the same regardless of where the grouping symbol is placed. This is not always the case in the use of the English language. Many phrases can have different meanings based on how the words are grouped. For each phrase, explain why the associative property would not hold.

(a) Cat fearing dog

(b) Hard test question

(c) Defective parts department

(d) Man eating animal

Write some phrases in which the associative property is satisfied.

73. Most maps contain legends that allow you to convert the distance between two points on the map to actual miles. For instance, if a map uses a legend that equates 1 in. to 5 mi and the distance between two towns is 4 in. on the map, then the towns are actually 20 mi apart.

(a) Obtain a map of your state and determine the shortest distance between any two major cities.

(b) Could you actually travel the route you measured in part **(a)**?

(c) Plan a trip between the two cities you selected in part **(a)** over established roads. Determine the distance that you actually travel using this route.

74. Calculate the product 378 · 215 in two ways.

Method 1: Round each factor to the nearest hundred and then multiply.

Method 2: Multiply first and then round the product to the nearest hundred.

(a) Compare your answers, and comment on the difference between the two results.

(b) List the advantages and disadvantages of each method.

(c) Describe situations in which each method is the preferred approach.

75. There are many different ways of rounding. One way used in computer applications is called **truncating.**

(a) Determine what rules would be used in truncating, and compare them to the rules used in rounding.

(b) Round 7473 to the hundreds place using truncating and rounding.

(c) State some possible problems that could occur in truncating.

76. Maria has been asked to estimate the number of pieces of paper in five large piles. She does not want to count every piece. Devise a plan to help her estimate the total number of pieces of paper.

77. Complete the following number cross.

Across	Down
1. 6 × 551	**1.** 5 × 7
5. 7 × 8	**2.** 9 × 41
6. 27 × 27	**3.** 67 × 100
7. 19 × 50	**4.** 2 × (49 + 100)
10. 3 × 67	**8.** 4 × 1301
12. 6 × 25	**9.** 100 + 10 + 1
13. 9 × 8	**11.** 2 × 87
15. 16 × 303	**14.** 25 + 3

ANSWERS

73. ______

74. ______

75. ______

76. ______

77. ______

Answers

1. 21 **3.** Factors, product **5.** 108 tiles **7.** 24 yd **9.** 24°C **11.** 18
13. 46 **15.** 4184 **17.** 19,293 **19.** 342,300 **21.** 2320 **23.** 816,300
25. 2128 **27.** 11,571 **29.** Commutative property of multiplication
31. Distributive property of multiplication over addition **33.** $(3 \cdot 2) + (3 \cdot 7)$
35. 272 cars **37.** 14,500 copies **39.** 420 stoves **41.** \$7504
43. \$6480 **45.** 60,000 **47.** 200,000 **49.** 6 ft^2 **51.** 187 $in.^2$
53. False **55.** False **57.** 36 yd^2 **59.** 56 $in.^2$ **61.** 153 $in.^2$
63. 6500 sleds **65.** \$280 **67.** \$252 **69.** **71.**
73. **75.**

77.

3	3	0	6	■	2
5	6	■	7	2	9
■	9	5	0	■	8
1	■	2	0	1	■
1	5	0	■	7	2
1	■	4	8	4	8

1.6 Division of Whole Numbers

1.6 OBJECTIVES

1. Use the language of division
2. Write a division problem as repeated subtraction
3. Divide whole numbers

Overcoming Math Anxiety

Learn to Take Useful Notes

Although some students find it easier to be organized than do other students, every student can become a better note taker. Below are some hints that can help you learn to take more useful notes. Good note taking begins with your preparation for class. Note that the first several items refer to your preparation for class.

1. **Read assigned material before the lecture.** This will help familiarize you with both the vocabulary and the concepts.
2. **Review your notes from the previous class meeting.** If you already have a concept in your lecture notes, and the idea is referred to again, you then simply jot a note to refer back to previous material in your notes.
3. **Get to class a few minutes early**—have your materials out and ready to go when your instructor walks in the door.
4. Be ready to listen as soon as the instructor walks in the door.
5. Have more than one pencil sharpened and ready to use.
6. Have a highlighter or colored pen available to mark particularly important segments of your notes.
7. **Know how to spell**—both plain old English words—and technical words that have already been presented in class. If you can spell them you won't have to waste time trying to figure out HOW to spell.
8. **Be aggressive in notetaking.** Don't wait for an idea to strike you—it's better to have too much material than too little.
9. If the professor repeats something, **WRITE IT DOWN**.
10. **Take notes, not dictation.** That means being able to develop your own form of shorthand.
11. **Develop abbreviations** for words that are used frequently in the course.
    - Real Numbers—R
    - Natural Numbers—N
12. Use symbols when you can
    - & for and
    - B for but
    - $\forall$ = for each
    - $\therefore$ = therefore
    - $\exists$ = there exists
    - $\ni$ = such that
    - $\in$ = is an element of
13. **Skip lines**—leave visual breaks between definitions, lists, or explanations.
14. If you miss something, **leave a blank in your notes.** You can fill it in later. If you try to copy it from your neighbor, both of you will lose more material.
15. **Get together with classmates** after lecture and pool your notes. That way you can be sure you have everything down. It will also help make sure you understand what you have written down. (That's how students survived before tape recorders were commonly used.)

We will now examine a fourth arithmetic operation, division. Just as multiplication was repeated addition, division is repeated subtraction. Division asks *how many times* one number is contained in another.

Example 1

Dividing by Using Subtraction

Joel needs to set up 48 chairs in the student union for a concert. If there is room for 8 chairs per row, how many rows will it take to set up all 48 chairs?

This problem can be solved by subtraction. Each row subtracts another 8 chairs.

$$\begin{array}{r} 48 \\ -8 \\ \hline 40 \end{array} \quad \begin{array}{r} 40 \\ -8 \\ \hline 32 \end{array} \quad \begin{array}{r} 32 \\ -8 \\ \hline 24 \end{array} \quad \begin{array}{r} 24 \\ -8 \\ \hline 16 \end{array} \quad \begin{array}{r} 16 \\ -8 \\ \hline 8 \end{array} \quad \begin{array}{r} 8 \\ -8 \\ \hline 0 \end{array}$$

Because 8 can be subtracted from 48 six times, there will be 6 rows.

This can also be seen as a division problem

$$48 \div 8 = 6 \qquad \text{or} \qquad 8\overline{)48}^{\,6} \qquad \text{or} \qquad \frac{48}{8} = 6$$

No matter which method we use, we call the 48 the **dividend,** the 8 the **divisor,** and the 6 the **quotient.**

CHECK YOURSELF 1

Carlotta is creating a garden path made of bricks. She has 72 bricks. Each row will have 6 bricks in it. How many rows can she make?

Units Analysis

When dividing a denominate number by an abstract number, the result will get the units of the denominate number. Here are a couple of examples.

76 trombones ÷ 4 = 19 trombones

$55 ÷ 11 = $5

When one denominate number is divided by another, the result will get the units of the dividend over the units of the divisor.

144 mi ÷ 6 gal = 24 mi/gal (which we read as "miles per gallon")

$120 ÷ 8 h = 15 dollars/h ("dollars per hour")

To solve a problem that requires division, you must first set up the problem as a division statement. Example 2 will illustrate this.

Example 2

Writing a Division Statement

Write a division statement that corresponds to the following situation. You need not do the division.

The staff at the Wok Inn Restaurant splits all tips at the end of each shift. Yesterday's evening shift collected a total of \$224. How much should each of the seven employees get in tips?

\$224 $\div$ 7 employees   Note that the units for the answer will be "dollars per employee."

CHECK YOURSELF 2

Write a division statement that corresponds to the following situation. You need not do the division.

All nine sections of basic math skills at SCC (Sum Community College) are full. There are a total of 315 students in the classes. How many students are in each class? What are the units for the answer?

In Section 1.5, we used a rectangular array of stars to represent multiplication. These same arrays can represent division. Just as $3 \cdot 4 = 12$ and $4 \cdot 3 = 12$, so is it true that $12 \div 3 = 4$ and $12 \div 4 = 3$.

```
* * *
* * *  } 4 • 3 = 12
* * *
* * *
```
or
$12 \div 3 = 4$

```
* * * *
* * * *  } 3 • 4 = 12
* * * *
```
or
$12 \div 4 = 3$

This relationship allows us to check our division results by doing multiplication.

Example 3

Checking Division by Using Multiplication

NOTE For a division problem to check, the *product* of the divisor and the quotient *must equal the dividend.*

(a) $7\overline{)21}$ with quotient 3   Check: $7 \cdot 3 = 21$

(b) $48 \div 6 = 8$ Check: $6 \cdot 8 = 48$

CHECK YOURSELF 3

Complete the division statements, and check your results.

(a) $9\overline{)45}$ **(b)** $28 \div 7 =$

NOTE Because $36 \div 9 = 4$, we say that 36 is *exactly divisible* by 9.

In our examples so far, the product of the divisor and the quotient has been equal to the dividend. This means that the dividend is *exactly divisible* by the divisor. That is not always the case. In Example 4, we are again using repeated subtraction.

Example 4

Dividing by Using Subtraction, Leaving a Remainder

How many times is 5 contained in 23?

NOTE Notice that the remainder must be smaller than the divisor or we could subtract again.

$$\begin{array}{r}23\\-\ 5\\\hline 18\end{array}\qquad\begin{array}{r}18\\-\ 5\\\hline 13\end{array}\qquad\begin{array}{r}13\\-\ 5\\\hline 8\end{array}\qquad\begin{array}{r}8\\-\ 5\\\hline 3\end{array}$$

We see that 5 is contained 4 times in 23, but 3 is "left over."

23 is not exactly divisible by 5. The "left over" 3 is called the **remainder** in the division. To check the division operation when a remainder is involved, we have a rule:

Definitions: Remainder

Dividend = divisor · quotient + remainder

CHECK YOURSELF 4

How many times is 7 contained in 38?

Example 5

Checking Division by a Single-Digit Number

Using the work of Example 4, we can write

$$\begin{array}{r}4\\5\overline{)23}\end{array}$$ with remainder 3

NOTE Another way to write the result is

$$\begin{array}{r}4\ \text{r}3\\5\overline{)23}\end{array}$$

The "r" stands for remainder.

To apply the remainder definition, we have

Divisor ↘ Quotient ↙

Dividend ⟶ $23 = 5 \cdot 4 + 3$ ⟵ Remainder

$23 = 20 + 3$

$23 = 23$ The division checks.

NOTE Notice that the multiplication is done before the 3 is added.

CHECK YOURSELF 5

Evaluate $7\overline{)38}$. *Check your answer.*

We must be careful when 0 is involved in a division problem. There are two special cases.

Rules and Properties: Division and Zero

1. 0 divided by any whole number (except 0) is 0.
2. Division by 0 is undefined.

The first case involving zero occurs when we are dividing into zero.

Example 6

Dividing into Zero

$0 \div 5 = 0$ because $0 = 5 \cdot 0$.

CHECK YOURSELF 6

(a) $0 \div 7 =$ **(b)** $0 \div 12 =$

Our second case illustrates what happens when 0 is the *divisor*. Here we have a special problem.

Example 7

Dividing by Zero

$8 \div 0 = ?$ This means that $8 = 0 \cdot ?$

Can 0 times some number ever be 8? From our multiplication facts, the answer is *no!* There is no answer to this problem, so we say that $8 \div 0$ is undefined.

CHECK YOURSELF 7

Decide whether each problem results in 0 *or is undefined.*

(a) $9 \div 0$ **(b)** $0 \div 9$ **(c)** $0 \div 15$ **(d)** $15 \div 0$

It is easy to divide when small whole numbers are involved, because much of the work can be done mentally. In working with larger numbers, we turn to a process called **long division.** This is a shorthand method for performing the steps of repeated subtraction.

To start, we can look at an example in which we subtract multiples of the divisor.

Example 8

Dividing by a Single-Digit Number

NOTE With larger numbers, repeated subtraction is just too time-consuming to be practical.

Divide 176 by 8.

Because 20 eights are 160, we know that there are at least 20 eights in 176.

Step 1 Write

$$\begin{array}{r} 20 \\ 8\overline{)176} \\ \underline{160} \\ 16 \end{array}$$

20 eights ⟶ 160

Subtracting 160 is just a shortcut for subtracting eight 20 times.

After subtracting the 20 eights, or 160, we are left with 16. There are 2 eights in 16, and so we continue.

Step 2

$$\begin{array}{r} 2 \\ 20 \\ 8\overline{)176} \\ \underline{160} \\ 16 \\ \underline{16} \\ 0 \end{array} \quad \left.\begin{array}{l} 2 \\ 20 \end{array}\right\} 22$$

Adding 20 and 2 gives us the quotient, 22.

2 eights ⟶ 16

Subtracting the 2 eights, we have a 0 remainder. So $176 \div 8 = 22$.

CHECK YOURSELF 8

Verify the results of Example 8, using multiplication.

The next step is to simplify this repeated-subtraction process one step further. The result will be the long-division method.

Example 9

Dividing by a Single-Digit Number

Divide 358 by 6.

The dividend is 358. We look at the first digit, 3. We cannot divide 6 into 3, and so we look at the *first two digits,* 35. There are 5 sixes in 35, and so we write 5 above the tens digit of the dividend.

$$\begin{array}{r} 5 \\ 6\overline{)358} \end{array}$$

When we place 5 as the tens digit, we really mean 5 tens, or 50.

Now multiply $5 \cdot 6$, place the product below 35, and subtract.

$$\begin{array}{r} 5 \\ 6\overline{)358} \\ \underline{30} \\ 5 \end{array}$$

We have actually subtracted 50 sixes (300) from 358.

Because the remainder, 5, is smaller than the divisor, 6, we bring down 8, the ones digit of the dividend.

$$\begin{array}{r} 5 \\ 6\overline{)358} \\ \underline{30\downarrow} \\ 58 \end{array}$$

Now divide 6 into 58. There are 9 sixes in 58, and so 9 is the ones digit of the quotient. Multiply $9 \cdot 6$ and subtract to complete the process.

NOTE Because the 4 is smaller than the divisor, we have a remainder of 4.

$$\begin{array}{r} 59 \\ 6\overline{)358} \\ \underline{30\downarrow} \\ 58 \\ \underline{54} \\ 4 \end{array}$$

We now have: $358 \div 6 = 59$ r4

NOTE Verify that this is true and that the division checks.

To check: $358 = 6 \cdot 59 + 4$

CHECK YOURSELF 9

Divide $7\overline{)453}$.

Long division becomes a bit more complicated when we have a two-digit divisor. It is now a matter of trial and error. We round the divisor and dividend to form a *trial divisor* and a *trial dividend.* We then estimate the proper quotient and must determine whether our estimate was correct.

Example 10

Dividing by a Two-Digit Number

Divide

$38\overline{)293}$

Round the divisor and dividend to the nearest ten. So 38 is rounded to 40, and 293 is rounded to 290. The trial divisor is then 40, and the trial dividend is 290.

NOTE Think: $\begin{array}{r} 7 \\ 4\overline{)29} \end{array}$

Now look at the nonzero digits in the trial divisor and dividend. They are 4 and 29. We know that there are 7 fours in 29, and so 7 is our first estimate of the quotient. Now we will see if 7 works.

$$\begin{array}{r} 7 \\ 38\overline{)293} \\ \underline{266} \\ 27 \end{array}$$

7 ⟵ Your estimate

Multiply $7 \cdot 38$. The product, 266, is less than 293, and so we can subtract.

The remainder, 27, is less than the divisor, 38, and so the process is complete.

$293 \div 38 = 7$ r27

Check: $293 = 38 \cdot 7 + 27$ You should verify that this statement is true.

CHECK YOURSELF 10

Divide.

$57\overline{)482}$

Because this process is based on estimation, we can't expect our first guess to always be right.

Example 11

Dividing by a Two-Digit Number

Divide

$54\overline{)428}$

NOTE Think: $\begin{array}{r} 8 \\ 5\overline{)43} \end{array}$

Rounding to the nearest ten, we have a trial divisor of 50 and a trial dividend of 430.

Looking at the nonzero digits, how many fives are in 43? There are 8. This is our first estimate.

$$\begin{array}{r} 8 \\ 54\overline{)428} \\ \underline{432} \end{array} \longleftarrow \text{Too large}$$

We multiply 8 · 54. Do you see what's wrong? The product, 432, is too large. We can't subtract. Our estimate of the quotient must be adjusted *downward.*

NOTE If we tried 6 as the quotient

$$\begin{array}{r} 6 \\ 54\overline{)428} \\ \underline{324} \\ 104 \end{array}$$

We have 104, which is too large to be a remainder.

We adjust the quotient downward to 7. We can now complete the division.

$$\begin{array}{r} 7 \\ 54\overline{)428} \\ \underline{378} \\ 50 \end{array}$$

We have

$428 \div 54 = 7 \text{ r}50$

Check: $428 = 54 \cdot 7 + 50$

CHECK YOURSELF 11

Divide.

$63\overline{)557}$

We have to be careful when a 0 appears as a digit in the quotient. Next, we look at an example in which this happens with a two-digit divisor.

Example 12

Dividing with Large Dividends

NOTE Our divisor, 32, will divide into 98, the first two digits of the dividend.

Divide

```
32)9871
```

Rounding to the nearest ten, we have a trial divisor of 30 and a trial dividend of 100. Think, "How many threes are in 10?" There are 3, and this is our first estimate of the quotient.

```
     3          Everything seems fine so far!
32)9871
   96
    2
```

Bring down 7, the next digit of the dividend.

```
     30
32)9871
   96↓      Now do you see the difficulty? We cannot divide
    27      32 into 27, and so we place 0 in the tens place
            of the quotient to indicate this fact.
```

We continue by multiplying by 0. After subtraction, we bring down 1, the last digit of the dividend.

```
     30
32)9871
   96 |
    27|
    00↓
    271
```

Another problem develops here. We round 32 to 30 for our trial divisor, and we round 271 to 270, which is the trial dividend at this point. Our estimate of the last digit of the quotient must be 9.

```
     309
32)9871
   96
    27
    00
    271
    288  ←—— Too large
```

We can't subtract. The trial quotient must be adjusted downward to 8. We can now complete the division.

```
     308
32)9871
   96
    27
    00
    271
    256
     15
```

$9871 \div 32 = 308 \text{ r}15$

Check: $9871 = 32 \cdot 308 + 15$

CHECK YOURSELF 12

Divide.

$43\overline{)8857}$

Because of the availability of the handheld calculator, it is rarely necessary that people find the exact answer when performing long division. On the other hand, it is frequently important that one be able to either estimate the result of long division, or confirm that a given answer (particularly from a calculator) is reasonable. As a result, the emphasis in this section will be to improve your estimation skills in division.

In Example 13, we divide a four-digit number by a two-digit number. Generally, we round the divisor to the nearest ten and the dividend to the nearest hundred.

Example 13

Estimating the Result of a Division Application

The Ramirez family took a trip of 2394 mi in their new car, using 77 gal of gas. Estimate their gas mileage (mi/gal).

Our estimate will be based on dividing 2400 by 80.

$$\begin{array}{r} 30 \\ 80\overline{)2400} \end{array}$$

They got approximately 30 mi/gal.

CHECK YOURSELF 13

Troy flew a light plane on a trip of 2844 mi that took 21 h. What was his approximate speed in miles per hour?

As before, we may have to combine operations to solve an application of the mathematics you have learned.

Example 14

Estimating the Result of a Division Application

Charles purchases a used car for \$8574. He agrees to make payments for 4 years. Interest charges will be \$978. Approximately what should his payments be?

First, we find the amount that Charles owes:

\$8574 + \$978 = \$9552

Now, to find the monthly payment, we divide that amount by 48 (months). To estimate the payment, we'll divide \$9600 by 50 months.

$$\begin{array}{r} 192 \\ 50\overline{)9600} \end{array}$$

The payments will be approximately \$192 per month.

CHECK YOURSELF 14

One \$10 bag of fertilizer will cover 310 ft^2. Approximately what would it cost to cover 2200 ft^2?

Using a Scientific Calculator to Divide

Of course, division is easily done using your calculator. However, as we will see, some special things come up when we use a calculator to divide. First we will outline the steps of division as it is done on a calculator.

Divide $35\overline{)2380}$.

Step 1 Enter the dividend. 2380

Step 2 Press the divide key. [÷]

Step 3 Enter the divisor. 35

Step 4 Press the equals key. [=] The desired quotient is now in your display.

NOTE Recall that a graphing calculator uses the [Enter] key rather than [=].

The display shows 68.

We have already mentioned some of the difficulties related to division with 0. We will experiment on the calculator.

Example 15

Using a Scientific Calculator to Divide

To find 0 ÷ 5, we use this sequence:

0 [÷] 5 [=]

Display 0

There is no problem with this. Zero divided by any whole number other than 0 is just 0.

CHECK YOURSELF 15

What is the result when you use your calculator to perform the given operation?

0 ÷ 17

We've seen what happens when dividing zero by another number, but what happens when we try to divide by zero? More importantly to this section, how does the calculator handle division by zero? Example 16 illustrates this concept.

Example 16

Using a Scientific Calculator to Divide

To find 5 ÷ 0, we use this sequence:

5 [÷] 0 [=]

Display Error

NOTE You may find that you must "clear" your calculator after trying this.

If we try this sequence, the calculator gives us an error! Do you see why? Division by 0 is not allowed. Try this on your calculator to see how this error is indicated.

CHECK YOURSELF 16

What is the result when you use your calculator to perform the given operation?

17 ÷ 0

Another special problem comes up when a remainder is involved in a division problem.

Example 17

Using a Scientific Calculator to Divide

Dividing 293 by 38 gives 7 with remainder 27.

NOTE Be aware that the calculator will not give you a remainder in the form we have been using in this chapter.

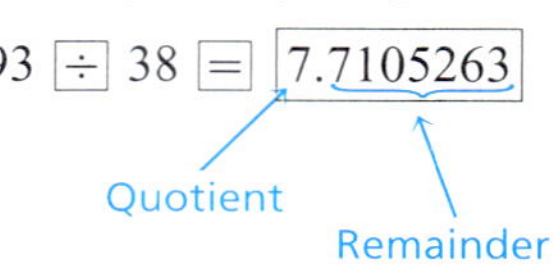

7 is the *whole-number part* of the quotient as before.

0.7105263 is the *decimal form of the remainder, 27, as a fraction of 38.*

CHECK YOURSELF 17

What is the result when you use your calculator to perform the given operation?

458 ÷ 36

The calculator can also help you combine division with other operations.

Example 18

Using a Scientific Calculator to Divide

To find 18 ÷ 2 + 3, use this sequence:

18 [÷] 2 [+] 3 [=]

Display 12

Do you see that the calculator has done the division as the first step?

CHECK YOURSELF 18

Use your calculator to compute.

15 ÷ 5 + 7

Example 19

Using a Scientific Calculator to Divide

To find 6 ÷ 3 · 2, use this sequence:

6 [÷] 3 [×] 2 [=]

Display 4

CHECK YOURSELF 19

Use your calculator to compute.

18 ÷ 6 · 5

CHECK YOURSELF ANSWERS

1. 12 **2.** 315 students ÷ 9 classes; students per class **3.** **(a)** 5; 9 · 5 = 45; **(b)** 4; 7 · 4 = 28 **4.** 5 **5.** 5 with remainder 3 **6.** **(a)** 0; **(b)** 0 **7.** **(a)** undefined; **(b)** 0; **(c)** 0; **(d)** undefined **8.** 8 · 22 = 176 **9.** 64 with remainder 5 **10.** 8 with remainder 26 **11.** 8 with remainder 53 **12.** 205 with remainder 42 **13.** 140 mi/h **14.** $70 **15.** 0 **16.** Error message **17.** 12.72222 **18.** 10 **19.** 15

Name ____________

Section ________ Date ________

1.6 Exercises

ANSWERS

1. ____
2. ____
3. ____ 4. ____
5. ____ 6. ____
7. ____ 8. ____
9. ____ 10. ____
11. ____ 12. ____
13. ____
14. ____
15. ____ 16. ____
17. ____ 18. ____
19. ____ 20. ____
21. ____
22. ____
23. ____ 24. ____
25. ____ 26. ____
27. ____ 28. ____
29. ____ 30. ____
31. ____ 32. ____
33. ____ 34. ____
35. ____ 36. ____
37. ____
38. ____

1. Given $48 \div 8 = 6$, 8 is the _______, 48 is the _______, and 6 is the _______.

2. In the statement $5\overline{)45}$ (quotient 9), 9 is the _______, 5 is the _______, and 45 is the _______.

3. Find $36 \div 9$ by repeated subtraction.

4. Find $40 \div 8$ by repeated subtraction.

5. Stefanie is planting rows of tomato plants. She wants to plant 63 plants with 9 plants per row. How many rows will she have?

6. Nick is designing a parking lot for a small office building. He must make room for 42 cars with 7 cars per row. How many rows should he plan for?

Divide, and identify the correct units for the quotient.

7. 36 pages ÷ 4

8. \$96 ÷ 8

9. 4900 km ÷ 7

10. 360 gal ÷ 18

11. 160 mi ÷ 4 h

12. 264 ft ÷ 3 s

13. 3720 h ÷ 5 months

14. 560 cal ÷ 7 g

Divide using long division, and check your work.

15. $5\overline{)43}$

16. $40 \div 9$

17. $9\overline{)65}$

18. $6\overline{)51}$

19. $57 \div 8$

20. $74 \div 8$

21. $0 \div 6$

22. $18 \div 0$

Divide.

23. $5\overline{)83}$

24. $9\overline{)78}$

25. $8\overline{)293}$

26. $7\overline{)346}$

27. $8\overline{)3136}$

28. $9\overline{)3527}$

29. $8\overline{)22,153}$

30. $5\overline{)43,287}$

31. $48\overline{)892}$

32. $54\overline{)372}$

33. $45\overline{)2367}$

34. $53\overline{)3480}$

35. $763\overline{)3071}$

36. $871\overline{)4321}$

In exercises 37 to 43, solve the applications.

37. Counting. Ramon bought 56 bags of candy. There were 8 bags in each box. How many boxes were there?

38. Capacity. There are 32 students who are taking a field trip. If each car can hold 4 students, how many cars will be needed for the field trip?

ANSWERS

39. ____________

40. ____________

41. ____________

42. ____________

43. ____________

44. ____________

39. Recreation. Ticket receipts for a play were \$552. If the tickets were \$4 each, how many tickets were purchased?

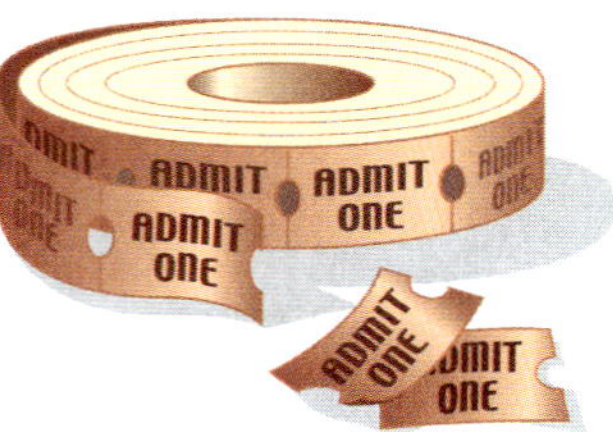

40. Construction. Construction of a fence section requires 8 boards. If you have 256 boards available, how many sections can you build?

41. Business. The homeowners along a street must share the \$2030 cost of new street lighting. If there are 14 homes, what amount will each owner pay?

42. Cost. A bookstore ordered 325 copies of a textbook at a cost of \$7800. What was the cost to the store for an individual textbook?

43. Bonuses. A company distributes \$16,488 in year-end bonuses. If each of the 36 employees receives the same amount, what bonus will each receive?

44. Complete the following number cross.

Across

1. $48 \div 4$
3. $1296 \div 8$
6. $2025 \div 5$
8. 4×5
9. 11×11
12. $15 \div 3 \times 111$
14. $144 \div (2 \times 6)$
16. $1404 \div 6$
18. $2500 \div 5$
19. 3×5

Down

1. $(12 + 16) \div 2$
2. 67×3
4. $744 \div 12$
5. $2600 \div 13$
7. $6300 \div 12$
10. $304 \div 2$
11. $5 \times (161 \div 7)$
13. $9027 \div 17$
15. $400 \div 20$
17. 9×5

Exponents and Order of Operations

OBJECTIVES

1. Use exponent notation
2. Evaluate expressions containing powers of whole numbers
3. Know the order of operations
4. Evaluate expressions that contain several operations

Overcoming Math Anxiety

Preparing for a Test

Preparation for a test really begins on the first day of class. Everything you have done in class and at home has been part of that preparation. However, there are a few things that you should focus on in the last few days before a scheduled test.

1. Plan your test preparation to end at least 24 hours before the test. The last 24 hours is too late, and besides, you will need some rest before the test.
2. Go over your homework and class notes with pencil and paper in hand. Write down all of the problem types, formulas, and definitions that you think might give you trouble on the test. Rework the class examples from your notes.
3. The day before the test, take the page(s) of notes from step 2, and transfer the most important ideas to a 3 by 5 card.
4. Just before the test, review the information on the card. You will be surprised at how much you remember about each concept.
5. Understand that, if you have been successful at completing your homework assignments, you can be successful on the test. This is an obstacle for many students, but it is an obstacle that can be overcome. Truly anxious students are often surprised that they scored as well as they did on a test. They tend to attribute this to blind luck. It is not. It is the first sign that you really do "get it." Enjoy the success.

NOTE Recall that

$3 + 3 + 3 + 3 = 4 \cdot 3$

Repeated addition was written as multiplication.

Earlier we described multiplication as a shorthand for repeated addition. There is also a shorthand for repeated multiplication. It uses **powers of a whole number.**

Example 1

Writing Repeated Multiplication as a Power

$3 \cdot 3 \cdot 3 \cdot 3$ can be written as 3^4. This is read as "3 to the fourth power."

In this case, repeated multiplication is written as the power of a number.

In this example, 3 is the **base** of the expression, and the raised number, 4, is the **exponent,** or **power.**

NOTE René Descartes, a French philosopher and mathematician, is generally credited with first introducing our modern exponent notation in about 1637.

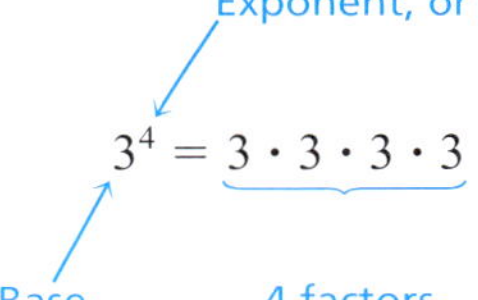

We count the factors and make this the power (or exponent) of the base.

CHECK YOURSELF 1

Write $2 \cdot 2 \cdot 2 \cdot 2 \cdot 2 \cdot 2$ *as a power of* 2.

Definitions: Exponents

The *exponent* tells us the number of times the base is to be used as a factor.

Example 2

Evaluating a Number Raised to a Power

2^5 is read "2 to the fifth power."

$$2^5 = \underbrace{2 \cdot 2 \cdot 2 \cdot 2 \cdot 2}_{\text{5 times}} = 32$$

2^5 tells us to use 2 as a factor 5 times. The result is 32.

Here 2 is the base, and 5 is the exponent.

CHECK YOURSELF 2

Read and evaluate 3^4.

Example 3

Evaluating a Number Raised to a Power

CAUTION

Be careful: 5^3 is *entirely different* from $5 \cdot 3$.

NOTE $5^3 = 125$ whereas $5 \cdot 3 = 15$.

Evaluate 8^2 and 5^3.

$8^2 = 8 \cdot 8 = 64$ Use 2 factors of 8.

And 8^2 is read "8 to the second power" or "8 squared."

$5^3 = 5 \cdot 5 \cdot 5 = 125$ Use 3 factors of 5.

5^3 is read "5 to the third power" or "5 cubed."

CHECK YOURSELF 3

Evaluate.

(a) 6^2 **(b)** 2^4

We need two special definitions for powers of whole numbers.

Rules and Properties: Raising a Number to the First Power

A whole number raised to the first power is just that number.

For example, $9^1 = 9$.

Rules and Properties: Raising a Number to the Zero Power

A whole number, other than 0, raised to the zero power is 1.

For example, $7^0 = 1$.

NOTE 0^0 is undefined.

Example 4

Evaluating Numbers Raised to the Power of Zero or One

(a) $8^0 = 1$ **(b)** $4^0 = 1$ **(c)** $5^1 = 5$ **(d)** $3^1 = 3$

CHECK YOURSELF 4

Evaluate.

(a) 9^0 **(b)** 7^1

We talked about *powers of 10* when we multiplied by numbers that end in 0. Because the powers of 10 have a special importance, we will list some of them.

$10^0 = 1$

$10^1 = 10$

$10^2 = 10 \cdot 10 = 100$

$10^3 = 10 \cdot 10 \cdot 10 = 1000$

$10^4 = 10 \cdot 10 \cdot 10 \cdot 10 = 10{,}000$

$10^5 = 10 \cdot 10 \cdot 10 \cdot 10 \cdot 10 = 100{,}000$

NOTE Notice that 10^3 is just a 1 followed by *three zeros.*

NOTE 10^5 is a 1 followed by *five zeros.*

Do you see why the powers of 10 are so important?

NOTE Archimedes (about 250 B.C.) reportedly estimated the number of grains of sand in the universe to be 10^{63}. This would be a 1 followed by 63 zeros!

Rules and Properties: Powers of 10

The powers of 10 correspond to the place values of our number system, ones, tens, hundreds, thousands, and so on.

This is what we meant earlier when we said that our number system was based on the number 10.

If multiplication is combined with addition or subtraction, you must know which operation to do first in finding the expression's value. We can easily illustrate this problem. How should we simplify this statement?

$3 + 4 \cdot 5 = ?$

Both multiplication and addition are involved in this expression, and we must decide which to do first to find the answer.

Step by Step: The Order of Operations

If multiplication, division, addition, and subtraction are involved in the same expression, do the operations in this order:

Step 1 Do all multiplication or division in order from left to right.
Step 2 Do all addition or subtraction in order from left to right.

Example 5

Using the Order of Operations

NOTE By this rule, we see that strategy (1) above was correct.

(a) $3 + 4 \cdot 5 = 3 + 20 = 23$ Multiply *first,* then add or subtract.
(b) $5 + 3 \cdot 6 = 5 + 18 = 23$
(c) $16 - 2 \cdot 3 = 16 - 6 = 10$
(d) $7 \cdot 8 - 20 = 56 - 20 = 36$
(e) $5 \cdot 6 + 4 \cdot 3 = 30 + 12 = 42$

CHECK YOURSELF 5

Evaluate.

(a) $8 + 3 \cdot 5$ **(b)** $15 \cdot 5 - 3$ **(c)** $4 \cdot 3 + 2 \cdot 6$

We now want to extend our rule for the order of operations.

NOTE Students sometimes remember this order by relating each step to part of the phrase

"Please	P
Excuse	E
My Dear	MD
Aunt Sally."	AS

Step by Step: The Order of Operations

Mixed operations in an expression should be done in this order:

Step 1 Do any operations inside parentheses.
Step 2 Evaluate any exponents.
Step 3 Do all multiplication or division in order from left to right.
Step 4 Do all addition or subtraction in order from left to right.

Example 6

Evaluating an Expression

Evaluate $4 \cdot 2^3$.

Step 1 There are no parentheses.

Step 2 Evaluate exponents.

$4 \cdot 2^3 = 4 \cdot 8$

Step 3 Multiply or divide.

$4 \cdot 8 = 32$

CHECK YOURSELF 6

Evaluate.

$3 \cdot 3^2$

Example 7

Evaluating an Expression

Evaluate $(2 + 3)^2 + 4 \cdot 3$.

Step 1 Do operations inside parentheses.

$(2 + 3)^2 + 4 \cdot 3 = 5^2 + 4 \cdot 3$

Step 2 Evaluate exponents.

$5^2 + 4 \cdot 3 = 25 + 4 \cdot 3$

Step 3 Multiply or divide.

$25 + 4 \cdot 3 = 25 + 12$

Step 4 Add or subtract.

$25 + 12 = 37$

CHECK YOURSELF 7

Evaluate.

$4 + (8 - 5)^2$

Example 8

Using the Order of Operations

(a) Evaluate $20 \div 2 \cdot 5$.

$$\underbrace{20 \div 2} \cdot 5$$
$$= 10 \cdot 5$$
$$= 50$$

So $20 \div 2 \cdot 5 = 50$.

Because the multiplication and division appear next to each other, work in order from left to right. Try it the other way and see what happens!

(b) Evaluate $(5 + 13) \div 6$.

$$\underbrace{(5 + 13)} \div 6$$
$$= 18 \div 6$$
$$= 3$$

So $(5 + 13) \div 6 = 3$.

Do the addition in the parentheses as the first step.

CHECK YOURSELF 8

Evaluate.

(a) $36 \div 4 \cdot 2$

(b) $(8 + 22) \div 5$

Using a Scientific Calculator to Evaluate an Expression

NOTE As we pointed out earlier, there are many differences among calculators. If any of the examples we consider in this section do not come out the same on the model you are using, check with your instructor.

Scientific calculators are designed to correctly apply the order of arithmetic operations. For each of the following two examples, be certain that you understand **why** the given answer is correct.

Example 9

Using a Scientific Calculator to Evaluate an Expression

To evaluate $6 + 3 \cdot 5$, use this sequence:

6 [+] 3 [×] 5 [=]

Display 21

Again the calculator follows the "order of operations." Try this with your calculator.

CHECK YOURSELF 9

Evaluate $9 + 3 \cdot 7$ *with your calculator.*

Most calculators have parentheses keys that will allow you to evaluate more complicated expressions easily.

Example 10

Using a Scientific Calculator to Evaluate an Expression

To evaluate $3 \cdot (4 + 5)$, use this sequence:

3 [×] [(]4 [+] 5[)] [=]

Display 27

Now the calculator does the addition in the parentheses as the first step. Then it does the multiplication.

CHECK YOURSELF 10

Evaluate $4 \cdot (2 + 9)$ *with your calculator.*

CHECK YOURSELF ANSWERS

1. 2^6 **2.** "Three to the fourth power" is 81 **3.** **(a)** 36; **(b)** 16 **4.** **(a)** 1; **(b)** 7 **5.** **(a)** 23; **(b)** 72; **(c)** 24 **6.** 27 **7.** 13 **8.** **(a)** 18; **(b)** 6 **9.** 30 **10.** 44

Name ______________

Section ________ Date ________

1.7 Exercises

Evaluate.

1. 3^2

2. 2^3

3. 2^4

4. 5^2

5. 5^1

6. 6^0

7. 9^0

8. 7^1

9. 10^3

10. 10^2

11. 10^6

12. 10^7

13. $2 \cdot 4^3$

14. $(2 \cdot 4)^3$

15. $5 + 2^2$

16. $(5 + 2)^2$

17. $(3 \cdot 2)^4$

18. $3 \cdot 2^4$

19. $2 \cdot 6^2$

20. $(2 \cdot 6)^2$

21. $14 - 3^2$

22. $12 + 4^2$

23. $(3 + 2)^3 - 20$

24. $5 + (9 - 5)^2$

25. $(7 - 4)^4 - 30$

26. $(5 + 2)^2 + 20$

27. $8 \div 4 + 2$

28. $3 \cdot 5 + 2$

29. $24 - 6 \div 3$

30. $3 + 9 \div 3$

31. $(24 - 6) \div 3$

32. $(3 + 9) \div 3$

33. $12 + 3 \div 3$

34. $6 \cdot 12 \div 3$

35. $18 \div 6 \cdot 3$

36. $30 \div 5 \cdot 2$

ANSWERS

1. ______ 2. ______
3. ______ 4. ______
5. ______ 6. ______
7. ______ 8. ______
9. ______ 10. ______
11. ______
12. ______
13. ______ 14. ______
15. ______ 16. ______
17. ______ 18. ______
19. ______ 20. ______
21. ______ 22. ______
23. ______ 24. ______
25. ______
26. ______
27. ______
28. ______
29. ______
30. ______
31. ______
32. ______
33. ______
34. ______
35. ______
36. ______

ANSWERS

37. __________

38. __________

39. __________

40. __________

41. __________

42. __________

43. __________

44. __________

45. __________

46. __________

47. __________

48. __________

49. __________

50. __________

51. __________

52. __________

53. __________

54. __________

55. __________

56. __________

57. __________

58. __________

37. $30 \div 6 - 12 \div 3$

38. $5 + 8 \div 4 - 3$

39. $4^2 \div 2$

40. $2 \cdot 4^3$

41. $5^2 \cdot 3$

42. $6^2 \div 3$

43. $3 \cdot 3^3$

44. $2^5 \cdot 3$

45. $(3^3 + 3) \div 10$

46. $(2^4 + 4) \div 5$

47. $15 \div (5 - 3 + 1)$

48. $20 \div (3 + 4 - 2)$

49. $27 \div (2^2 + 5)$

50. $48 \div (2^3 + 4)$

Numbers such as 3, 4, and 5 are called **Pythagorean triples,** after the Greek mathematician Pythagoras (sixth century B.C.), because

$3^2 + 4^2 = 5^2$

Which of the following sets of numbers are Pythagorean triples?

51. 6, 8, 10

52. 6, 11, 12

53. 5, 12, 13

54. 7, 24, 25

55. 8, 16, 18

56. 8, 15, 17

57. Is $(a + b)^P$ equal to $a^P + b^P$?

Try a few numbers and decide if you think this is true for all whole numbers, for some whole numbers, or never true. Write an explanation of your findings, and give examples.

58. Does $(a \cdot b)^P = a^P \cdot b^P$?

Try a few numbers and decide if you think this is true for all whole numbers, for some whole numbers, or never true. Write an explanation of your findings, and give examples.

Calculator Exercises

Multiply, using your calculator.

59. $\begin{array}{r} 256 \\ \times 508 \\ \hline \end{array}$

60. $\begin{array}{r} 18{,}569 \\ \times \quad 3286 \\ \hline \end{array}$

61. $78 \cdot 145 \cdot 36$

62. $37 \cdot 15 \cdot 42 \cdot 29$

Use your calculator to evaluate each of the following expressions.

63. $4 \cdot 5 - 7$

64. $6 \cdot 0 + 3$

65. $4 + 5 \cdot 0$

66. $8 \cdot (6 + 5)$

67. $5 \cdot 4 + 5 \cdot 7$

68. $8 \cdot 6 + 8 \cdot 5$

Solve the applications in exercises 69 and 70, using a calculator.

69. **Sales.** A car dealer kept a record of a month's sales. Complete the table.

Model	Number Sold	Profit per Sale	Monthly Profit
Subcompact	38	$528	_______
Compact	33	647	_______
Standard	19	912	_______
		Monthly Total Profit	_______

70. **Salary.** You take a job paying $1 the first day. On each following day your pay doubles. That is, on day 2 your pay is $2, on day 3 the pay is $4, and so on. Complete the table.

Day	Daily Pay	Total Pay
1	$1	$1
2	2	3
3	4	7
4	_______	_______
5	_______	_______
6	_______	_______
7	_______	_______
8	_______	_______
9	_______	_______
10	_______	_______

ANSWERS

59. _______
60. _______
61. _______
62. _______
63. _______
64. _______
65. _______
66. _______
67. _______
68. _______
69. _______
70. _______

Answers

1. 9 **3.** 16 **5.** 5 **7.** 1 **9.** 1000 **11.** 1,000,000 **13.** 128
15. 9 **17.** 1296 **19.** 72 **21.** 5 **23.** 105 **25.** 51 **27.** 4
29. 22 **31.** 6 **33.** 13 **35.** 9 **37.** 1 **39.** 8 **41.** 75
43. 81 **45.** 3 **47.** 5 **49.** 3 **51.** Yes **53.** Yes **55.** No
57. **59.** 130,048 **61.** 407,160 **63.** 13 **65.** 4 **67.** 55

69. Monthly profit:

$$\begin{array}{r} \$20{,}064 \\ 21{,}351 \\ \underline{17{,}328} \\ \$58{,}743 \end{array}$$

An Introduction to Equations

1. Identify expressions and equations
2. Determine whether an equation is true
3. Translate a sentence into an equation

Overcoming Math Anxiety

Taking a Test

Doing homework and asking questions are the best ways to prepare for a test. Once you are thoroughly prepared for the test, you must learn how to take it.

There is much to the psychology of anxiety that we can't readily address. There is, however, a physical aspect to anxiety that can be addressed rather easily. When people are in a stressful situation, they frequently start to panic. One symptom of the panic is shallow breathing. In a test situation, this starts a vicious cycle. If you breathe too shallowly, then not enough oxygen reaches the brain. When that happens, you are unable to think clearly. In a test situation, being unable to think clearly can cause you to panic. Hence we have a vicious cycle.

How do you break that cycle? It's pretty simple. Take a few deep breaths. We have seen students whose performance on math tests improved markedly after they got in the habit of writing "remember to breathe!" at the bottom of every test page. Try breathing; it will almost certainly improve your math test scores!

An ***expression*** is a number or a meaningful collection of operations ($+$, $-$, $\cdot$, $\div$) and numbers. Each of the following is an expression.

$14 \qquad 4 + 5 \cdot 2 \qquad 3 \cdot 5 - 1 \qquad 2^3 + 3 \cdot 2$

In Section 1.7, we used the order of operations to determine the value of an expression. Expressions are used to build equations.

An ***equation*** is two expressions connected by an equal sign. Each of the following is an equation.

$14 = 4 + 5 \cdot 2 \qquad 14 = 3 \cdot 5 - 1 \qquad 4 + 5 \cdot 2 = 3 \cdot 5 - 1$

Example 1

Identifying Expressions and Equations

Label each as an expression or an equation.

(a) $2 + 8 - 5$ is an expression.

(b) $2 \cdot 4 - 5 = 3$ is an equation.

(c) $2^3 + 4 = 12$ is an equation.

(d) $3^3 - 5 \cdot 4 - 6$ is an expression.

CHECK YOURSELF 1

Label each as an expression or an equation.

(a) $12 + 8 - 5 = 15$ **(b)** $3^3 - 5 = 22$

(c) $2^3 - 2 \cdot 4 + 5$

Did you notice that each of the equations in Example 1 was true? Unfortunately, that is not always the case. An important part of algebra is determining when an equation is true. Example 2 will help you to develop this skill.

Example 2

Determining Whether an Equation is True

An equation can also be called a statement. Label each equation as true or false.

(a) $17 + 8 - 9 = 16$ is a true statement.

(b) $2^3 - 5 = 1$ is a false statement.

(c) $3^2 - 2 = 2^2 + 3$ is a true statement because $3^2 - 2 = 9 - 2 = 7$ and $2^2 + 3 = 4 + 3 = 7$.

CHECK YOURSELF 2

Label each equation as true or false.

(a) $5 \cdot 4 - 7 = 13$

(b) $2^3 \cdot 4 = 32$

(c) $12 - 3^2 = 3^2 - 12$

In our final example in this chapter, we will translate sentences into equations.

Example 3

Translating Sentences into Equations

An equation can be read as a sentence. Given $2 \cdot 4 = 8$, we can read it as "two times four equals eight," or simply, "two times four is eight." For each sentence, write an equation.

(a) Three plus seven is two more than eight. This can be written as $3 + 7 = 8 + 2$. Notice that "two more than eight" is written as $8 + 2$.

(b) Four squared is the same as four times four. This can be written as $4^2 = 4 \cdot 4$.

(c) Three less than two squared is one. This is translated $2^2 - 3 = 1$. Notice here that "three less than two squared" is written as $2^2 - 3$.

CHECK YOURSELF 3

For each of the sentences, write an equation.

(a) Twenty-seven minus seven is two times ten.

(b) Four squared less than twenty is four.

(c) Thirteen more than two squared is seventeen.

CHECK YOURSELF ANSWERS

1. **(a)** Equation; **(b)** equation; **(c)** expression **2.** **(a)** True; **(b)** true; **(c)** false

3. **(a)** $27 - 7 = 2 \cdot 10$; **(b)** $20 - 4^2 = 4$; **(c)** $2^2 + 13 = 17$

Name ______________________

Section ________ Date ________

1.8 Exercises

Label each of the following as an expression or an equation.

1. $3 + 4 = 7$

2. $45 - 3 \cdot 5$

3. $36 \div 4 + 6 \cdot 3$

4. $2^2 + 15$

5. $2^2 + 1 = 5$

6. $27 - 13 + 6 = 2 \cdot 10$

7. $13 + 29 - 3$

8. $32 = 2^5$

Determine whether each equation is true or false.

9. $24 + 13 = 37$

10. $15 \cdot 3 = 35$

11. $2 \cdot 3 = 5$

12. $3 + 4 \cdot 2 = 11$

13. $12 - 3 \cdot 3 = 3$

14. $3^2 + 1 = 7$

15. $2 \cdot 3^2 = 18$

16. $2 \cdot 3^2 = 36$

17. $5 \cdot 2^2 = 100$

18. $5 \cdot 2^2 = 20$

Translate each phrase into an expression.

19. Three plus two minus five

20. Seven less than 12

21. Nine more than three

22. Six squared plus 5

Translate each expression into a phrase.

23. $5 + 3$

24. $2 \cdot 3$

25. $5 \cdot 2 - 1$

26. $3^2 + 1$

Translate each equation into a sentence.

27. $3 + 2 = 5$

28. $5 \cdot 2 = 10$

29. $15 - 3 = 12$

30. $3^2 + 2 = 11$

ANSWERS

1. ______
2. ______
3. ______
4. ______
5. ______
6. ______
7. ______
8. ______

9. ______ 10. ______
11. ______ 12. ______
13. ______ 14. ______
15. ______ 16. ______
17. ______ 18. ______
19. ______
20. ______ 21. ______
22. ______
23. ______
24. ______
25. ______
26. ______
27. ______
28. ______
29. ______
30. ______

ANSWERS

31. ______________________

32. ______________________

33. ______________________

34. ______________________

Translate each sentence into an equation.

31. Four more than seven is eleven.

32. Two more than three times five is seventeen.

33. Thirteen is five more than eight.

34. Six less than five squared is nineteen.

Answers

1. Equation **3.** Expression **5.** Equation **7.** Expression **9.** True
11. False **13.** True **15.** True **17.** False **19.** $3 + 2 - 5$
21. $3 + 9$ **23.** Five plus three **25.** One less than five times two
27. Two more than three is five **29.** Three less than fifteen is twelve
31. $7 + 4 = 11$ **33.** $13 = 8 + 5$

1 Summary

DEFINITION/PROCEDURE	EXAMPLE	REFERENCE
Introduction to Whole Numbers, Place Value		**Section 1.1**
Digits Digits are the basic symbols of the system.	0, 1, 2, 3, 4, 5, 6, 7, 8, and 9 are digits.	p. 3
Place Value The value of a digit in a number depends on its position or place.	7,352,589 — 9: Ones; 8: Tens; 5: Hundreds; 2: Thousands; 5: Ten thousands; 3: Hundred thousands; 7: Millions	p. 4
The **value of a number** is the sum of each digit multiplied by its place value.	$2345 = (2 \cdot 1000) + (3 \cdot 100) + (4 \cdot 10) + (5 \cdot 1)$	p. 5
Addition of Whole Numbers		**Section 1.2**
Addends The numbers that are being added. **Sum** The result of the addition.	$5 + 8$ Addends; 13 Sum	p. 12
The Properties of Addition **The Commutative Property** The order in which you add two whole numbers does not affect the sum.	$5 + 4 = 4 + 5$	p. 13
The Associative Property The way in which you group whole numbers in addition does not affect the final sum.	$(2 + 7) + 8 = 2 + (7 + 8)$	p. 14
The Additive Identity The sum of 0 and any whole number is just that whole number.	$6 + 0 = 0 + 6 = 6$	p. 14
Measuring Perimeter The perimeter is the total distance around the outside edge of a shape. The perimeter of a rectangle is $P = L + W + L + W$ (which can be written as $P = 2 \cdot L + 2 \cdot W$).	Rectangle: 6 ft, 2 ft, 6 ft, 2 ft $P = L + W + L + W$ $= 6\text{ ft} + 2\text{ ft} + 6\text{ ft} + 2\text{ ft}$ $= 16\text{ ft}$	p. 18

Continued

DEFINITION/PROCEDURE	EXAMPLE	REFERENCE
Subtraction of Whole Numbers		**Section 1.3**
Minuend The number we are subtracting from. **Subtrahend** The number that is being subtracted. **Difference** The result of the subtraction.	15 ⟵ Minuend − 9 ⟵ Subtrahend 6 ⟵ Difference	p. 29
Rounding, Estimation, and Ordering of Whole Numbers		**Section 1.4**
Step 1 To round a whole number to a certain decimal place, look at the digit to the *right* of that place. **Step 2** **a.** If that digit is 5 or more, that digit and all digits to the right become 0. The digit in the place you are rounding to is increased by 1. **b.** If that digit is less than 5, that digit and all digits to the right become 0. The digit in the place you are rounding to remains the same.	To the nearest hundred, 43,578 is rounded to 43,600. To the nearest thousand, 273,212 is rounded to 273,000.	p. 42
Order on the Whole Numbers For the numbers a and b, we can write **1.** $a < b$ (read "a is less than b") when a is *to the left* of b on the number line.	$8 < 12$ 8 9 10 11 12	p. 45
2. $a > b$ (read "a is greater than b") when a is *to the right* of b on the number line.	$15 > 10$ 10 11 12 13 14 15	p. 45
Multiplication of Whole Numbers		**Section 1.5**
Factors The numbers being multiplied. **Product** The result of the multiplication.	$7 \cdot 9 = 63$ ⟵ Product ↑ Factors	p. 51
The Properties of Multiplication **The Commutative Property** Multiplication, like addition, is a *commutative* operation. The order in which you multiply two whole numbers does not affect the product.	$7 \cdot 9 = 9 \cdot 7$	p. 51
The Distributive Property To multiply a factor by a sum of numbers, multiply the factor by each number inside the parentheses. Then add the products.	$2 \cdot (3 + 7) = (2 \cdot 3) + (2 \cdot 7)$	p. 53
Multiplicative Property of Zero The product of zero and any number is zero.	$3 \cdot 0 = 0 \cdot 3 = 0$	p. 54
Multiplicative Identity Property The product of one and any number is that number.	$3 \cdot 1 = 1 \cdot 3 = 3$	p. 54

DEFINITION/PROCEDURE	EXAMPLE	REFERENCE
Multiplication of Whole Numbers		**Section 1.5**
The Associative Property Multiplication is an *associative* operation. The way in which you group numbers in multiplication does not affect the final product.	$(3 \cdot 5) \cdot 6 = 3 \cdot (5 \cdot 6)$	p. 59
Finding the Area of a Rectangle The area of a rectangle is found using the formula $A = L \cdot W$.	6 ft, 2 ft $A = L \cdot W = 6\text{ ft} \cdot 2\text{ ft} = 12\text{ ft}^2$	p. 61
Division of Whole Numbers		**Section 1.6**
Divisor The number we are dividing by. **Dividend** The number being divided. **Quotient** The result of the division. **Remainder** The number "left over" after the division.	Divisor, Quotient 5 7)38 ← Dividend 35 3 ← Remainder	p. 72
Dividend = Divisor · Quotient + Remainder	$38 = 7 \cdot 5 + 3$	p. 74
The Role of 0 in Division Zero divided by any whole number (except 0) is 0.	$0 \div 7 = 0$	p. 75
Division by 0 is undefined.	$7 \div 0$ is undefined.	p. 75
Exponents and Order of Operations		**Section 1.7**
Using Exponents **Base** The number that is raised to a power. **Exponent** The exponent is written to the right and above the base. The exponent tells the number of times the base is to be used as a factor.	Exponent $5^3 = \underbrace{5 \cdot 5 \cdot 5}_{\text{Three factors}} = 125$ Base This is read "5 to the third power" or "5 cubed."	p. 87
The Order of Operations *Mixed operations* in an expression should be done in the following order: **Step 1** Do any operations inside parentheses. **Step 2** Evaluate any exponents. **Step 3** Do all multiplication and division in order from left to right. **Step 4** Do all addition and subtraction in order from left to right. Remember *P*lease *E*xcuse *M*y *D*ear *A*unt *S*ally	$4 \cdot (2 + 3)^2 - 7$ $= 4 \cdot 5^2 - 7$ $= 4 \cdot 25 - 7$ $= 100 - 7$ $= 93$	p. 90

Continued

DEFINITION/PROCEDURE	EXAMPLE	REFERENCE
An Introduction to Equations		**Section 1.8**
An **expression** is a number or a meaningful collection of operations (+, −, ·, ÷) and numbers. An **equation** is two expressions connected by an equal sign. An equation can be true or false.	$9 + 5 \cdot 2$ $2^3 - 5 \cdot 2$ $19 = 9 + 5 \cdot 2$ $14 - 3 \cdot 2 = 8$ is true $14 - 3 \cdot 2 = 22$ is false	p. 97

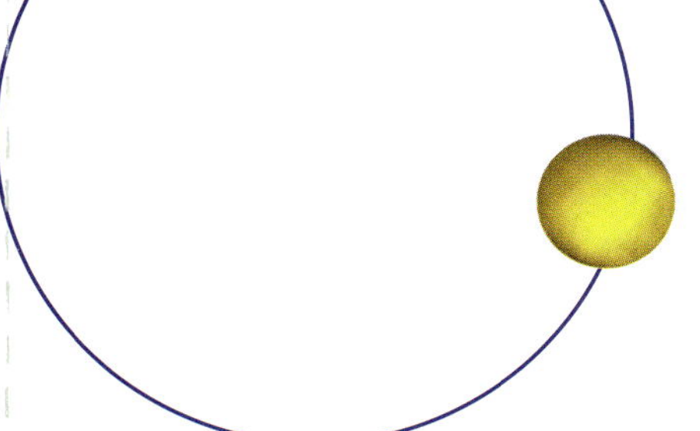

Summary and Review Exercises

You should now be reviewing the material in Chapter 1. These exercises will help in that process. Work all the exercises carefully. References are provided to the section for each exercise. If you have difficulty with an exercise, go back and review the related material.

[1.1] In exercises 1 and 2, give the place value of each of the indicated digits.

1. 6 in the number 5674

2. 5 in the number 543,400

In exercises 3 and 4, write each number in words.

3. 27,428

4. 200,305

Write each of the following as a number.

5. Thirty-seven thousand, five hundred eighty-three

6. Three hundred thousand, four hundred

[1.2] In exercises 7 and 8, name the property of addition that is illustrated.

7. $4 + 9 = 9 + 4$

8. $(4 + 5) + 9 = 4 + (5 + 9)$

In exercises 9 to 13, perform the indicated operations.

9.
$$\begin{array}{r} 784 \\ 385 \\ +\ 247 \\ \hline \end{array}$$

10.
$$\begin{array}{r} 2570 \\ 498 \\ 21{,}456 \\ +\quad 28 \\ \hline \end{array}$$

11.
$$\begin{array}{r} 367 \\ 289 \\ 1463 \\ +\ 2682 \\ \hline \end{array}$$

12.
$$\begin{array}{r} 6389 \\ 1567 \\ 315 \\ +\ 113{,}602 \\ \hline \end{array}$$

13. Find the value of 7 more than 4.

Find the perimeter of each figure.

14.

5 ft
2 ft
2 ft
2 ft
2 ft
5 ft

15.

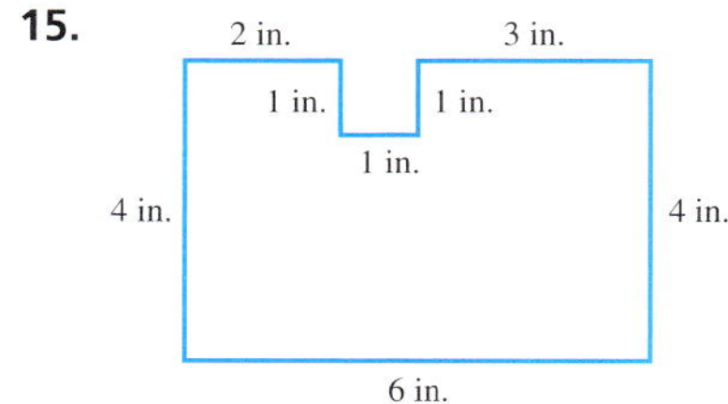

Solve the applications.

16. **Passenger count.** An airline had 173, 212, 185, 197, and 202 passengers on five morning flights between Washington, D.C., and New York. What was the total number of passengers?

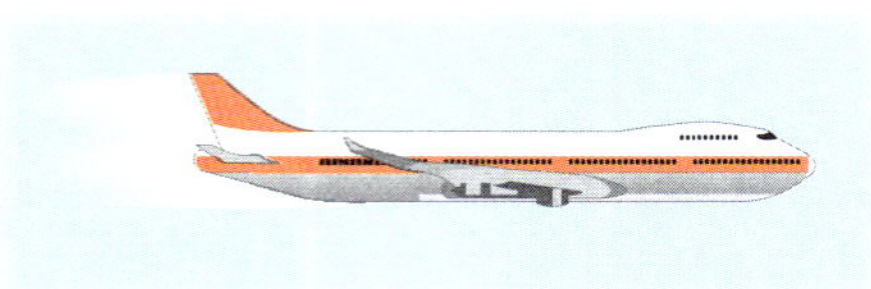

17. **Salaries.** The Future Stars summer camp employs five junior counselors. Their weekly salaries last week were \$108, \$135, \$81, \$135, and \$81. What was the total salary for the junior counselors?

[1.3] In exercises 18 to 23, perform the indicated operations.

18. $\begin{array}{r} 5325 \\ -\ 847 \\ \hline \end{array}$

19. $\begin{array}{r} 38{,}400 \\ -\ 19{,}600 \\ \hline \end{array}$

20. $\begin{array}{r} 86{,}000 \\ -\ 2169 \\ \hline \end{array}$

21. $\begin{array}{r} 2682 \\ -\ 108 \\ \hline \end{array}$

22. Find the difference of 7342 and 5579.

23. Find the value of 34 decreased by 7.

Solve the applications.

24. **Credit card payments.** Chuck owes \$795 on a credit card after a trip. He makes payments of \$75, \$125, and \$90. Interest of \$31 is charged. How much remains to be paid on the account?

25. **Total cost.** Juan bought a new car for \$16,785. The manufacturer offers a cash rebate of \$987. What was the cost after rebate?

[1.4] In exercises 26 to 28, round the numbers to the indicated place.

26. 6975 to the nearest hundred

27. 15,897 to the nearest thousand

28. 548,239 to the nearest ten thousand

In exercises 29 and 30, complete the statements by using the symbol $<$ or $>$.

29. 60 ____________ 70

30. 38 ____________ 35

[1.5] In exercises 31 to 34, name the property that is illustrated.

31. $7 \cdot 8 = 8 \cdot 7$

32. $3 \cdot (4 + 7) = 3 \cdot 4 + 3 \cdot 7$

33. $(8 \cdot 9) \cdot 4 = 8 \cdot (9 \cdot 4)$

34. $6 \cdot 1 = 6$

In exercises 35 to 37, perform the indicated operations.

35. $\begin{array}{r} 58 \\ \times\ 32 \\ \hline \end{array}$

36. $\begin{array}{r} 25 \\ \times\ 43 \\ \hline \end{array}$

37. $\begin{array}{r} 378 \\ \times\ 409 \\ \hline \end{array}$

Solve the application.

38. Costs. You wish to carpet a room that is 5 yd by 7 yd. The carpet costs \$18 per square yard. What will be the total cost of the materials?

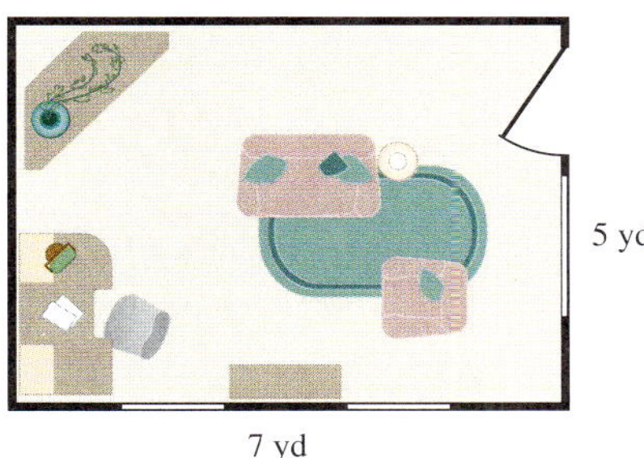

In exercise 39, perform the indicated operation.

39. $\begin{array}{r} 129 \\ \times\ 240 \\ \hline \end{array}$

Estimate the product by rounding each factor to the nearest hundred.

40. $\begin{array}{r} 1217 \\ \times\ 494 \\ \hline \end{array}$

Find the area of the given figures.

41. 6 in. / 3 in. / 3 in. / 6 in.

42.

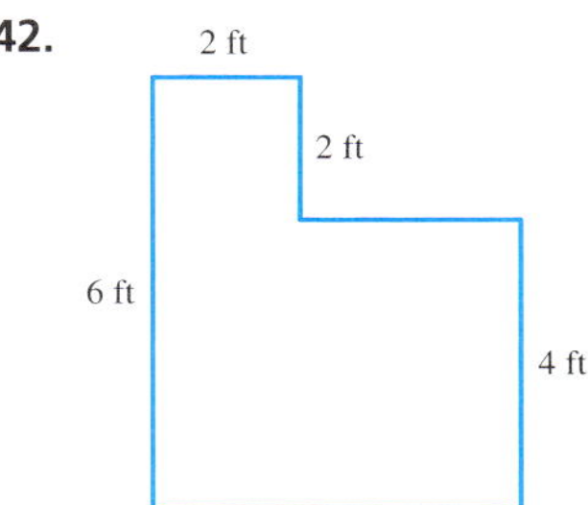

[1.6] In exercises 43 and 44, divide if possible.

43. $0 \div 8$

44. $5 \div 0$

In exercises 45 to 48, divide.

45. $8\overline{)2469}$

46. $39\overline{)2157}$

47. $64\overline{)31,809}$

48. $362\overline{)86,915}$

Solve the application.

49. **Mileage.** Hasina's odometer read 25,235 mi at the beginning of a trip and 26,215 mi at the end. If she used 35 gal of gas for the trip, what was her mileage (mi/gal)?

Estimate the following.

50. 356 divided by 37

51. 2125 divided by 28

Perform the indicated operations.

52. Find the value for the product of 9 and 5, divided by 3.

[1.7] In exercises 53 to 62, evaluate the expressions.

53. $5 \cdot 2^3$

54. $(5 \cdot 2)^3$

55. $4 + 8 \cdot 3$

56. $48 \div (2^3 + 4)$

57. $(4 + 8) \cdot 3$

58. $4 \cdot 3 + 8 \cdot 3$

59. $8 \div 4 \cdot 2 - 2 + 1$

60. $63 \cdot 2 \div 3 - 54 \div (12 \cdot 2 \div 4)$

61. $(3 \cdot 4)^2 - 100 \div 5 \cdot 6$

62. $(16 \cdot 2) \div 8 - (6 \div 3 \cdot 2)$

[1.8] Determine whether each is an expression or an equation.

63. $5 + 3 - 2 \cdot 2$

64. $7 - 2 \cdot 3 = 1$

65. $4^2 - 3 \cdot 2$

66. $2 \cdot 3^2 = 18$

Determine whether each equation is true or false.

67. $3 \cdot 2 + 1 = 7$

68. $5 + 3 \cdot 4 = 32$

69. $7 + 2 \cdot 2 = 11$

70. $3 \cdot 2^2 + 1 = 13$

Translate each of the sentences to an equation.

71. Five more than two squared is nine.

72. Eight less than four times three is four.

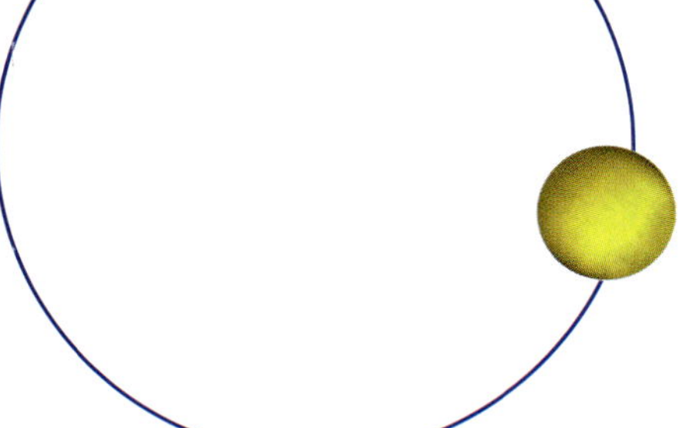

Chapter Test for Chapter 1

Name ______________________

Section ________ Date ________

ANSWERS

1. ______________
2. ______________
3. ______________
4. ______________
5. ______________
6. ______________
7. ______________
8. ______________
9. ______________
10. ______________
11. ______________
12. ______________
13. ______________
14. ______________
15. ______________

The purpose of the Chapter Test is to help you check your progress and review for a chapter test in class. When you are done, check your answers in the back of the book. If you missed any answers, be sure to go back and review the appropriate sections in the chapter.

1. Give the place value of 7 in 3,738,500.

2. Give the word name for 302,525.

3. Write two million, four hundred thirty thousand as a number.

In exercises 4 and 5, name the property of addition that is illustrated.

4. $5 + 12 = 12 + 5$

5. $(7 + 3) + 8 = 7 + (3 + 8)$

In exercises 6 and 7, perform the indicated operations.

6.
$$\begin{array}{r} 489 \\ 562 \\ 613 \\ +\ 254 \\ \hline \end{array}$$

7.
$$\begin{array}{r} 13 \\ 2543 \\ +\ 10{,}547 \\ \hline \end{array}$$

8. What is the total of 392, 95, 9237, and 11,972?

Solve the application.

9. **Game attendance.** The attendance for the games of a playoff series in basketball was 12,438, 14,325, 14,581, and 14,634. What was the total attendance for the series?

In exercises 10 to 13, subtract.

10. $289 - 54$

11. $53{,}294 - 41{,}074$

12. $32{,}345 - 1575$

13. $55{,}342 - 14{,}787$

14. The maximum load for a light plane with full gas tanks is 500 lb. Mr. Whitney weighs 215 lb, his wife 135 lb, and their daughter 78 lb. How much luggage can they take on a trip without exceeding the load limit?

In exercise 15, estimate the sum by rounding each addend to the nearest hundred.

15.
$$\begin{array}{r} 943 \\ 3281 \\ 778 \\ 2112 \\ +\ 570 \\ \hline \end{array}$$

ANSWERS

16. ________________

17. ________________

18. ________________

19. ________________

20. ________________

21. ________________

22. ________________

23. ________________

24. ________________

25. ________________

26. ________________

27. ________________

28. ________________

29. ________________

30. ________________

In exercises 16 and 17, complete the statements by using the symbol $<$ or $>$.

16. 49 ________________ 47

17. 80 ________________ 90

In exercises 18 to 20, name the property that is illustrated.

18. $3 \cdot (2 \cdot 7) = (3 \cdot 2) \cdot 7$

19. $4 \cdot (3 + 6) = (4 \cdot 3) + (4 \cdot 6)$

20. $5 \cdot 1 = 5$

Find the products.

21. $\begin{array}{r} 89 \\ \times\ 56 \\ \hline \end{array}$

22. $\begin{array}{r} 538 \\ \times\ 103 \\ \hline \end{array}$

23. A truck rental firm has ordered 25 new vans at a cost of \$12,350 per van. What will be the total cost of the order?

Divide using long division.

24. $28\overline{)2135}$

25. **Trip expenses.** Eight people estimate that the total expenses for a trip they are planning to take together will be \$1784. If each person pays an equal amount, what will be each person's share?

26. Evaluate the expression $15 - 12 \div 2^2 \cdot 3 + (12 \div 4 \cdot 3)$.

Find the perimeter of the figure shown.

27.

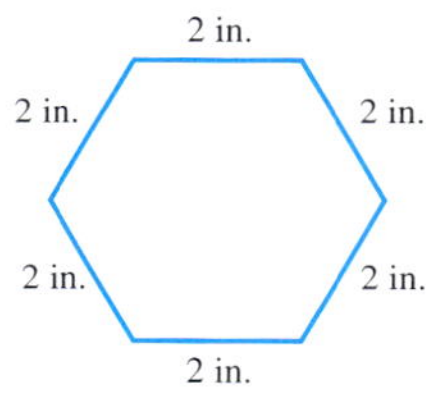

2 in.
2 in.
2 in.
2 in.
2 in.
2 in.

Find the area of the given figure.

28.

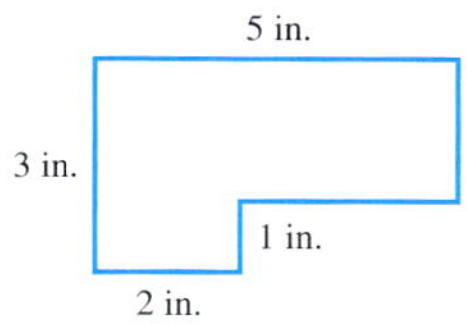

5 in.
3 in.
1 in.
2 in.

29. Is the equation $3 \cdot 5^2 = 75$ a true or false statement?

30. Translate into an equation: Two less than three squared is seven.

INTEGERS AND INTRODUCTION TO ALGEBRA

2

INTRODUCTION

Anthropologists and archaeologists investigate modern human cultures and societies as well as cultures that existed so long ago that their characteristics must be inferred from objects found buried in lost cities or villages. When some interesting object is found, such as the Rosetta stone, often the first questions that arise are "How old is this? When did this culture flourish?" With methods such as carbon dating, it has been established that large, organized cultures existed around 3000 B.C.E. in Egypt, 2800 B.C.E. in India, no later than 1500 B.C.E. in China, and around 1000 B.C.E. in the Americas.

How long ago was 1500 B.C.E.? Which is older, an object from 3000 B.C.E. or an object from A.D.* 500? Using the Christian notation for dates, we have to count A.D. years and B.C.E. years differently. An object from A.D. 500 is 2000 − 500 years old, or about 1500 years old. But an object from 3000 B.C.E. is 2000 + 3000 years old, or about 5000 years old. Why subtract in the first case but add in the second? Because of the way years are counted before the Christian era (B.C.E.) and after the birth of Christ (A.D.), the B.C.E. dates must be considered as *negative* numbers.

1000 B.C.E. = $^{-}1000$ A.D. 1000 = $^{+}1000$

Very early on, the Chinese accepted the idea that a number could be negative; they used red calculating rods for positive numbers and black for negative numbers. Hindu mathematicians in India worked out the arithmetic of negative numbers as long ago as A.D. 400, but western mathematicians did not recognize this idea until the sixteenth century. It would be difficult today to think of measuring things such as temperature, altitude, and money without using negative numbers.

*A.D. stands for the Latin *Anno Domini,* which means "in the year of the Lord."

Name ____________________

Section ________ Date ________

ANSWERS

1. ____________________
2. ____________________
3. ____________________
4. ____________________
5. ____________________
6. ____________________
7. ____________________
8. ____________________
9. ____________________
10. ____________________
11. ____________________
12. ____________________
13. ____________________
14. ____________________
15. ____________________
16. ____________________
17. ____________________
18. ____________________
19. ____________________
20. ________ 21. ________
22. ________ 23. ________
24. ________ 25. ________
26. ________ 27. ________
28. ____________________

Pre-Test Chapter 2

This pre-test will point out any difficulties you may be having with basic integer arithmetic and algebra. Do all the problems; then check your answers with those in the back of the book.

Represent the integers on the number line shown.

1. $6, -8, 4, -2, 10$

0

2. Place the following data set in ascending order: $5, -2, -4, 0, -1, 1$.

3. Determine the maximum and minimum of the following data set: $-4, 1, -5, 7, 3, 2$.

Evaluate:

4. $|-5|$ **5.** $|6|$ **6.** $|11 - 5|$

7. $|-11| - |5|$ **8.** $|4 + 5| - |6 - 3|$

Find the opposite of each integer.

9. -16 **10.** 23

Write each of the phrases using symbols.

11. 8 less than x

12. the quotient when w is divided by the product of x and 17

Identify which are expressions and which are not.

13. $7x - 5 = 11$ **14.** $3x - 2(x + 1)$

Perform the indicated operations.

15. $-7 + (-3)$ **16.** $8 + (-9)$ **17.** $(-3) + (-2)$

18. $8 - 11$ **19.** $-8 - 11$ **20.** $9 - (-3)$

21. $6 + (-6)$ **22.** $(-7)(-3)$ **23.** $\dfrac{-27 + 6}{-3}$

Evaluate each expression,

24. $5 - 4^2 \cdot 3 \div 6$ **25.** $(45 - 3 \cdot 5) + 5^2$

26. If $x = -2$, $y = 7$, and $w = -4$, evaluate the expression $\dfrac{x^2y}{w}$.

Combine like terms.

27. $5w^2t + 3w^2t$ **28.** $4a^2 - 3a + 5 + 7a - 2 - 5a^2$

Introduction to Integers

OBJECTIVES

1. Understand the meaning of a negative number
2. Recognize an integer
3. Represent an integer on a number line
4. Order integers
5. Evaluate expressions involving absolute value

When numbers are used to represent physical quantities (altitudes, temperatures, and amounts of money are examples), it may be necessary to distinguish between *positive* and *negative* quantities. It is convenient to represent these quantities with plus (+) or minus (−) signs. For instance, the altitude of Mount Whitney is 14,495 ft *above* sea level (+14,495).

Mount Whitney

The altitude of Death Valley is 282 ft *below* sea level (−282).

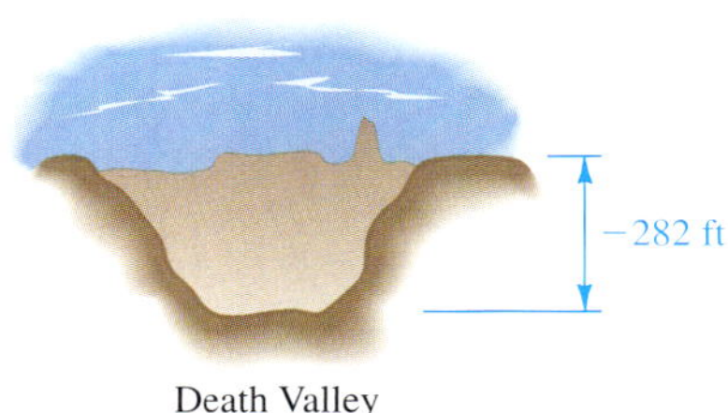

Death Valley

On a given day the temperature in Chicago is 10°F *below* zero (−10°).

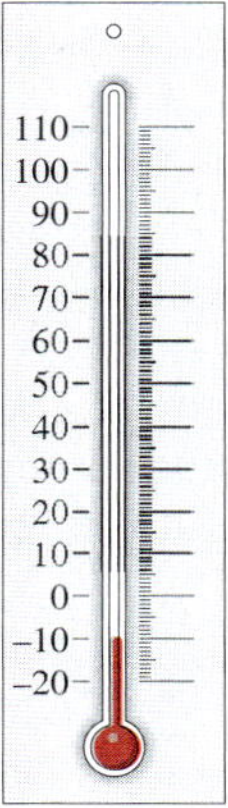

An account could show a *gain* of \$100 (+100), or a *loss* of \$100 (−100).

\$20
-\$25
-\$95
-\$100

These numbers suggest the need to extend the whole numbers to include both positive numbers (like +100) and negative numbers (like −282).

To represent the negative numbers, we extend the number line to the *left* of zero and name equally spaced points.

Numbers used to name points to the right of zero are positive numbers. They are written with a positive (+) sign or with no sign at all.

+6 and 9 are positive numbers

Numbers used to name points to the left of zero are negative numbers. They are always written with a negative (−) sign.

−3 and −20 are negative numbers

Read "negative 3."

Positive and negative numbers considered together are **signed numbers.**

Here is the number line extended to include both positive and negative numbers.

NOTE On the number line, we call zero the **origin.**

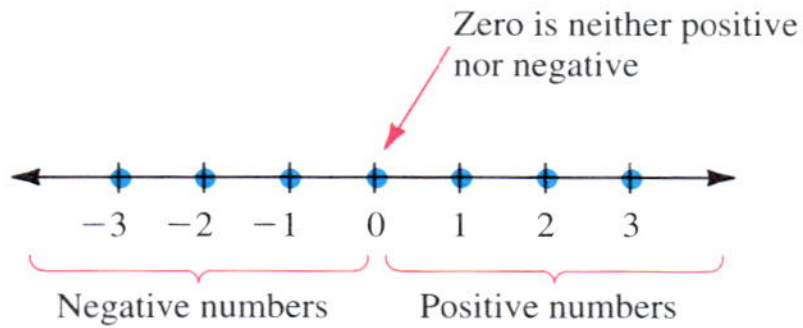

The numbers used to name the points shown on the number line are called the **integers.** The integers consist of the natural numbers, their negatives, and the number 0. We can represent the set of integers by

NOTE The dots are called *ellipses* and indicate that the pattern continues.

$$\{\ldots, -3, -2, -1, 0, 1, 2, 3, \ldots\}$$

Example 1

Representing Integers on the Number Line

Represent the integers on the number line shown.

$-3, -12, 8, 15, -7$

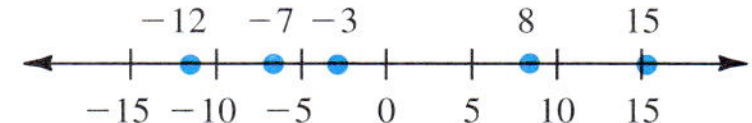

CHECK YOURSELF 1

Represent the integers on a number line.

$-1, -9, 4, -11, 8, 20$

−15 −10 −5 0 5 10 15 20

The set of numbers on the number line is *ordered.* The numbers get smaller moving to the left on the number line and larger moving to the right.

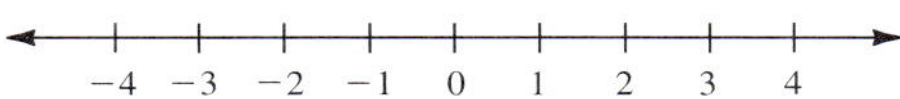

When a set of numbers is written from smallest to largest, the numbers are said to be in *ascending order.*

Example 2

Ordering Signed Numbers

Place each set of numbers in ascending order.

(a) $9, -5, -8, 3, 7$

From smallest to largest, the numbers are

$-8, -5, 3, 7, 9$

Note that this is the order in which the numbers appear on a number line as we move from left to right.

(b) $3, -2, 18, -20, -13$

From smallest to largest, the numbers are

$-20, -13, -2, 3, 18$

CHECK YOURSELF 2

Place each set of numbers in ascending order.

(a) $12, -13, 15, 2, -8, -3$

(b) $3, 6, -9, -3, 8$

The least and greatest numbers in a set are called the **extreme values.** The least element is called the **minimum,** and the greatest element is called the **maximum.**

Example 3

Labeling Extreme Values

For each set of numbers, determine the minimum and maximum values.

(a) 9, −5, −8, 3, 7

From our previous ordering of these numbers, we see that −8, the least element, is the minimum, and 9, the greatest element, is the maximum.

(b) 3, −2, 18, −20, −13

−20 is the minimum and 18 is the maximum.

CHECK YOURSELF 3

For each set of numbers, determine the minimum and maximum values.

(a) 12, −13, 15, 2, −8, −3 **(b)** 3, 6, −9, −3, 8

Integers are not the only kind of signed numbers. Decimals and fractions can also be thought of as signed numbers.

Example 4

Identifying Signed Numbers That Are Integers

Which of the signed numbers are also integers?

(a) 145 is an integer.

(b) −28 is an integer.

(c) 0.35 is not an integer.

(d) $-\frac{2}{3}$ is not an integer.

CHECK YOURSELF 4

Which of the signed numbers are also integers?

−23 1054 −0.23 0 −500 $-\frac{4}{5}$

Sometimes we refer to the negative of a number as its *opposite.* Example 5 illustrates this.

Example 5

Find the Opposite of Each Number

(a) 5 The opposite of 5 is -5.
(b) -9 The opposite of -9 is 9.

CHECK YOURSELF 5

Find the opposite of each number.

(a) 17 **(b)** -12

An important idea for our work in this chapter is the **absolute value** of a number. This represents the distance of the point named by the number from the origin on the number line.

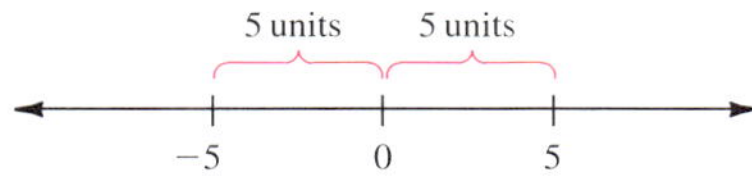

The absolute value of 5 is 5. The absolute value of -5 is also 5.

The absolute value of a positive number or zero is itself. The absolute value of a negative number is its opposite.

In symbols we write

$|5| = 5$ and $|-5| = 5$

Read "the absolute value of 5." Read "the absolute value of negative 5."

The absolute value of a number does *not* depend on whether the number is to the right or to the left of the origin, but on its *distance* from the origin.

Example 6

Simplifying Absolute Value Expressions

(a) $|7| = 7$
(b) $|-7| = 7$
(c) $-|-7| = -7$ This is the *negative,* or opposite, of the absolute value of negative 7.

CHECK YOURSELF 6

Evaluate.

(a) $|8|$ **(b)** $|-8|$ **(c)** $-|-8|$

To determine the order of operation for an expression that includes absolute values, note that the absolute value bars are a grouping symbol.

Example 7

Adding or Subtracting Absolute Values

(a) $|-10| + |10| = 10 + 10 = 20$

(b) $|8 - 3| = |5| = 5$ Absolute value bars, like parentheses, serve as a set of grouping symbols, so do the operation *inside* first.

(c) $|8| - |3| = 8 - 3 = 5$ Evaluate the absolute values, then subtract.

CHECK YOURSELF 7

Evaluate.

(a) $|-9| + |4|$ **(b)** $|9 - 4|$ **(c)** $|9| - |4|$

CHECK YOURSELF ANSWERS

1.

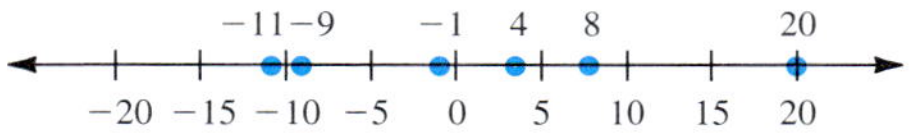

2. **(a)** $-13, -8, -3, 2, 12, 15$ **(b)** $-9, -3, 3, 6, 8$

3. **(a)** minimum is -13; maximum is 15 **(b)** minimum is -9; maximum is 8

4. -23, 1054, 0, and -500 **5.** **(a)** -17; **(b)** 12

6. **(a)** 8; **(b)** 8; **(c)** -8. **7.** **(a)** 13; **(b)** 5; **(c)** 5

Name ____________

Section ________ Date ________

2.1 Exercises

Represent each quantity with a signed number.

1. An altitude of 400 ft above sea level
2. An altitude of 80 ft below sea level
3. A loss of \$200
4. A profit of \$400
5. A decrease in population of 25,000
6. An increase in population of 12,500

Represent the integers on the number lines shown.

7. 5, −15, 18, −8, 3

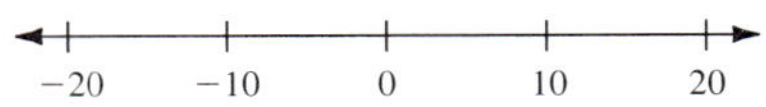

8. −18, 4, −5, 13, 9

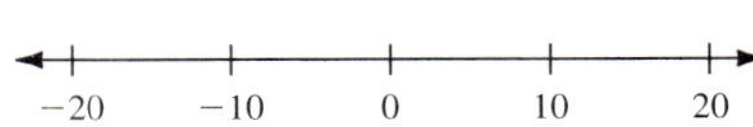

Which numbers in the sets are integers?

9. $\left\{5, -\frac{2}{9}, 175, -234, -0.64\right\}$
10. $\left\{-45, 0.35, \frac{3}{5}, 700, -26\right\}$

Place each of the sets in ascending order.

11. 3, −5, 2, 0, −7, −1, 8
12. −2, 7, 1, −8, 6, −1, 0
13. 9, −2, −11, 4, −6, 1, 5
14. 23, −18, −5, −11, −15, 14, 20
15. −6, 7, −7, 6, −3, 3
16. 12, −13, 14, −14, 15, −15

For each set, determine the maximum and minimum values.

17. 5, −6, 0, 10, −3, 15, 1, 8
18. 9, −1, 3, 11, −4, 2, 5, −2
19. 21, −15, 0, 7, −9, 16, −3, 11
20. −22, 0, 22, −31, 18, −5, 3
21. 3, 0, 1, −2, 5, 4, −1
22. 2, 7, −3, 5, −10, −5

Find the opposite of each number.

23. 15
24. 18

ANSWERS

1. ____________
2. ____________
3. ____________
4. ____________
5. ____________
6. ____________
7. ____________
8. ____________
9. ____________
10. ____________
11. ____________
12. ____________
13. ____________
14. ____________
15. ____________
16. ____________
17. ____________
18. ____________
19. ____________
20. ____________
21. ____________
22. ____________
23. ____________
24. ____________

ANSWERS

25. ______
26. ______
27. ______
28. ______
29. ______
30. ______
31. ______
32. ______
33. ______
34. ______
35. ______
36. ______
37. ______
38. ______
39. ______
40. ______
41. ______
42. ______
43. ______
44. ______
45. ______
46. ______
47. ______
48. ______
49. ______
50. ______
51. ______
52. ______

25. 11

26. 34

27. -19

28. -5

29. -7

30. -54

Evaluate.

31. $|17|$

32. $|28|$

33. $|-10|$

34. $|-7|$

35. $-|3|$

36. $-|5|$

37. $-|-8|$

38. $-|-13|$

39. $|-2| + |3|$

40. $|4| + |-3|$

41. $|-9| + |9|$

42. $|11| + |-11|$

43. $|4| - |-4|$

44. $|5| - |-5|$

45. $|15| - |8|$

46. $|11| - |3|$

47. $|15 - 8|$

48. $|11 - 3|$

49. $|-9| + |2|$

50. $|-7| + |4|$

51. $|-8| - |-7|$

52. $|-9| - |-4|$

ANSWERS

53. ______
54. ______
55. ______
56. ______
57. ______
58. ______
59. ______
60. ______
61. ______
62. ______
63. ______
64. ______
65. ______
66. ______
67. ______
68. ______

Label each statement as true or false.

53. All whole numbers are integers.

54. All nonzero integers are signed numbers.

55. All integers are whole numbers.

56. All signed numbers are integers.

57. All negative integers are whole numbers.

58. Zero is neither positive nor negative.

Place absolute value bars in the proper location on the left side of the expression so that the equation is true.

59. $(-6) + 2 = 8$

60. $(-8) + (-3) = 11$

61. $6 + (-2) = 8$

62. $8 + (-3) = 11$

Represent each quantity with a signed number.

63. Soil erosion. The erosion of 5 centimeters (cm) of topsoil from an Iowa cornfield.

64. Soil formation. The formation of 2.5 cm of new topsoil on the African savanna.

65. Checking accounts. The withdrawal of $50 from a checking account.

66. Saving accounts. The deposit of $200 in a savings account.

67. Temperature. The temperature change pictured.

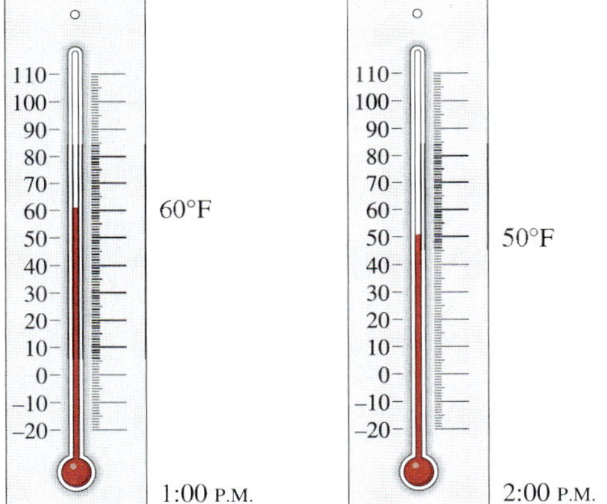

68. Stocks. An increase of 75 points in the Dow-Jones average.

ANSWERS

69. ____
70. ____
71. ____
72. ____
73. ____
74. ____
75. ____
76. ____
77. ____

69. Baseball. An eight-game losing streak by the local baseball team.

70. Population. An increase of 25,000 in the population of the city.

71. Positive trade balance. A country exported \$90,000,000 more than it imported, creating a positive trade balance.

72. Negative trade balance. A country exported \$60,000,000 less than it imported, creating a negative trade balance.

For each collection of numbers given in exercises 73 to 76, answer questions **(a)** to **(d)**:

(a) Which number is smallest?
(b) Which number lies farthest from the origin?
(c) Which number has the largest absolute value?
(d) Which number has the smallest absolute value?

73. $-6, 3, 8, 7, -2$

74. $-8, 3, -5, 4, 9$

75. $-2, 6, -1, 0, 2, 5$

76. $-9, 0, -2, 3, 6$

77. Simplify each of the following:

$-(-7) \qquad -(-(-7)) \qquad -(-(-(-7)))$

Based on your answers, generalize your results. 

Answers

1. 400 or $(+400)$ **3.** -200 **5.** $-25{,}000$

7. 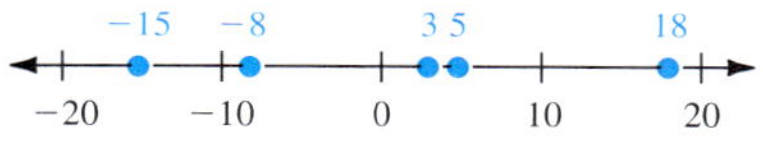 Number line with points at $-15, -8, 3, 5, 18$ (scale marked $-20, -10, 0, 10, 20$) **9.** 5, 175, -234

11. $-7, -5, -1, 0, 2, 3, 8$ **13.** $-11, -6, -2, 1, 4, 5, 9$
15. $-7, -6, -3, 3, 6, 7$ **17.** Max: 15; Min: -6 **19.** Max: 21; Min: -15
21. Max: 5; Min: -2 **23.** -15 **25.** -11 **27.** 19 **29.** 7 **31.** 17
33. 10 **35.** -3 **37.** -8 **39.** 5 **41.** 18 **43.** 0 **45.** 7
47. 7 **49.** 11 **51.** 1 **53.** True **55.** False **57.** False
59. $|-6| + 2 = 8$ **61.** $6 + |-2| = 8$ **63.** -5 **65.** -50
67. $-10°$ **69.** -8 **71.** $+90{,}000{,}000$ **73.** $-6; 8; 8; -2$
75. $-2; 6; 6; 0$ **77.**

Addition of Integers

OBJECTIVES

1. Use a number line to find the sum of two integers
2. Add two integers with the same sign
3. Add two integers with opposite signs
4. Solve applications involving integers

In Section 2.1 we introduced the idea of signed numbers. Now we will examine the four arithmetic operations (addition, subtraction, multiplication, and division) and see how those operations are performed when integers are involved. We start by considering addition.

An application may help. We will represent a gain of money as a positive number and a loss as a negative number.

If you gain \$3 and then gain \$4, the result is a gain of \$7:

$3 + 4 = 7$

If you lose \$3 and then lose \$4, the result is a loss of \$7:

$-3 + (-4) = -7$

If you gain \$3 and then lose \$4, the result is a loss of \$1:

$3 + (-4) = -1$

If you lose \$3 and then gain \$4, the result is a gain of \$1:

$-3 + 4 = 1$

The number line can be used to illustrate the addition of integers. Starting at the origin, we move to the *right* for positive integers and to the *left* for negative integers.

Example 1

Adding Integers

(a) Add 3 + 4.

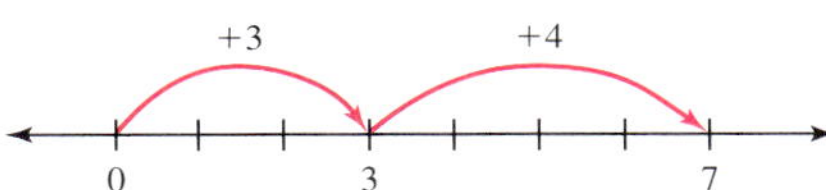

Start at the origin and move 3 units to the right. Then move 4 more units to the right to find the sum. From the number line we see that the sum is

$3 + 4 = 7$

(b) Add $(-3) + (-4)$.

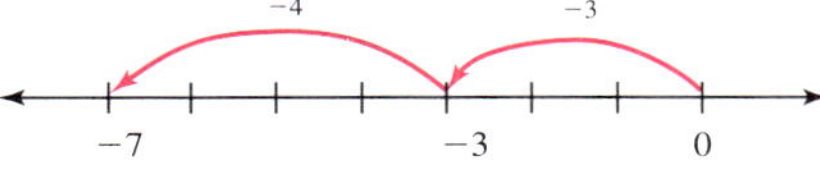

Start at the origin and move 3 units to the left. Then move 4 more units to the left to find the sum. From the number line we see that the sum is

$(-3) + (-4) = -7$

CHECK YOURSELF 1

Add.

(a) $(-4) + (-5)$
(b) $(-3) + (-7)$
(c) $(-5) + (-15)$
(d) $(-5) + (-3)$

You have probably noticed a helpful pattern in the previous example. This pattern will allow you to do the work mentally without having to use the number line. Look at the following rule.

Rules and Properties: Adding Integers Case 1: Same Sign

If two integers have the same sign, add their absolute values. Give the result the sign of the original integers.

NOTE This means that the sum of two positive integers is positive and the sum of two negative integers is negative. We first encountered absolute values in Section 2.1.

We can use the number line to illustrate the addition of two integers. This time the integers will have *different* signs.

Example 2

Adding Integers

(a) Add $3 + (-6)$.

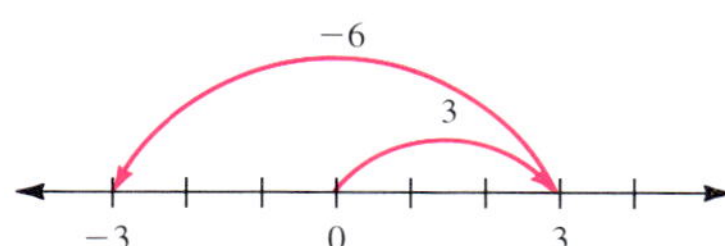

First move 3 units to the right of the origin. Then move 6 units to the left.

$3 + (-6) = -3$

(b) Add $-4 + 7$.

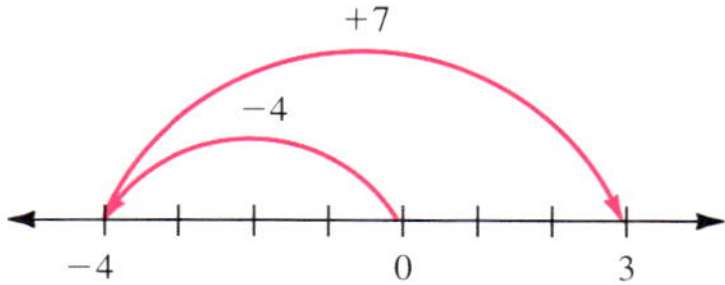

This time move 4 units to the left of the origin as the first step. Then move 7 units to the right.

$-4 + 7 = 3$

CHECK YOURSELF 2

Add.

(a) $7 + (-5)$
(b) $4 + (-8)$
(c) $-1 + 16$
(d) $-7 + 3$

You have no doubt noticed that, in adding a positive integer and a negative integer, sometimes the sum is positive and sometimes it is negative. This depends on which of the integers has the larger absolute value. This leads us to the second part of our addition rule.

Rules and Properties: Adding Integers Case 2: Different Signs

If two integers have different signs, subtract their absolute values, the smaller from the larger. Give the result the sign of the integer with the larger absolute value.

NOTE Again, we first encountered absolute values in Section 2.1.

Example 3

Adding Integers

(a) $7 + (-19) = -12$

Because the two integers have different signs, subtract the absolute values $(19 - 7 = 12)$. The sum of 7 and -19 has the sign $(-)$ of the integer with the larger absolute value, -19.

(b) $-13 + 7 = -6$

Subtract the absolute values $(13 - 7 = 6)$. The sum of -13 and 7 has the sign $(-)$ of the integer with the larger absolute value, -13.

CHECK YOURSELF 3

Add mentally.

(a) $5 + (-14)$ **(b)** $-7 + (-8)$

(c) $-8 + 15$ **(d)** $7 + (-8)$

In Section 1.2, we discussed the commutative, associative, and additive identity properties. There is another property of addition that we should mention.

Recall that every number has an *opposite*. It corresponds to a point the same distance from the origin as the given number, but in the opposite direction.

NOTE The opposite of a number is also called the **additive inverse** of that number.

NOTE 3 and -3 are opposites.

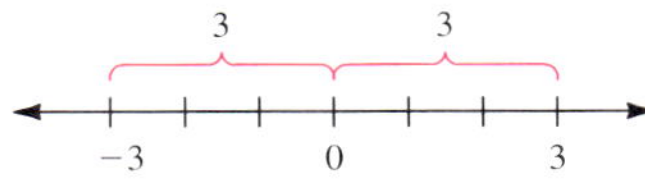

The opposite of 9 is -9.
The opposite of -15 is 15.

The additive inverse property states that the sum of any number and its opposite is 0.

Rules and Properties: Additive Inverse Property

For any number a, there exists a number $-a$ such that

$a + (-a) = (-a) + a = 0$

The sum of any number and its opposite, or additive inverse, is 0.

NOTE Here $-a$ represents the opposite of the number a. If a is positive, $-a$ is negative. If a is negative, $-a$ is positive.

Example 4

Adding Integers

(a) $9 + (-9) = 0$

(b) $-15 + 15 = 0$

Note that the opposite of 0 is 0 because $0 + 0 = 0$.

CHECK YOURSELF 4

Add.

(a) $(-17) + 17$ **(b)** $12 + (-12)$

When solving an application of integer arithmetic, the first step is to translate the phrase or statement using integers. Example 5 illustrates this step.

Example 5

An Application of the Addition of Integers

Shanique has \$250 in her checking account. She writes a check for \$120 and makes a deposit of \$90. What is the resulting balance?

First, translate the phrase using integers. Such problems will usually include something that is represented by negative integers and something that is represented by positive integers. In this case, a check can be represented as a negative integer and a deposit as a positive integer. We have

$250 + (-120) + 90$

This expression can now be evaluated.

$250 + (-120) + 90$

$= 130 + 90$

$= 220$

The resulting balance is \$220.

CHECK YOURSELF 5

Translate the first two sentences into an integer expression, and then answer the question.

When Kirin awoke, the temperature was twelve degrees below zero, Fahrenheit. Over the next six hours, the temperature increased by seventeen degrees. What was the temperature at that time?

CHECK YOURSELF ANSWERS

1. **(a)** -9; **(b)** -10; **(c)** -20; **(d)** -8 **2.** **(a)** 2; **(b)** -4; **(c)** 15; **(d)** -4
3. **(a)** -9; **(b)** -15; **(c)** 7; **(d)** -1 **4.** **(a)** 0; **(b)** 0
5. $-12 + 17$; the temperature was 5°F.

Name ____________

Section ______ Date ______

2.2 Exercises

Add.

1. $3 + 6$

2. $5 + 9$

3. $11 + 5$

4. $8 + 7$

5. $(-2) + (-3)$

6. $(-1) + (-9)$

7. $9 + (-3)$

8. $10 + (-4)$

9. $-9 + 0$

10. $-15 + 0$

11. $7 + (-7)$

12. $12 + (-12)$

13. $7 + (-9) + (-5) - 6$

14. $(-4) + 6 + (-3) + 0$

15. $7 + (-3) + 5 + (-11)$

16. $-6 + (-13) + 16$

In exercises 17 to 22, restate the problem using an expression involving integers and then answer the question.

17. **Checking account.** Amir has $100 in his checking account. He writes a check for $23 and makes a deposit of $51. What is his new balance?

18. **Checking account.** Olga has $250 in her checking account. She deposits $52 and then writes a check for $77. What is her new balance?

Bal: 250
Dep: 52
CK # 1111: 77

ANSWERS

1. ____
2. ____
3. ____
4. ____
5. ____
6. ____
7. ____
8. ____
9. ____
10. ____
11. ____
12. ____
13. ____
14. ____
15. ____
16. ____
17. ____
18. ____

ANSWERS

19. ____________

20. ____________

21. ____________

22. ____________

23. ____________

19. Football yardage. On four consecutive running plays, Marshall Faulk gained 23 yards, lost 5 yards, gained 15 yards, and lost 10 yards. What was his net yardage change for the series of plays?

20. Personal finance. Angelo owed his sister \$15. He later borrowed another \$10. What positive or negative number represents his current financial condition?

21. Education. A local community college had a decrease in enrollment of 750 students in the fall of 2001. In the spring of 2002, there was another decrease of 425 students. What was the total change in enrollment for both semesters?

22. Temperature. At 7 A.M., the temperature was $-15°F$. By 1 P.M., the temperature had increased by 18°F. What was the temperature at 1 P.M.?

23. In this chapter, it is stated that "Every number has an opposite." The opposite of 9 is -9. This corresponds to the idea of an opposite in English. In English an opposite is often expressed by a prefix, for example, *un*- or *ir*-.

(a) Write the opposite of these words: unmentionable, uninteresting, irredeemable, irregular, uncomfortable.

(b) What is the meaning of these expressions: not uninteresting, not irredeemable, not irregular, not unmentionable?

(c) Think of other prefixes that *negate* or change the meaning of a word to its *opposite.* Make a list of words formed with these prefixes, and write a sentence with three of the words you found. Make a sentence with two words and phrases from parts **(a)** and **(b)**.

What is the value of $-[-(-5)]$? What is the value of $-(-6)$? How does this relate to the given examples? Write a short description about this relationship.

Answers

1. 9 **3.** 16 **5.** -5 **7.** 6 **9.** -9 **11.** 0 **13.** -1
15. -2 **17.** \$128 **19.** 23 yards **21.** -1175 students **23.**

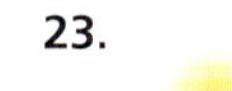

2.3 Subtraction of Integers

OBJECTIVES

1. Find the difference of two integers
2. Solve applications involving the subtraction of integers

To begin our discussion of subtraction when integers are involved, we can look back at a problem using natural numbers. Of course, we know that

$$8 - 5 = 3 \qquad (1)$$

From our work in adding integers, we know that it is also true that

$$8 + (-5) = 3 \qquad (2)$$

Comparing equations (1) and (2), we see that the results are the same. This leads us to an important pattern. Any subtraction problem can be written as a problem in addition. Subtracting 5 is the same as adding the opposite of 5, or -5. We can write this fact as follows:

$$8 - 5 = 8 + (-5) = 3$$

This leads us to the following rule for subtracting integers.

Rules and Properties: Subtracting Integers

1. To rewrite the subtraction problem as an addition problem:
   a. Change the subtraction operation to addition.
   b. Replace the integer being subtracted with its opposite.
2. Add the resulting integers as before.
   In symbols,

   $a - b = a + (-b)$

NOTE This is the *definition* of subtraction.

Example 1 illustrates the use of this definition while subtracting.

Example 1

Subtracting Integers

Subtraction *Addition*

Change the subtraction symbol ($-$) to an addition symbol ($+$).

(a) $15 - 7 = 15 + (-7)$

Replace 7 with its opposite, -7.

$= 8$

(b) $9 - 12 = 9 + (-12) = -3$

(c) $-6 - 7 = -6 + (-7) = -13$

(d) Subtract 5 from -2. We write the statement as $-2 - 5$ and proceed as before:

$-2 - 5 = -2 + (-5) = -7$

CHECK YOURSELF 1

Subtract.

(a) $18 - 7$ **(b)** $5 - 13$ **(c)** $-7 - 9$ **(d)** $-2 - 7$

The subtraction rule is used in the same way when the integer being subtracted is negative. Change the subtraction to addition. Replace the negative integer being subtracted with its opposite, which is positive. Example 2 illustrates this principle.

Example 2

Subtracting Integers

Subtraction *Addition*

Change the subtraction to an addition.

(a) $5 - (-2) = 5 + (+2) = 5 + 2 = 7$

Replace -2 with its opposite, $+2$ or 2.

(b) $7 - (-8) = 7 + (+8) = 7 + 8 = 15$

(c) $-9 - (-5) = -9 + 5 = -4$

(d) Subtract -4 from -5. We write

$-5 - (-4) = -5 + 4 = -1$

CHECK YOURSELF 2

Subtract.

(a) $8 - (-2)$ **(b)** $3 - (-10)$ **(c)** $-7 - (-2)$ **(d)** $7 - (-7)$

Example 3

An Application of the Subtraction of Integers

Susanna's checking account shows a balance of \$285. She has discovered that a deposit for \$47 was accidently recorded as a check for \$47. Write an integer expression that represents the correction on the balance. Then find the corrected balance.

$285 - (-47) + (47)$

Subtract the check and then add the deposit.

$285 - (-47) + (47) = 285 + 47 + 47 = 379$

The corrected balance is \$379.

CHECK YOURSELF 3

It appears that Marshal, a running back, gained 97 yards in the last game. A closer inspection of the statistics revealed that a 9-yard gain had been recorded as a 9-yard loss. Write an integer expression that represents the corrected yards gained, and then find that number.

Using Your Calculator to Add and Subtract Signed Integers

Your scientific (or graphing) calculator has a key that makes a number negative. This key is different from the "subtraction" key. The negative key is found on the bottom row of the calculator. It is marked either [+/−] or [(−)]. With a scientific calculator, this key is pressed *after* the number you wish to make negative is entered. All of the instructions in this section will assume that you have a scientific calculator.

Example 4

Entering a Negative Integer into the Calculator

Enter each of the following into your calculator.

(a) −24

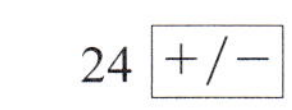

24 [+/−]

NOTE The 12 goes between positive and negative in the display. The final display is 12, because there are an even number of negative signs in front of the 12.

(b) −(−(−(−12)))

12 [+/−] [+/−] [+/−] [+/−]

CHECK YOURSELF 4

Enter each number into your calculator.

(a) −36 **(b)** −(−(−6))

Example 5

Adding Integers

Find the sum for each pair of integers.

(a) 256 + (−297)

256 [+] 297 [+/−] [=] −41

(b) −312 + (−569)

312 [+/−] [+] 569 [+/−] [=] −881

CHECK YOURSELF 5

Find the sum for each pair of integers.

(a) −368 + 547 **(b)** −596 + (−834)

Example 6

Subtracting Integers

Find the difference for −356 − (−469).

356 [+/−] [−] 469 [+/−] [=] 113

CHECK YOURSELF 6

Find the differences.

(a) $349 - (-49)$

(b) $-294 - (-137)$

CHECK YOURSELF ANSWERS

1. **(a)** 11; **(b)** −8; **(c)** −16; **(d)** −9 **2.** **(a)** 10; **(b)** 13; **(c)** −5; **(d)** 14
3. $97 - (-9) + 9 = 115$ yards **4.** **(a)** −36; **(b)** −6 **5.** **(a)** 179; **(b)** −1430
6. **(a)** 398; **(b)** −157

2.3 Exercises

Name ______________________

Section __________ Date __________

Subtract.

1. $21 - 13$
2. $36 - 22$
3. $82 - 45$
4. $103 - 56$
5. $8 - 10$
6. $14 - 19$
7. $24 - 45$
8. $136 - 352$
9. $-5 - 3$
10. $-15 - 8$
11. $-9 - 14$
12. $-8 - 12$
13. $5 - (-11)$
14. $7 - (-5)$
15. $7 - (-12)$
16. $3 - (-10)$
17. $-36 - (-24)$
18. $-28 - (-11)$
19. $-19 - (-27)$
20. $-11 - (-16)$

For exercises 21 to 23, write an integer expression that describes the situation. Then answer the question.

21. **Temperature.** The temperature at noon on a June day was 82°F. It fell by 12° in the next 4 h. What was the temperature at 4:00 P.M.?

22. Jason's checking account shows a balance of \$853. He has discovered that a deposit of \$70 was accidently recorded as a check for \$70. What is the corrected balance?

23. Ylena's checking account shows a balance of \$947. She has discovered that a check for \$86 was recorded as a deposit of \$86. What is the corrected balance?

24. How long ago was the year 1250 B.C.E.? What year was 3300 years ago? Make a number line and locate the following events, cultures, and objects on it. How long ago was each item in the list? Which two events are the closest to each other? You may want to learn more about some of the cultures in the list and the mathematics and science developed by that culture.

Inca culture in Peru—A.D. 1400
The *Ahmes Papyrus,* a mathematical text from Egypt—1650 B.C.E.
Babylonian arithmetic develops the use of a zero symbol—300 B.C.E.
First Olympic Games—776 B.C.E.
Pythagoras of Greece dies—500 B.C.E.
Mayans in Central America independently develop use of zero—A.D. 500
The *Chou Pei,* a mathematics classic from China—1000 B.C.E.
The *Aryabhatiya,* a mathematics work from India—A.D. 499
Trigonometry arrives in Europe via the Arabs and India—A.D. 1464
Arabs receive algebra from Greek, Hindu, and Babylonian sources and develop it into a new systematic form—A.D. 850
Development of calculus in Europe—A.D. 1670
Rise of abstract algebra—A.D. 1860
Growing importance of probability and development of statistics—A.D. 1902

ANSWERS

1. ______
2. ______
3. ______
4. ______
5. ______
6. ______
7. ______
8. ______
9. ______
10. ______
11. ______
12. ______
13. ______
14. ______
15. ______
16. ______
17. ______
18. ______
19. ______
20. ______
21. ______
22. ______
23. ______

24. ______

ANSWERS

25. ____________

26. ____________

27. ____________

28. ____________

29. ____________

30. ____________

31. ____________

32. ____________

33. ____________

34. ____________

35. ____________

36. ____________

37. ____________

38. ____________

25. Complete the following statement: "3 − (−7) is the same as ____ because . . ." Write a problem that might be answered by doing this subtraction.

26. Explain the difference between the two phrases: "a number subtracted from 5" and "a number less than 5." Use algebra and English to explain the meaning of these phrases. Write other ways to express subtraction in English. Which ones are confusing?

Calculator Exercises

Using your calculator, perform the following operations.

27. 345 + (−215)

28. 415 + (−123)

29. 679 + (−124)

30. −345 + (−215)

31. −789 + (−128)

32. −910 + (−567)

33. −349 + (−431)

34. −412 + (−367)

35. 47 − (−25)

36. 123 − (−219)

37. 234 − (−456)

38. 412 − (−123)

Answers

1. 8 **3.** 37 **5.** −2 **7.** −21 **9.** −8 **11.** −23 **13.** 16
15. 19 **17.** −12 **19.** 8 **21.** 70°F
23. 947 − 86 + (−86); $775 is the balance **25.** **27.** 130
29. 555 **31.** −917 **33.** −780 **35.** 72 **37.** 690

2.4 Multiplication of Integers

2.4 OBJECTIVES

1. Find the product of two or more integers
2. Use the order of operations with integers

When you first considered multiplication in arithmetic, it was thought of as repeated addition. Let's see what our work with the addition of integers can tell us about multiplication when integers are involved. For example,

$$3 \cdot 4 = \underbrace{4 + 4 + 4} = 12$$

We interpret multiplication as repeated addition to find the product, 12.

Now, consider the product $(3)(-4)$:

$$(3)(-4) = (-4) + (-4) + (-4) = -12$$

Looking at this product suggests the first portion of our rule for multiplying integers. The product of a positive integer and a negative integer is negative.

Rules and Properties: Multiplying Integers Case 1: Different Signs

The product of two integers with different signs is negative.

To use this rule in multiplying two integers with different signs, multiply their absolute values and attach a negative sign.

Example 1

Multiplying Integers

Multiply.

(a) $(5)(-6) = -30$

The product is negative.

(b) $(-10)(10) = -100$

(c) $(8)(-12) = -96$

CHECK YOURSELF 1

Multiply.

(a) $(-7)(5)$ **(b)** $(-12)(9)$ **(c)** $(-15)(8)$

The product of two negative integers is harder to visualize. The following pattern may help you see how we can determine the sign of the product.

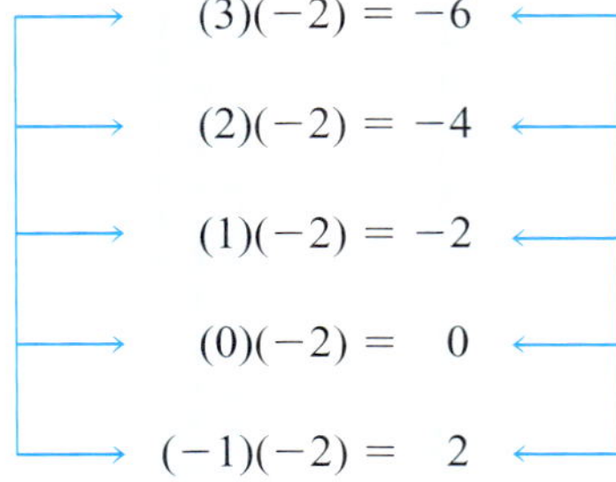

$(3)(-2) = -6$

$(2)(-2) = -4$

$(1)(-2) = -2$

$(0)(-2) = 0$

$(-1)(-2) = 2$

NOTE This number is decreasing by 1.

Do you see that the product is *increasing* by 2 each time?

NOTE $(-1)(-2)$ is the opposite of -2.

What should the product $(-2)(-2)$ be? Continuing the pattern shown, we see that

$(-2)(-2) = 4$

This suggests that the product of two negative integers is positive. That is the case. We can extend our multiplication rule.

Rules and Properties: Multiplying Integers Case 2: Same Sign

The product of two integers with the same sign is positive.

Example 2

Multiplying Integers

Multiply.

(a) $9 \cdot 7 = 63$ The product of two positive numbers (same sign, +) is positive.

(b) $(-8)(-5) = 40$ The product of two negative numbers (same sign, −) is positive.

CHECK YOURSELF 2

Multiply.

(a) $10 \cdot 12$ **(b)** $(-8)(-9)$

The Multiplicative Identity Property and Multiplicative Property of Zero studied in Section 1.5 can be applied to integers, as illustrated in Example 3.

Example 3

Multiplying Integers

Find each product.

(a) $(1)(-7) = -7$

(b) $(15)(1) = 15$

(c) $(-7)(0) = 0$

(d) $0 \cdot 12 = 0$

CHECK YOURSELF 3

Multiply.

(a) $(-10)(1)$ **(b)** $(0)(-17)$

We can now extend the rules for the order of operations learned in Section 1.7 to simplify expressions containing integers. First we will work with integers raised to a power.

Example 4

Integers with Exponents

Evaluate each expression.

(a) $(-3)^2 = (-3)(-3) = 9$

(b) $(-3)^3 = (-3)(-3)(-3) = -27$

(c) $-3^2 = -3 \cdot 3 = -9$ Note that the negative is *not* squared.

NOTE In part (b) of Example 4 we have a negative integer raised to a power.

In part (c) only the 3 is raised to a power. We have the opposite of 3 squared.

CHECK YOURSELF 4

Evaluate each expression.

(a) $(-4)^2$ **(b)** $(-4)^3$ **(c)** -4^2

In Example 5 we will apply the order of operations.

Example 5

Using Order of Operations with Integers

Evaluate each expression.

(a) $7(-9 + 12)$ Evaluate inside the parentheses first.

$= 7(3) = 21$

(b) $(-8)(-7) - 40$ Multiply first, then subtract.

$= 56 - 40$

$= 16$

(c) $(-5)^2 - 3$ Evaluate the power first.

$= (-5)(-5) - 3$ Note that $(-5)^2 = (-5)(-5) = 25$

$= 25 - 3$

$= 22$

(d) $-5^2 - 3$ Note that $-5^2 = -25$. The power applies *only* to the 5.

$= -25 - 3$

$= -28$

CHECK YOURSELF 5

Evaluate each expression.

(a) $8(-9 + 7)$ **(b)** $(-3)(-5) + 7$
(c) $(-4)^2 - (-4)$ **(d)** $-4^2 - (-4)$

CHECK YOURSELF ANSWERS

1. (a) -35; **(b)** -108; **(c)** -120 **2. (a)** 120; **(b)** 72 **3. (a)** -10; **(b)** 0
4. (a) 16; **(b)** -64; **(c)** -16 **5. (a)** -16; **(b)** 22; **(c)** 20; **(d)** -12

Name ______________________

Section ________ Date ________

2.4 Exercises

Multiply.

1. $4 \cdot 10$

2. $3 \cdot 14$

3. $(5)(-12)$

4. $(10)(-2)$

5. $(-8)(9)$

6. $(-12)(3)$

7. $(-8)(-7)$

8. $(-9)(-8)$

9. $(-5)(-12)$

10. $(-7)(-3)$

11. $(0)(-18)$

12. $(-17)(0)$

13. $(15)(0)$

14. $(0)(25)$

Do the indicated operations. Remember the rules for the order of operations.

15. $5(7 - 2)$

16. $7(8 - 5)$

17. $2(5 - 8)$

18. $6(14 - 16)$

19. $-3(9 - 7)$

20. $-6(12 - 9)$

21. $-3(-2 - 5)$

22. $-2(-7 - 3)$

23. $(-2)(3) - 5$

24. $(-6)(8) - 27$

25. $4(-7) - 5$

26. $(-3)(-9) - 11$

27. $(-5)(-2) - 12$

28. $(-7)(-3) - 25$

29. $(3)(-7) + 20$

30. $(2)(-6) + 8$

31. $-4 + (-3)(6)$

32. $-5 + (-2)(3)$

33. $7 - (-4)(-2)$

34. $9 - (-2)(-7)$

35. $(-7)^2 - 17$

36. $(-6)^2 - 20$

37. $(-5)^2 + 18$

38. $(-2)^2 + 10$

39. $-6^2 - 4$

40. $-5^2 - 3$

ANSWERS

1. ______
2. ______
3. ______
4. ______
5. ______
6. ______
7. ______
8. ______
9. ______
10. ______

11. ______	12. ______
13. ______	14. ______
15. ______	16. ______
17. ______	18. ______
19. ______	20. ______
21. ______	22. ______
23. ______	24. ______
25. ______	26. ______
27. ______	28. ______
29. ______	30. ______
31. ______	32. ______
33. ______	34. ______
35. ______	36. ______
37. ______	38. ______
39. ______	40. ______

ANSWERS

41. ______

42. ______

43. ______

44. ______

45. ______

46. ______

47. ______

48. ______

49. ______

50. ______

51. ______

52. ______

53. ______

41. $(-4)^2 - (-2)(-5)$

42. $(-3)^3 - (-8)(-2)$

43. $(-8)^2 - 5^2$

44. $(-6)^2 - 4^2$

45. $(-6)^2 - (-3)^2$

46. $(-8)^2 - (-4)^2$

47. $-8^2 - 5^2$

48. $-6^2 - 3^2$

49. $-8^2 - (-5)^2$

50. $-9^2 - (-6)^2$

51. **Basketball.** You score 23 points a game for 11 straight games. What is the total number of points that you scored?

52. **Gambling.** In Atlantic City, Nick played the slot machines for 12 h. He lost \$45 an hour. Use integers to represent the change in Nick's financial status at the end of the 12 h.

53. **Temperature.** The temperature is −6°F at 5:00 in the evening. If the temperature drops 2°F every hour, what is the temperature at 1:00 A.M.?

Answers

1. 40 **3.** −60 **5.** −72 **7.** 56 **9.** 60 **11.** 0 **13.** 0
15. 25 **17.** −6 **19.** −6 **21.** 21 **23.** −11 **25.** −33
27. −2 **29.** −1 **31.** −22 **33.** −1 **35.** 32 **37.** 43
39. −40 **41.** 6 **43.** 39 **45.** 27 **47.** −89 **49.** −89
51. 253 points **53.** −22°F

Division of Integers

OBJECTIVES

1. Find the quotient of two integers
2. Use the order of operations with integers

You know from your work in arithmetic that multiplication and division are related operations. We can use that fact, and our work of Section 2.4, to determine rules for the division of integers. Every division problem can be stated as an equivalent multiplication problem. For instance,

$$\frac{15}{5} = 3 \qquad \text{because} \qquad 15 = 5 \cdot 3$$

$$\frac{-24}{6} = -4 \qquad \text{because} \qquad -24 = (6)(-4)$$

$$\frac{-30}{-5} = 6 \qquad \text{because} \qquad -30 = (-5)(6)$$

These examples illustrate that because the two operations are related, the rule of signs that we stated in Section 2.4 for multiplication is also true for division.

Rules and Properties: Dividing Integers

1. The quotient of two integers with different signs is negative.
2. The quotient of two integers with the same sign is positive.

Again, the rule is easy to use. To divide two integers, divide their absolute values. Then attach the proper sign according to the rule.

Example 1

Dividing Integers

Divide.

(a) Positive → $\frac{28}{7} = 4$ ← Positive; Positive → (denominator)

(b) Negative → $\frac{-36}{-4} = 9$ ← Positive; Negative → (denominator)

(c) Negative → $\frac{-42}{7} = -6$ ← Negative; Positive → (denominator)

(d) Positive → $\frac{75}{-3} = -25$ ← Negative; Negative → (denominator)

CHECK YOURSELF 1

Divide.

(a) $\frac{-55}{11}$ **(b)** $\frac{80}{20}$ **(c)** $\frac{-48}{-8}$ **(d)** $\frac{144}{-12}$

As discussed in Section 1.6, we must be very careful when 0 is involved in a division problem. Remember that 0 divided by any nonzero number is just 0. This rule can be extended to include integers, so that

$$\frac{0}{-7} = 0 \qquad \text{because} \qquad 0 = (-7)(0)$$

However, if zero is the *divisor,* we have a special problem. Consider

$$\frac{-9}{0} = ?$$

This means that $-9 = 0 \cdot ?$.

Can 0 times a number ever be -9? No, so there is no solution.

Because $\frac{-9}{0}$ cannot be replaced by any number, we agree that *division by* 0 *is not allowed.* We say that division by 0 is *undefined.*

Example 2

Dividing Integers

Divide, if possible.

(a) $\frac{7}{0}$ is undefined.

(b) $\frac{-9}{0}$ is undefined.

(c) $\frac{0}{5} = 0$

(d) $\frac{0}{-8} = 0$

Note: The expression $\frac{0}{0}$ is called an **indeterminate form.** You will learn more about this in later mathematics classes.

CHECK YOURSELF 2

Divide, if possible.

(a) $\frac{0}{3}$ **(b)** $\frac{5}{0}$ **(c)** $\frac{-7}{0}$ **(d)** $\frac{0}{-9}$

Using Your Calculator to Multiply and Divide Integers

Finding the product of two integers using a calculator is relatively straightforward.

Example 3

Multiplying Integers

Find the product. $457 \cdot (-734)$

457 [×] 734 [+/−] [=] −335438

CHECK YOURSELF 3

Find the products.

(a) $36 \cdot (-91)$ **(b)** $-12 \cdot (-284)$

Finding the quotient of integers is also straightforward.

Example 4

Dividing Integers

Find the quotient. $\dfrac{-384}{16}$

384 [+/−] [÷] 16 [=] −24

CHECK YOURSELF 4

Find the quotient.

$-7865 \div -242$

We can also use the calculator to raise an integer to a power.

Example 5

Raising a Number to a Power

Evaluate.

$(-3)^6$

[(] 3 [)] [+/−] [y^x] 6 [=] 729

or, on some calculators

[(] [(−)] 3 [)] [^] 6 [Enter] 729

NOTE The parentheses ensure that the negative is attached to the 3 **before** it is raised to a power.

CHECK YOURSELF 5

Evaluate.

$(-2)^9$

The fraction bar, like parentheses and the absolute value bars, serves as a *grouping symbol*. This means that all operations in the numerator and denominator should be performed separately. Then the division is done as the last step. Example 6 illustrates this property.

Example 6

Using Order of Operations

Evaluate each expression.

(a) $\dfrac{(-6)(-7)}{3} = \dfrac{42}{3} = 14$

Multiply in the numerator, then divide.

(b) $\dfrac{3 + (-12)}{3} = \dfrac{-9}{3} = -3$

Add in the numerator, then divide.

(c) $\dfrac{-4 + (2)(-6)}{-6 - 2} = \dfrac{-4 + (-12)}{-6 - 2}$

Multiply in the numerator. Then add in the numerator and subtract in the denominator.

$= \dfrac{-16}{-8} = 2$

Divide as the last step.

CHECK YOURSELF 6

Evaluate each expression.

(a) $\dfrac{-4 + (-8)}{6}$ **(b)** $\dfrac{3 - (2)(-6)}{-5}$ **(c)** $\dfrac{(-2)(-4) - (-6)(-5)}{(-2)(11)}$

CHECK YOURSELF ANSWERS

1. **(a)** -5; **(b)** 4; **(c)** 6; **(d)** -12 **2.** **(a)** 0; **(b)** undefined; **(c)** undefined; **(d)** 0
3. **(a)** -3276; **(b)** 3408 **4.** 32.5 **5.** -512 **6.** **(a)** -2; **(b)** -3; **(c)** 1

Name ____________

Section ______ Date ______

2.5 Exercises

Divide.

1. $\frac{-20}{-4}$

2. $\frac{70}{14}$

3. $\frac{48}{6}$

4. $\frac{-24}{8}$

5. $\frac{50}{-5}$

6. $\frac{-32}{-8}$

7. $\frac{-52}{4}$

8. $\frac{56}{-7}$

9. $\frac{-75}{-3}$

10. $\frac{-60}{15}$

11. $\frac{0}{-8}$

12. $\frac{-125}{-25}$

13. $\frac{-9}{-1}$

14. $\frac{-10}{0}$

15. $\frac{-96}{-8}$

16. $\frac{-20}{2}$

17. $\frac{18}{0}$

18. $\frac{0}{8}$

19. $\frac{-17}{1}$

20. $\frac{-27}{-1}$

21. $\frac{-144}{-16}$

22. $\frac{-150}{6}$

Perform the indicated operations.

23. $\frac{(-6)(-3)}{2}$

24. $\frac{(-9)(5)}{-3}$

25. $\frac{(-8)(2)}{-4}$

26. $\frac{(7)(-8)}{-14}$

27. $\frac{24}{-4 - 8}$

28. $\frac{36}{-7 + 3}$

29. $\frac{-12 - 12}{-3}$

30. $\frac{-14 - 4}{-6}$

31. $\frac{55 - 19}{-12 - 6}$

32. $\frac{-11 - 7}{-14 + 8}$

33. $\frac{7 - 5}{2 - 2}$

34. $\frac{10 - 6}{4 - 4}$

ANSWERS

1. ____
2. ____
3. ____
4. ____
5. ____
6. ____
7. ____
8. ____
9. ____
10. ____
11. ____
12. ____
13. ____
14. ____
15. ____ 16. ____
17. ____
18. ____
19. ____ 20. ____
21. ____ 22. ____
23. ____ 24. ____
25. ____ 26. ____
27. ____ 28. ____
29. ____ 30. ____
31. ____ 32. ____
33. ____
34. ____

ANSWERS

35. ______

36. ______

37. ______

38. ______

39. ______

40. ______

41. ______

42. ______

43. ______

44. ______

45. ______

For exercises 35 to 37, use integers to write an expression that represents the situation. Then answer the question.

35. Mowing lawns. Patrick worked all day mowing lawns and was paid $9 per hour. If he had $125 at the end of a 9-h day, how much did he have before he started working?

36. Dieting. A woman lost 42 lb. If she lost 3 lb each week, how long has she been dieting?

37. Investment. Suppose that you and your two brothers bought equal shares of an investment for a total of $20,000 and sold it later for $16,232. How much did each person lose?

Calculator Exercises

Use your calculator to multiply and divide.

38. $15 \cdot (-45)$

39. $78 \cdot (-12)$

40. $(-56) \cdot 31$

41. $(-34) \cdot (-28)$

42. $(-71) \cdot (-19)$

43. $\dfrac{-28}{-14}$

44. $(-5)^4$

45. $(-4)^5$

Answers

1. 5 **3.** 8 **5.** -10 **7.** -13 **9.** 25 **11.** 0 **13.** 9 **15.** 12 **17.** Undefined **19.** -17 **21.** 9 **23.** 9 **25.** 4 **27.** -2 **29.** 8 **31.** -2 **33.** Undefined **35.** $125 - 9 \cdot 9 = \$44$ **37.** $\dfrac{20{,}000 - 16{,}232}{3} = \$1256$ **39.** -936 **41.** 952 **43.** 2 **45.** -1024

2.6 Introduction to Algebra: Variables and Expressions

OBJECTIVES

1. Represent addition, subtraction, multiplication, and division by using the symbols of algebra
2. Identify algebraic expressions

In arithmetic, you learned how to do calculations with numbers by using the basic operations of addition, subtraction, multiplication, and division.

In algebra, you will still use numbers and the same four operations. However, you will also use letters to represent numbers. Letters such as x, y, L, or W are called **variables** when they represent numerical values. If we need to represent the length and width of *any* rectangle, we can use the variables L and W.

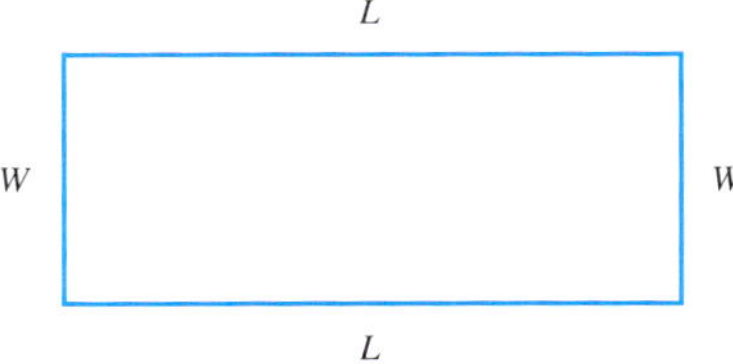

NOTE In arithmetic:
+ denotes addition
− denotes subtraction
· denotes multiplication
÷ denotes division.

You are familiar with the four symbols (+, −, ·, ÷) used to indicate the fundamental operations of arithmetic.

Next, we will look at how these operations are indicated in algebra. We begin by looking at addition.

Example 1

Writing Expressions That Indicate Addition

(a) *The sum of* a *and* 3 is written as $a + 3$.

(b) L *plus* W is written as $L + W$.

(c) 5 *more than* m is written as $m + 5$.

(d) x *increased by* 7 is written as $x + 7$.

CHECK YOURSELF 1

Write, using symbols.

(a) The sum of y and 4

(b) a plus b

(c) 3 more than x

(d) n increased by 6

In Example 2, we look at how subtraction is indicated in algebra.

Example 2

Writing Expressions That Indicate Subtraction

(a) r *minus* s is written as $r - s$.

(b) *The difference of* m *and* 5 is written as $m - 5$.

(c) x *decreased by* 8 is written as $x - 8$.

(d) 4 *less than* a is written as $a - 4$.

CHECK YOURSELF 2

Write, using symbols.

(a) w minus z

(b) The difference of a and 7

(c) y decreased by 3

(d) 5 less than b

You have seen that the operations of addition and subtraction are written exactly the same way in algebra as in arithmetic. This is not true in multiplication because the sign $\times$ looks like the letter x. So in algebra we use other symbols to show multiplication to avoid any confusion. Here are some ways to write multiplication.

NOTE x and y are called the **factors** of the product xy.

Definitions: Multiplication

A centered dot	$x \cdot y$	These all indicate the *product* of x and y or x times y.
Parentheses	$(x)(y)$	
Writing the letters next to each other	xy	

Example 3

Writing Expressions That Indicate Multiplication

NOTE You can place letters next to each other or numbers and letters next to each other to show multiplication. But you *cannot* place numbers side by side to show multiplication: 37 means the number "thirty-seven," not 3 times 7.

(a) The product of 5 and a is written as $5 \cdot a$, $(5)(a)$, or $5a$. The last expression, $5a$, is the shortest and the most common way of writing the product.

(b) 3 times 7 can be written as $3 \cdot 7$ or $(3)(7)$.

(c) Twice z is written as $2z$.

(d) The product of 2, s, and t is written as $2st$.

(e) 4 more than the product of 6 and x is written as $6x + 4$.

CHECK YOURSELF 3

Write, using symbols.

(a) m times n

(b) The product of h and b

(c) The product of 8 and 9

(d) The product of 5, w, and y

(e) 3 more than the product of 8 and a

Before we move on to division, we will look at how we can combine the symbols we have learned so far.

NOTE Not every collection of symbols is an expression.

Definitions: Expression

An **expression** is a meaningful collection of numbers, variables, and signs of operation.

Example 4

Identifying Expressions

(a) $2m + 3$ is an expression. It means that we multiply 2 and m, and then add 3.
(b) $x + \cdot + 3$ is not an expression. The three operations in a row have no meaning.
(c) $y = 2x - 1$ is not an expression. The equal sign is not an operation sign.
(d) $3a + 5b - 4c$ is an expression. Its meaning is clear.

CHECK YOURSELF 4

Identify which are expressions and which are not.

(a) $7 - \cdot x$ **(b)** $6 + y = 9$
(c) $a + b - c$ **(d)** $3x - 5yz$

To write more complicated products in algebra, we need some "punctuation marks." Parentheses ( ) mean that an expression is to be thought of as a single quantity. Brackets [ ] are used in exactly the same way as parentheses in algebra. Look at Example 5, which shows the use of these signs of grouping.

Example 5

Expressions with More Than One Operation

(a) 3 times the sum of a and b is written as

NOTE This can be read as "3 times the quantity a plus b."

$3(a + b)$

The sum of a and b is a single quantity, so it is enclosed in parentheses.

NOTE No parentheses are used here because the 3 multiplies *only* the a.

(b) The sum of 3 times a and b is written as $3a + b$.
(c) 2 times the difference of m and n is written as $2(m - n)$.
(d) The product of s plus t and s minus t is written as $(s + t)(s - t)$.
(e) The product of b and 3 less than b is written as $b(b - 3)$.

CHECK YOURSELF 5

Write, using symbols.

(a) Twice the sum of p and q
(b) The sum of twice p and q
(c) The product of a and the quantity $b - c$
(d) The product of x plus 2 and x minus 2
(e) The product of x and 4 more than x

NOTE In algebra the fraction form is usually used.

Now we will look at the operation of division.

Example 6

Writing Expressions That Indicate Division

(a) m divided by 3 is written as $\frac{m}{3}$.

(b) The quotient of a plus b and 5 is written as $\frac{a+b}{5}$.

(c) The sum p plus q divided by the difference p minus q is written as $\frac{p+q}{p-q}$.

CHECK YOURSELF 6

Write, using symbols.

(a) r divided by s

(b) The quotient when x minus y is divided by 7

(c) The difference a minus 2 divided by the sum a plus 2

Notice that we can use many different letters to represent variables. In Example 6 the letters m, a, b, p, and q represented different variables. We often choose a letter that reminds us of what it represents, for example, L for *length* or W for *width*.

Example 7

Writing Geometric Expressions

(a) *Length* times *width* is written LW.

(b) One-half of *altitude* times *base* is written $\frac{1}{2}ab$.

(c) *Length* times *width* times *height* is written LWH.

(d) Pi (π) times *diameter* is written πd.

CHECK YOURSELF 7

Write each geometric expression, using symbols.

(a) Two times *length* plus two times *width* **(b)** Two times pi (π) times *radius*

CHECK YOURSELF ANSWERS

1. (a) $y + 4$; **(b)** $a + b$; **(c)** $x + 3$; **(d)** $n + 6$ **2. (a)** $w - z$; **(b)** $a - 7$; **(c)** $y - 3$; **(d)** $b - 5$ **3. (a)** mn; **(b)** hb; **(c)** $8 \cdot 9$ or $(8)(9)$; **(d)** $5wy$; **(e)** $8a + 3$
4. (a) Not an expression; **(b)** not an expression; **(c)** an expression; **(d)** an expression
5. (a) $2(p + q)$; **(b)** $2p + q$; **(c)** $a(b - c)$; **(d)** $(x + 2)(x - 2)$; **(e)** $x(x + 4)$
6. (a) $\frac{r}{s}$; **(b)** $\frac{x-y}{7}$; **(c)** $\frac{a-2}{a+2}$ **7. (a)** $2L + 2W$; **(b)** $2\pi r$

Name ______________________

Section ________ Date ________

2.6 Exercises

Write each of the phrases, using symbols.

1. The sum of c and d
2. a plus 7
3. w plus z
4. The sum of m and n
5. x increased by 2
6. 3 more than b
7. 10 more than y
8. m increased by 4
9. a minus b
10. 5 less than s
11. b decreased by 7
12. r minus 3
13. 6 less than r
14. x decreased by 3
15. w times z
16. The product of 3 and c
17. The product of 5 and t
18. 8 times a
19. The product of 8, m, and n
20. The product of 7, r, and s
21. The product of 3 and the quantity p plus q
22. The product of 5 and the sum of a and b
23. Twice the sum of x and y
24. 3 times the sum of m and n
25. The sum of twice x and y

ANSWERS

1. ______
2. ______
3. ______
4. ______
5. ______
6. ______
7. ______
8. ______
9. ______
10. ______
11. ______
12. ______
13. ______
14. ______
15. ______
16. ______
17. ______
18. ______
19. ______
20. ______
21. ______
22. ______
23. ______
24. ______
25. ______

ANSWERS

26.
27.
28.
29.
30.
31.
32.
33.
34.
35.
36.
37.
38.
39.
40.
41.
42.
43.
44.
45.
46.
47.
48.
49.
50.
51.

26. The sum of 3 times m and n

27. Twice the difference of x and y

28. 3 times the difference of c and d

29. The quantity a plus b times the quantity a minus b

30. The product of x plus y and x minus y

31. The product of m and 3 less than m

32. The product of a and 7 more than a

33. x divided by 5

34. The quotient when b is divided by 8

35. The quotient of a plus b, and 7

36. The difference x minus y, divided by 9

37. The difference of p and q, divided by 4

38. The sum of a and 5, divided by 9

39. The sum of a and 3, divided by the difference of a and 3

40. The difference of m and n, divided by the sum of m and n

Write each of the phrases, using symbols. Use the variable x to represent the number in each case.

41. 5 more than a number

42. A number increased by 8

43. 7 less than a number

44. A number decreased by 10

45. 9 times a number

46. Twice a number

47. 6 more than 3 times a number

48. 5 times a number, decreased by 10

49. Twice the sum of a number and 5

50. 3 times the difference of a number and 4

51. The product of 2 more than a number and 2 less than that same number

ANSWERS

52. ______
53. ______
54. ______
55. ______
56. ______
57. ______
58. ______
59. ______
60. ______
61. ______
62. ______
63. ______
64. ______
65. ______
66. ______
67. ______
68. ______
69. ______
70. ______
71. ______
72. ______
73. ______
74. ______

52. The product of 5 less than a number and 5 more than that same number

53. The quotient of a number and 7

54. A number divided by 3

55. The sum of a number and 5, divided by 8

56. The quotient when 7 less than a number is divided by 3

57. 6 more than a number divided by 6 less than that same number

58. The quotient when 3 less than a number is divided by 3 more than that same number

Write each of the following geometric expressions using symbols.

59. Four times the length of a side (s)

60. $\frac{4}{3}$ times π times the cube of the radius (r)

61. The radius (r) squared times the height (h) times π

62. Twice the length (L) plus twice the width (W)

63. One-half the product of the height (h) and the sum of two unequal sides (b_1 and b_2)

64. Six times the length of a side (s) squared

Identify which are expressions and which are not.

65. $2(x + 5)$

66. $4 + (x - 3)$

67. $4 + \div m$

68. $6 + a = 7$

69. $2b = 6$

70. $x(y + 3)$

71. $2a + 5b$

72. $4x + \cdot 7$

73. Population growth. Earth's population has doubled in the last 40 years. If we let x represent Earth's population 40 years ago, what is the population today?

74. Species extinction. It is estimated that Earth is losing 4000 species of plants and animals every year. If S represents the number of species living last year, how many species are on Earth this year?

ANSWERS

75. ______

76. ______

77. ______

75. **Interest.** The simple interest (I) earned when a principal (P) is invested at a rate (r) for a time (t) is calculated by multiplying the principal times the rate times the time. Write a formula for the interest earned.

76. **Kinetic energy.** The kinetic energy (KE) of a particle of mass m is found by taking one-half of the product of the mass and the square of the velocity (v). Write a formula for the kinetic energy of a particle.

77. Rewrite each algebraic expression as an English phrase. Exchange papers with another student to edit your writing. Be sure the meaning in English is the same as in algebra. These expressions are not complete sentences, so your English does not have to be in complete sentences. Here is an example.

Algebra: $2(x - 1)$

English: We could write "One less than a number is doubled." Or we might write "A number is diminished by one and then multiplied by two."

(a) $n + 3$ **(b)** $\frac{x + 2}{5}$ **(c)** $3(5 + a)$ **(d)** $3 - 4n$ **(e)** $\frac{x + 6}{x - 1}$

Answers

1. $c + d$ **3.** $w + z$ **5.** $x + 2$ **7.** $y + 10$ **9.** $a - b$ **11.** $b - 7$ **13.** $r - 6$ **15.** wz **17.** $5t$ **19.** $8mn$ **21.** $3(p + q)$ **23.** $2(x + y)$ **25.** $2x + y$ **27.** $2(x - y)$ **29.** $(a + b)(a - b)$ **31.** $m(m - 3)$ **33.** $\frac{x}{5}$ **35.** $\frac{a + b}{7}$ **37.** $\frac{p - q}{4}$ **39.** $\frac{a + 3}{a - 3}$ **41.** $x + 5$ **43.** $x - 7$ **45.** $9x$ **47.** $3x + 6$ **49.** $2(x + 5)$ **51.** $(x + 2)(x - 2)$ **53.** $\frac{x}{7}$ **55.** $\frac{x + 5}{8}$ **57.** $\frac{x + 6}{x - 6}$ **59.** $4s$ **61.** $\pi r^2 h$ **63.** $\frac{1}{2}h(b_1 + b_2)$ **65.** Expression **67.** Not an expression **69.** Not an expression **71.** Expression **73.** $2x$ **75.** $I = Prt$ **77.**

2.7 Evaluating Algebraic Expressions

OBJECTIVES

1. Substitute integer values for variables in an expression and evaluate
2. Interpret summation notation

In applying algebra to problem solving, you will often want to find the value of an algebraic expression when you know certain values for the letters (or variables) in the expression. Finding the value of an expression is called *evaluating the expression* and uses the following steps.

Step by Step: To Evaluate an Algebraic Expression

Step 1 Replace each variable by the given number value.

Step 2 Do the necessary arithmetic operations, following the rules for order of operations.

Example 1

Evaluating Algebraic Expressions

Suppose that $a = 5$ and $b = 7$.

(a) To evaluate $a + b$, we replace a with 5 and b with 7.

$a + b = 5 + 7 = 12$

(b) To evaluate $3ab$, we again replace a with 5 and b with 7.

$3ab = 3 \cdot 5 \cdot 7 = 105$

CHECK YOURSELF 1

If $x = 6$ and $y = 7$, evaluate.

(a) $y - x$ **(b)** $5xy$

We are now ready to evaluate algebraic expressions that require following the rules for the order of operations.

Example 2

Evaluating Algebraic Expressions

Evaluate the expressions if $a = 2$, $b = 3$, $c = 4$, and $d = 5$.

(a) $5a + 7b = 5 \cdot 2 + 7 \cdot 3$ Multiply first.

$= 10 + 21 = 31$ Then add.

(b) $3c^2 = 3 \cdot 4^2$ Evaluate the power.

$= 3 \cdot 16 = 48$ Then multiply.

CAUTION

This is different from
$(3c)^2 = (3 \cdot 4)^2$
$= 12^2 = 144$

(c) $7(c + d) = 7(4 + 5)$ Add inside the parentheses.

$= 7 \cdot 9 = 63$

(d) $5a^4 - 2d^2 = 5 \cdot 2^4 - 2 \cdot 5^2$ Evaluate the powers.

$= 5 \cdot 16 - 2 \cdot 25$ Multiply.

$= 80 - 50 = 30$ Subtract.

NOTE Note that the parentheses attach the negative sign to the number **before** it is raised to a power.

(e) $(-a)^2 = (-2)^2$

$= (-2) \cdot (-2)$

$= 4$

(f) $-a^2 = -2^2$ Evaluate the powers.

$= -2 \cdot 2$

$= -4$

CHECK YOURSELF 2

If $x = 3$, $y = 2$, $z = 4$, and $w = 5$, evaluate the expressions.

(a) $4x^2 + 2$ (b) $5(z + w)$ (c) $7(z^2 - y^2)$ (d) $-x^2$ (e) $(-x)^2$

To evaluate algebraic expressions when a fraction bar is used, do the following: Start by doing all the work in the numerator, and then do the work in the denominator. Divide the numerator by the denominator as the last step.

Example 3

Evaluating Algebraic Expressions

If $p = 2$, $q = 3$, and $r = 4$, evaluate:

(a) $\frac{8p}{r}$

Replace p with 2 and r with 4.

NOTE As we mentioned in Section 2.5, the fraction bar is a grouping symbol, like parentheses. Work first in the numerator and then in the denominator.

$\frac{8p}{r} = \frac{8 \cdot 2}{4} = \frac{16}{4} = 4$ Divide as the last step.

(b) $\frac{7q + r}{p + q} = \frac{7 \cdot 3 + 4}{2 + 3}$ Now evaluate the top and bottom separately.

$= \frac{21 + 4}{2 + 3} = \frac{25}{5} = 5$

CHECK YOURSELF 3

Evaluate the expressions if $c = 5$, $d = 8$, and $e = 3$.

(a) $\frac{6c}{e}$ **(b)** $\frac{4d + e}{c}$ **(c)** $\frac{10d - e}{d + e}$

Example 4 shows how a scientific calculator can be used to evaluate algebraic expressions.

Example 4

Using a Calculator to Evaluate Expressions

Use a scientific calculator to evaluate the expressions.

(a) $\frac{4x + y}{z}$ if $x = 2$, $y = 1$, and $z = 3$

Replace x with 2, y with 1, and z with 3:

$$\frac{4x + y}{z} = \frac{4 \cdot 2 + 1}{3}$$

Now, use the following keystrokes:

[(] 4 [×] 2 [+] 1 [)] [÷] 3 [=]

The display will read 3.

(b) $\frac{7x - y}{3z - x}$ if $x = 2$, $y = 6$, and $z = 2$

$$\frac{7x - y}{3z - x} = \frac{7 \cdot 2 - 6}{3 \cdot 2 - 2}$$

Use the following keystrokes:

[(] 7 [×] 2 [−] 6 [)] [÷] [(] 3 [×] 2 [−] 2 [)] [=]

The display will read 2.

CHECK YOURSELF 4

Use a scientific calculator to evaluate the expressions if $x = 2$, $y = 6$, and $z = 5$.

(a) $\frac{2x + y}{z}$ **(b)** $\frac{4y - 2z}{x}$

Example 5

Evaluating Expressions

Evaluate $5a + 4b$ if $a = -2$ and $b = 3$.

Replace a with -2 and b with 3.

$$5a + 4b = 5(-2) + 4(3)$$
$$= -10 + 12$$
$$= 2$$

NOTE Remember the rules for the order of operations. Multiply first, then add.

CHECK YOURSELF 5

Evaluate $3x + 5y$ if $x = -2$ and $y = -5$.

We follow the same rules no matter how many variables are in the expression.

Example 6

Evaluating Expressions

Evaluate the expressions if $a = -4$, $b = 2$, $c = -5$, and $d = 6$.

This becomes $-(-20)$, or $+20$.

(a) $7a - 4c = 7(-4) - 4(-5)$
$$= -28 + 20$$
$$= -8$$

Evaluate the power first, then multiply by 7.

(b) $7c^2 = 7(-5)^2 = 7 \cdot 25$
$$= 175$$

(c) $b^2 - 4ac = 2^2 - 4(-4)(-5)$
$$= 4 - 4(-4)(-5)$$
$$= 4 - 80$$
$$= -76$$

Add inside the parentheses first.

(d) $b(a + d) = 2(-4 + 6)$
$$= 2(2)$$
$$= 4$$

CAUTION

When a squared variable is replaced by a negative number, square the negative.

$(-5)^2 = (-5)(-5) = 25$

The exponent applies to -5!

$-5^2 = -(5 \cdot 5) = -25$

The exponent applies only to 5!

CHECK YOURSELF 6

Evaluate the expressions if p = −4, q = 3, and r = −2.

(a) $5p - 3r$ **(b)** $2p^2 + q$ **(c)** $p(q + r)$
(d) $-q^2$ **(e)** $(-q)^2$

As mentioned earlier, the fraction bar is a grouping symbol. Example 7 further illustrates this concept.

Example 7

Evaluating Expressions

Evaluate the expressions if $x = 4$, $y = -5$, $z = 2$, and $w = -3$.

(a) $$\frac{z - 2y}{x} = \frac{2 - 2(-5)}{4}$$

$$= \frac{2 - (-10)}{4}$$

$$= \frac{12}{4} = 3$$

(b) $$\frac{3x - w}{2x + w} = \frac{3(4) - (-3)}{2(4) + (-3)} = \frac{12 + 3}{8 + (-3)}$$

$$= \frac{15}{5} = 3$$

CHECK YOURSELF 7

Evaluate the expressions if m = −6, n = 4, and p = −3.

(a) $\frac{m + 3n}{p}$ **(b)** $\frac{4m + n}{m + 4n}$

When an expression is evaluated by a calculator, the same order of cperations that we introduced in Section 1.7 is followed.

	Algebraic Notation	Calculator Notation
Addition	$6 + 2$	6 [+] 2
Subtraction	$4 - 8$	4 [−] 8
Multiplication	$(3)(-5)$	3 [×] [(−)] 5 or 3 [×] 5 [+/−]
Division	$\frac{8}{6}$	8 [÷] 6
Exponential	3^4	3 [^] 4 or 3 [y^x] 4

In many applications, you will need to find the sum of a set of numbers that you are working with. In mathematics, the shorthand symbol for "sum of" is the Greek letter Σ (capital sigma, the "S" of the Greek alphabet). The expression Σx, in which x refers to all the numbers in a given set, means the sum of all the numbers in that set.

Example 8

Summing a Set

Find Σx for the set of integers.

$-2, -6, 3, 5, -4$

$$\begin{aligned}\Sigma x &= -2 + (-6) + 3 + 5 + (-4)\\ &= (-8) + 3 + 5 + (-4)\\ &= (-8) + 8 + (-4)\\ &= -4\end{aligned}$$

CHECK YOURSELF 8

Find Σx for each set of integers.

(a) $-3, 4, -7, -9, 8$ **(b)** $-2, 6, -5, -3, 4, 7$

CHECK YOURSELF ANSWERS

1. (a) 1; **(b)** 210 **2. (a)** 38; **(b)** 45; **(c)** 84; **(d)** -9; **(e)** 9 **3. (a)** 10; **(b)** 7; **(c)** 7
4. (a) 2; **(b)** 7 **5.** -31 **6. (a)** -14; **(b)** 35; **(c)** -4; **(d)** -9; **(e)** 9
7. (a) -2; **(b)** -2 **8. (a)** -7; **(b)** 7

Name ______________________

Section ________ Date ________

Exercises

Evaluate each of the expressions if $a = -2$, $b = 5$, $c = -4$, and $d = 6$.

1. $3c - 2b$
2. $4c - 2b$
3. $8b + 2c$
4. $7a - 2c$
5. $-b^2 + b$
6. $(-b)^2 + b$
7. $3a^2$
8. $6c^2$
9. $c^2 - 2d$
10. $3a^2 + 4c$
11. $2a^2 + 3b^2$
12. $4b^2 - 2c^2$
13. $2(a + b)$
14. $5(b - c)$
15. $4(2a - d)$
16. $6(3c - d)$
17. $a(b + 3c)$
18. $c(3a - d)$
19. $\frac{6d}{c}$
20. $\frac{8b}{5c}$
21. $\frac{3d + 2c}{b}$
22. $\frac{2b + 3d}{2a}$

ANSWERS

1. ________
2. ________
3. ________
4. ________
5. ________
6. ________
7. ________
8. ________
9. ________
10. ________
11. ________
12. ________
13. ________
14. ________
15. ________
16. ________
17. ________
18. ________
19. ________
20. ________
21. ________
22. ________

ANSWERS

23. ______
24. ______
25. ______
26. ______
27. ______
28. ______
29. ______
30. ______
31. ______
32. ______
33. ______
34. ______
35. ______
36. ______
37. ______
38. ______
39. ______
40. ______
41. ______
42. ______
43. ______
44. ______
45. ______
46. ______
47. ______
48. ______
49. ______ 50. ______
51. ______ 52. ______

23. $\dfrac{2b - 3a}{c + 2d}$

24. $\dfrac{3d - 2b}{5a + d}$

25. $d^2 - b^2$

26. $c^2 - a^2$

27. $(d - b)^2$

28. $(c - a)^2$

29. $(d - b)(d + b)$

30. $(c - a)(c + a)$

31. $d^3 - b^3$

32. $c^3 + a^3$

33. $(d - b)^3$

34. $(c + a)^3$

35. $(d - b)(d^2 + db + b^2)$

36. $(c + a)(c^2 - ac + a^2)$

37. $b^2 + a^2$

38. $d^2 - a^2$

39. $(b + a)^2$

40. $(d - a)^2$

41. $a^2 + 2ad + d^2$

42. $b^2 - 2bc + c^2$

For each of the data sets, evaluate Σx.

43. 1, 2, 3, 7, 8, 9, 11

44. 2, 4, 5, 6, 10, 11, 12

45. −5, −3, −1, 2, 3, 4, 8

46. −4, −2, −1, 5, 7, 8, 10

47. 3, 2, −1, −4, −3, 8, 6

48. 3, −4, 2, −1, 2, −7, 9

For exercises 49 to 52, decide if the given values make the statement true or false.

49. $x - 7 = 2y + 5; x = 22, y = 5$

50. $3(x - y) = 6; x = 5, y = -3$

51. $2(x + y) = 2x + y; x = -4, y = -2$

52. $x^2 - y^2 = x - y; x = 4, y = -3$

53. Perimeter. The perimeter of a rectangle of length L and width W is sometimes given by the formula $P = 2L + 2W$. Find the perimeter when $L = 10$ in. and $W = 5$ in.

54. Using the equation given in exercise 53, find the perimeter of a sheet of paper that is 11 in. wide and 14 in. long.

55. A major-league pitcher throws a ball straight up into the air. The height of the ball (in feet) can be determined by substituting the number of seconds that have passed after the throw into the expression $110s - 16s^2$.

(a) Determine the height after 2 s. (Hint: Replace s with 2 in the expression, and then evaluate.)

(b) Find the height after 3 s.

56. Given the same equation used in exercise 55, find the height of the ball after 4 s. How does this compare to the answer for exercise 55? What has happened?

57. Is $a^n + b^n = (a + b)^n$? Try a few numbers and decide if you think this is true for all numbers, for some numbers, or never true. Write an explanation of your findings and give examples.

58. Enjoyment of patterns in art, music, and language is common to all cultures, and many cultures also delight in and draw spiritual significance from patterns in numbers. One such set of patterns is that of the "magic" square. One of these squares appears in a famous etching by Albrecht Dürer, who lived from 1471 to 1528 in Europe. He was one of the first artists in Europe to use geometry to give perspective, a feeling of three dimensions, in his work.

ANSWERS

53. ____________

54. ____________

55. ____________

56. ____________

57. ____________

58. ____________

The magic square in his work is this one:

16	3	2	13
5	10	11	8
9	6	7	12
4	15	14	1

Why is this square "magic"? It is magic because every row, every column, and both diagonals add to the same number. In this square there are 16 spaces for the numbers 1 through 16.

Part 1: What number does each row and column add to?

Write the square that you obtain by adding -17 to each number. Is this still a magic square? If so, what number does each column and row add to? If you add 5 to each number in the original magic square, do you still have a magic square? You have been studying the operations of addition, multiplication, subtraction, and division with integers and with rational numbers. What operations can you perform on this magic square and still have a magic square? Try to find something that will not work. Use algebra to help you decide what will work and what won't. Write a description of your work and explain your conclusions.

Part 2: Here is the oldest published magic square. It is from China, about 250 B.C.E. Legend has it that it was brought from the River Lo by a turtle to the Emperor Yii, who was a hydraulic engineer.

4	9	2
3	5	7
8	1	6

Check to make sure that this is a magic square. Work together to decide what operation might be done to every number in the magic square to make the sum of each row, column, and diagonal the *opposite* of what it is now. What would you do to every number to cause the sum of each row, column, and diagonal to equal zero?

Answers

1. -22 **3.** 32 **5.** -20 **7.** 12 **9.** 4 **11.** 83 **13.** 6
15. -40 **17.** 14 **19.** -9 **21.** 2 **23.** 2 **25.** 11 **27.** 1
29. 11 **31.** 91 **33.** 1 **35.** 91 **37.** 29 **39.** 9 **41.** 16
43. 41 **45.** 8 **47.** 11 **49.** True **51.** False **53.** 30 in.
55. **(a)** 156 ft; **(b)** 186 ft **57.**

2.8 Simplifying Algebraic Expressions

OBJECTIVES

1. Identify terms, like terms, and numerical coefficients
2. Simplify algebraic expressions by combining like terms

In Section 1.2, we found the perimeter of a rectangle by using the formula

$$P = L + W + L + W$$

There is another version of this formula that we can use. Because we are adding 2 times the length and 2 times the width, we can use the language of algebra to write this formula as

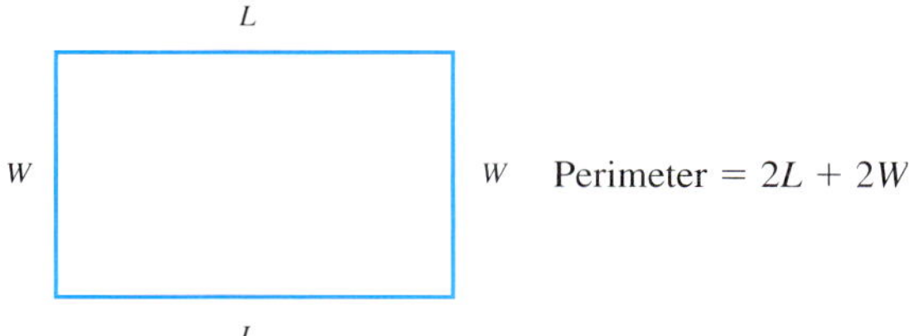

We call $2L + 2W$ an **algebraic expression,** or more simply an **expression.** Recall from Section 2.6 that an expression allows us to write a mathematical idea in symbols. It can be thought of as a meaningful collection of letters, numbers, and operation signs.

Some expressions are

$5x^2$ $\qquad$ $3a + 2b$ $\qquad$ $4x^3 + (-2y) + 1$

In algebraic expressions, the addition and subtraction signs break the expressions into smaller parts called *terms.*

Definitions: Term

A **term** is a number, or the product of a number and one or more variables, raised to a power.

In an expression, each sign (+ or −) is a part of the term that follows the sign.

Example 1

Identifying Terms

(a) $5x^2$ has one term.

(b) $\underbrace{3a}_{\text{Term}} + \underbrace{2b}_{\text{Term}}$ has two terms: $3a$ and $2b$.

(c) $\underbrace{4x^3}_{\text{Term}} + \underbrace{(-2y)}_{\text{Term}} + \underbrace{1}_{\text{Term}}$ has three terms: $4x^3$, $-2y$, and 1.

NOTE This could also be written as $4x^3 - 2y + 1$

CHECK YOURSELF 1

List the terms of each expression.

(a) $2b^4$ **(b)** $5m + 3n$ **(c)** $2s^2 - 3t - 6$

Note that a term in an expression may have any number of factors. For instance, $5xy$ is a term. It has factors of 5, x, and y. The number factor of a term is called the **numerical coefficient.** So for the term $5xy$, the numerical coefficient is 5.

Example 2

Identifying the Numerical Coefficient

(a) $4a$ has the numerical coefficient 4.

(b) $6a^3b^4c^2$ has the numerical coefficient 6.

(c) $-7m^2n^3$ has the numerical coefficient -7.

(d) Because $1 \cdot x = x$, the numerical coefficient of x is understood to be 1.

CHECK YOURSELF 2

Give the numerical coefficient for each term.

(a) $8a^2b$ **(b)** $-5m^3n^4$ **(c)** y

If terms contain exactly the *same letters* (or variables) raised to the *same powers,* they are called **like terms.**

Example 3

Identifying Like Terms

(a) The following are like terms.

$6a$ and $7a$

$5b^2$ and b^2

$10x^2y^3z$ and $-6x^2y^3z$

$-3m^2$ and m^2

Each pair of terms has the same letters, with each letter raised to the same power—the numerical coefficients can be any number.

(b) The following are *not* like terms.

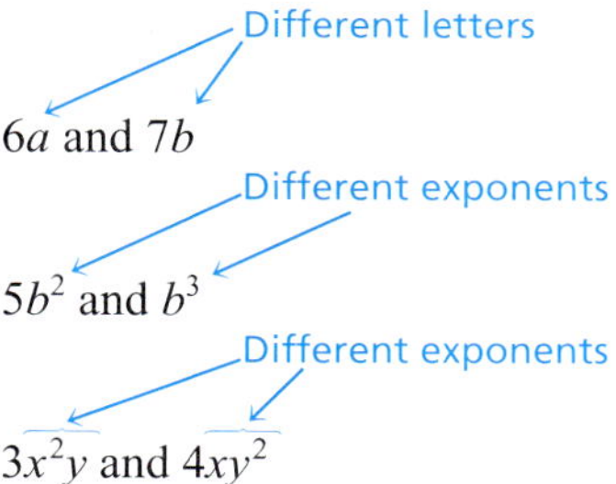

CHECK YOURSELF 3

Circle like terms.

$5a^2b \quad ab^2 \quad a^2b \quad -3a^2 \quad 4ab \quad 3b^2 \quad -7a^2b$

Like terms of an expression can always be combined into a single term. Look at the following:

$$\underbrace{x + x}_{2x} + \underbrace{x + x + x + x + x}_{5x} = \underbrace{x + x + x + x + x + x + x}_{7x}$$

Rather than having to write out all those x's, try

NOTE Here we use the distributive property from Section 1.5.

$$\begin{aligned} 2x + 5x &= (2 + 5)x \\ &= 7x \end{aligned}$$

In the same way,

NOTE You don't have to write all this out—just do it mentally!

$$\begin{aligned} 9b + 6b &= (9 + 6)b \\ &= 15b \end{aligned}$$

and

$$\begin{aligned} 10a - 4a &= 10a + (-4a) \\ &= (10 + (-4))a \\ &= 6a \end{aligned}$$

This leads us to a rule.

Step by Step: To Combine Like Terms

To combine like terms, use steps 1 and 2.

Step 1 Add or subtract the numerical coefficients.
Step 2 Attach the common variables.

Example 4

Combining Like Terms

Combine like terms.

(a) $8m + 5m = (8 + 5)m = 13m$

(b) $5pq^3 - 4pq^3 = 5pq^3 + (-4pq^3) = 1pq^3 = pq^3$

NOTE Remember that when any factor is multiplied by 0, the product is 0.

(c) $7a^3b^2 - 7a^3b^2 = 7a^3b^2 + (-7a^3b^2) = 0a^3b^2 = 0$

CHECK YOURSELF 4

Combine like terms.

(a) $6b + 8b$ **(b)** $12x^2 - 3x^2$

(c) $8xy^3 - 7xy^3$ **(d)** $9a^2b^4 - 9a^2b^4$

Now we will look at some expressions involving more than two terms. The idea is just the same.

Example 5

Combining Like Terms

Combine like terms.

(a) $5ab - 2ab + 3ab$

$= 5ab + (-2ab) + 3ab$

$= (5 + (-2) + 3)ab = 6ab$

NOTE The distributive property can be used over any number of like terms.

Only like terms can be combined.

(b) $8x - 2x + 5y$

$(8 + (-2))x + 5y$

$= 6x + 5y$

Like terms Like terms

(c) $5m + 8n + 4m - 3n$

$= (5m + 4m) + (8n + (-3n))$

$= 9m + 5n$

Here we have used the associative and commutative properties of addition.

NOTE With practice you won't be writing out these steps, but doing it mentally.

(d) $4x^2 + 2x - 3x^2 + x$

$= (4x^2 + (-3x^2)) + (2x + x)$

$= x^2 + 3x$

As these examples illustrate, combining like terms often means changing the grouping and the order in which the terms are written. Again all this is possible because of the properties of addition that we introduced in Section 1.2.

CHECK YOURSELF 5

Combine like terms.

(a) $4m^2 - 3m^2 + 8m^2$ **(b)** $9ab + 3a - 5ab$
(c) $4p + 7q + 5p - 3q$

As you have seen in arithmetic, subtraction can be performed directly. As this is the form used for most of mathematics, we will use that form throughout this text. Just remember, by using the additive inverse of numbers, you can always rewrite a subtraction problem as an addition problem.

Example 6

Combining Like Terms

Combine like terms.

(a) $2xy - 3xy + 5xy$

$= (2 - 3 + 5)xy$

$= 4xy$

(b) $5a - 2b + 7b - 8a$

$= 5a + (-2)b + 7b + (-8)a$

$= 5a + (-8)a + (-2)b + 7b$

$= -3a + 5b$

NOTE Note that x and x^2 are **not** like terms.

(c) $2x^2 + 3x + 4x^2 + x$

$= (2x^2 + 4x^2) + (3x + x)$

$= 6x^2 + 4x$

CHECK YOURSELF 6

Combine like terms.

(a) $4ab + 5ab - 3ab - 7ab$ **(b)** $2x - 7y - 8x - y$

The distributive property studied in Section 1.5 can be used to simplify expressions containing variables. This is illustrated in Example 7.

Example 7

Using the Distributive Property Before Combining Like Terms

Use the distributive property to remove the parentheses. Then simplify by combining like terms.

(a) $2(x - 5) - 8x$

$= 2(x) - 2(5) - 8x$

$= 2x - 10 - 8x$

$= 2x - 8x - 10$

$= (2 - 8)x - 10$

$= -6x - 10$

(b) $3a - 2b + 4(a + 3b)$

$= 3a - 2b + 4(a) + 4(3b)$

$= 3a - 2b + 4a + 12b$

$= 3a + 4a - 2b + 12b$

$= 7a + 10b$

CHECK YOURSELF 7

Use the distributive property to remove the parentheses. Then simplify by combining like terms.

(a) $7x + 3(x - 1)$ **(b)** $5(m + 2n) + 3(m - n)$

CHECK YOURSELF ANSWERS

1. **(a)** $2b^4$; **(b)** $5m$, $3n$; **(c)** $2s^2$, $-3t$, -6 **2.** **(a)** 8; **(b)** -5; **(c)** 1
3. The like terms are $5a^2b$, a^2b, and $-7a^2b$ **4.** **(a)** $14b$; **(b)** $9x^2$; **(c)** xy^3; **(d)** 0
5. **(a)** $9m^2$; **(b)** $4ab + 3a$; **(c)** $9p + 4q$ **6.** **(a)** $-ab$; **(b)** $-6x - 8y$
7. **(a)** $10x - 3$; **(b)** $8m + 7n$

Name ____________________

Section ________ Date ________

2.8 Exercises

List the terms of each expression.

1. $5a + 2$

2. $7a - 4b$

3. $4x^3$

4. $3x^2$

5. $3x^2 + 3x - 7$

6. $2a^3 - a^2 + a$

Circle the like terms in each group of terms.

7. $5ab$, $3b$, $3a$, $4ab$

8. $9m^2$, $8mn$, $5m^2$, $7m$

9. $4xy^2$, $2x^2y$, $5x^2$, $-3x^2y$, $5y$, $6x^2y$

10. $8a^2b$, $4a^2$, $3ab^2$, $-5a^2b$, $3ab$, $5a^2b$

Combine like terms.

11. $3m + 7m$

12. $6a^2 + 8a^2$

13. $7b^3 + 10b^3$

14. $7rs + 13rs$

15. $21xyz + 7xyz$

16. $4mn^2 + 15mn^2$

17. $9z^2 - 3z^2$

18. $7m - 6m$

19. $5a^3 - 5a^3$

20. $13xy - 9xy$

ANSWERS

1. ____________________
2. ____________________
3. ____________________
4. ____________________
5. ____________________
6. ____________________
7. ____________________
8. ____________________
9. ____________________
10. ____________________
11. ____________________
12. ____________________
13. ____________________
14. ____________________
15. ____________________
16. ____________________
17. ____________________
18. ____________________
19. ____________________
20. ____________________

ANSWERS

21. ____
22. ____
23. ____
24. ____
25. ____
26. ____
27. ____
28. ____
29. ____
30. ____
31. ____
32. ____
33. ____
34. ____
35. ____
36. ____
37. ____
38. ____
39. ____
40. ____

21. $19n^2 - 18n^2$

22. $7cd - 7cd$

23. $21p^2q - 6p^2q$

24. $17r^3s^2 - 8r^3s^2$

25. $10x^2 - 7x^2 + 3x^2$

26. $13uv + 5uv - 12uv$

27. $9a - 7a + 4b$

28. $5m^2 - 3m + 6m^2$

29. $7x + 5y - 4x - 4y$

30. $6a^2 + 11a + 7a^2 - 9a$

31. $4a + 7b + 3 - 2a + 3b - 2$

32. $5p^2 + 2p + 8 + 4p^2 + 5p - 6$

Perform the indicated operations.

33. Find the sum of $5a^4$ and $8a^4$.

34. Find the sum of $9p^2$ and $12p^2$.

35. Subtract $12a^3$ from $15a^3$.

36. Subtract $5m^3$ from $18m^3$.

37. Subtract $4x$ from the sum of $8x$ and $3x$.

38. Subtract $8ab$ from the sum of $7ab$ and $5ab$.

39. Subtract $3mn^2$ from the sum of $9mn^2$ and $5mn^2$.

40. Subtract $4x^2y$ from the sum of $6x^2y$ and $12x^2y$.

ANSWERS

41. ____

42. ____

43. ____

44. ____

45. ____

46. ____

47. ____

48. ____

49. ____

50. ____

51. ____

Use the distributive property to remove the parentheses in each expression. Then simplify by combining like terms.

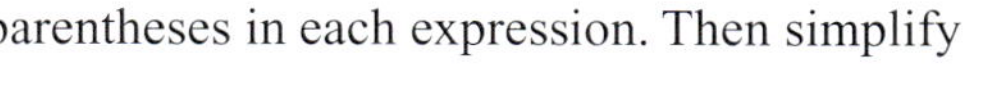

41. $2(3x + 2) + 4$

42. $3(4z + 5) - 9$

43. $5(6a - 2) + 12a$

44. $7(4w - 3) - 25w$

45. $4s + 2(s + 4) + 4$

46. $5p + 4(p + 3) - 8$

47. Write a paragraph explaining the difference between n^2 and $2n$.

48. Complete the explanation: "x^3 and $3x$ are not the same because . . ."

49. Complete the statement: "$x + 2$ and $2x$ are different because . . ."

50. Write an English phrase for each algebraic expression:

(a) $2x^3 + 5x$ **(b)** $(2x + 5)^3$ **(c)** $6(n + 4)^2$

51. Work with another student to complete this exercise. Place >, <, or = in the blank in these statements.

1^2____2^1

2^3____3^2

3^4____4^3

4^5____5^4

What happens as the table of numbers is extended? Try more examples

What sign seems to occur the most in your table? >, <, or =?

Write an algebraic statement for the pattern of numbers in this table. Do you think this is a pattern that continues? Add more lines to the table and extend the pattern to the general case by writing the pattern in algebraic notation. Write a short paragraph stating your conjecture.

Answers

1. $5a, 2$ **3.** $4x^3$ **5.** $3x^2, 3x, -7$ **7.** $5ab, 4ab$ **9.** $2x^2y, -3x^2y, 6x^2y$
11. $10m$ **13.** $17b^3$ **15.** $28xyz$ **17.** $6z^2$ **19.** 0 **21.** n^2
23. $15p^2q$ **25.** $6x^2$ **27.** $2a + 4b$ **29.** $3x + y$ **31.** $2a + 10b + 1$
33. $13a^4$ **35.** $3a^3$ **37.** $7x$ **39.** $11mn^2$ **41.** $6x + 8$ **43.** $42a - 10$
45. $6s + 12$ **47.** **49.** **51.**

2.9 Introduction to Linear Equations

2.9 OBJECTIVES

1. Use the language of equations
2. Determine whether a given number is a solution for an equation

In Chapter 1, we introduced one of the most important tools of mathematics, the equation. The ability to recognize and solve various types of equations is probably the most useful algebraic skill you will learn. We will continue to build upon the methods of this chapter throughout the remainder of the text. To start, we will review what we mean by an *equation.*

Definitions: Equation

An **equation** is a mathematical statement that two expressions are equal.

Some examples are $3 + 4 = 7$, $x + 3 = 5$, $P = 2L + 2W$.

Note that some of these equations contain letters. In algebra, we use letters to represent numerical values that we don't know or that could change. These letters are called variables. Recall that a **variable** is a letter used to represent a number. In the equation $x + 3 = 5$, x is a variable.

As you can see, an equal sign (=) separates the two equal expressions. These expressions are usually called the *left side* and the *right side* of the equation.

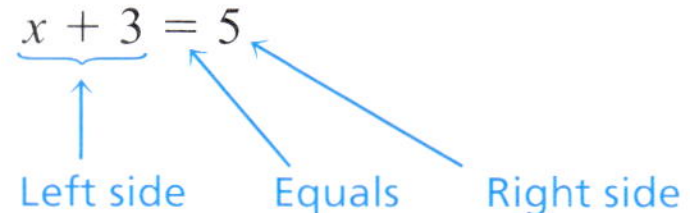

NOTE An equation such as $x + 3 = 5$ is called a **conditional equation** because it can be either true or false depending on the value given to the variable.

Just as the balance scale may be in balance or out of balance, an equation may be either true or false. For instance, $3 + 4 = 7$ is true because both sides name the same number. What about an equation such as $x + 3 = 5$ that has a letter or variable on one side? Any number can replace x in the equation. However, only one number will make this equation a true statement.

$$\text{If } x = \begin{cases} 1 & 1 + 3 = 5 \text{ is false} \\ 2 & 2 + 3 = 5 \text{ is true} \\ 3 & 3 + 3 = 5 \text{ is false} \end{cases}$$

The number 2 is called the **solution** (or *root*) of the equation $x + 3 = 5$ because substituting 2 for x gives a true statement.

Definitions: Solution

A **solution** for an equation is any value for the variable that makes the equation a true statement.

Example 1

Verifying a Solution

(a) Is 3 a solution for the equation $2x + 4 = 10$?

To find out, replace x with 3 and evaluate $2x + 4$ on the left.

Left side		*Right side*
$2 \cdot 3 + 4$	$\stackrel{?}{=}$	10
$6 + 4$	$\stackrel{?}{=}$	10
10	$=$	10

Because $10 = 10$ is a true statement, 3 is a solution of the equation.

(b) Is 5 a solution of the equation $3x - 2 = 2x + 1$?

To find out, replace x with 5 and evaluate each side separately.

NOTE Remember the rules for the order of operations. Multiply first; then add or subtract.

Left side		*Right side*
$3 \cdot 5 - 2$	$\stackrel{?}{=}$	$2 \cdot 5 + 1$
$15 - 2$	$\stackrel{?}{=}$	$10 + 1$
13	$\neq$	11

Because the two sides do not name the same number, we do not have a true statement, and 5 is not a solution.

CHECK YOURSELF 1

For the equation

$2x - 1 = x + 5$

(a) Is 4 a solution? **(b)** Is 6 a solution?

You may be wondering whether an equation can have more than one solution. It certainly can. For instance,

$$x^2 = 9$$

has two solutions. They are 3 and −3 because

$$3^2 = 9 \qquad \text{and} \qquad (-3)^2 = 9$$

In this chapter, however, we will always work with *linear equations in one variable.* These are equations that can be put into the form

$$ax + b = 0$$

in which the variable is x, a and b are any numbers, and a is not equal to 0. In a linear equation, the variable can appear only to the first power. No other power (x^2, x^3, etc.) can appear. Linear equations are also called **first-degree equations.** The degree of an equation in one variable is the highest power to which the variable appears.

Rules and Properties: Linear Equations

Linear equations in one variable are equations that can be written in the form

$$ax + b = 0 \qquad a \neq 0$$

Every such equation will have exactly one solution.

Example 2

Identifying Expressions and Equations

Label each as an expression, a linear equation, or an equation that is not linear.

(a) $4x + 5$ is an expression

(b) $2x + 8 = 0$ is a linear equation

(c) $3x^2 - 9 = 0$ is an equation that is not linear

(d) $5x = 15$ is a linear equation because this can be written as $5x - 15 = 0$.

CHECK YOURSELF 2

Label each as an expression, a linear equation, or an equation that is not linear.

(a) $2x^2 = 8$ **(b)** $2x - 3 = 0$

(c) $5x - 10$ **(d)** $2x + 1 = 7$

CHECK YOURSELF ANSWERS

1. **(a)** 4 is not a solution; **(b)** 6 is a solution
2. **(a)** Nonlinear equation; **(b)** linear equation; **(c)** expression; **(d)** linear equation

Name ______________________

Section ________ Date ________

2.9 Exercises

Is the number shown in parentheses a solution for the given equation?

1. $x + 4 = 9$ (5)

2. $x + 2 = 11$ (8)

3. $x - 15 = 6$ (−21)

4. $x - 11 = 5$ (16)

5. $5 - x = 2$ (4)

6. $10 - x = 7$ (3)

7. $4 - x = 6$ (−2)

8. $5 - x = 6$ (−3)

9. $3x + 4 = 13$ (8)

10. $5x + 6 = 31$ (5)

11. $4x - 5 = 7$ (2)

12. $2x - 5 = 1$ (3)

13. $5 - 2x = 7$ (−1)

14. $4 - 5x = 9$ (−2)

15. $4x - 5 = 2x + 3$ (4)

16. $5x + 4 = 2x + 10$ (4)

17. $x + 3 + 2x = 5 + x + 8$ (5)

18. $5x - 3 + 2x = 3 + x - 12$ (−2)

Label each as an expression or a linear equation.

19. $2x + 1 = 9$

20. $7x + 14$

21. $2x - 8$

22. $5x - 3 = 12$

23. $7x + 2x + 8 - 3$

24. $x + 5 = 13$

25. $2x - 8 = 3$

26. $12x - 5x + 2 + 5$

ANSWERS

1. ______
2. ______
3. ______
4. ______
5. ______
6. ______
7. ______
8. ______
9. ______
10. ______
11. ______
12. ______
13. ______
14. ______
15. ______
16. ______
17. ______
18. ______
19. ______
20. ______
21. ______
22. ______
23. ______
24. ______
25. ______
26. ______

ANSWERS

27. ______________________

28. ______________________

27. An algebraic equation is a complete sentence. It has a subject, a verb, and a predicate. For example, $x + 2 = 5$ can be written in English as "Two more than a number is five." Or, "A number added to two is five." Write an English version of the following equations. Be sure you write complete sentences and that the sentences express the same idea as the equations. Exchange sentences with another student, and see if your interpretation of each other's sentences result in the same equation.

(a) $2x - 5 = x + 1$

(b) $2(x + 2) = 14$

(c) $n + 5 = \frac{n}{2} - 6$

(d) $7 - 3a = 5 + a$

28. Complete the following explanation in your own words: "The difference between $3(x - 1) + 4 - 2x$ and $3(x - 1) + 4 = 2x$ is . . . ."

Answers

1. Yes **3.** No **5.** No **7.** Yes **9.** No **11.** No **13.** Yes **15.** Yes **17.** Yes **19.** Linear equation **21.** Expression **23.** Expression **25.** Linear equation **27.**

2.10 The Addition Property of Equality

OBJECTIVES

1. Use the addition property to solve an equation
2. Combine like terms while solving an equation
3. Use the distributive property while solving an equation

It is not difficult to find the solution for an equation such as $x + 3 = 8$ by guessing the answer to the question "What plus 3 is 8?" Here the answer to the question is 5, and that is also the solution for the equation. But for more complicated equations you are going to need something more than guesswork. A better method is to transform the given equation to an *equivalent equation* whose solution can be found by inspection. We can make a definition.

Definitions: Equivalent Equations

Equations that have the same solution are called **equivalent equations.**

These expressions are all equivalent equations:

$2x + 3 = 5 \qquad 2x = 2 \qquad \text{and} \qquad x = 1$

They all have the same solution, 1. We say that a linear equation is *solved* when it is transformed to an equivalent equation of the form

$x = \square$

The variable is alone on the left side.

The right side is some number, the solution.

NOTE In some cases we'll write the equation in the form

$\square = x$

The number will be our solution when the equation has the variable isolated on the left or on the right.

The addition property of equality is the first property you will need to transform an equation to an equivalent form.

Rules and Properties: The Addition Property of Equality

If $\quad a = b$

then $\quad a + c = b + c$

In words, adding the same quantity to both sides of an equation gives an equivalent equation.

REMEMBER An equation is a statement that the two sides are equal. Adding the same quantity to both sides does not change the equality or "balance."

Recall that we said that a true equation was like a scale in balance.

The addition property is equivalent to adding the same weight to both sides of the scale. It will remain in balance.

Example 1

Using the Addition Property to Solve an Equation

Solve.

$x - 3 = 9$

Remember that our goal is to isolate x on one side of the equation. Because 3 is being subtracted from x, we can add 3 to remove it. We must use the addition property to add 3 to both sides of the equation.

NOTE To check, replace x with 12 in the original equation:

$x - 3 = 9$

$12 - 3 \stackrel{?}{=} 9$

$9 = 9$

Because we have a true statement, 12 is the solution.

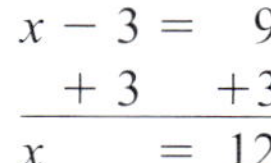

$$\begin{array}{rcr} x - 3 & = & 9 \\ +3 & & +3 \\ \hline x & = & 12 \end{array}$$

Adding 3 "undoes" the subtraction and leaves x alone on the left.

Because 12 is the solution for the equivalent equation $x = 12$, it is the solution for our original equation.

CHECK YOURSELF 1

Solve and check.

$x - 5 = 4$

The addition property also allows us to add a negative number to both sides of an equation. This is really the same as subtracting the same quantity from both sides.

Example 2

Using the Addition Property to Solve an Equation

NOTE Recall our comment that we could write an equation in the equivalent forms $x = \square$ or $\square = x$, in which $\square$ represents some number. Suppose we have an equation like

$12 = x + 7$

Adding -7 will isolate x *on the right:*

$$\begin{array}{rcl} 12 & = & x + 7 \\ -7 & & \quad -7 \\ \hline 5 & = & x \end{array}$$

and the solution is 5.

Solve.

$x + 5 = 9$

In this case, 5 is *added* to x on the left. We can use the addition property to add a -5 to both sides. Because $5 + (-5) = 0$, this will "undo" the addition and leave the variable x alone on one side of the equation.

$$\begin{array}{rcr} x + 5 & = & 9 \\ -5 & & -5 \\ \hline x & = & 4 \end{array}$$

The solution is 4. To check, replace x with 4 in the original equation.

$4 + 5 = 9$ (True)

CHECK YOURSELF 2

Solve and check.

$x + 6 = 13$

What if the equation has a variable term on both sides? You will have to use the addition property to add or subtract a term involving the variable to get the desired result.

Example 3

Using the Addition Property to Solve an Equation

Solve.

$5x = 4x + 7$

We will start by adding $-4x$ to both sides of the equation. Do you see why? Remember that an equation is solved when we have an equivalent equation of the form $x = \square$.

NOTE Recall that adding $-4x$ is identical to subtracting $4x$.

$$\begin{array}{rcr} 5x & = & 4x + 7 \\ -4x & & -4x \\ \hline x & = & 7 \end{array}$$

Adding $-4x$ to both sides *removes* $4x$ from the right.

To check: Because 7 is a solution for the equivalent equation $x = 7$, it should be a solution for the original equation. To find out, replace x with 7:

$$\begin{aligned} 5 \cdot 7 &\stackrel{?}{=} 4 \cdot 7 + 7 \\ 35 &\stackrel{?}{=} 28 + 7 \\ 35 &= 35 \quad \text{(True)} \end{aligned}$$

CHECK YOURSELF 3

Solve and check.

$7x = 6x + 3$

You may have to apply the addition property more than once to solve an equation. Look at Example 4.

Example 4

Using the Addition Property to Solve an Equation

Solve.

$7x - 8 = 6x$

We want all variables on *one* side of the equation. If we choose the left, we add $-6x$ to both sides of the equation. This will remove $6x$ from the right:

$$\begin{array}{rcr} 7x - 8 & = & 6x \\ -6x & & -6x \\ \hline x - 8 & = & 0 \end{array}$$

We want the variable alone, so we add 8 to both sides. This isolates x on the left.

$$\begin{array}{rcr} x - 8 & = & 0 \\ +8 & & +8 \\ \hline x & = & 8 \end{array}$$

The solution is 8. We'll leave it to you to check this result.

CHECK YOURSELF 4

Solve and check.

$9x + 3 = 8x$

Often an equation will have more than one variable term *and* more than one number. You will have to apply the addition property twice in solving these equations.

Example 5

Using the Addition Property to Solve an Equation

Solve.

$5x - 7 = 4x + 3$

We would like the variable terms on the left, so we start by adding $-4x$ to remove the $4x$ term from the right side of the equation:

$$\begin{array}{rcr} 5x - 7 &=& 4x + 3 \\ -4x \quad\; && -4x \quad\; \\ \hline x - 7 &=& 3 \end{array}$$

Now, to isolate the variable, we add 7 to both sides.

$$\begin{array}{rcr} x - 7 &=& 3 \\ +7 && +7 \\ \hline x \quad &=& 10 \end{array}$$

NOTE You could just as easily have added 7 to both sides and *then* added $-4x$. The result would be the same. In fact, some students prefer to combine the two steps.

The solution is 10. To check, replace x with 10 in the original equation:

$5 \cdot 10 - 7 \stackrel{?}{=} 4 \cdot 10 + 3$

$43 = 43$ (True)

CHECK YOURSELF 5

Solve and check.

(a) $4x - 5 = 3x + 2$ **(b)** $6x + 2 = 5x - 4$

NOTE Remember, by *simplify* we mean to combine all like terms.

In solving an equation, you should always simplify each side as much as possible before using the addition property.

Example 6

Combining Like Terms and Solving the Equation

Solve.

Like terms Like terms

$5 + 8x - 2 = 2x - 3 + 5x$

Because like terms appear on each side of the equation, we start by combining the numbers on the left (5 and -2). Then we combine the like terms ($2x$ and $5x$) on the right. We have

$3 + 8x = 7x - 3$

Now we can apply the addition property, as before:

$$\begin{array}{rcrl} 3 + 8x &=& 7x - 3 & \\ -\ 7x &=& -7x & \text{Add } -7x. \\ \hline 3 + \ x &=& -\ 3 & \\ -3 & & -\ 3 & \text{Add } -3. \\ \hline x &=& -\ 6 & \text{Isolate } x. \end{array}$$

The solution is -6. To check, always return to the original equation. That will catch any possible errors in simplifying. Replacing x with -6 gives

$$\begin{aligned} 5 + 8(-6) - 2 &\stackrel{?}{=} 2(-6) - 3 + 5(-6) \\ 5 - 48 - 2 &\stackrel{?}{=} -12 - 3 - 30 \\ -45 &= -45 \quad \text{(True)} \end{aligned}$$

CHECK YOURSELF 6

Solve and check.

(a) $3 + 6x + 4 = 8x - 3 - 3x$ **(b)** $5x + 21 + 3x = 20 + 7x - 2$

We may have to apply some of the properties discussed in Section 1.5 in solving equations. Example 7 illustrates the use of the distributive property to clear an equation of parentheses.

Example 7

Using the Distributive Property and Solving Equations

NOTE $2(3x + 4) = 2(3x) + 2(4) = 6x + 8$

Solve.

$$2(3x + 4) = 5x - 6$$

Applying the distributive property on the left, we have

$$6x + 8 = 5x - 6$$

We can then proceed as before:

$$\begin{array}{rcrl} 6x + 8 &=& 5x - 6 & \\ -5x & & -5x & \text{Add } -5x. \\ \hline x + 8 &=& -\ 6 & \\ -\ 8 & & -\ 8 & \text{Add } -8. \\ \hline x &=& -14 & \end{array}$$

NOTE Remember that $x = -14$ and $-14 = x$ are equivalent equations.

The solution is -14. We will leave the checking of this result to the reader.

Remember: Always return to the original equation to check.

CHECK YOURSELF 7

Solve and check each of the equations.

(a) $4(5x - 2) = 19x + 4$ **(b)** $3(5x + 1) = 2(7x - 3) - 4$

Given an expression such as

$$-2(x - 5)$$

the distributive property can be used to create the equivalent expression.

$$-2x + 10$$

The distribution of a negative number is used in Example 8.

Example 8

Distributing a Negative Number

Solve each of the equations.

(a) $-2(x - 5) = -3x + 2$

$$
\begin{array}{rcll}
-2x + 10 & = & -3x + 2 & \text{Distribute the } -2. \\
+3x & & +3x & \text{Add } 3x. \\
\hline
x + 10 & = & 2 & \\
-10 & = & -10 & \text{Add } -10. \\
\hline
x & = & -8 &
\end{array}
$$

(b) $-3(3x + 5) = -5(2x - 2)$

$$
\begin{array}{rcll}
-9x - 15 & = & -5(2x - 2) & \text{Distribute the } -3. \\
-9x - 15 & = & -10x + 10 & \text{Distribute the } -5. \\
+10x & & +10x & \text{Add } 10x. \\
\hline
x - 15 & = & 10 & \\
+15 & & +15 & \text{Add } 15. \\
\hline
x & = & 25 &
\end{array}
$$

CHECK YOURSELF 8

Solve each of the equations.

(a) $-2(x - 3) = -x + 5$ **(b)** $-4(2x - 1) = -3(3x + 2)$

When parentheses are preceded only by a negative, or by the minus sign, we say that we have a silent negative one. Example 9 illustrates this case.

Example 9

Distributing the Silent Negative One

Solve.

$-(2x + 3) = -3x + 7$

$-1(2x + 3) = -3x + 7$

$(-1)(2x) + (-1)(3) = -3x + 7$

$$
\begin{array}{rcll}
-2x - 3 & = & -3x + 7 & \\
+3x & & +3x & \text{Add } 3x. \\
\hline
x - 3 & = & 7 & \\
+3 & & +3 & \text{Add } 3. \\
\hline
x & = & 10 &
\end{array}
$$

CHECK YOURSELF 9

Solve.

$-(3x + 2) = -2x - 6$

CHECK YOURSELF ANSWERS

1. 9 **2.** 7 **3.** 3 **4.** -3 **5.** **(a)** 7; **(b)** -6 **6.** **(a)** -10; **(b)** -3
7. **(a)** 12; **(b)** -13 **8.** **(a)** 1; **(b)** -10 **9.** 4

Name ______________

Section ________ Date ________

2.10 Exercises

Solve and check each equation.

1. $x + 9 = 11$

2. $x - 4 = 6$

3. $x - 8 = 3$

4. $x + 11 = 15$

5. $x - 8 = -10$

6. $x + 5 = 2$

7. $x + 4 = -3$

8. $x - 5 = -4$

9. $11 = x + 5$

10. $x + 7 = 0$

11. $4x = 3x + 4$

12. $7x = 6x - 8$

13. $11x = 10x - 10$

14. $9x = 8x + 5$

15. $6x + 3 = 5x$

16. $12x - 6 = 11x$

17. $8x - 4 = 7x$

18. $9x - 7 = 8x$

19. $2x + 3 = x + 5$

20. $3x - 2 = 2x + 1$

21. $3x - 5 + 2x - 7 + x = 5x + 2$

22. $5x + 8 + 3x - x + 5 = 6x - 3$

23. $3(7x + 2) = 5(4x - 1) + 17$

24. $5(5x + 3) = 3(8x - 2) + 4$

25. Which equation is equivalent to the equation $5x - 7 = 4x - 12$?

(a) $9x = 19$ **(b)** $9x - 7 = -12$ **(c)** $x = -18$ **(d)** $x - 7 = -12$

26. Which equation is equivalent to the equation $12x - 6 = 8x + 14$?

(a) $4x - 6 = 14$ **(b)** $x = 20$ **(c)** $20x = 20$ **(d)** $4x = 8$

27. Which equation is equivalent to the equation $7x + 5 = 12x - 10$?

(a) $5x = -15$ **(b)** $7x - 5 = 12x$ **(c)** $-5 = 5x$ **(d)** $7x + 15 = 12x$

ANSWERS

1. ______
2. ______
3. ______
4. ______
5. ______
6. ______
7. ______
8. ______
9. ______
10. ______
11. ______
12. ______
13. ______
14. ______
15. ______
16. ______
17. ______
18. ______
19. ______
20. ______
21. ______
22. ______
23. ______
24. ______ 25. ______
26. ______ 27. ______

28. ______

29. ______

30. ______

True or false?

28. Every linear equation with one variable has exactly one solution.

29. Isolating the variable on the right side of the equation will result in a negative solution.

30. "Surprising Results!" Work with other students to try this experiment. Each person should do the following six steps mentally, not telling anyone else what their calculations are:

(a) Think of a number.
(b) Add 7.
(c) Multiply by 3.
(d) Add 3 more than the original number.
(e) Divide by 4.
(f) Subtract the original number.

What number do you end up with? Compare your answer with everyone else's. Does everyone have the same answer? Make sure that everyone followed the directions accurately. How do you explain the results? Algebra makes the explanation clear. Work together to do the problem again, using a variable for the number. Make up another series of computations that give "surprising results."

Answers

1. 2 **3.** 11 **5.** -2 **7.** -7 **9.** 6 **11.** 4 **13.** -10
15. -3 **17.** 4 **19.** 2 **21.** 14 **23.** 16 **25.** d **27.** d
29. False

2 Summary

DEFINITION/PROCEDURE	EXAMPLE	REFERENCE
Introduction to Integers		**Section 2.1**
Positive Integers Integers used to name whole numbers to the right of the origin on the number line. *Negative Integers* Integers used to name the opposites of whole numbers to the left of the origin on the number line. *Integers* Whole numbers and their opposites. The integers are $\{\ldots, -3, -2, -1, 0, 1, 2, 3, \ldots\}$	The origin $-3\ -2\ -1\ 0\ 1\ 2\ 3$ Negative integers Positive integers	p. 114
Absolute Value The distance (on the number line) between the point named by a signed number and the origin. The absolute value of x is written $\lvert x \rvert$.	$\lvert 7 \rvert = 7$ $\lvert -10 \rvert = 10$	p. 117
Addition of Integers		**Section 2.2**
Adding Integers **1.** If two integers have the same sign, add their absolute values. Give the result the sign of the original integers. **2.** If two integers have different signs, subtract their absolute values, the smaller from the larger. Give the result the sign of the integer with the larger absolute value.	$9 + 7 = 16$ $(-9) + (-7) = -16$ $15 + (-10) = 5$ $(-12) + 9 = -3$	p. 124 p. 125
Subtraction of Integers		**Section 2.3**
Subtracting Integers **1.** Rewrite the subtraction problem as an addition problem by **a.** Changing the subtraction symbol to an addition symbol **b.** Replacing the integer being subtracted with its opposite **2.** Add the resulting integers as before.	$16 - 8 = 16 + (-8) = 8$ $8 - 15 = 8 + (-15) = -7$ $-9 - (-7) = -9 + 7 = -2$	p. 129
Multiplication of Integers		**Section 2.4**
Multiplying Integers Multiply the absolute values of the two integers. **1.** If the integers have different signs, the product is negative. **2.** If the integers have the same sign, the product is positive.	$5(-7) = -35$ $(-10)(9) = -90$ $8 \cdot 7 = 56$ $(-9)(-8) = 72$ $(-2)^2 = (-2)(-2) = 4$ $-2^2 = -2 \cdot 2 = -4$	p. 135 p. 136

Continued

DEFINITION/PROCEDURE	EXAMPLE	REFERENCE
Division of Integers		**Section 2.5**
Dividing Integers Divide the absolute values of the two integers. **1.** If the integers have different signs, the quotient is negative. **2.** If the integers have the same sign, the quotient is positive.	$\frac{-32}{4} = -8$ $\frac{-18}{-9} = 2$	**p. 141**
Introduction to Algebra: Variables and Expressions		**Section 2.6**
Multiplication $\left.\begin{matrix} x \cdot y \\ (x)(y) \\ xy \end{matrix}\right\}$ These all mean the **product** of x **and** y or x **times** y.	The product of m and n is mn. The product of 2 and the sum of a and b is $2(a + b)$.	**p. 148**
Evaluating Algebraic Expressions		**Section 2.7**
Evaluating Algebraic Expressions To evaluate an algebraic expression: **1.** Replace each variable or letter with its number value. **2.** Do the necessary arithmetic, following the rules for the order of operations.	Evaluate $2x + 3y$ if $x = 5$ and $y = -2$. $2x + 3y$ $= 2 \cdot 5 + (3)(-2)$ $= 10 + (-6) = 4$	**p. 155**
Simplifying Algebraic Expressions		**Section 2.8**
Term A number or the product of a number and one or more variables.	$3xy$ is a term	**p. 165**
Combining Like Terms To combine like terms: **1.** Add or subtract the coefficients (the numbers multiplying the variables). **2.** Attach the common variable.	$5x + 2x = 7x$ $8a - 5a = 3a$	**p. 167**
Introduction to Linear Equations		**Section 2.9**
Equation A statement that two expressions are equal.	$2x - 3 = 5$ is an equation	**p. 175**
Solution A value for a variable that makes an equation a true statement.	4 is a solution for the above equation because $2(4) - 3 = 5$	**p. 176**
The Addition Property of Equality		**Section 2.10**
Equivalent Equations Equations that have exactly the same solutions.	$2x - 3 = 5$ and $x = 4$ are equivalent equations	**p. 181**
The Addition Property of Equality If $a = b$, then $a + c = b + c$	If $2x - 3 = 7$, then $2x - 3 + 3 = 7 + 3$	**p. 181**

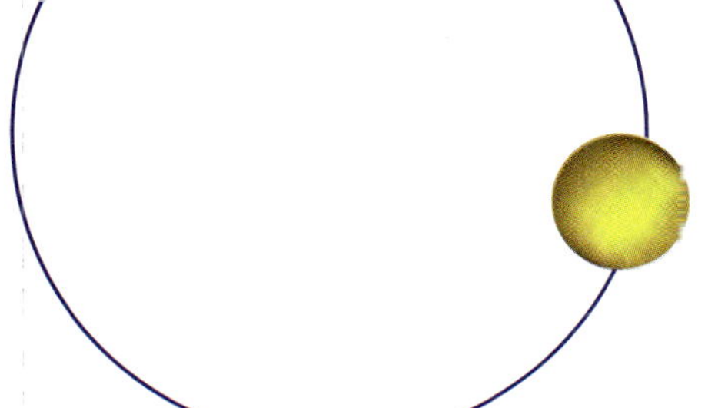

Summary and Review Exercises

This exercise set will give you practice with each of the objectives of the chapter. Each exercise is keyed to the appropriate chapter section. The answers are provided in the *Instructor's Manual.* Your instructor will give you guidelines on how to best use these exercises in your instructional setting.

[2.1] Represent the integers on the number line shown.

1. 6, −18, −3, 2, 15, −9

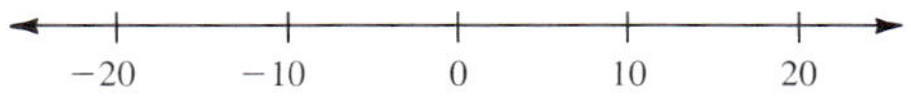

Place each of the sets in ascending order.

2. 4, −3, 6, −7, 0, 1, −2

3. −7, 8, −8, 1, 2, −3, 3, 0, 7

Find the opposite of each number.

4. 17

5. −63

Evaluate.

6. $|9|$

7. $|-9|$

8. $-|9|$

9. $-|-9|$

10. $|12 - 8|$

11. $|8| + |-12|$

12. $-|8 + 12|$

13. $|-18| - |-12|$

14. $|-7| - |-3|$

15. $|-9| + |-5|$

[2.2] Add.

16. $-3 + (-8)$

17. $10 + (-4)$

18. $6 + (-6)$

19. $-16 + (-16)$

20. $-18 + 0$

[2.3] Subtract.

21. $8 - 13$

22. $-7 - 10$

23. $10 - (-7)$

24. $-5 - (-1)$

25. $-9 - (-9)$

26. $0 - (-2)$

[2.4] Multiply.

27. $(10)(-7)$

28. $(-8)(-5)$

29. $(-3)(-15)$

30. $(1)(-15)$

31. $(0)(-8)$

Evaluate each of the expressions.

32. $18 - 3 \cdot 5$

33. $(18 - 3) \cdot 5$

34. $5 \cdot 4^2$

35. $(5 \cdot 4)^2$

36. $5 \cdot 3^2 - 4$

37. $5(3^2 - 4)$

38. $5(4 - 2)^2$

39. $5 \cdot 4 - 2^2$

40. $(5 \cdot 4 - 2)^2$

41. $3(5 - 2)^2$

42. $3 \cdot 5 - 2^2$

43. $(3 \cdot 5 - 2)^2$

[2.5] Divide.

44. $\frac{80}{16}$

45. $\frac{-63}{7}$

46. $\frac{-81}{-9}$

47. $\frac{0}{-5}$

48. $\frac{32}{-8}$

49. $\frac{-7}{0}$

Perform the indicated operations.

50. $\frac{-8 + 6}{-8 - (-10)}$

51. $\frac{-6 - 1}{5 - (-2)}$

52. $\frac{25 - 4}{-5 - (-2)}$

[2.6] Write, using symbols.

53. 5 more than y

54. c decreased by 10

55. The product of 8 and a

56. The quotient when y is divided by 3

57. 5 times the product of m and n

58. The product of a and 5 less than a

59. 3 more than the product of 17 and x

60. The quotient when a plus 2 is divided by a minus 2

Identify which are expressions and which are not.

61. $4(x + 3)$

62. $7 \div \cdot 8$

63. $y + 5 = 9$

64. $11 + 2(3x - 9)$

[2.7] Evaluate the expressions if $x = -3$, $y = 6$, $z = -4$, and $w = 2$.

65. $3x + w$

66. $5y - 4z$

67. $x + y - 3z$

68. $5z^2$

69. $3x^2 - 2w^2$

70. $3x^3$

71. $5(x^2 - w^2)$

72. $\dfrac{6z}{2w}$

73. $\dfrac{2x - 4z}{y - z}$

74. $\dfrac{3x - y}{w - x}$

75. $\dfrac{x(y^2 - z^2)}{(y + z)(y - z)}$

76. $\dfrac{y(x - w)^2}{x^2 - 2xw + w^2}$

[2.8] List the terms of the expressions.

77. $4a^3 - 3a^2$

78. $5x^2 - 7x + 3$

Circle like terms.

79. $5m^2, -3m, -4m^2, 5m^3, m^2$

80. $4ab^2, 3b^2, -5a, ab^2, 7a^2, -3ab^2, 4a^2b$

Combine like terms.

81. $5c + 7c$

82. $2x + 5x$

83. $4a - 2a$

84. $6c - 3c$

85. $9xy - 6xy$

86. $5ab^2 + 2ab^2$

87. $7a + 3b + 12a - 2b$

88. $6x - 2x + 5y - 3x$

89. $5x^3 + 17x^2 - 2x^3 - 8x^2$

90. $3a^3 + 5a^2 + 4a - 2a^3 - 3a^2 - a$

[2.9] Tell whether the number shown in parentheses is a solution for the given equation.

91. $7x + 2 = 16$ (2)

92. $5x - 8 = 3x + 2$ (4)

93. $7x - 2 = 2x + 8$ (2)

94. $4x + 3 = 2x - 11$ (−7)

95. $x + 5 + 3x = 2 + x + 23$ (6)

96. $\frac{1}{3}x + 2 = 10$ (21)

[2.10] Solve the equations and check your results.

97. $x + 5 = 7$

98. $x - 9 = 3$

99. $5x = 4x - 5$

100. $3x - 9 = 2x$

101. $5x - 3 = 4x + 2$

102. $9x + 2 = 8x - 7$

103. $7x - 5 = 6x - 4$

104. $3 + 4x - 1 = x - 7 + 2x$

105. $4(2x + 3) = 7x + 5$

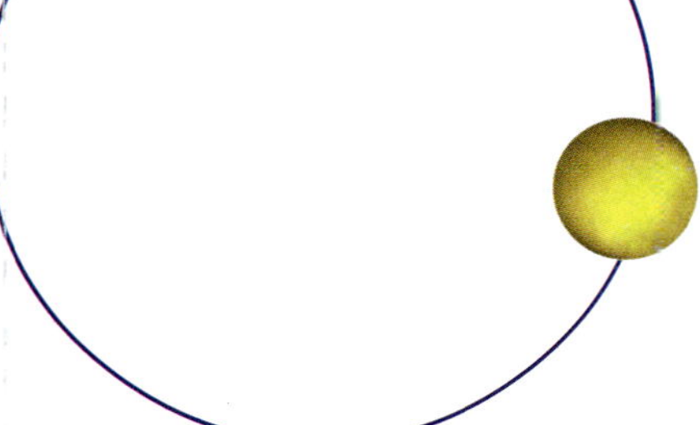

Chapter Test for Chapter 2

Name ______________________

Section __________ Date __________

ANSWERS

1. ______________
2. ______________
3. ______________
4. ______________
5. ______________
6. ______________
7. ______________
8. ______________
9. ______________
10. ______________
11. ______________
12. ______________
13. ______________
14. ______________
15. ______________
16. ______________
17. ______________
18. ______________
19. ______________

The purpose of the Chapter Test is to help you check your progress and review for a chapter test in class. When you are done, check your answers in the back of the book. If you missed any answers, be sure to go back and review the appropriate sections in the chapter.

Represent the integers on the number line shown.

1. 5, -12, 4, -7, 18, -17

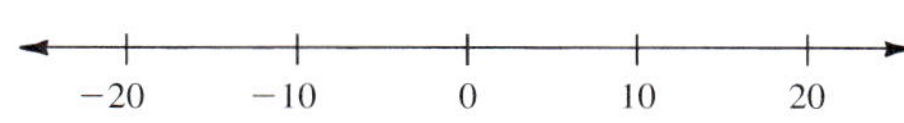

2. Place the following data set in ascending order: 4, -3, -6, 5, 0, 2, -2

Evaluate.

3. $|7|$

4. $|-7|$

5. $|18 - 7|$

6. $|18| - |-7|$

7. $-|24 - 5|$

Find the opposite of each integer.

8. 40

9. -19

Add.

10. $-8 + (-5)$

11. $6 + (-9)$

12. $(-9) + (-12)$

Subtract.

13. $9 - 15$

14. $-9 - 15$

15. $5 - (-4)$

16. $-7 - (-7)$

Multiply.

17. $(-8)(5)$

18. $(-9)(-7)$

19. $(6)(-4)$

ANSWERS

20. ____
21. ____
22. ____
23. ____
24. ____
25. ____
26. ____
27. ____
28. ____
29. ____
30. ____
31. ____
32. ____
33. ____
34. ____
35. ____
36. ____
37. ____
38. ____
39. ____
40. ____
41. ____
42. ____
43. ____
44. ____

Evaluate each expression.

20. $\frac{75}{-3}$

21. $\frac{-36 + 9}{-9}$

22. $\frac{(-15)(-3)}{-9}$

23. $\frac{9}{0}$

24. $23 - 4 \cdot 5$

25. $4 \cdot 5^2 - 35$

26. $4(2 + 4)^2$

27. $16 \div (-4) + (-5)$

28. If $x = 2$, $y = -1$, and $z = 3$, evaluate the expression $\frac{9x^2y}{3z}$.

Write, using symbols.

29. 5 less than a

30. The product of 6 and m

31. 4 times the sum of m and n

32. The quotient when the sum of a and b is divided by 3

Identify which are expressions and which are not.

33. $5x + 6 = 4$

34. $4 + (6 + x)$

Combine like terms.

35. $8a + 7a$

36. $10x + 8y + 9x - 3y$

37. Subtract $9a^2$ from the sum of $12a^2$ and $5a^2$.

38. Tom is 8 years younger than twice Moira's age. Write an expression for Tom's age. Let x represent Moira's age.

39. The length of a rectangle is 4 more than twice the width. Write an expression for the length of the rectangle.

Tell whether the number shown in parentheses is a solution for the given equation.

40. $7x - 3 = 25$ (5)

41. $8x - 3 = 5x + 9$ (4)

Solve the equations and check your results.

42. $x - 7 = 4$

43. $7x - 12 = 6x$

44. $9x - 2 = 8x + 5$

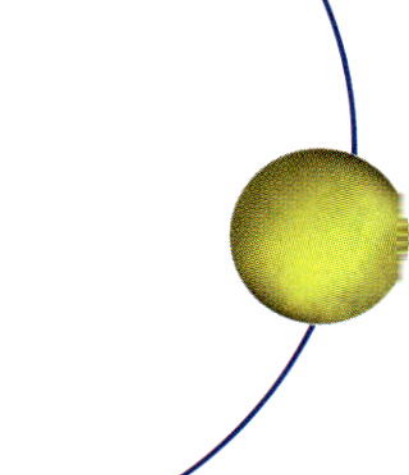

Cumulative Test for Chapters 1 and 2

Name ______________________

Section __________ Date __________

ANSWERS

1. ______________________

2. ______________________

3. ______________________

4. ______________________

5. ______________________

6. ______________________

7. ______________________

8. ______________________

9. ______________________

10. ______________________

11. ______________________

12. ______________________

13. ______________________

14. ______________________

This test is provided to help you in the process of reviewing Chapters 1 and 2. Answers are provided in the back of the book. If you missed any answers, be sure to go back and review the appropriate chapter sections.

1. Give the place value of 7 in 3,738,500.

2. Give the word name for 302,525.

3. Write two million, four hundred thirty thousand as a numeral.

In exercises 4 to 6, name the property of addition that is illustrated.

4. $5 + 12 = 12 + 5$

5. $9 + 0 = 9$

6. $(7 + 3) + 8 = 7 + (3 + 8)$

In exercises 7 and 8, perform the indicated operations.

7.
$$\begin{array}{r} 593 \\ 275 \\ +\ \ 98 \\ \hline \end{array}$$

8. Find the sum of 58, 673, 5325, and 17,295.

In exercises 9 and 10, round the numbers to the indicated place value.

9. 5873 to the nearest hundred

10. 953,150 to the nearest ten thousand

In exercise 11, estimate the sum by rounding to the nearest hundred.

11.
$$\begin{array}{r} 943 \\ 3281 \\ 778 \\ 2112 \\ +\ \ 570 \\ \hline \end{array}$$

Evaluate.

12. $|5 - 14|$

13. $|5| - |14|$

14. What is the opposite of -3?

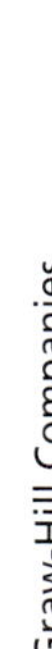

ANSWERS

15. ______
16. ______
17. ______
18. ______
19. ______
20. ______
21. ______
22. ______
23. ______
24. ______
25. ______
26. ______
27. ______
28. ______
29. ______
30. ______
31. ______
32. ______

In exercises 15 to 24, evaluate each expression.

15. $-12 + (-6)$

16. $-7 + 7$

17. $5 + (-7)$

18. $9 + (-4) + (-7)$

19. $-8 - (-8)$

20. $5 - (-5)$

21. $(-8)(-12)$

22. $(-6)(15)$

23. $14 \div (-7)$

24. $-25 \div 0$

In exercises 25 to 28, evaluate the expressions if $a = 5$, $b = -3$, $c = 4$, and $d = -2$.

25. $6ad$

26. $3b^2$

27. $3(c - 2d)$

28. $\frac{2a - 7d}{a - b}$

In exercises 29 and 30, combine like terms.

29. $6x + 14 - 3x + 5$

30. $4x + 8y - 2x - 7y$

In exercises 31 and 32, solve each equation and check your solution.

31. $x - 5 = 17$

32. $3x + 5 = 2x - 3$

FRACTIONS AND EQUATIONS

3

INTRODUCTION

The United States, along with the rest of the world, is trying to prepare for a steadily increasing population. One of the needs anticipated for a growing population is more housing. The solution is not as simple as just building more houses and apartment buildings. As the population increases, it also needs more farmland for food, more access room for highways and other forms of transit, and more recreation space.

Jennifer has worked as a land development architect for 15 years. She helps other developers come up with plans that meet the needs of both builders and communities. Her award-winning designs minimize the effect on the environment and maximize the usefulness of the space.

Name ______________________

Section ________ Date ________

ANSWERS

1. ______________

2. ______________

3. ______________

4. ______________

5. ______________

6. ______________

7. ______________

8. ______________

9. ______________

10. ______________

Pre-Test Chapter 3

This pre-test will point out any difficulties you may be having with fractions and the multiplication and division of fractions. Do all the problems. Then check your answers with those in the back of the book.

1. List all the factors of 42.

2. Write the prime factorization of 350.

3. Find the greatest common factor (GCF) of 24, 36, and 52.

4. Are $\frac{8}{12}$ and $\frac{20}{30}$ equivalent fractions?

5. Reduce $\frac{-18}{30}$ to lowest terms.

Carry out the indicated operations.

6. $\frac{-4}{5} \cdot \frac{5}{7}$

7. $\frac{6}{25} \div \frac{3}{10}$

8. $\frac{2}{3} \div 7$

Solve the following equations.

9. $7x + 3 = 2x - 2$

10. $\frac{x - 3}{8} = \frac{x - 2}{10}$

3.1 Introduction to Fractions

3.1 OBJECTIVES

1. Identify the numerator and denominator of a fraction
2. Use fractions to name parts of a whole
3. Identify proper and improper fractions

Earlier chapters dealt with integers and the operations that are performed on them. Now we will define *fraction,* learn the language of fractions, and learn to perform operations on fractions.

Definitions: Fraction

Whenever a unit or a whole quantity is divided into parts, we call those parts **fractions** of the unit.

NOTE Our word *fraction* comes from the Latin stem *fractio,* which means "breaking into pieces."

In Figure 1, the whole has been divided into five equivalent parts. We use the symbol $\frac{2}{5}$ to represent the shaded portion of the whole.

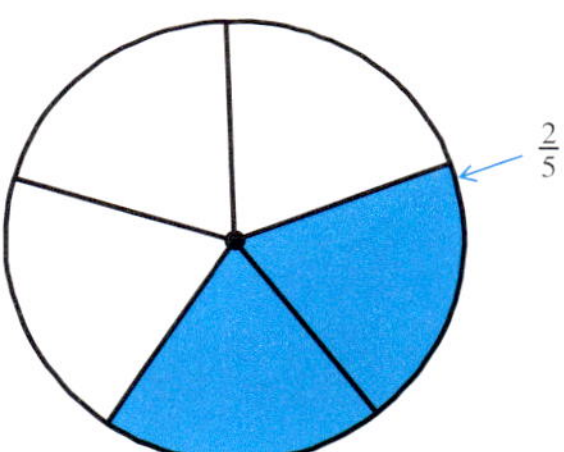

Figure 1

NOTE *Common fraction* is technically the correct term. We will normally just use *fraction* in these materials if there is no room for confusion.

The symbol $\frac{2}{5}$ is called a **common fraction,** or more simply a fraction. A fraction is written in the form $\frac{a}{b}$, in which a is an integer and b is a natural number.

We give the numbers a and b special names. The **denominator,** b, is the number on the bottom. This tells us into how many equal parts the unit or whole has been divided. The **numerator,** a, is the number on the top. This tells us how many parts of the unit are used.

In Figure 1, the *denominator* is 5; the unit or whole (the circle) has been divided into five equal parts. The *numerator* is 2. We have shaded two parts of the unit.

2 ⟵ Numerator

5 ⟵ Denominator

Example 1

Labeling Fraction Components

The fraction $\frac{4}{7}$ names the shaded part of the rectangle in Figure 2.

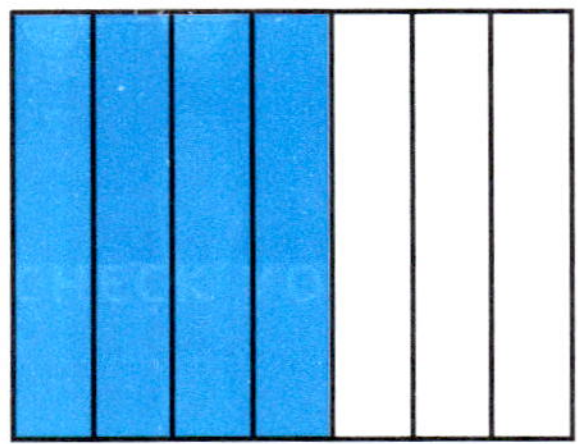

Figure 2

The unit or whole is divided into seven equivalent parts, so the denominator is 7. We have shaded four of those parts, and so we have a numerator of 4.

CHECK YOURSELF 1

What fraction names the shaded part of this diagram? Identify the numerator and denominator.

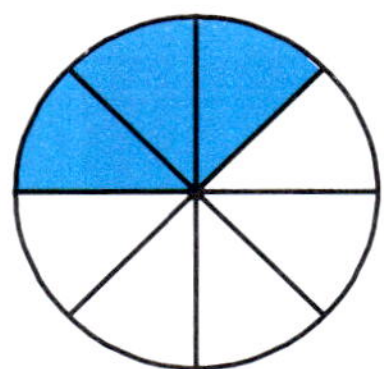

Fractions can also be used to name a part of a collection or a set of identical objects.

Example 2

Naming a Fractional Part

The fraction $\frac{5}{6}$ names the shaded part of Figure 3. We have shaded five of the six identical objects.

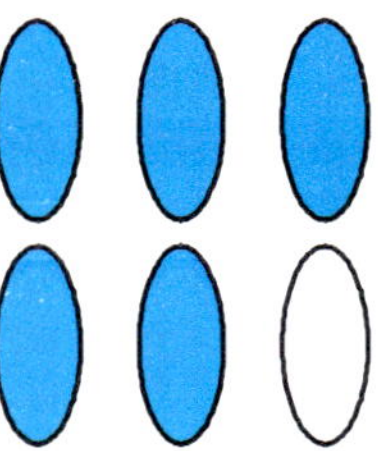

Figure 3

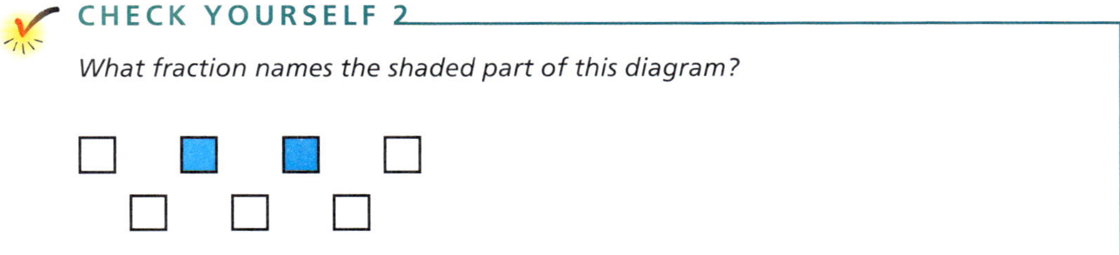

CHECK YOURSELF 2

What fraction names the shaded part of this diagram?

Example 3

Naming a Fractional Part

NOTE Of course, the fraction $\frac{8}{23}$ names the part of the class that is not women.

In a class of 23 students, 15 are women. We can name the part of the class that is women as $\frac{15}{23}$.

CHECK YOURSELF 3

Seven replacement parts out of a shipment of 50 were faulty. What fraction names the portion of the shipment that was faulty?

NOTE $\frac{a}{b}$ names the *quotient* when a is divided by b. Of course, b cannot be 0.

A fraction can also be thought of as indicating division. The symbol $\frac{a}{b}$ also means $a \div b$.

Example 4

Interpreting Division as a Fraction

$\frac{2}{3}$ names the quotient when 2 is divided by 3. So $\frac{2}{3} = 2 \div 3$.

Note: $\frac{2}{3}$ can be read as "two-thirds" or as "2 divided by 3."

CHECK YOURSELF 4

Using the numbers 5 and 9, write $\frac{5}{9}$ using division.

We can use the relative size of the numerator and denominator of a fraction to separate fractions into different categories.

Definitions: Proper Fraction

A fraction that names a number less than 1 (the numerator is *less than* the denominator) is called a **proper fraction.**

Definitions: Improper Fraction

A fraction that names a number greater than or equal to 1 (the numerator is *greater than* or *equal to* the denominator) is called an **improper fraction.**

Example 5

Categorizing Fractions

(a) $\frac{2}{3}$ is a proper fraction because the numerator is less than the denominator (Figure 4).

(b) $\frac{4}{3}$ is an improper fraction because the numerator is larger than the denominator (Figure 5).

(c) Also, $\frac{6}{6}$ is an improper fraction because it names exactly 1 unit; the numerator is equal to the denominator (Figure 6).

A proper fraction

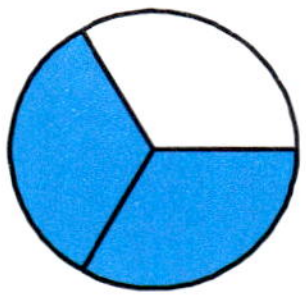

$\frac{2}{3}$ names less than 1 unit and $2 < 3$.

Numerator Denominator

Figure 4

An improper fraction

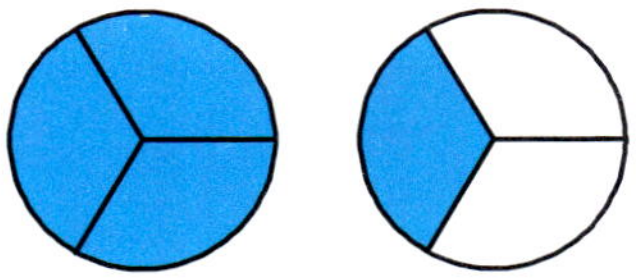

$\frac{4}{3}$ names more than 1 unit and $4 > 3$.

Numerator Denominator

Figure 5

An improper fraction

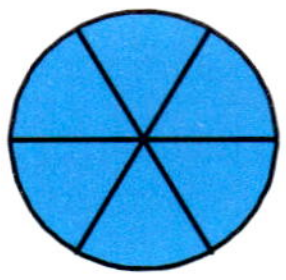

$\frac{6}{6} = 1$

Figure 6

CHECK YOURSELF 5

List the proper fractions and the improper fractions in the group:

$\frac{5}{4}, \frac{10}{11}, \frac{3}{4}, \frac{8}{5}, \frac{6}{6}, \frac{13}{10}, \frac{7}{8}, \frac{15}{8}$

One special kind of improper fraction should be mentioned at this point: a fraction with a denominator of 1.

Definitions: Fractions with a Denominator of 1

Any fraction with a denominator of 1 is equal to the numerator alone.

For example, $\frac{5}{1} = 5 \div 1 = 5$

CHECK YOURSELF ANSWERS

1. $\frac{3}{8}$ ⟵ Numerator, ⟵ Denominator

2. $\frac{2}{7}$ **3.** $\frac{7}{50}$ **4.** $5 \div 9$

5. Proper fractions: $\frac{10}{11}, \frac{3}{4}, \frac{7}{8}$ Improper fractions: $\frac{5}{4}, \frac{8}{5}, \frac{6}{6}, \frac{13}{10}, \frac{15}{8}$

Name ______

Section ______ Date ______

3.1 Exercises

Identify the numerator and denominator of each fraction.

1. $\frac{6}{11}$ **2.** $\frac{5}{12}$ **3.** $\frac{5}{3}$ **4.** $\frac{7}{2}$

What fraction names the shaded part of each of the figures?

5.

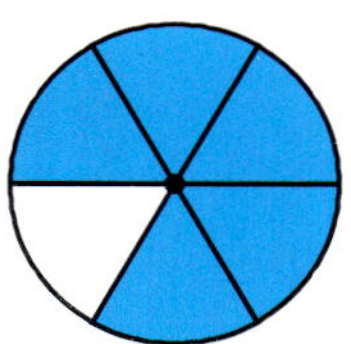

6.

7.

8.

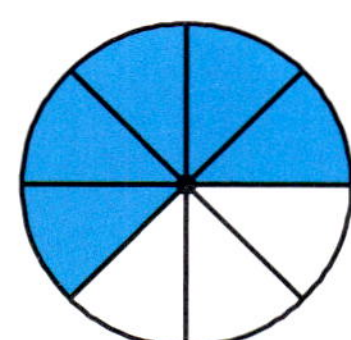

9.

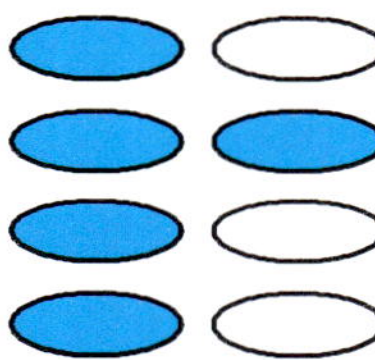

10.

Solve the applications.

11. Test scores. You missed 7 questions on a 20-question test. What fraction names the part you got correct? The part you got wrong?

12. Basketball. Of the five starters on a basketball team, two fouled out of a game. What fraction names the part of the starting team that fouled out?

13. Car sales. A used-car dealer sold 11 of the 17 cars in stock. What fraction names the portion sold? What fraction names the portion *not* sold?

14. Purchases. At lunch, five people out of a group of nine had hamburgers. What fraction names the part of the group who had hamburgers? What fraction names the part who did *not* have hamburgers?

ANSWERS

1. ______
2. ______
3. ______
4. ______
5. ______
6. ______
7. ______
8. ______
9. ______
10. ______
11. ______
12. ______
13. ______
14. ______

ANSWERS

15. __________

16. __________

17. __________

18. __________

19. __________

20. __________

21. __________

22. __________

23. __________

24. __________

25. __________

Identify each number as a proper fraction or an improper fraction.

15. $\frac{3}{5}$ **16.** $\frac{9}{5}$ **17.** $\frac{6}{6}$ **18.** $\frac{7}{9}$

19. $\frac{11}{8}$ **20.** $\frac{8}{8}$ **21.** $\frac{13}{17}$ **22.** $\frac{16}{15}$

23. Based on a 2000-calorie diet, consult with your local health officials and determine what fraction of your daily diet each of the following should contribute.

(a) Fat **(b)** Carbohydrate **(c)** Unsaturated fat **(d)** Protein

Using this information, work with your classmates to plan a daily menu that will satisfy the requirements.

24. Suppose the national debt had to be paid by individuals.

(a) How would the amount each individual owed be determined?
(b) Would this be a proper or improper fraction?

25. In statistics, we describe the probability of an event by dividing the number of ways the event can happen by the total number of possibilities. Using this definition, determine the probability of each of the following:

(a) Obtaining a head on a single toss of a coin
(b) Obtaining a prime number (in this case 2, 3, or 5) when a die is tossed

Answers

1. 6 is the numerator; 11 is the denominator
3. 5 is the numerator; 3 is the denominator **5.** $\frac{5}{6}$ **7.** $\frac{5}{5}$ **9.** $\frac{5}{8}$
11. $\frac{13}{20}, \frac{7}{20}$ **13.** $\frac{11}{17}, \frac{6}{17}$ **15.** Proper **17.** Improper **19.** Improper
21. Proper **23.** **25.** **(a)** $\frac{1}{2}$; **(b)** $\frac{3}{6} = \frac{1}{2}$

3.2 Prime Numbers and Factorization

3.2 OBJECTIVES

1. Find the factors of a whole number
2. Determine whether a number is prime, composite, or neither
3. Find the prime factorization for any number
4. Find the greatest common factor (GCF) of two or more numbers

Overcoming Math Anxiety

Working Together

How many of your classmates do you know? Whether you are by nature gregarious or shy, you have much to gain by getting to know your classmates.

1. It is important to have someone to call when you have missed class or if you are unclear on an assignment.
2. Working with another person is almost always beneficial to both people. If you don't understand something, it helps to have someone to ask about it. If you do understand something, nothing will cement that understanding more than explaining the idea to another person.
3. Sometimes we need to commiserate. If an assignment is particularly frustrating, it is reassuring to find that it is also frustrating for other students.
4. Have you ever thought you had the right answer, but it doesn't match the answer in the text? Frequently the answers are equivalent, but that's not always easy to see. A different perspective can help you see that. Occasionally there is an error in a textbook (here, we are talking about *other* textbooks). In such cases it is wonderfully reassuring to find that someone else has the same answer as you do.

NOTE 2 and 5 can also be called *divisors* of 10. They divide 10 exactly.

In Section 1.5 we said that because $2 \cdot 5 = 10$, we call 2 and 5 **factors** of 10.

Definitions: Factor

A **factor** of a natural number is another natural number that will *divide exactly* into that number. This means that the division will have a remainder of 0.

Example 1

Finding Factors

List all factors of 18.

NOTE This is a complete list of the factors. There are no other whole numbers that divide 18 exactly. Note that the factors of 18, except for 18 itself, are *smaller* than 18.

$1 \cdot 18 = 18$ 1 and 18 are factors of 18.

$2 \cdot 9 = 18$ 2 and 9 are also factors of 18.

$3 \cdot 6 = 18$ Because $3 \cdot 6 = 18$, 3 and 6 are factors (or divisors) of 18.

1, 2, 3, 6, 9, and 18 are all the factors of 18.

CHECK YOURSELF 1

List all the factors of 24.

Listing factors leads us to an important classification of whole numbers. Any whole number larger than 1 will be either a *prime* or a *composite* number. Look at the following definitions.

A whole number greater than 1 will always have itself and 1 as factors. Sometimes these will be the *only* factors. For instance, 1 and 3 are the only factors of 3.

Definitions: Prime Number

A **prime number** is any natural number (other than 1) that has exactly two factors, 1 and itself.

NOTE How large can a prime number be? There is no largest prime number. To date, the largest *known* prime is $2^{6972593} - 1$. This is a number with 2,098,960 digits, if you are curious. Of course, a computer had to be used to verify that a number of this size is prime. By the time you read this, someone may very well have found an even larger prime number.

As examples, 2, 3, 5, and 7 are prime numbers. Their only factors are 1 and themselves.

To check whether a number is prime, one approach is simply to divide the smaller primes, 2, 3, 5, 7, and so on, into the given number. If no factors other than 1 and the given number are found, the number is prime.

Here is the method known as the **sieve of Eratosthenes** for identifying prime numbers.

1. Write down a series of counting numbers, starting with the number 2. In this example, we stop at 50.
2. Start at the number 2. Delete every second number after the 2.
3. Move to the number 3. Delete every third number after 3 (some numbers will be deleted twice).
4. Continue this process, deleting every fourth number after 4, every fifth number after 5, and so on.
5. When you have finished, the undeleted numbers are the prime numbers.

	2	3	~~4~~	5	~~6~~	7	~~8~~	~~9~~	~~10~~
11	~~12~~	13	~~14~~	~~15~~	~~16~~	17	~~18~~	19	~~20~~
~~21~~	~~22~~	23	~~24~~	~~25~~	~~26~~	~~27~~	~~28~~	29	~~30~~
31	~~32~~	~~33~~	~~34~~	~~35~~	~~36~~	37	~~38~~	~~39~~	~~40~~
41	~~42~~	43	~~44~~	~~45~~	~~46~~	47	~~48~~	~~49~~	~~50~~

The prime numbers less than 50 are 2, 3, 5, 7, 11, 13, 17, 19, 23, 29, 31, 37, 41, 43, 47.

Example 2

Identifying Prime Numbers

Which of the numbers are prime?

17 is a prime number. 1 and 17 are the only factors.

29 is a prime number. 1 and 29 are the only factors.

33 is *not* prime. 1, 3, 11, and 33 are all factors of 33.

Note: If a number less than 100 is *not* a prime, it will have one or more of the numbers 2, 3, 5, or 7 as factors.

CHECK YOURSELF 2

Which of the following numbers are prime numbers?

2, 6, 9, 11, 15, 19, 23, 35, 41

We can now define a second class of whole numbers.

Definitions: Composite Number

A **composite number** is any natural number greater than 1 that is not prime. Every composite number has more than two factors.

NOTE This definition tells us that a composite number *does* have factors other than 1 and itself.

Example 3

Identifying Composite Numbers

Which of the numbers are composite?

18 is a composite number. 1, 2, 3, 6, 9, and 18 are all factors of 18.

23 is not a composite number. 1 and 23 are the only factors. This means that 23 is a *prime number.*

25 is a composite number. 1, 5, and 25 are factors.

38 is a composite number. 1, 2, 19, and 38 are factors.

CHECK YOURSELF 3

Which of the numbers are composite?

2, 6, 10, 13, 16, 17, 22, 27, 31, 35

By the definitions of prime and composite numbers:

Rules and Properties: One

The natural number 1 is neither prime nor composite.

This is simply a matter of the way in which prime and composite numbers are defined in mathematics. The number 1 is the *only* natural number that can't be classified as one or the other.

To **factor a number** means to write the number as a product of its natural-number factors.

Example 4

Factoring a Composite Number

Factor the number 10.

$10 = 2 \cdot 5$ The order in which you write the factors does not matter, so $10 = 5 \cdot 2$ would also be correct.

Of course, $10 = 10 \cdot 1$ is also a correct statement. However, in this section we are interested in factors other than 1 and the given number.

Factor the number 21.

$21 = 3 \cdot 7$

CHECK YOURSELF 4

Factor the number 35.

In writing composite numbers as a product of factors, there will be a number of different possible factorizations.

Example 5

Factoring a Composite Number

Find three ways to factor 72.

NOTE There have to be at least two different factorizations, because a composite number has factors other than 1 and itself.

$$
\begin{aligned}
72 &= 8 \cdot 9 \quad (1)\\
&= 6 \cdot 12 \quad (2)\\
&= 3 \cdot 24 \quad (3)
\end{aligned}
$$

CHECK YOURSELF 5

Find three ways to factor 42.

We now want to write composite numbers as a product of their **prime factors.** Look again at the first factored line of Example 5. The process of factoring can be continued until all the factors are prime numbers.

Example 6

Factoring a Composite Number

NOTE This is often called a **factor tree.**

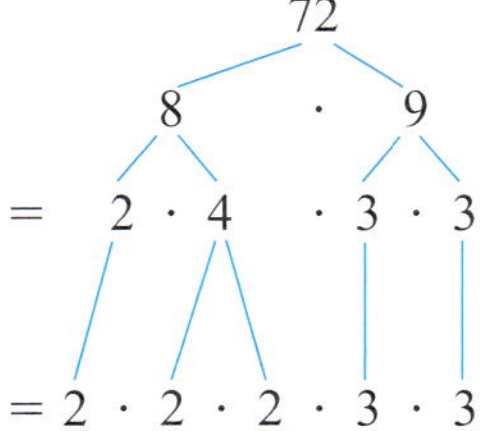

4 is still not prime, and so we continue by factoring 4.

72 is now written as a product of prime factors.

NOTE Finding the prime factorization of a number will be important in our later work in adding fractions.

When we write 72 as $2 \cdot 2 \cdot 2 \cdot 3 \cdot 3$, no further factorization is possible. This is called the *prime factorization* of 72.

Now, what if we start with the second factored line from the same example, $72 = 6 \cdot 12$?

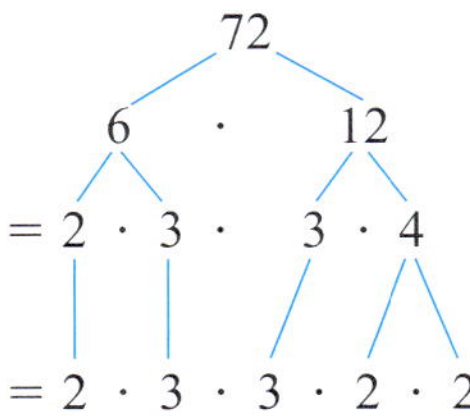

Continue to factor 6 and 12.

Continue again to factor 4. Other choices for the factors of 12 are possible. As we shall see, the end result will be the same.

No matter which pair of factors you start with, you will find the same prime factorization. In this case, there are three factors of 2 and two factors of 3. Because **multiplication is commutative,** the order in which we write the factors does not matter.

CHECK YOURSELF 6

We could also write

$72 = 2 \cdot 36$

Continue the factorization.

Rules and Properties: The Fundamental Theorem of Arithmetic

There is exactly one prime factorization for any composite number.

The method of Example 6 will always work. However, another method for factoring composite numbers exists. This method is particularly useful when numbers get large, in which case factoring with a number tree becomes unwieldy.

NOTE The prime factorization is then the product of all the prime divisors and the final quotient.

Rules and Properties: Factoring by Division

To find the prime factorization of a number, divide the number by a series of primes until the final quotient is a prime number.

Example 7

Finding Prime Factors

To write 60 as a product of prime factors, divide 2 into 60 for a quotient of 30. Continue to divide by 2 again for the quotient 15. Because 2 won't divide evenly into 15, we try 3. Because the quotient 5 is prime, we are done.

$2\overline{)60}$ = 30 → $2\overline{)30}$ = 15 → $3\overline{)15}$ = 5 Prime

Our factors are the prime divisors and the final quotient. We have

$60 = 2 \cdot 2 \cdot 3 \cdot 5$

CHECK YOURSELF 7

Complete the process to find the prime factorization of 90.

$2\overline{)90}$ = 45 → $?\overline{)45}$ = ?

Remember to continue until the final quotient is prime.

Writing composite numbers in their completely factored form can be simplified if we use a format called **continued division.**

Example 8

Finding Prime Factors Using Continued Division

Use the continued-division method to divide 60 by a series of prime numbers.

NOTE In each short division, we write the quotient *below* rather than above the dividend. This is just a convenience for the next division.

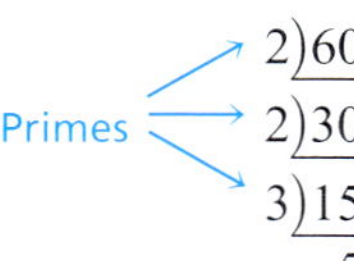

Stop when the final quotient is prime.

To write the factorization of 60, we list each divisor used and the final prime quotient. In our example, we have

$60 = 2 \cdot 2 \cdot 3 \cdot 5$

CHECK YOURSELF 8

Find the prime factorization of 234.

We know that a factor or a divisor of a whole number divides that number exactly.

The factors or divisors of 20 are

NOTE Again the factors of 20, other than 20 itself, are less than 20.

1, 2, 4, 5, 10, 20

Each of these numbers divides 20 exactly, that is, with no remainder.

Our work in this section involves common factors or divisors. A **common factor** or **divisor** for two numbers is any factor that divides both the numbers exactly.

Example 9

Finding Common Factors

Look at the numbers 20 and 30. Is there a common factor for the two numbers?

First, we list the factors. Then we circle the ones that appear in both lists.

Factors

20: (1), (2), 4, (5), (10), 20

30: (1), (2), 3, (5), 6, (10), 15, 30

We see that 1, 2, 5, and 10 are common factors of 20 and 30. Each of these numbers divides both 20 and 30 exactly.

Our later work with fractions will require that we find the greatest common factor (GCF) of a group of numbers.

Definitions: Greatest Common Factor

The **greatest common factor** (GCF) of a group of numbers is the *largest* number that will divide each of the given numbers exactly.

CHECK YOURSELF 9

List the factors of 30 *and* 36, *and then find the GCF.*

To find the GCF of a group of numbers, we use the prime factorization of each number. We will outline the process.

Step by Step: Finding the Greatest Common Factor

Step 1 Write the prime factorization for each of the numbers in the group.
Step 2 Locate the prime factors that are *common* to all the numbers.
Step 3 The greatest common factor will be the *product* of all the common prime factors.

NOTE If there are no common prime factors, the GCF is 1.

Example 10

Finding the Greatest Common Factor

Find the GCF of 20 and 30.

Step 1 Write the prime factorization of 20 and 30.

$20 = 2 \cdot 2 \cdot 5$

$30 = 2 \cdot 3 \cdot 5$

Step 2 Find the prime factors common to each number.

$20 = \textcircled{2} \cdot 2 \cdot \textcircled{5}$

$30 = \textcircled{2} \cdot 3 \cdot \textcircled{5}$

2 and 5 are the common prime factors.

Step 3 Form the product of the common prime factors.

$2 \cdot 5 = 10$

10 is the GCF.

CHECK YOURSELF 10

Find the GCF of 30 *and* 36.

To find the GCF of a group of more than two numbers, we use the same process.

Example 11

Finding the Greatest Common Factor

Find the GCF of 24, 30, and 36.

$24 = \textcircled{2} \cdot 2 \cdot 2 \cdot \textcircled{3}$

$30 = \textcircled{2} \cdot \textcircled{3} \cdot 5$

$36 = \textcircled{2} \cdot 2 \cdot \textcircled{3} \cdot 3$

2 and 3 are the prime factors common to *all three numbers.*

$2 \cdot 3 = 6$ is the GCF.

CHECK YOURSELF 11

Find the GCF of 15, 30, *and* 45.

Example 12

Finding the Greatest Common Factor

NOTE If two numbers, such as 15 and 28, have no common factor other than 1, they are called **relatively prime.**

Find the GCF of 15 and 28.

$15 = 3 \cdot 5$

$28 = 2 \cdot 2 \cdot 7$

There are no common prime factors listed. But remember that 1 is a factor of every whole number.

The GCF of 15 and 28 is 1.

CHECK YOURSELF 12

Find the GCF of 30 *and* 49.

CHECK YOURSELF ANSWERS

1. 1, 2, 3, 4, 6, 8, 12, and 24 **2.** 2, 11, 19, 23, and 41 are prime numbers
3. 6, 10, 16, 22, 27, and 35 are composite numbers **4.** $5 \cdot 7$
5. $2 \cdot 21, 3 \cdot 14, 6 \cdot 7$ **6.** $2 \cdot 2 \cdot 2 \cdot 3 \cdot 3$
7. $2\overline{)90}$ = 45 → $3\overline{)45}$ = 15 → $3\overline{)15}$ = 5 **8.** $2 \cdot 3 \cdot 3 \cdot 13$
$90 = 2 \cdot 3 \cdot 3 \cdot 5$
9. 30: (1), (2), (3), 5, (6), 10, 15, 30
36: (1), (2), (3), 4, (6), 9, 12, 18, 36
6 is the GCF.
10. $30 = (2) \cdot (3) \cdot 5$
$36 = (2) \cdot 2 \cdot (3) \cdot 3$
The GCF is $2 \cdot 3 = 6$.
11. 15 **12.** GCF is 1; 30 and 49 are relatively prime

Name ______________

Section ________ Date ________

Exercises

List the factors of each of the numbers.

1. 4
2. 6
3. 24
4. 32
5. 64
6. 66
7. 11
8. 37

Use the given list of numbers for exercises 9 and 10.

0, 1, 15, 19, 23, 31, 49, 55, 59, 87, 91, 97, 103, 105

9. Which of the given numbers are prime?

10. Which of the given numbers are composite?

Find the prime factorization of each number.

11. 18
12. 22
13. 30
14. 35
15. 70
16. 90
17. 66
18. 100
19. 130
20. 88
21. 315
22. 400

In the study of algebra, you often will want to find factors of a number with a given sum or difference. The exercises 23 to 26 use this technique.

23. Find two factors of 24 with a sum of 10.

24. Find two factors of 15 with a difference of 2.

25. Find two factors of 30 with a difference of 1.

26. Find two factors of 28 with a sum of 11.

ANSWERS

1. ______
2. ______
3. ______
4. ______
5. ______
6. ______
7. ______
8. ______
9. ______
10. ______
11. ______
12. ______
13. ______
14. ______
15. ______
16. ______
17. ______
18. ______
19. ______
20. ______
21. ______
22. ______
23. ______
24. ______
25. ______ 26. ______

ANSWERS

27. ______
28. ______
29. ______
30. ______
31. ______
32. ______
33. ______
34. ______
35. ______
36. ______
37. ______
38. ______
39. ______
40. ______
41. ______
42. ______

Find the GCF for each of the groups of numbers.

27. 4 and 6

28. 6 and 9

29. 20 and 21

30. 28 and 42

31. 18 and 24

32. 35 and 36

33. 18 and 54

34. 12 and 48

35. 12, 36, and 60

36. 15, 45, and 90

37. 25, 75, and 150

38. 36, 72, and 144

Solve the applications.

39. A natural number is said to be perfect if it is equal to the sum of its natural number divisors, except itself.

(a) Show that 28 is a perfect number.

(b) Identify another perfect number less than 28.

40. Find the smallest natural number that is divisible by all of the numbers. 2, 3, 4, 6, 8, 9.

41. Tom and Dick both work the night shift at the steel mill. Tom has every sixth night off, and Dick has every eighth night off. If they both have August 1 off, when will they both be off together again?

42. Mercury, Venus, and Earth revolve around the sun once every 3, 7, and 12 months, respectively. If the three planets are now in the same straight line, what is the smallest number of months that must pass before they line up again?

43. Use the *sieve of Eratosthenes* to determine all the prime numbers less than 100.

	2	3	4	5	6	7	8	9	10
11	12	13	14	15	16	17	18	19	20
21	22	23	24	25	26	27	28	29	30
31	32	33	34	35	36	37	38	39	40
41	42	43	44	45	46	47	48	49	50
51	52	53	54	55	56	57	58	59	60
61	62	63	64	65	66	67	68	69	70
71	72	73	74	75	76	77	78	79	80
81	82	83	84	85	86	87	88	89	90
91	92	93	94	95	96	97	98	99	100

44. Prime numbers that differ by 2 are called *twin primes.* Examples are 3 and 5, and 5 and 7. Find one pair of twin primes between 85 and 105.

45. Parts **(a)** to **(c)** refer to "twin primes" (see exercise 44).

(a) Search for, and make a list of several pairs of twin primes, in which the primes are greater than 3.

(b) What do you notice about each number that lies *between* a pair of twin primes?

(c) Write an explanation for your observation in part **(b)**.

46. Obtain (or imagine that you have) a quantity of square tiles. Six tiles can be arranged in the shape of a rectangle in two different ways:

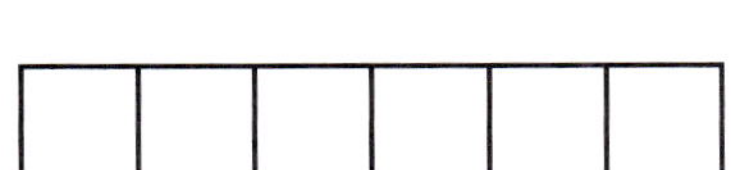

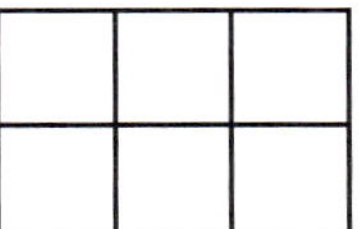

(a) Record the dimensions of the given rectangles.

(b) If you use 7 tiles, how many different rectangles can you form?

(c) If you use 10 tiles, how many different rectangles can you form?

(d) What kind of number (of tiles) permits *only one* arrangement into a rectangle? *More than one* arrangement?

47. The number 10 has 4 factors: 1, 2, 5, and 10. We can say that 10 has an even number of factors. Investigate several numbers to determine which numbers have an *even number* of factors and which numbers have an *odd number* of factors.

43. ______

44. ______

45. ______

46. ______

47. ______

48. ______________________

48. Suppose that a school has 1000 lockers and that they are all closed. A person passes through, opening every other locker, beginning with locker #2. Then another person passes through, changing every third locker (closing it if it is open, opening it if it is closed), starting with locker #3. Yet another person passes through, changing every fourth locker, beginning with locker #4. This process continues until 1000 people pass through.

(a) At the end of this process, which locker numbers are closed?

(b) Write an explanation for your answer to part **(a)**. (*Hint:* It may help to attempt exercise 47 first.)

Answers

1. 1, 2, 4 **3.** 1, 2, 3, 4, 6, 8, 12, 24 **5.** 1, 2, 4, 8, 16, 32, 64 **7.** 1, 11 **9.** 19, 23, 31, 59, 97, 103 **11.** $2 \cdot 3 \cdot 3$ **13.** $2 \cdot 3 \cdot 5$ **15.** $2 \cdot 5 \cdot 7$ **17.** $2 \cdot 3 \cdot 11$ **19.** $2\overline{)130}$, $5\overline{)65}$, 13; $130 = 2 \cdot 5 \cdot 13$ **21.** $3 \cdot 3 \cdot 5 \cdot 7$ **23.** 4, 6 **25.** 5, 6 **27.** 2 **29.** 1 **31.** 6 **33.** 18 **35.** 12 **37.** 25 **39.** **41.** August 25 **43.**  **45.**  **47.** 

3.3 Equivalent Fractions

3.3 OBJECTIVES

1. Determine whether two fractions are equivalent
2. Use the fundamental principle to simplify fractions
3. Use the fundamental principle to build fractions

It is possible to represent the same portion of the whole by different fractions. Look at Figure 1, representing $\frac{3}{6}$ and $\frac{1}{2}$. The two fractions are simply different names for the same number. They are called **equivalent fractions** for this reason.

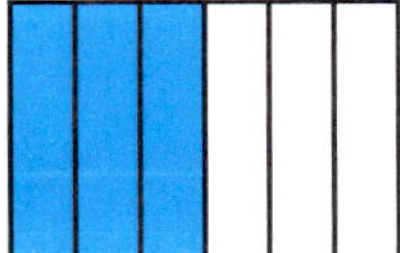

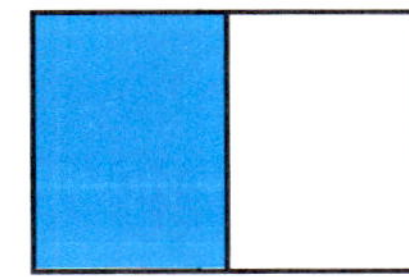

Figure 1

Any fraction has a large number of equivalent fractions. For instance, $\frac{2}{3}$, $\frac{4}{6}$, and $\frac{6}{9}$ are all equivalent fractions because they name the same part of a unit. This is illustrated in Figure 2.

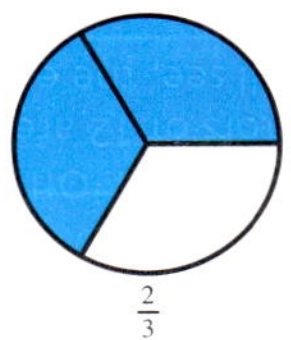

$\frac{2}{3}$

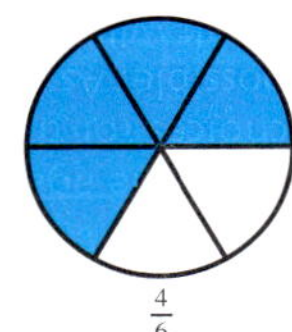

$\frac{4}{6}$

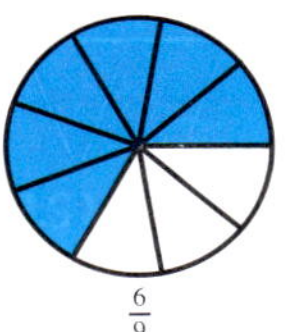

$\frac{6}{9}$

Figure 2

Many more fractions are equivalent to $\frac{2}{3}$. All these fractions can be used interchangeably. An easy way to find out if two fractions are equivalent is to use cross products.

$$\frac{a}{b} = \frac{c}{d}$$

We call $a \cdot d$ and $b \cdot c$ the **cross products.**

Rules and Properties: Testing for Equivalence

If the cross products for two fractions are equal, the two fractions are equivalent.

Example 1

Identifying Equivalent Fractions Using Cross Products

(a) Are $\frac{3}{24}$ and $\frac{4}{32}$ equivalent fractions?

The cross products are $3 \cdot 32$, or 96, and $24 \cdot 4$, or 96. Because the cross products are equal, the fractions are equivalent.

(b) Are $\frac{2}{5}$ and $\frac{3}{7}$ equivalent fractions?

The cross products are $2 \cdot 7$ and $5 \cdot 3$.

$2 \cdot 7 = 14$ and $5 \cdot 3 = 15$

Because $14 \neq 15$, the fractions are *not* equivalent.

NOTE Where would we find $\frac{-3}{11}$ on the number line? It is between 0 and −1.

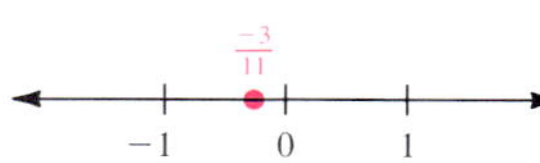

(c) Are $\frac{-3}{11}$ and $\frac{-6}{22}$ equivalent?

The cross products are

$(-3) \cdot 22 = -66$ and $11 \cdot (-6) = -66$

Because the cross products are equal, the fractions are equivalent.

CHECK YOURSELF 1

(a) Are $\frac{3}{8}$ and $\frac{9}{24}$ equivalent fractions? **(b)** Are $\frac{7}{8}$ and $\frac{8}{9}$ equivalent fractions?

In writing equivalent fractions, we use an important principle.

Rules and Properties: The Fundamental Principle of Fractions

For the fraction $\frac{a}{b}$ and any nonzero number c,

$$\frac{a}{b} = \frac{a \div c}{b \div c}$$

The Fundamental Principle of Fractions tells us that we can divide the numerator and denominator by the same nonzero number. The result is an equivalent fraction. For instance,

NOTE Divide the numerator and denominator by 2, 3, and 4, respectively.

$$\frac{2}{4} = \frac{2 \div 2}{4 \div 2} = \frac{1}{2} \qquad \frac{3}{6} = \frac{3 \div 3}{6 \div 3} = \frac{1}{2} \qquad \frac{4}{8} = \frac{4 \div 4}{8 \div 4} = \frac{1}{2}$$

NOTE Divide the numerator and denominator by 5, 6, and 7, respectively.

$$\frac{5}{10} = \frac{5 \div 5}{10 \div 5} = \frac{1}{2} \qquad \frac{6}{12} = \frac{6 \div 6}{12 \div 6} = \frac{1}{2} \qquad \frac{7}{14} = \frac{7 \div 7}{14 \div 7} = \frac{1}{2}$$

We will see how this is applied.

Simplifying a fraction or *reducing a fraction to lower terms* means finding an equivalent fraction with a *smaller* numerator and denominator than those of the original fraction. Dividing the numerator and denominator by the same nonzero number will do exactly that.

Consider Example 2.

Example 2

Simplifying Fractions

Simplify each fraction.

NOTE We apply the fundamental principle to divide the numerator and denominator by 5.

(a) $\frac{5}{15} = \frac{5 \div 5}{15 \div 5} = \frac{1}{3}$

$\frac{5}{15}$ and $\frac{1}{3}$ are equivalent fractions.

Check this by finding the cross products.

NOTE We divide the numerator and denominator by 2.

(b) $\frac{4}{8} = \frac{4 \div 2}{8 \div 2} = \frac{2}{4}$

$\frac{4}{8}$ and $\frac{2}{4}$ are equivalent fractions.

CHECK YOURSELF 2

Write two fractions that are equivalent to $\frac{30}{45}$.

(a) Divide the numerator and denominator by 5.
(b) Divide the numerator and denominator by 15.

We say that a fraction is in **simplest form,** or in **lowest terms,** if the numerator and denominator have no common factors other than 1. This means that the fraction has the smallest possible numerator and denominator.

In Example 2, $\frac{1}{3}$ is in simplest form because the numerator and denominator have no common factors other than 1. The fraction is in lowest terms.

NOTE In this case, the numerator and denominator are *not* as small as possible. The numerator and denominator have a common factor of 2.

$\frac{2}{4}$ is *not* in simplest form. Do you see that $\frac{2}{4}$ can also be written as $\frac{1}{2}$?

To write a fraction in simplest form or to *reduce a fraction to lowest terms,* divide the numerator and denominator by their GCF.

Example 3

Simplifying Fractions

Write $\frac{10}{15}$ in simplest form.

From our work in this chapter, we know that the GCF of 10 and 15 is 5. To write $\frac{10}{15}$ in simplest form, divide the numerator and denominator by 5.

$$\frac{10}{15} = \frac{10 \div 5}{15 \div 5} = \frac{2}{3}$$

The resulting fraction, $\frac{2}{3}$, is in lowest terms.

CHECK YOURSELF 3

Write $\frac{12}{18}$ *in simplest form by dividing the numerator and denominator by the GCF.*

Many students prefer another method of reducing fractions, which uses the prime factorizations of the numerator and denominator. Example 4 uses this method.

Example 4

Factoring to Simplify a Fraction

(a) Simplify $\frac{24}{42}$.

To simplify $\frac{24}{42}$, factor.

NOTE From the prime factorization of 24 and 42, we divide by the common factors of 2 and 3.

$$\frac{24}{42} = \frac{\overset{1}{\cancel{2}} \cdot 2 \cdot 2 \cdot \overset{1}{\cancel{3}}}{\underset{1}{\cancel{2}} \cdot \underset{1}{\cancel{3}} \cdot 7} = \frac{4}{7}$$

Note: The numerator of the simplified fraction is the *product* of the prime factors remaining in the numerator after the division by 2 and 3.

(b) Simplify $\frac{120}{180}$.

To reduce $\frac{120}{180}$ to lowest terms, write the prime factorizations of the numerator and denominator. Then divide by any common factors.

$$\frac{120}{180} = \frac{\overset{1}{\cancel{2}} \cdot \overset{1}{\cancel{2}} \cdot 2 \cdot \overset{1}{\cancel{3}} \cdot \overset{1}{\cancel{5}}}{\underset{1}{\cancel{2}} \cdot \underset{1}{\cancel{2}} \cdot \underset{1}{\cancel{3}} \cdot 3 \cdot \underset{1}{\cancel{5}}} = \frac{2}{3}$$

CHECK YOURSELF 4

Write each of the following fractions in simplest form.

(a) $\frac{60}{75}$ **(b)** $\frac{210}{252}$

There is another way to organize your work in simplifying fractions. It again uses the fundamental principle to divide the numerator and denominator by any common factors. We will illustrate with the fractions considered in Example 4.

Example 5

Using Common Factors to Simplify Fractions

(a) $\frac{24}{42} = \frac{\overset{12}{\cancel{24}}}{\underset{21}{\cancel{42}}} = \frac{\overset{4}{\cancel{12}}}{\underset{7}{\cancel{21}}} = \frac{4}{7}$

Divide by the common factor of 2.

Divide by the common factor of 3.

The original numerator and denominator are divisible by 2, and so we divide by that factor to arrive at $\frac{12}{21}$. A common factor of 3 still exists.

Divide again for the result $\frac{4}{7}$, which is in lowest terms.

Note: If we had seen the GCF of 6 at first, we could have divided by 6 and arrived at the same result in one step.

(b) $\frac{120}{180} = \frac{\overset{\overset{2}{\cancel{20}}}{\cancel{120}}}{\underset{\underset{3}{\cancel{30}}}{\cancel{180}}} = \frac{2}{3}$

Our first step is to divide by the common factor of 6. We then have $\frac{20}{30}$. There is still a common factor of 10, so we again divide.

Again, we could have divided by the GCF of 60 in one step if we had recognized it.

CHECK YOURSELF 5

Using the method of Example 5, write each of the fractions in simplest form.

(a) $\frac{60}{75}$ **(b)** $\frac{84}{196}$

Most of the fractions we have looked at to this point have been positive numbers. Fractions can certainly be negative numbers. The standard form for a negative fraction uses a negative number in the numerator and a positive number in the denominator.

Example 6

Writing Negative Fractions in Standard Form

Rewrite each fraction in standard form.

(a) $\dfrac{2}{-5}$

Recall that a positive number divided by a negative number results in a negative number. Here the fraction is negative. The standard form has the negative in the numerator, so we write $\dfrac{-2}{5}$.

(b) $\dfrac{-3}{-8}$

A negative number divided by a negative number results in a positive number. In standard form, we simply write this fraction as $\dfrac{3}{8}$.

CHECK YOURSELF 6

Rewrite each fraction in standard form.

(a) $\dfrac{-4}{-5}$ **(b)** $\dfrac{2}{-9}$ **(c)** $\dfrac{5}{-8}$

The fundamental principle is also used to simplify negative fractions.

Example 7

Simplifying Negative Fractions

Simplify $\dfrac{42}{-56}$.

First, we rewrite the fraction in standard form.

$$\frac{42}{-56} = \frac{-42}{56}$$

Then we divide by the common factor of 14 in the numerator and denominator.

$$\frac{-42}{56} = \frac{-3}{4}$$

CHECK YOURSELF 7

Simplify.

(a) $\dfrac{26}{-52}$ **(b)** $\dfrac{-18}{-64}$

To this point in this section, we have used the fundamental principle of fractions to simplify fractions. It can also be used to find an equivalent fraction with a larger denominator.

Rules and Properties: The Fundamental Principle of Fractions

For the fraction $\frac{a}{b}$ and any nonzero number c,

$$\frac{a}{b} = \frac{a \cdot c}{b \cdot c}$$

Example 8

Finding an Equivalent Fraction

Find the fraction equivalent to $\frac{2}{5}$ with a denominator of 25.

We are looking to find the new numerator where

$$\frac{2}{5} = \frac{?}{25}$$

Note that the original denominator, 5, must be multiplied by 5 to get the new denominator, 25. By the fundamental principle, we must multiply the numerator and denominator by the same number to get an equivalent fraction, so we multiply each by 5.

$$\frac{2}{5} = \frac{2 \cdot 5}{5 \cdot 5} = \frac{10}{25}$$

CHECK YOURSELF 8

Find the fraction equivalent to $\frac{5}{7}$ *with a denominator of* 42.

We can also use the fundamental principle to find equivalent negative fractions.

Example 9

Finding Equivalent Negative Fractions

Find the fraction equivalent to $\frac{-3}{7}$ with a denominator of 35.

We are looking to find the new numerator where

$$\frac{-3}{7} = \frac{?}{35}$$

Note that the original denominator, 7, must be multiplied by 5 to get the new denominator, 35. By the fundamental principle, we must multiply the numerator and denominator by the same number to get an equivalent fraction, so we multiply each by 5.

$$\frac{-3}{7} = \frac{-3 \cdot 5}{7 \cdot 5} = \frac{-15}{35}$$

CHECK YOURSELF 9

Find the fraction equivalent to $\frac{-2}{7}$ *with a denominator of* 21.

Using Your Calculator to Simplify Fractions

If you have a calculator that supports fraction arithmetic, you should learn to use it to check your work. We'll look at two different types of these calculators.

Scientific Calculator

Scientific calculators include the TI-34, the Casio fx-250 or fx-115, and the Sharp 506h or 509h.

Before doing Example 10, find the button on your scientific calculator that is labeled **a b/c**. This is the button that will be used to enter fractions.

Example 10

Using a Scientific Calculator to Simplify Fractions

Simplify the fraction $\frac{-24}{68}$.

There are four steps in simplifying fractions using a scientific calculator.

Step 1 Enter the numerator, 24 [+/−].

Step 2 Press the **a b/c** key.

Step 3 Enter the denominator, 68.

Step 4 Press [=].

The calculator will display the simplified fraction, $\frac{-6}{17}$.

CHECK YOURSELF 10

Simplify the fraction $\frac{-51}{81}$.

Graphing Calculator

Let's simplify the same fraction, $\frac{-24}{68}$, using a graphing calculator, such as the TI-82 or TI-83:

Step 1 Enter the fraction as a division problem: [(−)] 24 [÷] 68. The calculator will display $\frac{-24}{68}$.

Step 2 Press the **MATH** key.

Step 3 Select **1: ▶ Frac**.

Step 4 Press [Enter].

The calculator displays the simplified fraction, $\frac{-6}{17}$.

The graphing calculator is particularly useful for simplifying fractions with large values in the numerator and denominator.

NOTE Some scientific calculators cannot handle denominators larger than 999.

Example 11

Using a Graphing Calculator to Simplify Fractions

Simplify $\frac{-546}{637}$.

Using our calculator, we find that

$$\frac{-546}{637} = \frac{-6}{7}$$

CHECK YOURSELF 11

Simplify $\frac{-649}{885}$.

CHECK YOURSELF ANSWERS

1. **(a)** Yes; **(b)** No **2.** **(a)** $\frac{6}{9}$; **(b)** $\frac{2}{3}$

3. 6 is the GCF of 12 and 18, so $\frac{12}{18} = \frac{12 \div 6}{18 \div 6} = \frac{2}{3}$

4. **(a)** $\frac{60}{75} = \frac{2 \cdot 2 \cdot \cancel{3} \cdot \cancel{5}}{\cancel{3} \cdot \cancel{5} \cdot 5} = \frac{4}{5}$; **(b)** $\frac{210}{252} = \frac{\cancel{2} \cdot \cancel{3} \cdot 5 \cdot \cancel{7}}{\cancel{2} \cdot 2 \cdot 3 \cdot \cancel{3} \cdot \cancel{7}} = \frac{5}{6}$

5. **(a)** Divide by the common factors of 3 and 5, $\frac{60}{75} = \frac{4}{5}$;

(b) Divide by the common factors of 4 and 7, $\frac{84}{196} = \frac{3}{7}$

6. **(a)** $\frac{4}{5}$; **(b)** $\frac{-2}{9}$; **(c)** $\frac{-5}{8}$ **7.** **(a)** $\frac{-1}{2}$; **(b)** $\frac{9}{32}$ **8.** $\frac{30}{42}$ **9.** $\frac{-6}{21}$ **10.** $\frac{-17}{27}$

11. $\frac{-11}{15}$

Name ______________

Section ________ Date ________

3.3 Exercises

Are the pairs of fractions equivalent?

1. $\frac{1}{3}, \frac{3}{5}$

2. $\frac{3}{5}, \frac{9}{15}$

3. $\frac{1}{7}, \frac{4}{28}$

4. $\frac{2}{3}, \frac{3}{5}$

5. $\frac{-5}{6}, \frac{-15}{18}$

6. $\frac{-3}{4}, \frac{-16}{20}$

7. $\frac{2}{21}, \frac{4}{25}$

8. $\frac{20}{24}, \frac{5}{6}$

9. $\frac{-2}{7}, \frac{-3}{11}$

10. $\frac{-12}{15}, \frac{-36}{45}$

11. $\frac{16}{24}, \frac{40}{60}$

12. $\frac{15}{20}, \frac{20}{25}$

Solve the given applications.

13. **Test score.** On a test of 72 questions, Sam answered 54 correctly. On another test Sam answered 66 correct out of 88. Did Sam get the same portion of each test correct?

14. **Batting average.** Jeff Bagwell of the Houston Astros has 104 hits in 325 times at bat. Matt Williams of the Arizona Diamondbacks has 88 hits in 275 times at bat. Do they have the same batting average?

Write each fraction in simplest form.

15. $\frac{8}{12}$

16. $\frac{12}{15}$

17. $\frac{10}{14}$

18. $\frac{15}{50}$

ANSWERS

1. ______
2. ______
3. ______
4. ______
5. ______
6. ______
7. ______
8. ______
9. ______
10. ______
11. ______
12. ______
13. ______
14. ______
15. ______
16. ______
17. ______
18. ______

ANSWERS

19. ______
20. ______
21. ______
22. ______
23. ______
24. ______
25. ______
26. ______
27. ______
28. ______
29. ______
30. ______
31. ______
32. ______
33. ______
34. ______
35. ______
36. ______
37. ______ 38. ______
39. ______ 40. ______
41. ______ 42. ______

19. $\frac{-12}{18}$

20. $\frac{-28}{35}$

21. $\frac{35}{40}$

22. $\frac{21}{24}$

23. $\frac{-11}{-44}$

24. $\frac{-10}{-25}$

25. $\frac{12}{36}$

26. $\frac{18}{48}$

27. $\frac{24}{-27}$

28. $\frac{30}{-50}$

29. $\frac{32}{40}$

30. $\frac{17}{51}$

31. $\frac{-75}{105}$

32. $\frac{-62}{93}$

33. $\frac{48}{60}$

34. $\frac{48}{66}$

35. $\frac{-105}{-135}$

36. $\frac{-54}{-126}$

37. $\frac{66}{110}$

38. $\frac{280}{320}$

39. $\frac{16}{-21}$

40. $\frac{21}{-32}$

41. $\frac{31}{52}$

42. $\frac{42}{55}$

ANSWERS

43. ______
44. ______
45. ______
46. ______
47. ______
48. ______
49. ______
50. ______
51. ______
52. ______
53. ______
54. ______
55. ______
56. ______
57. ______
58. ______
59. ______
60. ______
61. ______
62. ______
63. ______
64. ______

Write each fraction as an equivalent fraction having the given denominator.

43. $\frac{1}{3}$; 9

44. $\frac{1}{5}$; 20

45. $\frac{1}{8}$; 128

46. $\frac{1}{10}$; 200

47. $\frac{-1}{4}$; 16

48. $\frac{-1}{6}$; 36

49. $\frac{-1}{7}$; 35

50. $\frac{-1}{9}$; 72

51. $\frac{2}{3}$; 27

52. $\frac{3}{5}$; 45

53. $\frac{4}{7}$; 28

54. $\frac{3}{20}$; 100

55. $\frac{-4}{5}$; 35

56. $\frac{-5}{9}$; 99

57. $\frac{-7}{24}$; 96

58. $\frac{-6}{17}$; 68

59. $\frac{27}{31}$; 93

60. $\frac{-17}{50}$; 200

Solve the applications.

61. Coins. A quarter is what fractional part of a dollar? Simplify your result.

62. Coins. A dime is what fractional part of a dollar? Simplify your result.

63. Time. What fractional part of an hour is 15 min? Simplify your result.

64. Time. What fractional part of a day is 6 h? Simplify your result.

ANSWERS

65. ______

66. ______

67. ______

68. ______

69. ______

70. ______

71. ______

65. Length. A meter is equal to 100 cm. What fractional part of a meter is 70 cm? Simplify your result.

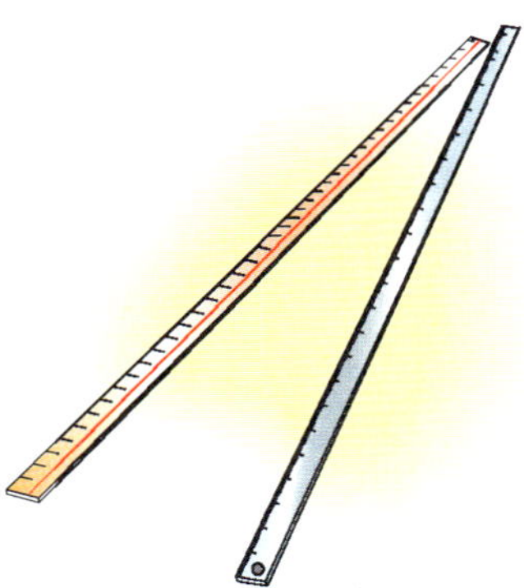

66. Length. A kilometer is equal to 1000 m. What fractional part of a kilometer is 300 m? Simplify your result.

67. Auto repairs. Susan did a tune-up on her automobile. She found that two of her eight spark plugs were fouled. What fraction represents the number of fouled plugs? Reduce to lowest terms.

68. Testing. Samantha answered 18 of 20 problems correctly on a test. What fractional part did she answer correctly? Reduce your answer to lowest terms.

69. Baseball. The local baseball team won 36 of the 58 games they played. What fractional part did they win? Reduce your answer to lowest terms.

70. Salary. Sharon earned \$250 at her after-school job. A new bike costs \$120. What fractional part of her money will remain after she purchases the bike? Reduce your answer to lowest terms.

71. A student is attempting to reduce the fraction $\frac{8}{12}$ to lowest terms. He produces the following argument:

$$\frac{8}{12} = \frac{4 + 4}{8 + 4} = \frac{4}{8} = \frac{1}{2}$$

What is the fallacy in this argument? What is the correct answer?

72. Can any of the following fractions be simplified?

(a) $\dfrac{824}{73}$ (b) $\dfrac{59}{11}$ (c) $\dfrac{135}{17}$

What characteristic do you notice about the denominator of each fraction? What rule would you make up based on your observations?

73. Consider Figure (a).

(a) Give the fraction that represents the shaded region.

(a)

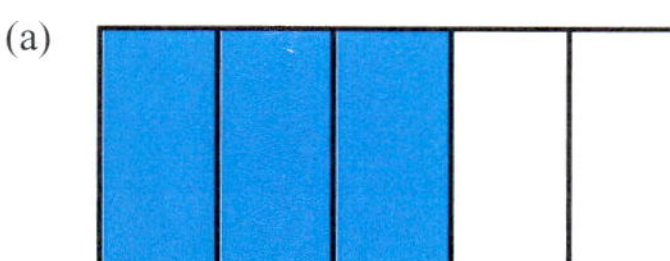

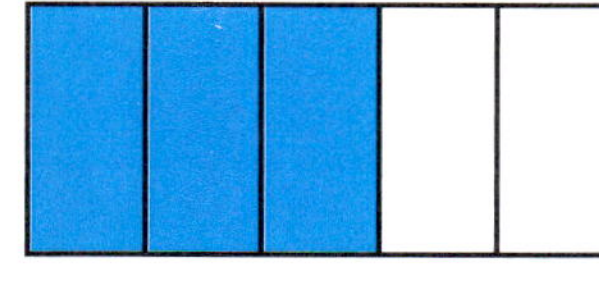

(b) Draw a horizontal line through the figure, as shown in Figure (b). Now give the fraction representing the shaded region.

(b)

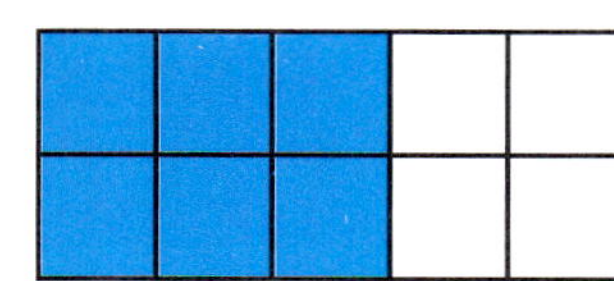

74. Repeat exercise 73 using these figures.

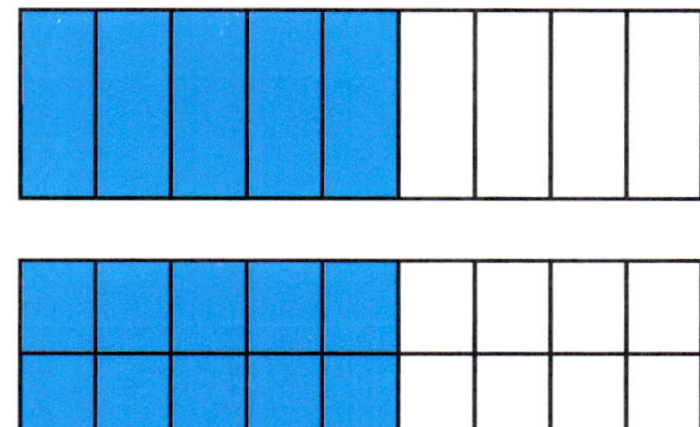

Calculator Exercises

Use your calculator to simplify the fractions.

75. $\dfrac{-28}{40}$

76. $\dfrac{-121}{132}$

77. $\dfrac{96}{-144}$

78. $\dfrac{385}{-605}$

79. $\dfrac{-445}{-623}$

80. $\dfrac{-153}{-255}$

81. $\dfrac{-299}{391}$

82. $\dfrac{-152}{209}$

83. $\dfrac{-289}{459}$

ANSWERS

72. ____________
73. ____________
74. ____________
75. ____________
76. ____________
77. ____________
78. ____________
79. ____________
80. ____________
81. ____________
82. ____________
83. ____________

Answers

1. $1 \cdot 5 = 5; 3 \cdot 3 = 9$. The fractions are not equivalent. **3.** Yes
5. Yes **7.** No **9.** No
11. $16 \cdot 60 = 960$, and $24 \cdot 40 = 960$. The fractions are equivalent. **13.** Yes
15. $\frac{2}{3}$ **17.** $\frac{5}{7}$ **19.** $\frac{-2}{3}$ **21.** $\frac{7}{8}$ **23.** $\frac{1}{4}$ **25.** $\frac{1}{3}$ **27.** $\frac{-8}{9}$
29. $\frac{4}{5}$ **31.** $\frac{-5}{7}$ **33.** $\frac{4}{5}$ **35.** $\frac{7}{9}$ **37.** $\frac{3}{5}$ **39.** $\frac{-16}{21}$ **41.** $\frac{31}{52}$
43. $\frac{3}{9}$ **45.** $\frac{16}{128}$ **47.** $\frac{-4}{16}$ **49.** $\frac{-5}{35}$ **51.** $\frac{18}{27}$ **53.** $\frac{16}{28}$ **55.** $\frac{-28}{35}$
57. $\frac{-28}{96}$ **59.** $\frac{81}{93}$ **61.** $\frac{1}{4}$ **63.** $\frac{1}{4}$ **65.** $\frac{7}{10}$ **67.** $\frac{1}{4}$ **69.** $\frac{18}{29}$
71. **73.** **(a)** $\frac{3}{5}$; **(b)** $\frac{6}{10}$ **75.** $\frac{-7}{10}$ **77.** $\frac{-2}{3}$ **79.** $\frac{5}{7}$
81. $\frac{-13}{17}$ **83.** $\frac{-17}{27}$

3.4 Multiplication and Division of Fractions

3.4 OBJECTIVES

1. Multiply two fractions
2. Determine the reciprocal of a number
3. Divide two fractions

Multiplication is the easiest of the four operations with fractions. We can illustrate multiplication by picturing fractions as parts of a whole or unit. Using this idea, we show the fractions $\frac{4}{5}$ and $\frac{2}{3}$.

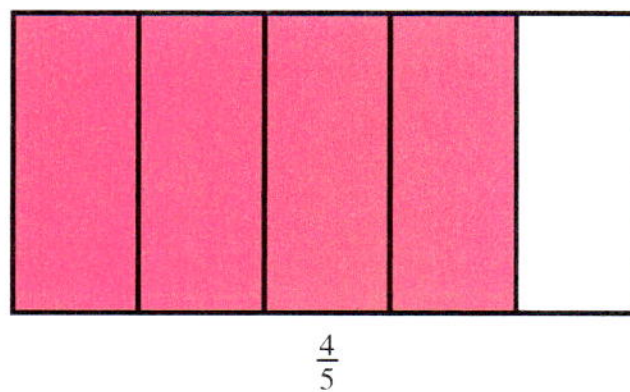

$\frac{4}{5}$

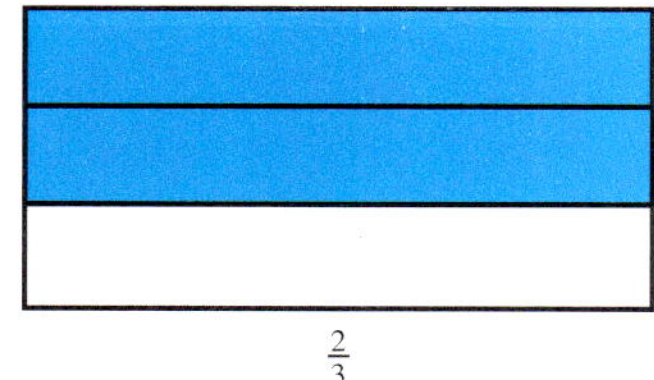

$\frac{2}{3}$

NOTE A fraction followed by the word *of* means that we want to multiply by that fraction.

Suppose now that we wish to find $\frac{2}{3}$ of $\frac{4}{5}$. We can combine the diagrams.

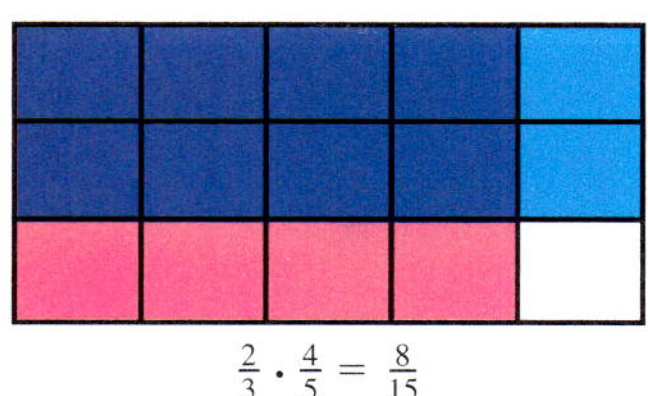

$\frac{2}{3} \cdot \frac{4}{5} = \frac{8}{15}$

The part of the whole representing the product $\frac{2}{3} \cdot \frac{4}{5}$ is the purple region in the figure. The unit has been divided into 15 parts and 8 of those parts are used, so $\frac{2}{3} \cdot \frac{4}{5}$ must be $\frac{8}{15}$.

The following rule is suggested by the diagrams.

Step by Step: To Multiply Fractions

Step 1 Multiply the numerators to find the numerator of the product.
Step 2 Multiply the denominators to find the denominator of the product.
Step 3 Simplify the resulting fraction if possible.

Example 1 will require using steps 1 and 2.

Example 1

Multiplying Two Fractions

NOTE We multiply fractions in this way *not* because it is easy, but because it works!

Multiply.

(a) $\frac{1}{4} \cdot \frac{1}{7} = \frac{1 \cdot 1}{4 \cdot 7} = \frac{1}{28}$

(b) $\frac{2}{3} \cdot \frac{4}{5} = \frac{2 \cdot 4}{3 \cdot 5} = \frac{8}{15}$

(c) $\frac{5}{8} \cdot \frac{7}{9} = \frac{5 \cdot 7}{8 \cdot 9} = \frac{35}{72}$

CHECK YOURSELF 1

Multiply.

(a) $\frac{7}{8} \cdot \frac{3}{10}$ **(b)** $\frac{5}{7} \cdot \frac{3}{4}$

Step 3 indicates that the product of fractions should always be simplified to lowest terms. Consider Example 2.

Example 2

Multiplying Two Fractions

Multiply and write the result in lowest terms.

(a) $\frac{3}{4} \cdot \frac{2}{9} = \frac{3 \cdot 2}{4 \cdot 9}$

$= \frac{6}{36}$

$= \frac{1}{6}$

Noting that $\frac{6}{36}$ is not in simplest form, we divide numerator and denominator by 6 to write the product in lowest terms.

(b) $\frac{4}{7} \cdot \frac{3}{8} = \frac{4 \cdot 3}{7 \cdot 8}$

$= \frac{12}{56}$

$= \frac{3}{14}$

CHECK YOURSELF 2

Multiply and write the result in lowest terms.

$\frac{5}{7} \cdot \frac{3}{10}$

To find the product of a fraction and a whole number, write the whole number as a fraction (the whole number divided by 1) and apply the multiplication rule as before. Example 3 illustrates this approach.

Example 3

Multiplying a Whole Number and a Fraction

Do the indicated multiplication.

Remember that $5 = \frac{5}{1}$.

(a) $5 \cdot \frac{3}{4} = \frac{5}{1} \cdot \frac{3}{4} = \frac{5 \cdot 3}{1 \cdot 4}$

$= \frac{15}{4}$

(b) $\frac{5}{12} \cdot 6 = \frac{5}{12} \cdot \frac{6}{1}$

$= \frac{5 \cdot 6}{12 \cdot 1}$

$= \frac{30}{12} = \frac{5}{2}$

NOTE Leave the product as an improper fraction unless instructed to write as a mixed number.

(c) $\frac{21}{2} \cdot \frac{4}{7} = \frac{84}{14}$

$= 6$

CHECK YOURSELF 3

Multiply.

(a) $\frac{3}{16} \cdot 8$ **(b)** $4 \cdot \frac{5}{7}$

When multiplying fractions, it is usually easier to simplify, that is, remove any common factors in the numerator and denominator, *before multiplying*. Remember that to simplify means to *divide* by the same common factor.

Example 4

Simplifying Before Multiplying Two Fractions

Simplify and then multiply.

NOTE Once again we are applying the fundamental principle to divide the numerator and denominator by 3.

NOTE Because we divide by any common factors before we multiply, the resulting product *is in simplest form.*

$$\frac{3}{5} \cdot \frac{4}{9} = \frac{\overset{1}{\cancel{3}} \cdot 4}{5 \cdot \underset{3}{\cancel{9}}}$$

$$= \frac{1 \cdot 4}{5 \cdot 3}$$

$$= \frac{4}{15}$$

To simplify, we divide the *numerator* and *denominator* by the common factor 3. Remember that $\overset{1}{\cancel{3}}$ means $3 \div 3 = 1$, and $\underset{3}{\cancel{9}}$ means $9 \div 3 = 3$.

CHECK YOURSELF 4

Simplify and then multiply.

$$\frac{7}{8} \cdot \frac{5}{21}$$

Our work in Example 4 leads to the following general rule about simplifying fractions in multiplication.

Rules and Properties: Simplifying Fractions Before Multiplying

In multiplying two or more fractions, we can divide any factor of the numerator and any factor of the denominator by the same nonzero number to simplify the product.

The same rule can be applied when we have more than two fractions, as in Example 5.

Example 5

Multiplying More Than Two Fractions

Find the product.

$$\frac{2}{5} \cdot \frac{5}{4} \cdot \frac{8}{9} = \frac{2 \cdot \overset{1}{\cancel{5}} \cdot \overset{2}{\cancel{8}}}{\underset{1}{\cancel{5}} \cdot \underset{1}{\cancel{4}} \cdot 9} = \frac{4}{9}$$

CHECK YOURSELF 5

Find the product.

$$\frac{3}{5} \cdot \frac{10}{13} \cdot \frac{26}{27}$$

If there are negative fractions involved in the product, it is best to first determine the sign of the product, then simplify.

Example 6

Multiplying More Than Two Signed Fractions

Find the product.

$$\frac{-3}{5} \cdot \frac{5}{7} \cdot \frac{-2}{9}$$

Note that the numerator will be positive (the product of two negatives). Because we will have a positive divided by a positive, our answer will be positive. We can now continue.

$$\frac{-3}{5} \cdot \frac{5}{7} \cdot \frac{-2}{9} = \frac{\overset{1}{\cancel{3}} \cdot \overset{1}{\cancel{5}} \cdot 2}{\underset{1}{\cancel{5}} \cdot 7 \cdot \underset{3}{\cancel{9}}} = \frac{2}{21}$$

CHECK YOURSELF 6

Find the product.

$$\frac{-3}{7} \cdot \frac{-12}{5} \cdot \frac{-2}{9}$$

We are now ready to look at the operation of division on fractions. Before we do so, we will need a new concept, the **reciprocal** of a fraction.

NOTE In general, the reciprocal of the fraction $\frac{a}{b}$ is $\frac{b}{a}$. Neither a nor b can be 0.

Rules and Properties: The Reciprocal of a Fraction

The product of any nonzero fraction and its **reciprocal** is 1. We invert, or interchange, the numerator and denominator of a fraction to write its reciprocal.

Example 7

Finding the Reciprocal of a Fraction

Find the reciprocal of (a) $\frac{3}{4}$ and (b) 5.

(a) The reciprocal of $\frac{3}{4}$ is $\frac{4}{3}$.

Just invert, or turn over, the fraction.

(b) The reciprocal of 5, or $\frac{5}{1}$, is $\frac{1}{5}$.

Write 5 as $\frac{5}{1}$ and then turn over the fraction.

CHECK YOURSELF 7

Find the reciprocal of (a) $\frac{5}{8}$ *and (b)* 3.

An important property relating a number and its reciprocal is given here.

Rules and Properties: Reciprocal Products

The product of any number and its reciprocal is 1. (Every number except zero has a reciprocal.)

We are now ready to use the reciprocal to find a rule for dividing fractions. Recall that we can represent the operation of division in several ways. We used the symbol ÷ earlier. Remember that a fraction also indicates division. For instance,

NOTE $3 \div 5$ and $\frac{3}{5}$ both mean "3 divided by 5."

$$3 \div 5 = \frac{3}{5}$$

In this statement, 5 is called the *divisor*. It follows the division sign ÷ and is written *below* the fraction bar.

Using this information, we can write a statement involving fractions and division as a *complex fraction,* which has a fraction as both its numerator and denominator, as Example 8 illustrates. We will study more about complex fractions in Section 3.7.

Example 8

Writing a Quotient as a Complex Fraction

Write $\frac{2}{5} \div \frac{3}{4}$ as a complex fraction.

The numerator is $\frac{2}{5}$.

$$\frac{\frac{2}{5}}{\frac{3}{4}}$$

A *complex fraction* is written by placing the dividend in the numerator and the divisor in the denominator.

The denominator is $\frac{3}{4}$.

CHECK YOURSELF 8

Write $\frac{2}{7} \div \frac{4}{5}$ *as a complex fraction.*

We will continue with the same division problem.

Example 9

Dividing Two Fractions

(1) $\frac{2}{5} \div \frac{3}{4} = \dfrac{\frac{2}{5}}{\frac{3}{4}}$ Write the original quotient as a complex fraction.

$= \dfrac{\frac{2}{5} \cdot \frac{4}{3}}{\frac{3}{4} \cdot \frac{4}{3}}$ Multiply the numerator and denominator by $\frac{4}{3}$, the reciprocal of the denominator. This does *not* change the value of the fraction.

$= \dfrac{\frac{2}{5} \cdot \frac{4}{3}}{1}$ The denominator becomes 1.

(2) $= \frac{2}{5} \cdot \frac{4}{3}$ Recall that a number divided by 1 is just that number.

We see from lines (1) and (2) that

$$\frac{2}{5} \div \frac{3}{4} = \frac{2}{5} \cdot \frac{4}{3}$$

NOTE Do you see a rule suggested?

We would certainly like to be able to divide fractions easily without all the work of Example 9. Look carefully at the example. A rule to divide fractions is suggested.

Rules and Properties: To Divide Fractions

To divide one fraction by another, multiply the dividend by the reciprocal of the divisor.

CHECK YOURSELF 9

Write $\frac{2}{7} \div \frac{4}{5}$ *as a multiplication problem.*

Example 10 applies the rule for dividing fractions.

Example 10

Dividing Two Fractions

Divide.

$\frac{1}{3} \div \frac{4}{7} = \frac{1}{3} \cdot \frac{7}{4}$ We invert the divisor, $\frac{4}{7}$, then multiply.

$= \frac{1 \cdot 7}{3 \cdot 4} = \frac{7}{12}$

NOTE *Remember,* the number inverted is the divisor. It *follows* the division sign.

CHECK YOURSELF 10

Divide.

$$\frac{2}{7} \div \frac{4}{5}$$

Now we will look at an example that contains signed numbers.

Example 11

Dividing Two Fractions

Divide.

$$\frac{-5}{8} \div \frac{3}{5} = \frac{-5}{8} \cdot \frac{5}{3} = \frac{-5 \cdot 5}{8 \cdot 3} = \frac{-25}{24}$$

NOTE If there is only one negative in a multiplication or division, the result will be negative.

CHECK YOURSELF 11

Divide.

$$\frac{-5}{6} \div \frac{3}{7}$$

Simplifying will also be useful in dividing fractions. Consider Example 12.

Example 12

Dividing Two Fractions

Divide.

NOTE Be careful! We must invert the divisor *before any simplification.*

$$\frac{3}{5} \div \frac{6}{7} = \frac{3}{5} \cdot \frac{7}{6}$$

Invert the divisor *first!* Then you can divide by the common factor of 3.

$$= \frac{\overset{1}{\cancel{3}} \cdot 7}{5 \cdot \underset{2}{\cancel{6}}} = \frac{7}{10}$$

CHECK YOURSELF 12

Divide.

$$\frac{4}{9} \div \frac{8}{15}$$

Using Your Calculator to Multiply and Divide Fractions

Scientific Calculator

To multiply fractions on a scientific calculator, you enter the first fraction, using the **a b/c** key, then press the multiplication sign, next enter the second fraction, and then press the equals sign.

Example 13

Multiplying Two Fractions

Find the product.

$$\frac{-7}{15} \cdot \frac{-5}{21}$$

The keystroke sequence is

7 [+/−] [**a b/c**] 15 [×] 5 [+/−] [**a b/c**] 21 [=]

The result is $\frac{1}{9}$.

CHECK YOURSELF 13

Find the product.

$$\frac{-24}{33} \cdot \frac{22}{39}$$

Graphing Calculator

When using a graphing calculator, you must choose the fraction option [**1:▶ Frac**] from the [**MATH**] menu before pressing [Enter].

For the fraction problem in Example 13, $\frac{-7}{15} \cdot \frac{-5}{21}$, the keystroke sequence is

[(−)] 7 [÷] 15 [×] [(−)] 5 [÷] 21 [**1:▶ Frac**] [Enter]

Again, the result will be $\frac{1}{9}$.

Scientific Calculator

Dividing fractions on a scientific calculator requires only that you enter the problem followed by the equal sign.

Example 14

Dividing Two Fractions

Find the quotient.

$$\frac{-24}{23} \div \frac{-16}{13}$$

The keystroke sequence is

24 [+/−] [a b/c] 23 [÷] 16 [+/−] [a b/c] 13 [=]

The result is $\frac{39}{46}$, a positive number (we divided two negative numbers).

CHECK YOURSELF 14

Find the quotient.

$$\frac{-17}{12} \div \frac{-16}{3}$$

Graphing Calculator

When using a graphing calculator, you must choose the fraction option [**1:▶ Frac**] from the [**MATH**] menu before pressing [Enter].

The keystroke sequence for the fraction problem in Example 14, $\frac{-24}{23} \div \frac{-16}{13}$, is

[(−)] 24 [÷] 23 [÷] [(] [(−)] 16 [÷] 13 [)] [**1:▶ Frac**] [Enter]

CHECK YOURSELF ANSWERS

1. **(a)** $\frac{7}{8} \cdot \frac{3}{10} = \frac{7 \cdot 3}{8 \cdot 10} = \frac{21}{80}$; **(b)** $\frac{5}{7} \cdot \frac{3}{4} = \frac{5 \cdot 3}{7 \cdot 4} = \frac{15}{28}$

2. $\frac{5}{7} \cdot \frac{3}{10} = \frac{5 \cdot 3}{7 \cdot 10} = \frac{15}{70} = \frac{3}{14}$ **3.** **(a)** $\frac{3}{2}$; **(b)** $\frac{20}{7}$

4. $\frac{7}{8} \cdot \frac{5}{21} = \frac{\overset{1}{\cancel{7}} \cdot 5}{8 \cdot \underset{3}{\cancel{21}}} = \frac{5}{24}$ **5.** $\frac{4}{9}$ **6.** $\frac{-8}{35}$ **7.** **(a)** $\frac{8}{5}$; **(b)** $\frac{1}{3}$

8. $\frac{\frac{2}{7}}{\frac{4}{5}}$ **9.** $\frac{2}{7} \cdot \frac{5}{4}$ **10.** $\frac{5}{14}$ **11.** $\frac{-35}{18}$

12. $\frac{4}{9} \div \frac{8}{15} = \frac{4}{9} \cdot \frac{15}{8} = \frac{\overset{1}{\cancel{4}} \cdot \overset{5}{\cancel{15}}}{\underset{3}{\cancel{9}} \cdot \underset{2}{\cancel{8}}} = \frac{5}{6}$ **13.** $\frac{-16}{39}$ **14.** $\frac{17}{64}$

Name ________________

Section ________ Date ________

3.4 Exercises

Multiply. Be sure to write each answer in simplest form.

1. $\frac{3}{4} \cdot \frac{5}{11}$

2. $\frac{2}{7} \cdot \frac{5}{9}$

3. $\frac{3}{4} \cdot \frac{7}{11}$

4. $\frac{2}{5} \cdot \frac{3}{7}$

5. $\frac{-3}{5} \cdot \frac{5}{7}$

6. $\frac{-6}{11} \cdot \frac{8}{6}$

7. $\frac{6}{13} \cdot \frac{4}{9}$

8. $\frac{5}{9} \cdot \frac{6}{11}$

9. $\frac{-3}{11} \cdot \frac{-7}{9}$

10. $\frac{-7}{9} \cdot \frac{-3}{5}$

11. $\frac{3}{10} \cdot \frac{5}{9}$

12. $\frac{5}{21} \cdot \frac{14}{25}$

13. $\frac{7}{9} \cdot \frac{6}{5}$

14. $\frac{8}{13} \cdot \frac{26}{5}$

15. $\frac{-3}{4} \cdot \frac{6}{7}$

16. $\frac{-3}{7} \cdot 14$

17. $9 \cdot \frac{5}{6}$

18. $15 \cdot \frac{5}{6}$

19. $\frac{-12}{25} \cdot \frac{-11}{18}$

20. $\frac{-10}{12} \cdot \frac{-16}{25}$

21. Find $\frac{2}{3}$ of $\frac{3}{7}$

22. What is $\frac{5}{6}$ of $\frac{9}{10}$?

Find the product.

23. $\frac{2}{3} \cdot \frac{3}{5} \cdot \frac{5}{7}$

24. $\frac{1}{7} \cdot \frac{7}{12} \cdot \frac{12}{13}$

25. $\frac{2}{5} \cdot \frac{10}{11} \cdot \frac{22}{25}$

26. $\frac{3}{8} \cdot \frac{1}{3} \cdot \frac{16}{17}$

27. $\frac{-1}{3} \cdot \frac{3}{8} \cdot \frac{12}{17}$

28. $\frac{4}{7} \cdot \frac{-3}{8} \cdot \frac{14}{15}$

29. $\frac{-1}{5} \cdot \frac{7}{9} \cdot \frac{-15}{14}$

30. $\frac{-2}{7} \cdot \frac{-5}{4} \cdot \frac{-14}{15}$

Find the quotient. Write each result in simplest form.

31. $\frac{1}{5} \div \frac{3}{4}$

32. $\frac{2}{5} \div \frac{1}{3}$

33. $\frac{-2}{5} \div \frac{3}{4}$

34. $\frac{-5}{8} \div \frac{3}{4}$

35. $\frac{8}{9} \div \frac{4}{3}$

36. $\frac{5}{9} \div \frac{8}{11}$

37. $\frac{-7}{10} \div \frac{-5}{9}$

38. $\frac{-8}{9} \div \frac{-11}{15}$

39. $\frac{8}{15} \div \frac{2}{5}$

ANSWERS

1. ________ 2. ________
3. ________ 4. ________
5. ________ 6. ________
7. ________ 8. ________
9. ________ 10. ________
11. ________ 12. ________
13. ________ 14. ________
15. ________ 16. ________
17. ________ 18. ________
19. ________ 20. ________
21. ________ 22. ________
23. ________ 24. ________
25. ________ 26. ________
27. ________ 28. ________
29. ________ 30. ________
31. ________ 32. ________
33. ________ 34. ________
35. ________ 36. ________
37. ________ 38. ________
39. ________

40. ______ 41. ______

42. ______ 43. ______

44. ______ 45. ______

46. ______

47. ______

48. ______

49. ______

50. ______

51. ______

52. ______

53. ______

54. ______

55. ______

56. ______

57. ______

58. ______

59. ______

60. ______

61. ______

62. ______

63. ______

64. ______

40. $\frac{5}{27} \div \frac{15}{54}$ **41.** $\frac{5}{27} \div \frac{25}{36}$ **42.** $\frac{9}{28} \div \frac{27}{35}$

43. $\frac{-4}{5} \div 4$ **44.** $-27 \div \frac{3}{7}$ **45.** $12 \div \frac{2}{3}$

46. $\frac{5}{8} \div 5$ **47.** $\frac{-12}{17} \div -6$ **48.** $\frac{-3}{4} \div -9$

Calculator Exercises

Find the products using your calculator.

49. $\frac{-15}{20} \cdot \frac{-8}{12}$ **50.** $\frac{-7}{8} \cdot \frac{-4}{21}$

51. $\frac{-18}{84} \cdot \frac{36}{27}$ **52.** $\frac{-6}{35} \cdot \frac{20}{12}$

53. $\frac{7}{12} \cdot \frac{-36}{63}$ **54.** $\frac{8}{27} \cdot \frac{-45}{64}$

55. $\frac{-27}{72} \cdot \frac{-24}{45}$ **56.** $\frac{-81}{136} \cdot \frac{-84}{135}$

Find the quotients using your calculator.

57. $\frac{1}{5} \div \frac{2}{15}$ **58.** $\frac{13}{17} \div \frac{39}{34}$

59. $\frac{20}{27} \div \frac{-35}{36}$ **60.** $\frac{13}{15} \div \frac{-39}{5}$

61. $\frac{15}{18} \div \frac{45}{27}$ **62.** $\frac{2}{3} \div \frac{10}{9}$

63. $\frac{-25}{45} \div \frac{-100}{135}$ **64.** $\frac{-19}{63} \div \frac{-38}{9}$

Answers

1. $\frac{15}{44}$ **3.** $\frac{21}{44}$ **5.** $\frac{-3}{7}$ **7.** $\frac{8}{39}$ **9.** $\frac{7}{33}$ **11.** $\frac{1}{6}$ **13.** $\frac{14}{15}$
15. $\frac{-9}{14}$ **17.** $\frac{15}{2}$ **19.** $\frac{22}{75}$ **21.** $\frac{2}{7}$ **23.** $\frac{2}{7}$ **25.** $\frac{8}{25}$ **27.** $\frac{-3}{34}$
29. $\frac{1}{6}$ **31.** $\frac{4}{15}$ **33.** $\frac{-8}{15}$ **35.** $\frac{2}{3}$ **37.** $\frac{63}{50}$ **39.** $\frac{4}{3}$ **41.** $\frac{4}{15}$
43. $\frac{-1}{5}$ **45.** 18 **47.** $\frac{2}{17}$ **49.** $\frac{1}{2}$ **51.** $\frac{-2}{7}$ **53.** $\frac{-1}{3}$ **55.** $\frac{1}{5}$
57. $\frac{3}{2}$ **59.** $\frac{-16}{21}$ **61.** $\frac{1}{2}$ **63.** $\frac{3}{4}$

3.5 The Multiplication Property of Equality

3.5 OBJECTIVE

1. Use the multiplication property of equality to solve an equation

In Chapter 2, we learned how to solve certain equations using the addition property. Now we will look at a different type of equation. For instance, what if we want to solve a multiplication equation? Here is an example:

$6x = 18$

Using the addition property won't help. We will need a second property for solving equations. First, we will revisit the idea of a reciprocal.

NOTE In general, the reciprocal of the fraction $\frac{a}{b}$ is $\frac{b}{a}$. Neither a nor b can be 0.

Example 1

Finding the Reciprocal of a Fraction

Find the reciprocal of (a) $\frac{4}{5}$ and (b) 7.

(a) The reciprocal of $\frac{4}{5}$ is $\frac{5}{4}$.

(b) The reciprocal of 7, or $\frac{7}{1}$, is $\frac{1}{7}$. Write 7 as $\frac{7}{1}$ and then turn over the fraction.

CHECK YOURSELF 1

Find the reciprocal of (a) $\frac{3}{7}$ *and (b)* 17.

The relationship of a number and its reciprocal is shown in Example 2.

Example 2

Multiplying a Number by Its Reciprocal

Multiply.

(a) $\left(\frac{3}{7}\right)\left(\frac{7}{3}\right) = 1$

(b) $(8)\left(\frac{1}{8}\right) = 1$

(c) $(-17)\left(\frac{-1}{17}\right) = 1$

CHECK YOURSELF 2

Find the products.

(a) $\left(\frac{1}{4}\right)(4)$

(b) $\left(\frac{-2}{11}\right)\left(\frac{-11}{2}\right)$

This relationship between a number and its reciprocal will be very important in applying the multiplication property of equality for solving equations.

NOTE The c cannot be zero. Multiplying by 0 gives $0 = 0$. We have lost the variable!

Rules and Properties: The Multiplication Property of Equality

If $a = b$ then $ac = bc$

In words, multiplying both sides of an equation by the same nonzero number gives an equivalent equation.

Again, we return to the image of the balance scale. We start with the assumption that a and b have the same weight.

The multiplication property tells us that the scale will be in balance as long as we have the same number of "a weights" as we have of "b weights."

We will work through some examples, using this second rule.

Example 3

Solving Equations by Using the Multiplication Property

Solve.

$6x = 18$

Here the variable x is multiplied by 6. So we apply the multiplication property and multiply both sides by the reciprocal of 6, $\frac{1}{6}$. Keep in mind that we want an equation of the form

$x = \square$

NOTE $\frac{1}{6}(6x) = \left(\frac{1}{6} \cdot 6\right)x$

$= 1 \cdot x$, or x

We then have x alone on the left, which is what we want.

$\frac{1}{6}(6x) = \frac{1}{6}(18)$

We can now simplify.

$1 \cdot x = 3$ or $x = 3$

The solution is 3. To check, replace x with 3:

$6 \cdot 3 \stackrel{?}{=} 18$

$18 = 18$ (True)

CHECK YOURSELF 3

Solve and check.

$8x = 32$

In Example 3 we solved the equation by multiplying both sides by the reciprocal of the coefficient of the variable.

Example 4 illustrates a slightly different approach to solving an equation by using the multiplication property.

Example 4

Solving Equations by Using the Multiplication Property

Solve.

$5x = -35$

NOTE Because division is defined in terms of multiplication, we can also divide both sides of an equation by the same nonzero number.

The variable x is multiplied by 5. We *divide* both sides by 5 to "undo" that multiplication:

$$\frac{5x}{5} = \frac{-35}{5}$$

$x = -7$ (Note that the right side reduces to −7. Be careful with the rules for signs.)

We will leave it to you to check the solution.

CHECK YOURSELF 4

Solve and check.

$7x = -42$

Example 5

Solving Equations by Using the Multiplication Property

Solve.

$-9x = 54$

In this case, x is multiplied by -9, so we divide both sides by -9 to isolate x on the left:

$$\frac{-9x}{-9} = \frac{54}{-9}$$

$$x = -6$$

The solution is -6. To check:

$$(-9)(-6) \stackrel{?}{=} 54$$

$$54 = 54 \quad \text{(True)}$$

CHECK YOURSELF 5

Solve and check.

$-10x = -60$

Example 6

Solving Equations by Using the Multiplication Property

Solve.

$$\frac{-1}{3}x = 6$$

Here x is multiplied by $\frac{-1}{3}$, so we multiply both sides by -3, the reciprocal of $\frac{-1}{3}$, to isolate x on the left:

$$(-3)\left(\frac{-1}{3}x\right) = (-3)(6)$$

We can now simplify.

$$1 \cdot x = -18 \qquad \text{or} \qquad x = -18$$

The solution is -18. To check, replace x with -18:

$$\frac{-1}{3}(-18) \stackrel{?}{=} 6$$

Rewriting the left side as $\frac{-18}{-3}$, we have

$$\frac{-18}{-3} \stackrel{?}{=} 6$$

$$6 = 6 \qquad \text{(True)}$$

CHECK YOURSELF 6

Solve and check.

$$\frac{1}{7}x = 9$$

You may sometimes have to simplify an equation before applying the methods of this section. Example 7 illustrates this property.

Example 7

Combining Like Terms and Solving Equations

Solve and check.

$$3x + 5x = 40$$

Using the distributive property, we can combine the like terms on the left to write

$$8x = 40$$

We can now proceed as before.

$$\frac{8x}{8} = \frac{40}{8} \qquad \text{Divide by 8.}$$

$$x = 5$$

The solution is 5. To check, we return to the original equation. Substituting 5 for x yields

$$3 \cdot 5 + 5 \cdot 5 \stackrel{?}{=} 40$$

$$15 + 25 \stackrel{?}{=} 40$$

$$40 = 40 \qquad \text{(True)}$$

The solution is verified.

CHECK YOURSELF 7

Solve and check.

$7x + 4x = -66$

CHECK YOURSELF ANSWERS

1. (a) $\frac{7}{3}$; **(b)** $\frac{1}{17}$ **2. (a)** 1; **(b)** 1 **3.** 4 **4.** -6 **5.** 6 **6.** 63 **7.** -6

Name ______________

Section ______ Date ______

3.5 Exercises

Find the reciprocal.

1. $\frac{7}{8}$

2. $\frac{9}{5}$

3. $\frac{1}{2}$

4. 6

5. $\frac{-2}{3}$

6. -5

7. 1

8. $\frac{1}{8}$

Find each product.

9. $\left(\frac{12}{5}\right)\left(\frac{5}{12}\right)$

10. $\left(\frac{-2}{3}\right)\left(\frac{-3}{2}\right)$

11. $(10)\left(\frac{1}{10}\right)$

12. $(-5)\left(\frac{-1}{5}\right)$

13. $\left(\frac{-1}{9}\right)(-9)$

14. $\left(\frac{1}{14}\right)(14)$

Solve for x and check your result.

15. $5x = 20$

16. $6x = 30$

17. $9x = 54$

18. $6x = -42$

19. $63 = 9x$

20. $66 = 6x$

21. $4x = -16$

22. $-3x = 27$

23. $-9x = 72$

24. $10x = -100$

25. $6x = -54$

26. $-7x = 49$

27. $-4x = -12$

28. $52 = -4x$

29. $-42 = 6x$

30. $-7x = -35$

31. $-6x = -54$

32. $-4x = -24$

ANSWERS

1. ______ 2. ______
3. ______ 4. ______
5. ______ 6. ______
7. ______ 8. ______
9. ______ 10. ______
11. ______ 12. ______
13. ______
14. ______
15. ______
16. ______
17. ______
18. ______
19. ______
20. ______
21. ______
22. ______
23. ______
24. ______
25. ______
26. ______
27. ______
28. ______
29. ______
30. ______
31. ______
32. ______

ANSWERS

33. ______
34. ______
35. ______
36. ______
37. ______
38. ______
39. ______
40. ______
41. ______
42. ______
43. ______
44. ______

33. $\frac{1}{2}x = 4$

34. $\frac{1}{3}x = 2$

35. $\frac{1}{5}x = 3$

36. $\frac{1}{8}x = 5$

37. $6 = \frac{1}{7}x$

38. $6 = \frac{1}{3}x$

39. $\frac{1}{5}x = -4$

40. $\frac{1}{7}x = -5$

41. $\frac{-1}{3}x = 8$

42. $5x + 4x = 36$

43. $16x - 9x = -35$

44. $4x - 2x + 7x = 36$

Answers

1. $\frac{8}{7}$ **3.** 2 **5.** $\frac{-3}{2}$ **7.** 1 **9.** 1 **11.** 1 **13.** 1 **15.** 4 **17.** 6 **19.** 7 **21.** -4 **23.** -8 **25.** -9 **27.** 3 **29.** -7 **31.** 9 **33.** 8 **35.** 15 **37.** 42 **39.** -20 **41.** -24 **43.** -5

3.6 Linear Equations in One Variable

3.6 OBJECTIVES

1. Combine the addition and multiplication properties to solve an equation
2. Use the order of operations when solving an equation
3. Recognize identities
4. Recognize equations with no solutions

In all our examples thus far, either the addition property or the multiplication property was used in solving an equation. Often, finding a solution will require the use of both properties.

Example 1

Solving Equations

(a) Solve.

$4x - 5 = 7$

Here x is *multiplied* by 4. The result, $4x$, then has 5 subtracted from it (or -5 added to it) on the left side of the equation. These two operations mean that both properties must be applied in solving the equation.

Because the variable term is already on the left, we start by adding 5 to both sides:

$$\begin{array}{rcr} 4x - 5 &=& 7 \\ +5 & & +5 \\ \hline 4x &=& 12 \end{array}$$

We now divide both sides by 4:

$$\frac{4x}{4} = \frac{12}{4}$$

$$x = 3$$

The solution is 3. To check, replace x with 3 in the original equation. Be careful to follow the rules for the order of operations.

$$4 \cdot 3 - 5 \stackrel{?}{=} 7$$

$$12 - 5 \stackrel{?}{=} 7$$

$$7 = 7 \quad \text{(True)}$$

(b) Solve.

$$\begin{array}{rcr} 3x + 8 &=& -4 \\ -8 & & -8 \\ \hline 3x &=& -12 \end{array} \qquad \text{Add } -8 \text{ to both sides.}$$

Now divide both sides by 3 to isolate x on the left.

$$\frac{3x}{3} = \frac{-12}{3}$$

$$x = -4$$

The solution is -4. We'll leave the check of this result to you.

CHECK YOURSELF 1

Solve and check.

(a) $6x + 9 = -15$ **(b)** $5x - 8 = 7$

The variable may appear in any position in an equation. Just apply the rules carefully as you try to write an equivalent equation, and you will find the solution. Example 2 illustrates this property.

Example 2

Solving Equations

Solve.

$$\begin{array}{rcr} 3 - 2x = & & 9 \\ -3 & & -3 \\ \hline -2x = & & 6 \end{array}$$

First add -3 to both sides.

NOTE $\frac{-2}{-2} = 1$, so we divide by -2 to isolate x on the left.

Now divide both sides by -2. This will leave x alone on the left.

$$\frac{-2x}{-2} = \frac{6}{-2}$$

$$x = -3$$

The solution is -3. We'll leave it to you to check this result.

CHECK YOURSELF 2

Solve and check.

$10 - 3x = 1$

You may also have to combine multiplication with addition or subtraction to solve an equation. Consider Example 3.

Example 3

Solving Equations

(a) Solve.

$$\frac{1}{5}x - 3 = 4$$

To get the x term alone, we first add 3 to both sides.

$$\begin{array}{rcr} \frac{1}{5}x - 3 = & & 4 \\ + 3 & & +3 \\ \hline \frac{1}{5}x \quad = & & 7 \end{array}$$

Now, multiply both sides of the equation by 5.

$$5\left(\frac{1}{5}x\right) = 5 \cdot 7$$

$$x = 35$$

The solution is 35. Just return to the original equation to check the result.

$$\frac{1}{5}(35) - 3 \stackrel{?}{=} 4$$

$$7 - 3 \stackrel{?}{=} 4$$

$$4 = 4 \qquad \text{(True)}$$

(b) Solve.

$$5 - \frac{1}{4}x = 2$$

To get the x term alone, we first add -5 to both sides.

$$\begin{array}{r} 5 - \frac{1}{4}x = \quad 2 \\ -5 \qquad\quad -5 \\ \hline \frac{-1}{4}x = -3 \end{array}$$

Now multiply both sides by -4, the reciprocal of $\frac{-1}{4}$.

$$(-4)\left(\frac{-1}{4}x\right) = (-4)(-3)$$

or

$$x = 12$$

The solution is 12. We'll leave it to you to check this result.

CHECK YOURSELF 3

Solve and check.

(a) $\frac{1}{6}x + 5 = 3$ **(b)** $-8 - \frac{1}{4}x = 10$

In Section 2.10, you learned how to solve certain equations when the variable appeared on both sides. Example 4 will show you how to extend that work by using the multiplication property of equality.

Example 4

Solving an Equation

Solve.

$$6x - 14 = 3x - 2$$

First add 14 to both sides. This will undo the subtraction on the left.

$$\begin{array}{r} 6x - 14 = 3x - \ \ 2 \\ +14 \qquad + 14 \\ \hline 6x \qquad = 3x + 12 \end{array}$$

Now add $-3x$ so that the terms in x will be on the left only.

$$\begin{array}{rcl} 6x &=& 3x + 12 \\ -3x & & -3x \\ \hline 3x &=& 12 \end{array}$$

Finally divide by 3.

$$\frac{3x}{3} = \frac{12}{3}$$

$$x = 4$$

Check:

$$6(4) - 14 \stackrel{?}{=} 3(4) - 2$$

$$24 - 14 \stackrel{?}{=} 12 - 2$$

$$10 = 10 \quad \text{(True)}$$

As you know, the basic idea is to use our two properties to form an equivalent equation with the x isolated. Here we added 14 and then subtracted $3x$. You can do these steps in either order. Try it for yourself the other way. In either case, the multiplication property is then used as the *last step* in finding the solution.

CHECK YOURSELF 4

Solve and check.

$7x - 5 = 3x + 15$

We will look at two approaches to solving equations in which the coefficient on the right side is greater than the coefficient on the left side.

Example 5

Solving an Equation (Two Methods)

Solve $4x - 8 = 7x + 7$.

Method 1

$$\begin{array}{rcl} 4x - 8 &=& 7x + 7 \\ -7x \quad & & -7x \\ \hline -3x - 8 &=& 7 \\ +8 & & +8 \\ \hline -3x &=& 15 \end{array}$$

Adding $-7x$ will get the variable terms on the left.

Adding 8 will leave the x term alone on the left.

$$\frac{-3x}{-3} = \frac{15}{-3}$$

Dividing by -3 will isolate x on the left.

$$x = -5$$

We'll let you check this result.

To avoid a negative coefficient (in this example, -3), some students prefer a different approach.

This time we'll work toward having the number on the *left* and the x term on the *right,* or

$\square = x.$

Method 2

NOTE It is usually easier to isolate the variable term on the side that will result in a positive coefficient.

$$\begin{array}{rcll} 4x - 8 &=& 7x + 7 & \\ -4x \quad &&-4x \quad & \text{Add } -4x \text{ to get the variables on the right.} \\ \hline -8 &=& 3x + 7 & \\ -7 &&-7 & \\ \hline -15 &=& 3x & \\ \dfrac{-15}{3} &=& \dfrac{3x}{3} & \text{Divide by 3 to isolate } x \text{ on the right.} \\ -5 &=& x & \end{array}$$

Because $-5 = x$ and $x = -5$ are equivalent equations, it really makes no difference; the solution is still -5! You can use whichever approach you prefer.

CHECK YOURSELF 5

Solve $5x + 3 = 9x - 21$ by finding equivalent equations of the form $x = \square$ and $\square = x$ to compare the two methods of finding the solution.

It may also be necessary to remove grouping symbols in solving an equation.

Example 6

Solving Equations That Contain Parentheses

NOTE
$5(x - 3)$
$= 5(x + (-3))$
$= 5x + 5(-3)$
$= 5x + (-15)$
$= 5x - 15$

Solve and check.

$$\begin{array}{rcll} 5(x - 3) - 2x &=& x + 7 & \text{First, apply the distributive property.} \\ 5x - 15 - 2x &=& x + 7 & \text{Combine like terms.} \\ 3x - 15 &=& x + 7 & \\ +15 &&+15 & \text{Add 15.} \\ \hline 3x &=& x + 22 & \\ -x &&-x & \text{Add } -x. \\ \hline 2x &=& 22 & \text{Divide by 2.} \\ x &=& 11 & \end{array}$$

The solution is 11. To check, substitute 11 for x in the original equation. Again note the use of our rules for the order of operations.

$$\begin{array}{rcll} 5(11 - 3) - 2 \cdot 11 &\stackrel{?}{=}& 11 + 7 & \text{Simplify terms in parentheses.} \\ 5 \cdot 8 - 2 \cdot 11 &\stackrel{?}{=}& 11 + 7 & \text{Multiply.} \\ 40 - 22 &\stackrel{?}{=}& 11 + 7 & \text{Add and subtract.} \\ 18 &=& 18 & \text{A true statement.} \end{array}$$

CHECK YOURSELF 6

Solve and check.

$$7(x + 5) - 3x = x - 7$$

An equation that is true for any value of x is called an **identity.**

Example 7

Solving an Equation

Solve the equation $2(x - 3) = 2x - 6$.

$$\begin{aligned} 2(x - 3) &= 2x - 6 \\ 2x - 6 &= 2x - 6 \\ -2x \quad &\quad -2x \\ \hline -6 &= -6 \end{aligned}$$

NOTE We could ask the question "For what values of x does $-6 = -6$?"

The statement $-6 = -6$ is true for any value of x. The original equation is an identity.

CHECK YOURSELF 7

Solve the equation $3(x - 4) - 2x = x - 12$.

There are also equations for which there are no solutions.

Example 8

Solving an Equation

Solve the equation $3(2x - 5) - 4x = 2x + 1$.

$$\begin{aligned} 3(2x - 5) - 4x &= 2x + 1 \\ 6x - 15 - 4x &= 2x + 1 \\ 2x - 15 &= 2x + 1 \\ -2x \quad &\quad -2x \\ \hline -15 &= 1 \end{aligned}$$

NOTE We could ask the question "For what values of x does $-15 = 1$?"

These two numbers are never equal. The original equation has no solutions.

CHECK YOURSELF 8

Solve the equation $2(x - 5) + x = 3x - 3$.

NOTE Such an outline of steps is sometimes called an **algorithm** for the process.

Step by Step: Solving Linear Equations

Step 1 Use the distributive property to remove any grouping symbols. Then simplify by combining like terms on each side of the equation.

Step 2 Add or subtract the same term on each side of the equation until the variable term is on one side and a number is on the other.

Step 3 Multiply or divide both sides of the equation by the same nonzero number so that the variable is alone on one side of the equation. If no variable remains, determine whether the original equation is an identity or whether it has no solutions.

Step 4 Check the solution in the original equation.

CHECK YOURSELF ANSWERS

1. **(a)** -4; **(b)** 3 **2.** 3 **3.** **(a)** -12; **(b)** -72 **4.** 5 **5.** 6 **6.** -14
7. The equation is an identity; x is any real number. **8.** There are no solutions.

Name ______________________

3.6 Exercises

Section ________ Date ________

Solve for x and check your result.

1. $2x + 1 = 9$
2. $3x - 1 = 17$
3. $3x - 2 = 7$
4. $5x + 3 = 23$
5. $4x + 7 = 35$
6. $7x - 8 = 13$
7. $2x + 9 = 5$
8. $6x + 25 = -5$
9. $4 - 7x = 18$
10. $8 - 5x = -7$
11. $3 - 4x = -9$
12. $5 - 4x = 25$
13. $\frac{1}{2}x + 1 = 5$
14. $\frac{1}{3}x - 2 = 3$
15. $\frac{1}{4}x - 5 = 3$
16. $\frac{1}{5}x + 3 = 8$
17. $5x = 2x + 9$
18. $7x = 18 - 2x$
19. $3x = 10 - 2x$
20. $11x = 7x + 20$
21. $9x + 2 = 3x + 38$
22. $8x - 3 = 4x + 17$
23. $4x - 8 = x - 14$
24. $6x - 5 = 3x - 29$
25. $7x - 3 = 9x + 5$
26. $5x - 2 = 8x - 11$

ANSWERS

1. ______
2. ______
3. ______
4. ______
5. ______
6. ______
7. ______
8. ______
9. ______
10. ______
11. ______
12. ______
13. ______
14. ______
15. ______
16. ______
17. ______
18. ______
19. ______
20. ______
21. ______
22. ______
23. ______
24. ______
25. ______
26. ______

ANSWERS

27. ______
28. ______
29. ______
30. ______
31. ______
32. ______
33. ______
34. ______
35. ______
36. ______
37. ______
38. ______
39. ______
40. ______
41. ______
42. ______
43. ______
44. ______
45. ______
46. ______
47. ______
48. ______

27. $5x + 4 = 7x - 8$

28. $2x + 23 = 6x - 5$

29. $2x - 3 + 5x = 7 + 4x + 2$

30. $8x - 7 - 2x = 2 + 4x - 5$

31. $6x + 7 - 4x = 8 + 7x - 26$

32. $7x - 2 - 3x = 5 + 8x + 13$

33. $9x - 2 + 7x + 13 = 10x - 13$

34. $5x + 3 + 6x - 11 = 8x + 25$

35. $7(2x - 1) - 5x = x + 25$

36. $9(3x + 2) - 10x = 12x - 7$

37. $4(x + 5) = 4x + 20$

38. $-3(2x - 4) - 12 = -6x$

39. $5(x + 1) - 4x = x - 5$

40. $-4(2x - 3) = -8x + 5$

41. $6x - 4x + 1 = 12 + 2x - 11$

42. $-2x + 5x - 9 = 3(x - 4) - 5$

43. $-4(x + 2) - 11 = 2(-2x - 3) - 13$

44. $4(-x - 2) + 5 = -2(2x + 7)$

45. Create an equation of the form $ax + b = c$ that has 2 as a solution.

46. Create an equation of the form $ax + b = c$ that has 7 as a solution.

47. The equation $3x = 3x + 5$ has no solution, whereas the equation $7x + 8 = 8$ has zero as a solution. Explain the difference between a solution of zero and no solution.

48. Construct an equation for which every real number is a solution.

Answers

1. 4 **3.** 3 **5.** 7 **7.** -2 **9.** -2 **11.** 3 **13.** 8 **15.** 32 **17.** 3 **19.** 2 **21.** 6 **23.** -2 **25.** -4 **27.** 6 **29.** 4 **31.** 5 **33.** -4 **35.** 4 **37.** Identity **39.** No solution **41.** Identity **43.** Identity **45.**

47.

3 Summary

DEFINITION/PROCEDURE	EXAMPLE	REFERENCE
Introduction to Fractions		**Section 3.1**
Fraction Fractions name a number of equal parts of a unit or whole. A fraction is written in the form $\frac{a}{b}$, in which a is an integer and b is a natural number.	$\frac{5}{8}$ is a fraction.	p. 201
Denominator The number of equal parts into which the whole is divided. **Numerator** The number of parts of the whole that are used.	$\frac{5}{8}$ (5 ← Numerator; 8 ← Denominator)	p. 201
Proper Fraction A fraction whose numerator is less than its denominator. It names a number less than 1.	$\frac{2}{3}$ and $\frac{11}{15}$ are proper fractions.	p. 203
Improper Fraction A fraction whose numerator is greater than or equal to its denominator It names a number greater than or equal to 1.	$\frac{7}{5}$, $\frac{21}{20}$, and $\frac{8}{8}$ are improper fractions.	p. 203
Prime Numbers and Factorization		**Section 3.2**
Prime Number Any whole number that has exactly two factors, 1 and itself.	7, 13, 29, and 73 are prime numbers.	p. 208
Composite Number Any whole number greater than 1 that is not prime.	8, 15, 42, and 65 are composite numbers.	p. 209
Prime Factorization To find the prime factorization of a number, divide the number by a series of primes until the final quotient is a prime number. The prime factors include each prime divisor and the final quotient.	$630 = 2 \cdot 3 \cdot 3 \cdot 5 \cdot 7.$	p. 210
Greatest Common Factor (GCF) The GCF is the *largest* number that is a factor of each of a group of numbers.		p. 212
To Find the GCF **Step 1** Write the prime factorization for each of the numbers in the group. **Step 2** Locate the prime factors that are common to all the numbers. **Step 3** The greatest common factor (GCF) will be the product of all the common prime factors. If there are no common prime factors, the GCF is 1.	To find the GCF of 24, 30, and 36: $24 = ⓶ \cdot 2 \cdot 2 \cdot ⓷$ $30 = ⓶ \cdot ⓷ \cdot 5$ $36 = ⓶ \cdot 2 \cdot ⓷ \cdot 3$ The GCF is $2 \cdot 3 = 6$	p. 213

Continued

DEFINITION/PROCEDURE	EXAMPLE	REFERENCE
Equivalent Fractions		**Section 3.3**
The Fundamental Principle of Fractions For the fraction $\frac{a}{b}$, and any nonzero number c, $$\frac{a}{b} = \frac{a \div c}{b \div c}$$ **In words:** We can divide the numerator and denominator of a fraction by the same nonzero number. The result will be an equivalent fraction.	$\frac{8}{12} = \frac{8 \div 4}{12 \div 4} = \frac{2}{3}$ $\frac{8}{12}$ and $\frac{2}{3}$ are equivalent fractions.	p. 220
Equivalent Fractions The fundamental principle can also be written as $$\frac{a}{b} = \frac{a \cdot c}{b \cdot c} \qquad c \neq 0$$ This is used to find equivalent fractions.	$\frac{2}{3} = \frac{2 \cdot 5}{3 \cdot 5} = \frac{10}{15}$	p. 224
Multiplication and Division of Fractions		**Section 3.4**
To Multiply Two Fractions **1.** Multiply numerator by numerator. This gives the numerator of the product. **2.** Multiply denominator by denominator. This gives the denominator of the product. **3.** Simplify the resulting fraction if possible. In multiplying fractions it is usually easiest to divide by any common factors in the numerator and denominator *before* multiplying.	$\frac{5}{8} \cdot \frac{3}{7} = \frac{5 \cdot 3}{8 \cdot 7} = \frac{15}{56}$ $\frac{5}{9} \cdot \frac{3}{10} = \frac{\overset{1}{\cancel{5}} \cdot \overset{1}{\cancel{3}}}{\underset{3}{\cancel{9}} \cdot \underset{2}{\cancel{10}}} = \frac{1}{6}$	p. 233
To Divide Two Fractions Replace the divisor by its reciprocal and multiply.	$\frac{3}{7} \div \frac{4}{5} = \frac{3}{7} \cdot \frac{5}{4} = \frac{15}{28}$	p. 238
The Multiplication Property of Equality		**Section 3.5**
The Multiplication Property of Equality If $a = b$ then $a \cdot c = b \cdot c$	If $\frac{1}{2}x = 7$ then $2\left(\frac{1}{2}x\right) = 2(7)$	p. 246
Linear Equations in One Variable		**Section 3.6**
Solving Linear Equations The steps of solving a linear equation are as follows: **1.** Use the distributive property to remove any grouping symbols. Then simplify by combining like terms. **2.** Add or subtract the same term on both sides of the equation until the variable term is on one side and a number is on the other. **3.** Multiply or divide both sides of the equation by the same nonzero number so that the variable is alone on one side of the equation. **4.** Check the solution in the original equation.	Solve: $3(x - 2) + 4x = 3x + 14$ $3x - 6 + 4x = 3x + 14$ $7x - 6 = 3x + 14$ $+6 \qquad +6$ $7x = 3x + 20$ $-3x \qquad -3x$ $4x = 20$ $\frac{4x}{4} = \frac{20}{4}$ $x = 5$	p. 258

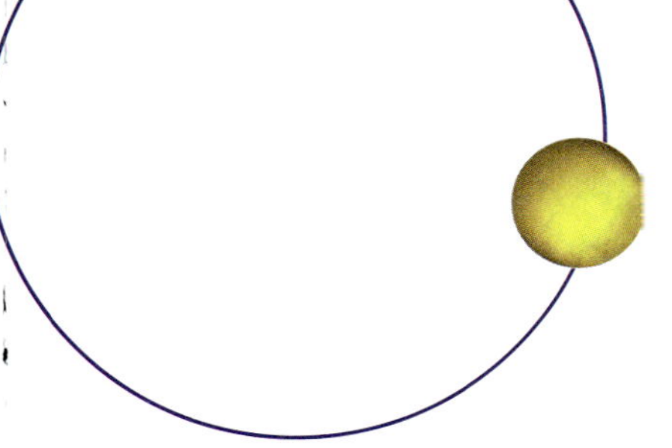

Summary and Review Exercises

This exercise set will give you practice with each of the objectives of the chapter. Each exercise is keyed to the appropriate chapter section. The answers are provided in the *Instructor's Manual.* Your instructor will give you guidelines on how to best use these exercises in your instructional setting.

[3.1] Identify the numerator and denominator of each fraction.

1. $\frac{5}{9}$

2. $\frac{17}{23}$

Give the fractions that name the shaded portions of the diagrams. Identify the numerator and the denominator.

3. 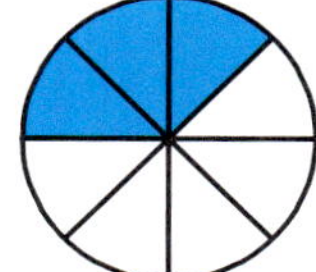

Fraction ____________

Numerator ____________

Denominator ____________

4. 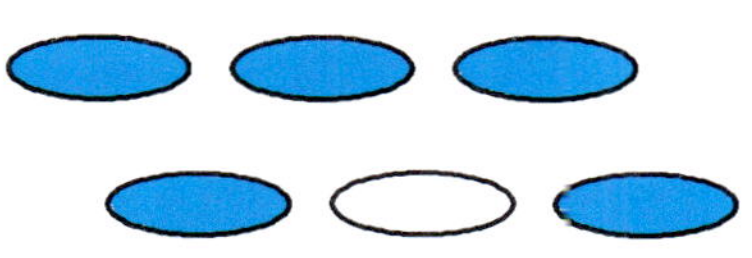

Fraction: ____________

Numerator: ____________

Denominator ____________

5. From the group of numbers:

$\frac{2}{3}, \frac{5}{4}, \frac{45}{8}, \frac{7}{7}, \frac{9}{1}, \frac{7}{10}, \frac{12}{5}$

List the proper fractions. ____________

List the improper fractions ____________

[3.2] In exercises 6 and 7, list all the factors of the given numbers.

6. 52

7. 41

In exercise 8, use the group of numbers 2, 5, 7, 11, 14, 17, 21, 23, 27, 39, and 43.

8. List the prime numbers; then list the composite numbers.

In exercises 9 and 10, determine which, if any, of the numbers 2, 3, and 5 are factors of the given numbers.

9. 2350

10. 33,451

In exercises 11 to 14, find the prime factorization for the given numbers.

11. 48

12. 420

13. 2640

14. 2250

In exercises 15 to 20, find the greatest common factor (GCF).

15. 15 and 20

16. 30 and 31

17. 24 and 40

18. 39 and 65

19. 49, 84, and 119

20. 77, 121, and 253

[3.3] Determine whether each of the pairs of fractions are equivalent.

21. $\frac{5}{8}, \frac{7}{12}$

22. $\frac{8}{15}, \frac{32}{60}$

Write each fraction in simplest form.

23. $\frac{24}{36}$

24. $\frac{45}{75}$

25. $\frac{140}{180}$

26. $\frac{16}{21}$

Find the missing numerators.

27. $\frac{15}{25} = \frac{?}{5}$

28. $\frac{32}{40} = \frac{?}{5}$

29. $\frac{2}{7} = \frac{?}{28}$

30. $\frac{4}{5} = \frac{?}{30}$

[3.4] Multiply.

31. $\frac{7}{15} \cdot \frac{-5}{21}$

32. $\frac{-10}{27} \cdot \frac{-9}{20}$

33. $4 \cdot \frac{3}{8}$

Divide.

34. $\frac{5}{12} \div \frac{5}{8}$

35. $\frac{-7}{15} \div \frac{14}{25}$

[3.5]–[3.6] Solve each equation.

36. $7x = 147$

37. $5x = 35$

38. $7x = -28$

39. $-6x = 24$

40. $-9x = -63$

41. $\frac{1}{4}x = 8$

42. $\frac{-1}{5}x = -3$

43. $5x - 3 = 12$

44. $4x + 3 = -13$

45. $7x + 8 = 3x$

46. $3 - 5x = -17$

47. $\frac{1}{3}x - 5 = 1$

48. $\frac{1}{4}x + 4 = 7$

49. $6x - 5 = 3x + 13$

50. $3x + 7 = x - 9$

51. $2x + 7 = 4x - 5$

52. $3x - 2 + 5x = 7 + 2x + 21$

53. $x - 5 - 6x = 3(9 + x)$

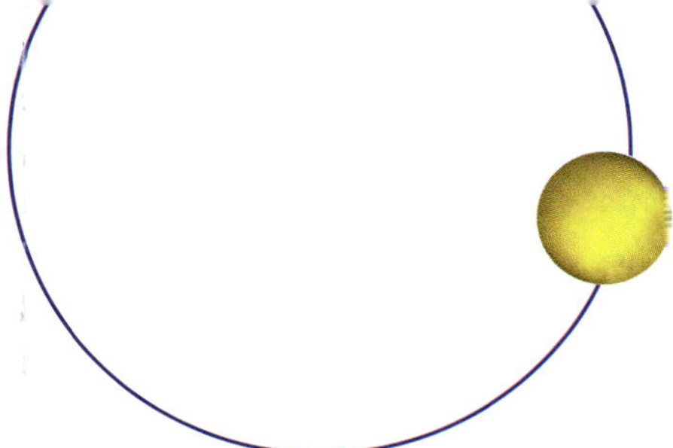

Chapter Test for Chapter 3

Name ____________

Section ______ Date ______

ANSWERS

1. ____________
2. ____________
3. ____________
4. ____________
5. ____________
6. ____________
7. ____________
8. ____________
9. ____________
10. ____________
11. ____________
12. ____________
13. ____________
14. ____________
15. ____________
16. ____________
17. ____________
18. ____________

The purpose of the Chapter Test is to help you check your progress and review for a chapter test in class. When you are done, check your answers in the back of the book. If you missed any answers, be sure to go back and review the appropriate sections in the chapter.

For exercises 1 to 3, what fraction names the shaded part of each diagram? Identify the numerator and denominator.

1. 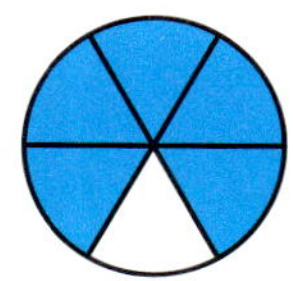

2. 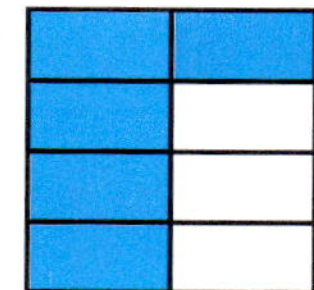

3. 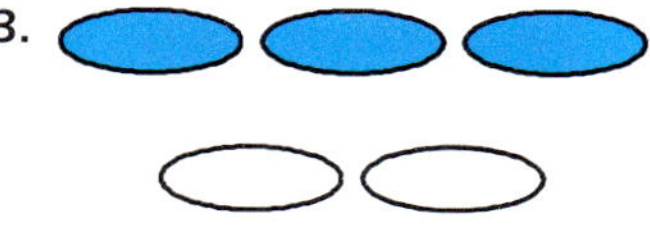

4. Identify the proper fractions and improper fractions in the following group.

$\frac{10}{11}, \frac{9}{5}, \frac{7}{7}, \frac{8}{1}, \frac{1}{8}$

Proper Improper

5. Which of the numbers 5, 9, 13, 17, 22, 27, 31, and 45 are prime numbers? Which are composite numbers?

6. Determine which, if any, of the numbers 2, 3, and 5 are factors of 54,204.

7. Find the prime factorization for 264.

In exercises 8 and 9, find the greatest common factor (GCF) for the given numbers.

8. 36 and 84

9. 16, 24, and 72

In exercises 10 to 12, use the cross-product method to determine whether the pair of fractions is equivalent.

10. $\frac{2}{7}, \frac{8}{28}$

11. $\frac{8}{20}, \frac{12}{30}$

12. $\frac{3}{20}, \frac{2}{15}$

In exercises 13 to 15, write the fractions in simplest form.

13. $\frac{21}{27}$

14. $\frac{36}{84}$

15. $\frac{8}{23}$

In exercises 16 to 18, find the missing numerators.

16. $\frac{28}{35} = \frac{?}{5}$

17. $\frac{42}{98} = \frac{?}{7}$

18. $\frac{105}{120} = \frac{?}{8}$

ANSWERS

19. ______
20. ______
21. ______
22. ______
23. ______
24. ______
25. ______
26. ______
27. ______
28. ______
29. ______
30. ______
31. ______
32. ______
33. ______
34. ______
35. ______

In exercises 19 and 20, multiply.

19. $\frac{9}{10} \cdot \frac{5}{8}$

20. $\frac{16}{35} \cdot \frac{-14}{24}$

In exercises 21 to 25, divide.

21. $\frac{7}{12} \div \frac{14}{15}$

22. $\left(\frac{-2}{5}\right) \div \frac{8}{3}$

23. $\left(\frac{-16}{15}\right) \div \left(\frac{-8}{3}\right)$

24. $\left(\frac{-7}{3}\right) \cdot \left(\frac{-3}{7}\right)$

25. $\left(\frac{-7}{3}\right) \div \left(\frac{-3}{7}\right)$

Solve the equations and check your results.

26. $7x = 49$

27. $\frac{1}{4}x = -3$

28. $\frac{-1}{5}x = -5$

29. $7x - 5 = 16$

30. $10 - 3x = -2$

31. $7x + 10 = 4x - 5$

32. $2x - 7 = 5x + 8$

33. $\frac{x}{3} - 5 = 2$

34. $\frac{2x}{5} + 1 = \frac{6}{5}$

35. $\frac{2x}{3} + 5 = \frac{1}{4}$

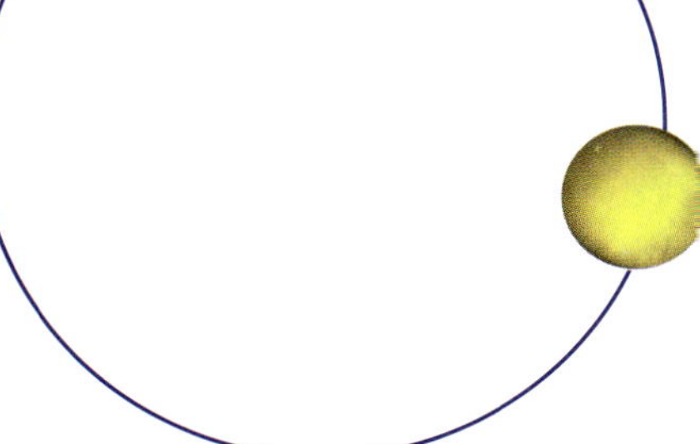

Cumulative Test for Chapters 1 to 3

Name ______________

Section ______ Date ______

ANSWERS

1. ______
2. ______
3. ______
4. ______
5. ______
6. ______
7. ______
8. ______
9. ______
10. ______
11. ______
12. ______
13. ______
14. ______
15. ______
16. ______
17. ______
18. ______
19. ______
20. ______
21. ______
22. ______
23. ______
24. ______

In exercises 1 to 15, perform the indicated operations.

1. $\begin{array}{r} 1369 \\ +\ 5804 \\ \hline \end{array}$

2. $\begin{array}{r} 489 \\ 562 \\ 613 \\ +\ 254 \\ \hline \end{array}$

3. $\begin{array}{r} 357 \\ 28 \\ +\ 2346 \\ \hline \end{array}$

4. $\begin{array}{r} 13 \\ 2543 \\ +\ 10{,}547 \\ \hline \end{array}$

5. $-289 + 54$

6. $53{,}294 - 41{,}074$

7. $503 - 74$

8. $5731 - 2492$

9. $58 \cdot (-3)$

10. Find the product of -273 and -7.

11. $\begin{array}{r} 89 \\ \times\ 56 \\ \hline \end{array}$

12. $\begin{array}{r} 538 \\ \times\ 103 \\ \hline \end{array}$

13. $281\overline{)6935}$

14. $571\overline{)12{,}583}$

15. $293\overline{)61{,}382}$

In exercises 16 to 21, evaluate each of the expressions.

16. $12 \div 6 + 3$

17. $4 + 12 \div 4$

18. $3^3 \div 9$

19. $28 \div 7 \cdot 4$

20. $5 \cdot 8 \div 2$

21. $36 \div (3^2 + 3)$

In exercise 22, identify the proper fractions and improper fractions from the group.

$\frac{5}{7}, \frac{15}{9}, \frac{8}{8}, \frac{11}{1}, \frac{2}{5}$

22. Proper: ___ Improper: _______

In exercises 23 and 24, find out whether the pair of fractions is equivalent.

23. $\frac{8}{32}, \frac{9}{36}$

24. $\frac{6}{11}, \frac{7}{9}$

ANSWERS

25. ____________________
26. ____________________
27. ____________________
28. ____________________
29. ____________________
30. ____________________
31. ____________________
32. ____________________
33. ____________________
34. ____________________
35. ____________________
36. ____________________
37. ____________________
38. ____________________
39. ____________________
40. ____________________

Perform the indicated operations.

25. $\frac{7}{15} \cdot \frac{5}{21}$

26. $\frac{-10}{27} \cdot \frac{-9}{20}$

27. $-4 \cdot \frac{3}{8}$

28. $3 \cdot \frac{-5}{8}$

29. $\frac{-5}{6} \div \frac{-1}{3}$

30. $\frac{12}{13} \div (-4)$

Simplify each expression.

31. $3x - 5x + 7x$

32. $4(x + 2) - 5(x - 3)$

33. $7 - 2(x + 3)$

34. $2x - 5y - 4x + 8y$

35. $3(x - y) + 5(2x + 3y)$

Solve the equations and check your results.

36. $9x - 5 = 8x$

37. $\frac{-3}{4}x = 18$

38. $6x - 8 = 2x - 3$

39. $2x + 3 = 7x + 5$

40. $\frac{4}{3}x - 6 = 4 - \frac{2}{3}x$

4 APPLICATIONS OF FRACTIONS AND EQUATIONS

INTRODUCTION

When Charity applied for the job at the Gnu Deli, she was surprised when they asked her how comfortable she was in dealing with fractions. She assumed her employer was talking about reading the scale when weighing meats and cheeses. Once she started the job, she discovered why fraction skills were important.

At the Gnu Deli, they sell a lot of chunks of cheese. Customers usually order $\frac{1}{2}$ or $\frac{1}{4}$ lb of one of the cheeses. Unfortunately, the cheeses are of very different densities, so $\frac{1}{2}$ lb of Parmesan is a very different size from $\frac{1}{2}$ lb of Gruyère. To closely approximate the size of the chunk to be cut off, Charity learned a neat trick that has really sharpened her fraction skills. Assume that she is asked to cut off a $\frac{1}{2}$-lb chunk of Havarti cheese. She first weighs the piece of cheese in the deli case. Then she figures out what fraction of that weight $\frac{1}{2}$ lb would be. This tells her how much of the cheese to cut off.

When she started the job, she had to have a pencil and paper handy every time somebody ordered a chunk of cheese. Now that she has been practicing this skill for a couple of months, she is amazed at how quickly she can do the math in her head!

Name ______________________

Section ________ Date ________

ANSWERS

1. ______________________

2. ______________________

3. ______________________

4. ______________________

5. ______________________

6. ______________________

7. ______________________

8. ______________________

9. ______________________

10. ______________________

11. ______________________

12. ______________________

13. ______________________

14. ______________________

15. ______________________

16. ______________________

Pre-Test Chapter 4

This pre-test will point out any difficulties you may be having with applications of fractions and equations. Do all the problems. Then check your answers with those in the back of the book.

1. **(a)** $\frac{3}{7} + \frac{2}{7}$ **(b)** $\frac{8}{9} - \frac{3}{9}$

2. Find the least common denominator for fractions with the denominators 16 and 24.

3. Find the least common multiple of each group of numbers.

(a) 25 and 40 **(b)** 20, 24, and 30

Add or subtract. Simplify when possible.

4. $\frac{4}{5} + \frac{7}{8} + \frac{9}{20}$

5. $\frac{7}{24} + \frac{-2}{9}$

6. Convert $\frac{37}{4}$ to a mixed number.

7. Convert $6\frac{4}{7}$ to an improper fraction.

Carry out the indicated operation.

8. $2\frac{2}{5} \cdot 1\frac{3}{4}$

9. $\frac{2}{3} \div 7$

10. $3\frac{5}{6} + 2\frac{7}{8}$

11. $2\frac{1}{9} - 1\frac{1}{12}$

12. Simplify. $\frac{\frac{3}{8}}{\frac{5}{6}}$

13. A piece of wood that is $15\frac{3}{4}$ in. long is to be cut into blocks $1\frac{1}{8}$ in. long. How many blocks can be cut?

14. A house plan calls for $12\frac{3}{4}$ yd^2 of carpeting in the living room and $5\frac{1}{2}$ yd^2 in the hallway. How much carpeting will be needed?

15. Solve for x. $\frac{x}{4} - \frac{x}{5} = 2$

16. If two-fifths of a number is added to one-half of that number, the sum is 27. Find the number.

Addition and Subtraction of Fractions

OBJECTIVES

1. Add and subtract fractions with like denominators
2. Find a least common multiple (LCM)
3. Order fractions with unlike denominators
4. Add and subtract fractions with unlike denominators

Recall from our work in Chapter 1 that adding can be thought of as combining groups of the *same kind* of objects. This is also true when you think about adding fractions.

Fractions can be added only if they name a number of the *same parts* of a whole. This means that we can add fractions only when they are **like fractions,** that is, when they have the *same* (*a common*) denominator.

As long as we are dealing with like fractions, addition is an easy matter. Just use this rule:

NOTE For instance, we can add two nickels and three nickels to get five nickels. We *cannot* directly add two nickels and three dimes!

Step by Step: To Add Like Fractions

Step 1 Add the numerators.
Step 2 Place the sum over the common denominator.
Step 3 Simplify the resulting fraction when necessary.

Example 1 illustrates the use of this rule.

Example 1

Adding Like Fractions

Add.

$$\frac{1}{5} + \frac{3}{5}$$

Step 1 Add the numerators.

$1 + 3 = 4$

Step 2 Write that sum over the common denominator, 5. We are done at this point because the answer, $\frac{4}{5}$, is in the simplest possible form.

$$\frac{1}{5} + \frac{3}{5} = \underbrace{\frac{1+3}{5}}_{\text{Step 1}} = \underbrace{\frac{4}{5}}_{\text{Step 2}}$$

We will illustrate with a diagram.

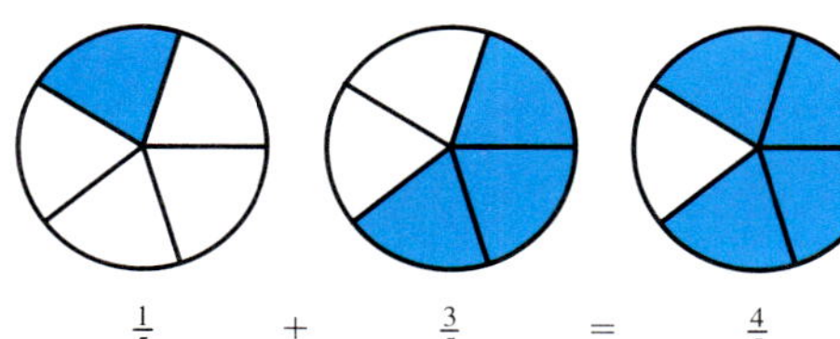

NOTE Combining 1 of the 5 parts with 3 of the 5 parts gives a total of 4 of the 5 equal parts.

CHECK YOURSELF 1

Add.

$$\frac{2}{9} + \frac{5}{9}$$

CAUTION

Be Careful! In adding fractions, *do not* follow the rule for multiplying fractions. To multiply $\frac{1}{5} \cdot \frac{3}{5}$, you would multiply both the numerators and the denominators:

$$\frac{1}{5} \cdot \frac{3}{5} = \frac{3}{25}$$

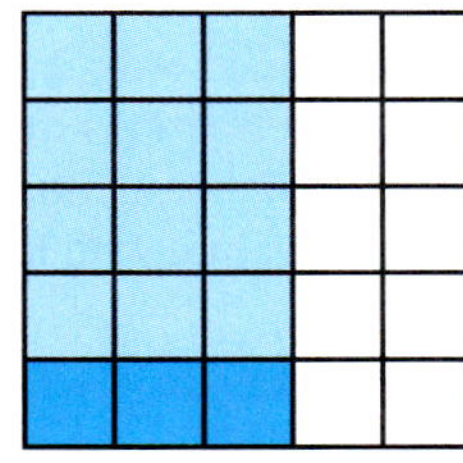

$\frac{1}{5}$ of $\frac{3}{5} = \frac{3}{25}$

But to add fractions, you must have a common denominator that becomes the denominator of the sum.

$$\frac{1}{5} + \frac{3}{5} \neq \frac{4}{10}$$

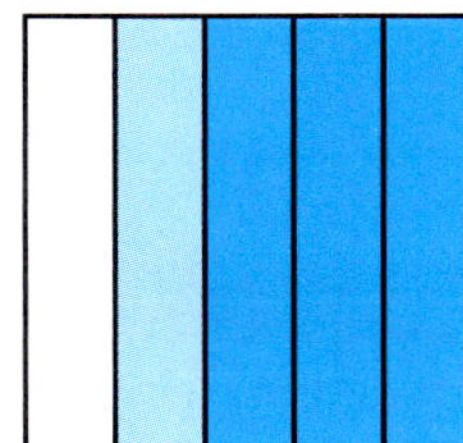

$\frac{1}{5} + \frac{3}{5} = \frac{4}{5}$

Step 3 of the addition rule for like fractions tells us to *simplify* the sum. Sums of fractions are usually written in lowest terms. Consider Example 2.

Example 2

Adding Like Fractions That Require Simplifying

Add and simplify.

Step 3

$$\frac{3}{12} + \frac{5}{12} = \frac{8}{12} = \frac{2}{3}$$

The sum $\frac{8}{12}$ is *not* in lowest terms.
Divide the numerator and denominator by 4 to simplify the result.

CHECK YOURSELF 2

Add.

$$\frac{4}{15} + \frac{6}{15}$$

We can also easily extend our addition rule to find the sum of more than two fractions as long as they all have the same denominator. This is shown in Example 3.

Example 3

Adding a Group of Like Fractions

Add.

$$\frac{2}{7} + \frac{3}{7} + \frac{6}{7} = \frac{11}{7}$$ Add the numerators: $2 + 3 + 6 = 11$.

CHECK YOURSELF 3

Add.

$$\frac{1}{8} + \frac{3}{8} + \frac{5}{8}$$

NOTE Like fractions have the same denominator.

If a problem involves like fractions, then subtraction, like addition, is not difficult.

NOTE Note the similarity to our rule for adding like fractions.

Step by Step: To Subtract Like Fractions

Step 1 Subtract the numerators.
Step 2 Place the difference over the common denominator.
Step 3 Simplify the resulting fraction when necessary.

Example 4

Subtracting Like Fractions

Subtract.

(a) $\frac{4}{5} - \frac{2}{5} = \frac{4 - 2}{5} = \frac{2}{5}$ (Step 1: $\frac{4-2}{5}$; Step 2: $\frac{2}{5}$)

Subtract the numerators: $4 - 2 = 2$. Write the difference over the common denominator, 5. Step 3 is not necessary because the difference is in simplest form.

Illustrating with a diagram:

NOTE Subtracting 2 of the 5 parts from 4 of the 5 parts leaves 2 of the 5 parts.

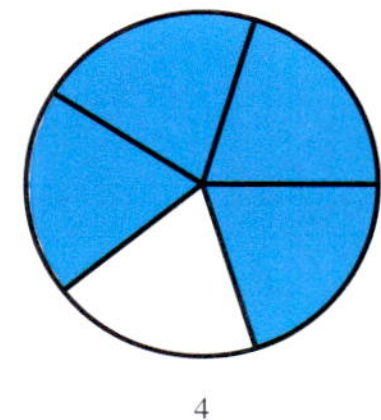

$\frac{4}{5}$ −

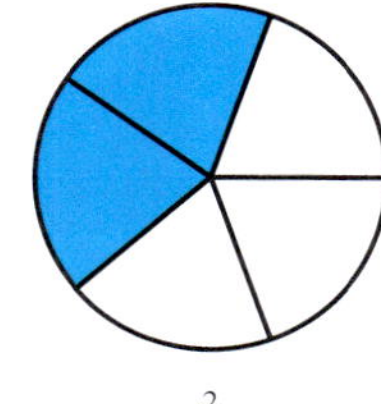

$\frac{2}{5}$ =

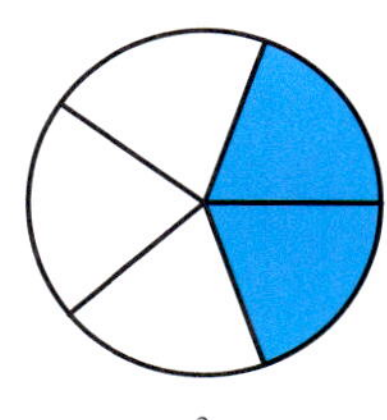

$\frac{2}{5}$

NOTE Always write the result in lowest terms.

(b) $\frac{5}{8} - \frac{3}{8} = \frac{5 - 3}{8} = \frac{2}{8} = \frac{1}{4}$

CHECK YOURSELF 4

Subtract.

$$\frac{11}{12} - \frac{5}{12}$$

As was the case with integers, we should determine the sign of the result first when adding signed fractions.

Example 5

Adding Two Signed Fractions

Find the sum.

(a) $\frac{-3}{5} + \frac{2}{5}$

NOTE You may think of the magnitude of a number as its distance from 0 on the number line.

Because the signs are opposite and $\frac{-3}{5}$ has greater magnitude than $\frac{2}{5}$, the sum will be negative.

$$\frac{-3}{5} + \frac{2}{5} = \frac{-1}{5}$$

(b) $\frac{-3}{5} + \frac{-2}{5}$

The signs are the same (both are negative), so we know that the sum will be negative.

$$\frac{-3}{5} + \frac{-2}{5} = \frac{-5}{5} = -1$$

CHECK YOURSELF 5

Find the sum.

(a) $\frac{-4}{9} + \frac{7}{9}$

(b) $\frac{-6}{35} + \frac{-12}{35}$

Recall that, when subtracting integers, we changed the exercise to an equivalent addition exercise. We will do the same thing with fractions when negative numbers are involved.

Example 6

Subtracting Two Signed Fractions

Find the difference.

(a) $\frac{-3}{7} - \frac{2}{7}$

First, we change the operation to addition, also replacing $\frac{2}{7}$ with its opposite $\left(\frac{-2}{7}\right)$.

$$\frac{-3}{7} - \frac{2}{7} = \frac{-3}{7} + \frac{-2}{7} = \frac{-5}{7}$$

(b) $\frac{-3}{7} - \frac{-4}{7}$

Again, we change to an addition exercise.

(c) $\frac{-3}{7} - \frac{-4}{7} = \frac{-3}{7} + \frac{4}{7}$

We are now finding the sum of two numbers with opposite signs. First we find the difference of $\frac{4}{7}$ and $\frac{3}{7}$. Because $\frac{4}{7}$ has greater magnitude than $\frac{-3}{7}$, the result is positive.

$$\frac{-3}{7} + \frac{4}{7} = \frac{1}{7}$$

CHECK YOURSELF 6

Find the difference.

(a) $\frac{-4}{11} - \frac{7}{11}$

(b) $\frac{-6}{7} - \frac{-12}{7}$

In this section, we are discussing the process used for adding or subtracting two fractions. One of the most important concepts we use in the addition and subtraction of fractions is that of **multiples.**

Definitions: Multiples

The *multiples* of a number are the product of that number with the natural numbers 1, 2, 3, 4, 5, ...

Example 7

Listing Multiples

List the multiples of 3.

The multiples of 3 are

$3 \cdot 1, 3 \cdot 2, 3 \cdot 3, 3 \cdot 4, \ldots$

or

$3, 6, 9, 12, \ldots$ The three dots indicate that the list continues indefinitely.

NOTE Notice that the multiples, except for 3 itself, are *larger* than 3.

An easy way of listing the multiples of 3 is to think of *counting by threes.*

CHECK YOURSELF 7

List the first seven multiples of 4.

Sometimes we need to find common multiples of two or more numbers.

Definitions: Common Multiples

If a number is a multiple of each of a group of numbers, it is called a *common multiple* of the numbers; that is, it is a number that is evenly divisible by all the numbers in the group.

Example 8

Finding Common Multiples

NOTE 15, 30, 45, and 60 are multiples of *both* 3 and 5.

Find four common multiples of 3 and 5.

Some multiples of 3 are

3, 6, 9, 12, 15, 18, 21, 24, 27, 30, 33, 36, 39, 42, 45

Some multiples of 5 are

5, 10, 15, 20, 25, 30, 35, 40, 45, 50, 55, 60

So, common multiples of 3 and 5 include

15, 30, 45

CHECK YOURSELF 8

List the first six multiples of 6. *Then look at your list from Check Yourself* 7 *and list some common multiples of* 4 *and* 6.

For our later work, we will use the *least common multiple* of a group of numbers.

Definitions: Least Common Multiple

The **least common multiple** (LCM) of a group of numbers is the *smallest* number that is a multiple of each number in the group.

It is possible to simply list the multiples of each number and then find the LCM by inspection.

Example 9

Finding the Least Common Multiple

Find the least common multiple of 6 and 8.

Multiples

6: 6, 12, 18, (24), 30, 36, 42, 48, . . .

8: 8, 16, (24), 32, 40, 48, . . .

NOTE 48 is also a common multiple of 6 and 8, but we are looking for the *smallest* such number.

We see that 24 is the smallest number common to both lists. So 24 is the LCM of 6 and 8.

CHECK YOURSELF 9

Find the least common multiple of 20 *and* 30 *by listing the multiples of each number.*

The technique of Example 9 will work for any group of numbers. However, it becomes tedious for larger numbers. We can outline a different approach.

Step by Step: Finding the Least Common Multiple

Step 1 Write the prime factorization for each of the numbers in the group.

Step 2 Find all the prime factors that appear in any one of the prime factorizations.

Step 3 Form the product of those prime factors, using each factor the greatest number of times it occurs in any one factorization.

NOTE For instance, if a number appears three times in the factorization of a number, it must be included at least three times in forming the least common multiple.

Some students prefer a slightly different method of lining up the factors to help in remembering the process of finding the LCM of a group of numbers.

Example 10

Finding the Least Common Multiple

To find the LCM of 10 and 18, factor:

NOTE Line up the *like* factors vertically.

$$10 = 2 \qquad \cdot 5$$

$$18 = \underline{2 \cdot 3 \cdot 3}$$

$$2 \cdot 3 \cdot 3 \cdot 5 \qquad \text{Bring down the factors.}$$

2 and 5 appear, at most, one time in any one factorization. And 3 appears two times in one factorization.

$$2 \cdot 3 \cdot 3 \cdot 5 = 90$$

So 90 is the LCM of 10 and 18.

CHECK YOURSELF 10

Use the method of Example 10 *to find the LCM of* 24 *and* 36.

The procedure is the same for a group of more than two numbers.

Example 11

Finding the Least Common Multiple

To find the LCM of 12, 18, and 20, we factor:

NOTE The different factors that appear are 2, 3, and 5.

$$12 = 2 \cdot 2 \cdot 3$$

$$18 = 2 \qquad \cdot 3 \cdot 3$$

$$20 = \underline{2 \cdot 2 \qquad\quad \cdot 5}$$

$$2 \cdot 2 \cdot 3 \cdot 3 \cdot 5$$

2 and 3 appear twice in one factorization, and 5 appears just once.

$$2 \cdot 2 \cdot 3 \cdot 3 \cdot 5 = 180$$

So 180 is the LCM of 12, 18, and 20.

CHECK YOURSELF 11

Find the LCM of 3, 4, *and* 6.

The process of finding the least common multiple is very useful when adding, subtracting, or comparing unlike fractions (fractions with different denominators).

Suppose you are asked to compare the sizes of the fractions $\frac{3}{7}$ and $\frac{4}{7}$. Because each unit in the diagram is divided into seven parts, it is easy to see that $\frac{4}{7}$ is larger than $\frac{3}{7}$.

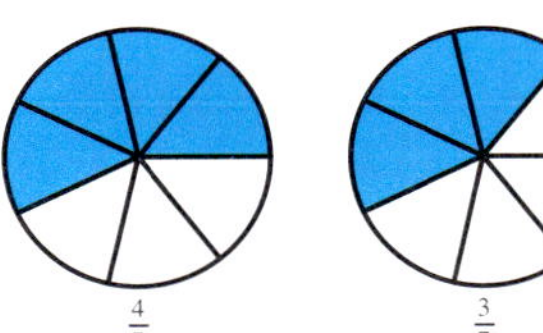

Four parts of seven are a greater portion than three parts. Now compare the size of the fractions $\frac{2}{5}$ and $\frac{3}{7}$.

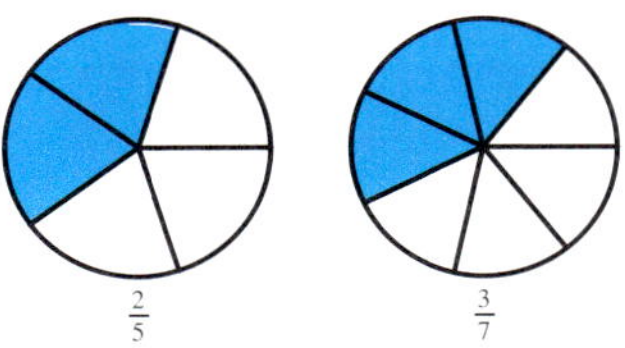

We *cannot* compare fifths with sevenths! $\frac{2}{5}$ and $\frac{3}{7}$ are *not* like fractions. Because they name different ways of dividing the whole, deciding which fraction is larger is not nearly so easy.

To compare the sizes of fractions, we change them to equivalent fractions having a *common denominator*. This common denominator must be a common multiple of the original denominators.

Example 12

Finding Common Denominators to Order Fractions

Compare the sizes of $\frac{2}{5}$ and $\frac{3}{7}$.

The original denominators are 5 and 7. Because 35 is a common multiple of 5 and 7, we will use 35 as our common denominator.

NOTE $\frac{2}{5}$ and $\frac{14}{35}$ are **equivalent fractions.** They name the same part of a whole.

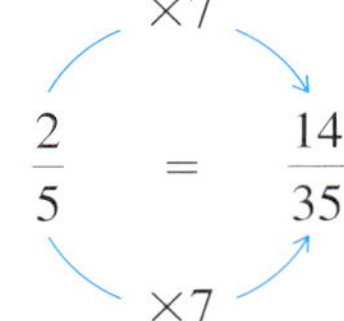

Think, "What must we multiply 5 by to get 35?" The answer is 7. Multiply the numerator and denominator by that number.

$$\frac{3}{7} = \frac{15}{35}$$ ($\times 5$ numerator and denominator)

Multiply the numerator and denominator by 5.

NOTE 15 of 35 parts represents a greater portion of the whole than 14 of 35 parts.

Because $\frac{2}{5} = \frac{14}{35}$ and $\frac{3}{7} = \frac{15}{35}$, we see that $\frac{3}{7}$ is larger than $\frac{2}{5}$.

CHECK YOURSELF 12

Which is larger, $\frac{5}{9}$ *or* $\frac{4}{7}$*?*

Now we will consider an example that uses the inequality notation.

Example 13

Using an Inequality Symbol with Two Fractions

NOTE The inequality symbol "points" to the smaller quantity.

Use the inequality symbol $<$ or $>$ to complete the statement.

$$\frac{5}{8} \underline{\qquad} \frac{3}{5}$$

Once again we must compare the sizes of the two fractions, and this is done by converting the fractions to equivalent fractions with a common denominator. Here we will use 40 as that denominator.

NOTE The LCM of 8 and 5 is 40.

$$\frac{5}{8} \;\; \frac{25}{40} \qquad \frac{3}{5} \;\; \frac{24}{40}$$ ($\times 5$ and $\times 8$ respectively)

Because $\frac{5}{8}\left(\text{or } \frac{25}{40}\right)$ is larger than $\frac{3}{5}\left(\text{or } \frac{24}{40}\right)$, we write

$$\frac{5}{8} > \frac{3}{5}$$

CHECK YOURSELF 13

Use the symbol < or > to complete the statement.

$\frac{5}{9}$ ___ $\frac{6}{11}$

Thus far, we have dealt only with finding sums and differences of like fractions (fractions with a common denominator). What about a sum that deals with **unlike fractions,** such as $\frac{1}{3} + \frac{1}{4}$?

NOTE Only *like* fractions can be added.

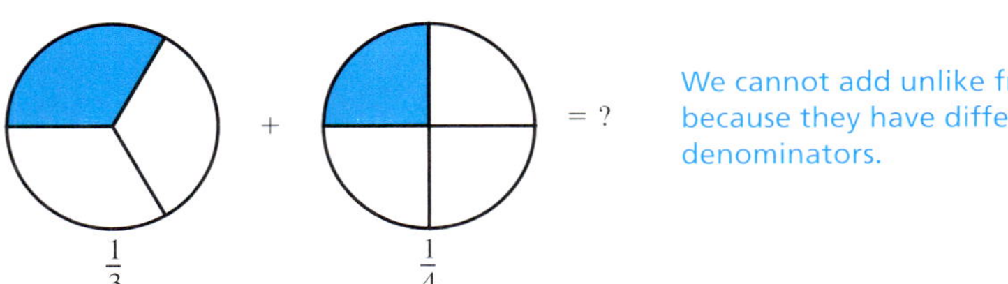

We cannot add unlike fractions because they have different denominators.

To add unlike fractions, write them as equivalent fractions with a common denominator. In this case, we will use 12 as the denominator.

NOTE We can now add because we have like fractions.

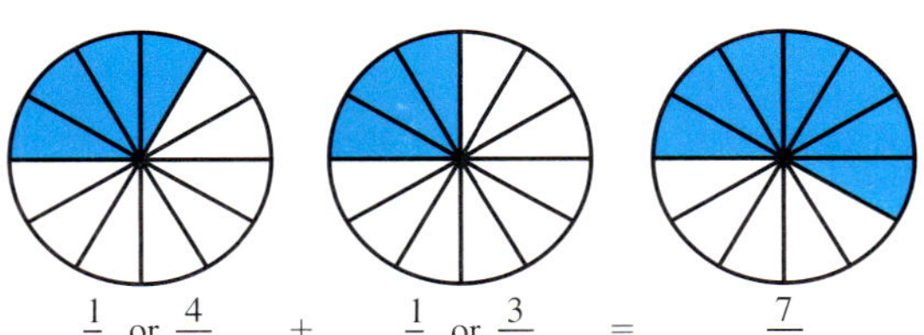

We have chosen 12 because it is a *multiple* of 3 and 4.

$\frac{1}{3}$ is equivalent to $\frac{4}{12}$.

$\frac{1}{4}$ is equivalent to $\frac{3}{12}$.

Any common multiple of the denominators will work in forming equivalent fractions. For instance, we can write $\frac{1}{3}$ as $\frac{8}{24}$ and $\frac{1}{4}$ as $\frac{6}{24}$. Our work is simplest, however, if we use the smallest possible number for the common denominator. This is called the **least common denominator (LCD).**

The LCD is the least common multiple of the denominators of the fractions. This is the *smallest* number that is a multiple of all the denominators. For example, the LCD for $\frac{1}{3}$ and $\frac{1}{4}$ is 12, *not* 24.

NOTE This is virtually identical to the Step by Step on page 277 for finding the LCM.

Step by Step: To Find the Least Common Denominator

Step 1 Write the prime factorization for each of the denominators.

Step 2 Find all the prime factors that appear in any one of the prime factorizations.

Step 3 Form the product of those prime factors, using each factor the greatest number of times it occurs in any one factorization.

We are now ready to add unlike fractions. In this case, the fractions must be renamed as equivalent fractions that have the same denominator. We will use this rule:

Step by Step: To Add Unlike Fractions

Step 1 Find the LCD of the fractions.
Step 2 Change each unlike fraction to an equivalent fraction with the LCD as a common denominator.
Step 3 Add the resulting like fractions as before.

Example 14 shows the use of this rule.

Example 14

Adding Unlike Fractions

Add the fractions $\frac{1}{6}$ and $\frac{3}{8}$.

Step 1 We find that the LCD for fractions with denominators of 6 and 8 is 24.

Step 2 Convert the fractions so that they have the denominator 24.

$$\frac{1}{6} \overset{\times 4}{=} \frac{4}{24}$$

How many sixes are in 24? There are 4. So multiply the numerator and denominator by 4.

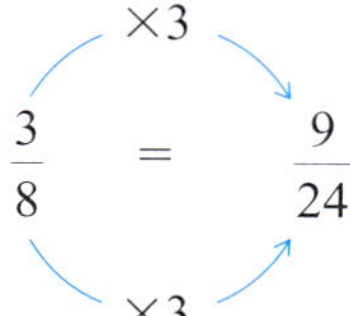

$$\frac{3}{8} \overset{\times 3}{=} \frac{9}{24}$$

How many eights are in 24? There are 3. So multiply the numerator and denominator by 3.

Step 3 We can now add the equivalent like fractions.

$$\frac{1}{6} + \frac{3}{8} = \frac{4}{24} + \frac{9}{24} = \frac{13}{24}$$

Add the numerators and place that sum over the common denominator.

CHECK YOURSELF 14

Add.

$$\frac{3}{5} + \frac{1}{3}$$

Here is a similar example. Remember that the sum should always be written in simplest form.

Example 15

Adding Unlike Fractions That Require Simplifying

Add the fractions $\frac{7}{10}$ and $\frac{2}{15}$.

Step 1 The LCD for fractions with denominators of 10 and 15 is 30.

Step 2 $\frac{7}{10} = \frac{21}{30}$

$\frac{2}{15} = \frac{4}{30}$

Do you see how the equivalent fractions are formed?

Step 3 $\frac{7}{10} + \frac{2}{15} = \frac{21}{30} + \frac{4}{30}$

$= \frac{25}{30} = \frac{5}{6}$

Add the resulting like fractions. Be sure the sum is in simplest form.

CHECK YOURSELF 15

Add.

$\frac{1}{6} + \frac{7}{12}$

We can easily add more than two fractions by using the same procedure. Example 16 illustrates this approach.

Example 16

Adding a Group of Unlike Fractions

Add $\frac{5}{6} + \frac{2}{9} + \frac{4}{15}$.

Step 1 The LCD is 90.

Step 2 $\frac{5}{6} = \frac{75}{90}$ Multiply the numerator and denominator by 15.

$\frac{2}{9} = \frac{20}{90}$ Multiply the numerator and denominator by 10.

$\frac{4}{15} = \frac{24}{90}$ Multiply the numerator and denominator by 6.

Step 3 $\frac{75}{90} + \frac{20}{90} + \frac{24}{90} = \frac{119}{90}$ Now add.

CHECK YOURSELF 16

Add.

$$\frac{2}{5} + \frac{3}{8} + \frac{7}{20}$$

To subtract unlike fractions, which are fractions that do not have the same denominator, we have this rule:

NOTE Of course, this is the same as our rule for adding fractions. We just subtract instead of add!

Step by Step: To Subtract Unlike Fractions

Step 1 Find the LCD of the fractions.
Step 2 Change each unlike fraction to an equivalent fraction with the LCD as a common denominator.
Step 3 Subtract the resulting like fractions as before.

Example 17

Subtracting Unlike Fractions

Subtract $\frac{5}{8} - \frac{1}{6}$.

Step 1 The LCD is 24.

Step 2 Convert the fractions so that they have the common denominator 24.

$$\frac{5}{8} = \frac{15}{24}$$

$$\frac{1}{6} = \frac{4}{24}$$

The first two steps are exactly the same as if we were adding.

NOTE You can use your calculator to check your result.

Step 3 Subtract the equivalent like fractions.

$$\frac{5}{8} - \frac{1}{6} = \frac{15}{24} - \frac{4}{24} = \frac{11}{24}$$

Be Careful! You *cannot* subtract the numerators and subtract the denominators.

$$\frac{5}{8} - \frac{1}{6} \quad \text{is } \textit{not} \quad \frac{4}{2}$$

CHECK YOURSELF 17

Subtract.

$$\frac{7}{10} - \frac{1}{4}$$

Using Your Calculator to Add and Subtract Fractions

Adding and subtracting fractions on the calculator is very much like the multiplication and division you did in Chapter 3. The only thing that changes is the operation.

Scientific Calculator

Here's where the fraction calculator is a great tool for checking your work. No muss, no fuss, no searching for a common denominator. Just enter the fractions and get the right answer!

Example 18

Adding Fractions

Find the sum or difference.

(a) $\frac{-3}{14} + \frac{-7}{12}$

The keystroke sequence is

3 [+/−] **[a b/c]** 14 [+] 7 [+/−] **[a b/c]** 12 [=]

The result is $\frac{-67}{84}$.

(b) $\frac{-5}{8} - \frac{-7}{18}$

5 [+/−] **[a b/c]** 8 [−] 7 [+/−] **[a b/c]** 18 [=]

The result is $\frac{-17}{72}$.

Graphing Calculator

Use your graphing calculator to find the sum.

$\frac{-7}{24} + \frac{-17}{42}$

The keystroke sequence is

[(−)] 7 [÷] 24 [+] [(−)] 17 [÷] 42 [Math] **[1: ▶ Frac]** [Enter]

The result is $\frac{-39}{56}$.

CHECK YOURSELF 18

Find the sum or difference.

(a) $\frac{-5}{24} + \frac{-11}{18}$ **(b)** $\frac{-9}{11} - \frac{-1}{3}$

CHECK YOURSELF ANSWERS

1. $\frac{2}{9} + \frac{5}{9} = \frac{2+5}{9} = \frac{7}{9}$ **2.** $\frac{2}{3}$ **3.** $\frac{9}{8}$ **4.** $\frac{1}{2}$ **5.** **(a)** $\frac{3}{9} = \frac{1}{3}$; **(b)** $\frac{-18}{35}$

6. **(a)** $\frac{-11}{11} = -1$; **(b)** $\frac{6}{7}$

7. The first seven multiples of 4 are 4, 8, 12, 16, 20, 24, and 28.

8. 6, 12, 18, 24, 30, 36; some common multiples of 4 and 6 are 12, 24, and 36.

9. The multiples of 20 are 20, 40, 60, 80, 100, 120, . . . ; the multiples of 30 are 30, 60, 90, 120, 150, . . . ; the least common multiple of 20 and 30 is 60, the smallest number common to both lists.

10. $2 \cdot 2 \cdot 2 \cdot 3 \cdot 3 = 72$ **11.** 12 **12.** $\frac{4}{7}$ is larger **13.** $\frac{5}{9} > \frac{6}{11}$

14. $\frac{14}{15}$ **15.** $\frac{1}{6} + \frac{7}{12} = \frac{2}{12} + \frac{7}{12} = \frac{9}{12} = \frac{3}{4}$ **16.** $\frac{9}{8}$ **17.** $\frac{9}{20}$

18. **(a)** $\frac{-59}{72}$; **(b)** $\frac{-16}{33}$

Name ____________

Section ______ Date ______

4.1 Exercises

Add or subtract as indicated. Write all answers in lowest terms.

1. $\frac{3}{5} + \frac{1}{5}$

2. $\frac{4}{7} + \frac{1}{7}$

3. $\frac{4}{11} + \frac{6}{11}$

4. $\frac{5}{16} + \frac{4}{16}$

5. $\frac{2}{10} + \frac{3}{10}$

6. $\frac{5}{12} + \frac{1}{12}$

7. $\frac{3}{7} + \frac{4}{7}$

8. $\frac{3}{20} + \frac{7}{20}$

9. $\frac{-9}{30} + \frac{-11}{30}$

10. $\frac{-4}{9} + \frac{-5}{9}$

11. $\frac{-13}{48} + \frac{-23}{48}$

12. $\frac{-17}{60} + \frac{-31}{60}$

13. $\frac{3}{7} + \frac{6}{7}$

14. $\frac{3}{5} + \frac{4}{5}$

15. $\frac{7}{10} - \frac{-9}{10}$

16. $\frac{5}{8} - \frac{-7}{8}$

17. $\frac{11}{12} - \frac{-10}{12}$

18. $\frac{13}{18} - \frac{-11}{18}$

19. $\frac{1}{8} + \frac{1}{8} + \frac{3}{8}$

20. $\frac{1}{10} + \frac{3}{10} + \frac{3}{10}$

21. $\frac{1}{9} + \frac{4}{9} + \frac{5}{9}$

22. $\frac{7}{12} + \frac{11}{12} + \frac{1}{12}$

23. $\frac{3}{5} + \frac{-1}{5}$

24. $\frac{5}{7} + \frac{-2}{7}$

25. $\frac{7}{9} + \frac{-4}{9}$

26. $\frac{7}{10} + \frac{-3}{10}$

27. $\frac{13}{20} - \frac{3}{20}$

28. $\frac{19}{30} - \frac{17}{30}$

ANSWERS

1. ____
2. ____
3. ____
4. ____
5. ____
6. ____
7. ____
8. ____
9. ____
10. ____
11. ____
12. ____
13. ____ 14. ____
15. ____ 16. ____
17. ____ 18. ____
19. ____ 20. ____
21. ____ 22. ____
23. ____ 24. ____
25. ____ 26. ____
27. ____ 28. ____

29. ______
30. ______
31. ______
32. ______
33. ______
34. ______
35. ______
36. ______
37. ______
38. ______
39. ______
40. ______
41. ______
42. ______
43. ______
44. ______
45. ______
46. ______
47. ______
48. ______
49. ______
50. ______
51. ______
52. ______

29. $\frac{19}{24} - \frac{5}{24}$

30. $\frac{25}{36} - \frac{13}{36}$

31. $\frac{11}{12} - \frac{7}{12}$

32. $\frac{9}{10} - \frac{6}{10}$

33. $\frac{-8}{9} - \frac{-3}{9}$

34. $\frac{-5}{8} - \frac{-1}{8}$

Evaluate. Write all answers in lowest terms.

35. $\frac{7}{12} - \frac{4}{12} + \frac{3}{12}$

36. $\frac{8}{9} + \frac{3}{9} - \frac{5}{9}$

37. $\frac{6}{13} - \frac{3}{13} + \frac{11}{13}$

38. $\frac{9}{11} - \frac{3}{11} + \frac{7}{11}$

39. $\frac{18}{23} - \frac{13}{23} - \frac{3}{23}$

40. $\frac{17}{18} - \frac{11}{18} - \frac{5}{18}$

Find the least common multiple for each group of numbers. Use whichever method you wish.

41. 2 and 3

42. 3 and 5

43. 4 and 6

44. 6 and 9

45. 10 and 20

46. 12 and 36

47. 9 and 12

48. 20 and 30

49. 18, 21, and 28

50. 8, 15, and 20

51. 20, 30, and 45

52. 12, 20, and 35

ANSWERS

Arrange the given fractions from smallest to largest.

53. $\frac{11}{12}, \frac{4}{5}, \frac{5}{6}$

54. $\frac{5}{8}, \frac{9}{16}, \frac{13}{32}$

Complete the statements, using the symbol < or >.

55. $\frac{5}{6}$ ____ $\frac{2}{5}$

56. $\frac{3}{4}$ ____ $\frac{10}{11}$

57. $\frac{5}{16}$ ____ $\frac{7}{20}$

58. $\frac{7}{12}$ ____ $\frac{9}{15}$

Find the least common denominator for fractions with the given denominators.

59. 48 and 80

60. 60 and 84

61. 3, 4, and 5

62. 3, 4, and 6

63. 8, 10, and 15

64. 6, 22, and 33

Add or subtract as indicated.

65. $\frac{2}{3} + \frac{1}{4}$

66. $\frac{-3}{5} + \frac{-1}{3}$

67. $\frac{-1}{5} + \frac{-3}{10}$

68. $\frac{1}{3} + \frac{1}{18}$

69. $\frac{5}{8} - \frac{-1}{12}$

70. $\frac{5}{12} - \frac{-3}{10}$

71. $\frac{1}{5} + \frac{7}{10} + \frac{4}{15}$

72. $\frac{2}{3} + \frac{1}{4} + \frac{3}{8}$

73. $\frac{4}{5} + \frac{-1}{3}$

74. $\frac{7}{9} + \frac{-1}{6}$

75. $\frac{11}{15} - \frac{3}{5}$

76. $\frac{5}{6} - \frac{2}{7}$

77. $\frac{-3}{8} - \frac{-1}{4}$

78. $\frac{-9}{10} - \frac{-4}{5}$

79. $\frac{15}{16} + \frac{5}{8} - \frac{1}{4}$

80. $\frac{9}{10} - \frac{1}{5} + \frac{1}{2}$

53. ____
54. ____
55. ____
56. ____
57. ____
58. ____
59. ____
60. ____
61. ____
62. ____
63. ____
64. ____
65. ____
66. ____
67. ____
68. ____
69. ____
70. ____
71. ____
72. ____
73. ____ 74. ____
75. ____ 76. ____
77. ____ 78. ____
79. ____ 80. ____

ANSWERS

81. ______

82. ______

83. ______

84. ______

85. ______

86. ______

87. ______

88. ______

89. ______

90. ______

Calculator Exercises

Find the following sums or differences using your calculator.

81. $\frac{2}{3} + \frac{-1}{2}$

82. $\frac{11}{12} + \frac{-5}{6}$

83. $\frac{3}{4} + \frac{-7}{9}$

84. $\frac{7}{11} + \frac{-5}{6}$

85. $\frac{5}{12} - \frac{-1}{6}$

86. $\frac{2}{7} - \frac{-3}{8}$

87. $\frac{15}{17} + \frac{-9}{11}$

88. $\frac{31}{42} + \frac{-18}{51}$

89. $\frac{-4}{9} - \frac{-2}{5}$

90. $\frac{-11}{13} - \frac{-2}{3}$

Answers

1. $\frac{4}{5}$ **3.** $\frac{10}{11}$ **5.** $\frac{2}{10} + \frac{3}{10} = \frac{5}{10} = \frac{1}{2}$ **7.** $\frac{3}{7} + \frac{4}{7} = \frac{7}{7} = 1$
9. $\frac{-2}{3}$ **11.** $\frac{-3}{4}$ **13.** $\frac{9}{7}$ **15.** $\frac{7}{10} + \frac{9}{10} = \frac{16}{10} = \frac{8}{5}$ **17.** $\frac{7}{4}$
19. $\frac{5}{8}$ **21.** $\frac{10}{9}$ **23.** $\frac{2}{5}$ **25.** $\frac{1}{3}$ **27.** $\frac{1}{2}$ **29.** $\frac{7}{12}$ **31.** $\frac{1}{3}$
33. $\frac{-5}{9}$ **35.** $\frac{1}{2}$ **37.** $\frac{14}{13}$ **39.** $\frac{2}{23}$ **41.** 6 **43.** 12 **45.** 20
47. 36 **49.** 252 **51.** 180 **53.** $\frac{4}{5}, \frac{5}{6}, \frac{11}{12}$ **55.** $>$ **57.** $<$
59. 240 **61.** 60 **63.** 120 **65.** $\frac{11}{12}$ **67.** $\frac{-1}{2}$ **69.** $\frac{17}{24}$
71. $\frac{1}{5} + \frac{7}{10} + \frac{4}{15} = \frac{6}{30} + \frac{21}{30} + \frac{8}{30} = \frac{35}{30} = \frac{7}{6}$
73. $\frac{4}{5} - \frac{1}{3} = \frac{12}{15} - \frac{5}{15} = \frac{7}{15}$ **75.** $\frac{2}{15}$ **77.** $\frac{-1}{8}$ **79.** $\frac{21}{16}$
81. $\frac{1}{6}$ **83.** $\frac{-1}{36}$ **85.** $\frac{7}{12}$ **87.** $\frac{12}{187}$ **89.** $\frac{-2}{45}$

4.2 Operations on Mixed Numbers

4.2 OBJECTIVES

1. Write improper fractions as mixed numbers
2. Write mixed numbers as improper fractions
3. Multiply two mixed numbers
4. Estimate products by rounding
5. Divide two mixed numbers
6. Add two mixed numbers
7. Subtract two mixed numbers

Another way to write a fraction whose absolute value is larger than 1 is called a **mixed number.**

Definitions: Mixed Number

A **mixed number** is the sum of a whole number and a proper fraction.

Example 1

Identifying a Mixed Number

NOTE $2\frac{3}{4}$ means $2 + \frac{3}{4}$. In fact, we read the mixed number as "two *and* three-fourths." The addition sign is usually not written.

$2\frac{3}{4}$ is a mixed number. It represents the sum of the whole number 2 and the fraction $\frac{3}{4}$. Look at the diagram, which represents $2\frac{3}{4}$.

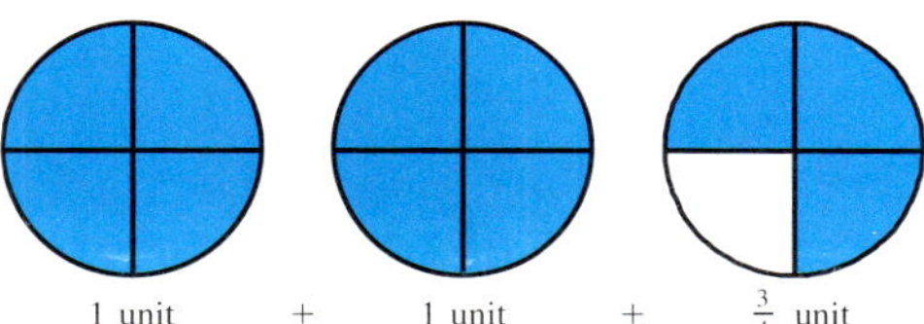

Notice that there are eleven *quarters,* and hence the improper fraction $\frac{11}{4}$.

CHECK YOURSELF 1

Give the mixed number that names the shaded portion of the given diagram.

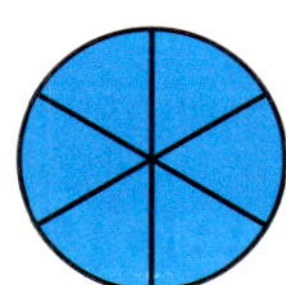
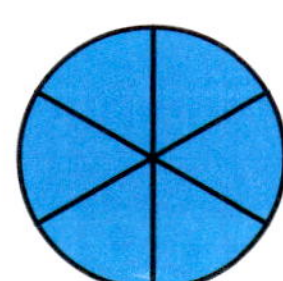
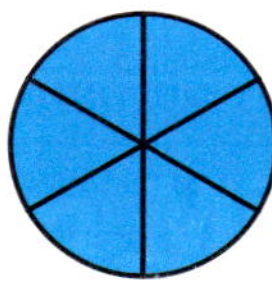
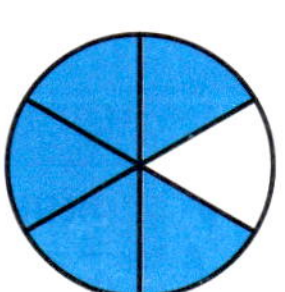

For our later work it will be important to be able to change back and forth between improper fractions and mixed numbers. Because an improper fraction represents a number with absolute value greater than or equal to 1, we have this rule:

NOTE In subsequent courses, you will find that improper fractions are preferred to mixed numbers.

Rules and Properties: Improper Fractions to Mixed Numbers

An improper fraction can always be written as either a mixed number or as a whole number.

NOTE You can write the fraction $\frac{7}{5}$ as $7 \div 5$. We divide the numerator by the denominator.

To do this, remember that you can think of a fraction as indicating division. The numerator is divided by the denominator. This leads us to the next rule:

Step by Step: To Change an Improper Fraction to a Mixed Number

Step 1 Divide the numerator by the denominator.

Step 2 If there is a remainder, write the remainder over the original denominator.

NOTE In step 1, the quotient gives the whole-number portion of the mixed number. Step 2 gives the fractional portion of the mixed number.

Example 2

Converting a Fraction to a Mixed Number

Convert $\frac{17}{5}$ to a mixed number.

Divide 17 by 5.

$$\begin{array}{r} 3 \\ 5\overline{)17} \\ \underline{15} \\ 2 \end{array} \qquad \frac{17}{5} = 3\frac{2}{5}$$

Remainder (the 2); Original denominator (the 5); Quotient (the 3)

In diagram form:

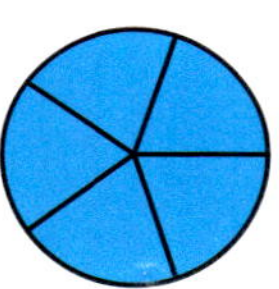
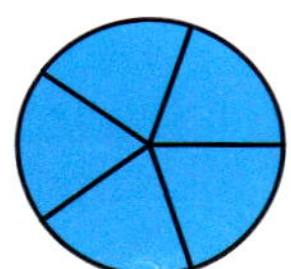
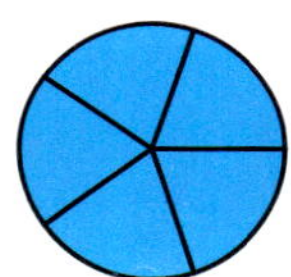
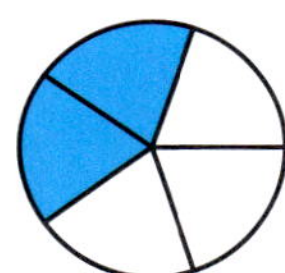

$$\frac{17}{5} = 3\frac{2}{5}$$

CHECK YOURSELF 2

Convert $\frac{32}{5}$ to a mixed number.

Example 3

Converting a Fraction to a Mixed Number

Convert $\frac{21}{7}$ to a mixed or a whole number.

Divide 21 by 7.

NOTE If there is *no* remainder, the improper fraction is equal to some whole number, in this case the number 3.

$$\begin{array}{r} 3 \\ 7\overline{)21} \\ \underline{21} \\ 0 \end{array} \qquad \frac{21}{7} = 3$$

CHECK YOURSELF 3

Convert $\frac{48}{6}$ to a mixed or a whole number.

It is also easy to convert mixed numbers to improper fractions. Just use this rule:

Step by Step: To Change a Mixed Number to an Improper Fraction

Step 1 Multiply the denominator of the fraction by the whole-number portion of the mixed number.

Step 2 Add the numerator of the fraction to that product.

Step 3 Write that sum over the original denominator to form the improper fraction.

Example 4

Converting Mixed Numbers to Improper Fractions

(a) Convert $3\frac{2}{5}$ to an improper fraction.

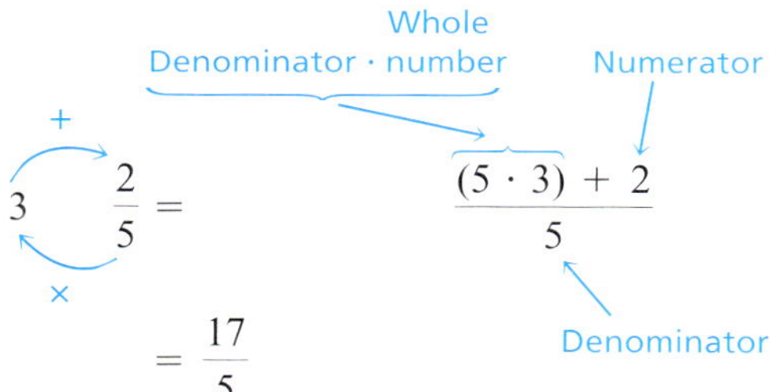

$$3\frac{2}{5} = \frac{(5 \cdot 3) + 2}{5}$$

Multiply the denominator by the whole number (5 · 3 = 15). Add the numerator. We now have 17.

$$= \frac{17}{5}$$

Write 17 over the original denominator.

In diagram form:

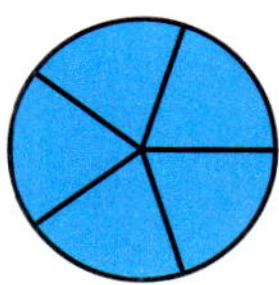 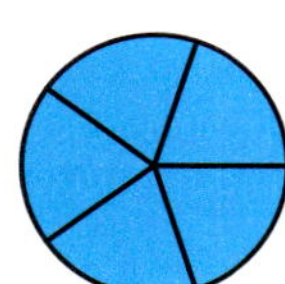 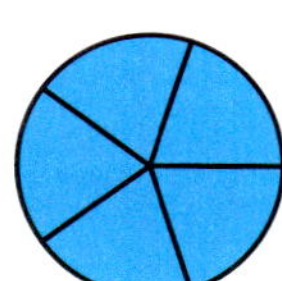 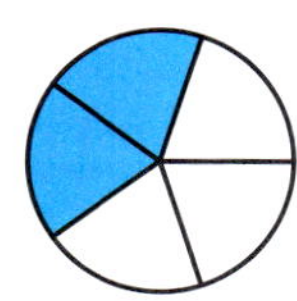

Each of the three units has 5 fifths, so the whole-number portion is 5 · 3, or 15, fifths. Then add the 2 fifths from the fractional portion for 17 fifths.

(b) Convert $4\frac{5}{7}$ to an improper fraction.

NOTE Multiply the denominator, 7, by the whole number, 4, and add the numerator, 5.

$$4\frac{5}{7} = \frac{(7 \cdot 4) + 5}{7} = \frac{33}{7}$$

CHECK YOURSELF 4

Convert $5\frac{3}{8}$ to an improper fraction.

When mixed numbers are involved in multiplication, the problem requires an additional step. First, change any mixed numbers to improper fractions. Then apply our multiplication rule for fractions.

Example 5

Multiplying a Mixed Number and a Fraction

$1\frac{1}{2} \cdot \frac{3}{4} = \frac{3}{2} \cdot \frac{3}{4}$ Change the mixed number to an improper fraction.

Here $1\frac{1}{2} = \frac{3}{2}$.

$= \frac{3 \cdot 3}{2 \cdot 4}$ Multiply as before.

$= \frac{9}{8} = 1\frac{1}{8}$ When the original problem includes mixed numbers, the product is usually written in mixed-number form.

CHECK YOURSELF 5

Multiply.

$\frac{5}{8} \cdot 3\frac{1}{2}$

If two mixed numbers are involved, change both of the mixed numbers to improper fractions. Example 6 illustrates this.

Example 6

Multiplying Two Mixed Numbers

Multiply.

$3\frac{2}{3} \cdot 2\frac{1}{2} = \frac{11}{3} \cdot \frac{5}{2}$ Change the mixed numbers to improper fractions.

$= \frac{11 \cdot 5}{3 \cdot 2} = \frac{55}{6} = 9\frac{1}{6}$

CAUTION

NOTE This incorrect pattern leads to an answer of $6\frac{1}{3}$. The correct answer is $9\frac{1}{6}$.

Be Careful! Students sometimes think of

$3\frac{2}{3} \cdot 2\frac{1}{2}$ as $(3 \cdot 2) + \left(\frac{2}{3} \cdot \frac{1}{2}\right)$

This is *not* the correct multiplication pattern. You must first change the mixed numbers to improper fractions.

CHECK YOURSELF 6

Multiply.

$2\frac{1}{3} \cdot 3\frac{1}{2}$

The method of simplifying before multiplying two fractions that we studied in Section 3.4 can also be used when mixed numbers are involved. Consider Example 7.

Example 7

Simplifying Before Multiplying Two Mixed Numbers

Multiply.

$$2\frac{2}{3} \cdot 2\frac{1}{4} = \frac{8}{3} \cdot \frac{9}{4}$$ First, convert the mixed numbers to improper fractions.

$$= \frac{\overset{2}{\cancel{8}} \cdot \overset{3}{\cancel{9}}}{\underset{1}{\cancel{3}} \cdot \underset{1}{\cancel{4}}}$$ To simplify, divide by the common factors of 3 and 4.

$$= \frac{2 \cdot 3}{1 \cdot 1}$$ Multiply as before.

$$= \frac{6}{1} = 6$$

CHECK YOURSELF 7

Simplify and then multiply.

$$3\frac{1}{3} \cdot 2\frac{2}{5}$$

The ideas of Examples 1 to 7 will also allow us to find the product of more than two fractions.

Example 8

Simplifying Before Multiplying Three Numbers

Simplify and then multiply.

NOTE Remember our earlier rule: We can divide *any* factor of the numerator and *any* factor of the denominator by the same nonzero number.

$$\frac{2}{3} \cdot 1\frac{4}{5} \cdot \frac{5}{8} = \frac{2}{3} \cdot \frac{9}{5} \cdot \frac{5}{8}$$ Write any mixed or whole numbers as improper fractions.

$$= \frac{\overset{1}{\cancel{2}} \cdot \overset{3}{\cancel{9}} \cdot \overset{1}{\cancel{5}}}{\underset{1}{\cancel{3}} \cdot \underset{1}{\cancel{5}} \cdot \underset{4}{\cancel{8}}}$$ To simplify, divide by the common factors in the numerator and denominator.

$$= \frac{3}{4}$$

CHECK YOURSELF 8

Simplify and then multiply.

$$\frac{5}{8} \cdot 4\frac{4}{5} \cdot \frac{1}{6}$$

We encountered estimation by rounding in our earlier work with whole numbers. Estimation can also be used to check the reasonableness of an answer when we are working with fractions or mixed numbers.

Example 9

Estimating the Product of Two Mixed Numbers

Estimate the product of

$$3\frac{1}{8} \cdot 5\frac{5}{6}$$

Round each mixed number to the nearest whole number.

$$3\frac{1}{8} \rightarrow 3$$

$$5\frac{5}{6} \rightarrow 6$$

Our estimate of the product is then

$$3 \cdot 6 = 18$$

Note: The actual product in this case is $18\frac{11}{48}$, which certainly seems reasonable in view of our estimate.

CHECK YOURSELF 9

Estimate the product.

$$2\frac{7}{8} \cdot 8\frac{1}{3}$$

We learned how to divide fractions in Section 3.4. When mixed or whole numbers are involved, the process is similar. Simply change the mixed or whole numbers to improper fractions as the first step. Then proceed with the division rule. Example 10 illustrates this approach.

Example 10

Dividing Two Mixed Numbers

Divide.

$$2\frac{3}{8} \div 1\frac{3}{4} = \frac{19}{8} \div \frac{7}{4}$$ Write the mixed numbers as improper fractions.

$$= \frac{19}{\underset{2}{\cancel{8}}} \cdot \frac{\overset{1}{\cancel{4}}}{7}$$ Invert the divisor and multiply as before.

$$= \frac{19}{14} = 1\frac{5}{14}$$

CHECK YOURSELF 10

Divide.

$$3\frac{1}{5} \div 2\frac{2}{5}$$

Example 11 illustrates the division process when a whole number is involved.

Example 11

Dividing a Mixed Number and a Whole Number

Divide and simplify.

NOTE Write the whole number 6 as $\frac{6}{1}$.

$$1\frac{4}{5} \div 6 = \frac{9}{5} \div \frac{6}{1}$$

$$= \frac{\overset{3}{\cancel{9}}}{5} \cdot \frac{1}{\underset{2}{\cancel{6}}}$$ Invert the divisor, then divide by the common factor of 3.

$$= \frac{3}{10}$$

CHECK YOURSELF 11

Divide.

$$8 \div 4\frac{4}{5}$$

We learned how to add and subtract fractions in Section 4.1. Once you know how to add fractions, adding mixed numbers should be no problem if you keep in mind that addition involves combining groups of the *same kind* of objects. Because mixed numbers consist of two parts—a whole number and a fraction—we could work with the whole numbers and the fractions separately. Generally, it is easier to rewrite mixed numbers as improper fractions, and then do the addition.

Consider $1\frac{1}{5} + 2\frac{2}{5}$.

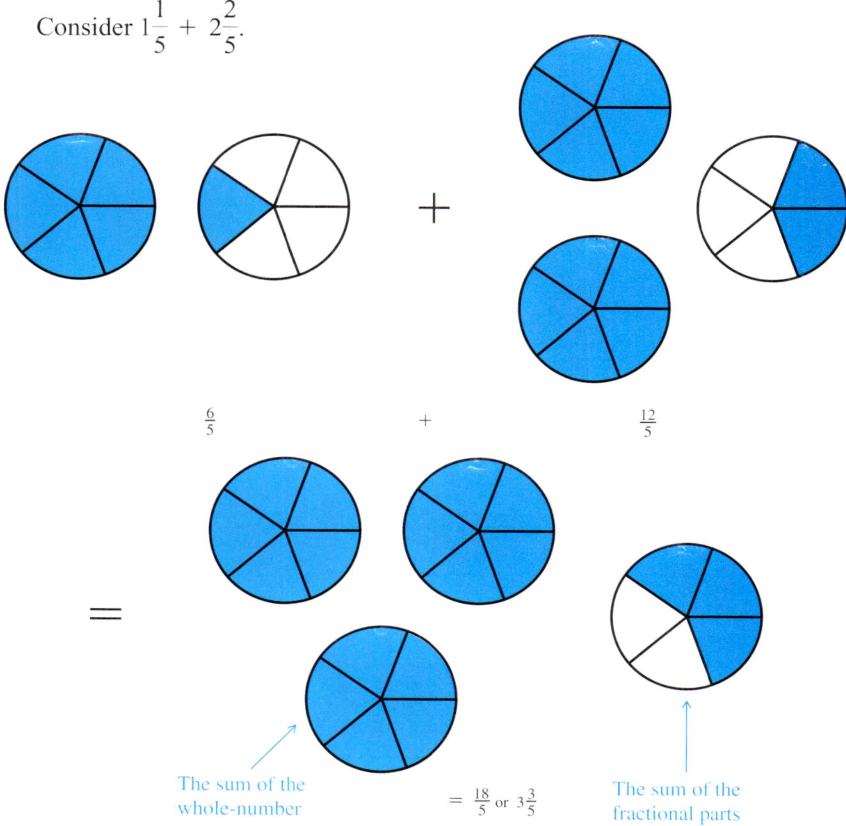

This suggests a general rule.

Step by Step: To Add Mixed Numbers

NOTE Step 2 requires that the fractional parts have the same denominator.

Step 1 Change the mixed numbers to improper fractions.
Step 2 Add the fractions.
Step 3 Rewrite the result as a mixed number if required.

Example 12 illustrates the use of this rule.

Example 12

Adding Mixed Numbers

Add and write the result as a mixed number.

$$3\frac{1}{5} + 4\frac{2}{5} = \frac{16}{5} + \frac{22}{5}$$ Rewrite as improper fractions.

$$= \frac{38}{5}$$ Add the numerators.

$$= 7\frac{3}{5}$$ Rewrite as a mixed number.

CHECK YOURSELF 12

Add $2\frac{3}{10} + 3\frac{4}{10}$. *Write the result as a mixed number.*

When the fractional portions of the mixed numbers have different denominators, we must rename these fractions as equivalent fractions with the least common denominator to perform the addition in step 2. Consider Example 13.

Example 13

Adding Mixed Numbers with Different Denominators

Add, and write the result as a mixed number.

$$3\frac{1}{6} + 2\frac{3}{8} = \frac{19}{6} + \frac{19}{8}$$ The LCD of the fractions is 24. Rename them with that denominator.

$$= \frac{76}{24} + \frac{57}{24}$$ Then add as before.

$$= \frac{133}{24}$$

$$= 5\frac{13}{24}$$

NOTE
$$\begin{array}{r} 5 \\ 24\overline{)133} \\ \underline{120} \\ 13 \end{array}$$

CHECK YOURSELF 13

Add $5\frac{7}{10} + 3\frac{5}{6}$. *Write the result as a mixed number.*

You follow the same procedure if more than two mixed numbers are involved in the problem.

Example 14

Adding Mixed Numbers with Different Denominators

Add.

NOTE The LCD of the three fractions is 40. Convert to equivalent fractions.

$$2\frac{1}{5} + 3\frac{3}{4} + 4\frac{1}{8} = \frac{11}{5} + \frac{15}{4} + \frac{33}{8}$$
$$= \frac{88}{40} + \frac{150}{40} + \frac{165}{40}$$
$$= \frac{403}{40}$$
$$= 10\frac{3}{40}$$

CHECK YOURSELF 14

Add $5\frac{1}{2} + 4\frac{2}{3} + 3\frac{3}{4}$.

We can use a similar technique for *subtracting* mixed numbers. The rule is similar to that stated earlier for adding mixed numbers.

Step by Step: To Subtract Mixed Numbers

Step 1 Change the mixed numbers to improper fractions.
Step 2 Subtract the fractions.
Step 3 Rewrite the result as a mixed number if required.

Example 15 illustrates the use of this rule.

Example 15

Subtracting Mixed Numbers with Like Denominators

Subtract.

$$5\frac{7}{12} - 3\frac{5}{12} = \frac{67}{12} - \frac{41}{12}$$
$$= \frac{26}{12}$$
$$= \frac{13}{6}$$
$$= 2\frac{1}{6}$$

CHECK YOURSELF 15

Subtract $8\frac{7}{8} - 5\frac{3}{8}$.

Again, we must rename the fractions if different denominators are involved. This approach is shown in Example 16.

Example 16

Subtracting Mixed Numbers with Different Denominators

Subtract.

$$8\frac{7}{10} - 3\frac{3}{8} = \frac{87}{10} - \frac{27}{8}$$ Write the fractions with denominator 40.

$$= \frac{348}{40} - \frac{135}{40}$$ Subtract as before.

$$= \frac{213}{40}$$

$$= 5\frac{13}{40}$$

CHECK YOURSELF 16

Subtract $7\frac{11}{12} - 3\frac{5}{8}$.

To subtract a mixed number from a whole number, we use the same techniques.

Example 17

Subtracting Mixed Numbers

Subtract.

$$6 - 2\frac{3}{4}$$

$$6 - 2\frac{3}{4} = \frac{24}{4} - \frac{11}{4}$$ Write both the whole number and the mixed number as improper fractions with a common denominator.

$$= \frac{13}{4}$$

$$= 3\frac{1}{4}$$

NOTE

$6 = \frac{6}{1} = \frac{24}{4}$

Multiply the numerator and denominator by 4 to form a common denominator.

CHECK YOURSELF 17

Subtract $7 - 3\frac{2}{5}$.

Using Your Calculator to Add and Subtract Mixed Numbers

We have already seen how to add, subtract, multiply, and divide fractions using our calculators. Now we will use our calculators to add and subtract mixed numbers.

Scientific Calculator

To enter a mixed number on a scientific calculator, press the fraction key between both the whole number and the numerator and denominator. For example, to enter $3\frac{7}{12}$, press

3 [a b/c] 7 [a b/c] 12

Example 18

Adding Mixed Numbers

Add.

$$-3\frac{7}{12} + 2\frac{11}{16}$$

The keystroke sequence is

3 [a b/c] 7 [a b/c] 12 [+/−] [+] 2 [a b/c] 11 [a b/c] 16 [=]

The result is $\frac{-43}{48}$.

Graphing Calculator

As with multiplying and dividing fractions, when using a graphing calculator, you must choose the fraction option from the math menu before pressing [Enter].

For the problem in Example 18, $-3\frac{7}{12} + 2\frac{11}{16}$, the keystroke sequence is

[(−)] [(] 3 [+] 7 [÷] 12 [)] [+] 2 [+] 11 [÷] 16 [Math] [►Frac] [Enter]

The display will read $\frac{-43}{48}$.

NOTE When representing a negative mixed number, the number must be enclosed in parentheses. Do you see the difference between $-3 + \frac{7}{12}$ and $-\left(3 + \frac{7}{12}\right)$?

CHECK YOURSELF 18

Find the sum.

$$-8\frac{3}{7} + 4\frac{5}{6}$$

In this section, we learned to work with mixed numbers. In actual practice, and throughout the remainder of this text, mixed numbers are used almost exclusively when working with applications. For example, if, through the use of some formula, we discover that we need $\frac{13}{3}$ gal of paint, it makes far more sense to express that as $4\frac{1}{3}$ gal. On the other hand, if we solve an algebraic equation and find that $x = \frac{13}{3}$, we would leave the result as an improper fraction.

CHECK YOURSELF ANSWERS

1. $3\frac{5}{6}$ **2.** $6\frac{2}{5}$ **3.** 8 **4.** $\frac{43}{8}$ **5.** $\frac{5}{8} \cdot 3\frac{1}{2} = \frac{5}{8} \cdot \frac{7}{2} = \frac{35}{16} = 2\frac{3}{16}$ **6.** $8\frac{1}{6}$

7. $3\frac{1}{3} \cdot 2\frac{2}{5} = \frac{10}{3} \cdot \frac{12}{5} = \frac{\overset{2}{\cancel{10}} \cdot \overset{4}{\cancel{12}}}{\underset{1}{\cancel{3}} \cdot \underset{1}{\cancel{5}}} = \frac{8}{1} = 8$ **8.** $\frac{1}{2}$ **9.** 24

10. $3\frac{1}{5} \div 2\frac{2}{5} = \frac{16}{5} \div \frac{12}{5} = \frac{\overset{4}{\cancel{16}}}{\underset{1}{\cancel{5}}} \cdot \frac{\overset{1}{\cancel{5}}}{\underset{3}{\cancel{12}}} = \frac{4}{3} = 1\frac{1}{3}$ **11.** $1\frac{2}{3}$

12. $5\frac{7}{10}$ **13.** $5\frac{7}{10} + 3\frac{5}{6} = \frac{57}{10} + \frac{23}{6} = \frac{171}{30} + \frac{115}{30} = \frac{286}{30} = 9\frac{16}{30} = 9\frac{8}{15}$

14. $13\frac{11}{12}$ **15.** $3\frac{1}{2}$ **16.** $7\frac{11}{12} - 3\frac{5}{8} = \frac{95}{12} - \frac{29}{8} = \frac{190}{24} - \frac{87}{24} = \frac{103}{24} = 4\frac{7}{24}$

17. $3\frac{3}{5}$ **18.** $-3\frac{25}{42}$

Exercises

Name ______________

Section ________ Date ________

Identify each number as a proper fraction, an improper fraction, or a mixed number.

1. $\frac{3}{5}$

2. $\frac{-9}{5}$

3. $2\frac{3}{5}$

4. $\frac{7}{9}$

5. $\frac{6}{6}$

6. $-1\frac{1}{5}$

7. $\frac{11}{8}$

8. $\frac{8}{8}$

9. $4\frac{5}{7}$

10. $-5\frac{3}{7}$

11. $\frac{-13}{17}$

12. $\frac{16}{15}$

Give the mixed number that names the shaded portion of each diagram.

13. 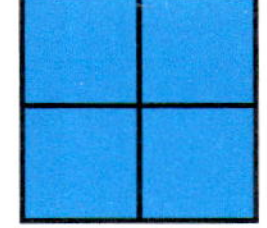 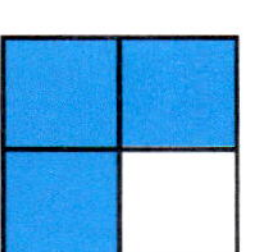

14. 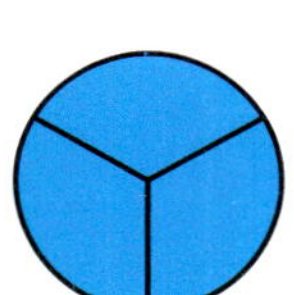 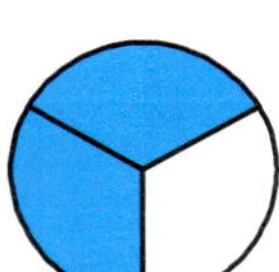

15. 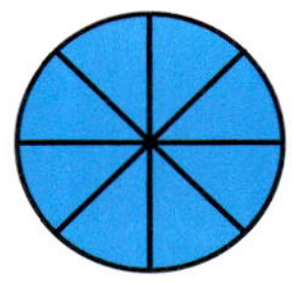 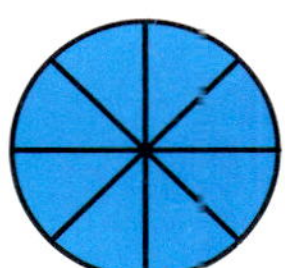

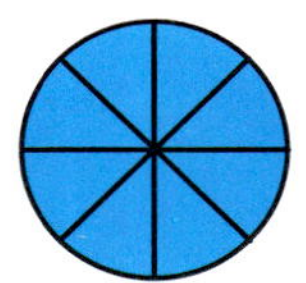 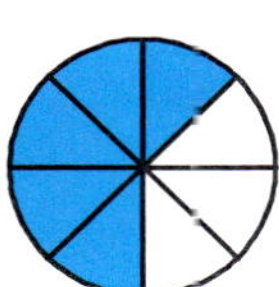

16.  

Change to a mixed or whole number.

17. $\frac{22}{5}$

18. $\frac{27}{8}$

19. $\frac{-34}{5}$

20. $\frac{-25}{6}$

ANSWERS

1. ______________
2. ______________
3. ______________
4. ______________
5. ______________
6. ______________
7. ______________
8. ______________
9. ______________
10. ______________
11. ______________
12. ______________
13. ______________
14. ______________
15. ______________
16. ______________
17. ______________
18. ______________
19. ______________
20. ______________

ANSWERS

21. ____
22. ____
23. ____
24. ____
25. ____
26. ____
27. ____
28. ____
29. ____
30. ____
31. ____
32. ____
33. ____
34. ____
35. ____
36. ____
37. ____
38. ____
39. ____ 40. ____
41. ____ 42. ____
43. ____ 44. ____
45. ____ 46. ____
47. ____ 48. ____

Change to an improper fraction.

21. 8

22. $-4\frac{5}{8}$

23. $-6\frac{2}{9}$

24. 7

25. $3\frac{3}{7}$

26. $2\frac{2}{9}$

Multiply. Write each answer as a mixed number, whole number, or proper fraction.

27. $\frac{4}{9} \cdot 3\frac{3}{5}$

28. $5\frac{1}{3} \cdot \frac{7}{8}$

29. $\frac{-10}{27} \cdot 3\frac{3}{5}$

30. $-1\frac{1}{3} \cdot 1\frac{1}{5}$

31. $4\frac{1}{5} \cdot \frac{10}{21} \cdot \frac{9}{20}$

32. $\frac{7}{8} \cdot 5\frac{1}{3} \cdot \frac{-5}{14}$

Estimate the products.

33. $3\frac{1}{5} \cdot 4\frac{2}{3}$

34. $5\frac{1}{7} \cdot 2\frac{2}{13}$

35. $11\frac{3}{4} \cdot 5\frac{1}{4}$

36. $3\frac{4}{5} \cdot 5\frac{6}{7}$

37. $8\frac{2}{9} \cdot 7\frac{11}{12}$

38. $\frac{9}{10} \cdot 2\frac{2}{7}$

Find the reciprocal.

39. $2\frac{1}{3}$

40. $4\frac{3}{5}$

41. $9\frac{3}{4}$

42. $5\frac{1}{8}$

Divide. Write each result in simplest form.

43. $5\frac{3}{5} \div \frac{7}{15}$

44. $\frac{-7}{18} \div 5\frac{5}{6}$

45. $-1\frac{1}{3} \div 1\frac{1}{7}$

46. $3\frac{1}{2} \div 2\frac{4}{5}$

47. $3\frac{3}{4} \div 1\frac{3}{8}$

48. $5\frac{1}{3} \div 2\frac{2}{5}$

ANSWERS

49. ______ 50. ______
51. ______ 52. ______
53. ______ 54. ______
55. ______ 56. ______
57. ______ 58. ______
59. ______ 60. ______
61. ______
62. ______
63. ______
64. ______
65. ______
66. ______
67. ______
68. ______
69. ______
70. ______
71. ______
72. ______
73. ______
74. ______
75. ______
76. ______

Do the indicated operations.

49. $2\frac{2}{9} + 3\frac{5}{9}$

50. $5\frac{2}{9} + 6\frac{4}{9}$

51. $-7\frac{7}{8} + 3\frac{3}{8}$

52. $-3\frac{5}{6} + 1\frac{1}{6}$

53. $3\frac{2}{5} - 1\frac{4}{5}$

54. $5\frac{3}{7} - 2\frac{1}{7}$

55. $3\frac{2}{3} - 2\frac{1}{4}$

56. $5\frac{4}{5} - 1\frac{1}{6}$

57. $-7\frac{5}{12} + 3\frac{11}{18}$

58. $-9\frac{3}{7} + 2\frac{13}{21}$

59. $5 - 2\frac{1}{4}$

60. $4 - 1\frac{2}{3}$

61. $3\frac{3}{4} + 5\frac{1}{2} - 2\frac{3}{8}$

62. $1\frac{5}{6} + 3\frac{5}{12} - 2\frac{1}{4}$

63. $2\frac{3}{8} + 2\frac{1}{4} - 1\frac{5}{6}$

64. $1\frac{1}{15} + 3\frac{3}{10} - 2\frac{4}{5}$

Calculator Exercises

Add or subtract.

65. $3\frac{2}{3} + 2\frac{1}{4}$

66. $6\frac{1}{6} + 8\frac{2}{3}$

67. $-5\frac{4}{9} + \left(-2\frac{2}{3}\right)$

68. $-2\frac{3}{7} + \left(-4\frac{9}{14}\right)$

69. $11\frac{2}{3} + 5\frac{1}{4}$

70. $6\frac{3}{8} + 14\frac{5}{12}$

71. $14\frac{13}{18} + 22\frac{23}{27}$

72. $3\frac{41}{45} + 5\frac{25}{27}$

73. $-4\frac{7}{9} + 2\frac{11}{18}$

74. $-7\frac{8}{11} + 4\frac{13}{22}$

75. $5\frac{11}{16} - 2\frac{5}{12}$

76. $18\frac{5}{24} - 11\frac{3}{40}$

ANSWERS

77. ______

78. ______

79. ______

80. ______

77. $-6\frac{2}{3} + 1\frac{5}{6}$

78. $-131\frac{43}{45} + 99\frac{27}{60}$

79. $10\frac{2}{3} + 4\frac{1}{5} + 7\frac{2}{15}$

80. $7\frac{1}{5} + 3\frac{2}{3} + 1\frac{1}{5}$

Answers

1. Proper **3.** Mixed number **5.** Improper **7.** Improper **9.** Mixed number **11.** Proper **13.** $1\frac{3}{4}$ **15.** $3\frac{5}{8}$ **17.** $4\frac{2}{5}$ **19.** $-6\frac{4}{5}$ **21.** $\frac{8}{1}$ **23.** $\frac{-56}{9}$ **25.** $3\frac{3}{7} = \frac{(7 \cdot 3) + 3}{7} = \frac{24}{7}$ **27.** $1\frac{3}{5}$ **29.** $-1\frac{1}{3}$ **31.** $\frac{9}{10}$ **33.** 15 **35.** 60 **37.** 64 **39.** $\frac{3}{7}$ **41.** $\frac{4}{39}$ **43.** 12 **45.** $-1\frac{1}{6}$ **47.** $2\frac{8}{11}$ **49.** $5\frac{7}{9}$ **51.** $-4\frac{1}{2}$ **53.** $1\frac{3}{5}$ **55.** $1\frac{5}{12}$ **57.** $-3\frac{29}{36}$ **59.** $2\frac{3}{4}$ **61.** $6\frac{7}{8}$ **63.** $2\frac{19}{24}$ **65.** $5\frac{11}{12}$ **67.** $-8\frac{1}{9}$ **69.** $16\frac{11}{12}$ **71.** $37\frac{31}{54}$ **73.** $-2\frac{1}{6}$ **75.** $3\frac{13}{48}$ **77.** $-4\frac{5}{6}$ **79.** 22

4.3 Complex Fractions

4.3 OBJECTIVE

1. Simplify complex fractions

In Section 3.4, we learned to write a quotient such as $\frac{2}{3} \div \frac{4}{5}$ as the complex fraction $\frac{\frac{2}{3}}{\frac{4}{5}}$. A **complex fraction** has a fraction in its numerator, in its denominator, or in both. We then learned to rewrite the original quotient as the product of the reciprocal, so $\frac{2}{3} \div \frac{4}{5} = \frac{2}{3} \cdot \frac{5}{4}$. To simplify a complex fraction, multiply the numerator by the reciprocal of the denominator.

Example 1

Simplifying Complex Fractions

NOTE Unless we are solving an application problem, we write improper fractions rather than mixed numbers.

$$\frac{\frac{3}{4}}{\frac{5}{8}} = \frac{3}{4} \div \frac{5}{8} = \frac{3}{\cancel{4}_1} \cdot \frac{\cancel{8}^2}{5} = \frac{6}{5}$$

CHECK YOURSELF 1

Simplify.

(a) $\frac{\frac{4}{7}}{\frac{3}{7}}$ (b) $\frac{\frac{3}{8}}{\frac{5}{6}}$

Like parentheses, the fraction bar acts as a grouping symbol. Operations in the numerator or denominator should be applied before the fraction can be simplified.

Example 2

Simplifying a Complex Fraction

Simplify $\frac{2 - \frac{3}{5}}{4 - \frac{1}{3}}$.

To apply the operations in the numerator and denominator, we first find common denominators.

$$\frac{2 - \frac{3}{5}}{4 - \frac{1}{3}} = \frac{\frac{10}{5} - \frac{3}{5}}{\frac{12}{3} - \frac{1}{3}}$$

Now we continue with the subtractions.

$$\frac{\frac{10}{5} - \frac{3}{5}}{\frac{12}{3} - \frac{1}{3}} = \frac{\frac{7}{5}}{\frac{11}{3}}$$

$$\frac{\frac{7}{5}}{\frac{11}{3}} = \frac{7}{5} \div \frac{11}{3} = \frac{7}{5} \cdot \frac{3}{11} = \frac{21}{55}$$

CHECK YOURSELF 2

Simplify.

$$\frac{5 + \frac{3}{7}}{7 - \frac{2}{3}}$$

Consider the complex fraction we just studied in Example 3. It may be easier to simplify *without* applying the operations first. We simply must remember to correctly apply the distributive property. This approach is shown in Example 4.

Example 3

Simplifying a Complex Fraction

Simplify $\frac{2 - \frac{3}{5}}{4 - \frac{1}{3}}$.

First we note that the LCD of all the fractions that appear is 15. Then we multiply the main numerator and main denominator by 15.

$$\frac{15 \cdot \left(2 - \frac{3}{5}\right)}{15 \cdot \left(4 - \frac{1}{3}\right)} = \frac{15 \cdot 2 - 15 \cdot \frac{3}{5}}{15 \cdot 4 - 15 \cdot \frac{1}{3}} = \frac{30 - 9}{60 - 5} = \frac{21}{55}$$

CHECK YOURSELF 3

Simplify $\frac{5 + \frac{3}{7}}{7 - \frac{2}{3}}$ *using the approach shown in Example 4.*

CHECK YOURSELF ANSWERS

1. **(a)** $\frac{4}{3}$; **(b)** $\frac{9}{20}$ **2.** $\frac{6}{7}$ **3.** $\frac{6}{7}$

Exercises

Name ______________

Section ________ Date ________

Simplify each complex fraction.

1. $\dfrac{\frac{2}{3}}{\frac{6}{8}}$

2. $\dfrac{\frac{5}{6}}{\frac{10}{15}}$

3. $\dfrac{\frac{1}{2}}{\frac{1}{4}}$

4. $\dfrac{\frac{3}{4}}{\frac{1}{8}}$

5. $\dfrac{\frac{4}{5}}{\frac{9}{10}}$

6. $\dfrac{\frac{5}{8}}{\frac{6}{10}}$

7. $\dfrac{\frac{6}{19}}{\frac{5}{38}}$

8. $\dfrac{\frac{8}{5}}{\frac{23}{25}}$

9. $\dfrac{6\frac{1}{2}}{\frac{2}{3}}$

10. $\dfrac{5\frac{1}{3}}{\frac{7}{6}}$

11. $\dfrac{4\frac{1}{2}}{5\frac{1}{4}}$

12. $\dfrac{8\frac{1}{3}}{9\frac{1}{6}}$

13. $\dfrac{2 - \frac{1}{2}}{2 + \frac{1}{4}}$

14. $\dfrac{3 + \frac{5}{6}}{7 - \frac{2}{3}}$

15. $\dfrac{3 + \frac{1}{8}}{4 - \frac{1}{4}}$

16. $\dfrac{5 - \frac{1}{3}}{5 + \frac{2}{3}}$

17. $\dfrac{2 + \frac{1}{6}}{5 - \frac{1}{3}}$

18. $\dfrac{4 + \frac{1}{5}}{6 + \frac{3}{20}}$

19. $\dfrac{8 + \frac{1}{2}}{4 - \frac{1}{4}}$

20. $\dfrac{5 + \frac{1}{2}}{2 - \frac{1}{10}}$

21. $\dfrac{4 - \frac{1}{3}}{3 + \frac{1}{2}}$

ANSWERS

1. ______________
2. ______________
3. ______________
4. ______________
5. ______________
6. ______________
7. ______________
8. ______________
9. ______________
10. ______________
11. ______________
12. ______________
13. ______________
14. ______________
15. ______________
16. ______________
17. ______________
18. ______________
19. ______________
20. ______________
21. ______________

ANSWERS

22. ______

23. ______

24. ______

22. $\dfrac{2 - \dfrac{1}{4}}{5 + \dfrac{2}{3}}$

23. $\dfrac{5 + \dfrac{3}{4}}{7 - \dfrac{1}{6}}$

24. $\dfrac{3 - \dfrac{2}{3}}{5 + \dfrac{1}{5}}$

Answers

1. $\dfrac{8}{9}$ 3. 2 5. $\dfrac{8}{9}$ 7. $\dfrac{12}{5}$ 9. $\dfrac{39}{4}$ 11. $\dfrac{6}{7}$ 13. $\dfrac{2}{3}$ 15. $\dfrac{5}{6}$

17. $\dfrac{13}{28}$ 19. $\dfrac{34}{15}$ 21. $\dfrac{22}{21}$ 23. $\dfrac{69}{82}$

4.4 Applications Involving Fractions

4.4 OBJECTIVE

1. Solve applications involving fractions

Our knowledge of fractions is critical in solving many applied problems. For example, you probably do many conversions between mixed and whole numbers without even thinking about the process that you follow, as Example 1 illustrates.

Example 1

Converting Quarter Dollars to Dollars

Maritza has 53 quarters in her bank. How many dollars does she have?

Because there are 4 quarters in each dollar, 53 quarters can be written as

$$\frac{53}{4}$$

Converting the amount to dollars is the same as rewriting it as a mixed number.

$$\frac{53}{4} = 13\frac{1}{4}$$

She has $13\frac{1}{4}$ dollars, which you would probably write as \$13.25. (*Note:* We will discuss decimal point usage in Chapter 5.)

CHECK YOURSELF 1

Kevin is doing the inventory in the convenience store in which he works. He finds there are 11 half gallons of milk. Write the amount of milk as a mixed number of gallons.

NOTE Recall that a denominate number has a unit of measure attached to it.

Units Analysis

When dividing two denominate numbers, the units are also divided. This yields a unit in fraction form.

Examples

$$250 \text{ mi} \div 10 \text{ gal} = \frac{250 \text{ mi}}{10 \text{ gal}} = \frac{25 \text{ mi}}{1 \text{ gal}} = 25 \frac{\text{mi}}{\text{gal}} \text{ (read "miles per gallon")}$$

$$360 \text{ ft} \div 30 \text{ s} = \frac{360 \text{ ft}}{30 \text{ s}} = 12 \frac{\text{ft}}{\text{s}} \text{ ("feet per second")}$$

When we multiply denominate numbers that have these units in fraction form, they behave just like fractions.

Examples

$$25 \frac{\text{mi}}{\text{gal}} \cdot 12 \text{ gal} = \frac{25 \text{ mi}}{1 \not{\text{gal}}} \cdot \frac{12 \not{\text{gal}}}{1} = 300 \text{ mi}$$

(If we look at the units, we see that the gallons essentially "cancel" when one is in the numerator and the other is in the denominator.)

$$12 \frac{\text{ft}}{\text{s}} \cdot 60 \frac{\text{s}}{\text{min}} = \frac{12 \text{ ft}}{1 \not{\text{s}}} \cdot \frac{60 \not{\text{s}}}{1 \text{ min}} = \frac{720 \text{ ft}}{1 \text{ min}} = 720 \frac{\text{ft}}{\text{min}}$$

(Again, the seconds cancel, leaving feet in the numerator and minutes in the denominator.)

Now we will look at some applications of our work with fraction operations. In solving these word problems, we will use the same approach we used earlier with whole numbers. We will review the four-step process introduced in Section 1.2.

Step by Step: Solving Applications Involving Fractions

Step 1 Read the problem carefully to determine the given information and what you are asked to find.
Step 2 Decide upon the operation or operations to be used.
Step 3 Write down the complete statement necessary to solve the problem and do the calculations.
Step 4 Check to make sure that you have answered the question of the problem and that your answer seems reasonable.

Now we will work through some examples, using these steps.

Example 2

An Application Involving Multiplication

Lisa worked $10\frac{1}{4}\ \frac{\text{h}}{\text{day}}$ for 5 days. How many hours did she work?

Step 1 We are looking for the total hours Lisa worked.

Step 2 We will multiply the hours per day by the days.

Step 3 $10\frac{1}{4}\ \frac{\text{h}}{\text{day}} \cdot 5 \text{ days} = \frac{41}{4}\ \frac{\text{h}}{\text{day}} \cdot 5 \text{ days} = \frac{205}{4}\text{ h} = 51\frac{1}{4}\text{ h}$

Step 4 Note the days cancel, leaving only the unit *hours*. The units should always be compared to the desired units from step 1. The answer also seems reasonable. An answer like 5 h or 500 h would not seem reasonable.

CHECK YOURSELF 2

Carlos worked $8\frac{1}{2}$ *h per day for 6 days. How many hours did he work?*

In Example 3, we will follow the four steps for solving applications, but we won't label the steps. You should still think about these steps as we solve the problem.

Example 3

An Application Involving the Multiplication of Mixed Numbers

A sheet of notepaper is $6\frac{3}{4}$ in. wide by $8\frac{2}{3}$ in. long. Find the area of the paper.

Multiply the given length by the width. This will give the desired area. First, we will estimate the area.

$7 \text{ in.} \cdot 9 \text{ in.} = 63 \text{ in.}^2$

Now, we will find the exact area.

$$8\frac{2}{3}\text{ in.}\cdot 6\frac{3}{4}\text{ in.} = \frac{26}{3}\text{ in.}\cdot\frac{27}{4}\text{ in.}$$
$$= \frac{117}{2}\text{ in.}^2$$
$$= 58\frac{1}{2}\text{ in.}^2$$

NOTE Recall that the area of a rectangle is the product of its length and its width.

The units (square inches) are units of area. Note from our estimate that the result is reasonable.

CHECK YOURSELF 3

A window is $4\frac{1}{2}$ ft high by $2\frac{1}{3}$ ft wide. What is its area?

Example 4

An Application Involving the Multiplication of a Mixed Number and a Fraction

A state park contains $38\frac{2}{3}$ acres. According to the plan for the park, $\frac{3}{4}$ of the park is to be left as a wildlife preserve. How many acres will this be?

NOTE The word *of* indicates multiplication.

We want to find $\frac{3}{4}$ of $38\frac{2}{3}$ acres. We then multiply as shown:

$$\frac{3}{4}\cdot 38\frac{2}{3} = \frac{\overset{1}{\cancel{3}}}{\underset{1}{\cancel{4}}}\cdot\frac{\overset{29}{\cancel{116}}}{\underset{1}{\cancel{3}}}\text{ acres} = 29\text{ acres}$$

CHECK YOURSELF 4

A backyard has $25\frac{3}{4}$ yd^2 of open space. If Patrick wants to build a vegetable garden covering $\frac{2}{3}$ of the open space, how many square yards will this be?

Units Analysis

When dividing by denominate numbers that have fractional units, we multiply by the reciprocal of the number *and its units.*

Examples

$$500\text{ mi} \div \frac{25\text{ mi}}{1\text{ gal}} = 500\text{ mi}\cdot\frac{1\text{ gal}}{25\text{ mi}} = 20\text{ gal}$$

$$\$24{,}000 \div \frac{\$400}{1\text{ yr}} = \$24{,}000\cdot\frac{1\text{ yr}}{\$400} = 60\text{ yr}$$

(As always, note that in each case, the arithmetic of the units produces the final units.)

As was the case with multiplication, our work with the division of fractions will be used in the solution of a variety of applications. The steps of the problem-solving process remain the same.

Example 5

An Application Involving the Division of Mixed Numbers

NOTE A kilometer, abbreviated km, is a metric unit of distance. It is about $\frac{3}{5}$ mi.

Jack traveled 140 km in $2\frac{1}{3}$ h. What was his average speed?

NOTE The important formula is Speed = distance ÷ time.

Distance → 140 km; Time → $2\frac{1}{3}$ h

$$\text{Speed} = 140 \text{ km} \div 2\frac{1}{3} \text{ h}$$

We know the distance traveled and the time for that travel. To find the *average* speed, we must use division. Do you remember why?

$$= \frac{140}{1} \text{ km} \div \frac{7}{3} \text{ h}$$

$$= \frac{\overset{20}{\cancel{140}}}{1} \cdot \frac{3}{\underset{1}{\cancel{7}}} \frac{\text{km}}{\text{h}}$$

$\frac{\text{km}}{\text{h}}$ is read "kilometers per hour." This is a unit of speed.

$$= 60 \frac{\text{km}}{\text{h}}$$

CHECK YOURSELF 5

A light plane flew 280 *mi in* $1\frac{3}{4}$ *h. What was its average speed?*

Example 6

An Application Involving the Division of Mixed Numbers

An electrician needs pieces of wire $2\frac{3}{5}$ in. long. If she has a $20\frac{4}{5}$-in. piece of wire, how many of the shorter pieces can she cut?

NOTE We must divide the length of the longer piece by the desired length of the shorter piece.

$$20\frac{4}{5} \div 2\frac{3}{5} = \frac{104}{5} \div \frac{13}{5}$$

$$= \frac{\overset{8}{\cancel{104}}}{\underset{1}{\cancel{5}}} \cdot \frac{\overset{1}{\cancel{5}}}{\underset{1}{\cancel{13}}}$$

$$= 8 \text{ pieces}$$

CHECK YOURSELF 6

A piece of plastic water pipe 63 *in. long is to be cut into lengths of* $3\frac{1}{2}$ *in. How many of the shorter pieces can be cut?*

Many applications can be solved by adding fractions.

Example 7

An Application Involving the Adding of Like Fractions

Noel walked $\frac{9}{10}$ mi to Jensen's house and then walked $\frac{7}{10}$ mi to school. How far did Noel walk?

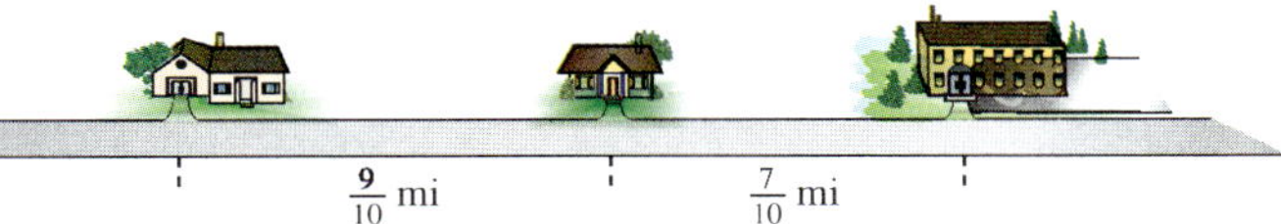

To find the total distance Noel walked, add the two distances.

$$\frac{9}{10} + \frac{7}{10} = \frac{16}{10} = 1\frac{6}{10} = 1\frac{3}{5}$$

Noel walked $1\frac{3}{5}$ mi.

CHECK YOURSELF 7

Emir bought $\frac{7}{16}$ lb of candy at one store and $\frac{11}{16}$ lb at another store. How much candy did Emir buy?

Many of the measurements you deal with in everyday life involve fractions. Now we will look at some typical situations.

Example 8

An Application Involving the Addition of Unlike Fractions

Jack's doctor wants him to run at least 2 mi per week. Jack ran $\frac{1}{2}$ mi on Monday, $\frac{2}{3}$ mi on Wednesday, and $\frac{3}{4}$ mi on Friday. How far did he run during the week? Did he meet the goal set by his doctor?

The three distances that Jack ran are the given information in the problem. We want to find a total distance, so we must add for the solution.

$$\frac{1}{2} + \frac{2}{3} + \frac{3}{4} = \frac{6}{12} + \frac{8}{12} + \frac{9}{12}$$

$$= \frac{23}{12} = 1\frac{11}{12} \text{ mi}$$

Because we have no common denominator, we must convert to equivalent fractions before we can add.

Jack ran $1\frac{11}{12}$ mi during the week. He did not quite meet his goal.

CHECK YOURSELF 8

Susan is designing an office complex. She needs $\frac{2}{5}$ acre for buildings, $\frac{1}{3}$ acre for driveways and parking, and $\frac{1}{6}$ acre for walks and landscaping. How much land does she need?

Example 9

An Application Involving the Addition of Unlike Fractions

Sam bought three packages of spices weighing $\frac{1}{4}$, $\frac{5}{8}$, and $\frac{1}{2}$ lb. What was the total weight?

We need to find the total weight, so we must add.

NOTE The abbreviation for pounds is *lb* from the Latin *libra*, meaning "balance" or "scales."

$$\frac{1}{4} + \frac{5}{8} + \frac{1}{2} = \frac{2}{8} + \frac{5}{8} + \frac{4}{8}$$

Write each fraction with the denominator 8.

$$= \frac{11}{8} = 1\frac{3}{8} \text{ lb}$$

The total weight was $1\frac{3}{8}$ lb.

CHECK YOURSELF 9

For three different recipes, Max needs $\frac{3}{8}$, $\frac{1}{2}$, and $\frac{5}{8}$ gal tomato sauce. How many gallons should he buy altogether?

We will next look at an example that applies our work in subtracting unlike fractions.

Example 10

An Application Involving the Subtraction of Unlike Fractions

You have $\frac{7}{8}$ yd of a handwoven linen. A pattern for a placemat calls for $\frac{1}{2}$ yd. Will you have enough left for two napkins that will use $\frac{1}{3}$ yd?

First, find out how much fabric is left over after the placemat is made.

$$\frac{7}{8} \text{ yd} - \frac{1}{2} \text{ yd} = \frac{7}{8} \text{ yd} - \frac{4}{8} \text{ yd} = \frac{3}{8} \text{ yd}$$

NOTE Remember that $\frac{3}{8}$ yd is left over and that $\frac{1}{3}$ yd is needed.

Now compare the size of $\frac{1}{3}$ and $\frac{3}{8}$.

$$\frac{3}{8} \text{ yd} = \frac{9}{24} \text{ yd} \quad \text{and} \quad \frac{1}{3} \text{ yd} = \frac{8}{24} \text{ yd}$$

Because $\frac{3}{8}$ yd is *more than* the $\frac{1}{3}$ yd that is needed, there is enough material for the place-mat *and* two napkins.

CHECK YOURSELF 10

A concrete walk will require $\frac{3}{4}$ yd^3 of concrete. If you have mixed $\frac{8}{9}$ yd^3, will enough concrete remain to do a project that will use $\frac{1}{6}$ yd^3? (Note: yd^3 is a measure of volume.)

Our next application involves measurement in inches. Note that on a ruler or yardstick, the marks divide each inch into $\frac{1}{2}$-in., $\frac{1}{4}$-in., and $\frac{1}{8}$-in. sections, and on some rulers, $\frac{1}{16}$-in. sections. We will use denominators of 2, 4, 8, and 16 in our measurement applications.

Example 11

An Application Involving the Subtraction of Unlike Fractions

Alexei is cutting two slats that are each to be $\frac{3}{16}$ in. in width from a piece of wood that is $1\frac{3}{4}$ in. across. How much will be left?

The two $\frac{3}{16}$-in. pieces will total

$$2 \cdot \frac{3}{16} = \frac{6}{16} = \frac{3}{8} \text{ in.}$$

$$1\frac{3}{4} = \frac{7}{4} = \frac{14}{8}$$

$$\frac{14}{8} - \frac{3}{8} = \frac{11}{8}$$

The remaining strip will be $\frac{11}{8}$ in. or $1\frac{3}{8}$ in. wide.

CHECK YOURSELF 11

Ricardo is cutting three strips from a piece of metal with a width of 1 in. Each strip has a width of $\frac{3}{16}$ in. How much metal will remain after the cuts?

Often we will have to use more than one operation to find the solution to a problem. Consider Example 12.

Example 12

An Application Involving Mixed Numbers

A rectangular poster is to have a total length of $12\frac{1}{4}$ in. We want a $1\frac{3}{8}$-in. border on the top and a 2-in. border on the bottom. What is the length of the printed part of the poster?

First, we will draw a sketch of the poster:

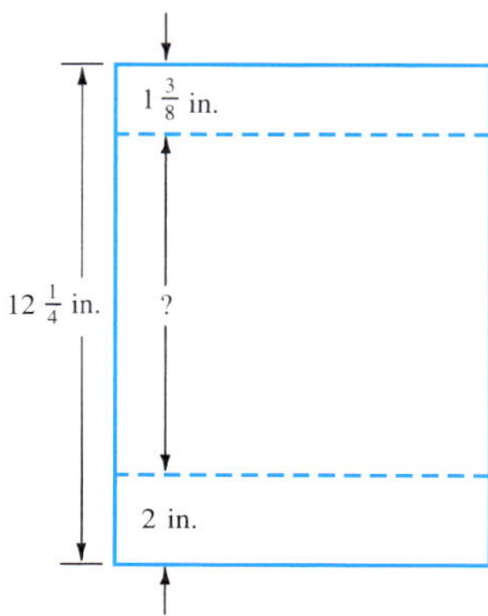

Now, we will use that sketch to find the total width of the top and bottom borders.

$$1\frac{3}{8} + 2 = \frac{11}{8} + \frac{16}{8} = \frac{27}{8} \text{ in.}$$

Now *subtract* that sum (the top and bottom borders) from the total length of the poster.

$$12\frac{1}{4} - \frac{27}{8} = \frac{49}{4} - \frac{27}{8} = \frac{98}{8} - \frac{27}{8}$$

$$= \frac{71}{8} = 8\frac{7}{8} \text{ in.}$$

The length of the printed part is $8\frac{7}{8}$ in.

CHECK YOURSELF 12

You cut one shelf $3\frac{3}{4}$ ft long and one $4\frac{1}{2}$ ft long from a 12-ft piece of lumber. Can you cut another shelf 4 ft long?

CHECK YOURSELF ANSWERS

1. $\frac{11}{2} = 5\frac{1}{2}$ gal **2.** 51 h **3.** $10\frac{1}{2}$ ft^2 **4.** $17\frac{1}{6}$ yd^2 **5.** $160\ \frac{\text{mi}}{\text{h}}$

6. 18 pieces **7.** $1\frac{1}{8}$ lb **8.** $\frac{9}{10}$ acre **9.** $1\frac{1}{2}$ gal

10. $\frac{5}{36}$ yd^3 will remain. You do *not* have enough concrete for both projects.

11. $\frac{7}{16}$ in. **12.** No, only $3\frac{3}{4}$ ft is "left over."

4.4 Exercises

Name ______________________

Section ________ Date ________

Solve the applications.

1. **Savings.** Clayton has 64 quarters in his bank. How many dollars does he have?

2. **Savings.** Amy has 19 quarters in her purse. How many dollars does she have?

3. **Inventory.** Manuel counted 35 half gallons of orange juice in his store. Write the amount of orange juice as a mixed number of gallons.

4. **Inventory.** Sarah has 19 half gallons of turpentine in her paint store. Write the amount of turpentine as a mixed number of gallons.

5. **Recipes.** A recipe calls for $\frac{2}{3}$ cup of sugar for each serving. How much sugar is needed for six servings?

6. **Recipes.** Mom's French toast requires $\frac{3}{4}$ cup of batter for each serving. If five people are expected for breakfast, how much batter is needed?

7. **Gardening.** A patch of dirt needs $3\frac{5}{6}$ ft^2 of sod to cover it. If Nick decides to cover only $\frac{3}{4}$ of the dirt, how much sod does he need?

8. **Construction.** A driveway requires $4\frac{5}{6}$ yd^3 of concrete to cover it. If Sheila wants to enlarge her driveway to $2\frac{1}{2}$ times its current size, how much concrete will she need?

ANSWERS

1. ______________________
2. ______________________
3. ______________________
4. ______________________
5. ______________________
6. ______________________
7. ______________________
8. ______________________

ANSWERS

9. __________

10. __________

11. __________

12. __________

13. __________

14. __________

15. __________

16. __________

17. __________

9. Map scales. The scale on a map is 1 in. = 200 mi. What actual distance, in miles, does $\frac{3}{8}$ in. represent?

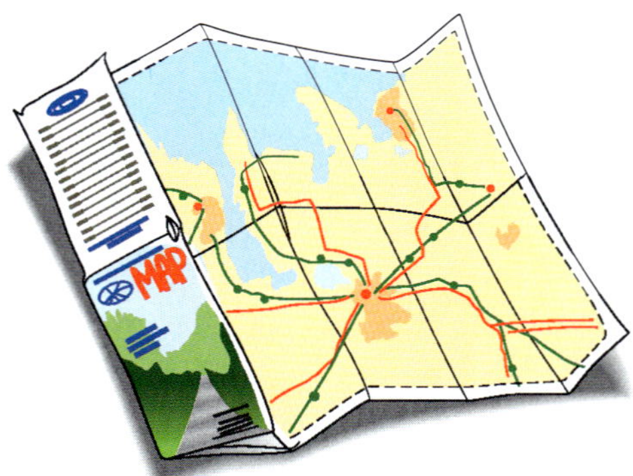

10. Salary. You make \$90 a day on a job. What will you receive for working $\frac{3}{4}$ of a day?

11. Size. A lumberyard has a stack of 80 sheets of plywood. If each sheet is $\frac{3}{4}$ in. thick, how high will the stack be?

12. Family budget. A family uses $\frac{2}{5}$ of its monthly income for housing and utilities on average. If the family's monthly income is \$1,750, what is spent for housing and utilities? What amount remains?

13. Elections. Of the eligible voters in an election, $\frac{3}{4}$ were registered. Of those registered, $\frac{5}{9}$ actually voted. What fraction of those people who were eligible voted?

14. Surveys. A survey has found that $\frac{7}{10}$ of the people in a city own pets. Of those who own pets, $\frac{2}{3}$ have dogs. What fraction of those surveyed own dogs?

15. Area. A kitchen has dimensions $3\frac{1}{3}$ yd by $3\frac{3}{4}$ yd. How many square yards of linoleum must be bought to cover the floor?

16. Distance. If you drive at an average speed of $52\ \frac{\text{mi}}{\text{h}}$ for $1\frac{3}{4}$ h, how far will you travel?

17. Distance. A jet flew at an average speed of $540\ \frac{\text{mi}}{\text{h}}$ on a $4\frac{2}{3}$-h flight. What was the distance flown?

18. **Area.** A piece of land that has $11\frac{2}{3}$ acres is being subdivided for home lots. It is estimated that $\frac{2}{7}$ of the area will be used for roads. What amount remains to be used for lots?

19. **Circumference.** To find the approximate circumference or distance around a circle, we multiply its diameter by $\frac{22}{7}$. What is the circumference of a circle with a diameter of 21 in.?

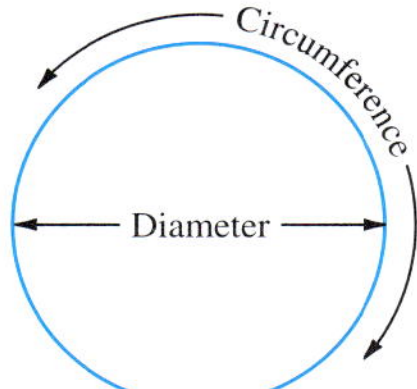

20. **Area.** The length of a rectangle is $\frac{6}{7}$ yd, and its width is $\frac{21}{26}$ yd. What is its area in square yards?

The formula for the area of a triangle is

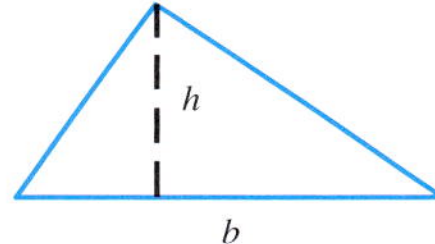

$$A = \frac{1}{2} \cdot h \cdot b$$

in which h is the height of the triangle and b is the base.

21. Find the area of a triangle with a height of $2\frac{2}{5}$ in. and a base of $3\frac{1}{3}$ in.

22. Find the area of a triangle with a height of $1\frac{7}{8}$ in. and a base of $2\frac{2}{5}$ in.

23. A recipe calls for the following ingredients:

$\frac{7}{8}$ cup of flour, $\frac{3}{4}$ cup of sugar, $\frac{2}{3}$ cup of milk, and $\frac{5}{6}$ teaspoon of salt. This recipe makes eight servings. What amount of each quantity would you use if you wanted to serve two people?

24. Obtain a map of your state and, using the legend provided, determine the distance between your state capital and any other city. Would this be the actual distance you would travel by car if you made the journey? Why or why not?

ANSWERS

18. ____________

19. ____________

20. ____________

21. ____________

22. ____________

23. ____________

24. ____________

ANSWERS

25. ______________

26. ______________

27. ______________

28. ______________

29. ______________

30. ______________

31. ______________

32. ______________

33. ______________

34. ______________

35. ______________

25. Wire cutting. A wire $5\frac{1}{4}$ ft long is to be cut into 7 pieces of the same length. How long will each piece be?

26. Quantity. A potter uses $\frac{2}{3}$ lb of clay in making a bowl. How many bowls can be made from 16 lb of clay?

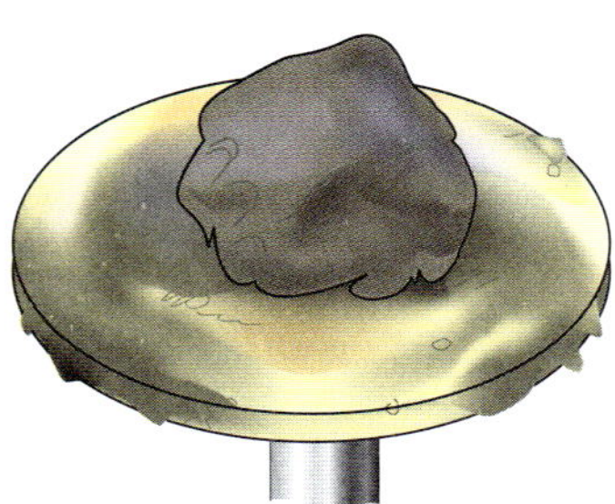

27. Speed. Virginia made a trip of 95 mi in $1\frac{1}{4}$ h. What was her average speed?

28. Cost. A piece of land measures $3\frac{3}{4}$ acres and is for sale at $60,000. What is the price per acre?

29. Number of servings. A roast weighs $3\frac{1}{4}$ lb. How many $\frac{1}{4}$-lb servings will the roast provide?

30. Number of books. A bookshelf is 55 in. long. If the books have an average thickness of $1\frac{1}{4}$ in., how many books can be put on the shelf?

31. Quantity. A butcher wants to wrap $\frac{3}{8}$-lb packages of ground beef from a cut of meat weighing $19\frac{1}{8}$ lb. How many packages can be prepared?

32. Quantity. A manufacturer has $45\frac{1}{2}$ yd of imported cotton fabric. A shirt pattern uses $1\frac{3}{4}$ yd. How many shirts can be made?

33. Number of pieces. A stack of $\frac{3}{4}$-in.-thick plywood is 48 in. high. How many sheets of plywood are in the stack?

34. Area. A landfill occupies land that measures $10\frac{2}{3}$ mi by $6\frac{3}{4}$ mi. If there are 144 cells of equal area in the landfill, what is the area of each cell?

35. Manuel has $7\frac{1}{2}$ yd of cloth. He wants to cut it into strips $1\frac{3}{4}$ yd long. How many strips will he have? How much cloth remains, if any?

36. Evette has $41\frac{1}{2}$ ft of string. She wants to cut it into pieces $3\frac{3}{4}$ ft long. How many pieces of string will she have? How much string remains, if any?

37. In squeezing oranges for fresh juice, three oranges yield about $\frac{1}{3}$ of a cup.

(a) How much juice could you expect to obtain from a bag containing 24 oranges?

(b) If you needed 8 cups of orange juice, how many bags of oranges should you buy?

38. A farmer died and left 17 cows to be divided among three workers. The first worker was to receive $\frac{1}{2}$ of the cows, the second worker was to receive $\frac{1}{3}$ of the cows, and the third worker was to receive $\frac{1}{9}$ of the cows. The executor of the farmer's estate realized that 17 cows could not be divided into halves, thirds, or ninths and so added a neighbor's cow to the farmer's. With 18 cows, the executor gave 9 cows to the first worker, 6 cows to the second worker, and 2 cows to the third worker. This accounted for the 17 cows, so the executor returned the borrowed cow to the neighbor. Explain why this works.

39. Division of fractions is not commutative.

For example, $\frac{3}{4} \div \frac{5}{6} \neq \frac{5}{6} \div \frac{3}{4}$.

There could be an exception. Can you think of a situation in which division of fractions would be commutative?

40. Josephine's boss tells her that her salary is to be divided by $\frac{1}{3}$. Should she quit?

41. Compare the English phrases: "divide in half" and "divide by one-half." Do they say the same thing? Create examples to support your answer.

ANSWERS

36. ____________

37. ____________

38. ____________

39. ____________

40. ____________

41. ____________

42. ______

43. ______

44. ______

45. ______

46. ______

47. ______

48. ______

42. **(a)** Compute: $5 \div \frac{1}{10}$; $5 \div \frac{1}{100}$; $5 \div \frac{1}{1000}$; $5 \div \frac{1}{10000}$.

(b) As the divisor gets smaller (approaches 0), what happens to the quotient?

(c) What does this say about the answer to $5 \div 0$?

Solve the applications. Write each answer in lowest terms.

43. **Money.** You collect 3 dimes, 2 dimes, and then 4 dimes. How much money do you have as a fraction of a dollar?

44. **Money.** You collect 7 nickels, 4 nickels, and then 5 nickels. How much money do you have as a fraction of a dollar?

45. **Work.** You work 7 h one day, 5 h the second day, and 6 h the third day. How long did you work, as a fraction of a 24-h day?

46. **Time.** One task took 7 min, a second task took 12 min, and a third task took 21 min. How long did the three tasks take, as a fraction of an hour?

47. **Perimeter.** What is the perimeter of a rectangle if the length is $\frac{7}{10}$ in. and the width is $\frac{2}{10}$ in.?

48. **Perimeter.** Find the perimeter of a rectangular picture frame if the width is $\frac{7}{9}$ yd and the length is $\frac{5}{9}$ yd.

ANSWERS

49. ______

50. ______

51. ______

52. ______

53. ______

54. ______

55. ______

56. ______

57. ______

49. Athletics. Patrick spent $\frac{4}{9}$ of an hour in the batting cages on Friday, and $\frac{7}{9}$ of an hour on Saturday. He wants to spend 2 h over 3 days. How much time should he spend on Sunday to accomplish this goal?

50. Quality control. Maria, a road inspector, must inspect $\frac{17}{30}$ of a mile of road. If she has already inspected $\frac{11}{30}$ of a mile, how much more does she need to inspect?

51. Perimeter. Find the perimeter of the figure.

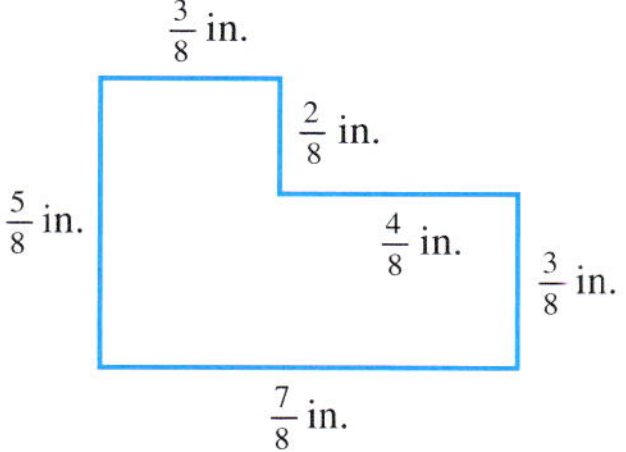

Find the perimeters of the triangles.

52.

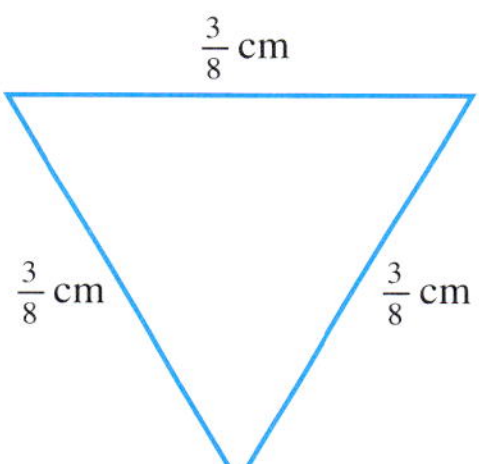

53.

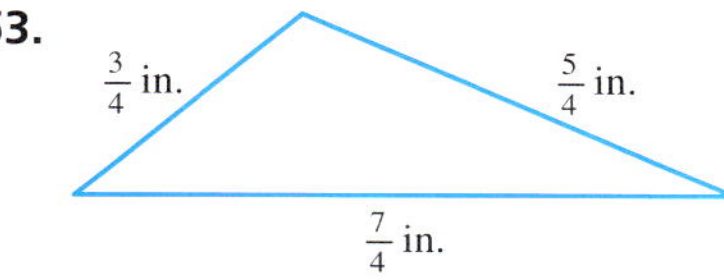

54.

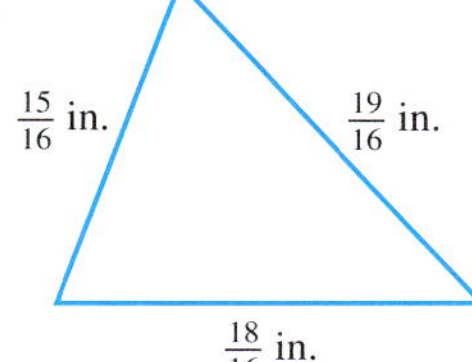

55.

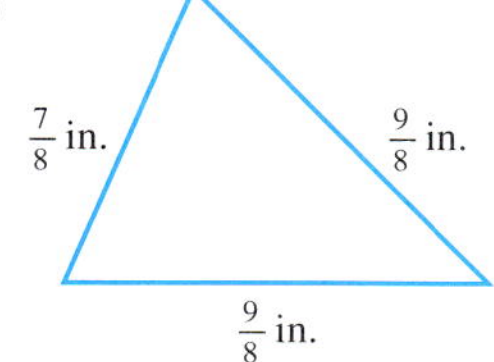

Find the perimeters of the polygons.

56.

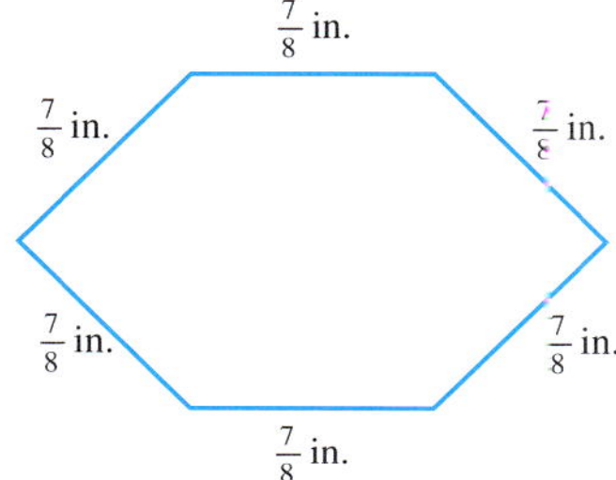

57.

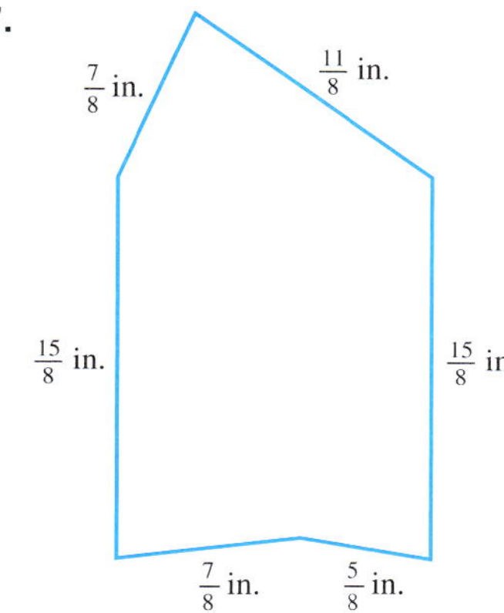

ANSWERS

58. ______

59. ______

60. ______

61. ______

62. ______

63. ______

64. ______

65. ______

58. Bolt size. Bolts can be purchased with diameters of $\frac{3}{8}$, $\frac{1}{4}$, or $\frac{3}{16}$ in. Which is smallest?

59. Plywood size. Plywood comes in thicknesses of $\frac{5}{8}$, $\frac{3}{4}$, $\frac{1}{2}$, and $\frac{3}{8}$ in. Which size is thickest?

60. Doweling. Doweling is sold with diameters of $\frac{1}{2}$, $\frac{9}{16}$, $\frac{5}{8}$, and $\frac{3}{8}$ in. Which size is smallest?

61. Elian is asked to create a fraction equivalent to $\frac{1}{4}$. His answer is $\frac{4}{7}$. What did he do wrong? What would be a correct answer?

62. A sign on a busy highway says Exit 5A is $\frac{3}{4}$ mi away and Exit 5B is $\frac{5}{8}$ mi away. Which exit is first?

63. Complete the following cross number puzzle.

ACROSS

2. The LCM of 11 and 13

4. The GCF of 120 and 300

7. The GCF of 13 and 52

8. The GCF of 360 and 540

DOWN

1. The LCM of 8, 14, and 21

3. The LCM of 16 and 12

5. The LCM of 2, 5, and 13

6. The GCF of 54 and 90

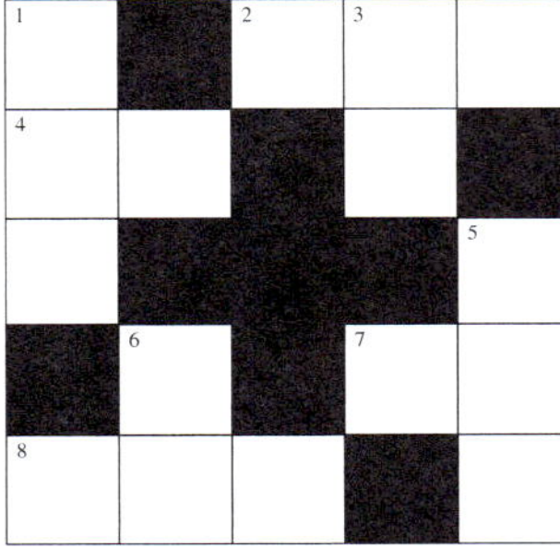

64. Countertop thickness. A countertop consists of a board $\frac{3}{4}$ in. thick and tile $\frac{3}{8}$ in. thick. What is the overall thickness?

65. Budgets. Amy budgets $\frac{2}{5}$ of her income for housing and $\frac{1}{6}$ of her income for food. What fraction of her income is budgeted for these two purposes? What fraction of her income remains?

66. ______
67. ______
68. ______
69. ______
70. ______
71. ______
72. ______

66. Daily schedule. A person spends $\frac{3}{8}$ day at work and $\frac{1}{3}$ day sleeping. What fraction of a day do these two activities use? What fraction of the day remains?

67. Distance. Jose walked $\frac{3}{4}$ mi to the store, $\frac{1}{2}$ mi to a friend's house, and then $\frac{2}{3}$ mi home. How far did he walk?

68. Perimeter. Find the perimeter of, or the distance around, the figure.

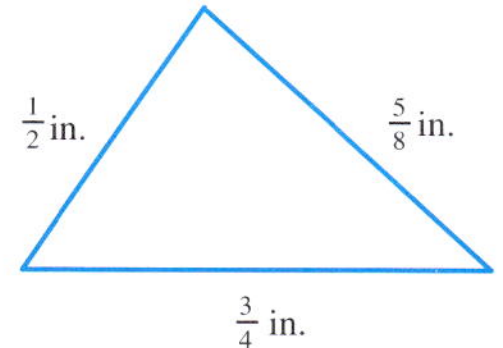

69. Budgeting. A budget guide states that you should spend $\frac{1}{4}$ of your salary for housing, $\frac{3}{16}$ for food, $\frac{1}{16}$ for clothing, and $\frac{1}{8}$ for transportation. What total portion of your salary will these four expenses account for?

70. Salary. Deductions from your paycheck are made roughly as follows: $\frac{1}{8}$ for federal tax, $\frac{1}{20}$ for state tax, $\frac{1}{20}$ for social security, and $\frac{1}{40}$ for a savings withholding plan. What portion of your pay is deducted?

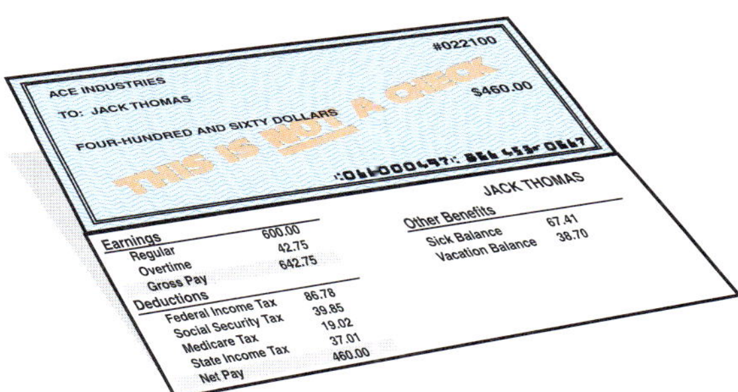

For exercises 71 and 72, find the missing dimension (?) in the given figure.

71.

$\frac{7}{16}$ in. ?

$\frac{3}{4}$ in.

72.

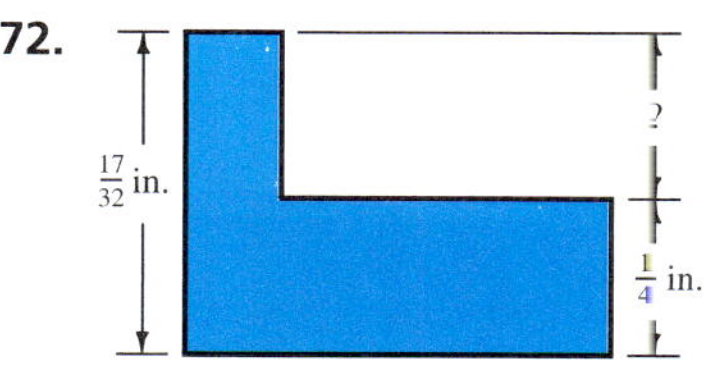

73. ______

74. ______

75. ______

76. ______

77. ______

78. ______

79. ______

73. Quantity of material. A roll of paper contains $30\frac{1}{4}$ yd. If $16\frac{7}{8}$ yd is cut from the roll, how much paper remains?

74. Geometry. Find the missing dimension in the figure.

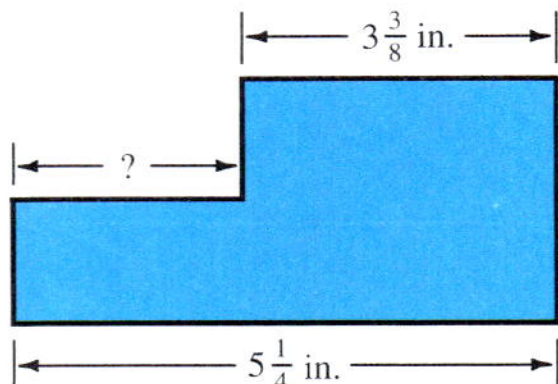

75. Carpentry. A $4\frac{1}{4}$-in. bolt is placed through a board that is $3\frac{1}{2}$ in. thick. How far does the bolt extend beyond the board?

76. Working hours. Ben can work 20 h per week on a part-time job. He works $5\frac{1}{2}$ h on Monday and $3\frac{3}{4}$ h on Tuesday. How many more hours can he work during the week?

77. Geometry. Find the missing dimension in the figure.

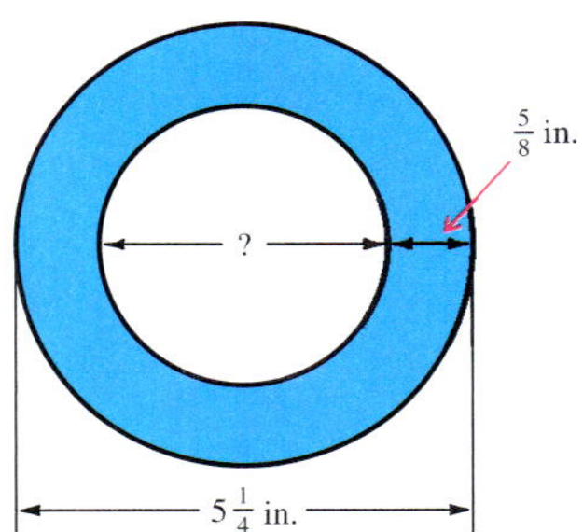

78. Carpeting. The Whites used $20\frac{3}{4}$ yd^2 of carpet for their living room, $15\frac{1}{2}$ yd^2 for the dining room, and $6\frac{1}{4}$ yd^2 for a hallway. How much will remain if they began with a 50-yd^2 roll of carpeting?

79. Construction. A construction company has bids for paving roads of $1\frac{1}{2}$, $\frac{3}{4}$, and $3\frac{1}{3}$ mi for the month of July. With their present equipment, they can pave 8 mi in 1 month. How much more work can they take on in July?

ANSWERS

80. ________

81. ________

82. ________

83. ________

84. ________

80. **Travel.** On an 8-h trip, Jack drives $2\frac{3}{4}$ h and Pat drives $2\frac{1}{2}$ h. How many hours are left to drive?

81. **Distance.** A runner has told herself that she will run 20 mi each week. She runs $5\frac{1}{2}$ mi on Sunday, $4\frac{1}{4}$ mi on Tuesday, $4\frac{3}{4}$ mi on Wednesday, and $2\frac{1}{8}$ mi on Friday. How far must she run on Saturday to meet her goal?

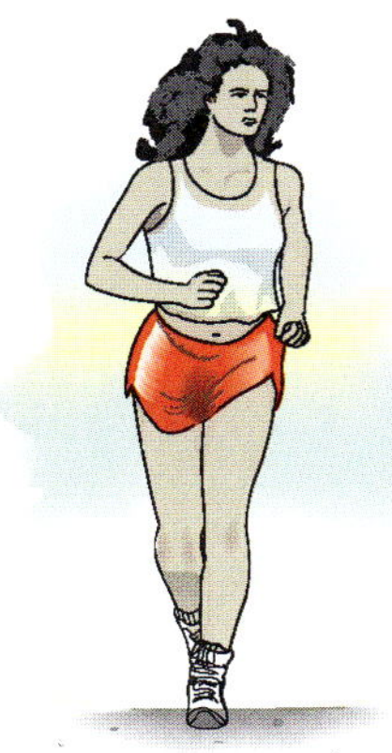

82. **Environment.** If paper takes up $\frac{1}{2}$ of the space in a landfill and plastic takes up $\frac{1}{10}$ of the space, how much of the landfill is used for other materials?

83. **Environment.** If paper takes up $\frac{1}{2}$ of the space in a landfill and organic waste takes up $\frac{1}{8}$ of the space, how much of the landfill is used for other materials?

84. **Interest.** The interest rate on an auto loan in May was $12\frac{3}{8}\%$. By September the rate was up to $14\frac{1}{4}\%$. How much did the interest rate increase over the period?

Answers

1. \$16 **3.** $17\frac{1}{2}$ gal **5.** 4 cups **7.** $2\frac{7}{8}$ ft^2 **9.** 75 mi **11.** 60 in. **13.** $\frac{5}{12}$ **15.** $12\frac{1}{2}$ yd^2 **17.** 2520 mi **19.** 66 in. **21.** 4 in.^{2} **23.**  **25.** $\frac{3}{4}$ ft **27.** 76 $\frac{\text{mi}}{\text{h}}$ **29.** 13 servings **31.** 51 packages **33.** 64 sheets **35.** 4; $\frac{1}{2}$ yd **37.** $2\frac{2}{3}$ cups; 3 bags **39.**

41. 

43. $\frac{9}{10}$ of a dollar

45. $\frac{3}{4}$ day

47. $\frac{9}{5}$ in. or $1\frac{4}{5}$ in.

49. $\frac{7}{9}$ h

51. 3 in.

53. $\frac{15}{4}$ in. $= 3\frac{3}{4}$ in.

55. $\frac{25}{8}$ in. $= 3\frac{1}{8}$ in.

57. $\frac{60}{8}$ in. $= 7\frac{1}{2}$ in.

59. $\frac{3}{4}$ in.

61. 

63.

1		1	4	3
6	0		8	
8				1
	1		1	3
1	8	0		0

65. $\frac{17}{30}, \frac{13}{30}$

67. $1\frac{11}{12}$ mi

69. $\frac{5}{8}$

71. $\frac{5}{16}$ in.

73. $13\frac{3}{8}$ yd

75. $\frac{3}{4}$ in.

77. 4 in.

79. $2\frac{5}{12}$ mi

81. $3\frac{3}{8}$ mi

83. $\frac{3}{8}$

4.5 Equations Containing Fractions

OBJECTIVES

1. Solve equations containing fractions
2. Distinguish between solving fractional equations and simplifying fractional expressions

Recall from Section 3.6 that to solve an equation such as $\frac{1}{7}x = 9$ we multiply both sides of the equation by 7, the reciprocal of $\frac{1}{7}$. From our work with fractions, we know that $\frac{1}{7}x = 9$ is equivalent to the equation $\frac{x}{7} = 9$ since $\frac{1}{7}x = \frac{1}{7} \cdot \frac{x}{1} = \frac{x}{7}$. This means that we may solve the equation $\frac{x}{7} = 9$ in the same way we solved $\frac{1}{7}x = 9$, by multiplying both sides of the equation by the reciprocal of $\frac{1}{7}$ which is 7.

If we rewrite $\frac{x}{7} = 9$ as $\frac{x}{7} = \frac{9}{1}$, we observe that 7, the number by which we multiplied both sides of the equation, is the LCD of the fractions $\frac{x}{7}$ and $\frac{9}{1}$. This observation leads us to a method for solving **fractional equations,** which are equations that contain fractions as one or more of their terms.

To solve a fractional equation, we multiply each term of the equation by the LCD of all of the fractions. The resulting equation should be equivalent to the original equation and be cleared of all fractions.

Example 1

Solving Fractional Equations

(a) Solve.

$$\frac{x}{3} = 6$$

$\frac{x}{3} = 6$ is equivalent to $\frac{x}{3} = \frac{6}{1}$. The LCD for $\frac{x}{3}$ and $\frac{6}{1}$ is 3. Multiply both sides of the equation by 3.

$$3\left(\frac{x}{3}\right) = 3 \cdot 6$$

$$x = 18$$

This leaves x alone on the left because

$3\left(\frac{x}{3}\right) = \frac{3}{1} \cdot \frac{x}{3} = \frac{x}{1} = x$

The solution is 18. To check, replace x with 18 in the *original* equation:

$$\frac{18}{3} \stackrel{?}{=} 6$$

$$6 = 6 \quad \text{(True)}$$

The solution is verified.

(b) Solve.

$$\frac{x}{5} = -9$$

$\frac{x}{5} = -9$ is equivalent to $\frac{x}{5} = \frac{-9}{1}$. The LCD for $\frac{x}{5}$ and $\frac{-9}{1}$ is 5. Multiply both sides of the equation by 5.

$$5\left(\frac{x}{5}\right) = 5(-9)$$

$$x = -45$$

The solution is -45. To check, we replace x with -45:

$$\frac{-45}{5} \stackrel{?}{=} -9$$

$$-9 = -9 \quad \text{(True)}$$

The solution is verified.

CHECK YOURSELF 1

Solve and check.

(a) $\frac{x}{7} = 3$ **(b)** $\frac{x}{4} = -8$

When the variable is multiplied by a fraction that has a numerator other than 1, there are two approaches to finding the solution.

Example 2

Solving Fractional Equations

Solve.

NOTE Remember that, $\frac{3}{5}x$ can be written as $\frac{3x}{5}$.

$$\frac{3}{5}x = 9$$

One approach is to multiply by 5, the LCD of $\frac{3x}{5}$ and $\frac{9}{1}$, as the first step.

$$5\left(\frac{3}{5}x\right) = 5 \cdot 9$$

$$3x = 45$$

Now we divide by 3.

$$\frac{3x}{3} = \frac{45}{3}$$

$$x = 15$$

To check:

$$\frac{3}{5} \cdot 15 \stackrel{?}{=} 9$$

$$9 = 9 \quad \text{(True)}$$

The solution is verified.

A second approach uses our knowledge of reciprocals and is generally a bit more efficient. We multiply both sides of the equation by $\frac{5}{3}$.

NOTE Recall that $\frac{5}{3}$ is the *reciprocal* of $\frac{3}{5}$, and the product of a number and its reciprocal is just 1! So

$$\left(\frac{5}{3}\right)\left(\frac{3}{5}\right) = 1$$

$$\frac{5}{3}\left(\frac{3}{5}x\right) = \frac{5}{3} \cdot 9$$

$$x = \frac{5}{\cancel{3}_1} \cdot \frac{\cancel{9}^3}{1} = 15$$

So $x = 15$, as before.

CHECK YOURSELF 2

Solve and check.

$$\frac{2}{3}x = 18$$

When an equation has more than one fraction, we first multiply by the LCD to clear the denominators.

Example 3

Solving an Equation with Two Fractions

Solve each equation.

(a) $\frac{x}{3} = \frac{5}{8}$

The LCD is 24. Multiplying both sides of the equation by 24, we get

$$\frac{x}{3} \cdot 24 = \frac{5}{8} \cdot 24$$

$$8x = 15$$

$$x = \frac{15}{8}$$

NOTE For practice you should verify that $\dfrac{\frac{15}{8}}{3} = \frac{5}{8}$

(b) $\frac{3x}{10} = \frac{7}{12}$

The LCD is 60. Multiplying both sides by 60, we get

$$\frac{3x}{10} \cdot 60 = \frac{7}{12} \cdot 60$$

$$18x = 35$$

$$x = \frac{35}{18}$$

NOTE Again, verify that $\frac{3\left(\frac{35}{18}\right)}{10} = \frac{7}{12}$

CHECK YOURSELF 3

Solve each equation.

(a) $\frac{x}{10} = \frac{7}{2}$ **(b)** $\frac{2x}{6} = \frac{3}{15}$

Caution! Remember that, when we are solving an equation, we are looking for a value for the variable (usually x) that makes the equation true. When we are simplifying an expression, we are finding an equivalent expression.

Example 4

Identifying Equations

For each, decide whether you are given an equation or an expression. Then state whether you could look for a solution.

(a) $\frac{3x}{5} = \frac{3}{12}$ This is an equation. You can look for a solution.

(b) $\frac{3x}{10} + \frac{7}{12}$ This is an expression. You cannot solve an expression!

(c) $\frac{x}{3} - \frac{1}{5}$ This is an expression. You cannot solve an expression.

(d) $\frac{x}{6} + 6 = \frac{4}{9}$ This is an equation. We will look at solving equations of this form in the remainder of this section.

CHECK YOURSELF 4

For each, decide whether you are given an equation or an expression. Then state whether you could look for a solution.

(a) $\frac{x}{15} - \frac{3}{12}$ **(b)** $\frac{3x}{10} = \frac{7}{12}$

(c) $\frac{x}{2} - 6$ **(d)** $\frac{x}{5} + 6 = \frac{4}{9}$

Our method of multiplying by the LCD to clear fractions will be applied to solving an equation with more than one term on one side of the equation in Example 5. We will see that the distributive property plays an important role.

Example 5

Solving Fractional Equations

Solve.

NOTE This equation has three terms: $\frac{x}{2}$, $-\frac{1}{3}$, and $\frac{2x+3}{6}$. The sign of the term is not used to find the LCD.

$$\frac{x}{2} - \frac{1}{3} = \frac{2x+3}{6}$$

The LCD for $\frac{x}{2}, \frac{1}{3}$, and $\frac{2x+3}{6}$ is 6. Multiply both sides of the equation by 6. Using the distributive property, we multiply *each* term by 6.

NOTE By the multiplication property of equality, this equation is equivalent to the original equation.

$$6 \cdot \frac{x}{2} - 6 \cdot \frac{1}{3} = 6\left(\frac{2x+3}{6}\right) \qquad \text{or} \qquad 3x - 2 = 2x + 3$$

Solving as before, we have

$$3x - 2x = 3 + 2 \qquad \text{or} \qquad x = 5$$

To check, substitute 5 for x in the *original* equation:

$$\frac{5}{2} - \frac{1}{3} \stackrel{?}{=} \frac{2 \cdot 5 + 3}{6}$$

$$\frac{15}{6} - \frac{2}{6} \stackrel{?}{=} \frac{10+3}{6}$$

$$\frac{13}{6} = \frac{13}{6} \qquad \text{(True)}$$

The solution is verified.

CHECK YOURSELF 5

Solve and check.

$$\frac{x}{4} - \frac{1}{6} = \frac{4x-5}{12}$$

CAUTION

Caution! Sometimes, one or more of the terms in an equation will be integers. It is important to remember that every term must be multiplied by the LCD. That includes these integer terms.

Example 6

Solving Equations Involving Both Fractions and Integers

Solve the equation.

$$\frac{x}{2} + 6 = \frac{3}{4}$$

The LCD is 4. Multiplying both sides by 4, we get

$$4 \cdot \left(\frac{x}{2} + 6\right) = \frac{3}{4} \cdot 4$$

$$4 \cdot \frac{x}{2} + 4 \cdot 6 = \frac{3}{4} \cdot 4$$

$$2x + 24 = 3$$

$$2x = -21$$

$$x = \frac{-21}{2}$$

CHECK YOURSELF 6

Solve the equation.

$$\frac{x}{3} - 4 = \frac{1}{4}$$

CHECK YOURSELF ANSWERS

1. **(a)** 21; **(b)** -32 **2.** 27 **3.** **(a)** $x = 35$; **(b)** $x = \frac{3}{5}$

4. **(a)** An expression, can't be solved; **(b)** an equation, can be solved; **(c)** an expression, can't be solved; **(d)** an equation, can be solved.

5. 3 **6.** $x = \frac{51}{4}$

Exercises

Name ______________________

Section __________ Date __________

Solve for x and check your result.

1. $\frac{x}{4} = 8$

2. $\frac{x}{3} = 12$

3. $\frac{x}{8} = 5$

4. $\frac{x}{7} = 2$

5. $\frac{x}{6} = -3$

6. $\frac{x}{2} = -20$

7. $\frac{x}{5} = -4$

8. $\frac{x}{9} = -3$

9. $\frac{3}{4}x = 15$

10. $\frac{4}{5}x = 12$

11. $\frac{2}{3}x = 16$

12. $\frac{2}{5}x = 8$

13. $\frac{1}{6}x = -3$

14. $\frac{1}{4}x = 2$

15. $\frac{1}{5}x = 5$

16. $\frac{1}{3}x = -12$

17. $\frac{3}{2}x = 24$

18. $\frac{4}{3}x = 12$

19. $\frac{5}{3}x = -15$

20. $\frac{5}{2}x = -20$

ANSWERS

1. ______________________
2. ______________________
3. ______________________
4. ______________________
5. ______________________
6. ______________________
7. ______________________
8. ______________________
9. ______________________
10. ______________________
11. ______________________
12. ______________________
13. ______________________
14. ______________________
15. ______________________
16. ______________________
17. ______________________
18. ______________________
19. ______________________
20. ______________________

ANSWERS

21. ______
22. ______
23. ______
24. ______
25. ______
26. ______
27. ______
28. ______
29. ______
30. ______
31. ______
32. ______
33. ______
34. ______
35. ______
36. ______
37. ______
38. ______
39. ______
40. ______
41. ______
42. ______

In exercises 21 to 34, determine the smallest multiplier to use in order to clear the equation of fractions. Do not solve.

21. $\frac{x}{6} = \frac{5}{3}$

22. $\frac{x}{4} = \frac{7}{8}$

23. $\frac{x}{2} = \frac{-7}{3}$

24. $\frac{x}{4} = \frac{-5}{12}$

25. $\frac{x}{5} = \frac{2}{3}$

26. $\frac{x}{7} = \frac{3}{4}$

27. $\frac{x}{6} = \frac{-9}{4}$

28. $\frac{x}{8} = \frac{-7}{12}$

29. $\frac{x}{12} = \frac{5}{9}$

30. $\frac{x}{15} = \frac{3}{10}$

31. $\frac{2x}{5} = \frac{3}{8}$

32. $\frac{4x}{3} = \frac{2}{5}$

33. $\frac{3x}{8} = \frac{-1}{6}$

34. $\frac{5x}{12} = \frac{-4}{9}$

Solve for x and check your result.

35. $\frac{x}{6} = \frac{5}{3}$

36. $\frac{x}{4} = \frac{7}{8}$

37. $\frac{x}{2} = \frac{-7}{3}$

38. $\frac{x}{4} = \frac{-5}{12}$

39. $\frac{x}{5} = \frac{2}{3}$

40. $\frac{x}{7} = \frac{3}{4}$

41. $\frac{x}{6} = \frac{-9}{4}$

42. $\frac{x}{8} = \frac{-7}{12}$

43. $\frac{x}{12} = \frac{5}{9}$

44. $\frac{x}{15} = \frac{3}{10}$

45. $\frac{2x}{5} = \frac{3}{8}$

46. $\frac{4x}{3} = \frac{2}{5}$

47. $\frac{3x}{8} = \frac{-1}{6}$

48. $\frac{5x}{12} = \frac{-4}{9}$

For exercises 49 to 56, decide whether you are given an equation or an expression. Then state whether you could look for a solution.

49. $\frac{x}{6} = \frac{7}{3}$

50. $\frac{x}{5} + \frac{3}{8}$

51. $\frac{2x}{5} - \frac{7}{8}$

52. $\frac{3x}{4} = \frac{12}{5}$

53. $\frac{x}{2} - 3$

54. $\frac{x}{8} + 4 = 0$

55. $\frac{3}{5}x - 2 = 6$

56. $\frac{2}{3}x + 4 - \frac{5}{6}$

Solve for x and check your result.

57. $\frac{x}{5} - \frac{1}{3} = \frac{x - 7}{3}$

58. $\frac{x}{6} + \frac{3}{4} = \frac{x - 1}{4}$

59. $\frac{x}{4} - \frac{1}{5} = \frac{4x + 3}{20}$

60. $\frac{x}{12} - \frac{1}{6} = \frac{2x - 7}{12}$

61. $\frac{x}{3} + 4 = \frac{5}{2}$

62. $\frac{x}{5} + 2 = \frac{7}{10}$

63. $\frac{x}{6} - 5 = \frac{3}{4}$

64. $\frac{x}{4} - 3 = \frac{5}{3}$

43. ______
44. ______
45. ______
46. ______
47. ______
48. ______
49. ______
50. ______
51. ______
52. ______
53. ______
54. ______
55. ______
56. ______
57. ______
58. ______
59. ______
60. ______
61. ______
62. ______
63. ______
64. ______

Answers

1. $x = 32$ **3.** $x = 40$ **5.** $x = -18$ **7.** $x = -20$ **9.** $x = 20$
11. $x = 24$ **13.** $x = -18$ **15.** $x = 25$ **17.** $x = 16$ **19.** $x = -9$
21. 6 **23.** 6 **25.** 15 **27.** 12 **29.** 36 **31.** 40 **33.** 24
35. $x = 10$ **37.** $x = \frac{-14}{3}$ **39.** $x = \frac{10}{3}$ **41.** $x = \frac{-27}{2}$ **43.** $x = \frac{20}{3}$
45. $x = \frac{15}{16}$ **47.** $x = \frac{-4}{9}$ **49.** Equation; yes **51.** Expression; no
53. Expression; no **55.** Equation; yes **57.** $x = 15$ **59.** $x = 7$
61. $x = \frac{-9}{2}$ **63.** $x = \frac{69}{2}$

4.6 Applications of Linear Equations in One Variable

4.6 OBJECTIVES

1. Solve applications involving linear equations
2. Solve applications involving perimeter

The main reason for learning how to set up and solve algebraic equations is so that we can use them to solve word problems. In fact, algebraic equations were *invented* to make solving word problems much easier. The first word problems that we know about are over 4,000 years old. They were literally "written in stone," on Babylonian tablets, about 500 years before the first algebraic equation made its appearance.

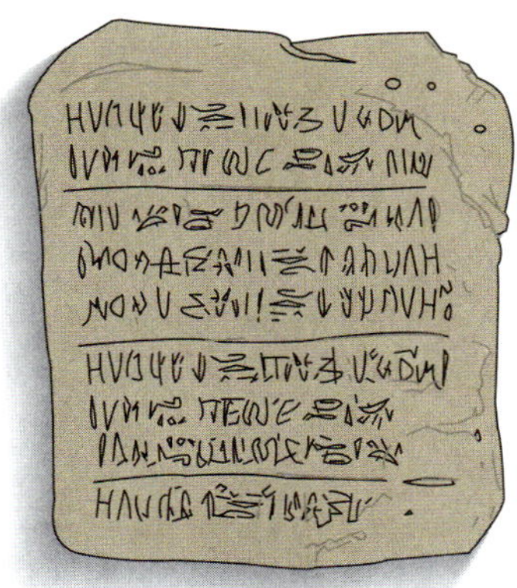

Before algebra, people solved word problems primarily by **substitution,** which is a method of finding unknown numbers by using trial and error in a logical way. Example 1 shows how to solve a word problem using substitution.

Example 1

Solving a Word Problem by Substitution

The sum of two consecutive integers is 37. Find the two integers.

If the two integers were 20 and 21, their sum would be 41. Because that's more than 37, the integers must be smaller. If the integers were 15 and 16, the sum would be 31. More trials yield that the sum of 18 and 19 is 37.

CHECK YOURSELF 1

The sum of two consecutive integers is 91. *Find the two integers.*

Most word problems are not so easily solved by substitution. For more complicated word problems, a five-step procedure is used. Using this step-by-step approach will, with practice, allow you to organize your work. Organization is the key to solving word problems. Here are the five steps.

Step by Step: Using Equations to Solve Word Problems

Step 1 Read the problem carefully. Then reread it to decide what you are asked to find.

Step 2 Choose a letter to represent one of the unknowns in the problem. Then represent all other unknowns of the problem with expressions that use the same letter.

Step 3 Translate the problem to the language of algebra to form an equation.

Step 4 Solve the equation and answer the question of the original problem.

Step 5 Check your solution by returning to the original problem.

NOTE We discussed these translations in Section 2.6. You might find it helpful to review that section before going on.

The third step is usually the hardest part. We must translate words to the language of algebra. Before we look at a complete example, the table may help you review that translation step.

Translating Words to Algebra

Words	Algebra
The sum of x and y	$x + y$
3 plus a	$3 + a$ or $a + 3$
5 more than m	$m + 5$
b increased by 7	$b + 7$
The difference of x and y	$x - y$
4 less than a	$a - 4$
s decreased by 8	$s - 8$
The product of x and y	$x \cdot y$ or xy
5 times a	$5 \cdot a$ or $5a$
Twice m	$2m$
The quotient of x and y	$\frac{x}{y}$
a divided by 6	$\frac{a}{6}$
One-half of b	$\frac{b}{2}$ or $\frac{1}{2}b$

Now we will look at some typical examples of translating phrases to algebra.

Example 2

Translating Statements

Translate each statement to an algebraic expression.

(a) The sum of a and 2 times b $a + 2b$

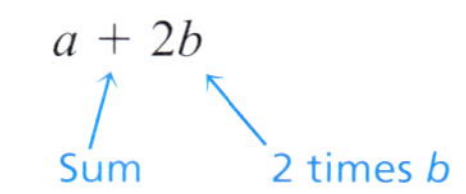

(b) 5 times m, increased by 1 $5m + 1$ (5 times m; Increased by 1)

(c) 5 less than 3 times x $3x - 5$ (3 times x; 5 less than)

(d) The product of x and y, divided by 3 $\frac{xy}{3}$ (The product of x and y; Divided by 3)

CHECK YOURSELF 2

Translate to algebra.

(a) 2 more than twice x
(b) 4 less than 5 times n
(c) The product of twice a and b
(d) The sum of s and t, divided by 5

Now we will work through a complete example. Although this problem could be solved by substitution, it is presented here to help you practice the five-step approach.

Example 3

Solving an Application

The sum of a number and 5 is 17. What is the number?

Step 1 *Read carefully.* You must find the unknown number.

Step 2 *Choose letters or variables.* Let x represent the unknown number. There are no other unknowns.

Step 3 *Translate.*

$$x + 5 = 17$$

(The sum of: $x + 5$; is: $=$)

NOTE Always return to the *original problem* to check your result and *not* to the equation of step 3. This will prevent possible errors!

Step 4 *Solve.*

$$\begin{aligned} x + 5 &= 17 \\ -5 \quad &\;\; -5 \\ \hline x &= 12 \end{aligned}$$

Add −5.

So the number is 12.

Step 5 *Check.* Is the sum of 12 and 5 equal to 17? Yes ($12 + 5 = 17$). We have checked our solution.

CHECK YOURSELF 3

The sum of a number and 8 *is* 35. *What is the number?*

Step by Step: Representing Consecutive Integers

Consecutive integers are integers that follow one another. To represent them in algebra:

If x is an integer, then $x + 1$ is the next consecutive integer, $x + 2$ is the next, and so on.

We'll need this idea in Example 4.

Example 4

Solving an Application

REMEMBER THE STEPS!
Read the problem carefully. What do you need to find?
Assign letters to the unknown or unknowns.
Write an equation.

The sum of two consecutive integers is 41. What are the two integers?

Step 1 We want to find the two consecutive integers.

Step 2 Let x be the first integer. Then $x + 1$ must be the next.

Step 3

The first integer — The second integer

$$x + \overbrace{x + 1} = 41$$

The sum — Is

NOTE Solve the equation.

Step 4

$$x + x + 1 = 41$$
$$2x + 1 = 41$$
$$2x = 40$$
$$x = 20$$

The first integer (x) is 20, and the next integer ($x + 1$) is 21.

NOTE Check.

Step 5 The sum of the two integers 20 and 21 is 41.

CHECK YOURSELF 4

The sum of three consecutive integers is 51. *What are the three integers?*

Sometimes algebra is used to reconstruct missing information. Example 5 does just that with some election information.

Example 5

Solving an Application

There were 55 more yes votes than no votes on an election measure. If 735 votes were cast in all, how many yes votes were there? How many no votes?

NOTE What do you need to find?

Step 1 We want to find the number of yes votes and the number of no votes.

NOTE Assign letters to the unknowns.

Step 2 Let x be the number of no votes. Then

$$\underbrace{x + 55}_{\text{55 more than } x}$$

is the number of yes votes.

NOTE Write an equation.

Step 3

$$\underset{\text{No votes}}{x} + \underbrace{x + 55}_{\text{Yes votes}} = 735$$

NOTE Solve the equation.

Step 4

$$x + x + 55 = 735$$
$$2x + 55 = 735$$
$$2x = 680$$
$$x = 340$$
$$\text{No votes } (x) = 340$$
$$\text{Yes votes } (x + 55) = 395$$

NOTE Check.

Step 5 Thus 340 no votes plus 395 yes votes equals 735 total votes. The solution checks.

CHECK YOURSELF 5

Francine earns $120 per month more than Rob. If they earn a total of $2,680 per month, what are their monthly salaries?

Similar methods will allow you to solve a variety of word problems. Example 6 includes three unknown quantities but uses the same basic solution steps.

Example 6

Solving an Application

Juan worked twice as many hours as Jerry. Marcia worked 3 more hours than Jerry. If they worked a total of 31 h, find out how many hours each worked.

Step 1 We want to find the hours each worked, so there are three unknowns.

Step 2 Let x be the hours that Jerry worked.

Twice Jerry's hours

Then $2x$ is Juan's hours worked

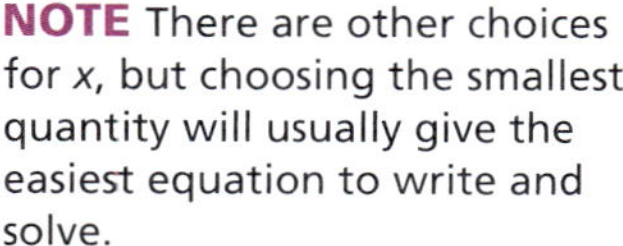

NOTE There are other choices for x, but choosing the smallest quantity will usually give the easiest equation to write and solve.

3 more hours than Jerry worked

and $x + 3$ is Marcia's hours.

Step 3

Jerry, Juan, Marcia

$$x + 2x + x + 3 = 31$$

Sum of their hours

Step 4

$$x + 2x + x + 3 = 31$$

$$4x + 3 = 31$$

$$4x = 28$$

$$x = 7$$

Jerry's hours $(x) = 7$

Juan's hours $(2x) = 14$

Marcia's hours $(x + 3) = 10$

Step 5 The sum of their hours $(7 + 14 + 10)$ is 31, and the solution is verified.

CHECK YOURSELF 6

Lucy jogged twice as many miles as Paul but 3 less than Isaac. If the three ran a total of 23 mi, how far did each person run?

The solutions for many problems from geometry will also yield linear equations. Consider Example 7.

Example 7

Solving a Geometry Application

NOTE Whenever you are working on an application involving geometric figures, you should draw a sketch of the problem, including the labels assigned in step 2.

The length of a rectangle is 1 cm less than 3 times the width. If the perimeter is 54 cm, find the dimensions of the rectangle.

Step 1 You want to find the dimensions (the width and length).

Step 2 Let x be the width.

Then $3x - 1$ is the length.

(3 times the width; 1 less than)

Step 3 To write an equation, we'll use this formula for the perimeter of a rectangle:

$$P = 2W + 2L$$

So

$$2x + 2(3x - 1) = 54$$

(Twice the width; Twice the length; Perimeter)

Step 4 Solve the equation.

$$2x + 2(3x - 1) = 54$$
$$2x + 6x - 2 = 54$$
$$8x = 56$$
$$x = 7$$

The width x is 7 cm, and the length, $3x - 1$, is 20 cm. We leave step 5, the check, to you.

NOTE Be sure to return to the original statement of the problem when checking your result.

CHECK YOURSELF 7

The length of a rectangle is 5 in. more than twice the width. If the perimeter of the rectangle is 76 in., what are the dimensions of the rectangle?

CHECK YOURSELF ANSWERS

1. 45 and 46 **2.** **(a)** $2x + 2$; **(b)** $5n - 4$; **(c)** $2ab$; **(d)** $\frac{s + t}{5}$

3. The equation is $x + 8 = 35$. The number is 27.

4. The equation is $x + x + 1 + x + 2 = 51$. The integers are 16, 17, and 18.

5. The equation is $x + x + 120 = 2{,}680$. Rob's salary is $1,280, and Francine's is $1,400.

6. Paul: 4 mi; Lucy: 8 mi; Isaac: 11 mi

7. The width is 11 in.; the length is 27 in.

Name ______________________

Section ________ Date ________

Exercises

Translate each statement to an algebraic equation. Let x represent the number in each case. Do not solve.

1. 3 more than a number is 7.

2. 5 less than a number is 12.

3. 7 less than 3 times a number is twice that same number.

4. 4 more than 5 times a number is 6 times that same number.

5. 2 times the sum of a number and 5 is 18 more than that same number.

6. 3 times the sum of a number and 7 is 4 times that same number.

7. 3 more than twice a number is 7.

8. 5 less than 3 times a number is 25.

9. 7 less than 4 times a number is 41.

10. 10 more than twice a number is 44.

11. 3 times a number is 12 more than that number.

12. 5 times a number is 8 less than that number.

Solve the word problems. Be sure to label the unknowns and to show the equation you use for the solution.

13. Number problem. The sum of a number and 7 is 33. What is the number?

14. Number problem. The sum of a number and 15 is 22. What is the number?

15. Number problem. The sum of a number and -15 is 7. What is the number?

16. Number problem. The sum of a number and -8 is 17. What is the number?

17. Number of votes cast. In an election, the winning candidate has 1,840 votes. If the total number of votes cast was 3,260, how many votes did the losing candidate receive?

18. Monthly earnings. Mike and Stefanie work at the same company and make a total of $2,760 per month. If Stefanie makes $1,400 per month, how much does Mike earn every month?

ANSWERS

1. ______
2. ______
3. ______
4. ______
5. ______
6. ______
7. ______
8. ______
9. ______
10. ______
11. ______
12. ______
13. ______
14. ______
15. ______
16. ______
17. ______
18. ______

ANSWERS

19. ______
20. ______
21. ______
22. ______
23. ______
24. ______
25. ______
26. ______
27. ______
28. ______
29. ______
30. ______
31. ______
32. ______
33. ______
34. ______

19. Number addition. The sum of twice a number and 7 is 33. What is the number?

20. Number addition. 3 times a number, increased by 8, is 50. Find the number.

21. Number subtraction. 5 times a number, minus 12, is 78. Find the number.

22. Number subtraction. 4 times a number, decreased by 20, is 44. What is the number?

23. Consecutive integers. The sum of two consecutive integers is 71. Find the two integers.

24. Consecutive integers. The sum of two consecutive integers is 145. Find the two integers.

25. Consecutive integers. The sum of three consecutive integers is 63. What are the three integers?

26. Consecutive integers. If the sum of three consecutive integers is 93, find the three integers.

27. Even integers. The sum of two consecutive even integers is 66. What are the two integers? (*Hint:* Consecutive even integers such as 10, 12, and 14 can be represented by x, $x + 2$, $x + 4$, and so on.)

28. Even integers. If the sum of two consecutive even integers is 86, find the two integers.

29. Odd integers. If the sum of two consecutive odd integers is 52, what are the two integers? (*Hint:* Consecutive odd integers such as 21, 23, and 25 can be represented by x, $x + 2$, $x + 4$, and so on.)

30. Odd integers. The sum of two consecutive odd integers is 88. Find the two integers.

31. Odd integers. The sum of three consecutive odd integers is 105. What are the three integers?

32. Even integers. The sum of three consecutive even integers is 126. What are the three integers?

33. Consecutive integers. The sum of four consecutive integers is 86. What are the four integers?

34. Consecutive integers. The sum of four consecutive integers is 62. What are the four integers?

35. Consecutive integers. 4 times an integer is 9 more than 3 times the next consecutive integer. What are the two integers?

36. Even integers. 4 times an even integer is 30 less than 5 times the next consecutive even integer. Find the two integers.

37. Election votes. In an election, the winning candidate had 160 more votes than the loser. If the total number of votes cast was 3,260, how many votes did each candidate receive?

38. Monthly salaries. Jody earns $140 more per month than Frank. If their monthly salaries total $2,760, what amount does each earn?

39. Appliance costs. A washer-dryer combination costs $650. If the washer costs $70 more than the dryer, what does each appliance cost?

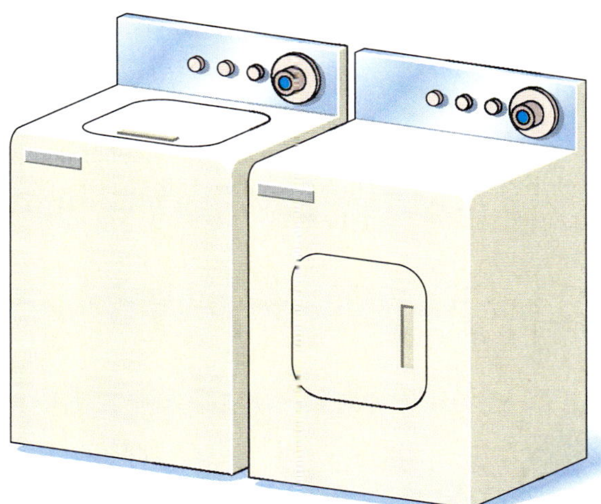

40. Length of materials. Yuri has a board that is 98 in. long. He wishes to cut the board into two pieces so that one piece will be 10 in. longer than the other. What should be the length of each piece?

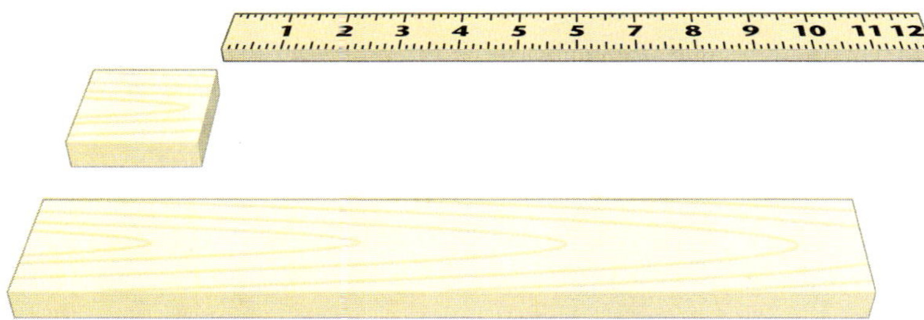

41. Age. Yan Ling is 1 year less than twice as old as his sister. If the sum of their ages is 14 years, how old is Yan Ling?

42. Age. Diane is twice as old as her brother Dan. If the sum of their ages is 27 years, how old are Diane and her brother?

ANSWERS

35. ______

36. ______

37. ______

38. ______

39. ______

40. ______

41. ______

42. ______

ANSWERS

43. ______

44. ______

45. ______

46. ______

47. ______

48. ______

49. ______

50. ______

51. ______

52. ______

43. Age. Maritza is 3 years less than 4 times as old as her daughter. If the sum of their ages is 37, how old is Maritza?

44. Age. Mrs. Jackson is 2 years more than 3 times as old as her son. If the difference between their ages is 22 years, how old is Mrs. Jackson?

45. Airfare costs. On her vacation in Europe, Jovita's expenses for food and lodging were \$60 less than twice as much as her airfare. If she spent \$2,400 in all, what was her airfare?

46. Earnings. Rachel earns \$6,000 less than twice as much as Tom. If their two incomes total \$48,000, how much does each earn?

47. Number of students. There are 99 students registered in three sections of algebra. There are twice as many students in the 10 A.M. section as the 8 A.M. section and 7 more students at 12 P.M. than at 8 A.M. How many students are in each section?

48. Gallons of oil. The Randolphs used 12 more gallons of fuel oil in October than in September and twice as much oil in November as in September. If they used 132 gal for the 3 months, how much was used during each month?

In exercises 49 to 52, find the length of each side of the figure for the given perimeter.

49.

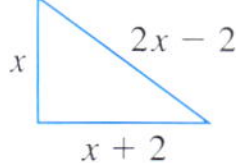

$P = 24$ in.

50.

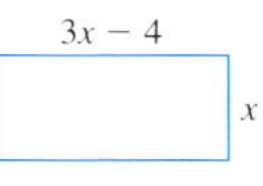

$P = 32$ cm

51.

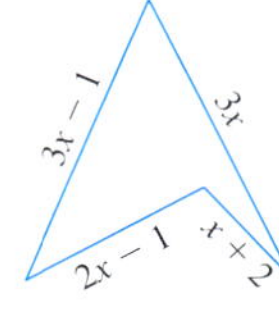

$P = 90$ in.

52.

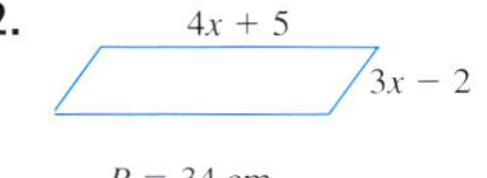

$P = 34$ cm

ANSWERS

53. ____

54. ____

55. ____

56. ____

57. ____

58. ____

53. "I make \$2.50 an hour more in my new job." If x = the amount I used to make per hour and y = the amount I now make, which equation(s) say the same thing as the given statement? Explain your choice(s) by translating the equation into English and comparing with the original statement.

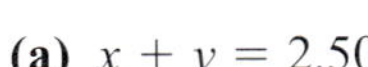

(a) $x + y = 2.50$ **(b)** $x - y = 2.50$

(c) $x + 2.50 = y$ **(d)** $2.50 + y = x$

(e) $y - x = 2.50$ **(f)** $2.50 - x = y$

54. "The river rose 4 ft above flood stage last night." If a = the river's height at flood stage, b = the river's height last night, which equations say the same thing as the given statement? Explain your choices by translating the equations into English and comparing the meaning with the original statement.

(a) $a + b = 4$ **(b)** $b - 4 = a$

(c) $a - 4 = b$ **(d)** $a + 4 = b$

(e) $b + 4 = b$ **(f)** $b - a = 4$

55. Maxine lives in Pittsburgh, Pennsylvania, and pays $8\frac{33}{100}$ cents per kilowatt hour (kWh) for electricity. During the 6 months of cold winter weather, her household uses about 1,500 kWh of electric power per month. During the two hottest summer months, the usage is also high because the family uses electricity to run an air conditioner. During these summer months, the usage is 1,200 kWh per month; the rest of the year, usage averages 900 kWh per month.

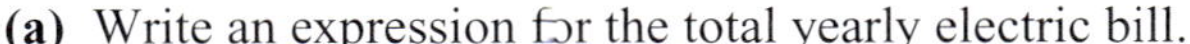

(a) Write an expression for the total yearly electric bill.

(b) Maxine is considering spending \$2,000 for more insulation for her home so that it is less expensive to heat and to cool. The insulation company claims that "with proper installation the insulation will reduce your heating and cooling bills by 25%." If Maxine invests the money in insulation, how long will it take her to get her money back in saving on her electric bill? Write to her about what information she needs to answer this question. Give her your opinion about how long it will take to save \$2,000 on heating and cooling bills, and explain your reasoning. What is your advice to Maxine?

56. Subtracting numbers. If one-third of a number is subtracted from three-fourths of that number, the difference is 15. What is the number?

57. Subtracting numbers. If one-fourth of a number is subtracted from two-fifths of the number, the difference is 3. Find the number.

58. Adding numbers. If five-sixths of a number is added to one-fifth of the number, the sum is 31. What is the number?

Answers

1. $x + 3 = 7$ **3.** $3x - 7 = 2x$ **5.** $2(x + 5) = x + 18$ **7.** $2x + 3 = 7$ **9.** $4x - 7 = 41$ **11.** $3x = x + 12$ **13.** 26 **15.** 22 **17.** 1,420 **19.** 13 **21.** 18 **23.** 35, 36 **25.** 20, 21, 22 **27.** 32, 34 **29.** 25, 27 **31.** 33, 35, 37 **33.** 20, 21, 22, 23 **35.** 12, 13 **37.** 1,710; 1,550 **39.** Washer, \$360; dryer, \$290 **41.** 9 years old **43.** 29 years old **45.** \$820 **47.** 8 A.M.: 23; 10 A.M.: 46; 12 P.M.: 30 **49.** 6 in., 8 in., 10 in. **51.** 12 in., 19 in., 29 in., 30 in. **53.** 
55.   **57.** 20

4 Summary

DEFINITION/PROCEDURE	EXAMPLE	REFERENCE
Addition and Subtraction of Fractions		**Section 4.1**
To Add (Subtract) Like Fractions 1. Add (subtract) the numerators. 2. Place the sum (difference) over the common denominator. 3. Simplify the resulting fraction if necessary.	$\frac{5}{18} + \frac{7}{18} = \frac{12}{18} = \frac{2}{3}$	p. 271, 273
Least Common Multiple (LCM) The LCM is the *smallest* number that is a multiple of each of a group of numbers.		p. 277
To Find the LCD of a Group of Fractions 1. Write the prime factorization for each of the denominators. 2. Find all the prime factors that appear in any one of the prime factorizations. 3. Form the product of those prime factors, using each factor the greatest number of times it occurs in any one factorization.	To find the LCD of fractions with denominators 4, 6, and 15: $4 = 2 \cdot 2$ $6 = 2 \quad \cdot 3$ $15 = \quad 3 \cdot 5$ $2 \cdot 2 \cdot 3 \cdot 5$ The LCD $= 2 \cdot 2 \cdot 3 \cdot 5$, or 60.	p. 281
To Add (Subtract) Unlike Fractions 1. Find the LCD of the fractions. 2. Change each fraction to an equivalent fraction with the LCD as a common denominator. 3. Add (subtract) the resulting like fractions as before.	$\frac{3}{4} + \frac{7}{10} = \frac{15}{20} + \frac{14}{20}$ $= \frac{29}{20}$	p. 282, 284
Operations on Mixed Numbers		**Section 4.2**
Mixed Number The sum of a whole number and a proper fraction.	$2\frac{1}{3}$ and $5\frac{7}{8}$ are mixed numbers. Note that $2\frac{1}{3}$ means $2 + \frac{1}{3}$.	p. 291
To Change an Improper Fraction into a Mixed Number 1. Divide the numerator by the denominator. The quotient is the whole-number portion of the mixed number. 2. If there is a remainder, write the remainder over the original denominator. This gives the fractional portion of the mixed number.	$\frac{22}{5} = 4\frac{2}{5}$ $5\overline{)22}$ quotient 4 ← Quotient $\underline{20}$ 2 ← Remainder	p. 292
To Change a Mixed Number to an Improper Fraction 1. Multiply the denominator of the fraction by the whole-number portion of the mixed number. 2. Add the numerator of the fraction to that product. 3. Write that sum over the original denominator to form the improper fraction.	Denominator, Whole number, Numerator $5\frac{3}{4} = \frac{(4 \cdot 5) + 3}{4} = \frac{23}{4}$ Denominator	p. 293

Continued

DEFINITION/PROCEDURE	EXAMPLE	REFERENCE
Multiplying or Dividing Mixed Numbers Convert any mixed or whole numbers to improper fractions. Then multiply or divide the fractions as before.	$6\frac{2}{3} \cdot 3\frac{1}{5} = \frac{\overset{4}{\cancel{20}}}{3} \cdot \frac{16}{\underset{1}{\cancel{5}}}$ $= \frac{64}{3} = 21\frac{1}{3}$	p. 296
To Add or Subtract Mixed Numbers **1.** Rewrite as improper fractions. **2.** Add or subtract the fractions. **3.** Rewrite the results as a mixed number if required.	$5\frac{1}{2} - 3\frac{3}{4} = \frac{11}{2} - \frac{15}{4}$ $= \frac{22}{4} - \frac{15}{4}$ $= \frac{7}{4}$ $= 1\frac{3}{4}$	p. 298
Complex Fractions		**Section 4.3**
A complex fraction has a fraction in its numerator or denominator (or both).	$\dfrac{\frac{2}{3}}{\frac{5}{6}}$ is a complex fraction.	p. 307
To simplify a complex fraction, multiply the numerator and denominator by the LCD of the fractions within the complex fraction.	$\dfrac{\frac{2}{3}}{\frac{5}{6}} = \dfrac{\frac{2}{3} \cdot 6}{\frac{5}{6} \cdot 6} = \frac{4}{5}$	p. 308
Equations Containing Fractions		**Section 4.5**
To solve an equation containing one or more fractions: **1.** Find the LCD of the denominators. **2.** Multiply *every term* by the LCD. **3.** Solve the resulting equation as before.	To solve $\frac{x}{5} + 1 = \frac{1}{2}$: **1.** The LCD is 10. **2.** $10\left(\frac{x}{5} + 1\right) = 10\left(\frac{1}{2}\right)$ **3.** $2x + 10 = 5$ $2x = -5$ $x = \frac{-5}{2}$	p. 333
Applications of Linear Equations in One Variable		**Section 4.6**
To use an equation to solve a word problem: **1.** Read the problem carefully to decide what you are asked to find. **2.** Choose a letter to represent one of the unknowns. Then represent all other unknowns with expressions using that same letter. **3.** Translate the problem to algebra to form an equation. **4.** Solve the equation, and answer the original question. **5.** Check your solution by returning to the original problem.		p. 342
Consecutive integers If x is an integer, then $x + 1$ is the next consecutive integer, $x + 2$ is the next, and so on.	If 20 is an integer, $20 + 1 = 21$ is the next consecutive integer.	p. 344

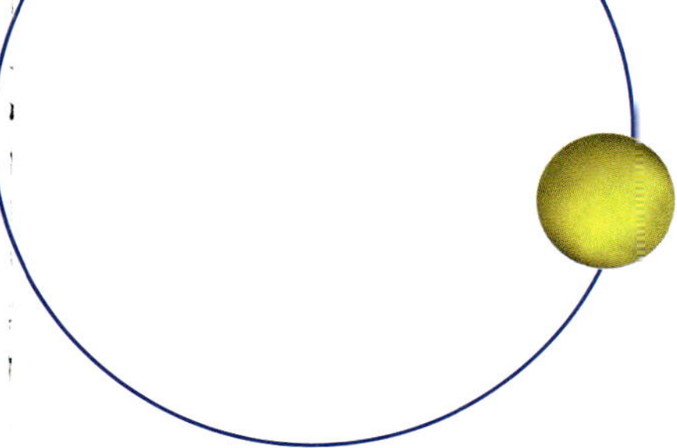

Summary and Review Exercises

[4.1] Add. Simplify when possible.

1. $\frac{8}{15} + \frac{2}{15}$

2. $\frac{-4}{7} + \frac{-3}{7}$

3. $\frac{2}{9} + \frac{5}{9} + \frac{4}{9}$

4. $\frac{4}{15} + \frac{7}{15} + \frac{7}{15}$

Find the LCM for each group of numbers.

5. 9, 12, and 24

6. 14, 21, and 28

Arrange the fractions in order from smallest to largest.

7. $\frac{5}{8}, \frac{7}{12}$

8. $\frac{5}{6}, \frac{4}{5}, \frac{7}{10}$

Complete the statements using the symbol $<$, $=$, or $>$.

9. $\frac{5}{12}$ ____ $\frac{3}{8}$

10. $\frac{3}{7}$ ____ $\frac{9}{21}$

11. $\frac{9}{16}$ ____ $\frac{7}{12}$

Write as equivalent fractions with the LCD as a common denominator.

12. $\frac{1}{6}, \frac{7}{8}$

13. $\frac{3}{10}, \frac{5}{8}, \frac{7}{12}$

Find the LCD for fractions with the given denominators.

14. 6 and 24

15. 12 and 18

16. 2, 5, and 8

17. 3, 6, and 8

Add or subtract as indicated.

18. $\frac{-3}{10} + \frac{-7}{12}$

19. $\frac{3}{8} + \frac{5}{12}$

20. $\frac{5}{36} + \frac{7}{24}$

21. $\frac{-2}{15} + \frac{-9}{20}$

22. $\frac{3}{8} + \frac{5}{12} + \frac{7}{18}$

23. $\frac{5}{6} + \frac{8}{15} + \frac{9}{20}$

24. $\frac{8}{9} + \frac{-3}{9}$

25. $\frac{9}{10} - \frac{6}{10}$

26. $\frac{5}{8} - \frac{1}{8}$

27. $\frac{11}{12} + \frac{-7}{12}$

28. $\frac{7}{8} - \frac{2}{3}$

29. $\frac{5}{6} - \frac{3}{5}$

30. $\frac{11}{18} + \frac{-2}{9}$

31. $\frac{5}{6} + \frac{-1}{4}$

32. $\frac{11}{12} - \frac{1}{4} - \frac{1}{3}$

33. $\frac{13}{15} + \frac{2}{3} - \frac{3}{5}$

[4.2] Convert to mixed or whole numbers.

34. $\frac{41}{6}$

35. $\frac{-32}{8}$

36. $\frac{23}{3}$

37. $\frac{47}{4}$

Convert to improper fractions.

38. $7\frac{5}{8}$

39. $-4\frac{3}{10}$

40. $5\frac{2}{7}$

41. $-12\frac{8}{13}$

Multiply.

42. $-5\frac{1}{3} \cdot 1\frac{4}{5}$

43. $1\frac{5}{12} \cdot 8$

44. $3\frac{1}{5} \cdot \frac{7}{8} \cdot 2\frac{6}{7}$

Divide.

45. $3\frac{3}{8} \div 2\frac{1}{4}$

46. $3\frac{3}{7} \div 8$

Perform the indicated operations.

47. $6\frac{5}{7} + 3\frac{4}{7}$

48. $-5\frac{7}{10} + \left(-3\frac{11}{12}\right)$

49. $-7\frac{7}{9} + 3\frac{4}{9}$

50. $2\frac{1}{3} + 5\frac{1}{6} - 2\frac{4}{5}$

[4.3] Simplify.

51. $\dfrac{\frac{2}{3}}{\frac{3}{5}}$

52. $\dfrac{\frac{5}{8}}{\frac{3}{4}}$

53. $\dfrac{6 - \frac{2}{3}}{3 + \frac{1}{6}}$

54. $\dfrac{5 + \frac{1}{5}}{17 - \frac{3}{10}}$

[4.4] Solve the applications.

55. Distance. The scale on a map is 1 in. = 80 mi. If two cities are $2\frac{3}{4}$ in. apart on the map, what is the actual distance between the cities?

56. Cost of linoleum. A kitchen measures $5\frac{1}{3}$ yd by $4\frac{1}{4}$ yd. If you purchase linoleum costing \$9 per square yard, what will it cost to cover the floor?

57. Cost of carpet. Your living room measures $6\frac{2}{3}$ yd by $4\frac{1}{2}$ yd. If you purchase carpeting at \$18 per square yard, what will it cost to carpet the room?

58. Area. A living room has dimensions $5\frac{2}{3}$ yd by $4\frac{1}{2}$ yd. How much carpeting must be purchased to cover the room?

59. Speed. If you drive 126 mi in $2\frac{1}{4}$ h, what is your average speed?

60. Average speed. If you drive 117 mi in $2\frac{1}{4}$ h, what is your average speed?

61. Number of lots. An 18-acre piece of land is to be subdivided into home lots that are each $\frac{3}{8}$ acre. How many lots can be formed?

62. Baking. A recipe calls for $\frac{1}{3}$ cup of milk. You have $\frac{3}{4}$ cup. How much milk will be left over?

63. Length. Bradley needs two shelves, one $32\frac{3}{8}$ in. long and the other $36\frac{11}{16}$ in. long. What is the total length of shelving that is needed?

64. Perimeter. Find the perimeter of the triangle.

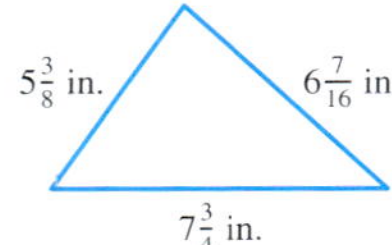

65. Plywood thickness. A sheet of plywood consists of two outer sections that are $\frac{3}{16}$ in. thick and a center section that is $\frac{3}{8}$ in. thick. How thick is the plywood overall?

[4.5] Solve the equations.

66. $\frac{x}{4} = 8$

67. $-\frac{x}{5} = -3$

68. $\frac{2}{3}x = 18$

69. $\frac{3}{4}x = 24$

70. $\frac{x}{3} - 5 = 1$

71. $\frac{3}{4}x - 2 = 7$

72. $\frac{x}{11} = \frac{12}{33}$

73. $\frac{x}{10} = \frac{9}{30}$

74. $\frac{x+1}{5} = \frac{20}{25}$

75. $\frac{2}{5} = \frac{x-2}{20}$

[4.6] Solve the word problems. Be sure to label the unknowns and to show the equation you used.

76. The sum of 3 times a number and 7 is 25. What is the number?

77. 5 times a number, decreased by 8, is 32. Find the number?

78. If the sum of two consecutive integers is 85, find the two integers?

79. The sum of three consecutive odd integers is 57. What are the three integers?

80. Rafael earns $35 more per week than Andrew. If their weekly salaries total $715, what amount does each earn?

81. Larry is 2 years older than Susan, and Nathan is twice as old as Susan. If the sum of their ages is 30 years, find each of their ages.

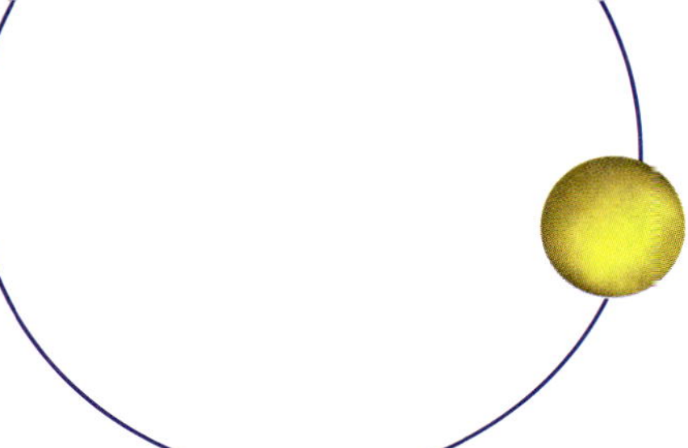

Chapter Test for Chapter 4

Name ______________

Section ________ Date ________

ANSWERS

1. ______
2. ______
3. ______
4. ______
5. ______
6. ______
7. ______
8. ______
9. ______
10. ______
11. ______
12. ______
13. ______
14. ______
15. ______
16. ______
17. ______
18. ______
19. ______
20. ______
21. ______

The purpose of the Chapter Test is to help you check your progress and review for a chapter test in class. When you are done, check your answers in the back of the book. If you missed any answers, be sure to go back and review the appropriate sections in the chapter.

In exercises 1 and 2, add.

1. $\frac{3}{10} + \frac{6}{10}$

2. $\frac{5}{12} + \frac{3}{12}$

3. Find the least common multiple of 18, 24, and 36.

In exercises 4 and 5, find the least common denominator for fractions with the given denominators.

4. 12 and 15

5. 3, 4, and 18

In exercises 6 to 8, add.

6. $\frac{2}{5} + \frac{4}{10}$

7. $\frac{-1}{6} + \frac{-3}{7}$

8. $\frac{1}{4} + \frac{5}{8} + \frac{7}{10}$

In exercises 9 to 11, subtract.

9. $\frac{7}{9} - \frac{-4}{9}$

10. $\frac{-7}{18} - \frac{5}{18}$

11. $\frac{11}{12} - \frac{3}{20}$

12. Convert $\frac{17}{4}$ to a mixed number.

13. Convert $8\frac{2}{9}$ to an improper fraction.

Perform the indicated operation.

14. $2\frac{2}{3} \cdot 1\frac{2}{7}$

15. $5\frac{1}{3} \cdot \frac{3}{4}$

16. $5\frac{3}{5} \div 2\frac{1}{10}$

17. $4\frac{1}{6} + 3\frac{3}{4}$

18. $7\frac{3}{8} - 5\frac{5}{8}$

19. $7 - 5\frac{7}{15}$

Simplify each complex fraction.

20. $\dfrac{\frac{2}{3}}{\frac{5}{6}}$

21. $\dfrac{2 - \frac{1}{5}}{4 + \frac{3}{10}}$

ANSWERS

22. ______

23. ______

24. ______

25. ______

26. ______

27. ______

28. ______

29. ______

30. ______

Solve the applications.

22. **Number of homes.** A $31\frac{1}{3}$-acre piece of land is subdivided into home lots. Each home lot is to be $\frac{2}{3}$ acre. How many homes can be built?

23. **Number of books.** A bookshelf is 66 in. long. If the thickness of each book on the shelf is $1\frac{3}{8}$ in., how many books can be placed on the shelf?

24. The average person drinks about $3\frac{1}{5}$ cups of coffee per workday. If a person works 5 days a week for 50 weeks every year, estimate how many cups of coffee that person will drink in a working lifetime of $51\frac{3}{4}$ years.

Solve for x and check your result.

25. $\frac{x}{4} = -8$

26. $\frac{2x}{3} = \frac{-8}{9}$

27. $\frac{2}{5}x - 3 = 17$

28. $\frac{x}{2} - 3 = \frac{5}{3}$

Solve the applications.

29. 3 times an integer is 41 less than 5 times the next consecutive integer. Find the two integers.

30. On a shopping trip, Caitlin spent \$10 less than twice the amount that Ben spent. Together they spent \$95. How much did Ben spend?

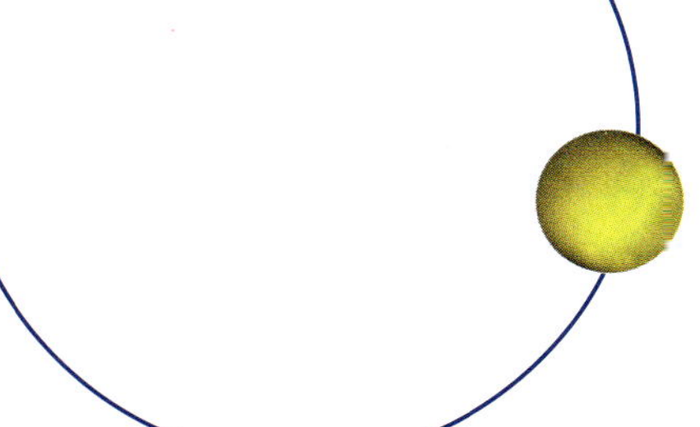

Cumulative Test for Chapters 1 to 4

Name ____________

Section ______ Date ______

ANSWERS

1. ______
2. ______
3. ______
4. ______
5. ______
6. ______
7. ______
8. ______
9. ______
10. ______
11. ______ 12. ______
13. ______ 14. ______
15. ______ 16. ______
17. ______ 18. ______
19. ______
20. ______
21. ______
22. ______

This test covers selected topics from Chapters 1 through 4.

In exercises 1 to 14, perform the indicated operations.

1. $8 + (-4)$

2. $-7 + (-5)$

3. $\frac{7}{3} + \frac{11}{3}$

4. $\frac{4}{5} - \frac{2}{3}$

5. $(-6)(3)$

6. $\frac{1}{3} \cdot \frac{3}{5}$

7. $\frac{3}{7} \cdot \frac{-2}{3}$

8. $(-50) \div (-5)$

9. $\frac{-2}{5} \div \frac{3}{10}$

10. $15 \div 0$

11. $0 \div \frac{1}{2}$

12. $\frac{1}{5} + \frac{3}{4} - \frac{1}{3}$

13. $(-3)(-9)$

14. $\frac{3}{5} \cdot \frac{1}{3} \cdot \frac{5}{7}$

Convert to mixed numbers.

15. $\frac{16}{9}$

16. $\frac{36}{5}$

Convert to improper fractions.

17. $5\frac{3}{4}$

18. $6\frac{1}{9}$

Identify each as an expression or an equation.

19. $2x + 1$

20. $3x - 4 = 2$

21. $\frac{2}{3}x + \frac{1}{5}$

22. $\frac{1}{4}x - 3 = 5$

ANSWERS

23. ________
24. ________
25. ________
26. ________
27. ________
28. ________
29. ________
30. ________
31. ________
32. ________
33. ________
34. ________
35. ________
36. ________
37. ________
38. ________
39. ________
40. ________

In exercises 23 to 30, perform the indicated operations.

23. $3\frac{2}{5} \cdot \frac{-5}{8}$

24. $1\frac{5}{12} \cdot 8$

25. $3\frac{5}{7} + 2\frac{4}{7}$

26. $-8\frac{1}{9} + 3\frac{5}{9}$

27. $9 - 5\frac{3}{8}$

28. $3\frac{1}{6} + 3\frac{1}{4} - 2\frac{7}{8}$

29. $2\frac{1}{4} \div (-3)$

30. $-5 \div 3\frac{1}{3}$

Solve each application.

31. A $6\frac{1}{2}$-in. bolt is placed through a wall that is $5\frac{7}{8}$ in. thick. How far does the bolt extend beyond the wall?

32. On a 6-h trip, Carlos drove $1\frac{3}{4}$ h and then Maria drove for another $2\frac{1}{3}$ h. How many hours remained on the trip?

Solve for x and check your result.

33. $3x - 2 = 5x + 4$

34. $\frac{3}{4}x = \frac{x + 4}{2}$

Solve the word problems. Be sure to show the equation used for the solution.

35. If 4 times a number decreased by 7 is 45, find that number.

36. The sum of two consecutive integers is 93. What are those two integers?

37. If 3 times an odd integer is 12 more than the next consecutive odd integer, what is that integer?

38. Michelle earns $120 more per week than Dimitri. If their weekly salaries total $720, how much does Michelle earn?

39. The length of a rectangle is 2 cm more than 3 times its width. If the perimeter of the rectangle is 44 cm, what are the dimensions of the rectangle?

40. One side of a triangle is 5 in. longer than the shortest side. The third side is twice the length of the shortest side. If the triangle perimeter is 37 in., find the length of each leg.

5

DECIMALS

INTRODUCTION

When you look into the cockpit of a plane, you have to be impressed with the number of gauges that face the pilot. It is remarkable that the pilot can keep all that information straight. It is even more remarkable to realize that the pilot has to know how to calculate with pencil and paper much of the information available.

When Gwen decided to become a pilot, she went to the local airport to sign up for classes. Upon registration, she received a packet that described the test she was going to have to pass to get a permit. She was amazed to see that the test was almost entirely made up of math questions. It took Gwen a month of reviewing to prepare for the test. Now that she has her pilot's license, she understands why she needed to be able to do all that math!

Name ____________

Section ________ Date ________

ANSWERS

1. ____________
2. ____________
3. ____________
4. ____________
5. ____________
6. ____________
7. ____________
8. ____________
9. ____________
10. ____________
11. ____________
12. ____________
13. ____________
14. ____________
15. ____________

Pre-Test Chapter 5

This pre-test will point out any difficulties you may be having working with decimals. Do all the exercises. Then check your answers with those in the back of the book.

1. Give the place value of 5 in the decimal 13.4658.

2. Write $2\frac{371}{1000}$ in decimal form and in words.

3. **(a)** $56 + (-5.16) + 1.8 + (-0.33)$ **(b)** $4.6 - 2.225$

4. **Consumer spending.** You have \$20 in cash and make purchases of \$6.89 and \$10.75. How much cash do you have left?

5. **(a)** $0.357 \cdot (-2.41)$ **(b)** $0.5362 \cdot 1{,}000$

6. Round 2.35878 to the nearest hundredth.

7. **Fuel costs.** You fill up your car with 9.2 gal of fuel at \$1.299 per gallon. What is the cost of the fill-up (to the nearest cent)?

8. Solve $4.2 + 1.5x = -3.3$.

9. Solve $2.38 - (6.8 + 4.5x) = -3.25x$.

10. Divide $57\overline{)242.25}$.

11. Divide $1.6\overline{)3.896}$.

12. **Salary.** Manny worked 27.5 h in a week and earned \$209. What was his hourly rate of pay?

13. Divide $53.4 \div 1{,}000$.

14. Find the decimal equivalent of each of the following.

    (a) $\frac{3}{8}$ **(b)** $\frac{7}{24}$ (to the nearest hundredth)

15. Use the Pythagorean theorem to find the length of the hypotenuse for $\triangle ABC$.

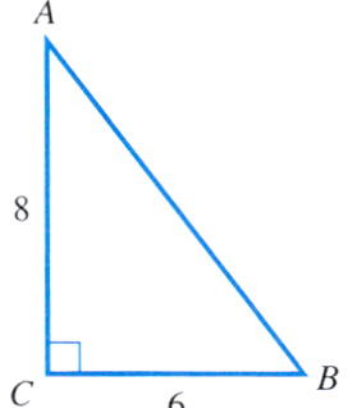

5.1 Introduction to Decimals, Place Value, and Rounding

OBJECTIVES

1. Identify place value in a decimal fraction
2. Write a decimal in words
3. Write a decimal as a fraction or mixed number
4. Compare the size of several decimals
5. Round a decimal to any specified decimal place

In Chapter 4, we looked at common fractions. We will turn now to a special kind of fraction, a **decimal fraction.** A decimal fraction is a fraction whose denominator is a *power of 10.* Some examples of decimal fractions are $\frac{3}{10}$, $\frac{45}{100}$, and $\frac{123}{1000}$.

In Chapter 1, we talked about the idea of place value. Recall that in our decimal place-value system, each place has *one-tenth* the value of the place to its left.

NOTE Remember that the powers of 10 are 1, 10, 100, 1,000, and so on. You might want to review Section 1.7 before going on.

Example 1

Identifying Place Values

Label the place values for the number 538.

5	3	8
↑	↑	↑
Hundreds	Tens	Ones

The ones place value is one-tenth of the tens place value; the tens place value is one-tenth of the hundreds place value; and so on.

CHECK YOURSELF 1

Label the place values for the number 2,793.

We now want to extend this idea *to the right* of the ones place. Write a period to the *right* of the ones place. This is called the **decimal point.** Each digit to the right of that decimal point will represent a fraction whose denominator is a power of 10. The first place to the right of the decimal point is the tenths place:

NOTE The decimal point separates the whole-number part and the fractional part of a decimal fraction.

$$0.1 = \frac{1}{10}$$

Example 2

Writing a Number in Decimal Form

Write the mixed number $3\frac{2}{10}$ in decimal form.

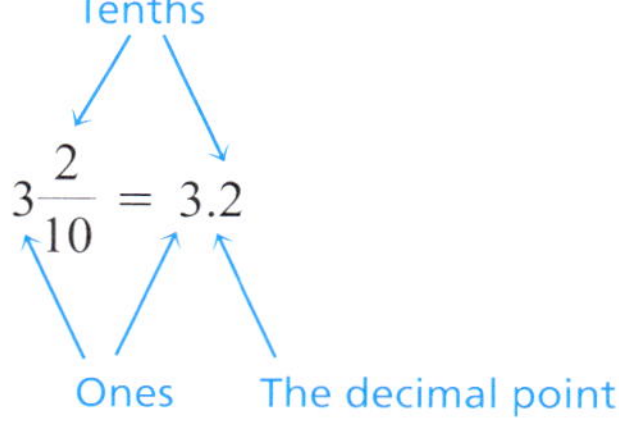

CHECK YOURSELF 2

Write $5\frac{3}{10}$ *in decimal form.*

As you move farther to the *right,* each place value must be one-tenth of the value before it. The second place value is hundredths $\left(0.01 = \frac{1}{100}\right)$. The next place is thousandths, the fourth position is the ten thousandths place, and so on. The figure illustrates the value of each position as we move to the right of the decimal point.

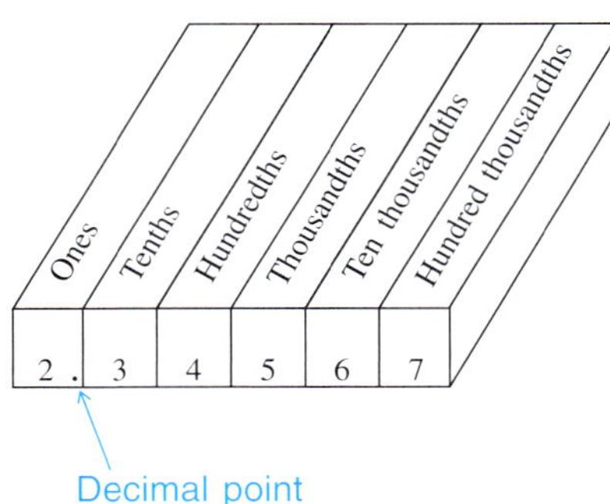

Example 3

Identifying Place Values

NOTE For convenience we will shorten the term "decimal fraction" to "decimal" from this point on.

What are the place values for 4 and 6 in the decimal 2.34567? The place value of 4 is hundredths, and the place value of 6 is ten thousandths.

CHECK YOURSELF 3

What is the place value of 5 *in the decimal of Example* 3?

Understanding place values will allow you to read and write decimals by using these steps.

NOTE If there are *no* nonzero digits to the left of the decimal point, start directly with step 3.

Step by Step: Reading or Writing Decimals in Words

Step 1 Read the digits *to the left* of the decimal point as a whole number.
Step 2 Read the decimal point as the word *and.*
Step 3 Read the digits *to the right* of the decimal point as a whole number followed by the place value of the rightmost digit.

Example 4

Writing a Decimal Number in Words

Write each decimal number in words.

5.03 is read "five and three hundredths."

Hundredths — The rightmost digit, 3, is in the hundredths position.

12.057 is read "twelve and fifty-seven thousandths."

Thousandths — The rightmost digit, 7, is in the thousandths position.

NOTE An informal way of reading decimals is to simply read the digits in order and use the word *point* to indicate the decimal point. 2.58 can be read "two point five eight." 0.689 can be read "zero point six eight nine."

0.5321 is read "five thousand three hundred twenty-one ten thousandths."

When the decimal has no whole-number part, we have chosen to write a 0 to the left of the decimal point. This simply makes sure that you don't miss the decimal point. However, both 0.5321 and .5321 are correct.

CHECK YOURSELF 4

Write 2.58 *in words.*

NOTE The number of digits to the right of the decimal point is called the number of **decimal places** in a decimal number. So, 0.35 has two decimal places.

One quick way to write a decimal as a common fraction is to remember that the number of decimal places must be the same as the number of zeros in the denominator of the common fraction.

Example 5

Writing a Decimal Number as a Mixed Number

Write each decimal as a common fraction or mixed number.

$$0.35 = \frac{35}{100}$$

Two places — Two zeros

This fraction can then be simplified to $\frac{7}{20}$.

The same method can be used with decimals that are greater than 1. Here the result will be a mixed number.

NOTE The 0 to the right of the decimal point is a "placeholder" that is not needed in the common-fraction form.

$$2.058 = 2\frac{58}{1000}$$

Three places — Three zeros

This mixed number can be simplified to $2\frac{29}{500}$.

CHECK YOURSELF 5

Write as common fractions or mixed numbers.

(a) 0.528 **(b)** 5.08

REMEMBER: By the fundamental principle of fractions, multiplying the numerator and denominator of a fraction by the same nonzero number does not change the value of the fraction.

It is often useful to compare the sizes of two decimal fractions. One approach to comparing decimals uses this fact: Writing zeros to the right *does not change* the value of a decimal. 0.53 is the same as 0.530. Look at the fractional form:

$$\frac{53}{100} = \frac{530}{1000}$$

The fractions are equivalent. We have multiplied the numerator and denominator by 10.

We will see how this is used to compare decimals in Example 6.

Example 6

Comparing the Sizes of Two Decimal Numbers

Which is larger?

0.84 or 0.842

Write 0.84 as 0.840. Then we see that 0.842 (or 842 thousandths) is greater than 0.840 (or 840 thousandths), and we can write

$$0.842 > 0.84$$

CHECK YOURSELF 6

Complete the statement, using the symbol $<$ or $>$.

0.588 ______ 0.59

Whenever a decimal represents a measurement made by some instrument (a rule or a scale), the decimals are not exact. They are accurate only to a certain number of places and are called **approximate numbers.** Usually, we want to make all decimals in a particular problem accurate to a specified decimal place or tolerance. This will require **rounding** the decimals. We can picture the process on a number line.

Example 7

Rounding to the Nearest Tenth

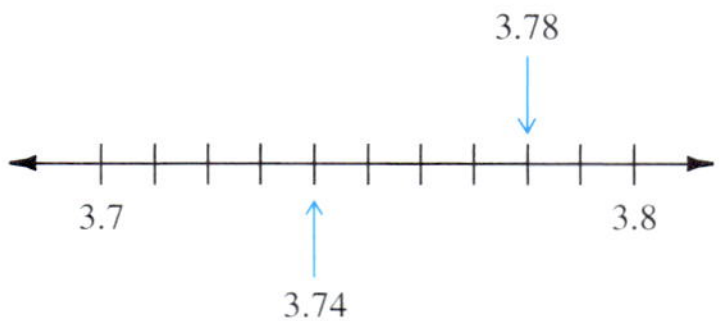

NOTE 3.74 is closer to 3.7 than it is to 3.8. 3.78 is closer to 3.8.

3.74 is rounded down to the nearest tenth, 3.7. 3.78 is rounded up to 3.8.

CHECK YOURSELF 7

Use the number line in Example 7 to round 3.77 to the nearest tenth.

Rather than using the number line, this rule can be applied.

Step by Step: To Round a Decimal

Step 1 Find the place to which the decimal is to be rounded.
Step 2 If the next digit to the right is 5 or more, increase the digit in the place you are rounding by 1. Discard remaining digits to the right.
Step 3 If the next digit to the right is less than 5, just discard that digit and any remaining digits to the right.

Example 8

Rounding to the Nearest Tenth

Round 34.58 to the nearest tenth.

NOTE Many students find it easiest to mark this digit with an arrow.

34.58   Locate the digit you are rounding to. The 5 is in the tenths place.

Because the next digit to the right, (8), is 5 or more, increase the tenths digit by 1. Then discard the remaining digits.

34.58 is rounded to 34.6.

CHECK YOURSELF 8

Round 48.82 to the nearest tenth.

Example 9

Rounding to the Nearest Hundredth

Round 5.673 to the nearest hundredth.

5.673   The 7 is in the hundredths place.

The next digit to the right, (3), is less than 5. Leave the hundredths digit as it is, and discard the remaining digits to the right.

5.673 is rounded to 5.67.

CHECK YOURSELF 9

Round 29.247 to the nearest hundredth.

Example 10

Rounding to a Specified Decimal Place

Round 3.14159 to four decimal places.

NOTE The fourth place to the *right* of the decimal point is the ten thousandths place.

3.14159 The 5 is in the ten thousandths place.

The next digit to the right, (9), is 5 or more, so increase the digit you are rounding to by 1. Discard the remaining digits to the right.

3.14159 is rounded to 3.1416.

CHECK YOURSELF 10

Round 0.8235 *to three decimal places.*

CHECK YOURSELF ANSWERS

1. 2 7 9 3 (2: Thousands; 7: Hundreds; 9: Tens; 3: Ones) **2.** $5\frac{3}{10} = 5.3$ **3.** Thousandths

4. Two and fifty-eight hundredths **5.** **(a)** $\frac{528}{1000}$; **(b)** $5\frac{8}{100}$ **6.** $0.588 < 0.59$

7. 3.8 **8.** 48.8 **9.** 29.25 **10.** 0.824

Name ____________

Section ________ Date ________

5.1 Exercises

For the decimal 8.57932:

1. What is the place value of 7?

2. What is the place value of 5?

3. What is the place value of 3?

4. What is the place value of 2?

Write in decimal form.

5. $\frac{23}{100}$

6. $\frac{371}{1000}$

7. $\frac{209}{10000}$

8. $3\frac{5}{10}$

9. $23\frac{56}{1000}$

10. $7\frac{431}{10000}$

Write in words.

11. 0.23

12. 0.371

13. 0.071

14. 0.0251

15. 12.07

16. 23.056

Write in decimal form.

17. Fifty-one thousandths

18. Two hundred fifty-three ten thousandths

19. Seven and three tenths

20. Twelve and two hundred forty-five thousandths

Write each as a common fraction or mixed number.

21. 0.65

22. 0.00765

23. 5.231

24. 4.0171

ANSWERS

1. ____________
2. ____________
3. ____________
4. ____________
5. ____________
6. ____________
7. ____________
8. ____________
9. ____________
10. ____________
11. ____________
12. ____________
13. ____________
14. ____________
15. ____________
16. ____________
17. ____________
18. ____________
19. ____________
20. ____________
21. ______ 22. ______
23. ______ 24. ______

ANSWERS

25. ____________
26. ____________
27. ____________
28. ____________
29. ____________
30. ____________
31. ____________
32. ____________
33. ____________
34. ____________
35. ____________
36. ____________
37. ____________
38. ____________
39. ____________
40. ____________
41. ____________
42. ____________

Complete each statement, using the symbol $<$, $=$, or $>$.

25. 0.69 __________ 0.689

26. 0.75 __________ 0.752

27. 1.23 __________ 1.230

28. 2.451 __________ 2.45

29. 10 __________ 9.9

30. 4.98 __________ 5

31. 1.459 __________ 1.46

32. 0.235 __________ 0.2350

33. Arrange in order from smallest to largest.

0.71, 0.072, $\frac{7}{10}$, 0.007, 0.0069

$\frac{7}{100}$, 0.0701, 0.0619, 0.0712

34. Arrange in order from smallest to largest.

2.05, $\frac{25}{10}$, 2.0513, 2.059

$\frac{251}{100}$, 2.0515, 2.052, 2.051

Round to the indicated place.

35. 53.48 tenths

36. 6.785 hundredths

37. 21.534 hundredths

38. 5.842 tenths

39. 0.342 hundredths

40. 2.3576 thousandths

41. 2.71828 thousandths

42. 1.543 tenths

43. 0.0475 tenths

44. 0.85356 ten thousandths

45. 4.85344 ten thousandths

46. 52.8728 thousandths

47. 6.734 two decimal places

48. 12.5467 three decimal places

49. 6.58739 four decimal places

50. 503.824 two decimal places

Round 56.35829 to the nearest:

51. Tenth

52. Ten thousandth

53. Thousandth

54. Hundredth

In exercises 55 to 60, determine the decimal that corresponds to the shaded portion of each "decimal square." Note that the total value of a decimal square is 1.

55.

56.

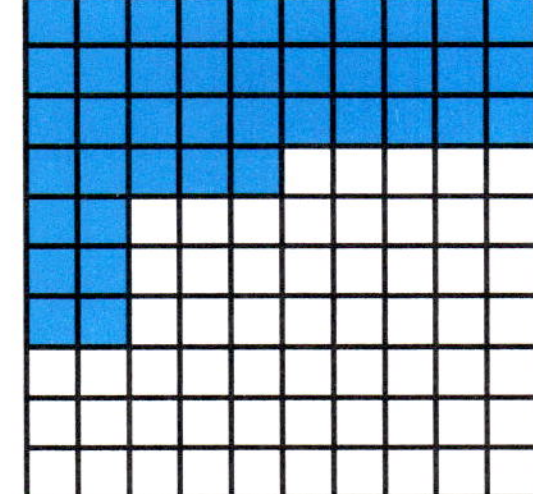

57.

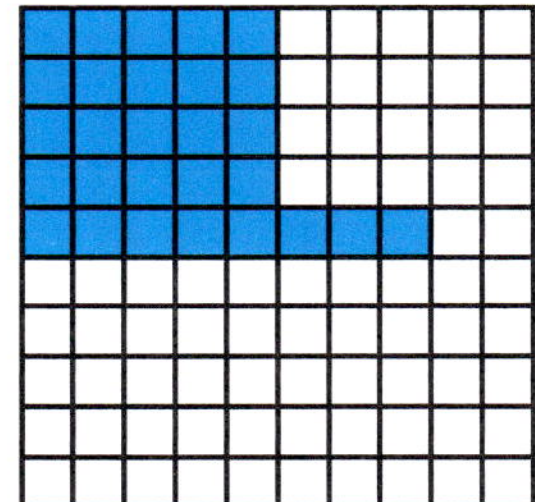

58.

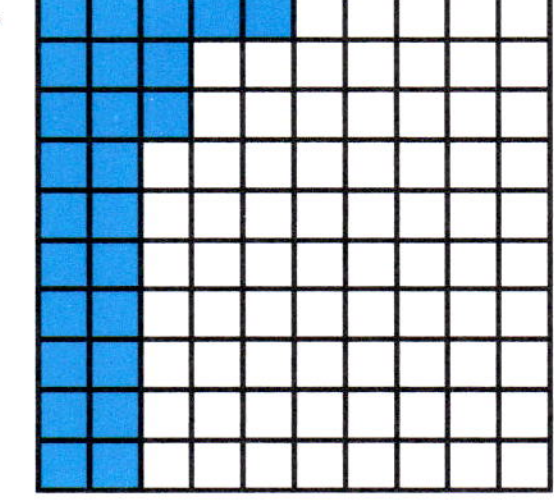

ANSWERS

43. __________

44. __________

45. __________

46. __________

47. __________

48. __________

49. __________

50. __________

51. __________

52. __________

53. __________

54. __________

55. __________

56. __________

57. __________

58. __________

ANSWERS

59. ____________

60. ____________

61. ____________

62. ____________

63. ____________

64. ____________

65. ____________

66. ____________

59.

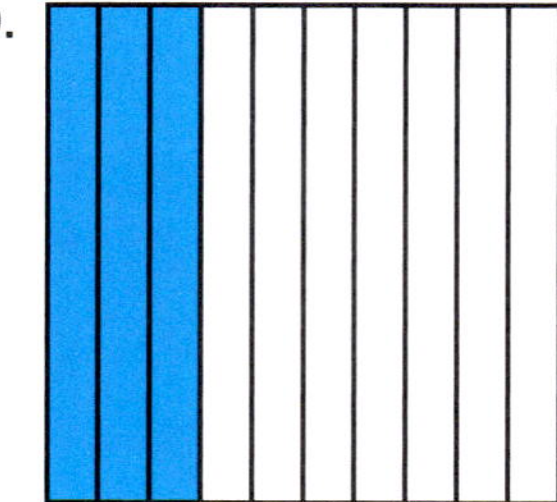

60.

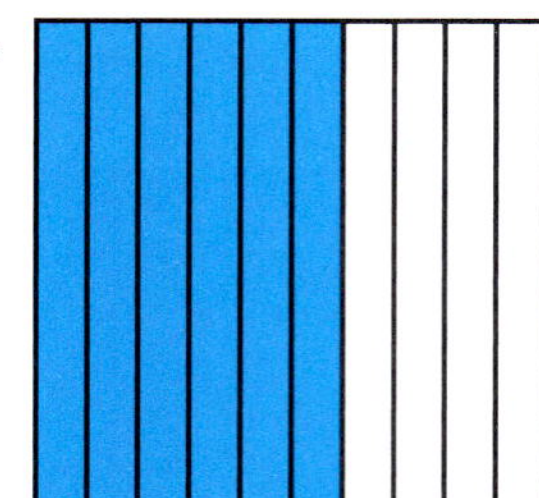

In exercises 61 to 64, shade the portion of the square that is indicated by the given decimal.

61. 0.23

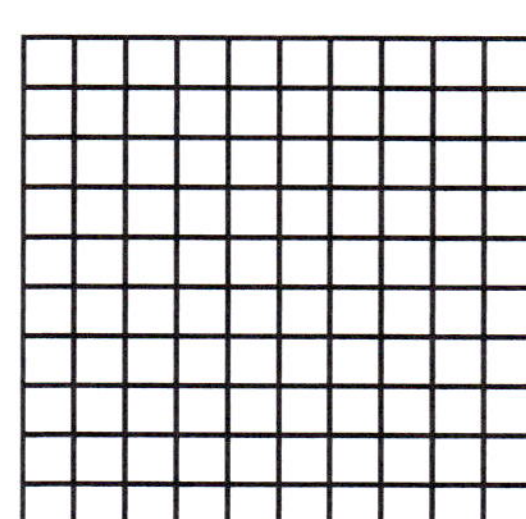

62. 0.89

63. 0.3

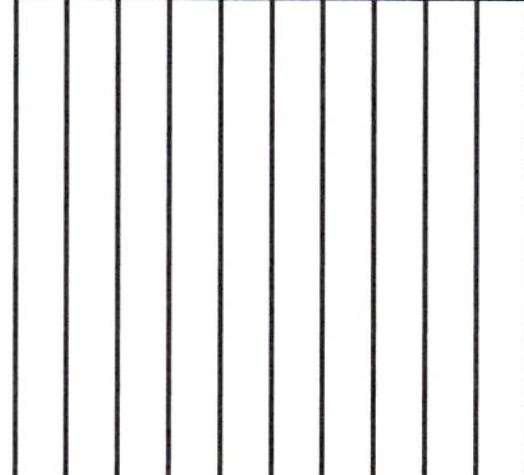

64. 0.30

65. Plot (draw a dot) 3.2 and 3.7 on the number line. Then estimate the location for 3.62.

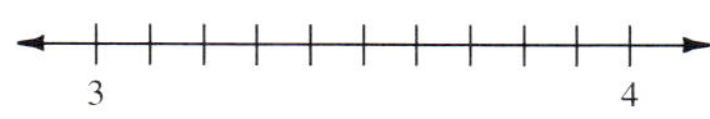

66. Plot 12.51 and 12.58 on the number line. Then estimate the location for 12.537.

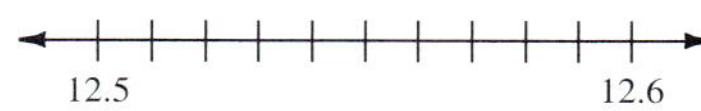

67. Plot 7.124 and 7.127 on the number line. Then estimate the location of 7.1253.

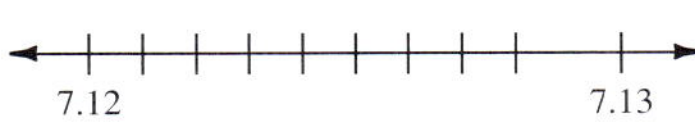

68. Plot 5.73 and 5.74 on the number line. Then estimate the location for 5.782.

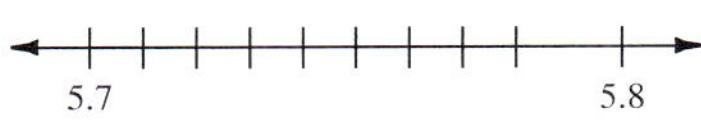

69. Estimate, to the tenth of a degree, the reading of the Fahrenheit thermometer shown.

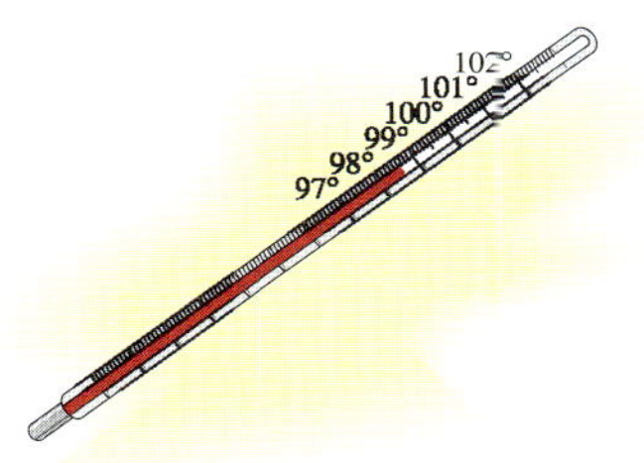

70. Estimate, to the tenth of a centimeter, the length of the pencil shown.

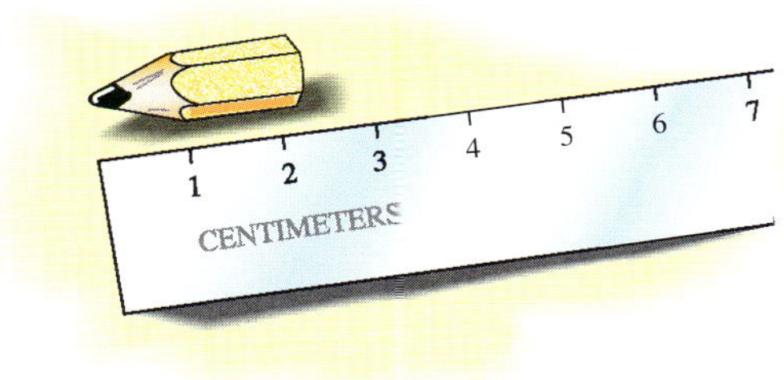

71. **(a)** What is the difference in these values: 0.120, 0.1200, and 0.12000?

(b) Explain in your own words why placing zeros to the right of a decimal point does not change the value of the number.

72. Lula wants to round 76.24491 to the nearest hundredth. She first rounds 76.24491 to 76.245 and then rounds 76.245 to 76.25 and claims that this is the final answer. What is wrong with this approach?

ANSWERS

67. ______

68. ______

69. ______

70. ______

71. ______

72. ______

Answers

1. Hundredths **3.** Ten thousandths **5.** 0.23 **7.** 0.0209 **9.** 23.056

11. Twenty-three hundredths **13.** Seventy-one thousandths

15. Twelve and seven hundredths **17.** 0.051 **19.** 7.3 **21.** $\frac{65}{100}\left(\text{or } \frac{13}{20}\right)$

23. $5\frac{231}{1000}$ **25.** $0.69 > 0.689$ **27.** $1.23 = 1.230$ **29.** $10 > 9.9$

31. $1.459 < 1.46$ **33.** 0.0069, 0.007, 0.0619, $\frac{7}{100}$, 0.0701, 0.0712, 0.072, $\frac{7}{10}$, 0.71

35. 53.5 **37.** 21.53 **39.** 0.34 **41.** 2.718 **43.** 0.0 **45.** 4.8534

47. 6.73 **49.** 6.5874 **51.** 56.4 **53.** 56.358 **55.** 0.44 **57.** 0.28

59. 0.3 **61.**

63.

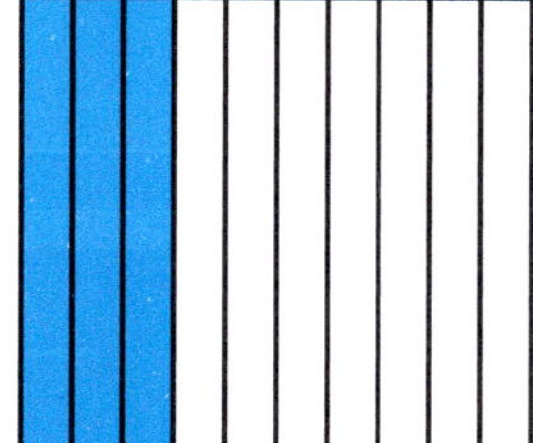

65.

3 4

67.

7.12 7.13

69. 98.6°F **71.**

5.2 Addition and Subtraction of Decimals

OBJECTIVES

1. Add two or more decimals
2. Subtract one decimal from another

Working with decimals rather than common fractions makes the basic operations much easier. We will start by looking at addition. One method for adding decimals is to write the decimals as common fractions, add, and then change the sum back to a decimal.

$$0.34 + 0.52 = \frac{34}{100} + \frac{52}{100} = \frac{86}{100} = 0.86$$

It is much more efficient to leave the numbers in decimal form and perform the addition in the same way as we did with whole numbers. You can use this rule.

Step by Step: To Add Decimals

Step 1 Write the numbers being added in column form *with their decimal points aligned vertically.*

Step 2 Add just as you would with whole numbers.

Step 3 Place the decimal point of the sum in line with the decimal points of the addends.

Example 1 illustrates the use of this rule.

Example 1

Adding Decimals

Add 0.13, 0.42, and 0.31.

NOTE Placing the decimal points in a vertical line ensures that we are adding digits of the same place value.

$$\begin{array}{r} 0.13 \\ 0.42 \\ +\ 0.31 \\ \hline 0.86 \end{array}$$

CHECK YOURSELF 1

Add 0.23, 0.15, *and* 0.41.

In adding decimals, you can use the *carrying process* just as you did in adding whole numbers. Consider Example 2.

Example 2

Adding Decimals Involving Carrying

Add 0.35, 1.58, and 0.67.

$$\begin{array}{r} {\scriptstyle 1\ 2} \\ 0.35 \\ 1.58 \\ +\ 0.67 \\ \hline 2.60 \end{array}$$

← Carries

In the hundredths column:
5 + 8 + 7 = 20
Write 0 and carry 2 to the tenths column.
In the tenths column:
2 + 3 + 5 + 6 = 16
Write 6 and carry 1 to the ones column.

Note: The carrying process works with decimals, just as it did with whole numbers, because each place value is again *one-tenth* the value of the place to its left.

CHECK YOURSELF 2

Add 23.546, 0.489, 2.312, *and* 6.135.

In adding decimals, the numbers may not have the same number of decimal places. Just fill in as many zeros as needed so that all of the numbers added have the same number of decimal places.

Recall that adding zeros to the right *does not change* the value of a decimal. 0.53 is the same as 0.530.

We will see how this is used in Example 3.

Example 3

Adding Decimals

Add 0.53, 4, 2.7, and 3.234.

NOTE Be sure that the decimal points are in a vertical line.

$$\begin{array}{r} 0.53 \\ 4. \\ 2.7 \\ +\ 3.234 \\ \hline \end{array}$$

Note that for a whole number, the decimal is understood to be to its right. So 4 = 4.

Now fill in the missing zeros, and add as before.

$$\begin{array}{r} 0.530 \\ 4.000 \\ 2.700 \\ +\ 3.234 \\ \hline 10.464 \end{array}$$

Now all the numbers being added have *three* decimal places.

CHECK YOURSELF 3

Add 6, 2.583, 4.7, *and* 2.54.

The addition of signed decimals follows the same rules as the addition of signed integers or fractions.

1. If the signs are the same, find the sum of the magnitudes (absolute values) of the numbers and give the result the sign of the numbers.
2. If the signs are opposite, find the difference of the absolute value of the numbers and give the result the sign of the number with greater magnitude (absolute value).

Example 4

Adding Signed Decimals

Find each sum.

(a) $-2.34 + (-15.7)$

The signs are the same, so we find the sum of the absolute values and give the result a negative sign.

$-2.34 + (-15.7) = -18.04$

(b) $-3.56 + 2.14$

The signs are opposite, so we find the difference of the absolute values and give the result the sign of the number with greater magnitude. The -3.56 has greater magnitude, so the result will be negative.

$-3.56 + 2.14 = -1.42$

CHECK YOURSELF 4

Find each sum.

(a) $-9.4 + (-19.26)$ **(b)** $-12.3 + 7.2$

Much of what we have said about adding decimals is also true of subtraction. To subtract decimals, we use this rule:

Step by Step: To Subtract Decimals

Step 1 Write the numbers being subtracted in column form *with their decimal points aligned vertically.*

Step 2 Subtract just as you would with whole numbers.

Step 3 Place the decimal point of the difference in line with the decimal points of the numbers being subtracted.

Example 5 illustrates the use of this rule.

Example 5

Subtracting a Decimal

Subtract 1.23 from 3.58.

$$\begin{array}{r} 3.58 \\ -\ 1.23 \\ \hline 2.35 \end{array}$$

Subtract in the hundredths, the tenths, and then the ones columns.

CHECK YOURSELF 5

Subtract 9.87 − 5.45.

Because each place value is one-tenth the value of the place to its left, borrowing, when you are subtracting decimals, works just as it did in subtracting whole numbers.

Example 6

Subtraction of a Decimal That Involves Borrowing

Subtract 1.86 from 6.54.

$$\begin{array}{r} {}^{5}\,{}^{1}4\,{}_{1} \\ \not{6}.\not{5}4 \\ -\ 1.86 \\ \hline 4.68 \end{array}$$

Here, borrow from the tenths and ones places to do the subtraction.

CHECK YOURSELF 6

Subtract 35.35 − 13.89.

In subtracting decimals, as in adding, we can write zeros to the right of the decimal point so that both decimals have the same number of decimal places.

Example 7

Subtracting a Decimal

(a) Subtract 2.36 from 7.5.

NOTE When you are subtracting, align the decimal points, then write zeros to the right to align the digits.

$$\begin{array}{r} \overset{4\ 1}{7.\not{5}0} \\ -\ 2.36 \\ \hline 5.14 \end{array}$$

We have written a 0 at the end of 7.5. Next, borrow 1 tenth from the 5 tenths in the minuend.

(b) Subtract 3.657 from 9.

NOTE 9 has been rewritten as 9.000.

$$\begin{array}{r} \overset{8\ 99}{\not{9}.\not{0}\not{0}0} \\ -\ 3.657 \\ \hline 5.343 \end{array}$$

In this case, move left to the ones place to begin the borrowing process.

CHECK YOURSELF 7

Subtract 5 − 2.345.

When subtracting signed decimals, change the exercise to an addition problem and follow the rules stated before Example 4.

Example 8

Subtracting Signed Decimals

Perform each subtraction.

(a) Subtract 4.31 from −6.55.

First, translate the statement into an expression.

$-6.55 - 4.31$

Rewrite as an addition problem.

$-6.55 + (-4.31)$

The signs are the same, so add the absolute values and make the result negative.

$-6.55 + (-4.31) = -10.86$

(b) Subtract −12.4 from −7.2.

Translate the statement into an expression.

$-7.2 - (-12.4)$

Rewrite as an addition problem.

$-7.2 + 12.4$

The signs are opposite, so find the difference of the absolute values and make the answer positive (12.4 has greater magnitude).

$-7.2 + 12.4 = 5.2$

CHECK YOURSELF 8

(a) *Subtract* 6.39 *from* −5.12.
(b) *Subtract* −12.34 *from* 1.5.
(c) *Subtract* −8.1 *from* −15.62.

Using Your Calculator to Add and Subtract Decimals

REMEMBER: The reason for this book is to help you review the basic skills of arithmetic. We are using these calculator sections to show you how the calculator can be helpful as a tool. Unless your instructor says otherwise, you should be using your calculator *only on the problems in these special sections.*

Entering decimals in your calculator is similar to entering whole numbers. There is just one difference: The decimal point key [•] is used to place the decimal point as you enter the number.

Example 9

Entering a Decimal Number into a Calculator

To enter 12.345, press

[1] [2] [•] [3] [4] [5]

Display 12.345

CHECK YOURSELF 9

Enter 14.367 *on your calculator.*

Example 10

Entering a Decimal Number into a Calculator

To enter 0.678, press

NOTE You don't have to press the 0 key for the digit 0 to the left of the decimal point.

[•] [6] [7] [8]

Display 0.678

CHECK YOURSELF 10

Enter 0.398 *on your calculator.*

The process of adding and subtracting signed decimals on your calculator is the same as we saw in Chapter 1 when we were adding and subtracting integers.

Example 11

Adding Decimals

To add 2.567 + (−0.89), enter

2.567 [+] 0.89 [+/−] [=]

or, on a graphing calculator,

2.567 [+] [(−)] 0.89 [Enter]

Display 1.677

CHECK YOURSELF 11

Add on your calculator.

$5.39 + (-9.7)$

Subtraction of decimals on the calculator is similar.

Example 12

Subtracting a Signed Decimal

To subtract $-4.2 - (-2.875)$, enter

4.2 [+/−] [−] 2.875 [+/−] [=]   or   [(−)] 4.2 [−] [(−)] 2.875 [Enter]

Display −1.325

CHECK YOURSELF 12

Subtract on your calculator.

$-16.3 - (-7.895)$

Often both addition and subtraction are involved in a calculation. In this case, just enter the decimals and the operation signs, + or −, as they appear in the problem.

Example 13

Adding and Subtracting Decimals

NOTE Again there are differences in the operation of various calculators. Try this problem on yours to check that its operation sequence is correct.

To find $23.7 - 5.2 + 3.87 - 2.341$, enter

23.7 [−] 5.2 [+] 3.87 [−] 2.341 [=]

Display 20.029

CHECK YOURSELF 13

Use your calculator to find

$52.8 - 36.9 + 15.87 - 9.36$

CHECK YOURSELF ANSWERS

1. 0.79 **2.** 32.482 **3.**
$$\begin{array}{r} 6.000 \\ 2.583 \\ 4.700 \\ +\ 2.540 \\ \hline 15.823 \end{array}$$
4. **(a)** −28.66; **(b)** −5.1 **5.** 4.42
6. 21.46 **7.** 2.655 **8.** **(a)** −11.51; **(b)** 13.84; **(c)** −7.52 **9.** 14.367
10. 0.398 **11.** −4.31 **12.** −8.405 **13.** 22.41

Exercises

Name ____________

Section ________ Date ________

Add.

1. $\begin{array}{r} 0.28 \\ +\ 0.79 \\ \hline \end{array}$

2. $\begin{array}{r} 2.59 \\ +\ 0.63 \\ \hline \end{array}$

3. $-1.045 + (-0.23)$

4. $-2.485 + (-1.25)$

5. $\begin{array}{r} 0.62 \\ 4.23 \\ +\ 12.5 \\ \hline \end{array}$

6. $\begin{array}{r} 0.50 \\ 2.99 \\ +\ 24.8 \\ \hline \end{array}$

7. $\begin{array}{r} 5.28 \\ +\ 19.455 \\ \hline \end{array}$

8. $\begin{array}{r} 23.845 \\ +\ 7.29 \\ \hline \end{array}$

9. $\begin{array}{r} 13.58 \\ 7.239 \\ +\ 1.5 \\ \hline \end{array}$

10. $-8.625 + (-2.45) + (-12.6)$

11. $\begin{array}{r} 25.3582 \\ 6.5 \\ 1.898 \\ +\ 0.69 \\ \hline \end{array}$

12. $\begin{array}{r} 1.336 \\ 15.6857 \\ 7.9 \\ +\ 0.85 \\ \hline \end{array}$

13. $0.43 + 0.8 + 0.561$

14. $1.25 + 0.7 + 0.259$

15. $5 + 23.7 + 8.7 + 9.85$

16. $28.3 + 6 + 8.76 + 3.8$

17. $-25.83 + (-5.62)$

18. $-32.59 + (-9.56)$

19. $42.731 + 1.058 + 103.24$

20. $27.4 + 213.321 + 39.38$

ANSWERS

1. ____________
2. ____________
3. ____________
4. ____________
5. ____________
6. ____________
7. ____________
8. ____________
9. ____________
10. ____________
11. ____________
12. ____________
13. ____________
14. ____________
15. ____________
16. ____________
17. ____________
18. ____________
19. ____________
20. ____________

ANSWERS

21. ______

22. ______

23. ______

24. ______

In exercises 21 to 24, use decimal square shading to represent the addition process. Shade each square and the total.

21.

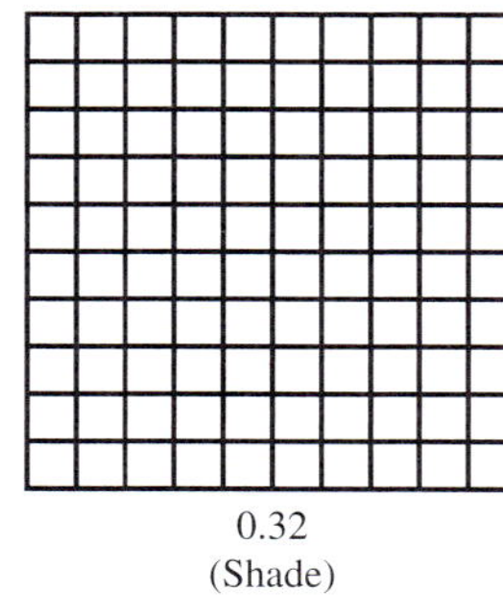
0.32
(Shade)

+

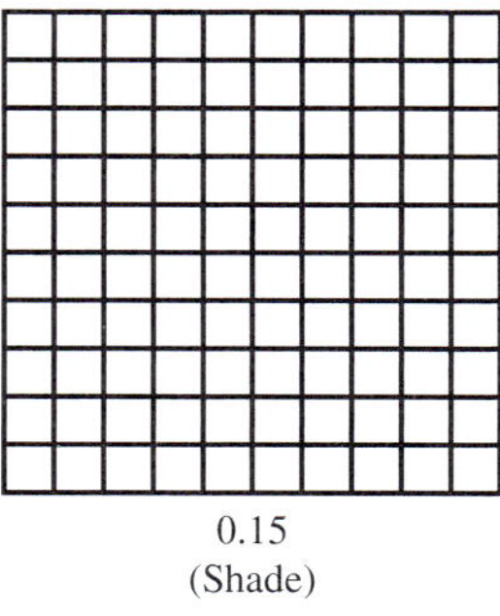
0.15
(Shade)

=

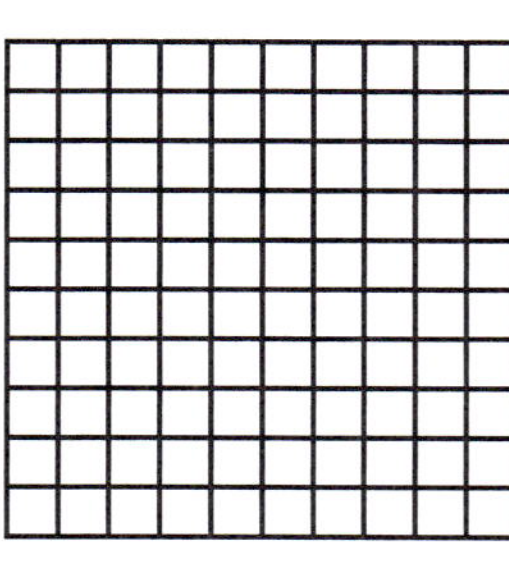
(Shade Total)

22.

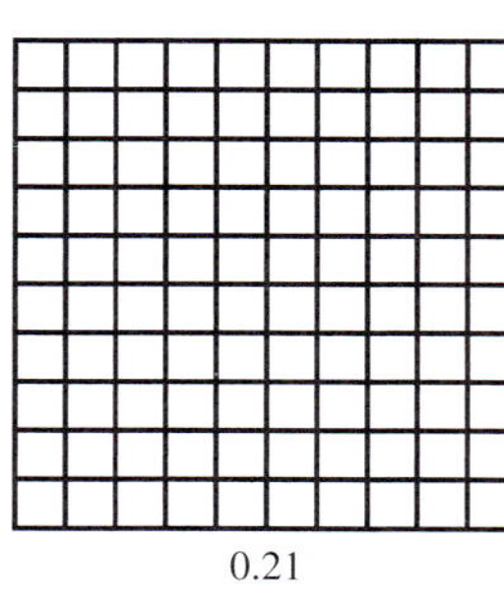
0.21

+

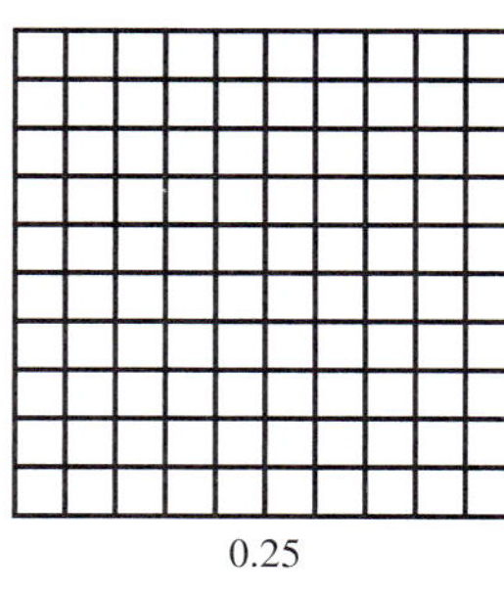
0.25

=

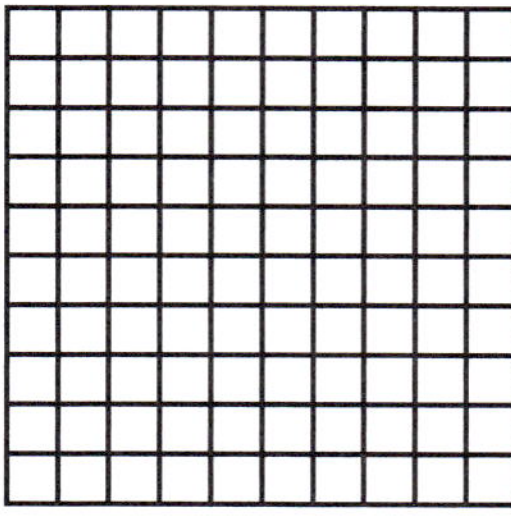

23.

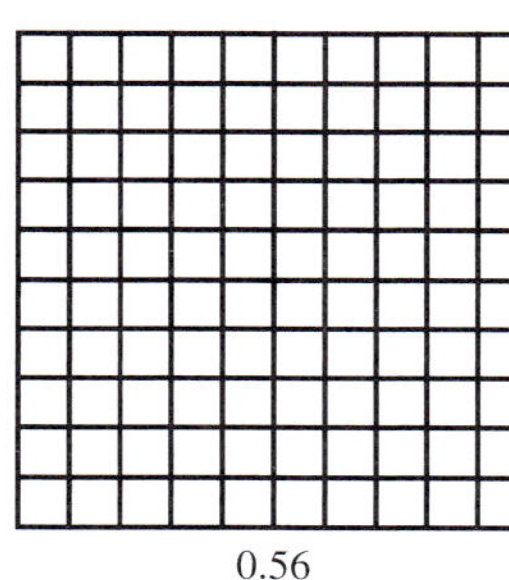
0.56

+

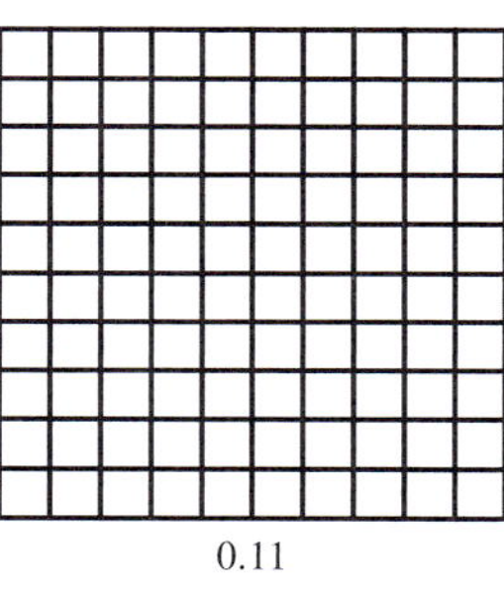
0.11

=

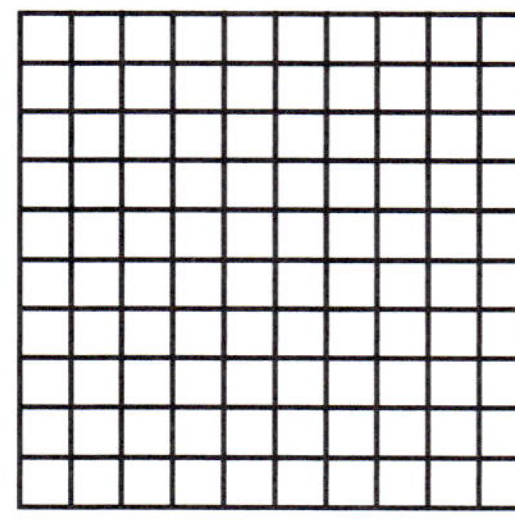

24.

0.43

+

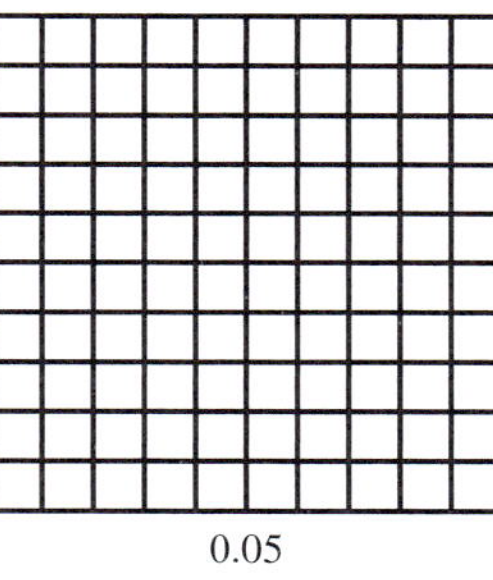
0.05

=

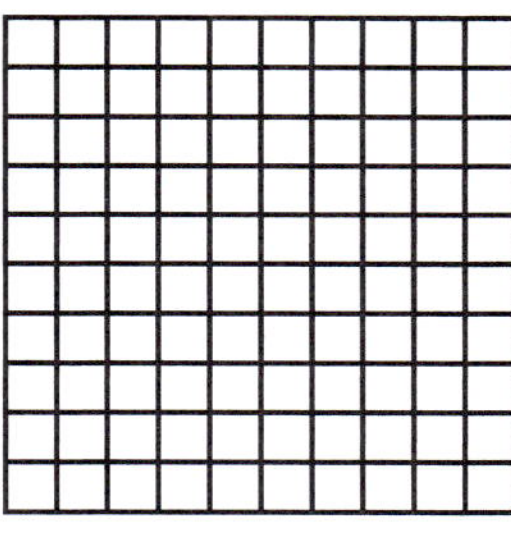

ANSWERS

25. ______
26. ______
27. ______
28. ______
29. ______
30. ______
31. ______
32. ______
33. ______
34. ______
35. ______
36. ______
37. ______
38. ______
39. ______
40. ______
41. ______
42. ______
43. ______
44. ______

Add or subtract as indicated.

25. $\begin{array}{r} 0.85 \\ -\ 0.59 \\ \hline \end{array}$

26. $\begin{array}{r} 5.68 \\ -\ 2.65 \\ \hline \end{array}$

27. $23.81 + (-6.57)$

28. $48.03 + (-19.95)$

29. $\begin{array}{r} 17.134 \\ -\ \ 3.502 \\ \hline \end{array}$

30. $\begin{array}{r} 40.092 \\ -\ 21.595 \\ \hline \end{array}$

31. $-35.8 + 7.45$

32. $-7.83 + 5.2$

33. $\begin{array}{r} 3.82 \\ -\ 1.565 \\ \hline \end{array}$

34. $\begin{array}{r} 8.59 \\ -\ 5.6 \\ \hline \end{array}$

35. $\begin{array}{r} 7.02 \\ -\ 4.7 \\ \hline \end{array}$

36. $\begin{array}{r} 45.6 \\ -\ \ 8.75 \\ \hline \end{array}$

37. $-12 - (-5.35)$

38. $-15 - (-8.85)$

39. Subtract 2.87 from 6.84.

40. Subtract 3.69 from 10.57.

41. Subtract −7.75 from 9.4.

42. Subtract 5.82 from 12.

43. Subtract 0.24 from 5.

44. Subtract −8.7 from 16.32.

ANSWERS

45. ______

46. ______

47. ______

48. ______

49. ______

50. ______

51. ______

52. ______

53. ______

54. ______

Calculator Exercises

Solve the exercises using your calculator.

45. $5.87 + 3.6 + 9.25$

46. $3.456 + 10 + 2.8 + 5.62$

47. $-28.21 + (-387.6) + (-3{,}935.21)$

48. $-10{,}345.2 + (-2{,}308.35) + (-153.58)$

49. $-4.59 - (-2.389)$

50. $-19.375 - (-14.2)$

51. $27.85 - 3.45 - 2.8$

52. $8.8 - 4.59 - 2.325 + 8.5$

53. $14 + 3.2 - 9.35 - 3.375$

54. $8.7675 + 2.8 - 3.375 - 6$

Answers

1. 1.07 **3.** -1.275 **5.** 17.35 **7.** 24.735 **9.** 22.319 **11.** 34.4462

13. 1.791 **15.**

$$\begin{array}{r} {}^{22}\\ 5.00 \\ 23.70 \\ 8.70 \\ +\ 9.85 \\ \hline 47.25 \end{array}$$

17. -31.45 **19.** 147.029 **21.** 0.47

23. 0.67 **25.** 0.26 **27.**

$$\begin{array}{r} {}^{1\,1\ \ 7\,1}\\ 23.81 \\ -\ 6.57 \\ \hline 17.24 \end{array}$$

29. 13.632 **31.** -28.35

33. 2.255 **35.** 2.32 **37.** -6.65 **39.** 3.97 **41.** 17.15 **43.** 4.76

45. 18.72 **47.** $-4{,}351.02$ **49.** -2.201 **51.** 21.6 **53.** 4.475

5.3 Multiplication of Decimals

OBJECTIVES

1. Multiply two or more decimals
2. Estimate the product of decimals
3. Multiply a decimal by a power of ten

To start our discussion of the multiplication of decimals, we will write the decimals in common-fraction form and then multiply.

Example 1

Multiplying Two Decimals

$$0.32 \cdot 0.2 = \frac{32}{100} \cdot \frac{2}{10} = \frac{64}{1000} = 0.064$$

Here 0.32 has *two* decimal places, and 0.2 has *one* decimal place. The product 0.064 has *three* decimal places.

Note:

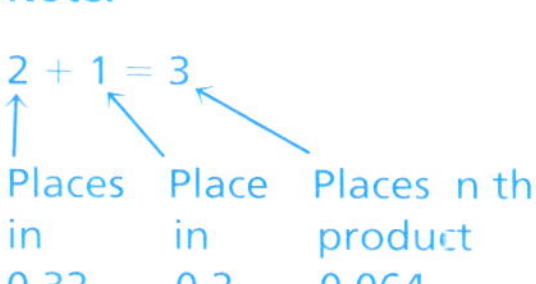

CHECK YOURSELF 1

Find the product and the number of decimal places.

$0.14 \cdot 0.054$

You do not need to write decimals as common fractions to multiply. Our work suggests this rule.

Step by Step: To Multiply Decimals

Step 1 Multiply the decimals as though they were whole numbers.
Step 2 Add the number of decimal places in the numbers being multiplied.
Step 3 Place the decimal point in the product so that the number of decimal places in the product is the sum of the number of decimal places in the factors.

Example 2 illustrates this rule.

Example 2

Multiplying Two Decimals

Multiply 0.23 by 0.7.

$$\begin{array}{r} 0.23 \\ \times\ 0.7 \\ \hline 0.161 \end{array}$$

0.23 ⟵ Two places
0.7 ⟵ One place
0.161 ⟵ Three places

CHECK YOURSELF 2

Multiply 0.36 · 1.52.

You may have to affix zeros to the left in the product to place the decimal point. Consider Example 3.

Example 3

Multiplying Two Decimals

Multiply.

$$\begin{array}{r} 0.136 \\ \times \quad 0.28 \\ \hline 1088 \\ 272 \\ \hline 0.03808 \end{array}$$

0.136 ⟵ Three places
0.28 ⟵ Two places
3 + 2 = 5

0.03808 ⟵ Five places
Insert 0

Insert a 0 to mark off five decimal places.

CHECK YOURSELF 3

Multiply 0.234 · 0.24.

Estimation is also helpful in multiplying decimals.

Example 4

Estimating the Product of Two Decimals

Estimate the product 24.3 · 5.8.

$$\begin{array}{r} 24.3 \\ \times \quad 5.8 \\ \hline \end{array} \quad \xrightarrow{\text{Round}} \quad \begin{array}{r} 24 \\ \times \quad 6 \\ \hline 144 \end{array}$$

Multiply for the estimate.

CHECK YOURSELF 4

Estimate the product.

17.95 · 8.17

When multiplying signed decimals, recall these rules:

1. The product of two numbers with the same sign will be positive.
2. The product of two numbers with opposite signs will be negative.

Example 5

Multiplying Signed Decimals

Find each product.

(a) $2.5 \cdot (-1.4)$

The signs are opposite, so the product will be negative.

$$\begin{array}{r} 1.4 \\ \times\ 2.5 \\ \hline 70 \\ 28 \\ \hline 3.50 \end{array}$$

so $2.5 \cdot (-1.4) = -3.5$.

(b) $-4.6 \cdot (-1.3)$

The signs are the same, so the product will be positive.

$$\begin{array}{r} 1.3 \\ \times\ 4.6 \\ \hline 78 \\ 52 \\ \hline 5.98 \end{array}$$

so $-4.6 \cdot (-1.3) = 5.98$.

CHECK YOURSELF 5

Find each product.

(a) $-1.7 \cdot 4.2$ **(b)** $-2.6 \cdot (-2.3)$ **(c)** $6.8 \cdot (-1.6)$

We will study applications involving operations with decimals in Section 5.8. For now, we will note that there are enough applications involving multiplication by the powers of 10 to make it worthwhile to develop a special rule so you can do such operations quickly and easily. Look at the patterns in some of these special multiplications.

$$\begin{array}{r} 0.679 \\ \times\quad 10 \\ \hline 6.790 \end{array}\text{, or } 6.79 \qquad \begin{array}{r} 23.58 \\ \times\quad 10 \\ \hline 235.80 \end{array}\text{, or } 235.8$$

Do you see that multiplying by 10 has moved the decimal point *one place to the right?* Now we will look at what happens when we multiply by 100.

NOTE The rule will be used to multiply by 10, 100, 1,000, and so on.

$$\begin{array}{r} 0.892 \\ \times\ 100 \\ \hline 89.200 \end{array}\text{, or } 89.2 \qquad \begin{array}{r} 5.74 \\ \times\ 100 \\ \hline 574.00 \end{array}\text{, or } 574$$

NOTE The digits remain the same. Only the *position* of the decimal point is changed.

NOTE Multiplying by 10, 100, or any other larger power of 10 makes the number *larger*. Move the decimal point *to the right.*

Multiplying by 100 shifts the decimal point *two places to the right.* The pattern of these examples gives us this rule:

Rules and Properties: To Multiply by a Power of 10

Move the decimal point to the right the same number of places as there are zeros in the power of 10.

Example 6

Multiplying by Powers of Ten

$2.356 \cdot 10 = 23.56$

One zero — The decimal point has moved one place to the right.

$3.67 \cdot 1{,}000 = 3{,}670.$

Three zeros — The decimal point has moved three places to the right. Note that we inserted a 0 to place the decimal point correctly.

NOTE Remember that 10^5 is just a 1 followed by five zeros.

$0.005672 \cdot 10^5 = 567.2$

Five zeros — The decimal point has moved five places to the right.

CHECK YOURSELF 6

Multiply.

(a) $43.875 \cdot 100$ **(b)** $0.0083 \cdot 10^3$

The steps for finding the product of decimals on a calculator are similar to the ones we used for multiplying whole numbers. To find the product of a group of decimals, just extend the process.

Example 7

Multiplying a Group of Decimals

To multiply $2.8 \cdot 3.45 \cdot 3.725$, enter

2.8 [×] 3.45 [×] 3.725 [=]

Display 35.9835 — Again, look at the original expression to see if this answer is reasonable.

CHECK YOURSELF 7

Multiply $3.1 \cdot 5.72 \cdot 6.475$.

You can also easily find powers of decimals with your calculator by using a procedure similar to that in Example 7.

Example 8

Finding the Power of a Decimal Number

REMEMBER:
$(2.35)^3 = 2.35 \cdot 2.35 \cdot 2.35$

Find $(2.35)^3$.
Enter

2.35 [×] 2.35 [×] 2.35 [=]

Display 12.977875

CHECK YOURSELF 8

Evaluate $(6.2)^4$.

As we stated earlier, some calculators have keys that will find powers more quickly. Look for keys marked [x^2] or [y^x]. Other calculators have a power key marked [∧].

Example 9

Finding the Power of a Decimal Number Using Power Keys

Find $(2.35)^3$.
Enter

2.35 [∧] 3 [Enter] or 2.35 [y^x] 3 [=]

The result is 12.977875.

CHECK YOURSELF 9

Find $(6.2)^4$.

How many places can your calculator display? Most calculators can display either 8, 9, or 10 digits. To find the display capability of your calculator, just enter digits until the calculator can accept no more numbers. For example, try entering

1 [−] 0.226592266 [=]

Does your calculator display 10 digits? Now turn the calculator upside down. What does it say? (It may take a little imagination to see it.)

What happens when your calculator wants to display an answer that is too big to fit in the display? We will try an experiment to see. Enter

10 [×] 10 [=]

Now continue to multiply this answer by 10. Many calculators will let you do this by simply pressing [=]. Others require you to enter "[×] 10" for each calculation. Multiply by 10 until the display is no longer a 1 followed by a series of zeros. The new display represents the power of 10 of the answer. It will be displayed as either

[1^{10}]

(which looks like 1 to the tenth power, but means 1 times 10^{10}) or

[1 E 10]

(which also means 1 times 10^{10}).

Answers that are displayed in this way are said to be in **scientific notation.** This is a topic that you will study in your next math course. In this text we will avoid exercises with answers that are too large to display in the decimal notation that you already know. If you do get such an answer, you should go back and check your work. Do not be afraid to try experimenting with your calculator. It is amazing how much math you can (accidently) learn while playing!

Example 10

Multiplying by a Power of Ten Using the Power Key on a Calculator

Find the product $3.485 \cdot 10^4$.

Use your calculator to enter

3.485 [×] 10 [y^x] 4 [=]   or   3.485 [×] 10 [∧] 4 [Enter]

The result will be 34,850. Note that the decimal point has moved four places (the power of 10) to the right.

CHECK YOURSELF 10

Find the product $8.755 \cdot 10^6$.

CHECK YOURSELF ANSWERS

1. 0.00756, 5 decimal places **2.** 0.5472 **3.** 0.05616 **4.** 144 **5.** **(a)** −7.14; **(b)** 5.98; **(c)** −10.88 **6.** **(a)** 4,387.5; **(b)** 8.3 **7.** 114.8147 **8.** 1,477.6336 **9.** 1,477.6336 **10.** 8,755,000

Name ______

Section ______ Date ______

5.3 Exercises

Multiply.

1. $\begin{array}{r} 2.3 \\ \times 3.4 \\ \hline \end{array}$

2. $\begin{array}{r} 6.5 \\ \times 4.3 \\ \hline \end{array}$

3. $\begin{array}{r} 8.4 \\ \times 5.2 \\ \hline \end{array}$

4. $\begin{array}{r} 9.2 \\ \times 4.6 \\ \hline \end{array}$

5. $\begin{array}{r} 2.56 \\ \times \ 72 \\ \hline \end{array}$

6. $\begin{array}{r} 56.7 \\ \times \ 35 \\ \hline \end{array}$

7. $\begin{array}{r} 0.78 \\ \times \ 2.3 \\ \hline \end{array}$

8. $\begin{array}{r} 9.5 \\ \times 0.45 \\ \hline \end{array}$

9. $\begin{array}{r} 15.7 \\ \times 2.35 \\ \hline \end{array}$

10. $\begin{array}{r} 28.3 \\ \times 0.59 \\ \hline \end{array}$

11. $\begin{array}{r} 0.354 \\ \times \ 0.8 \\ \hline \end{array}$

12. $\begin{array}{r} 0.624 \\ \times \ 0.85 \\ \hline \end{array}$

13. $3.28 \cdot (-5.07)$

14. $0.582 \cdot (-6.3)$

15. $5.238 \cdot 0.48$

16. $0.372 \cdot 58$

17. $-1.053 \cdot (-0.552)$

18. $-2.375 \cdot 0.28$

19. $0.0056 \cdot (-0.082)$

20. $-1.008 \cdot (-0.046)$

21. $0.8 \cdot 2.376$

22. $3.52 \cdot 58$

23. $0.3085 \cdot 4.5$

24. $0.028 \cdot 0.685$

Multiply.

25. $5.89 \cdot 10$

26. $-0.895 \cdot 100$

27. $-23.79 \cdot 100$

28. $2.41 \cdot 10$

29. $0.045 \cdot 10$

30. $5.8 \cdot 100$

31. $0.431 \cdot (-100)$

32. $0.025 \cdot (-10)$

33. $0.471 \cdot 100$

34. $0.95 \cdot 10{,}000$

35. $-0.7125 \cdot (-1{,}000)$

36. $-23.42 \cdot (-1{,}000)$

ANSWERS

1. ______ 2. ______
3. ______ 4. ______
5. ______ 6. ______
7. ______ 8. ______
9. ______ 10. ______
11. ______ 12. ______
13. ______
14. ______
15. ______
16. ______
17. ______
18. ______
19. ______
20. ______
21. ______
22. ______
23. ______
24. ______
25. ______
26. ______ 27. ______
28. ______ 29. ______
30. ______ 31. ______
32. ______ 33. ______
34. ______ 35. ______
36. ______

ANSWERS

37. ____________
38. ____________
39. ____________
40. ____________
41. ____________
42. ____________
43. ____________
44. ____________
45. ____________
46. ____________
47. ____________
48. ____________
49. ____________
50. ____________
51. ____________
52. ____________
53. ____________
54. ____________
55. ____________
56. ____________
57. ____________
58. ____________
59. ____________
60. ____________

37. $4.25 \cdot 10^2$

38. $0.36 \cdot 10^3$

39. $3.45 \cdot 10^4$

40. $0.058 \cdot 10^5$

Calculator Exercises

Compute.

41. $0.08 \cdot (-7.375)$

42. $21.34 \cdot (-0.005)$

43. $-21.38 \cdot (-13.75)$

44. $-58.05 \cdot (-13.02)$

45. $-127.85 \cdot 0.055 \cdot 15.84$

46. $-18.28 \cdot 143.45 \cdot (-0.075)$

47. $(2.65)^2$

48. $(0.08)^3$

49. $(3.95)^3$

50. $(0.521)^2$

Find the products using your calculator.

51. $3.365 \cdot 10^3$

52. $4.128 \cdot 10^3$

53. $4.316 \cdot 10^5$

54. $8.163 \cdot 10^6$

55. $7.236 \cdot 10^8$

56. $5.234 \cdot 10^7$

57. $32.136 \cdot 10^5$

58. $41.234 \cdot 10^4$

59. $31.789 \cdot 10^4$

60. $61.356 \cdot 10^3$

Answers

1. 7.82 **3.** 43.68 **5.** 184.32 **7.** 1.794 **9.** 36.895 **11.** 0.2832 **13.** −16.6296 **15.** 2.51424 **17.** 0.581256 **19.** −0.0004592 **21.** 1.9008 **23.** 1.38825 **25.** 58.9 **27.** −2,379 **29.** 0.45 **31.** −43.1 **33.** 47.1 **35.** 712.5 **37.** 425 **39.** 34,500 **41.** −0.59 **43.** 293.975 **45.** −111.38292 **47.** 7.0225 **49.** 61.629875 **51.** 3,365 **53.** 431,600 **55.** 723,600,000 **57.** 3,213,600 **59.** 317,890

5.4 Division of Decimals

5.4 OBJECTIVES

1. Divide a decimal by a whole number
2. Divide a decimal by a decimal
3. Divide a decimal by a power of ten
4. Use order of operations with decimals

The division of decimals is very similar to our earlier work with dividing whole numbers. The only difference is in learning to place the decimal point in the quotient. We will start with the case of dividing a decimal by a whole number. Here, placing the decimal point is easy. You can apply this rule.

Step by Step: To Divide a Decimal by a Whole Number

Step 1 Place the decimal point in the quotient *directly above* the decimal point of the dividend.

Step 2 Divide as you would with whole numbers.

Example 1

Dividing a Decimal by a Whole Number

Divide 29.21 by 23.

NOTE Do the division just as if you were dealing with whole numbers. Just remember to place the decimal point in the quotient *directly above* the one in the dividend.

```
     1.27
23)29.21
   23
    6 2
    4 6
    1 61
    1 61
       0
```

The quotient is 1.27. You can check by multiplying $1.27 \cdot 23$.

CHECK YOURSELF 1

Divide 80.24 *by* 34.

We will look at another example of dividing a decimal by a whole number.

Example 2

Dividing a Decimal by a Whole Number

Divide 122.2 by 52.

NOTE Again place the decimal point of the quotient above that of the dividend.

```
      2.3
52)122.2
   104
    18 2
    15 6
     2 6
```

We normally do not use a remainder when dealing with decimals. Instead, we write an extra 0 in the dividend and continue.

NOTE Remember that writing a 0 does not change the value of the dividend. It simply allows us to complete the division process in this case.

```
      2.35
52)122.20   ← Add a 0.
   104
    18 2
    15 6
     2 60
     2 60
        0
```

So $122.2 \div 52 = 2.35$. The quotient is 2.35.

CHECK YOURSELF 2

Divide 234.6 *by* 68.

Often you will be asked to give a quotient to a certain place value. In this case, continue the division process to *one digit past* the indicated place value. Then round the result back to the desired accuracy.

When working with money, for instance, we normally give the quotient to the nearest hundredth of a dollar (the nearest cent). This means carrying the division out to the thousandths place and then rounding back.

Example 3

Dividing a Decimal by a Whole Number and Rounding the Result

NOTE Find the quotient to *one place past* the desired place, and then round the result.

Find the quotient of $25.75 \div 15$ to the nearest hundredth.

```
     1.716
15)25.750   ← Write a 0 to carry the division
   15          to the thousandths place.
   10 7
   10 5
      25
      15
      100
       90
       10
```

So $25.75 \div 15 = 1.72$ (to the nearest hundredth).

CHECK YOURSELF 3

Find $99.26 \div 35$ *to the nearest hundredth.*

We want now to look at division *by* decimals. Here is an example using a fractional form.

Example 4

Rewriting a Problem That Requires Dividing by a Decimal

Rewrite the division problem so that the divisor is a whole number.

$$2.57 \div 3.4 = \frac{2.57}{3.4}$$ Write the division as a fraction.

$$= \frac{2.57 \cdot 10}{3.4 \cdot 10}$$ We multiply the numerator and denominator by 10 so the divisor is a whole number. This *does not change* the value of the fraction.

$$= \frac{25.7}{34}$$ Multiplying by 10, shift the decimal point in the numerator and denominator *one place to the right.*

$$= 25.7 \div 34$$ Our division problem is rewritten so that the divisor is a whole number.

NOTE It's always easier to rewrite a division problem so that you're dividing by a whole number. Dividing by a whole number makes it easy to place the decimal point in the quotient.

So

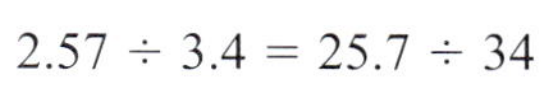

$$2.57 \div 3.4 = 25.7 \div 34$$ After we multiply the numerator and denominator by 10, we see that $2.57 \div 3.4$ is the same as $25.7 \div 34$.

CHECK YOURSELF 4

Rewrite the division problem so that the divisor is a whole number.

$3.42 \div 2.5$

NOTE Of course, multiplying by any whole-number power of 10 greater than 1 is just a matter of shifting the decimal point to the right.

Do you see the rule suggested by Example 4? We multiplied the numerator and the denominator (the dividend and the divisor) by 10. We made the divisor a whole number without altering the actual digits involved. All we did was shift the decimal point in the divisor and dividend the same number of places. This leads us to the next rule.

Step by Step: To Divide by a Decimal

Step 1 Move the decimal point in the divisor *to the right,* making the divisor a whole number.

Step 2 Move the decimal point in the dividend to the right *the same number of places.* Add zeros if necessary.

Step 3 Place the decimal point in the quotient directly above the decimal point of the dividend.

Step 4 Divide as you would with whole numbers.

We will now look at an example of the use of our division rule.

Example 5

Rounding the Result of Dividing by a Decimal

Divide 1.573 by 0.48 and give the quotient to the nearest tenth.

Write

0.48‸)1.57‸3 Shift the decimal points two places to the right to make the divisor a whole number.

Now divide:

NOTE Once the division statement is rewritten, place the decimal point in the quotient above that in the dividend.

```
      3.27
48)157.30
   144
    13 3
     9 6
     3 70
     3 36
       34
```

Note that we add a 0 to carry the division to the hundredths place. In this case, we want to find the quotient to the nearest tenth.

Round 3.27 to 3.3. So

$1.573 \div 0.48 = 3.3$ (to the nearest tenth)

CHECK YOURSELF 5

Divide, rounding the quotient to the nearest tenth.

$3.4 \div 1.24$

Recall that you can multiply decimals by powers of 10 by simply shifting the decimal point to the right. A similar approach will work for division by powers of 10.

Example 6

Dividing a Decimal by a Power of 10

(a) Divide 35.3 by 10.

```
     3.53
10)35.30
   30
    5 3
    5 0
      30
      30
       0
```

The dividend is 35.3. The quotient is 3.53. The decimal point has been shifted *one place to the left.* Note also that the divisor, 10, has *one* zero.

(b) Divide 378.5 by 100.

```
      3.785
100)378.500
    300
     78 5
     70 0
      8 50
      8 00
        500
        500
          0
```

Here the dividend is 378.5, whereas the quotient is 3.785. The decimal point is now shifted *two places to the left.* In this case the divisor, 100, has *two* zeros.

CHECK YOURSELF 6

Perform each of the divisions.

(a) $52.6 \div 10$ **(b)** $267.9 \div 100$

Example 6 suggests this rule.

Rules and Properties: To Divide a Decimal by a Power of 10

Move the decimal point *to the left* the same number of places as there are zeros in the power of 10.

Example 7

Dividing a Decimal by a Power of 10

Divide.

(a) $27.3 \div 10 = 2_\wedge 7.3$ Shift one place to the left.

$= 2.73$

(b) $57.53 \div 100 = 0_\wedge 57.53$ Shift two places to the left.

$= 0.5753$

NOTE As you can see, we may have to insert zeros to correctly place the decimal point.

(c) $39.75 \div 1{,}000 = 0_\wedge 039.75$ Shift three places to the left.

$= 0.03975$

(d) $85 \div 1{,}000 = 0_\wedge 085.$ The decimal after the 85 is implied.

$= 0.085$

REMEMBER: 10^4 is a 1 followed by *four zeros*.

(e) $235.72 \div 10^4 = 0_\wedge 0235.72$ Shift four places to the left.

$= 0.023572$

CHECK YOURSELF 7

Divide.

(a) $3.84 \div 10$ **(b)** $27.3 \div 1{,}000$

Recall the rules associated with the division of signed numbers:

1. The quotient of two numbers with the same sign is positive.
2. The quotient of two numbers with opposite signs is negative.

Example 8

Dividing Signed Decimals

Find each quotient.

(a) $-1.61 \div 2.3$

The signs are opposite, so the quotient will be negative. Dividing 1.61 by 2.3, we find

$$2.3\overline{)1.61}$$

$$= 23\overline{)16.1} \quad .7$$

$$\begin{array}{r} 16\ 1 \\ \hline 0 \end{array}$$

so, $-1.61 \div 2.3 = -0.7$.

(b) $-5.13 \div (-6.84)$

The signs are the same, so the result will be positive. Dividing 5.13 by 6.84, we find

$$6.84\overline{)5.13}$$

$$= 684\overline{)513.00} \quad .75$$

$$\begin{array}{r} 478\ 8 \\ \hline 34\ 20 \\ 34\ 20 \\ \hline 0 \end{array}$$

so, $-5.13 \div (-6.84) = 0.75$.

CHECK YOURSELF 8

Find each quotient.

(a) $-3.612 \div 2.58$ **(b)** $-2.224 \div (-2.78)$

Recall that the order of operations is always used to simplify a mathematical expression with several operations. You should recall the order of operations as given here.

Rules and Properties: The Order of Operations

1. Perform any operations enclosed in **parentheses.**
2. Apply any **exponents.**
3. Do any **multiplication** and **division,** moving from left to right.
4. Do any **addition** and **subtraction,** moving from left to right.

Example 9

Applying the Order of Operations

Simplify each expression.

(a) $4.6 + (0.5 \cdot 4.4)^2 - 3.93$

$= 4.6 + (2.2)^2 - 3.93$ parentheses

$= 4.6 + 4.84 - 3.93$ exponent

$= 9.44 - 3.93$ add (left of the subtraction)

$= 5.51$ subtract

(b) $16.5 - (2.8 + 0.2)^2 + 4.1 \cdot 2$

$= 16.5 - (3)^2 + 4.1 \cdot 2$ parentheses

$= 16.5 - 9 + 4.1 \cdot 2$ exponent

$= 16.5 - 9 + 8.2$ multiply

$= 7.5 + 8.2$ subtraction (left of the addition)

$= 15.7$ add

(c) $4.8 + 6(8.9 - 10.9)^2 - 8.5 \cdot 4$

$= 4.8 + 6(-2)^2 - 8.5 \cdot 4$ parentheses

$= 4.8 + 6 \cdot 4 - 8.5 \cdot 4$ exponent

$= 4.8 + 24 - 34$ multiply

$= 28.8 - 34$ add (left of the subtraction)

$= -5.2$ subtract

CHECK YOURSELF 9

Simplify each expression.

(a) $6.35 + (0.2 \cdot 8.5)^2 - 3.7$ **(b)** $2.5^2 - (3.57 - 2.14) + 3.2 \cdot 1.5$

(c) $2.7 + 5 \cdot (4.75 - 9.75)^2 - 5^3$

Using Your Calculator to Divide Decimals

It would be most surprising if you had reached this point without using your calculator to divide decimals. It is a good way to check your work, and a reasonable way to solve applications.

Example 10

Dividing Decimals

Use your calculator to find the quotient.

$211.56 \div (-82)$

Enter the problem in the calculator.

211.56 [÷] 82 [+/−] [=]

We find that the answer is -2.58.

CHECK YOURSELF 10

Use your calculator to find the quotient.

$-304.32 \div 9.6$

CHECK YOURSELF ANSWERS

1. 2.36 **2.** 3.45 **3.** 2.84 **4.** $34.2 \div 25$ **5.** 2.7
6. **(a)** 5.26; **(b)** 2.679 **7.** **(a)** 0.384; **(b)** 0.0273 **8.** **(a)** -1.4; **(b)** 0.8
9. **(a)** 5.54; **(b)** 9.62; **(c)** 2.7 **10.** -31.7

Name ____________

Section ________ Date ________

5.4 Exercises

Divide.

1. $16.68 \div 6$
2. $43.92 \div 8$
3. $1.92 \div (-4)$
4. $-5.52 \div (-6)$
5. $-5.48 \div (-8)$
6. $-2.76 \div 8$
7. $13.89 \div 6$
8. $21.92 \div 5$
9. $185.6 \div (-32)$
10. $-165.6 \div (-36)$
11. $-79.9 \div (-34)$
12. $179.3 \div (-55)$
13. $52\overline{)13.78}$
14. $76\overline{)26.22}$
15. $0.6\overline{)11.07}$
16. $0.8\overline{)10.84}$
17. $3.8\overline{)7.22}$
18. $2.9\overline{)13.34}$
19. $5.2\overline{)11.622}$
20. $6.4\overline{)3.616}$
21. $0.27\overline{)1.8495}$
22. $0.038\overline{)0.8132}$
23. $0.046\overline{)1.587}$
24. $0.52\overline{)3.2318}$
25. $-0.658 \div 2.8$
26. $0.882 \div (-0.36)$

Divide by moving the decimal point.

27. $5.8 \div 10$
28. $5.1 \div 10$
29. $-4.568 \div 100$
30. $-3.817 \div 100$
31. $24.39 \div 1{,}000$
32. $8.41 \div 100$
33. $-6.9 \div 1{,}000$
34. $-7.2 \div 1{,}000$
35. $7.8 \div 10^2$
36. $3.6 \div 10^3$
37. $-45.2 \div 10^5$
38. $-57.3 \div 10^4$

Divide and round the quotient to the indicated decimal place.

39. $23.8 \div 9$ tenths
40. $5.27 \div 8$ hundredths
41. $38.48 \div 46$ hundredths
42. $3.36 \div 36$ thousandths

ANSWERS

1. ______ 2. ______
3. ______ 4. ______
5. ______ 6. ______
7. ______ 8. ______
9. ______ 10. ______
11. ______ 12. ______
13. ______ 14. ______
15. ______ 16. ______
17. ______ 18. ______
19. ______ 20. ______
21. ______ 22. ______
23. ______ 24. ______
25. ______ 26. ______
27. ______ 28. ______
29. ____________
30. ____________
31. ____________
32. ____________
33. ____________
34. ____________
35. ______ 36. ______
37. ____________
38. ____________
39. ______ 40. ______
41. ______ 42. ______

ANSWERS

43. ______
44. ______
45. ______
46. ______
47. ______
48. ______
49. ______
50. ______
51. ______
52. ______
53. ______
54. ______
55. ______
56. ______
57. ______
58. ______
59. ______
60. ______
61. ______
62. ______
63. ______
64. ______
65. ______
66. ______

43. $125.4 \div 52$ tenths

44. $2.563 \div 54$ thousandths

45. $0.7\overline{)1.642}$ hundredths

46. $0.6\overline{)7.695}$ tenths

47. $4.5\overline{)8.415}$ tenths

48. $5.8\overline{)16}$ hundredths

49. $3.12\overline{)4.75}$ hundredths

50. $64.2\overline{)16.3}$ thousandths

Simplify each expression.

51. $4.2 - 3.1 \cdot 1.5 + (3.1 + 0.4)^2$

52. $150 + 4.1 \cdot 1.5 - (2.5 \cdot 1.6)^3 \cdot 2.4$

53. $17.9 \cdot 1.1 - (2.3 \cdot 1.1)^2 + (13.4 - 2.1 \cdot 4.6)$

54. $6.89^2 - 3.14 \cdot 2.5 + (4.1 - 3.2 \cdot 1.6)^2$

Calculator Exercises

Divide and check.

55. $8.901 \div (-2.58)$

56. $-16.848 \div (-0.288)$

57. $-99.705 \div (-34.5)$

58. $-171.25 \div 2.74$

59. $-0.01372 \div 0.056$

60. $0.200754 \div (-0.00855)$

Divide and round to the indicated place.

61. $2.546 \div 1.38$ hundredths

62. $45.8 \div 9.4$ tenths

63. $0.5782 \div 1.236$ thousandths

64. $1.25 \div 0.785$ hundredths

65. $1.34 \div 2.63$ two decimal places

66. $12.364 \div 4.361$ three decimal places

Answers

1. 2.78 **3.** −0.48 **5.** 0.685 **7.** 2.315 **9.** −5.8 **11.** 2.35 **13.** 0.265 **15.** 18.45 **17.** 1.9 **19.** 2.235 **21.** 6.85 **23.** 34.5 **25.** −0.235 **27.** 0.58 **29.** −0.04568 **31.** 0.02439 **33.** −0.0069 **35.** 0.078 **37.** −0.000452 **39.** 2.6 **41.** 0.84 **43.** 2.4 **45.** 2.35 **47.** 1.9 **49.** 1.52 **51.** 11.8 **53.** 17.0291 **55.** −3.45 **57.** 2.89 **59.** −0.245 **61.** 1.84 **63.** 0.468 **65.** 0.51

5.5 Fractions and Decimals

OBJECTIVES

1. Convert a common fraction to a decimal
2. Convert a common fraction to a repeating decimal
3. Convert a mixed number to a decimal
4. Convert a decimal to a common fraction
5. Compare the sizes of fractions and decimals

Because a common fraction can be interpreted as division, you can divide the numerator of the common fraction by its denominator to convert a common fraction to a decimal. The result is called a **decimal equivalent.**

Example 1

Converting a Fraction to a Decimal Equivalent

Write $\frac{5}{8}$ as a decimal.

NOTE Remember that 5 can be written as 5.0, 5.00, 5.000, and so on. In this case, we continue the division by adding zeros to the dividend until a 0 remainder is reached.

$$\begin{array}{r} 0.625 \\ 8\overline{)5.000} \\ \underline{4\,8} \\ 20 \\ \underline{16} \\ 40 \\ \underline{40} \\ 0 \end{array}$$

Because $\frac{5}{8}$ means 5 ÷ 8, divide 8 into 5.

We see that $\frac{5}{8} = 0.625$; 0.625 is the decimal equivalent of $\frac{5}{8}$.

CHECK YOURSELF 1

Find the decimal equivalent of $\frac{7}{8}$.

Some fractions are used so often that we have listed their decimal equivalents for your reference.

NOTE The division used to find these decimal equivalents stops when a 0 remainder is reached. The equivalents are called **terminating decimals.**

Some Common Decimal Equivalents

$\frac{1}{2} = 0.5$	$\frac{1}{4} = 0.25$	$\frac{1}{5} = 0.2$	$\frac{1}{8} = 0.125$
	$\frac{3}{4} = 0.75$	$\frac{2}{5} = 0.4$	$\frac{3}{8} = 0.375$
		$\frac{3}{5} = 0.6$	$\frac{5}{8} = 0.625$
		$\frac{4}{5} = 0.8$	$\frac{7}{8} = 0.875$

If a decimal equivalent does not terminate, you can round the result to approximate the fraction to some specified number of decimal places. Consider Example 2.

Example 2

Converting a Fraction to a Decimal Equivalent

Write $\frac{3}{7}$ as a decimal. Round the answer to the nearest thousandth.

$$\begin{array}{r} 0.4285 \\ 7\overline{)3.0000} \\ \underline{2\,8} \\ 20 \\ \underline{14} \\ 60 \\ \underline{56} \\ 40 \\ \underline{35} \\ 5 \end{array}$$

In this example, we are choosing to round to three decimal places, so we must add enough zeros to carry the division to four decimal places.

NOTE Multiply 0.429 times 7 and then multiply 0.428 times 7. Which product is closer to 3?

So $\frac{3}{7} = 0.429$ (to the nearest thousandth).

CHECK YOURSELF 2

Find the decimal equivalent of $\frac{5}{11}$ to the nearest thousandth.

If a fraction's decimal equivalent does *not* terminate, it will *repeat* a sequence of digits. These decimals are called **repeating decimals.**

Example 3

Converting a Fraction to a Repeating Decimal

(a) Write $\frac{1}{3}$ as a decimal.

$$\begin{array}{r} 0.333 \\ 3\overline{)1.000} \\ \underline{9} \\ 10 \\ \underline{9} \\ 10 \\ \underline{9} \end{array}$$

The digit 3 will just repeat itself indefinitely because each new remainder will be 1.

Adding more zeros and going on will simply lead to more threes in the quotient.

We can say $\frac{1}{3} = 0.333 \cdots$

The three dots mean "and so on" and tell us that 3 will repeat itself indefinitely.

(b) Write $\frac{5}{12}$ as a decimal.

$$
\begin{array}{r}
0.4166\cdots \\
12\overline{)5.0000} \\
\underline{4\,8} \\
20 \\
\underline{12} \\
80 \\
\underline{72} \\
80 \\
\underline{72} \\
8
\end{array}
$$

In this example, the digit 6 will just repeat itself because the remainder, 8, will keep occurring if we add more zeros and continue the division.

CHECK YOURSELF 3

Find the decimal equivalent of each fraction.

(a) $\frac{2}{3}$ **(b)** $\frac{7}{12}$

Some important decimal equivalents (rounded to the nearest thousandth) are given here for reference.

$\frac{1}{6} = 0.167$ $\frac{1}{3} = 0.333$ $\frac{2}{3} = 0.667$ $\frac{5}{6} = 0.833$

Another way to write a repeating decimal is with a bar placed over the digit or digits that repeat. For example, we can write

$0.37373737\cdots$

as

$0.\overline{37}$

The bar placed over the digits indicates that "37" repeats indefinitely.

$\frac{1}{6} = 0.1\overline{6}$

The bar placed over the digit 6 indicates that only this 6 repeats.

Example 4

Converting a Fraction to a Repeating Decimal

Write $\frac{5}{11}$ as a decimal.

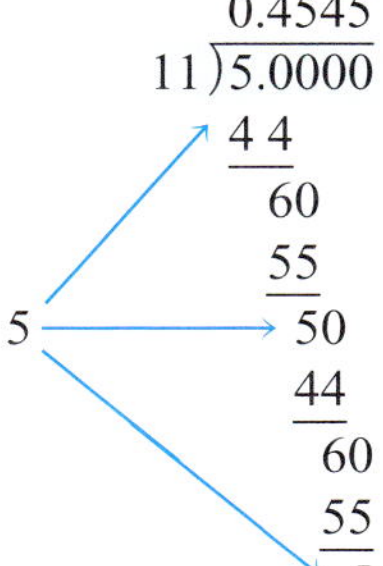

$$
\begin{array}{r}
0.4545 \\
11\overline{)5.0000} \\
\underline{4\,4} \\
60 \\
\underline{55} \\
50 \\
\underline{44} \\
60 \\
\underline{55} \\
5
\end{array}
$$

As soon as a remainder repeats itself, as 5 does here, the pattern of digits will repeat in the quotient.

$\frac{5}{11} = 0.\overline{45}$

$= 0.4545\cdots$

CHECK YOURSELF 4

Use the bar notation to write the decimal equivalent of $\frac{5}{7}$. (Be patient. You'll have to divide for a while to find the repeating pattern.)

You can find the decimal equivalents for mixed numbers in a similar way. Find the decimal equivalent of the fractional part of the mixed number, and then combine that with the whole-number part. Example 5 illustrates this approach.

Example 5

Converting a Mixed Number to a Decimal Equivalent

NOTE Recall that $3\frac{5}{16} = 3 + \frac{5}{16}$

Find the decimal equivalent of $3\frac{5}{16}$.

$\frac{5}{16} = 0.3125$ First find the equivalent of $\frac{5}{16}$ by division.

$3\frac{5}{16} = 3.3125$ Add 3 to the result.

CHECK YOURSELF 5

Find the decimal equivalent of $2\frac{5}{8}$.

To this point in the section, we have seen that to find the decimal equivalent of a fraction, we use long division. Because the remainder must be less than the divisor, the remainder must *either repeat or become 0.* Thus *every common fraction* will have a *repeating* or a *terminating* decimal as its decimal equivalent.

Using what we have learned about place values, you can easily write decimals as common fractions. Use this rule.

Step by Step: To Convert a Terminating Decimal Less Than 1 to a Common Fraction

Step 1 Write the digits of the decimal without the decimal point. This will be the numerator of the common fraction.

Step 2 The denominator of the fraction is a 1 followed by as many zeros as there are places in the decimal.

Example 6

Converting a Decimal to a Common Fraction

$0.7 = \frac{7}{10}$ (One place; One zero)

$0.09 = \frac{9}{100}$ (Two places; Two zeros)

$0.257 = \frac{257}{1000}$ (Three places; Three zeros)

CHECK YOURSELF 6

Write as common fractions.

(a) 0.3 **(b)** 0.311

When a decimal is converted to a common fraction, the common fraction that results should be written in lowest terms.

Example 7

Converting a Decimal to a Common Fraction

Convert 0.395 to a fraction and write the result in lowest terms.

NOTE Divide the numerator and denominator by 5.

$$0.395 = \frac{395}{1000} = \frac{79}{200}$$

CHECK YOURSELF 7

Write 0.275 *as a common fraction.*

If the decimal has a whole-number portion, write the digits to the right of the decimal point as a proper fraction and then form a mixed number for your result.

Example 8

Converting a Decimal to a Mixed Number

NOTE Repeating decimals can also be written as common fractions, although the process is more complicated. We will limit ourselves to the conversion of terminating decimals in this textbook.

Write 12.277 as a mixed number.

$$0.277 = \frac{277}{1000} \quad \text{so} \quad 12.277 = 12\frac{277}{1000}$$

CHECK YOURSELF 8

Write 32.433 *as a mixed number.*

Comparing the sizes of common fractions and decimals requires finding the decimal equivalent of the common fraction and then comparing the resulting decimals.

Example 9

Comparing the Sizes of Common Fractions and Decimals

Which is larger, $\frac{3}{8}$ or 0.38?

Write the decimal equivalent of $\frac{3}{8}$. That decimal is 0.375. Now comparing 0.375 and 0.38, we see that 0.38 is the larger of the numbers:

$$0.38 > \frac{3}{8}$$

CHECK YOURSELF 9

Which is larger, $\frac{3}{4}$ *or* 0.8*?*

A calculator is very useful in converting common fractions to decimals. Just divide the numerator by the denominator, and the decimal equivalent will be in the display.

Example 10

Converting Fractions to Decimals

Find the decimal equivalent of $\frac{7}{16}$.

7 [÷] 16 [=]

Display 0.4375   0.4375 is the decimal equivalent of $\frac{7}{16}$.

CHECK YOURSELF 10

Find the decimal equivalent of $\frac{5}{16}$.

Often, you will want to round the result in the display.

Example 11

Converting Fractions to Decimals

NOTE Some calculators show the 0 to the left of the decimal point and seven digits to the right. Others omit the 0 and show eight digits to the right. Check yours with this example.

Find the decimal equivalent of $\frac{5}{24}$ to the nearest hundredth.

5 [÷] 24 [=]

Display 0.2083333

$\frac{5}{24} = 0.21$ (nearest hundredth)

CHECK YOURSELF 11

Find the decimal equivalent of $\frac{7}{29}$ to the nearest hundredth.

To find the decimal equivalent of a mixed number, use the sequence given in Example 12.

Example 12

Converting Mixed Numbers to Decimals

NOTE There are several ways to do this, depending on the calculator you are using. For example,

7 [+] 5 [÷] 8 [=]

will work on most scientific calculators. Try it on yours.

Change $7\frac{5}{8}$ to a decimal.

5 [÷] 8 [+] 7 [=]

Display 7.625

CHECK YOURSELF 12

Find the decimal equivalent of $3\frac{3}{8}$.

Depending on the calculator you are using, the result may be rounded at its last displayed digit.

Example 13

Converting Fractions to Decimals

For $\frac{5}{9}$:

5 $\boxed{\div}$ 9 $\boxed{=}$

Display 0.555555555 or 0.555555556 ⟵ Rounded display, actually $\frac{5}{9} = 0.\overline{5}$

CHECK YOURSELF 13

Find the decimal equivalent of $\frac{7}{11}$.

CHECK YOURSELF ANSWERS

1. 0.875 **2.** $\frac{5}{11} = 0.455$ (to the nearest thousandth)

3. **(a)** 0.666 · · · ; **(b)** $\frac{7}{12} = 0.583 \cdots$ The digit 3 will continue indefinitely.

4. $\frac{5}{7} = 0.\overline{714285}$ **5.** 2.625 **6.** **(a)** $\frac{3}{10}$; **(b)** $\frac{311}{1000}$ **7.** $0.275 = \frac{11}{40}$

8. $32\frac{433}{1000}$ **9.** $0.8 > \frac{3}{4}$ **10.** 0.3125 **11.** 0.24 **12.** 3.375

13. 0.63636363 or 0.63636364

Name ____________

Section ________ Date ________

5.5 Exercises

Find the decimal equivalents for each fraction.

1. $\frac{3}{4}$

2. $\frac{4}{5}$

3. $\frac{9}{20}$

4. $\frac{3}{10}$

5. $\frac{1}{5}$

6. $\frac{1}{8}$

7. $\frac{5}{16}$

8. $\frac{11}{20}$

9. $\frac{7}{10}$

10. $\frac{7}{16}$

11. $\frac{27}{40}$

12. $\frac{17}{32}$

Find the decimal equivalents rounded to the indicated place.

13. $\frac{5}{6}$ thousandths

14. $\frac{7}{12}$ hundredths

15. $\frac{4}{15}$ thousandths

Write the decimal equivalents, using the bar notation.

16. $\frac{1}{18}$

17. $\frac{4}{9}$

18. $\frac{3}{11}$

Find the decimal equivalents for each of the mixed numbers.

19. $5\frac{3}{5}$

20. $7\frac{3}{4}$

21. $4\frac{7}{16}$

Find the decimal equivalent for each fraction. Use the bar notation.

22. $\frac{1}{11}$

23. $\frac{1}{111}$

24. $\frac{1}{1111}$

25. From the pattern of exercises 22 to 24, can you guess the decimal representation for $\frac{1}{11111}$?

ANSWERS

1. ____________
2. ____________
3. ____________
4. ____________
5. ____________
6. ____________
7. ____________
8. ____________
9. ____________
10. ____________
11. ____________
12. ____________
13. ____________
14. ____________
15. ____________
16. ____________
17. ____________
18. ____________
19. ____________
20. ____________
21. ____________
22. ____________
23. ____________
24. ____________
25. ____________

ANSWERS

26.

27.

28.

29.

30.

31.

32.

33.

34.

35.

36.

37.

38.

39.

40.

41. 42.

43. 44.

45. 46.

47. 48.

49. 50.

51. 52.

53. 54.

55. 56.

Insert > or < to form a true statement.

26. $\frac{18}{21}$ __ 0.863

27. $\frac{31}{34}$ __ 0.9118

28. $\frac{21}{37}$ __ 0.5664

29. $\frac{13}{17}$ __ 0.7657

Write each decimal as a common fraction or mixed number. Write your answer in lowest terms.

30. 0.3

31. 0.8

32. 0.6

33. 0.37

34. 0.97

35. 0.587

36. 0.379

37. 0.48

38. 0.75

39. 0.58

40. 0.65

41. 0.425

42. 0.116

43. 0.375

44. 0.225

45. 0.136

46. 0.575

47. 0.059

48. 0.067

49. 0.0625

50. 0.0425

51. 6.3

52. 5.7

53. 2.17

54. 3.31

55. 5.28

56. 15.35

Complete each statement, using the symbol < or >.

57. $\frac{7}{8}$ ______ 0.87

58. $\frac{5}{16}$ ______ 0.313

59. $\frac{9}{25}$ ______ 0.4

60. $\frac{11}{17}$ ______ 0.638

Calculator Exercises

Find the decimal equivalents.

61. $\frac{7}{8}$

62. $\frac{11}{16}$

63. $\frac{9}{16}$

64. $\frac{7}{24}$ hundredth

65. $\frac{5}{32}$ thousandth

66. $\frac{11}{75}$ thousandth

67. $\frac{3}{11}$ use bar notation

68. $\frac{7}{11}$ use bar notation

69. $\frac{16}{33}$ use bar notation

70. $3\frac{4}{5}$

71. $3\frac{7}{8}$

72. $8\frac{3}{16}$

Convert each decimal to a fraction.

73. 0.3

74. 0.55

75. 0.305

76. 0.1

77. 0.875

78. 0.125

ANSWERS

57. ______
58. ______
59. ______
60. ______
61. ______
62. ______
63. ______
64. ______
65. ______
66. ______
67. ______
68. ______
69. ______
70. ______
71. ______
72. ______
73. ______
74. ______
75. ______
76. ______
77. ______
78. ______

Answers

1. 0.75 **3.** 0.45 **5.** 0.2 **7.** 0.3125 **9.** 0.7 **11.** 0.675
13. 0.833 **15.** 0.267 **17.** $0.\overline{4}$ **19.** 5.6 **21.** 4.4375 **23.** $0.\overline{009}$
25. $0.\overline{00009}$ **27.** $<$ **29.** $<$ **31.** $\frac{4}{5}$ **33.** $\frac{37}{100}$ **35.** $\frac{587}{1000}$
37. $\frac{12}{25}$ **39.** $\frac{29}{50}$ **41.** $\frac{17}{40}$ **43.** $\frac{3}{8}$ **45.** $\frac{17}{125}$ **47.** $\frac{59}{1000}$
49. $\frac{1}{16}$ **51.** $6\frac{3}{10}$ **53.** $2\frac{17}{100}$ **55.** $5\frac{7}{25}$ **57.** $>$ **59.** $<$
61. 0.875 **63.** 0.5625 **65.** 0.156 **67.** $0.\overline{27}$ **69.** $0.\overline{48}$ **71.** 3.875
73. $\frac{3}{10}$ **75.** $\frac{61}{200}$ **77.** $\frac{7}{8}$

5.6 Equations Containing Decimals

5.6 OBJECTIVE

1. Solve equations containing decimals

Equations involving decimals can be solved by the methods we studied in Chapter 4. For instance, to solve $2.3x = 6.9$, we simply use the multiplication property of equality to divide both sides of the equation by 2.3. This will isolate the variable on the left as desired. This is illustrated in Example 1.

Example 1

Solving an Equation Involving Decimals

Solve the equation.

$2.3x = 6.9$

Dividing both sides by 2.3, we get

$$\frac{2.3x}{2.3} = \frac{6.9}{2.3} = \frac{69}{23}$$

or $x = 3$.

CHECK YOURSELF 1

Solve the equation.

$4.1x = 20.5$

The addition property of equality is also used to solve equations containing decimals.

Example 2

Solving an Equation Involving Decimals

Solve the equation.

$1.7x - 1.68 = 2.5x + 5.8$

First, we use the addition property to collect variable terms on the left and constant terms on the right.

$$\begin{array}{rcr} 1.7x - 1.68 & = & 2.5x + 5.8 \\ + 1.68 & & + 1.68 \\ \hline 1.7x & = & 2.5x + 7.48 \\ -2.5x & & -2.5x \\ \hline -0.8x & = & 7.48 \end{array}$$

We can now divide each side of the equation by the -0.8.

$$\frac{-0.8x}{-0.8} = \frac{7.48}{-0.8}$$

$$x = -9.35$$

CHECK YOURSELF 2

Solve the equation.

$5.3x - 3.46 = 7.1x + 5.09$

Care must be taken when an equation contains parentheses.

Example 3

Solving an Equation Containing Parentheses

Solve the equation.

$3.1x - (x - 4.3) = 1.3x - 3.2$

First, we rewrite the equation so that subtraction is changed to the addition of the opposite. To accomplish this, we distribute the negative over the expression $x - 4.3$.

$3.1x + (-x + 4.3) = 1.3x + (-3.2)$

or

$3.1x + (-x) + 4.3 = 1.3x + (-3.2)$

Collecting like terms on the left,

$$\begin{array}{rcr} 2.1x + 4.3 & = & 1.3x + (-3.2) \\ -\ 4.3 & & -4.3 \\ \hline 2.1x & = & 1.3x + (-7.5) \\ -1.3x & & -1.3x \\ \hline 0.8x & = & -7.5 \end{array}$$

Subtract 4.3 from each side.

Dividing both sides by 0.8, we find

$x = -9.375$

CHECK YOURSELF 3

Solve the equation.

$2.3x + 13.2 = 1.52 - (x - 4.75)$

CHECK YOURSELF ANSWERS

1. $x = 5$ **2.** $x = -4.75$ **3.** $x = -2.1$

Name ______________

Section ________ Date ________

5.6 Exercises

Solve each equation for x.

1. $3.2x = 12.8$

2. $5.1x = -15.3$

3. $-4.5x = 13.5$

4. $-8.2x = -32.8$

5. $1.3x + 2.8x = 12.3$

6. $2.7x + 5.4x = -16.2$

7. $9.3x + (-6.2x) = 12.4$

8. $12.5x + (-7.2x) = -21.2$

9. $5.3x + (-7) = 2.3x + 5$

10. $9.8x + 2 = 3.8x + 20$

11. $3x - 0.54 = 2(x - 0.15)$

12. $7x + 0.125 = 6x - 0.289$

13. $6x + 3(x - 0.2789) = 4(2x + 0.3912)$

14. $9x - 2(3x - 0.124) = 2x + 0.965$

15. $5x - (0.345 - x) = 5x + 0.8713$

16. $-3(0.234 - x) = 2(x + 0.974)$

17. $2.3x - 4.25 = 3.3x + 2.15$

18. $5.7x + 3.84 = 6.7x + 5.26$

19. $7.1x - 14 = 4.3x - 8.54$

20. $5.6x + 7.5 = 2.9x + 0.885$

21. $3.2x + 8.36 = 5x + 13.94$

22. $4.8x - 2.35 = 7x - 8.51$

23. $7.4(x - 1.2) = 8.4(x - 0.5)$

24. $4.8(x + 3.5) = 5.8(x - 2.5)$

ANSWERS

1. ______________
2. ______________
3. ______________
4. ______________
5. ______________
6. ______________
7. ______________
8. ______________
9. ______________
10. ______________
11. ______________
12. ______________
13. ______________
14. ______________
15. ______________
16. ______________
17. ______________
18. ______________
19. ______________
20. ______________
21. ______________
22. ______________
23. ______________
24. ______________

ANSWERS

25. ____________

26. ____________

27. ____________

28. ____________

25. $3.5(x - 1.4) = 1.3x - 3.14$

26. $2.6x - 12.6 = 5.9(x - 1.8)$

27. $5 - 1.6(x + 2) = 2(x - 1.3) - 0.1$

28. $3 - 2.4(x - 5) = 4(x + 1.5) - 7.64$

Answers

1. 4 **3.** -3 **5.** 3 **7.** 4 **9.** 4 **11.** 0.24 **13.** 2.4015 **15.** 1.2163 **17.** -6.4 **19.** 1.95 **21.** -3.1 **23.** -4.68 **25.** 0.8 **27.** 1.25

5.7 Square Roots and the Pythagorean Theorem

OBJECTIVES

1. Find the square root of a perfect square
2. Apply the Pythagorean theorem
3. Approximate the square root of a number
4. Use a calculator to find a square root

Some numbers can be written as the product of two identical factors, for example,

$9 = 3 \cdot 3$

Either factor is called a **square root** of the number. The symbol $\sqrt{\ }$ (called a **radical sign**) is used to indicate a square root. Thus $\sqrt{9} = 3$ because $3 \cdot 3 = 9$. The term *square root* is used because three **squared** is nine.

$3^2 = 9 \qquad \sqrt{9} = 3$

Example 1

Finding the Square Root

Find the square root of 49 and of 16.

(a) $\sqrt{49} = 7$ Because $7 \cdot 7 = 7^2 = 49$

(b) $\sqrt{16} = 4$ Because $4 \cdot 4 = 4^2 = 16$

CHECK YOURSELF 1

Find each square root.

(a) $\sqrt{121}$ **(b)** $\sqrt{36}$

When the square root of a whole number is itself a whole number, the original number is called a perfect square. In Example 1, we see that 49 and 16 are perfect squares.

The most frequently used theorem in geometry is undoubtedly the Pythagorean theorem. In this section you will use that theorem. You will also learn a little about the history of the theorem. It is a theorem that applies only to right triangles.

The side opposite the right angle of a right triangle is called the **hypotenuse.** The hypotenuse will always be the largest side of a right triangle.

Example 2

Identifying the Hypotenuse

In the right triangle shown, the side labeled c is the hypotenuse.

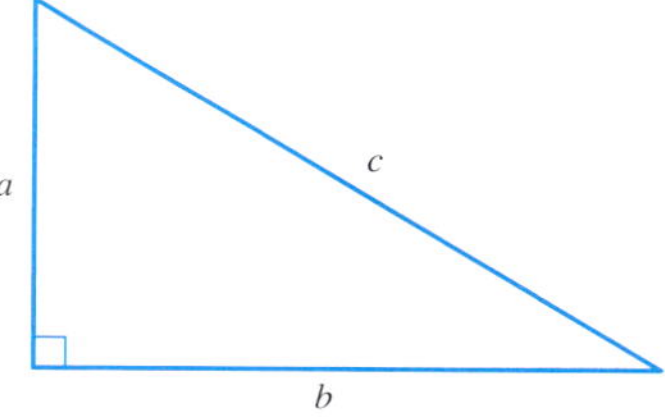

CHECK YOURSELF 2

Which side represents the hypotenuse of the given right triangle?

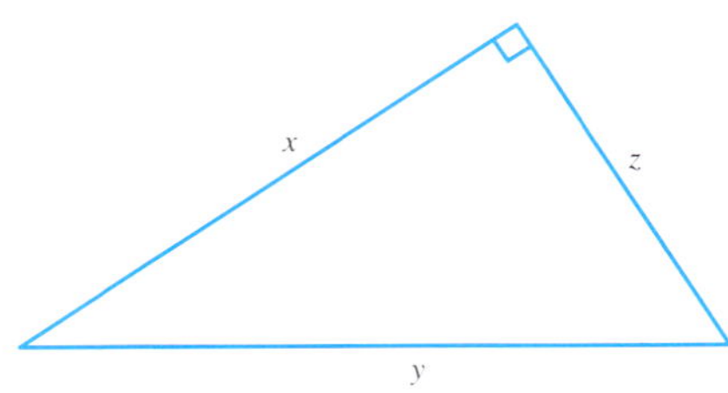

NOTE Such a triple is also called a **Pythagorean triple.**

The numbers 3, 4, and 5 have a special relationship. Together they are called a **perfect triple,** which means that when you square all three numbers, the sum of the smaller squares equals the squared value of the larger number.

Example 3

Identifying Perfect Triples

Show that each group of numbers is a perfect triple.

(a) 3, 4, and 5

$3^2 = 9, \qquad 4^2 = 16, \qquad 5^2 = 25$

and $9 + 16 = 25$, so we can say that $3^2 + 4^2 = 5^2$.

(b) 7, 24, and 25

$7^2 = 49, \qquad 24^2 = 576, \qquad 25^2 = 625$

and $49 + 576 = 625$, so we can say that $7^2 + 24^2 = 25^2$.

CHECK YOURSELF 3

Show that each group of numbers is a perfect triple.

(a) 5, 12, and 13 **(b)** 6, 8, and 10

All the triples that you have seen, and many more, were known by the Babylonians more than 4,000 years ago. Stone tablets that had dozens of perfect triples carved into them have been found. The basis of the Pythagorean theorem was understood long before the time of Pythagoras (ca. 540 B.C.). The Babylonians not only understood perfect triples but also knew how triples related to a right triangle.

Rules and Properties: The Pythagorean Theorem (Version 1)

If the lengths of the three sides of a right triangle are all integers, they will form a perfect triple.

There are two other forms in which the Pythagorean theorem is regularly presented. It is important that you see the connection between the three forms.

Rules and Properties: The Pythagorean Theorem (Version 2)

The square of the hypotenuse of a right triangle is equal to the sum of the squares of the other two sides.

NOTE This is the version that you will refer to in your algebra classes.

Rules and Properties: The Pythagorean Theorem (Version 3)

Given a right triangle with sides a and b and hypotenuse c, it is always true that

$c^2 = a^2 + b^2$

Example 4

Finding the Length of a Side of a Right Triangle

Find the missing integer length for each right triangle.

(a)

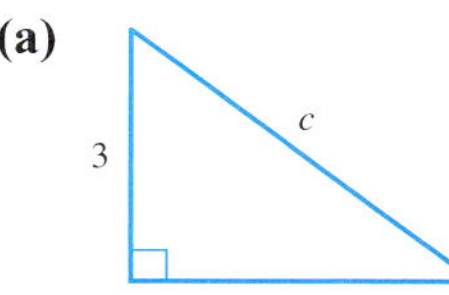

$$c^2 = 3^2 + 4^2$$
$$= 9 + 16$$
$$= 25$$

We need the square root of 25.

$c = 5$

(b)

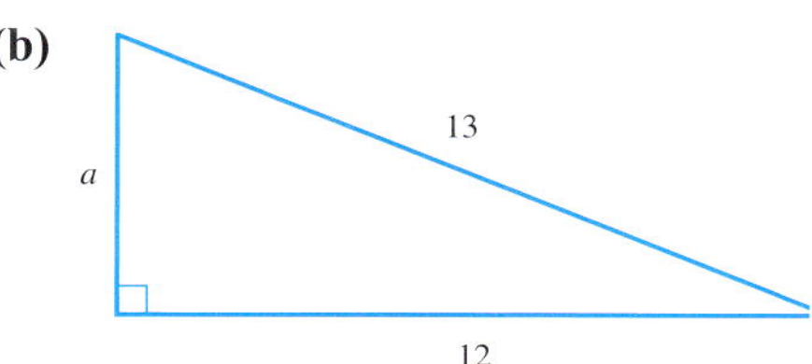

$$13^2 = a^2 + 12^2$$

$$\begin{array}{r} 169 = a^2 + 144 \\ -144 \quad\quad -144 \\ \hline 25 = a^2 \\ 5 = a \end{array}$$

CHECK YOURSELF 4

Find the missing integer length for each right triangle.

(a)

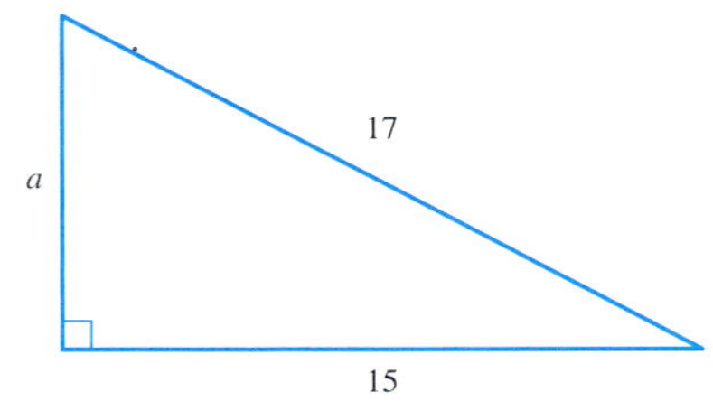

(b)

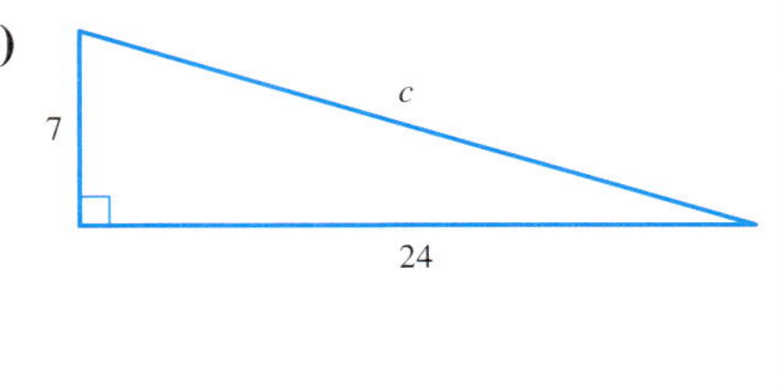

Example 5

Using the Pythagorean Theorem

NOTE The triangle has sides 6, 8, and 10.

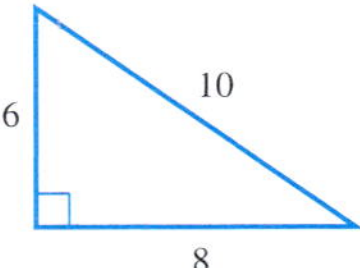

If the lengths of two sides of a right triangle are 6 and 8, find the length of the hypotenuse.

$c^2 = a^2 + b^2$ The value of the hypotenuse is found from the Pythagorean theorem with $a = 6$ and $b = 8$.

$c^2 = (6)^2 + (8)^2 = 36 + 64 = 100$

$c = \sqrt{100} = 10$ The length of the hypotenuse is 10 (because $10^2 = 100$).

CHECK YOURSELF 5

Find the hypotenuse of a right triangle whose sides measure 9 *and* 12.

In some right triangles, the lengths of the hypotenuse and one side are given and we are asked to find the length of the missing side.

Example 6

Using the Pythagorean Theorem

Find the missing length.

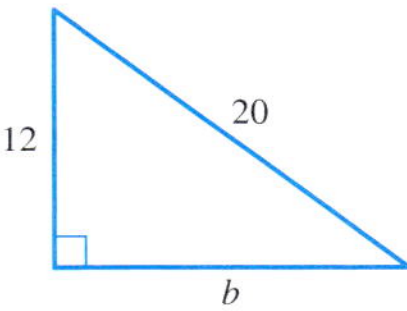

$a^2 + b^2 = c^2$ Use the Pythagorean theorem with $a = 12$ and $c = 20$.

$(12)^2 + b^2 = (20)^2$

$144 + b^2 = 400$

$b^2 = 400 - 144 = 256$

$b = \sqrt{256} = 16$ The missing side is 16.

CHECK YOURSELF 6

Find the missing length for a right triangle with one leg measuring 8 *cm and the hypotenuse measuring* 10 *cm.*

Not every square root is a whole number. In fact, there are only 10 whole-number square roots for the numbers from 1 to 100. They are the square roots of 1, 4, 9, 16, 25, 36, 49, 64, 81, and 100. However, we can approximate square roots that are not whole numbers. For example, we know that the square root of 12 is not a whole number. We also know that its value must lie somewhere between the square root of 9 ($\sqrt{9} = 3$) and the square root of 16 ($\sqrt{16} = 4$). That is, $\sqrt{12}$ is between 3 and 4.

Example 7

Approximating Square Roots

Approximate $\sqrt{29}$.

$\sqrt{25} = 5$ and $\sqrt{36} = 6$, so $\sqrt{29}$ must be between 5 and 6.

CHECK YOURSELF 7

$\sqrt{19}$ *is between which two numbers?*

(a) 4 and 5 **(b)** 5 and 6 **(c)** 6 and 7

To find a square root on your scientific calculator, you use the square root key. On some calculators, you simply enter the number, and then press the square root key. With others, you must use the second function on the $\boxed{x^2}$ (or $\boxed{y^x}$) key and specify the root you wish to find.

Example 8

Finding a Square Root Using the Calculator

Find the square root of 256.

256 $\boxed{\sqrt{\ }}$

Display 16

or

256 $\boxed{\text{2nd}}$ $\boxed{\sqrt[x]{y} \; y^x}$ 2 $\boxed{=}$

Display 16 The "2" is entered for the 2nd (square) root.

NOTE Recall that, to use the radical key ($\boxed{\sqrt{\ }}$) with a scientific calculator, first enter the 49 and then press the radical key. With a graphing calculator, press the radical key first, and then enter the 49 and a closing parenthesis.

CHECK YOURSELF 8

Find the square root of 361.

As we saw in Example 7, not every square root is a whole number. Your calculator can help give you the *approximate* square root of any number.

Example 9

Finding an Approximate Square Root Using the Calculator

Approximate the square root of 29. Round your answer to the nearest tenth.

Enter

29 [√]

Your calculator display will read something like this:

Display 5.385164807

This is an *approximation* of the square root. It is rounded to the nearest billionth. The calculator cannot display the exact answer because there is no end to the sequence of digits (and also no pattern.) If the square root of a whole number is not another whole number, then the answer has an infinite number of digits.

CAUTION

If we square 5.4, we will not get exactly 29. Try this.

To find the approximate square root, we round to the nearest tenth. Our approximation for the square root of 29 is 5.4.

CHECK YOURSELF 9

Approximate the square root of 19. *Round your answer to the nearest tenth.*

Example 10

Evaluating Expressions Using a Calculator

Use a scientific calculator to approximate the value of $\sqrt{177}$.

$\sqrt{177}$ Using the calculator, you find $\sqrt{177} = 13.3041\cdots$ To the nearest hundredth, $\sqrt{177} = 13.30$.

CHECK YOURSELF 10

Use a scientific calculator to approximate the value of each expression. Round your answer to the nearest hundredth.

(a) $\sqrt{357}$ **(b)** $7(\sqrt{71})$

CHECK YOURSELF ANSWERS

1. **(a)** 11; **(b)** 6 **2.** Side y **3.** **(a)** $5^2 + 12^2 = 25 + 144 = 169$, $13^2 = 169$, so $5^2 + 12^2 = 13^2$; **(b)** $6^2 + 8^2 = 36 + 64 = 100$, $10^2 = 100$, so $6^2 + 8^2 = 10^2$ **4.** **(a)** 8; **(b)** 25 **5.** 15 **6.** 6 cm **7.** **(a)** 4 and 5 **8.** 19 **9.** 4.4 **10.** **(a)** 18.89; **(b)** 58.98

Name ______________________

Section ________ Date ________

5.7 Exercises

In exercises 1 to 4, find the square root.

1. $\sqrt{64}$
2. $\sqrt{121}$
3. $\sqrt{169}$
4. $\sqrt{196}$

Identify the hypotenuse of the given triangles by giving its letter.

5.

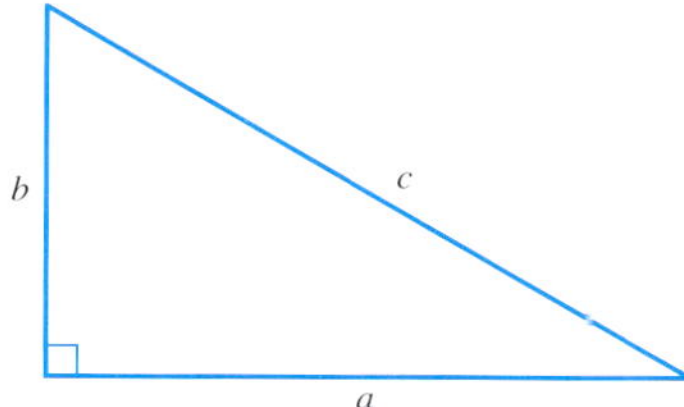

6.

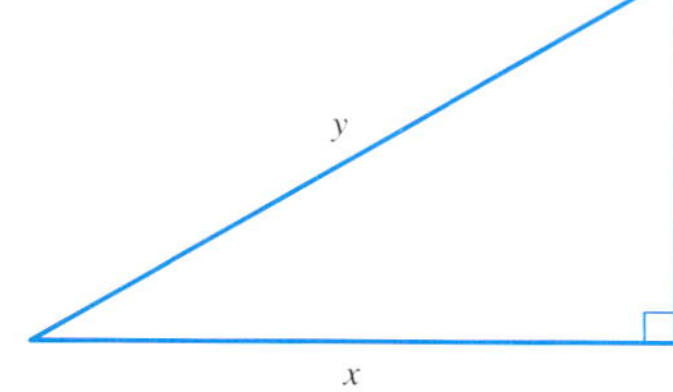

For exercises 7 to 12, identify which numbers are perfect triples.

7. 3, 4, 5
8. 4, 5, 6
9. 7, 12, 13
10. 5, 12, 13
11. 8, 15, 17
12. 9, 12, 15

For exercises 13 to 16, find the missing length for each right triangle.

13.

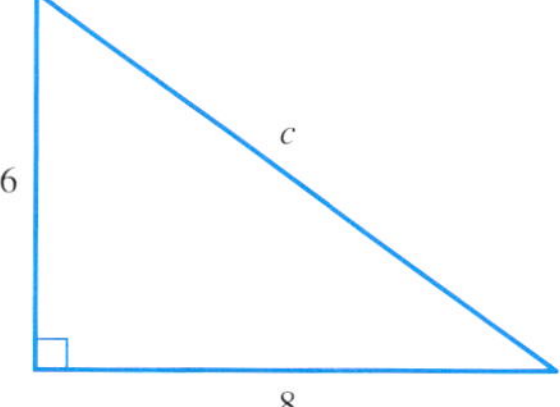

14.

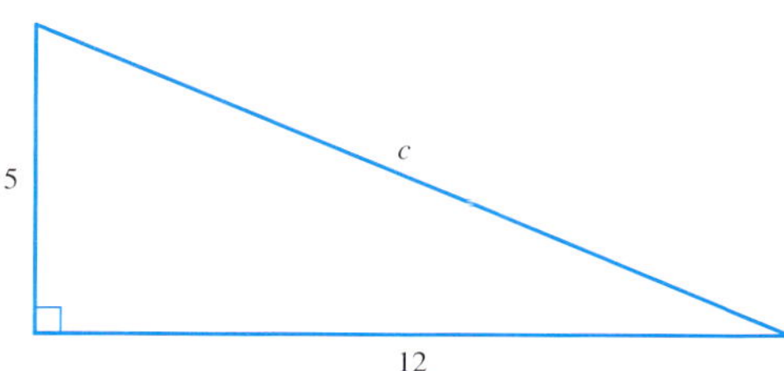

15.

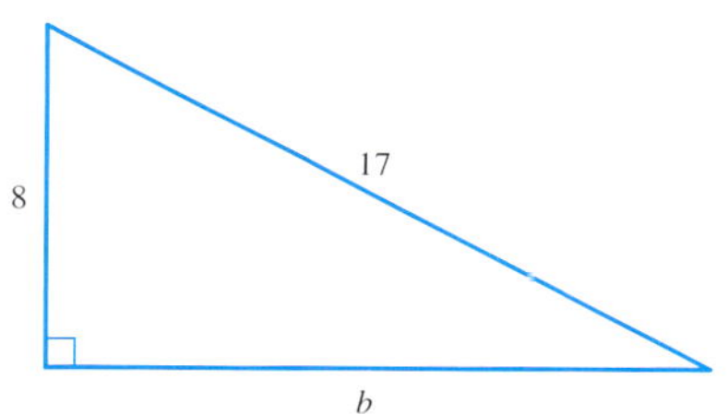

ANSWERS

1. ______
2. ______
3. ______
4. ______
5. ______
6. ______
7. ______
8. ______
9. ______
10. ______
11. ______
12. ______
13. ______
14. ______
15. ______

ANSWERS

16. ______

17. ______

18. ______

19. ______

20. ______

21. ______

22. ______

23. ______

24. ______

25. ______

16.

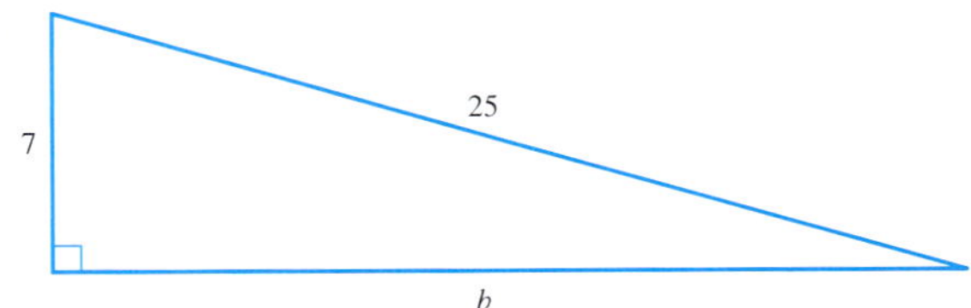

Select the correct approximation.

17. Is $\sqrt{23}$ between **(a)** 3 and 4, **(b)** 4 and 5, or **(c)** 5 and 6?

18. Is $\sqrt{15}$ between **(a)** 1 and 2, **(b)** 2 and 3, or **(c)** 3 and 4?

19. Is $\sqrt{44}$ between **(a)** 6 and 7, **(b)** 7 and 8, or **(c)** 8 and 9?

20. Is $\sqrt{31}$ between **(a)** 3 and 4, **(b)** 4 and 5, or **(c)** 5 and 6?

In exercises 21 to 24, find the perimeter of each triangle shown. (*Hint:* First find the missing side.)

21.

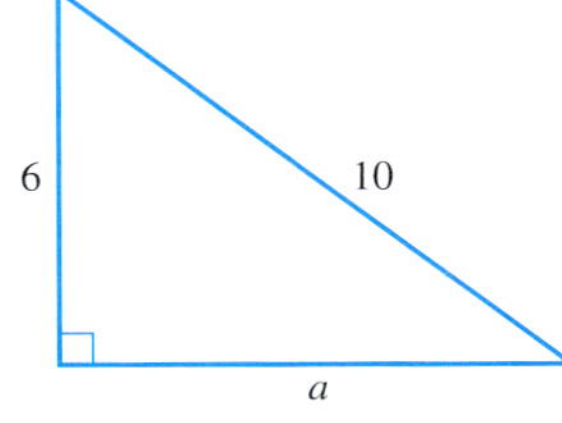

22.

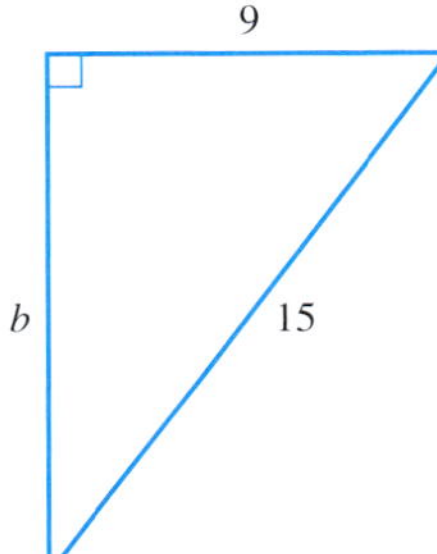

23.

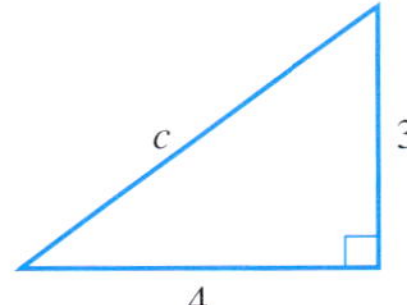

24.

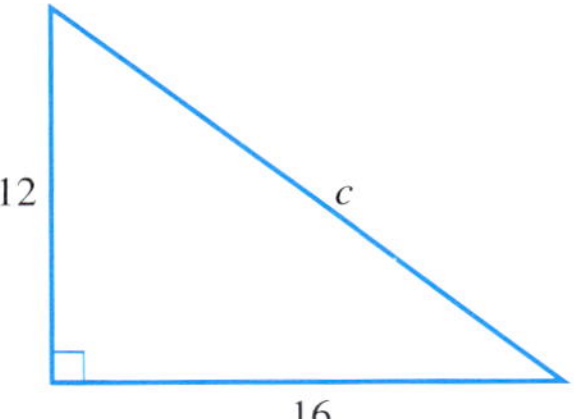

25. Find the altitude, h, of the triangle shown.

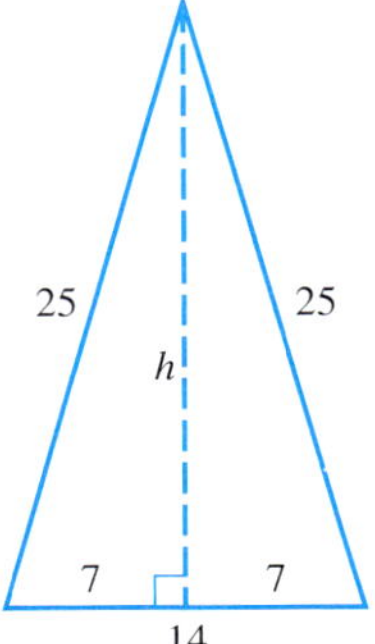

26. Find the altitude of the triangle shown.

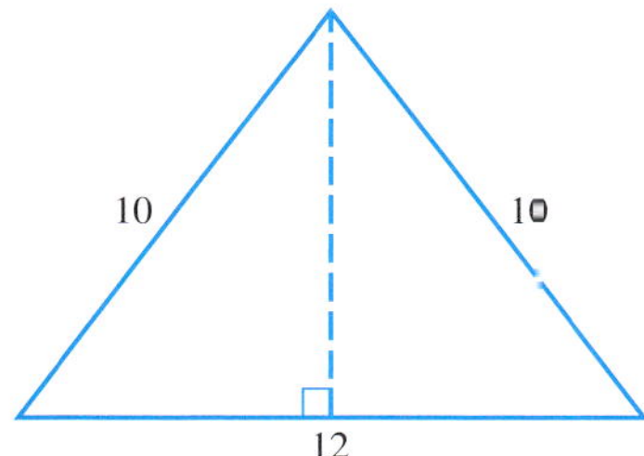

In exercises 27 and 28, find the length of the diagonal of each rectangle.

27.

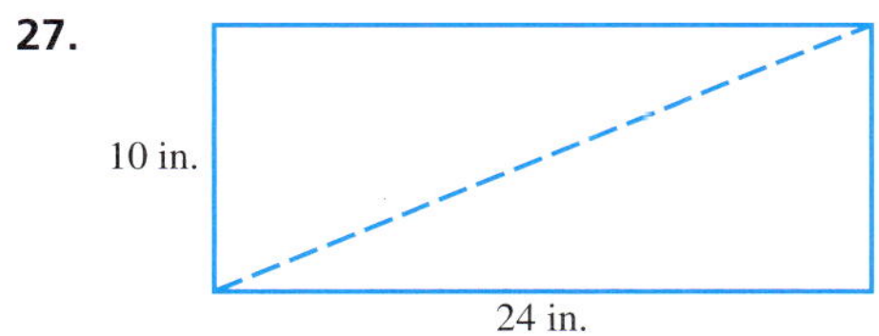

28.

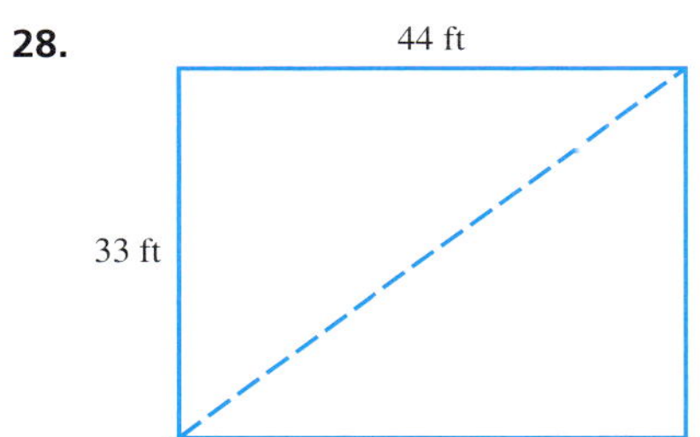

29. A castle wall, 24 ft high, is surrounded by a moat 7 ft across. Will a 26-ft ladder, placed at the edge of the moat, be long enough to reach the top of the wall?

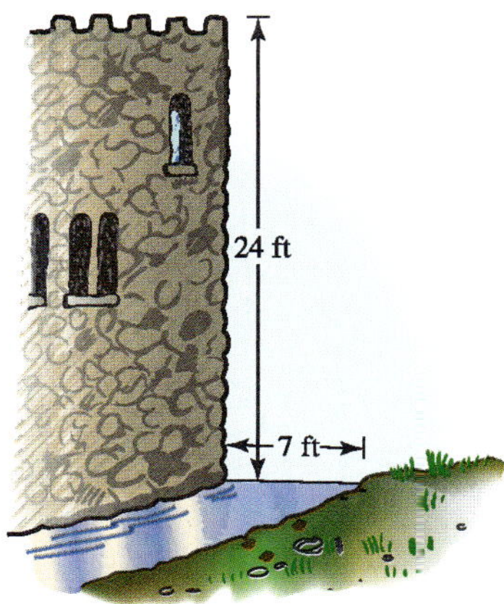

ANSWERS

26. ____________

27. ____________

28. ____________

29. ____________

ANSWERS

30. ______________________

31. ______________________

32. ______________________

33. ______________________

34. ______________________

35. ______________________

36. ______________________

37. ______________________

38. ______________________

39. ______________________

40. ______________________

41. ______________________

42. ______________________

43. ______________________

44. ______________________

45. ______________________

46. ______________________

30. A baseball diamond is the shape of a square that has sides of length 90 ft. Find the distance from home plate to second base. Round your answer to the nearest hundredth.

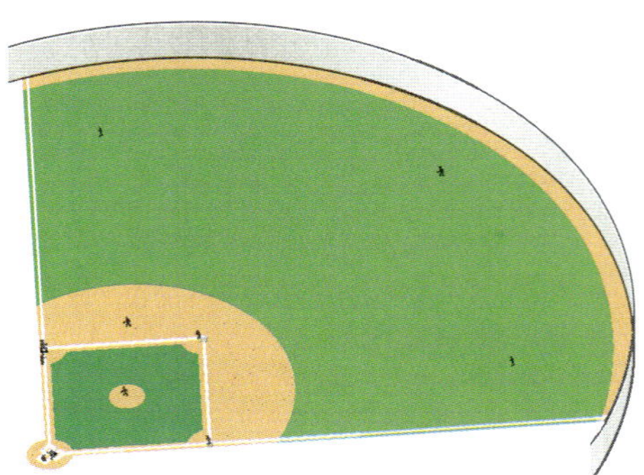

Calculator Exercises

Use your calculator to find the square root of each number.

31. 64

32. 144

33. 289

34. 1,024

35. 1,849

36. 784

37. 8,649

38. 5,329

39. 3,844

40. 3,364

Use your calculator to approximate each square root. Round to the nearest tenth.

41. $\sqrt{23}$

42. $\sqrt{31}$

43. $\sqrt{51}$

44. $\sqrt{42}$

45. $\sqrt{134}$

46. $\sqrt{251}$

Answers

1. 8 **3.** 13 **5.** *c* **7.** Yes **9.** No **11.** Yes **13.** 10 **15.** 15 **17.** b **19.** a **21.** 24 **23.** 12 **25.** 24 **27.** 26 in. **29.** Yes **31.** 8 **33.** 17 **35.** 43 **37.** 93 **39.** 62 **41.** 4.8 **43.** 7.1 **45.** 11.6

5.8 Applications

5.8 OBJECTIVES

1. Use addition of decimals to solve application problems
2. Use subtraction of decimals to solve application problems
3. Use multiplication of decimals to solve application problems
4. Use multiplication of decimals by a power of 10 to solve application problems
5. Use division of decimals to solve application problems
6. Use division of decimals by a power of 10 to solve application problems
7. Use the Pythagorean theorem to solve application problems

Many applied problems require working with decimals. For instance, filling up at a gas station means reading decimal amounts.

Example 1

An Application of the Addition of Decimals

On a trip the Chang family kept track of their gas purchases. If they bought 12.3, 14.2, 10.7, and 13.8 gal, how much gas did they use on the trip?

NOTE Because we want a total amount, addition is used for the solution.

$$\begin{array}{r} 12.3 \\ 14.2 \\ 10.7 \\ +\ 13.8 \\ \hline 51.0 \text{ gal} \end{array}$$

CHECK YOURSELF 1

The Higueras kept track of the gasoline they purchased on a recent trip. If they bought 12.4, 13.6, 9.7, 11.8, and 8.3 gal, how much gas did they buy on the trip?

Every day you deal with amounts of money. Because our system of money is a decimal system, most problems involving money also involve operations with decimals.

Example 2

An Application of the Addition of Decimals

Andre makes deposits of \$3.24, \$15.73, \$50, \$28.79, and \$124.38 during May. What is the total of his deposits for the month?

$$\begin{array}{r} \$\quad 3.24 \\ 15.73 \\ 50.00 \\ 28.79 \\ +\ 124.38 \\ \hline \$222.14 \end{array}$$

Simply add the amounts of money deposited as decimals. Note that we write \$50 as \$50.00.

← The total of deposits for May

CHECK YOURSELF 2

Your textbooks for the fall term cost \$63.50, \$78.95, \$43.15, \$82, *and* \$85.85. *What was the total cost of textbooks for the term?*

In Chapter 1, we defined *perimeter* as the distance around the outside of a straight-edged shape. Finding the perimeter often requires that we add decimal numbers.

Example 3

An Application Involving the Addition of Decimals

Rachel is going to put a fence around the perimeter of her farm. The figure shows a picture of the land, measured in kilometers. How much fence does she need to buy?

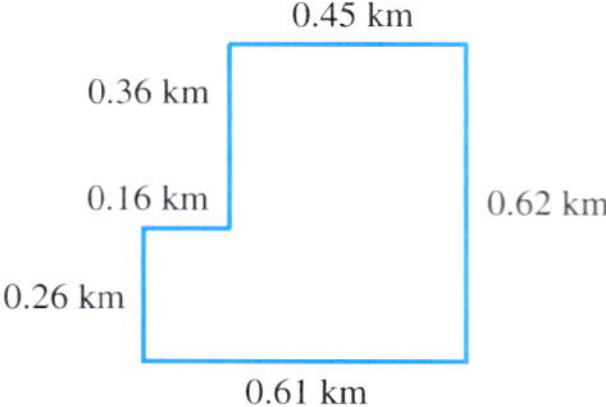

The perimeter is the sum of the lengths of the sides, so we add those lengths to find the total fencing needed.

$0.16 + 0.36 + 0.45 + 0.62 + 0.61 + 0.26 = 2.46$

Rachel needs 2.46 km of fence for the perimeter of her farm.

CHECK YOURSELF 3

Manuel intends to build a walkway around the perimeter of his garden (shown in the figure). What will the total length of the walkway be?

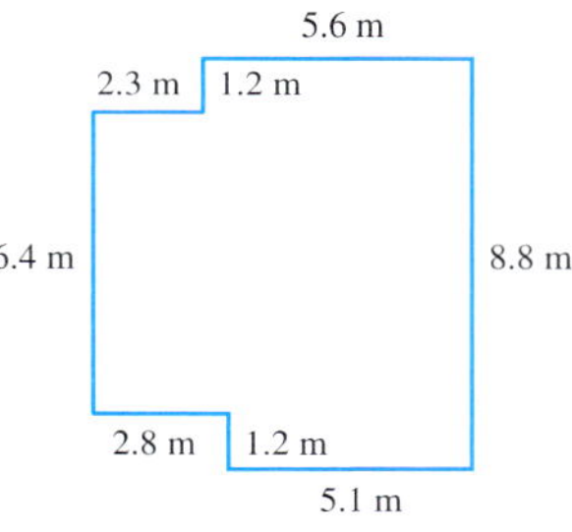

We can apply the subtraction methods of Section 5.2 in solving applications involving decimals.

Example 4

An Application of the Subtraction of a Decimal Number

Jonathan was 98.3 cm tall on his sixth birthday. On his seventh birthday he was 104.2 cm. How much did he grow during the year?

NOTE We want to find the difference between the two measurements, so we subtract.

$$\begin{array}{r} 104.2 \text{ cm} \\ -\ \ 98.3 \text{ cm} \\ \hline 5.9 \text{ cm} \end{array}$$

Jonathan grew 5.9 cm during the year.

CHECK YOURSELF 4

A car's highway mileage before a tune-up was 28.8 *miles per gallon* $\left(\frac{mi}{gal}\right)$. *After the tune-up, it measured* $30.1 \frac{mi}{gal}$. *What was the increase in mileage?*

The same method can be used in working with money.

Example 5

An Application of the Subtraction of a Decimal Number

At the grocery store, Sally buys a roast that is marked \$12.37. She pays for her purchase with a \$20 bill. How much change does she get?

NOTE Sally's change will be the *difference* between the price of the roast and the \$20 paid. We must use subtraction for the solution.

$$\begin{array}{r} \$20.00 \\ -\ 12.37 \\ \hline \$\ 7.63 \end{array}$$

Add zeros to write \$20 as \$20.00. Then subtract as before.

Sally will receive \$7.63 in change after her purchase.

CHECK YOURSELF 5

A stereo system that normally sells for \$549.50 *is discounted (or marked down) to* \$499.95 *for a sale. What is the savings?*

Keeping your checkbook balanced requires addition and subtraction of decimal numbers.

Example 6

An Application Involving the Addition and Subtraction of Decimals

For the check register, find the running balance.

Beginning balance	\$234.15
Check # 301	23.88
Balance	______
Check # 302	38.98
Balance	______
Check # 303	114.66
Balance	______
Deposit	175.75
Balance	______
Check # 304	212.55
Ending balance	______

To keep a running balance, we add the deposits and subtract the checks.

Beginning balance	$234.15	
Check # 301	23.88	Subtract
Balance	210.27	
Check # 302	38.98	Subtract
Balance	171.29	
Check # 303	114.66	Subtract
Balance	56.63	
Deposit	175.75	Add
Balance	232.38	
Check # 304	212.55	Subtract
Ending balance	19.83	

CHECK YOURSELF 6

For the check register, add the deposit amounts and subtract the check amounts to find the balance.

	Beginning balance	$398.00
	Check # 401	19.75
(a)	Balance	______
	Check # 402	56.88
(b)	Balance	______
	Check # 403	117.59
(c)	Balance	______
	Deposit	224.67
(d)	Balance	______
	Check # 404	411.48
(e)	Ending balance	______

Now we will look at some applications of our work in multiplying decimals.

Example 7

An Application Involving the Multiplication of Two Decimals

A sheet of paper has dimensions 27.5 cm by 21.5 cm. What is its area?

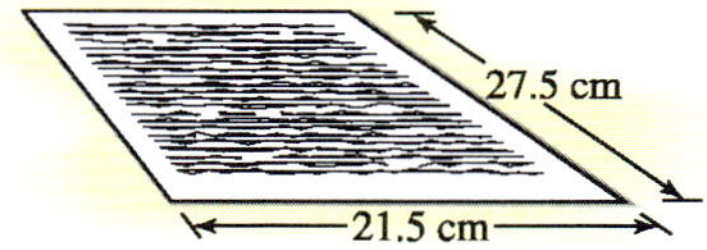

We multiply to find the required area.

NOTE Recall that area is length times width, so multiplication is the necessary operation.

$$\begin{array}{r} 27.5 \text{ cm} \\ \times\ 21.5 \text{ cm} \\ \hline 137\ 5 \\ 275\ \ \\ 550\ \ \ \\ \hline 591.25 \text{ cm}^2 \end{array}$$

The area of the paper is 591.25 cm^2.

CHECK YOURSELF 7

If 1 kg is 2.2 lb, how many pounds equal 5.3 kg?

Example 8

An Application Involving the Multiplication of Two Decimals

NOTE Usually in problems dealing with money we round the result to the nearest cent (hundredth of a dollar).

Jack buys 8.7 gal of kerosene at 98.9 cents per gallon. Find the cost of the kerosene.

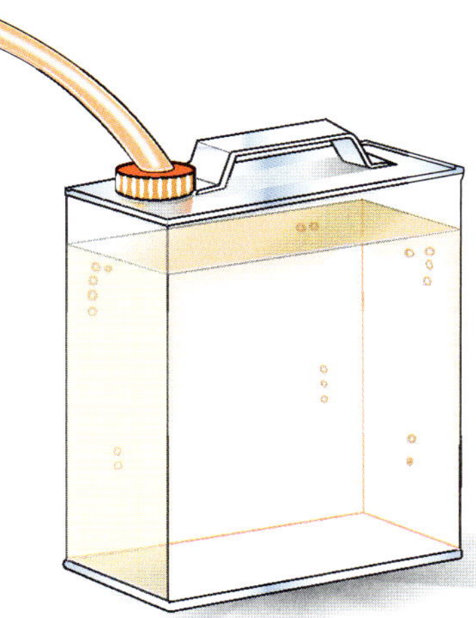

We multiply the cost per gallon by the number of gallons. Then we round the result to the nearest cent. Note that the units of the answer will be cents.

$$\begin{array}{r} 98.9 \\ \times\ 8.7 \\ \hline 69\ 23 \\ 791\ 2 \\ \hline 860.43 \end{array}$$

The product 860.43 (cents) is rounded to 860 (cents), or \$8.60.

The cost of Jack's kerosene will be \$8.60.

CHECK YOURSELF 8

One liter (L) is approximately 0.265 gal. On a trip to Europe, the Bernards purchased 88.4 L of gas for their rental car. How many gallons of gas did they purchase, to the nearest tenth of a gallon?

Sometimes we will have to use more than one operation for a solution, as Example 9 shows.

Example 9

An Application Involving Two Operations

Steve purchased a television set for \$299.50. He agreed to pay for the set by making payments of \$27.70 for 12 months. How much extra did he pay on the installment plan?

First we multiply to find the amount actually paid.

$$\begin{array}{r} \$\ 27.70 \\ \times\ \quad 12 \\ \hline 55\ 40 \\ 277\ 0 \\ \hline \$332.40 \end{array}$$

$\longleftarrow$ Amount paid

Now subtract the listed price. The difference will give the extra amount Steve paid.

$$\begin{array}{r} \$332.40 \\ -\ \ 299.50 \\ \hline \$\ \ 32.90 \end{array} \longleftarrow \text{Extra amount}$$

Steve will pay an additional \$32.90 on the installment plan.

CHECK YOURSELF 9

Sandy's new car had a list price of \$10,985. She paid \$1,500 down and will pay \$305.35 per month for 36 months on the balance. How much extra will she pay with this loan arrangement?

Example 10 is just one of many applications that require multiplying by a power of 10.

Example 10

An Application Involving Multiplication by a Power of 10

NOTE There are 1,000 meters in a kilometer.

To convert from kilometers to meters, multiply by 1,000. Find the number of meters in 2.45 km.

NOTE If the result is a whole number, there is no need to write the decimal point.

2.45 km = 2450. m

Just move the decimal point three places to the right to make the conversion. Note that we added a zero to place the decimal point correctly.

CHECK YOURSELF 10

To convert from kilograms to grams, multiply by 1,000. Find the number of grams in 5.23 kg.

Example 11

An Application Involving the Division of a Decimal by a Whole Number

A carton of 144 items costs \$56.10. What is the price per item to the nearest cent?

To find the price per item, divide the total price by 144.

NOTE You might want to review the rules for rounding decimals in Section 5.1.

$$\begin{array}{r} 0.389 \\ 144\overline{)56.100} \\ \underline{43\ 2} \\ 12\ 90 \\ \underline{11\ 52} \\ 1\ 380 \\ \underline{1\ 296} \\ 84 \end{array}$$

Carry the division to the thousandths place and then round back.

The cost per item is rounded to \$0.39, or 39¢.

CHECK YOURSELF 11

An office paid \$26.55 for 72 pens. What was the cost per pen to the nearest cent?

Example 12

Solving an Application Involving the Division of Decimals

Andrea worked 41.5 h in a week and earned $239.87. What was her hourly rate of pay?

To find the hourly rate of pay we must use division. We divide the number of hours worked into the total pay.

NOTE Notice that we must add a zero to the dividend to complete the division process.

```
           5.78
41.5 )239.8 70
      207 5
       32 3 7
       29 0 5
          3 3 20
          3 3 20
               0
```

Andrea's hourly rate of pay was $5.78.

CHECK YOURSELF 12

A developer wants to subdivide a 12.6-acre piece of land into 0.45-acre lots. How many lots are possible?

Example 13

Solving an Application Involving the Division of Decimals

At the start of a trip the odometer read 34,563. At the end of the trip, it read 36,235. If 86.7 gal of gas were used, find the number of miles per gallon (to the nearest tenth).

First, find the number of miles traveled by subtracting the initial reading from the final reading.

```
  36,235    Final reading
− 34,563    Initial reading
   1,672    Miles covered
```

Next, divide the miles traveled by the number of gallons used. This will give us the miles per gallon.

```
             1 9.28
86.7 )1,672.0 00
        867
        805 0
        780 3
         24 7 0
         17 3 4
            7 3 60
            6 9 36
              4 24
```

Round 19.28 to 19.3 $\frac{\text{mi}}{\text{gal}}$.

CHECK YOURSELF 13

John starts his trip with an odometer reading of 15,436 and ends with a reading of 16,238. If he used 45.9 gal of gas, find the number of miles per gallon (to the nearest tenth).

Now we will look at an application of our work in dividing by powers of 10.

Example 14

Solving an Application Involving a Power of 10

To convert from millimeters to meters, we divide by 1,000. How many meters does 3,450 mm equal?

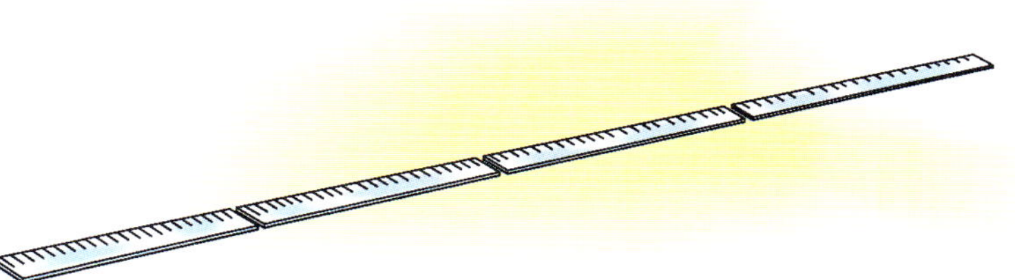

$$3{,}450 \text{ mm} = 3\ 450. \text{ m}$$ Shift three places to the left to divide by 1,000.

$$= 3.450 \text{ m}$$

CHECK YOURSELF 14

A shipment of 1,000 notebooks cost a stationery store \$658. What was the cost per notebook to the nearest cent?

Example 15

An Application Involving the Division of Decimals Using the Calculator

Omar drove 256.3 mi on a tank of gas. When he filled up the tank, it took 9.1 gal. What was his gas mileage?

Here's where students get into trouble when they use a calculator. Entering these values, you may be tempted to answer "$28.16483516 \frac{\text{mi}}{\text{gal}}$." The difficulty is that there is no way you can compute gas mileage to the nearest hundred-millionth mile. How do you decide where to round off the answer that the calculator gives you? A good rule of thumb when multiplication or division is involved is to report in your answer the least number of digits given in the problem. In this case, you were given a number with four digits and another with two digits. Your answer should not have more than two digits. Instead of 28.16483516, the answer could be $28 \frac{\text{mi}}{\text{gal}}$. Think about the question. If you were asked for gas mileage, how precise an answer would you give? The best answer to this question would be to give the nearest whole number of miles per gallon: $28 \frac{\text{mi}}{\text{gal}}$.

CHECK YOURSELF 15

Emmet gained a total of 857 yards in 209 times that he carried the football. How many yards did he average for each time he carried the ball?

The Pythagorean theorem can be applied to solve a variety of geometric problems.

Example 16

Solving for the Length of the Diagonal

Find, to the nearest tenth, the length of the diagonal of a rectangle that is 8 cm long and 5 cm wide. Let x be the unknown length of the diagonal:

NOTE Always draw and label a sketch showing the information from a problem when geometric figures are involved.

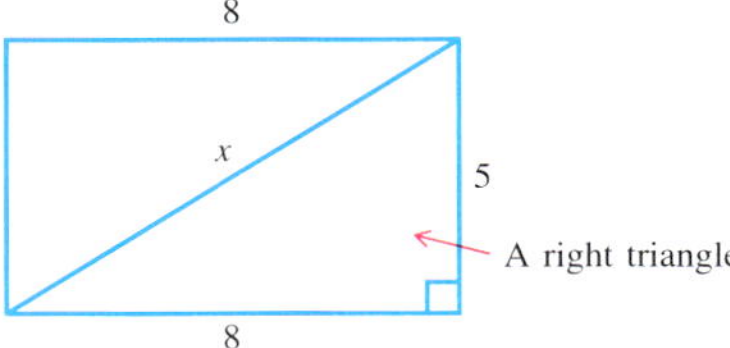

So

$$x^2 = 5^2 + 8^2$$
$$= 25 + 64$$
$$= 89$$
$$x = \sqrt{89}$$

Thus

$$x \approx 9.4 \text{ cm}$$

CHECK YOURSELF 16

The diagonal of a rectangle is 12 in. and its width is 6 in. Find its length to the nearest tenth.

The next application also makes use of the Pythagorean theorem.

Example 17

Solving an Application

How long must a guywire be to reach from the top of a 30-ft pole to a point on the ground 20 ft from the base of the pole? Round your answer to the nearest foot.

Again be sure to draw a sketch of the problem.

NOTE Always check to see if your final answer is reasonable.

x
30 ft
20 ft

$$x^2 = 20^2 + 30^2$$
$$= 400 + 900$$
$$= 1{,}300$$
$$x = \sqrt{1{,}300}$$
$$\approx 36 \text{ ft}$$

CHECK YOURSELF 17

A 16.0-ft ladder leans against a wall with its base 4.00 ft from the wall. How far off the floor is the top of the ladder?

Example 18

Approximating Length with a Calculator

Approximate the length of the diagonal of a rectangle. The diagonal forms the hypotenuse of a triangle with legs 12.2 in. and 15.7 in. The length of the diagonal would be $\sqrt{12.2^2 + 15.7^2} = \sqrt{395.33} \approx 19.9$ in.

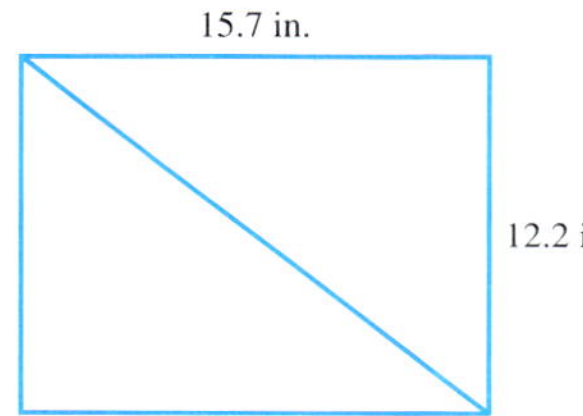

CHECK YOURSELF 18

Approximate the length of the diagonal of the rectangle to the nearest tenth.

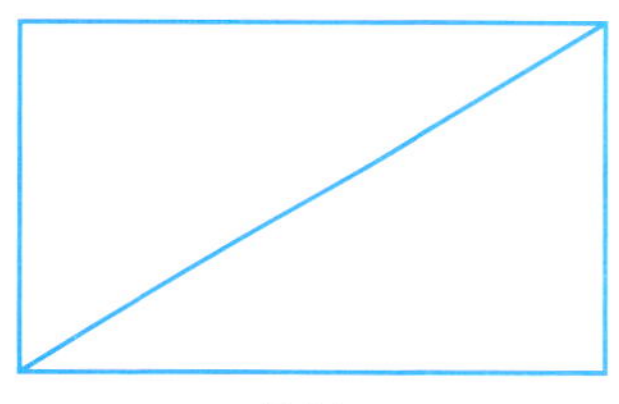

CHECK YOURSELF ANSWERS

1. 55.8 gal **2.** \$353.45 **3.** 33.4 m **4.** $1.3\ \frac{\text{mi}}{\text{gal}}$ **5.** \$49.55
6. **(a)** \$378.25; **(b)** \$321.37; **(c)** \$203.78; **(d)** \$428.45; **(e)** \$16.97 **7.** 11.66 lb
8. 23.4 gal **9.** \$1,507.60 **10.** 5,230 g **11.** \$0.37, or 37¢ **12.** 28 lots
13. $17.5\ \frac{\text{mi}}{\text{gal}}$ **14.** 66¢ **15.** 4.10 yd **16.** Length is approximately 10.4 in.
17. Height is approximately 15.5 ft. **18.** Length is approximately 24.0 in.

5.8 Exercises

Name ______________

Section ________ Date ________

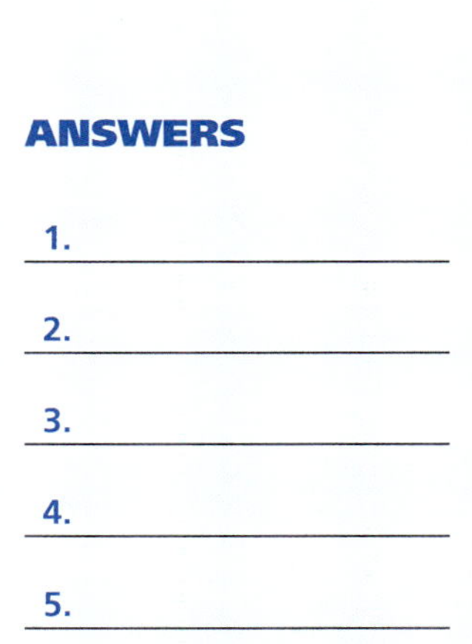

ANSWERS

1. ______________
2. ______________
3. ______________
4. ______________
5. ______________
6. ______________
7. ______________
8. ______________
9. ______________

Perform the additions.

1. Add twenty-three hundredths, five tenths, and two hundred sixty-eight thousandths.

2. Add seven tenths, four hundred fifty-eight thousandths, and fifty-six hundredths.

3. Add five and three tenths, seventy-five hundredths, twenty and thirteen hundredths, and twelve and seven tenths.

4. Add thirty-eight and nine tenths, five and fifty-eight hundredths, seven, and fifteen and eight tenths.

Solve the applications.

5. **Gas purchase.** On a 3-day trip, Dien bought 12.7, 15.9, and 13.8 gal of gas. How many gallons of gas did he buy?

6. **Distance.** Felix ran 2.7 mi on Monday, 1.9 mi on Wednesday, and 3.6 mi on Friday. How far did he run during the week?

7. Rainfall was recorded in centimeters during the winter months as indicated on the bar graph.

   (a) How much rain fell during those months?

   (b) How much more rain fell in December than in February?

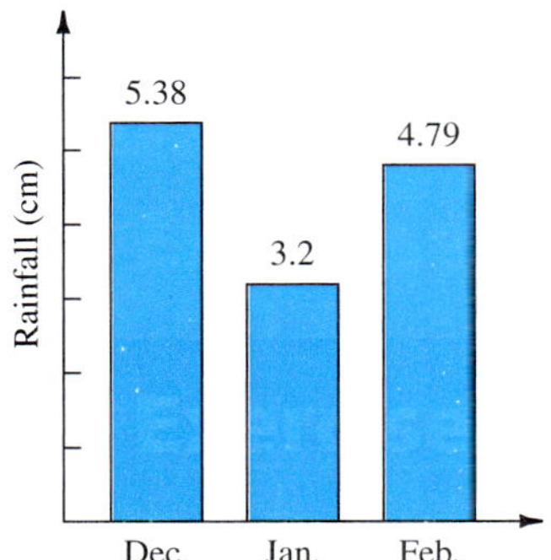

8. **Total length.** A metal fitting has three sections, with lengths 2.5, 1.775, and 1.45 in. What is the total length of the fitting?

9. **Total expenses.** Nicole had the following expenses on a business trip: gas, \$45.69; food, \$123; lodging, \$95.60; and parking and tolls, \$8.65. What were her total expenses during the trip?

ANSWERS

10. ______________

11. ______________

12. ______________

13. ______________

14. ______________

15. ______________

16. ______________

10. Textbook costs. Hok Sum's textbooks for one term cost \$29.95, \$47, \$52.85, \$33.35, and \$10. What was his total cost for textbooks?

11. Checking. Jordan wrote checks of \$50, \$11.38, \$112.57, and \$9.73 during a single week. What was the total amount of the checks he wrote?

12. The deposit slip shown indicates the amounts that made up a deposit Peter Rabbit made. What was the total amount of his deposit?

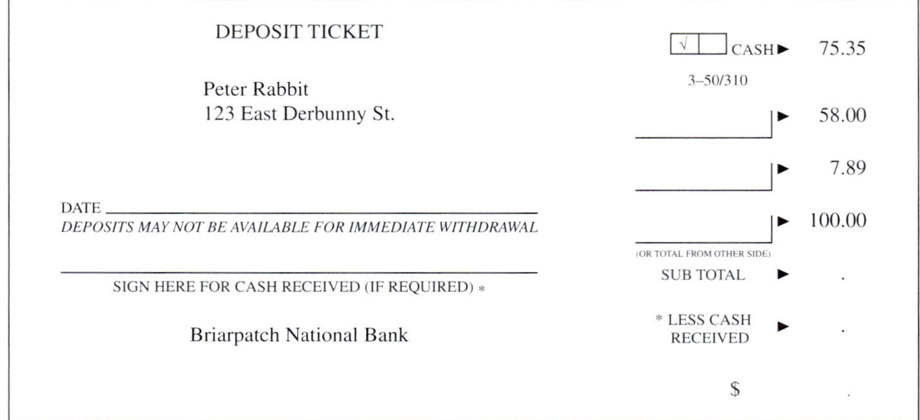
DEPOSIT TICKET

Peter Rabbit
123 East Derbunny St.

DATE ______________
DEPOSITS MAY NOT BE AVAILABLE FOR IMMEDIATE WITHDRAWAL

SIGN HERE FOR CASH RECEIVED (IF REQUIRED) *

Briarpatch National Bank

CASH ►	75.35
3–50/310 ►	58.00
►	7.89
►	100.00
(OR TOTAL FROM OTHER SIDE) SUB TOTAL ►	.
* LESS CASH RECEIVED ►	.
\$	.

13. Perimeter. Lupe is putting a fence around her yard. Her yard is rectangular and measures 8.16 yd long and 12.68 yd wide. How much fence should Lupe purchase?

14. Perimeter. Find the perimeter of the given figure.

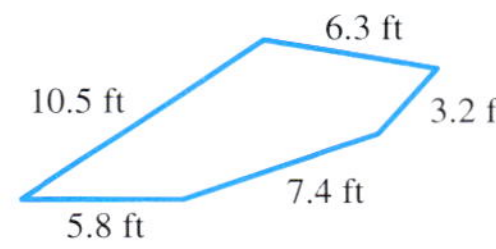

15. Fencing. The figure gives the distance in miles of the boundary sections around a ranch. How much fencing is needed for the property?

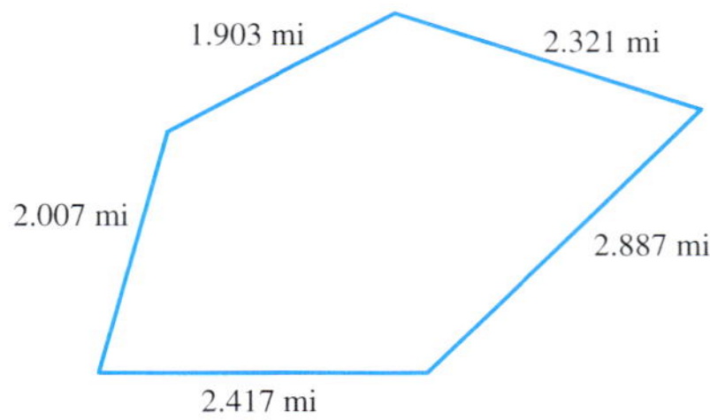

16. Discounts. A television set selling for \$399.50 is discounted (or marked down) to \$365.75. What is the savings?

17. ______

18. ______

19. ______

20. ______

21. ______

22. ______

17. **Checkbook balance.** For the check register, find the running balance.

Beginning balance	$456.00
Check # 601	$199.29
Balance	______
Service charge	$ 18.00
Balance	______
Check # 602	$ 85.78
Balance	______
Deposit	$250.45
Balance	______
Check # 603	$201.24
Ending balance	______

18. **Checkbook balance.** For the check register, find the running balance.

Beginning balance	$589.21
Check # 678	$175.63
Balance	______
Check # 679	$ 56.92
Balance	______
Deposit	$121.12
Balance	______
Check # 680	$345.99
Ending balance	______

19. **Car maintenance.** Your bill for a car tune-up includes $7.80 for oil, $5.90 for a filter, $3.40 for spark plugs, $4.10 for points, and $28.70 for labor. Estimate your total cost by rounding each amount to the nearest dollar.

20. **Payroll.** The payroll at a car repair shop for 1 week was $456.73, utilities were $123.89, advertising was $212.05, and payments to distributors were $415.78. Estimate the amount spent in 1 week by rounding each amount to the nearest dollar.

21. **Expenses.** On a recent business trip your expenses were $343.78 for airfare, $412.78 for lodging, $148.89 for food, and $102.15 for other items. Estimate your total expenses by rounding each amount to the nearest dollar.

22. Here are charges on a credit card account:
$8.97, $32.75, $15.95, $67.32, $215.78, $74.95, $83.90, and $257.28

(a) Estimate the total bill for the charges by rounding each number to the nearest dollar and adding the results.

(b) Estimate the total bill by adding the charges and then rounding to the nearest dollar.

(c) What are the advantages and disadvantages of the methods in **(a)** and **(b)**?

ANSWERS

23. ______________________

24. ______________________

25. ______________________

26. ______________________

27. ______________________

28. ______________________

29. ______________________

30. ______________________

31. ______________________

23. Find the next number in the sequence: 3.125, 3.375, 3.625, . . .

Recall that a magic square is one in which the sum of every row, column, and diagonal is the same. Complete the magic squares in exercises 24 and 25.

24.

1.6		1.2
	1	
0.8		

25.

2.4		7.2
10.8		
4.8		

26. Find the next two numbers in each of the sequences:

(a) 0.75   0.62   0.5   0.39

(b) 1.0   1.5   0.9   3.5   0.8

27. Total cost. Kurt bought four shirts on sale as pictured. What was the total cost of the purchase?

28. Fuel Consumption. A light plane uses 5.8 $\frac{\text{gal}}{\text{h}}$ of fuel. How much fuel is used on a flight of 3.2 h? Give your answer to the nearest tenth of a gallon.

29. Cost. The Hallstons select a carpet costing \$15.49 per square yard. If they need 7.8 yd^2 of carpet, what is the cost to the nearest cent?

30. Car payment. Maureen's car payment is \$242.38 per month for 4 years. How much will she pay altogether?

31. Area. A classroom is 7.9 m wide and 11.2 m long. Estimate its area.

32. Cost. You buy a roast that weighs 6.2 lb and costs \$3.89 per pound. Estimate the cost of the roast.

33. Cost. A store purchases 100 items at a cost of \$1.38 each. Find the total cost of the order.

34. Conversion. To convert from meters to centimeters, multiply by 100. How many centimeters are there in 5.3 m?

35. Conversion. How many grams are there in 2.2 kg? Multiply by 1,000 to make the conversion.

36. Cost. An office purchases 1,000 pens at a cost of 17.8 cents each. What is the cost of the purchase in dollars?

Meyer's Office Supply

371 Maple Dr., Treynor IA 50001

Item	Quantity	Item Price	Total
Pens	1000	\$0.178	

37. Label making. We have 91.25 in. of plastic labeling tape and wish to make labels that are 1.25 in. long. How many labels can be made?

38. Wages. Alberto worked 32.5 h, earning \$306.15. How much did he make per hour?

39. Cost per pound. A roast weighing 5.3 lb sold for \$14.89. Find the cost per pound to the nearest cent.

40. Weight. One nail weighs 0.025 oz. How many nails are there in 1 lb? (1 lb is 16 oz.)

41. Mileage. A family drove 1,390 mi, stopping for gas three times. If they purchased 15.5, 16.2, and 10.8 gal of gas, find the number of miles per gallon (the mileage) to the nearest tenth.

42. Mileage. On a trip an odometer changed from 36,213 to 38,319. If 136 gal of gas were used, find the number of miles per gallon (to the nearest tenth).

43. Conversion. To convert from millimeters to inches, we can divide by 25.4. If film is 35 mm wide, find the width to the nearest hundredth of an inch.

44. Conversion. To convert from centimeters to inches, we can divide by 2.54. The rainfall in Paris was 11.8 cm during 1 week. What was that rainfall to the nearest hundredth of an inch?

45. The blood alcohol content (BAC) of a person who has been drinking is determined by Widmark's formula:

$$\text{BAC} = \frac{\text{total oz} \cdot \text{percent alcohol} \cdot 1.055}{\text{body weight} \cdot 0.68} - (\text{hours drinking} \cdot 0.017)$$

A 125-lb person is driving and is stopped by a policewoman on suspicion of driving under the influence (DUI). The driver claims that in the past 2 h he consumed only six 12-oz bottles of 3.9% beer. If he undergoes a breathalyzer test, what will his BAC be? Will this amount be under the legal limit for your state?

ANSWERS

32. ______
33. ______
34. ______
35. ______
36. ______
37. ______
38. ______
39. ______
40. ______
41. ______
42. ______
43. ______
44. ______

45. ______

ANSWERS

46. ______

47. ______

48. ______

49. ______

50. ______

51. ______

52. ______

46. Four brands of soap are available in a local store.

Brand	Ounces	Total Price	Unit Price
Squeaky Clean	5.5	$0.36	
Smell Fresh	7.5	0.41	
Feel Nice	4.5	0.31	
Look Bright	6.5	0.44	

Compute the unit price, and decide which brand is the best buy.

47. Sophie is a quality control expert. She inspects boxes of #2 pencils. Each pencil weighs 4.4 g. The contents of a box of pencils weigh 66.6 g. If a box is labeled CONTENTS: 16 PENCILS, should Sophie approve the box as meeting specifications? Explain your answer.

48. Write a plan to determine the number of miles per gallon your car (or your family car) gets. Use this plan to determine your car's actual miles per gallon.

49. Express the width and length of a $1 bill in centimeters. Then express the same dimensions in millimeters.

50. If the perimeter of a square is 19.2 cm, how long is each side?

$P = 19.2$ cm

51. If the perimeter of an equilateral triangle (all sides have equal length) is 16.8 cm, how long is each side?

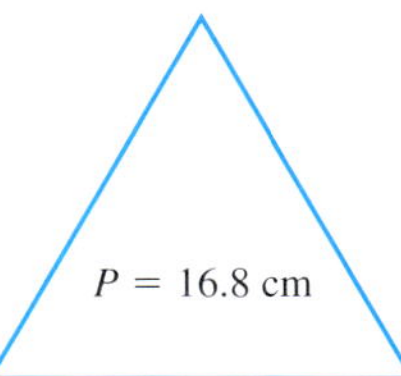

52. If the perimeter of a regular pentagon (all sides have equal length) is 23.5 in., how long is each side?

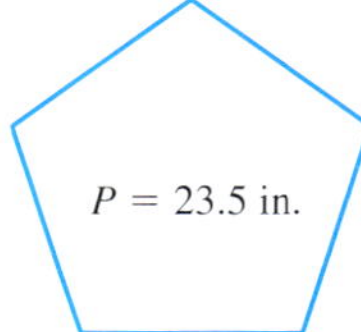

ANSWERS

53. ______

54. ______

55. ______

56. ______

57. ______

58. ______

59. ______

60. ______

61. ______

In exercises 53 to 58, express your answer to the nearest thousandth.

53. Length of a diagonal. Find the length of the diagonal of a rectangle with a length of 10 cm and a width of 7 cm.

54. Length of a diagonal. Find the length of the diagonal of a rectangle with 5 in. width and 7 in. length.

55. Width of a rectangle. Find the width of a rectangle whose diagonal is 12 ft and whose length is 10 ft.

56. Length of a rectangle. Find the length of a rectangle whose diagonal is 9 in. and whose width is 6 in.

57. Length of a wire. How long must a guywire be to run from the top of a 20-ft pole to a point on the ground 8 ft from the base of the pole?

58. Height of a ladder. The base of a 15-ft ladder is 5 ft away from a wall. How high from the floor is the top of the ladder?

Calculator Exercises

Solve the applications using your calculator.

59. Checking balance. Your checking account has a balance of $532.89. You write checks of $50, $27.54, and $134.75 and make a deposit of $50. What is your ending balance?

60. Checking balance. Your checking account has a balance of $278.45. You make deposits of $200 and $135.46. You write checks for $389.34, $249, and $53.21. What is your ending balance? Be careful with this problem. A negative balance means that your account is overdrawn

61. Dr. Rogers is concerned with the increasing cost of making photocopies. She wants to examine alternatives to the current financing plan. The office currently leases a copy machine for $110 per month and $0.025 per copy. A 3-year payment plan is available that costs $125 per month and $0.015 per copy.

(a) If the office expects to run 100,000 copies per year, which is the better plan?

(b) How much money will the better plan save over the other plan?

ANSWERS

62. ______

63. ______

64. ______

65. ______

62. In a bottling company, a machine can fill a 2-L bottle in 0.5 s and move the next bottle into place in 0.1 s. How many 2-L bottles can be filled by the machine in 2 h?

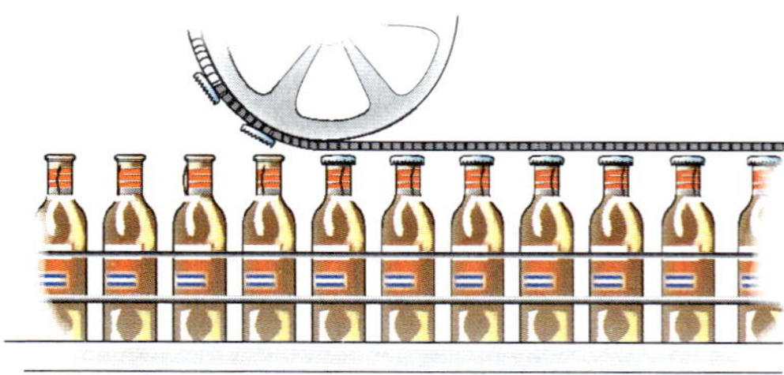

63. The owner of a bakery sells a finished cake for $8.99. The cost of baking 16 cakes is $75.63. Write a plan to find out how much profit the baker can make on each cake sold.

64. Area. An 80.5-acre piece of land is being subdivided into 0.35-acre lots. How many lots are possible in the subdivision?

65. Salary. In 1 week, Tom earned $178.30 by working 36.25 h. What was his hourly rate of pay to the nearest cent?

Answers

1. 0.998 **3.**
$$\begin{array}{r} \overset{1}{5.30} \\ 0.75 \\ 20.13 \\ +\ 12.70 \\ \hline 38.88 \end{array}$$
5. 42.4 gal **7. (a)** 13.37 cm; **(b)** 0.59 cm

9. $272.94 **11.** $183.68 **13.** 41.68 yd **15.** 11.535 mi

17. End balance: $202.14 **19.** $50

21. $1,008 **23.** **25.**

2.4	8.4	7.2
10.8	6	1.2
4.8	3.6	9.6

27. $39.92

29. $120.82 **31.** 88 m^2 **33.** $138 **35.** 2,200 g **37.** 73 labels

39. $2.81 **41.** $32.7\ \dfrac{\text{mi}}{\text{gal}}$ **43.** 1.38 in. **45.**  **47.**

49.   **51.** 5.6 cm **53.** Approximately 12.207 cm

55. Approximately 6.633 ft **57.** Approximately 21.541 ft **59.** $370.60

61. (a) Current plan: $11,460; 3-year lease: $9,000; **(b)** Savings: $2,460

63. **65.** $4.92

5 Summary

DEFINITION/PROCEDURE	EXAMPLE	REFERENCE
Introduction to Decimals, Place Value, and Rounding		**Section 5.1**
Decimal Fraction A fraction whose denominator is a power of 10. We call decimal fractions *decimals.*	$\frac{7}{10}$ and $\frac{47}{100}$ are decimal fractions.	p. 367
Decimal Place Each position for a digit to the right of the decimal point. Each decimal place has a place value that is one-tenth the value of the place to its left.	2.3456 — Ten thousandths, Thousandths, Hundredths, Tenths	p. 368
Reading and Writing Decimals in Words 1. Read the digits *to the left* of the decimal point as a whole number. 2. Read the decimal point as the word *and.* 3. Read the digits *to the right* of the decimal point as a whole number followed by the place value of the rightmost digit.	Hundredths 8.15 is read "eight and fifteen hundredths."	p. 368
Rounding Decimals 1. Find the place to which the decimal is to be rounded. 2. If the next digit to the right is 5 or more, increase the digit in the place you are rounding to by 1. Discard any remaining digits to the right. 3. If the next digit to the right is less than 5, just discard that digit and any remaining digits to the right.	To round 5.87 to the nearest tenth: ↓ 5.87 is rounded to 5.9 To round 12.3454 to the nearest thousandth: ↓ 12.3454 is rounded to 12.345.	p. 370
Addition and Subtraction of Decimals		**Section 5.2**
To Add or Subtract Decimals 1. Write the numbers being added (or subtracted) in column form with their decimal points in a vertical line. You may have to place zeros to the right of the existing digits. 2. Add (or subtract) just as you would with whole numbers. 3. Place the decimal point of the sum in line with the decimal points of the addends.	To subtract 5.875 from 8.5: 8.500 − 5.875 2.625	p. 379, 381
Multiplication of Decimals		**Section 5.3**
To Multiply Decimals 1. Multiply the decimals as though they were whole numbers. 2. Add the number of decimal places in the factors. 3. Place the decimal point in the product so that the number of decimal places in the product is the sum of the number of decimal places in the factors.	To multiply 2.85 · 0.045: 2.85 ⟵ Two places × 0.045 ⟵ Three places 1425 1140 0.12825 ⟵ Five places	p. 389
Multiplying by Powers of 10 Move the decimal point to the right the same number of places as there are zeros in the power of 10.	$2.37 \cdot 10 = 23.7$ $0.567 \cdot 1000 = 567$	p. 392

Continued

DEFINITION/PROCEDURE	EXAMPLE	REFERENCE
Division of Decimals		**Section 5.4**
To Divide by a Decimal 1. Move the decimal point to the right, making the divisor a whole number. 2. Move the decimal point in the dividend to the right the same number of places. Add zeros if necessary. 3. Place the decimal point in the quotient directly above the decimal point of the dividend. 4. Divide as you would with whole numbers.	To divide 16.5 by 5.5, move the decimal points: $\begin{array}{r} 3 \\ 5.5_{\wedge}\overline{)16.5_{\wedge}} \\ \underline{16\ 5} \\ 0 \end{array}$	**p. 399**
To Divide by a Power of 10 Move the decimal point to the left the same number of places as there are zeros in the power of 10.	$25.8 \div 10 = 2_{\wedge}5.8 = 2.58$	**p. 400**
Fractions and Decimals		**Section 5.5**
To Convert a Common Fraction to a Decimal 1. Divide the numerator of the common fraction by its denominator. 2. The quotient is the decimal equivalent of the common fraction.	To convert $\frac{1}{2}$ to a decimal: $\begin{array}{r} 0.5 \\ 2\overline{)1.0} \\ \underline{1\ 0} \\ 0 \end{array}$	**p. 408**
To Convert a Terminating Decimal Less Than 1 to a Common Fraction 1. Write the digits of the decimal without the decimal point. This will be the numerator of the common fraction. 2. The denominator of the fraction is a 1 followed by as many zeros as there are places in the decimal.	To convert 0.275 to a common fraction: $0.275 = \frac{275}{1000} = \frac{11}{40}$	**p. 410**
Equations Containing Decimals		**Section 5.6**
To solve an equation that contains decimals, use the same procedure used for solving other linear equations.	$1.2x + 3.9x = 2.7x - 3.2$ $5.1x = 2.7x - 3.2$ $2.4x = -3.2$ $x = \frac{-3.2}{2.4}$ $x = -1.\overline{33}$	**p. 419**
Square Roots and the Pythagorean Theorem		**Section 5.7**
The square root of a number is a value that, when squared, gives us that number. The length of the three sides of a right triangle will form a perfect triple.	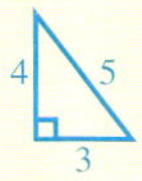 $3^2 + 4^2 = 5^2$	**p. 423**
The Pythagorean theorem is usually written as $c^2 = a^2 + b^2$	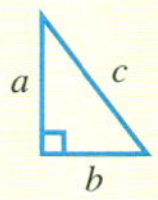	**p. 425**

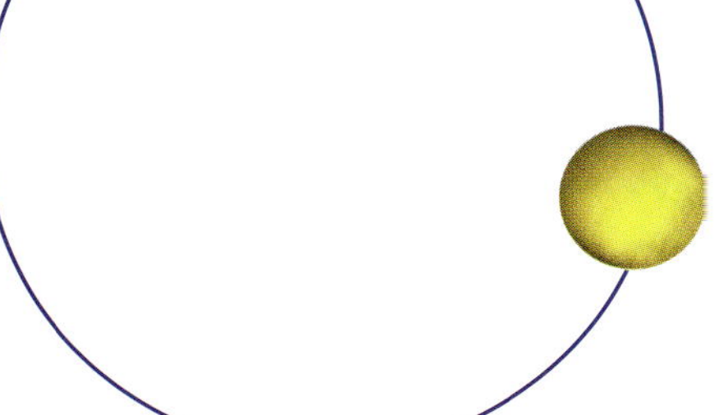

Summary and Review Exercises

You should now be reviewing the material in Chapter 5. These exercises will help in that process. Work all the exercises carefully. References are provided to the section for each exercise. If you made an error, go back and review the related material.

[5.1] In exercises 1 and 2, find the indicated place values.

1. 7 in 3.5742

2. 3 in 0.5273

In exercises 3 and 4, write the fractions in decimal form.

3. $\frac{37}{100}$

4. $\frac{307}{10000}$

In exercises 5 and 6, write the decimals in words.

5. 0.071

6. 12.39

In exercises 7 and 8, write the fractions in decimal form.

7. Four and five tenths

8. Four hundred and thirty-seven thousandths

In exercises 9 to 12, complete each statement using the symbol $<$, $=$, or $>$.

9. 0.79 ______ 0.785

10. 1.25 ______ 1.250

11. 12.8 ______ 13

12. 0.832 ______ 0.83

In exercises 13 to 15, round to the indicated place.

13. 5.837 hundredths

14. 9.5723 thousandths

15. 4.87625 three decimal places

[5.2] In exercises 16 to 19, add.

16. $\begin{array}{r} 2.58 \\ +\ 0.89 \\ \hline \end{array}$

17. $\begin{array}{r} 3.14 \\ 0.8 \\ 2.912 \\ +\ 12 \\ \hline \end{array}$

18. -1.3, 25, -5.27, and 6.158

19. Add eight, forty-three thousandths, five and nineteen hundredths, and seven and three tenths.

In exercises 20 to 23, subtract.

20. $\begin{array}{r} 29.21 \\ -\ 5.89 \\ \hline \end{array}$

21. $\begin{array}{r} 6.73 \\ -\ 2.485 \\ \hline \end{array}$

22. 1.735 from −2.81

23. −12.38 from 19

[5.3] In exercises 24 to 29, multiply.

24. $\begin{array}{r} 22.8 \\ \times\ 0.72 \\ \hline \end{array}$

25. $\begin{array}{r} 0.0045 \\ \times\ 0.058 \\ \hline \end{array}$

26. $-1.24 \cdot 56$

27. $-0.0025 \cdot (-0.491)$

28. $0.052 \cdot 1{,}000$

29. $0.045 \cdot 10^4$

[5.4] In exercises 30 to 32, divide. Round answers to the nearest hundredth.

30. $8\overline{)3.08}$

31. $58\overline{)269.7}$

32. $55\overline{)17.69}$

In exercises 33 to 36, divide. Round answers to the nearest thousandth.

33. $0.7\overline{)1.865}$

34. $-3.042 \div (-0.37)$

35. $5.3\overline{)6.748}$

36. $0.2549 \div (-2.87)$

In exercises 37 to 39, divide.

37. $7.6 \div 10$

38. $80.7 \div 1{,}000$

39. $457 \div 10^4$

[5.5] In exercises 40 to 43, find the decimal equivalents.

40. $\frac{7}{16}$

41. $\frac{3}{7}$ (round to the thousandth)

42. $\frac{4}{15}$ (use bar notation)

43. $3\frac{3}{4}$

In exercises 44 to 48, write as common fractions or mixed numbers. Simplify your answers.

44. 0.21

45. 0.084

46. 5.28

47. 0.0067

48. 21.857

[5.6] Solve the equations and check your results.

49. $3.7x + 8 = 1.7x + 16$

50. $5.4x + (-3) = 8.4x + 9$

51. $2.9x = 4.9x - 3.3$

52. $1.4x = 4.4x + 9.75$

53. $2(x - 1.8) = 4.2(x + 0.9) + 3.18$

54. $3(x + 5.8) = 6.5(x - 1.2) + 12.25$

[5.7] In exercises 55 and 56, find the square root.

55. $\sqrt{324}$

56. $\sqrt{784}$

57. Find the hypotenuse of the triangle whose sides are 33 and 44.

[5.8] In exercises 58 to 71, solve the applications.

58. Geometry. Find the perimeter (to the nearest hundredth of a centimeter) of a rectangle that has dimensions 5.37 cm by 8.64 cm.

59. Distance. Janice ran 4.8 mi on Sunday, 5.3 mi on Tuesday, 3.9 mi on Thursday, and 8.2 mi on Saturday. How far did she run during the week?

60. Dimensions. Find dimension *a* in the figure.

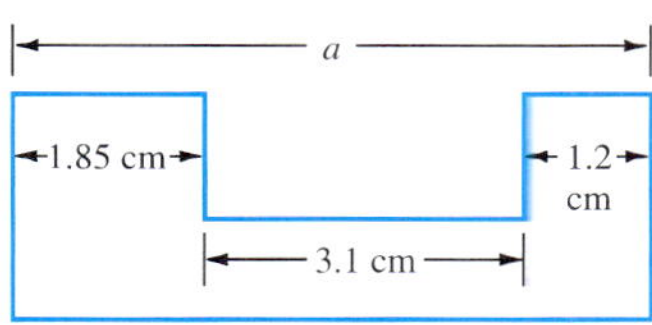

61. Savings. A stereo system that normally sells for $499.50 is discounted (or marked down) to $437.75 for a sale. Find the savings.

62. Cash remaining. If you cash a $50 check and make purchases of $8.71, $12.53, and $9.83, how much money do you have left?

63. Earnings. Neal worked for 37.4 h during a week. If his hourly rate of pay was $7.25, how much did he earn?

64. Interest. To find the simple interest on a loan at $11\frac{1}{2}\%$ for 1 year, we must multiply the amount of the loan by 0.115. Find the simple interest on a $2,500 loan at $11\frac{1}{2}\%$ for 1 year.

65. Installment plan costs. A television set has an advertised price of $499.50. You buy the set and agree to make payments of $27.15 per month for 2 years. How much extra are you paying by buying on this installment plan?

66. **Total cost.** A stereo dealer buys 100 portable radios for a promotion sale. If she pays $57.42 per radio, what is her total cost?

67. **Employee donation.** During a charity fund-raising drive 37 employees of a company donated a total of $867.65. What was the average donation per employee?

68. **Mileage.** In six readings, Faith's gas mileage was 38.9, 35.3, 39.0, 41.2, 40.5, and 40.8 $\frac{\text{mi}}{\text{gal}}$. What was the average mileage to the nearest tenth of a mile per gallon? (*Hint:* First find the sum of the mileages. Then divide the sum by 6, because there are 6 mileages.)

69. **Lot quantity.** A developer is planning to subdivide an 18.5-acre piece of land. She estimates that 5 acres will be used for roads and wants individual lots of 0.25 acre. How many lots are possible?

70. **Mileage.** Paul drives 949 mi using 31.8 gal of gas. What is his mileage for the trip (to the nearest tenth of a mile per gallon)?

71. **Tape cost.** A shipment of 1,000 videotapes cost a dealer $7,090. What was the cost per tape to the dealer?

Solve each of the applications. Approximate your answer to one decimal place where necessary.

72. Find the length of the diagonal of a rectangle whose length is 12 in. and whose width is 9 in.

73. Find the length of a rectangle whose diagonal has a length of 10 cm and whose width is 5 cm.

74. How long must a guywire be to run from the top of an 18-ft pole to a point on level ground 16 ft away from the base of the pole?

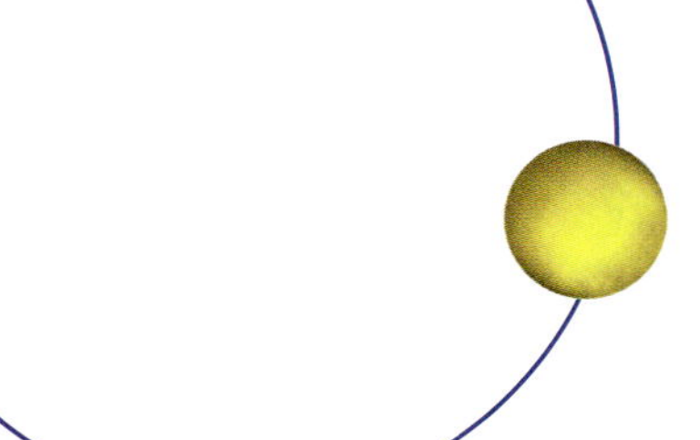

Chapter Test for Chapter 5

Name ______________

Section ________ Date ________

ANSWERS

1. ______
2. ______
3. ______
4. ______
5. ______
6. ______
7. ______
8. ______
9. ______
10. ______
11. ______
12. ______
13. ______
14. ______
15. ______
16. ______
17. ______
18. ______
19. ______
20. ______ 21. ______
22. ______ 23. ______
24. ______ 25. ______
26. ______ 27. ______
28. ______

This test is provided to help you in the process of reviewing Chapter 5. Answers are provided in the back of the book. If you missed any answers, be sure to go back and review the appropriate chapter sections.

1. Find the place value of 8 in 0.5248.

2. Write $\frac{49}{1000}$ in decimal form.

3. Write 2.53 in words.

4. Write twelve and seventeen thousandths in decimal form.

In exercises 5 and 6, complete the statement, using the symbol $<$ or $>$.

5. 0.889 _______ 0.89

6. 0.531 _______ 0.53

In exercises 7 to 9, add.

7.
$$\begin{array}{r} 3.45 \\ 0.6 \\ +\ 12.59 \\ \hline \end{array}$$

8. 2.4, 35, -4.73, and -5.123.

9. Seven, seventy-nine hundredths, and five and thirteen thousandths.

In exercises 10 and 11, round to the indicated place.

10. 0.5977 thousandths

11. 23.5724 two decimal places

In exercises 12 to 14, subtract.

12.
$$\begin{array}{r} 18.32 \\ -\ 7.78 \\ \hline \end{array}$$

13.
$$\begin{array}{r} 40 \\ -\ 15.625 \\ \hline \end{array}$$

14. -1.742 from -5.63

In exercises 15 to 19, multiply.

15.
$$\begin{array}{r} 32.9 \\ \times\ 0.53 \\ \hline \end{array}$$

16.
$$\begin{array}{r} 0.049 \\ \times\ 0.57 \\ \hline \end{array}$$

17. $2.75 \cdot (-0.53)$

18. $0.735 \cdot 1{,}000$

19. $1.257 \cdot 10^4$

In exercises 20 to 26, divide. When indicated, round to the given place value.

20. $8\overline{)3.72}$

21. $27\overline{)63.45}$

22. $2.72 \div 53$ thousandths

23. $4.1\overline{)10.455}$

24. $0.6\overline{)1.431}$

25. $-3.969 \div 0.54$

26. $0.263 \div 3.91$ three decimal places

In exercises 27 and 28, divide.

27. $4.983 \div 1{,}000$

28. $523 \div 10^5$

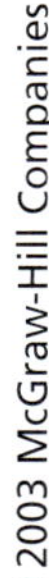

ANSWERS

29. ______
30. ______
31. ______
32. ______
33. ______
34. ______
35. ______
36. ______
37. ______
38. ______
39. ______
40. ______
41. ______
42. ______
43. ______
44. ______
45. ______
46. ______
47. ______
48. ______
49. ______

In exercises 29 to 31, find the decimal equivalents of the common fractions. When indicated, round to the given place value.

29. $\frac{9}{16}$ **30.** $\frac{4}{7}$ (thousandths) **31.** $\frac{7}{11}$ (use bar notation)

In exercises 32 and 33, write the decimals as common fractions or mixed numbers. Simplify your answer.

32. 0.072 **33.** 4.44

34. Insert $<$ or $>$ to form a true statement.

$0.168 \underline{\qquad} \frac{3}{25}$

35. A baseball team has a winning percentage of 0.458. Write this as a fraction in simplest form.

In exercises 36 to 40, solve each equation.

36. $5.2x = 7.54$ **37.** $2.3x = -8.28$

38. $1.83 + 2x = 5.04 + 5x$ **39.** $6.5(x - 1.4) = 4.4(x - 3.2) + 9.495$

40. $1.3 + 2(x + 6.8) = 3x + 20.2$

41. Find the square root of 441.

42. The legs of a right triangle are 39 m and 52 m in length. Find the length of the hypotenuse.

43. Find the perimeter of the triangle shown.

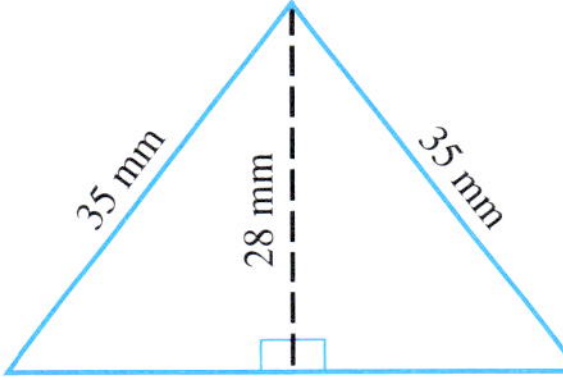

44. Gasoline purchased. On a business trip, Martin bought the following amounts of gasoline: 14.4, 12, 13.8, and 10 gal. How much gasoline did he purchase on the trip?

45. Cash remaining. You pay for purchases of $13.99, $18.75, $9.20, and $5 with a $50 bill. How much cash will you have left?

46. Find the area of a rectangle with length 3.5 in. and width 2.15 in.

47. Total costs. A college bookstore purchases 1,000 pens at a cost of 54.3 cents per pen. Find the total cost of the order in dollars.

48. Number of lots. A 14-acre piece of land is being developed into home lots. If 2.8 acres of land will be used for roads and each home site is to be 0.35 acre, how many lots can be formed?

49. Costs to families. A street improvement project will cost $57,340 and that cost is to be divided among the 100 families in the area. What will be the cost to each individual family?

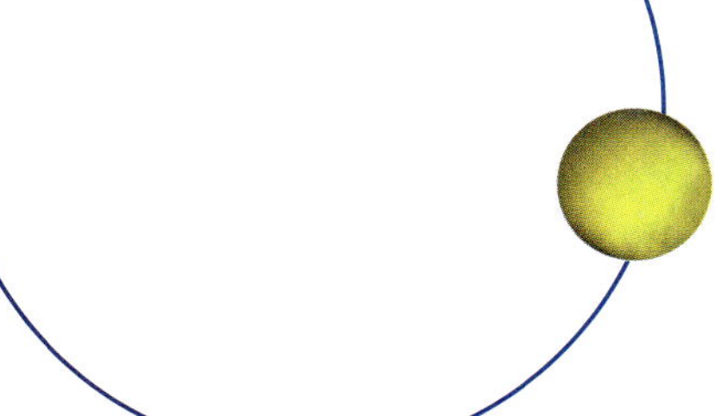

Cumulative Test for Chapters 1 to 5

Name ____________

Section ________ Date ________

ANSWERS

1. ____
2. ____
3. ____
4. ____
5. ____
6. ____
7. ____
8. ____
9. ____
10. ____
11. ____
12. ____
13. ____
14. ____
15. ____
16. ____
17. ____
18. ____
19. ____
20. ____
21. ____
22. ____
23. ____ 24. ____

This review covers selected topics from Chapters 1 through 5.

Perform the indicated operations.

1. $9 - (-3)$

2. $(-4)(-5)$

3. $0 \div (-7)$

4. $27 - 3 \cdot 2^2$

5. $8(-9 + 7)$

6. $\frac{2}{5} \cdot \frac{15}{8}$

7. $\frac{-3}{4} + \frac{-1}{3}$

8. $\frac{2}{5} \div \frac{-3}{10}$

9. $12.3 + 8.52$

10. $19.3 - 8.47$

11. $0.03 \cdot 425$

12. $0.0042 \cdot 1{,}000$

13. $53.26 \div 100$

Round to the indicated place.

14. 6.82148 thousandths

15. 5.982 tenths

Complete each statement using the symbol $<$, $=$, or $>$.

16. 15.6295 ________ 15.631

17. 7.04 ________ 7.040

Write as a common fraction.

18. 0.15

19. 0.08

Write as a decimal.

20. $\frac{5}{8}$

21. $\frac{6}{7}$ round to the hundredths

22. Write the prime factorization of 210.

23. Find the greatest common factor of 30 and 48.

24. Find the least common multiple of 15 and 18.

ANSWERS

25. ____
26. ____
27. ____
28. ____
29. ____
30. ____
31. ____
32. ____
33. ____
34. ____
35. ____
36. ____
37. ____
38. ____
39. ____
40. ____
41. ____
42. ____
43. ____
44. ____

Identify each as an expression or an equation.

25. $5x - 3 = 8$

26. $\frac{3}{4}x - \frac{1}{4}$

Evaluate each expression if $x = -3$, $y = 6$, $z = -4$, and $w = 2$.

27. $x + 3y - 3z$

28. $\frac{3x - y}{w - x}$

Combine like terms.

29. $7x + 5y - 4x - 6y$

30. $\frac{13}{5}x + 2 - \frac{3}{5}x + 5$

Solve each equation and check your result.

31. $4x + 3 = 5x + 7$

32. $2x - 6 = 5x + 9$

33. $\frac{x}{4} = 12$

34. $\frac{2}{5}x = -10$

35. $-1.2x = 3$

36. $4.2x - 5 = 1.2x - 29$

In exercises 37 and 38, use the figure:

37. Find the perimeter of the figure shown.

38. Find the area of the figure shown.

Solve each application.

39. If a number increased by 8 is 23, find that number.

40. A dresser costs \$200 more than a bed frame. Together the two pieces cost \$900. How much does the dresser cost alone?

41. Francisco drove 273 mi using 15 gal of gas. What was his mileage (miles per gallon) for the trip?

42. The length of a rectangle is 4 cm less than 3 times its width. The perimeter of the rectangle is 88 cm. What are the dimensions of the rectangle?

43. Find the square root of 289.

44. The legs of a right triangle are 13 in. and 18 in. in length. Find the length of the hypotenuse. Round to the nearest tenth.

Ratio, Rate, and Proportion

6

INTRODUCTION

Bookkeeping and accounting used to be done exclusively with pencils and ledgers. Now, because of the popularity of computers and accounting software, many people count on their computers to do their accounting.

When Jean became an accountant, she expected that she would work with the books of a single, large company. Instead, she has almost 50 clients for whom she does accounting. Many of these clients enter the information into their computers and bring only the disks to Jean. She says that many of the errors she finds could be avoided if people had better estimating skills. Jean learned early in her math classes that no answer (especially one obtained by use of a calculator or computer) should be accepted without checking to see if it is reasonable.

Name ____________________

Section ________ Date ________

ANSWERS

1. ____________________

2. ____________________

3. ____________________

4. ____________________

5. ____________________

6. ____________________

7. ____________________

8. ____________________

9. ____________________

10. ____________________

11. ____________________

12. ____________________

13. ____________________

14. ____________________

Pre-Test Chapter 6

This pre-test will point out any difficulties you may be having with ratios and proportions. Do all the problems, and then check your answers with those in the back of the book.

1. Write the ratio of 7 to 10.

2. Write the ratio of 20 to 15 in lowest terms.

3. Find the rate equivalent to $\frac{551 \text{ mi}}{19 \text{ gal}}$.

4. Find the unit price given that a dozen cans of cat food cost $9.48.

5. Is $\frac{4}{7} = \frac{12}{21}$ a true proportion?

6. Is $\frac{5}{9} = \frac{9}{16}$ a true proportion?

7. Solve for x: $\frac{x}{4} = \frac{5}{2}$

8. Solve for a: $\frac{5}{a} = \frac{7}{21}$

9. Solve for n: $\frac{\frac{1}{2}}{2} = \frac{3}{n}$

10. Use a proportion to find the unknown side, labeled x, given the following pair of similar triangles.

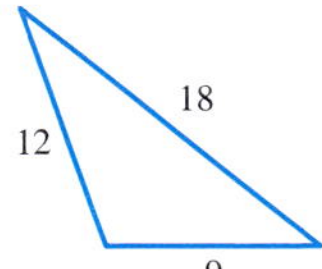

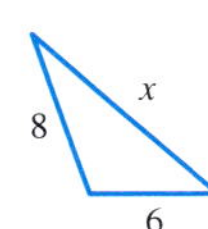

11. A 6-ft-tall man casts a 10-ft shadow. Find the height of a building casting a 220-ft shadow.

12. **Pricing.** Cans of tomato juice are marked 2 for $1.05. At this price, what will 12 cans cost?

13. **Car sales.** The ratio of compact cars to larger model cars sold during a month was 9 to 4. If 72 compact cars were sold during that period, how many larger cars were sold?

14. **Cost.** If 2 gal of paint will cover 450 ft^2, how many gallons will be needed to paint a room with 2,475 ft^2 of wall surface?

6.1 Ratios

OBJECTIVES

1. Write the ratio of two numbers in simplest form
2. Write the ratio of two quantities in simplest form

In Chapter 3, you saw two meanings for a fraction:

1. A fraction can name a certain number of parts of a whole. For example, $\frac{3}{5}$ names 3 parts of a whole that has been divided into 5 equal parts.

2. A fraction can indicate division. The same $\frac{3}{5}$ can be thought of as $3 \div 5$.

We now want to turn to a third meaning for a fraction:

3. A fraction can be a ratio. A **ratio** is a means of comparing two numbers or quantities.

NOTE Another way of writing the ratio of 3 to 5 is 3:5. We have chosen to use only the fraction notation for a ratio in this textbook.

Example 1

Writing a Ratio as a Fraction

Write the ratio 3 to 5 as a fraction.

To compare 3 to 5, we write the ratio of 3 to 5 as $\frac{3}{5}$. So $\frac{3}{5}$ also means "the ratio of 3 to 5."

CHECK YOURSELF 1

Write the ratio of 7 to 12 as a fraction.

Example 2 illustrates the use of a ratio in comparing *like quantities,* which means we're comparing inches to inches, cm to cm, apples to apples, etc.

Example 2

Applying the Concept of Ratio

The width of a rectangle is 7 cm and its length is 19 cm. Write the ratio of its width to its length as a fraction.

$$\frac{7 \text{ cm}}{19 \text{ cm}} = \frac{7}{19}$$

We are comparing centimeters to centimeters, so the units "cancel."

NOTE A ratio fraction can be greater than 1.

NOTE In this case the ratio is *never* written as a mixed number. It is left as an improper fraction.

The ratio of its length to its width is

$$\frac{19 \text{ cm}}{7 \text{ cm}} = \frac{19}{7}$$

CHECK YOURSELF 2

A basketball team wins 17 of its 29 games in a season.

(a) Write the ratio of wins to games played. **(b)** Write the ratio of wins to losses.

Because a ratio is a fraction, we can reduce it to simplest form. Consider Example 3.

Example 3

Writing a Ratio in Simplest Form

NOTE When simplifying a fraction, you are actually multiplying by one.

$\frac{20 \div 10}{30 \div 10}$ is the same as

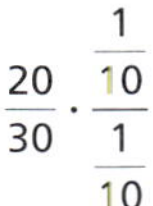

$\frac{20}{30} \cdot \frac{\frac{1}{10}}{\frac{1}{10}}$

This is another application of the fundamental rule of fractions.

Write the ratio of 20 to 30 in lowest terms.

$$\frac{20}{30} = \frac{2}{3}$$

Divide the numerator and denominator by the common factor of 10.

CHECK YOURSELF 3

Write the ratio of 24 to 32 in lowest terms.

Example 4 relates to an application of ratios.

Example 4

Simplifying the Ratio of Two Dimensions

A common size for a movie screen is 32 ft by 18 ft. Write this as a ratio in simplest form.

$$\frac{32 \text{ ft}}{18 \text{ ft}} = \frac{32}{18} = \frac{16}{9}$$

CHECK YOURSELF 4

A common computer display mode is 640 pixels (picture elements) by 480 pixels. Write this as a ratio in simplest form.

Some ratios include fractions or decimals, as in Examples 5 and 6.

Example 5

Simplifying a Ratio Involving a Fraction

Loren sank a $22\frac{1}{2}$-ft putt, and Carrie sank a 30-ft putt. Express the ratio of the two distances as a ratio of whole numbers.

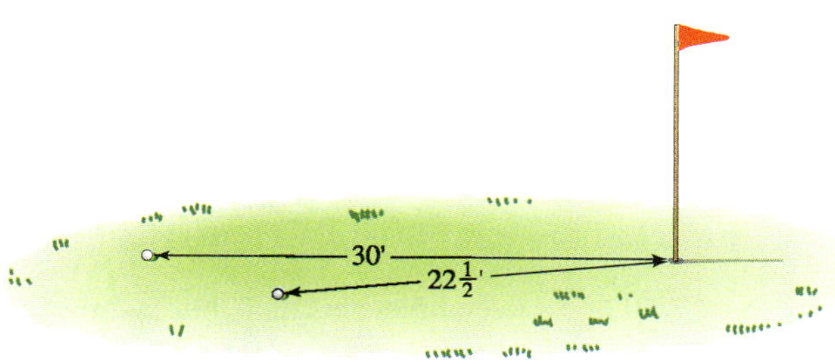

$$\frac{22\frac{1}{2}}{30} = \frac{\frac{45}{2}}{30} = \frac{\frac{45}{2}}{\frac{30}{1}}$$

Because we are dividing a fraction by a fraction, we invert and multiply.

$$\frac{45}{2} \div \frac{30}{1} = \frac{45}{2} \cdot \frac{1}{30} = \frac{3}{4}$$

The ratio $22\frac{1}{2}$ to 30 is equivalent to the ratio 3 to 4.

CHECK YOURSELF 5

Rita jogged $3\frac{1}{2}$ mi this morning, and Yi jogged $4\frac{1}{4}$ mi. Express the ratio of the two distances as a ratio of whole numbers.

Example 6 simplifies a ratio involving decimals.

Example 6

Simplifying a Ratio Involving Decimals

The diameter of a 20-oz bottle is 2.8 in., and the diameter of a 2-L bottle is 5.25 in. Express the ratio of the two diameters as a ratio of whole numbers.

$$\frac{2.8}{5.25} = \frac{2.8 \cdot 100}{5.25 \cdot 100} = \frac{280}{525}$$

$$\frac{280}{525} = \frac{8}{15}$$

The ratio of the diameters 2.8 to 5.25 is equivalent to the ratio 8 to 15.

CHECK YOURSELF 6

The width of a standard newspaper column is 2.625 in., and the length of a standard column is 19.5 in. Express the ratio of the two measurements as a ratio of whole numbers.

In Example 7, we will see that sometimes, to find a ratio, we must rewrite two denominate numbers so that they have the same dimensions.

Example 7

Rewriting Denominate Numbers to Find a Ratio

Joe took 2 full hours to complete his final, but Jaymie finished hers in 75 min. Find the ratio of the two times.

To find a ratio, both numbers must have the same units. If we first convert the 2 h to 120 min, both units are minutes.

NOTE

$$\frac{2\ \cancel{h}}{1} \cdot \frac{60\text{ min}}{1\ \cancel{h}} = \frac{120\text{ min}}{1}$$

$$\frac{2\text{ h}}{75\text{ min}} = \frac{120\text{ min}}{75\text{ min}} = \frac{8}{5}$$

CHECK YOURSELF 7

Find the ratio of whole numbers that is equivalent to the ratio of 16 ft to 10 yd.

CHECK YOURSELF ANSWERS

1. $\frac{7}{12}$ **2.** **(a)** $\frac{17}{29}$; **(b)** $\frac{17}{12}$ (the team lost 12 games) **3.** $\frac{3}{4}$ **4.** $\frac{4}{3}$ **5.** $\frac{14}{17}$

6. $\frac{7}{52}$ **7.** $\frac{8}{15}$

6.1 Exercises

Name ____________

Section ________ Date ________

Write each of the ratios in simplest form.

1. The ratio of 9 to 13
2. The ratio of 5 to 4
3. The ratio of 9 to 4
4. The ratio of 5 to 12
5. The ratio of 10 to 15
6. The ratio of 16 to 12
7. The ratio of $3\frac{1}{2}$ to 14
8. The ratio of $5\frac{3}{5}$ to $2\frac{1}{10}$
9. The ratio of 10.5 to 2.7
10. The ratio of 2.2 to 0.6
11. The ratio of 12 mi to 18 mi
12. The ratio of 100 cm to 90 cm
13. The ratio of 40 ft to 65 ft
14. The ratio of 12 oz to 18 oz
15. The ratio of $48 to $42
16. The ratio of 20 ft to 24 ft
17. The ratio of 75 s to 3 min
18. The ratio of 7 oz to 3 lb
19. The ratio of 4 nickels to 5 dimes
20. The ratio of 8 in. to 3 ft
21. The ratio of 2 days to 10 h
22. The ratio of 4 ft to 4 yd
23. The ratio of 5 gal to 12 quarts (qt)
24. The ratio of 7 dimes to 3 quarters

ANSWERS

1. ____________
2. ____________
3. ____________
4. ____________
5. ____________
6. ____________
7. ____________
8. ____________
9. ____________
10. ____________
11. ____________
12. ____________
13. ____________
14. ____________
15. ____________
16. ____________

17. ______ 18. ______

19. ______ 20. ______

21. ______ 22. ______

23. ______ 24. ______

ANSWERS

25. ____________

26. ____________

27. ____________

28. ____________

29. ____________

30. ____________

31. ____________

32. ____________

33. ____________

Solve the applications.

25. Class make-up ratio. An algebra class has 7 men and 13 women. Write the ratio of men to women. Write the ratio of women to men.

26. Football ratio. A football team wins 9 of its 16 games with no ties. Write the ratio of wins to games played. Write the ratio of wins to losses.

27. Election ratio. In a school election 4,500 yes votes were cast, and 3,000 no votes were cast. Write the ratio of yes to no votes.

28. Basketball ratio. A basketball player made 42 of the 70 shots taken in a tournament. Write the ratio of shots made to shots taken.

29. Carla walked $2\frac{1}{4}$ mi this afternoon and Mario walked $5\frac{3}{8}$ mi this afternoon. Express the ratio of the two distances as a ratio of whole numbers.

30. One car has an $11\frac{1}{2}$-gal tank and another has a $17\frac{3}{4}$-gal tank. Express the ratio of the capacities as a ratio of whole numbers.

31. One refrigerator holds $2\frac{2}{3}$ ft^3 of food and another holds $5\frac{3}{4}$ ft^3 of food. Express the ratio of the capacities as a ratio of whole numbers.

32. The price of an antibiotic in one drugstore is \$12.50. The price of the same antibiotic in another drugstore is \$8.75. Write the ratio of the prices as a ratio of whole numbers.

33. The width of a notebook is 3.5 in. and the length is 6.75 in. Write the ratio of length to width as a ratio of whole numbers.

ANSWERS

34. ______

35. ______

36. ______

37. ______

34. Marc took 3 h to mow a lawn. Angelina took 150 min to mow the same lawn a week earlier. Write the ratio of Marc's time to Angelina's time as a ratio of whole numbers.

35. Employment ratio. A company employs 24 women and 18 men. Write the ratio of men to women employed by the company.

36. Measurement ratio. If a room is 30 ft long and 6 yd wide, write the ratio of the length to the width of the room.

37. (a) Buy a 1.69-oz (medium size) bag of M&Ms. Determine the ratio of the number of M&Ms of each color (yellow, red, blue, orange, brown, and green) to the total number of M&Ms in the bag.

(b) Compare your ratios to those your classmates obtain.

(c) Use the information from parts **(a)** and **(b)** to determine a ratio for all the different colors in a bag of M&Ms.

(d) E-mail the manufacturer of M&Ms (Mars, Inc.) at www.m-ms.com and see if they use fixed ratios to determine the distribution of the colors in a bag. If they do, compare these ratios to yours.

38. ______________________

38. Two pencils are shown. Write the ratio of the length of the smaller pencil to the larger pencil.

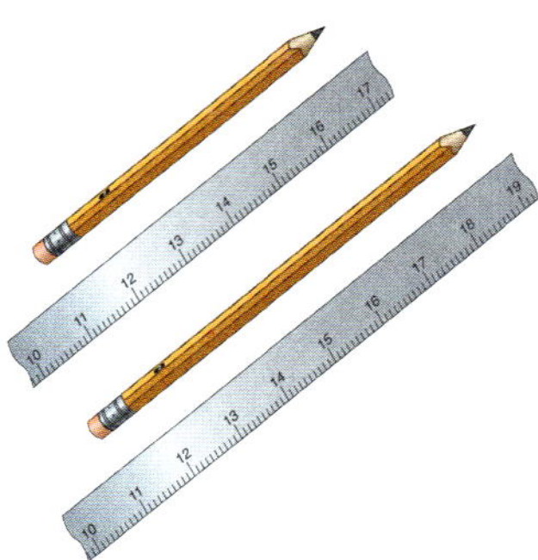

Answers

1. $\frac{9}{13}$ **3.** $\frac{9}{4}$ **5.** $\frac{2}{3}$ **7.** $\frac{1}{4}$ **9.** $\frac{35}{9}$ **11.** $\frac{2}{3}$ **13.** $\frac{8}{13}$ **15.** $\frac{8}{7}$
17. $\frac{5}{12}$ **19.** $\frac{2}{5}$ **21.** $\frac{24}{5}$ **23.** $\frac{5}{3}$ **25.** $\frac{7}{13}$; $\frac{13}{7}$ **27.** $\frac{3}{2}$ **29.** $\frac{18}{43}$
31. $\frac{32}{69}$ **33.** $\frac{27}{14}$ **35.** $\frac{3}{4}$ **37.** 

6.2 Rates

6.2 OBJECTIVES

1. Write a rate as a fraction
2. Find unit rates
3. Find unit prices

A ratio compares two denominate numbers. Examples of a ratio include

$$\frac{9 \text{ s}}{12 \text{ s}} = \frac{9}{12} = \frac{3}{4}$$

or

$$\frac{5 \text{ mi}}{7 \text{ mi}} = \frac{5}{7}$$

In these examples, the units of the numerator and denominator are identical. When we compare denominate numbers with different units, we get a **rate.**

Example 1

Finding a Rate

Find each rate.

(a) $\frac{12 \text{ ft}}{16 \text{ s}} = \frac{12}{16} \frac{\text{ft}}{\text{s}} = \frac{3}{4} \frac{\text{ft}}{\text{s}}$

(b) $\frac{200 \text{ mi}}{10 \text{ gal}} = \frac{200}{10} \frac{\text{mi}}{\text{gal}} = 20 \frac{\text{mi}}{\text{gal}}$

(c) $\frac{10 \text{ gal}}{200 \text{ mi}} = \frac{10}{200} \frac{\text{gal}}{\text{mi}} = \frac{1}{20} \frac{\text{gal}}{\text{mi}}$

CHECK YOURSELF 1

Find each rate.

(a) $\frac{250 \text{ mi}}{10 \text{ h}}$ **(b)** $\frac{\$60{,}000}{2 \text{ yr}}$ **(c)** $\frac{2 \text{ yr}}{\$60{,}000}$

Sometimes, we need to find the appropriate rate within a statement, as in Example 2.

Example 2

Finding a Rate

Randy Johnson had 320 strikeouts in 280 innings. What was his strikeout per inning rate?

NOTE Recall that in Section 6.1 we stated that mixed numbers were inappropriate for ratios. When we write a rate, a mixed number, usually rewritten as a decimal, is not just appropriate, it is preferred.

$$\frac{320 \text{ strikeouts}}{280 \text{ innings}} = \frac{320}{280} \frac{\text{strikeouts}}{\text{inning}}$$

$$= 1\frac{1}{7} \frac{\text{strikeouts}}{\text{inning}}$$

To rewrite this as a decimal, we approximate. The approximation is indicated by the symbol $\approx$.

$$1\frac{1}{7} \frac{\text{strikeouts}}{\text{inning}} \approx 1.14 \frac{\text{strikeouts}}{\text{inning}}$$

CHECK YOURSELF 2

Chamique Holdsclaw scored 450 points in 15 games. What was her $\frac{\text{points}}{\text{game}}$ *rate?*

One purpose for computing a rate is for comparison. Example 3 illustrates.

Example 3

Comparing Rates

Player A scores 50 points in 9 games, and player B scores 260 points in 36 games. Which player scored at a higher rate?

Player A's rate was $\frac{50 \text{ points}}{9 \text{ games}} = \frac{50}{9} \frac{\text{points}}{\text{game}}$

$$\approx 5.56 \frac{\text{points}}{\text{game}}$$

Player B's rate was $\frac{260 \text{ points}}{36 \text{ games}} = \frac{260}{36} \frac{\text{points}}{\text{game}}$

$$= \frac{65}{9} \frac{\text{points}}{\text{game}}$$

$$\approx 7.22 \frac{\text{points}}{\text{game}}$$

Player B scored at a higher rate.

CHECK YOURSELF 3

Hassan scored 25 goals in 8 games, and Lee scored 52 goals in 18 games. Which player scored at a higher rate?

One type of rate, seen commonly in supermarkets, is unit price.

Definitions: Unit Price

The **unit price** (or unit rate) relates a price to some common unit.

NOTE A unit price is a price *per unit.* The unit used may be ounces, pints, pounds, or some other unit.

Example 4

Finding a Unit Price

Find the unit price for each item.

(a) 8 oz of cream cost $1.53

$$\frac{\$1.53}{8 \text{ oz}} = \frac{153 \text{ cents}}{8 \text{ oz}} = \frac{153}{8} \frac{\text{cents}}{\text{oz}}$$

$$\approx 19 \frac{\text{cents}}{\text{oz}}$$

(b) 20 lb of potatoes cost $3.98

$$\frac{\$3.98}{20 \text{ lb}} = \frac{398 \text{ cents}}{20 \text{ lb}} = \frac{398}{20} \frac{\text{cents}}{\text{lb}}$$

$$\approx 20 \frac{\text{cents}}{\text{lb}}$$

CHECK YOURSELF 4

Find the unit price for each item.

(a) 12 soda cans cost \$2.98 **(b)** 25 lb of dog food cost \$9.99

CHECK YOURSELF ANSWERS

1. **(a)** $25\ \frac{\text{mi}}{\text{h}}$; **(b)** $30{,}000\ \frac{\text{dollars}}{\text{yr}}$; **(c)** $\frac{1}{30{,}000}\ \frac{\text{yr}}{\text{dollar}}$ **2.** $30\ \frac{\text{points}}{\text{game}}$

3. Hassan had a higher rate **4.** **(a)** $\approx 25\ \frac{\text{cents}}{\text{can}}$; **(b)** $\approx 40\ \frac{\text{cents}}{\text{lb}}$

Name ______________

Section ________ Date ________

6.2 Exercises

Find each rate.

1. $\frac{300 \text{ mi}}{4 \text{ h}}$

2. $\frac{95 \text{ cents}}{5 \text{ pencils}}$

3. $\frac{69 \text{ ft}}{3 \text{ s}}$

4. $\frac{3 \text{ s}}{69 \text{ ft}}$

5. $\frac{\$10{,}000}{5 \text{ yr}}$

6. $\frac{5 \text{ yr}}{\$10{,}000}$

7. $\frac{680 \text{ ft}}{17 \text{ s}}$

8. $\frac{480 \text{ mi}}{15 \text{ gal}}$

9. $\frac{15 \text{ gal}}{480 \text{ mi}}$

10. $\frac{7200 \text{ revolutions}}{16 \text{ mi}}$

11. $\frac{57 \text{ oz}}{3 \text{ cans}}$

12. $\frac{\$2{,}000{,}000}{4 \text{ yr}}$

13. $\frac{150 \text{ calories}}{3 \text{ oz}}$

14. $\frac{240 \text{ lb of fertilizer}}{6 \text{ lawns}}$

15. $\frac{192 \text{ diapers}}{32 \text{ babies}}$

16. $\frac{657{,}200 \text{ library books}}{5200 \text{ students}}$

Solve the applications.

17. Trac drives 256 mi using 3 gal of gasoline. How many miles per gallon does his car get?

18. Seven pounds of fertilizer cover fourteen hundred square feet. How many square feet are covered by one pound of fertilizer?

19. A local college has 6,000 registered vehicles for 2,400 campus parking spaces. How many vehicles are there for each parking space?

ANSWERS

1. ______________
2. ______________
3. ______________
4. ______________
5. ______________
6. ______________
7. ______________
8. ______________
9. ______________
10. ______________
11. ______________
12. ______________
13. ______________
14. ______________
15. ______________
16. ______________
17. ______________
18. ______________
19. ______________

ANSWERS

20. ____________

21. ____________

22. ____________

23. ____________

24. ____________

25. ____________

26. ____________

27. ____________

28. ____________

20. Curt Schilling has 141 strikeouts in 163 innings. What was his strikeout per inning rate?

21. A water pump can produce 280 gal in 24 h. How many gallons per hour is this?

22. If 214 shares of stock cost \$5,992, what was the cost per share?

23. A printer produces 4 pages in 6 s. How many pages are produced per second?

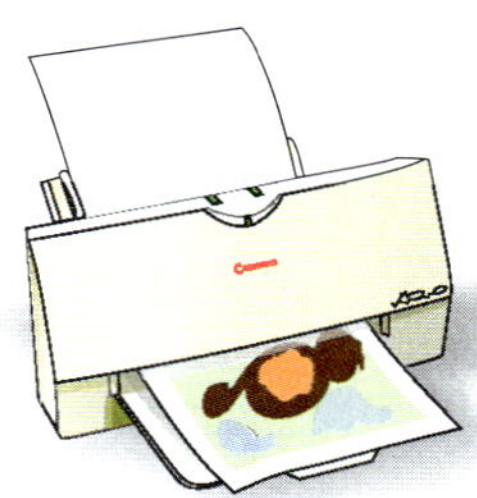

24. Augie eats 12 hamburgers in 48 min. How many minutes does it take Augie to eat one hamburger?

25. A 12-oz can of tuna costs \$4.80. What is the cost of tuna per ounce?

26. The fabric for a dress costs \$76.45 for 9 yd. What is the cost per yard?

27. Gerry laid 634 bricks in 35 min, and his friend Matt laid 515 bricks in 27 min. Who is the faster bricklayer?

28. Mike drove 135 mi in 2.5 h. Sam drove 91 mi in 1.75 h. Who drives at the faster speed?

ANSWERS

29. ______
30. ______
31. ______
32. ______
33. ______
34. ______
35. ______
36. ______
37. ______
38. ______

29. Luis Gonzalez has 137 hits in 387 at bats. Larry Walker has 119 hits in 324 at bats. Who has the higher batting average?

30. What is the better buy—5 lb of sugar for $4.75 or 20 lb of sugar for $19.92?

Find the unit price for each item.

31. $57.50 for 5 shirts

32. $104.93 for 7 CDs

33. $5.16 for a dozen oranges

34. $10.44 for 18 bottles of water

Find the best buy in each of the exercises.

35. Dishwashing liquid:

(a) 12 oz for 79¢
(b) 22 oz for $1.29

36. Canned corn:

(a) 10 oz for 21¢
(b) 17 oz for 39¢

37. Syrup:

(a) 12 oz for 99¢
(b) 24 oz for $1.59
(c) 36 oz for $2.19

38. Shampoo:

(a) 4 oz for $1.16
(b) 7 oz for $1.52
(c) 15 oz for $3.39

ANSWERS

39. ____________

40. ____________

41. ____________

42. ____________

39. Salad oil (1 qt is 32 oz):

(a) 18 oz for 89¢
(b) 1 qt for \$1.39
(c) 1 qt 16 oz for \$2.19

40. Tomato juice [1 pint (pt) is 16 oz]:

(a) 8 oz for 37¢
(b) 1 pt 10 oz for \$1.19
(c) 1 qt 14 oz for \$1.99

41. Peanut butter (1 lb is 16 oz):

(a) 12 oz for \$1.25
(b) 18 oz for \$1.72
(c) 1 lb 12 oz for \$2.59
(d) 2 lb 8 oz for \$3.76

42. Laundry detergent:

(a) 1 lb 2 oz for \$1.99
(b) 1 lb 12 oz for \$2.89
(c) 2 lb 8 oz for \$4.19
(d) 5 lb for \$7.99

Answers

1. $75 \frac{\text{mi}}{\text{h}}$ **3.** $23 \frac{\text{ft}}{\text{s}}$ **5.** $2{,}000 \frac{\text{dollars}}{\text{yr}}$ **7.** $40 \frac{\text{ft}}{\text{s}}$ **9.** $\approx 0.03 \frac{\text{gal}}{\text{mi}}$
11. $19 \frac{\text{oz}}{\text{can}}$ **13.** $50 \frac{\text{calories}}{\text{oz}}$ **15.** $6 \frac{\text{diapers}}{\text{baby}}$ **17.** $32 \frac{\text{mi}}{\text{gal}}$
19. 2.5 vehicles per space **21.** $\approx 11.67 \frac{\text{gal}}{\text{h}}$ **23.** $\frac{2}{3} \frac{\text{page}}{\text{s}}$ **25.** $40 \frac{\text{cents}}{\text{oz}}$
27. Matt **29.** Larry Walker **31.** $\frac{\$11.50}{\text{shirt}}$ **33.** $\frac{\$0.43}{\text{orange}}$ **35.** b
37. c **39.** b **41.** c

6.3 Proportions

6.3 OBJECTIVES

1. Write a proportion
2. Determine whether two fractions are equivalent
3. Determine whether two rates are proportional
4. Solve a proportion for an unknown value

Definitions: Proportion

A **proportion** is an equation that compares two equal fractions (or rates).

NOTE This is the same as saying the fractions are equivalent. They name the same number.

NOTE Remember that we call a letter representing an unknown value a *variable*. Here *a*, *b*, *c*, and *d* are variables. We could have chosen any letters.

Because the ratio of 1 to 3 is equal to the ratio of 2 to 6, we can write the proportion

$$\frac{1}{3} = \frac{2}{6}$$

The proportion $\frac{a}{b} = \frac{c}{d}$ is read "*a* is to *b* as *c* is to *d*." We read the proportion $\frac{1}{3} = \frac{2}{6}$ as "one is to three as two is to six."

Example 1

Writing a Proportion

Write the proportion 3 is to 7 as 9 is to 21.

$$\frac{3}{7} = \frac{9}{21}$$

CHECK YOURSELF 1

Write the proportion 4 *is to* 12 *as* 6 *is to* 18.

When you write a proportion for two rates, placement is important.

Example 2

Writing a Proportion with Two Rates

Write a proportion that is equivalent to the statement: If it takes 3 h to mow 4 acres of grass, it will take 6 h to mow 8 acres.

$$\frac{3 \text{ h}}{4 \text{ acres}} = \frac{6 \text{ h}}{8 \text{ acres}}$$

Note that, in both fractions, the hours units are in the numerator and the acres units are in the denominator.

CHECK YOURSELF 2

Write a proportion that is equivalent to the statement: If it takes 5 *rolls of wallpaper to cover* 400 *ft*^{2}, *it will take* 7 *rolls to cover* 560 *ft*^{2}.

If two fractions are equal, they form a **true proportion.**

Rules and Properties: The Proportion Rule

If $\frac{a}{b} = \frac{c}{d}$, then $a \cdot d = b \cdot c$.

We say that the fractions $\frac{a}{b}$ and $\frac{c}{d}$ are equivalent.

Example 3

Determining Whether Two Fractions Are Equivalent

Determine whether each pair of fractions is equivalent.

(a) $\frac{5}{6} \stackrel{?}{=} \frac{10}{12}$

Multiply:

$6 \cdot 10 = 60$
$5 \cdot 12 = 60$
Equal

Because $b \cdot c = a \cdot d$, $\frac{5}{6} = \frac{10}{12}$ is a proportion. It can be called a *true proportion* or a *true equation*.

(b) $\frac{3}{7} \stackrel{?}{=} \frac{4}{9}$

Multiply:

$7 \cdot 4 = 28$
$3 \cdot 9 = 27$
Not equal

The products are not equal, so $\frac{3}{7} = \frac{4}{9}$ is not a proportion. It is a *false proportion.*

CHECK YOURSELF 3

Determine whether each pair of fractions is equivalent.

(a) $\frac{5}{8} \stackrel{?}{=} \frac{20}{32}$ **(b)** $\frac{7}{9} \stackrel{?}{=} \frac{3}{4}$

If two fractions are equivalent, we say that they are **proportional.**

Example 4

Verifying a Proportion

Determine whether each pair of fractions is proportional.

(a) $\dfrac{3}{\frac{1}{2}} \stackrel{?}{=} \dfrac{30}{5}$

$\dfrac{1}{2} \cdot 30 = 15$

$3 \cdot 5 = 15$

Because the products are equal, the fractions are proportional.

(b) $\dfrac{0.4}{20} \stackrel{?}{=} \dfrac{3}{100}$

$20 \cdot 3 = 60$

$0.4 \cdot 100 = 40$

Because the products are *not* equal, the fractions are not proportional.

CHECK YOURSELF 4

Determine whether each pair of fractions is proportional.

(a) $\dfrac{0.5}{8} \stackrel{?}{=} \dfrac{3}{48}$

(b) $\dfrac{\frac{1}{4}}{6} \stackrel{?}{=} \dfrac{3}{80}$

The proportion rule can also be used to verify that rates are proportional.

Example 5

Verifying a Proportion Involving Rates

Is the rate $\dfrac{\text{5 U.S. dollars}}{\text{15,000 colones}}$ equivalent to the rate $\dfrac{\text{27 U.S. dollars}}{\text{81,000 colones}}$?

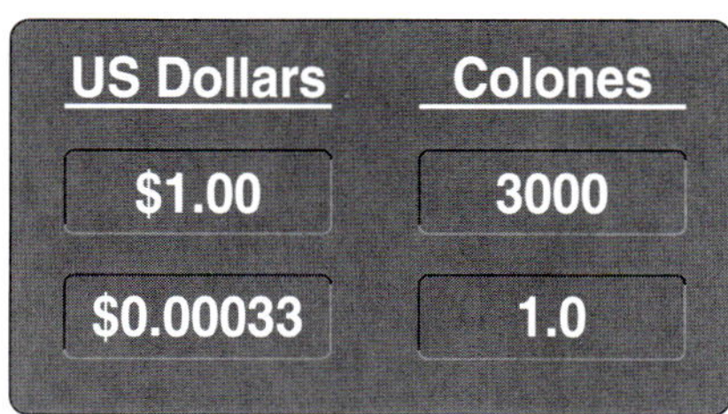

We want to know if the equation

$$\frac{5}{15{,}000} \stackrel{?}{=} \frac{27}{81{,}000}$$

is true.

$$5 \cdot 81{,}000 = 405{,}000$$

$$27 \cdot 15{,}000 = 405{,}000$$

The rates are equivalent.

CHECK YOURSELF 5

Is the rate $\frac{50 \text{ pages}}{45 \text{ min}}$ *equivalent to the rate* $\frac{30 \text{ pages}}{25 \text{ min}}$*?*

We can use what we have learned so far about proportions and equations in order to **solve a proportion.** A proportion consists of four values. If three of the four values of a proportion are known, you can always find the missing or unknown value.

In the proportion $\frac{a}{3} = \frac{10}{15}$, the first value is unknown. We have chosen to represent the unknown value with the letter a. Using the proportion rule, we can proceed.

NOTE $\frac{?}{3} = \frac{10}{15}$ is a proportion in which the first value is unknown. Our work in this section will be learning how to find that unknown value.

$$\frac{a}{3} = \frac{10}{15}$$

$$15 \cdot a = 3 \cdot 10 \qquad \text{or} \qquad 15 \cdot a = 30$$

The equals-sign tells us that $15 \cdot a$ and 30 are just different names for the same number. In Section 2.9, we learned that this type of statement is called an **equation.**

Recall that we can divide both sides of an equation by the same nonzero number.

$$15 \cdot a = 30$$ Divide both sides by 15.

$$\frac{15 \cdot a}{15} = \frac{30}{15}$$

$$\frac{\overset{1}{\cancel{15}} \cdot a}{\underset{1}{\cancel{15}}} = \frac{\overset{2}{\cancel{30}}}{\underset{1}{\cancel{15}}}$$ Divide by the coefficient of the variable. Do you see why we divided by 15? It leaves our unknown a by itself in the left term.

$$a = 2$$

You should always check your result. It is easy in this case. We found a value of 2 for a. Replace the unknown a with that value. Then verify that the fractions are proportional. We started with $\frac{a}{3} = \frac{10}{15}$ and found a value of 2 for a. So we write

NOTE Replace a with 2 and multiply.

$$\frac{2}{3} \stackrel{?}{=} \frac{10}{15}$$

$$3 \cdot 10 \stackrel{?}{=} 2 \cdot 15$$

$$30 = 30$$

The value of 2 for a is correct.

The procedure for solving a proportion is summarized here.

Step by Step: To Solve a Proportion

Step 1 Use the proportion rule to write the equivalent equation $a \cdot d = b \cdot c$.
Step 2 Divide both terms of the equation by the coefficient of the variable.
Step 3 Use the value found to replace the unknown in the original proportion. Check that the ratios or the rates are proportional.

NOTE This gives us the unknown value. Now check the result.

Example 6

Solving Proportions for Unknown Values

Find the unknown value.

$$\frac{8}{x} = \frac{6}{9}$$

Step 1 Using the proportion rule, we have

$$6 \cdot x = 8 \cdot 9$$

or $$6x = 72$$

Step 2 Locate the coefficient of the variable, 6, and divide both sides of the equation by that coefficient.

$$\frac{\overset{1}{\cancel{6}}x}{\underset{1}{\cancel{6}}} = \frac{\overset{12}{\cancel{72}}}{\underset{1}{\cancel{6}}}$$

$$x = 12$$

Step 3 To check, replace x with 12 in the original proportion.

$$\frac{8}{12} \stackrel{?}{=} \frac{6}{9}$$

Multiply:

$$12 \cdot 6 \stackrel{?}{=} 8 \cdot 9$$

$$72 = 72$$ The value of 12 for x checks.

CHECK YOURSELF 6

Solve the proportions for n. Check your result.

(a) $\frac{4}{5} = \frac{n}{25}$ **(b)** $\frac{7}{9} = \frac{42}{n}$

In solving for a missing term in a proportion, we may find an equation involving fractions or decimals. Example 7 involves finding the unknown value in such cases.

Example 7

Solving Proportions for Unknown Values

(a) Solve the proportion for x.

$$\frac{\frac{1}{4}}{3} = \frac{4}{x}$$

$$\frac{1}{4}x = 12$$

$$4 \cdot \frac{1}{4}x = 4 \cdot 12$$

$$x = 48$$

To check, replace x with 48 in the original proportion.

$$\frac{\frac{1}{4}}{3} \stackrel{?}{=} \frac{4}{48}$$

$$3 \cdot 4 \stackrel{?}{=} \frac{1}{4} \cdot 48$$

$$12 = 12$$

(b) Solve the proportion for a.

NOTE Here we must divide 6 by 0.5 to find the unknown value. The steps of that division are shown here for review.

$$\begin{array}{r} 12. \\ 0.5\overline{)6.0} \\ 5 \\ \hline 1\,0 \\ 1\,0 \\ \hline 0 \end{array}$$

$$\frac{0.5}{2} = \frac{3}{a}$$

$$0.5a = 6$$

$$\frac{0.5a}{0.5} = \frac{6}{0.5} \qquad \text{Divide by the coefficient, 0.5.}$$

$$a = 12$$

We will leave it to you to confirm that $0.5 \cdot 12 = 2 \cdot 3$.

CHECK YOURSELF 7

(a) Solve for a.

$$\frac{\frac{1}{2}}{5} = \frac{3}{a}$$

(b) Solve for x.

$$\frac{0.4}{x} = \frac{2}{30}$$

CHECK YOURSELF ANSWERS

1. $\frac{4}{12} = \frac{6}{18}$ **2.** $\frac{5 \text{ rolls}}{400 \text{ ft}^2} = \frac{7 \text{ rolls}}{560 \text{ ft}^2}$ **3. (a)** Yes; **(b)** no **4. (a)** Yes; **(b)** no

5. No

6. (a) $5n = 100$; $\frac{5n}{5} = \frac{100}{5}$; $n = 20$. To check: $\frac{4}{5} \stackrel{?}{=} \frac{20}{25}$; $5 \cdot 20 \stackrel{?}{=} 4 \cdot 25$; $100 = 100$

(b) $7n = 42 \cdot 9$; $\frac{7n}{7} = \frac{42 \cdot 9}{7}$; $n = 6 \cdot 9$; $n = 54$. To check: $\frac{7}{9} \stackrel{?}{=} \frac{42}{54}$; $7 \cdot 54 \stackrel{?}{=} 9 \cdot 42$; $378 = 378$

7. (a) 30; **(b)** 6

Name ____________

Section ______ Date ______

6.3 Exercises

Write each as a proportion.

1. 4 is to 9 as 8 is to 18.

2. 6 is to 11 as 18 is to 33.

3. 2 is to 9 as 8 is to 36.

4. 10 is to 15 as 20 is to 30.

5. 3 is to 5 as 15 is to 25.

6. 8 is to 11 as 16 is to 22.

7. 9 is to 13 as 27 is to 39.

8. 15 is to 21 as 60 is to 84.

In exercises 9 to 14, write the proportion that is equivalent to the given statement.

9. If 15 lb of string beans cost \$4, then 45 lb will cost \$12.

10. If Maria hit 8 home runs in 15 softball games, then she should hit 24 home runs in 45 games.

11. If 3 credit hours at Bucks County Community College cost \$216, then 12 credits will cost \$864.

12. If 16 lb of fertilizer cover 1,520 ft^2, then 21 lb should cover 1,995 ft^2.

13. If Audrey travels 180 mi on interstate I-95 in 3 h, then he should travel 300 mi in 5 h.

14. If 2 vans can transport 18 people, then 5 vans can transport 45 people.

ANSWERS

1. ____________
2. ____________
3. ____________
4. ____________
5. ____________
6. ____________
7. ____________
8. ____________
9. ____________
10. ____________
11. ____________
12. ____________
13. ____________
14. ____________

ANSWERS

15. ____________

16. ____________

17. ____________

18. ____________

19. ____________

20. ____________

21. ____________

22. ____________

23. ____________

24. ____________

25. ____________

26. ____________

27. ____________

28. ____________

29. ____________

30. ____________

31. ____________

32. ____________

33. ____________

34. ____________

35. ____________

36. ____________

37. ____________

38. ____________

39. ____________

40. ____________

41. ____________

Determine whether each pair of fractions is proportional.

15. $\frac{3}{4} \stackrel{?}{=} \frac{9}{12}$

16. $\frac{6}{7} \stackrel{?}{=} \frac{18}{21}$

17. $\frac{3}{4} \stackrel{?}{=} \frac{15}{20}$

18. $\frac{3}{5} \stackrel{?}{=} \frac{6}{10}$

19. $\frac{11}{15} \stackrel{?}{=} \frac{9}{13}$

20. $\frac{9}{10} \stackrel{?}{=} \frac{2}{7}$

21. $\frac{8}{3} \stackrel{?}{=} \frac{24}{9}$

22. $\frac{5}{8} \stackrel{?}{=} \frac{15}{24}$

23. $\frac{6}{17} \stackrel{?}{=} \frac{9}{11}$

24. $\frac{5}{12} \stackrel{?}{=} \frac{8}{20}$

25. $\frac{7}{16} \stackrel{?}{=} \frac{21}{48}$

26. $\frac{2}{5} \stackrel{?}{=} \frac{7}{9}$

27. $\frac{10}{3} \stackrel{?}{=} \frac{150}{50}$

28. $\frac{5}{8} \stackrel{?}{=} \frac{75}{120}$

29. $\frac{3}{7} \stackrel{?}{=} \frac{18}{42}$

30. $\frac{12}{7} \stackrel{?}{=} \frac{96}{50}$

31. $\frac{7}{15} \stackrel{?}{=} \frac{84}{180}$

32. $\frac{76}{24} \stackrel{?}{=} \frac{19}{6}$

33. $\frac{60}{36} \stackrel{?}{=} \frac{25}{15}$

34. $\frac{\frac{1}{2}}{4} \stackrel{?}{=} \frac{5}{40}$

35. $\frac{3}{\frac{1}{5}} \stackrel{?}{=} \frac{30}{6}$

36. $\frac{\frac{2}{3}}{6} \stackrel{?}{=} \frac{1}{12}$

37. $\frac{\frac{3}{4}}{12} \stackrel{?}{=} \frac{1}{16}$

38. $\frac{0.3}{4} \stackrel{?}{=} \frac{1}{20}$

39. $\frac{3}{60} \stackrel{?}{=} \frac{0.3}{6}$

40. $\frac{0.6}{0.12} \stackrel{?}{=} \frac{2}{0.4}$

41. $\frac{0.6}{15} \stackrel{?}{=} \frac{2}{75}$

ANSWERS

42. ________
43. ________
44. ________
45. ________
46. ________
47. ________
48. ________
49. ________
50. ________
51. ________
52. ________
53. ________
54. ________
55. ________
56. ________
57. ________
58. ________
59. ________
60. ________
61. ________
62. ________

In exercises 42 to 50, determine if the given rates are equivalent.

42. $\dfrac{\text{7 cups of flour}}{\text{4 loaves of bread}} \stackrel{?}{=} \dfrac{\text{4 cups of flour}}{\text{3 loaves of bread}}$

43. $\dfrac{\text{6 American dollars}}{\text{50 krone}} \stackrel{?}{=} \dfrac{\text{15 American dollars}}{\text{125 krone}}$

44. $\dfrac{\text{22 mi}}{\text{15 gal}} \stackrel{?}{=} \dfrac{\text{55 mi}}{\text{35 gal}}$

45. $\dfrac{\text{46 pages}}{\text{30 min}} \stackrel{?}{=} \dfrac{\text{18 pages}}{\text{8 min}}$

46. $\dfrac{\text{9 in.}}{\text{57 mi}} \stackrel{?}{=} \dfrac{\text{6 in.}}{\text{38 mi}}$

47. $\dfrac{\text{12 yen}}{\text{5 pesos}} \stackrel{?}{=} \dfrac{\text{108 yen}}{\text{45 pesos}}$

48. $\dfrac{\text{12 gal of paint}}{\text{8329 ft}^2} \stackrel{?}{=} \dfrac{\text{9 gal of paint}}{\text{1240 ft}^2}$

49. $\dfrac{\text{12 in. of snow}}{\text{1.4 in. of rain}} \stackrel{?}{=} \dfrac{\text{36 in. of snow}}{\text{7 in. of rain}}$

50. $\dfrac{\text{9 people}}{\text{2 cars}} \stackrel{?}{=} \dfrac{\text{11 people}}{\text{3 cars}}$

Solve for the unknown in each of the proportions.

51. $\dfrac{x}{3} = \dfrac{6}{9}$

52. $\dfrac{x}{6} = \dfrac{3}{9}$

53. $\dfrac{10}{n} = \dfrac{15}{6}$

54. $\dfrac{x}{8} = \dfrac{15}{24}$

55. $\dfrac{a}{42} = \dfrac{5}{7}$

56. $\dfrac{7}{12} = \dfrac{m}{24}$

57. $\dfrac{18}{12} = \dfrac{12}{p}$

58. $\dfrac{x}{32} = \dfrac{7}{8}$

59. $\dfrac{x}{18} = \dfrac{64}{72}$

60. $\dfrac{20}{15} = \dfrac{100}{a}$

61. $\dfrac{6}{n} = \dfrac{75}{100}$

62. $\dfrac{36}{x} = \dfrac{8}{6}$

ANSWERS

63. ______
64. ______
65. ______
66. ______
67. ______
68. ______
69. ______
70. ______
71. ______
72. ______
73. ______
74. ______
75. ______
76. ______
77. ______
78. ______
79. ______
80. ______

63. $\frac{5}{35} = \frac{a}{28}$

64. $\frac{20}{24} = \frac{p}{18}$

65. $\frac{12}{100} = \frac{3}{x}$

66. $\frac{b}{7} = \frac{21}{49}$

67. $\frac{p}{24} = \frac{25}{120}$

68. $\frac{5}{x} = \frac{20}{88}$

69. $\frac{\frac{1}{2}}{2} = \frac{3}{a}$

70. $\frac{x}{5} = \frac{2}{\frac{1}{3}}$

71. $\frac{\frac{1}{4}}{12} = \frac{m}{96}$

72. $\frac{12}{\frac{1}{3}} = \frac{108}{y}$

73. $\frac{\frac{2}{5}}{8} = \frac{1.2}{n}$

74. $\frac{4}{a} = \frac{\frac{2}{5}}{10}$

75. $\frac{0.2}{2} = \frac{1.2}{a}$

76. $\frac{n}{3} = \frac{6}{0.5}$

77. $\frac{p}{7} = \frac{8}{0.7}$

78. $\frac{y}{12} = \frac{5}{0.6}$

79. $\frac{x}{3.3} = \frac{1.1}{6.6}$

80. $\frac{0.5}{a} = \frac{1.25}{5}$

Answers

1. $\frac{4}{9} = \frac{8}{18}$ **3.** $\frac{2}{9} = \frac{8}{36}$ **5.** $\frac{3}{5} = \frac{15}{25}$ **7.** $\frac{9}{13} = \frac{27}{39}$ **9.** $\frac{15 \text{ lb}}{\$4} = \frac{45 \text{ lb}}{\$12}$ **11.** $\frac{3 \text{ credits}}{\$216} = \frac{12 \text{ credits}}{\$864}$ **13.** $\frac{180 \text{ mi}}{3 \text{ h}} = \frac{300 \text{ mi}}{5 \text{ h}}$ **15.** Yes **17.** Yes **19.** No **21.** Yes **23.** No **25.** Yes **27.** No **29.** Yes **31.** Yes **33.** Yes **35.** No **37.** Yes **39.** Yes **41.** No **43.** Yes **45.** No **47.** Yes **49.** No **51.** $9x = 18; x = 2$ **53.** 4 **55.** 30 **57.** $18p = 144; p = 8$ **59.** 16 **61.** 8 **63.** $35a = 140; a = 4$ **65.** 25 **67.** 5 **69.** 12 **71.** 2 **73.** 24 **75.** 12 **77.** 80 **79.** 0.55

6.4 Similar Triangles and Proportions

6.4 OBJECTIVES

1. Use proportions to find a missing measurement from similar triangles
2. Solve an application involving similar triangles

Have you ever wondered how tall a certain tree was? In this section, you will learn how to find the height of a tree without climbing it! First we must introduce the idea of similar right triangles. Recall that a triangle in which two of the sides are perpendicular is called a right triangle.

NOTE The square always indicates a right triangle.

Right triangles

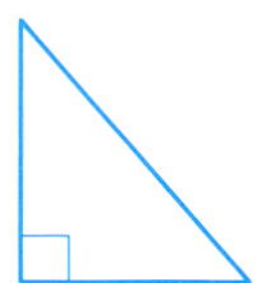

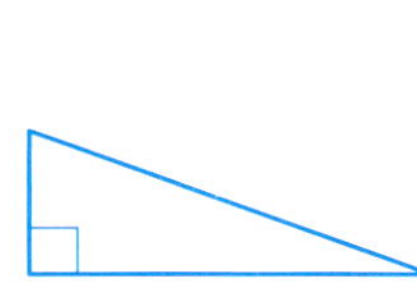

Other triangles

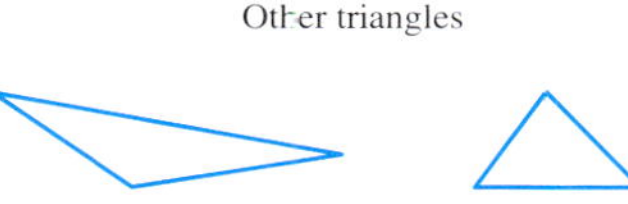

Definitions: Similar Right Triangles

Two right triangles are similar if the ratios of corresponding sides are equivalent.

Example 1

Confirming Similar Triangles

Show that the two right triangles are similar.

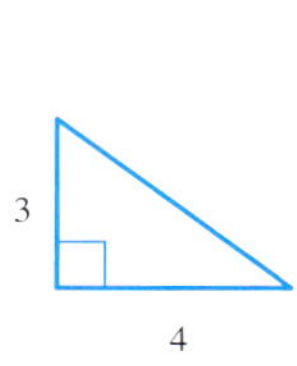

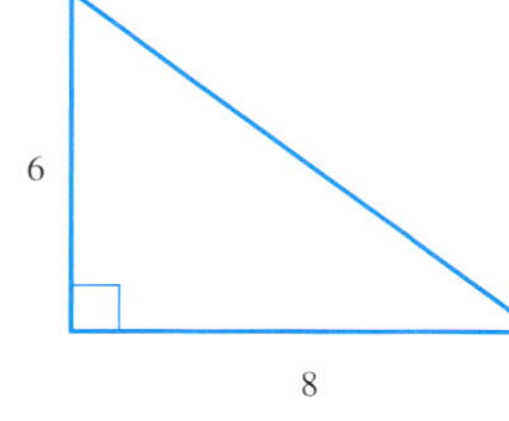

The 3 and the 6 are corresponding sides, as are the 4 and the 8.

We need to use two different pairs of sides to show that two triangles are similar. In this case,

$$\frac{3}{6} \stackrel{?}{=} \frac{4}{8}$$

is a true proportion, so the triangles are similar.

CHECK YOURSELF 1

Show that the two triangles are similar.

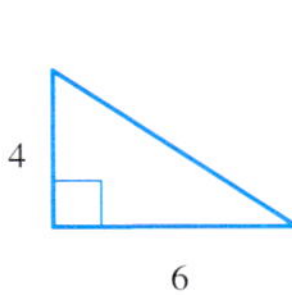

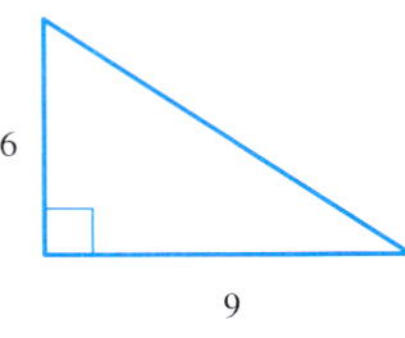

If we know that two triangles (even if they do not happen to be right triangles) are similar, we can find the length of a missing side.

Example 2

Using Similar Triangles to Find a Missing Length

Given the two similar triangles, find the length of the side marked x.

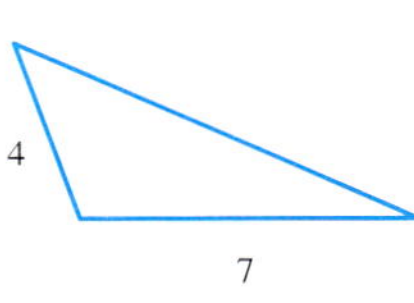

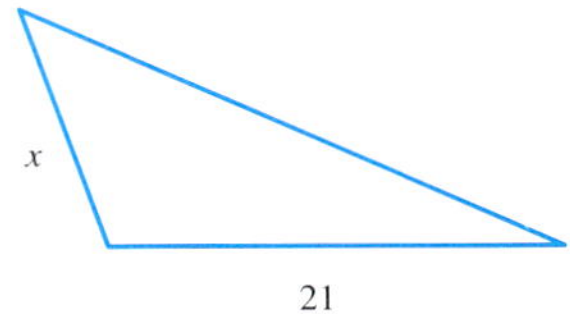

The triangles are similar, so the sides must be proportional.

$$\frac{4}{7} = \frac{x}{21}$$

$$4 \cdot 21 = 7 \cdot x$$

$$84 = 7x$$

$$12 = x$$

CHECK YOURSELF 2

Find the length of the side marked x if the triangles are similar.

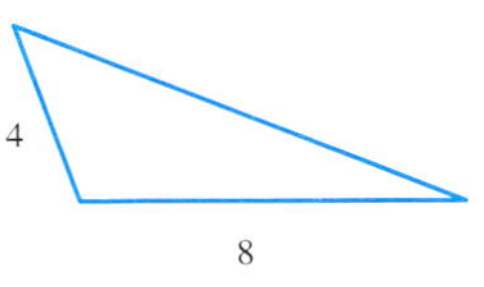

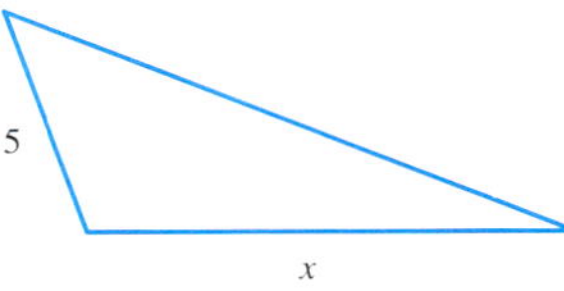

Proportions can also be used to find the length of the long side (called the **hypotenuse**) of a right triangle.

Example 3

Finding a Missing Length

Find the length of the side marked with an x, given that the triangles are similar.

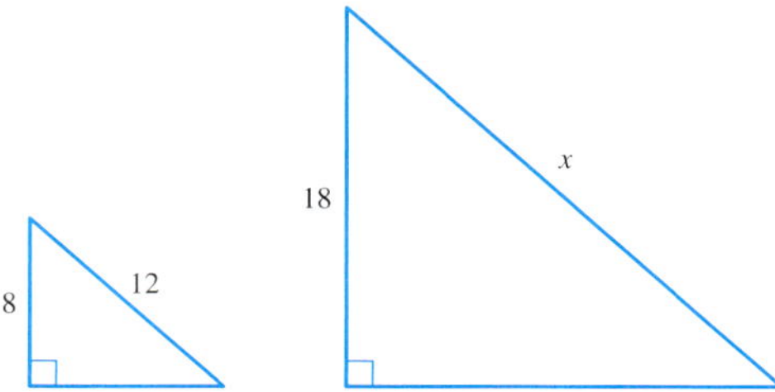

Again, we use proportions

$$\frac{8}{12} = \frac{18}{x}$$

$$8 \cdot x = 12 \cdot 18$$

$$8x = 216$$

$$x = 27$$

CHECK YOURSELF 3

Find the length of the side marked with an x, given that the triangles are similar.

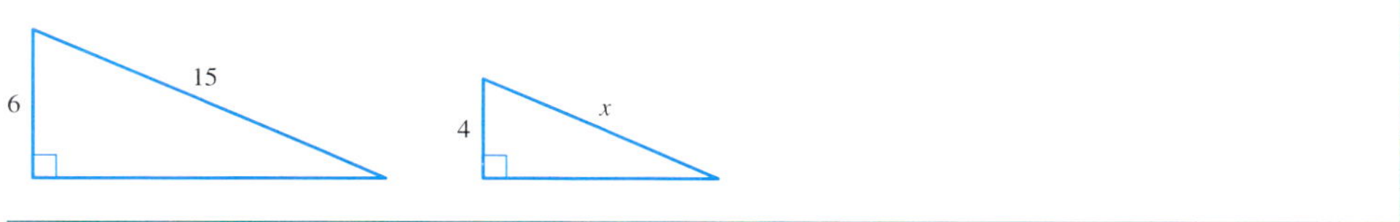

A common application of similar triangles occurs when one is looking for the height of a tall object. Example 4 illustrates this application of similar triangles.

Example 4

Solving an Application Using Similar Triangles

If a 6-ft-tall man casts a shadow that is 10 ft long, how tall is a tree that casts a shadow that is 70 ft long?

We can look at a picture of the two triangles involved.

NOTE Connect the top of the tree to the end of the shadow to create a triangle. Connecting the top of the man to the end of his shadow creates a similar triangle.

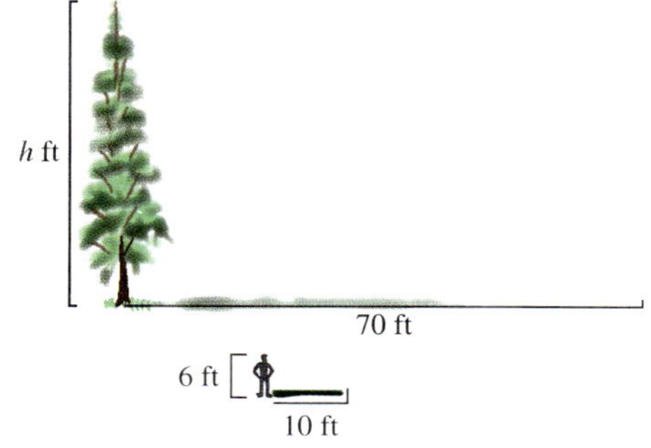

From the similar triangles, we have the proportion

$$\frac{6}{10} = \frac{h}{70}$$

Using the proportion rule, we have $6 \cdot 70 = 10 \cdot h$

$$10 \cdot h = 420$$

$$\frac{10 \cdot h}{10} = \frac{420}{10}$$

$$h = 42$$

The tree is 42 ft tall.

CHECK YOURSELF 4

If a woman who is $5\frac{1}{2}$ ft tall casts a shadow that is 3 ft long, how tall is a building that casts a shadow that is 90 ft long?

CHECK YOURSELF ANSWERS

1. $\frac{4}{6} = \frac{6}{9}$ **2.** 10 **3.** 10 **4.** 165 ft tall

Name ______________________

Section ________ Date ________

6.4 Exercises

Use a proportion to find the unknown side, labeled x, in each of the pairs of similar figures.

1.

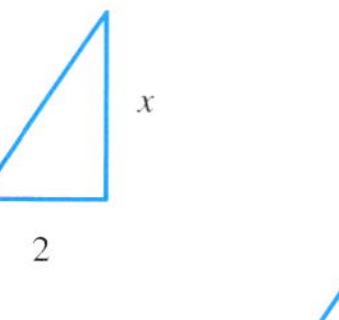

2.

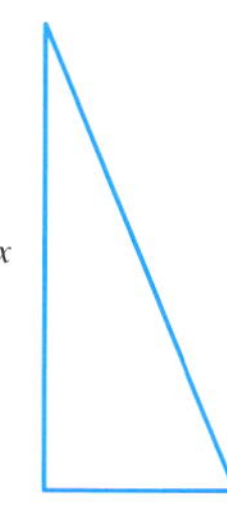

3.

4.

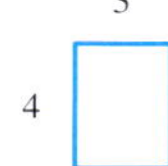

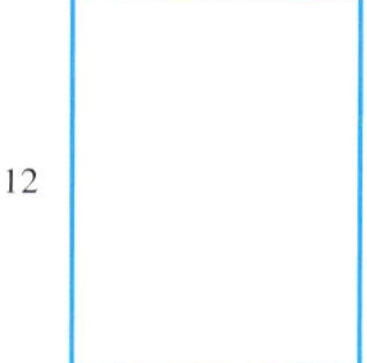

In exercises 5 to 10, the two triangles shown are similar. Find the indicated side.

5. Find v.

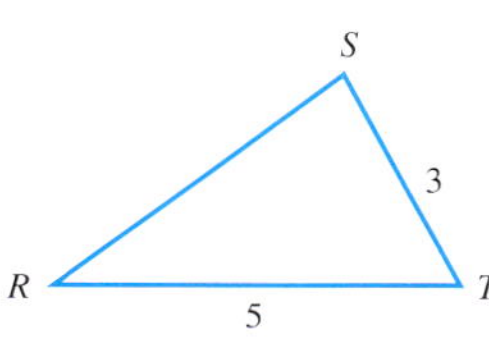

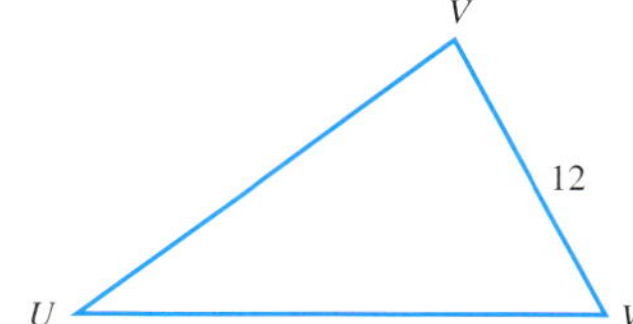

6. Find f.

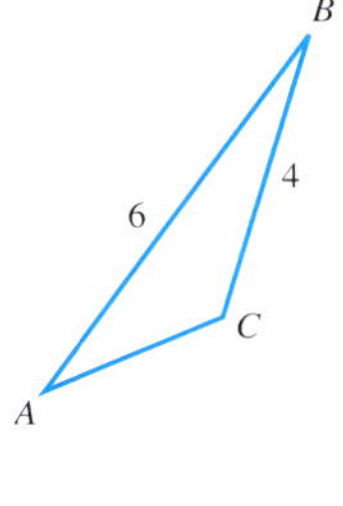

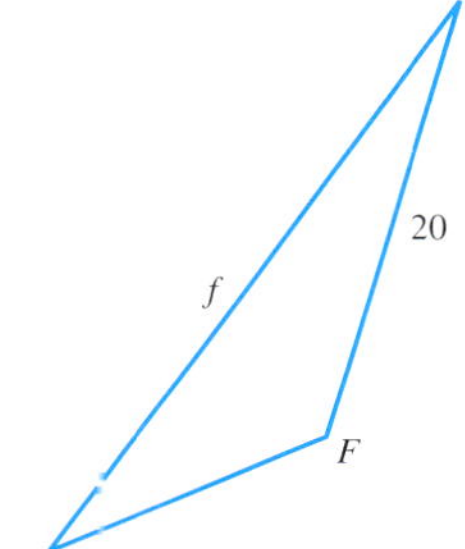

ANSWERS

1. ______________________
2. ______________________
3. ______________________
4. ______________________
5. ______________________
6. ______________________

ANSWERS

7. ______

8. ______

9. ______

10. ______

7. Find g.

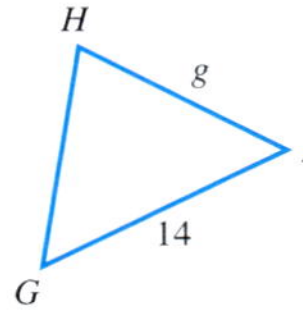

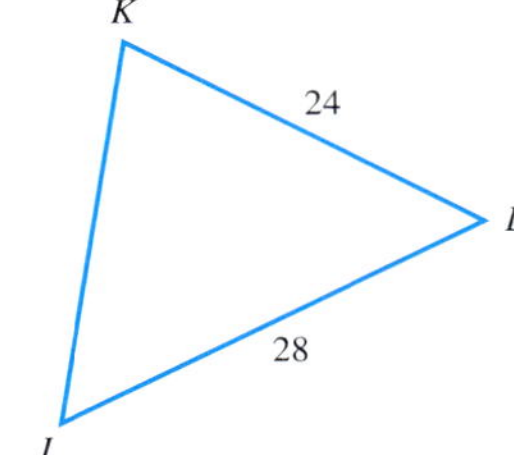

8. Find m.

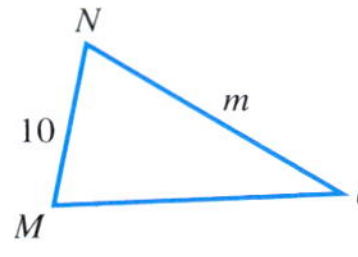

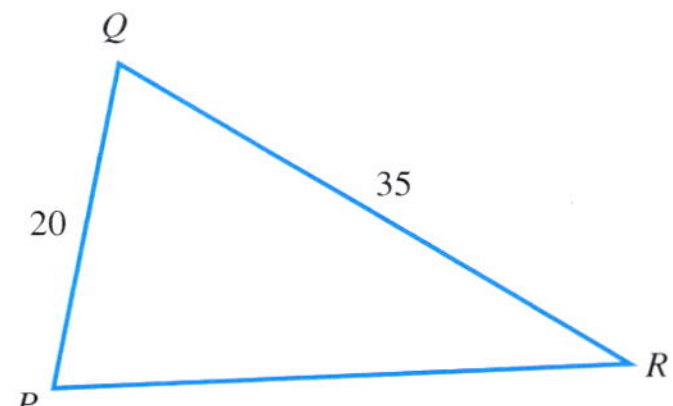

9. Find t.

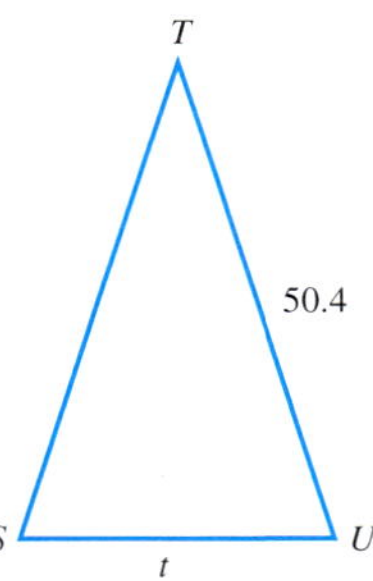

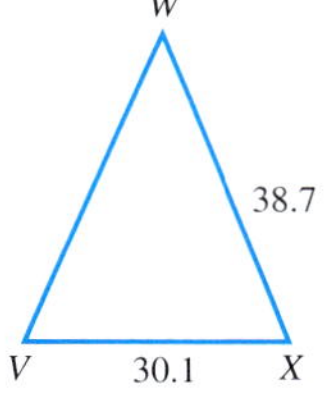

10. Find e.

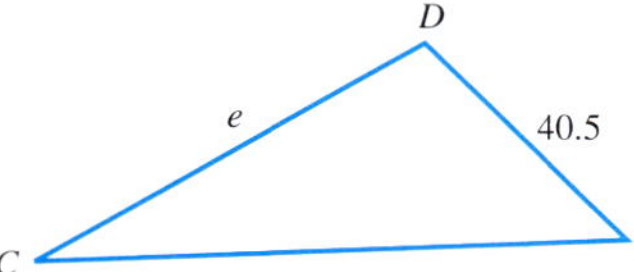

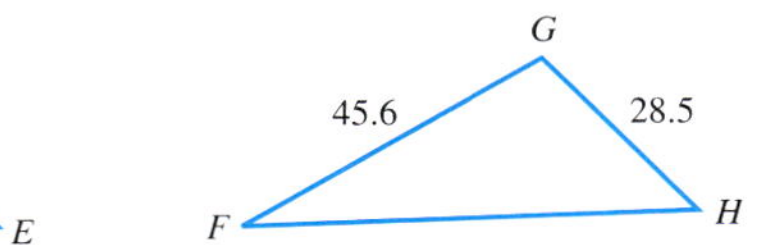

ANSWERS

11. ______

12. ______

13. ______

14. ______

Solve the applications.

11. Lighting. A 9-ft light pole casts a 15-ft shadow. Find the height of a nearby tree that is casting a 40-ft shadow.

12. Fencing. A 6-ft fence post casts a 9-ft shadow. How tall is a nearby pole that casts a 15-ft shadow?

13. A light pole casts a shadow that measures 4 ft. At the same time, a yardstick casts a shadow that is 9 in. long. How tall is the pole?

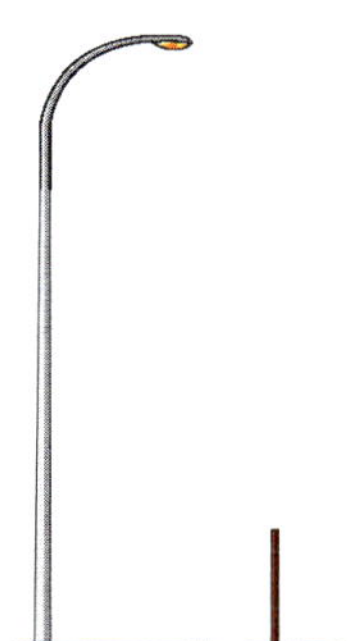

14. A tree casts a shadow that measures 5 m. At the same time, a meterstick casts a shadow that is 0.4 m long. How tall is the tree?

ANSWERS

15. ________________

15. Use the ideas of similar triangles to determine the height of a pole or tree on your campus. Work with one or two partners.

Answers

1. 3 **3.** 6 **5.** 20 **7.** 12 **9.** 39.2 **11.** 24 ft **13.** 16 ft
15.

6.5 More Applications of Proportion

OBJECTIVES

1. Solve an application that involves a proportion
2. Solve an application involving unit pricing

Now that you have learned how to find an unknown value in a proportion, and to set up and solve proportions involving similar triangles, this can be used in the solution of other types of applications.

Step by Step: Solving Applications of Proportions

Step 1 Read the problem carefully to determine the given information.

Step 2 Write the proportion necessary to solve the problem. Use a letter to represent the unknown quantity. Be sure to include the units in writing the proportion.

Step 3 Solve, answer the question of the original problem, and check the proportion as before.

Example 1

Using a Proportion to Find an Unknown Value

In a shipment of 400 parts, 14 are found to be defective. How many defective parts should be expected in a shipment of 1,000?

Assume that the ratio of defective parts to the total number remains the same.

$$\frac{14 \text{ defective}}{400 \text{ total}} = \frac{x \text{ defective}}{1000 \text{ total}}$$

We have decided to let x be the unknown number of defective parts.

Multiply:

$400x = 14{,}000$

Divide by the coefficient, 400.

$x = 35$

35 defective parts should be expected in the shipment.

Checking the original proportion, we get

$14 \cdot 1{,}000 \stackrel{?}{=} 400 \cdot 35$

$14{,}000 = 14{,}000$

CHECK YOURSELF 1

An investment of \$3,000 *earned* \$330 *for* 1 *year. How much will an investment of* \$10,000 *earn at the same rate for* 1 *year?*

Now we will look at an application involving fractions in the proportion

Example 2

Using Proportions to Find an Unknown Value

The scale on a map is given as $\frac{1}{4}$ in. = 3 mi. The distance between two towns is 4 in. on the map. How far apart are the towns in miles?

NOTE We could divide both sides by $\frac{1}{4}$:

$$\frac{\frac{1}{4} \cdot x}{\frac{1}{4}} = \frac{3 \cdot 4}{\frac{1}{4}}$$

$$x = \frac{3 \cdot 4}{\frac{1}{4}}$$

$$x = \frac{12}{\frac{1}{4}}$$

and then invert and multiply

$$x = \frac{12}{1} \cdot \frac{4}{1}$$

$$= 48$$

For this solution we use the fact that the ratio of inches (on the map) to miles remains the same.

$$\frac{\frac{1}{4} \text{ in.}}{3 \text{ mi}} = \frac{4 \text{ in.}}{x \text{ mi}}$$

$$\frac{1}{4} \cdot x = 3 \cdot 4 = 12$$

$$4\left(\frac{1}{4}\right) \cdot x = 4 \cdot 12$$

$$1 \cdot x = 4 \cdot 12$$

$$x = 48 \text{ (mi)}$$

CHECK YOURSELF 2

Jack drives 125 mi in $2\frac{1}{2}$ h. At the same rate, how far will he be able to travel in 4 h? (Hint: Write $2\frac{1}{2}$ as an improper fraction.)

We may also find decimals in the solution of an application.

Example 3

Using Proportions to Find an Unknown Value

Jill works 4.2 h and receives \$21. How much will she get if she works 10 h?

The ratio of hours worked to the amount of pay remains the same.

$$\frac{4.2 \text{ h}}{\$21} = \frac{10 \text{ h}}{\$a}$$

Let a be the unknown amount of pay.

$$4.2a = 210$$

$$\frac{4.2a}{4.2} = \frac{210}{4.2}$$ Divide both sides by 4.2.

$$a = \$50$$

CHECK YOURSELF 3

A piece of cable 8.5 *cm long weighs* 68 *g. What will a* 10-*cm length of the same cable weigh?*

In Example 4, we must convert the units stated in the problem.

Example 4

Using Proportions to Find an Unknown Value

A machine produces 15 cans in 2 min. At this rate how many cans can it make in an 8-h period?

In writing a proportion for this problem, we must write the times involved in terms of the same units.

$$\frac{15 \text{ cans}}{2 \text{ min}} = \frac{x \text{ cans}}{480 \text{ min}}$$ Because 1 h is 60 min, convert 8 h to 480 min.

$$2x = 15 \cdot 480$$

or $$2x = 7{,}200$$

$$x = 3{,}600 \text{ cans}$$

CHECK YOURSELF 4

Instructions on a can of film developer call for 2 *oz of concentrate to* 1 *qt of water. How much of the concentrate is needed to mix with* 1 *gal of water? (*4 *qt* = 1 *gal)*

In practical applications, you may have to round the result after using your calculator in the solution of a proportion. Example 5 shows such a situation.

Example 5

Using Proportions to Find an Unknown Value

Micki drives 278 mi on 13.6 gal of gas. If the gas tank of her car holds 21 gal, how far can she travel on a full tank of gas?

We can write the proportion

$$\frac{278 \text{ mi}}{13.6 \text{ gal}} = \frac{x \text{ mi}}{21 \text{ gal}}$$

Multiply.

$$13.6\,x = 278 \cdot 21$$

Now divide both terms by 13.6.

$$\frac{\overset{1}{\cancel{13.6}}x}{\underset{1}{\cancel{13.6}}} = \frac{278 \cdot 21}{13.6}$$

Now to find x, we must multiply 278 by 21 and then divide by 13.6. On the calculator,

278 [×] 21 [÷] 13.6 [=] 429.26471

We will round the result to the nearest mile; Micki can drive about 429 mi on a full tank of gas.

CHECK YOURSELF 5

Life insurance costs $4.37 for each $1,000 of insurance. How much does a $25,000 policy cost?

Your calculator can be very handy for comparing prices at the grocery store. As we said in Section 6.2, to find the unit price, just divide the cost of the item by the number of units.

Example 6

Finding Unit Prices

A dishwashing liquid comes in three sizes:

(a) 12 oz for 77¢
(b) 22 oz for $1.33
(c) 32 oz for $1.85

Which is the best buy?

For each size, let's find the unit price in cents per ounce.

(a) For the first size (77¢ for 12 oz), using your calculator, divide.

77 [÷] 12 [=] 6.4166667

$$\frac{77¢}{12 \text{ oz}} \approx 6.4¢ \text{ per ounce}$$

We have chosen to round to the nearest tenth of a cent.

(b) To find the unit price for the second size, divide again.

NOTE Notice that we consider $1.33 as 133¢ to find the rate "cents per ounce."

133 [÷] 22 [=] 6.0454545

$$\frac{\$1.33}{22 \text{ oz}} \approx 6.0¢ \text{ per ounce}$$

Again round to the nearest tenth of a cent.

(c) For the third size,

185 [÷] 32 [=] 5.78125

$$\frac{\$1.85}{32 \text{ oz}} \approx 5.8¢ \text{ per ounce}$$

Comparing the three unit prices, we see that the 32-oz size of dishwashing liquid, at 5.8¢ per ounce, is the best buy.

CHECK YOURSELF 6

A floor cleaner comes in three sizes:

32 oz for \$2.89
48 oz for \$3.43
70 oz for \$4.96

Which is the best buy?

All rates used must be in terms of the same units. If quantities involve different units, they must be converted.

Example 7

Finding Unit Prices

Vegetable oil is sold in the following quantities:

(a) 16 oz for \$1.27
(b) 1 pt 8 oz for \$1.79
(c) 1 qt 6 oz for \$2.89

Which is the best buy?

(a) $\frac{\$1.27}{16 \text{ oz}} \approx 7.9¢$ per ounce

(b) Because 1 pt is 16 oz, 1 pt 8 oz is (16 + 8) oz, or 24 oz. So we write 1 pt 8 oz as 24 oz.

$\frac{\$1.79}{24 \text{ oz}} \approx 7.5¢$ per ounce

(c) Because 1 qt is 32 oz, 1 qt 6 oz is (32 + 6) oz, or 38 oz. Write 1 qt 6 oz as 38 oz.

$\frac{\$2.89}{38 \text{ oz}} \approx 7.6¢$ per ounce

In this case, by comparing the unit prices we see that the 1-pt 8-oz size is the best buy.

CHECK YOURSELF 7

Ketchup is sold in the following quantities:

12 oz for \$0.68
1 pt 5 oz for \$1.05
1 qt 7 oz for \$1.89

Which is the best buy?

CHECK YOURSELF ANSWERS

1. \$1,100 **2.** $\dfrac{125 \text{ mi}}{\frac{5}{2} \text{ h}} = \dfrac{x \text{ mi}}{4 \text{ h}}$

$$\frac{5}{2}x = 500$$ Divide both sides by $\frac{5}{2}$.

$$x = 200 \text{ mi}$$

3. 80 g **4.** 8 oz **5.** \$109.25 **6.** 70-oz size at 7.1¢ per oz
7. 1 qt 7 oz for 4.8¢ per oz

6.5 Exercises

Name ____________

Section ________ Date ________

Solve the applications.

1. **Book purchases.** If 12 books are purchased for $40, how much will you pay for 18 books at the same rate?

2. **Construction.** If an 8-ft two-by-four costs 96¢, what should a 12-ft two-by-four cost?

3. **Consumer affairs.** A box of 18 tea bags is marked 90¢. At that price, what should a box of 48 tea bags cost?

4. **Consumer affairs.** Cans of orange juice are marked 2 for 93¢. What would the price of a case of 24 cans be?

5. **Workload.** A worker can complete the assembly of 15 tape players in 6 h. At this rate, how many can the worker complete in a 40-h workweek?

6. **Consumer affairs.** If 3 lb of apples cost 90¢, what will 10 lb cost?

7. **Elections.** The ratio of yes to no votes in an election was 3 to 2. How many no votes were cast if there were 2,880 yes votes?

8. **College enrollment.** The ratio of men to women at a college is 7 to 5. How many women students are there if there are 3,500 men?

9. **Photography.** A photograph 5 in. wide by 6 in. high is to be enlarged so that the new width is 15 in. What will be the height of the enlargement?

10. **Shift work.** Meg's job is assembling lawn chairs. She can put together 55 chairs in 4 h. At this rate, how many chairs can she assemble in an 8-h shift?

11. **Distance.** Christy can travel 110 mi in her new car on 5 gal of gas. How far can she travel on a full tank, which has 12 usable gallons?

ANSWERS

1. ____________
2. ____________
3. ____________
4. ____________
5. ____________
6. ____________
7. ____________
8. ____________
9. ____________
10. ____________
11. ____________

ANSWERS

12. ______________________

13. ______________________

14. ______________________

15. ______________________

16. ______________________

17. ______________________

12. Property taxes. The Changs purchased an $80,000 home, and the property taxes were $1,400. If they make improvements and the house is now valued at $120,000, what will the new property tax be?

13. Distance. A car travels 165 mi in 3 h. How far will it travel in 8 h if it continues at the same speed?

14. Consumer affairs. A battery pack is on sale at 2 for $3. At this rate, how much will 7 packs cost?

15. Manufacturing. The ratio of teeth on a smaller gear to those on a larger gear is 3 to 7. If the smaller gear has 15 teeth, how many teeth does the larger gear have?

16. Consumer affairs. A store has T-shirts on sale at 2 for $5.50. At this rate, what will five shirts cost?

Using the given map, find the distances between the cities named in exercises 17 to 20.

17. Find the distance from Harrisburg to Philadelphia.

ANSWERS

18. ______

19. ______

20. ______

21. ______

22. ______

23. ______

24. ______

25. ______

26. ______

27. ______

18. Find the distance from Punxsutawney (home of the groundhog) to State College (home of the Nittany Lions).

19. Find the distance from Gettysburg to Meadville.

20. Find the distance from Scranton to Waynesburg.

21. Manufacturing. An inspection reveals 30 defective parts in a shipment of 500. How many defective parts should be expected in a shipment of 1,200?

22. Investments. You invest \$4,000 in a stock that pays a \$180 dividend in 1 year. At the same rate, how much will you need to invest to earn \$270?

23. Football. A football back ran 212 yd in the first two games of the season. If he continues at the same pace, how many yards should he gain in the 11-game season?

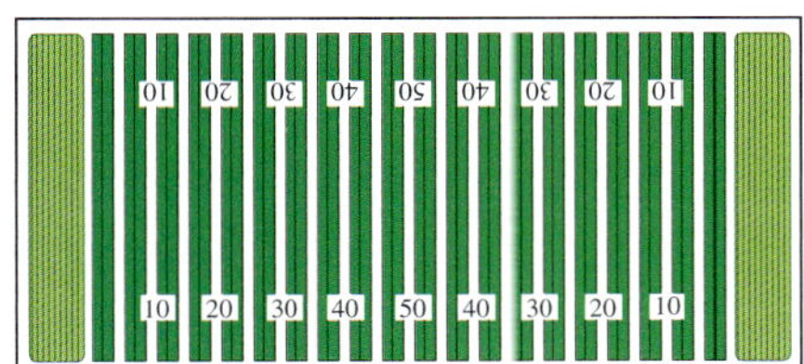

24. Cooking. A 6-lb roast will serve 14 people. What size roast is needed to serve 21 people?

25. Lawn care. A 2-lb box of grass seed is supposed to cover 2,500 ft^2 of lawn. How much seed will you need for 8,750 ft^2 of lawn?

26. Construction. On the blueprint of the Wilsons' new home, the scale is 5 in. equals 7 ft. What will be the actual length of a bedroom if it measures 10 in. long on the blueprint?

27. Distance. The scale on a map is $\frac{1}{2}$ in. = 50 mi. If the distance between two towns on the map is 6 in., how far apart are they in miles?

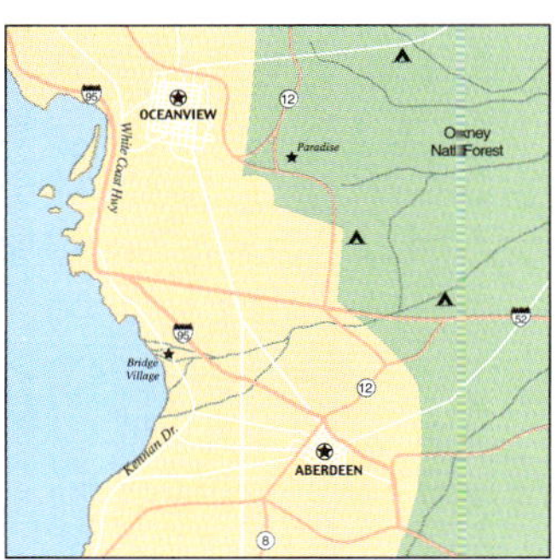

SCALE: 1/2 in. = 50 mi. 0 50 100 200

28. ______

29. ______

30. ______

31. ______

32. ______

33. ______

34. ______

35. ______

36. ______

37. ______

28. Science. A metal bar expands $\frac{1}{4}$ in. for each 12°F rise in temperature. How much will it expand if the temperature rises 48°F?

29. Car maintenance. Your car burns $2\frac{1}{2}$ qt of oil on a trip of 5,000 mi. How many quarts should you expect to use when driving 7,200 mi?

30. Lighting. A 6-ft person casts a $7\frac{1}{2}$-ft shadow. If the shadow of a nearby pole is 30 ft long, how tall is the pole?

31. Manufacturing. A piece of tubing 10.5 cm long weighs 35 g. What is the weight of a piece of the same tubing that is 15 cm long?

32. Salary. Jane works 7.75 h and receives $38.75 pay. What will she receive at the same rate if she works 12 h?

33. Sales tax. The sales tax on an item costing $80 is $5.20. What will the tax be for an item costing $150?

34. Conversion. If 8 km is approximately 4.8 mi, how many kilometers will equal 12 mi?

35. Timing. You find that your watch gains 2 min in 6 h. How much will it gain in 3 days?

36. Painting. If 2 qt of paint will cover 225 ft^2, how many square feet will 2 gal cover? (1 gal = 4 qt)

37. Construction. Directions on a box of 4 cups of wallpaper paste are to mix the contents with 5 qt of water. To mix a smaller batch using 1 cup of paste, how much water (in ounces) should be added? (1 qt = 32 oz)

ANSWERS

38. ______

39. ______

40. ______

41. ______

42. ______

43. ______

44. ______

45. ______

46. ______

38. Film processing. A film processing machine can develop three rolls of film every 8 min. At this rate, how many rolls can be developed in a 4-h period?

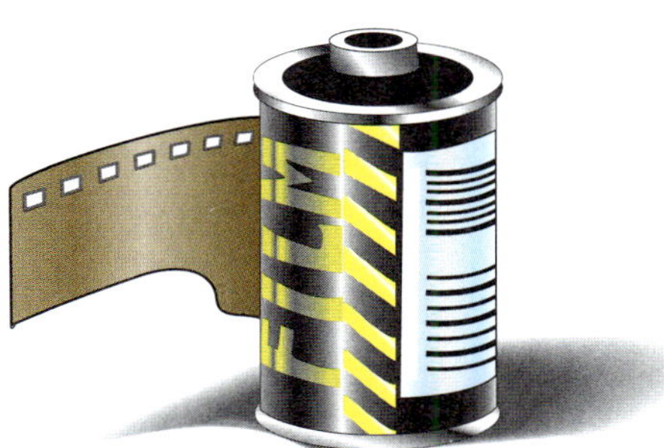

39. Carpooling. Approximately 7 out of every 10 people in the U.S. workforce drive to work alone. During morning rush hour there are 115,000 cars on the streets of a medium-sized city. How many of these cars have one person in them?

40. Carpooling. Approximately 15 out of every 100 people in the U.S. workforce carpool to work. There are an estimated 320,000 people in the workforce of a given city. How many of these people are in car pools?

41. A recipe for 12 servings lists the following ingredients:

12 cups ziti	7 cups spaghetti sauce	4 cups ricotta cheese
$\frac{1}{2}$ cup parsley	1 teaspoon garlic powder	$\frac{1}{2}$ teaspoon pepper
4 cups mozzarella cheese	2 tablespoons parmesan cheese	

Determine the amount of ingredients necessary to serve 5 people.

Calculator Exercises

Solve for the unknown.

42. $\frac{770}{1988} = \frac{n}{71}$

43. $\frac{x}{4.7} = \frac{11.8}{16.9}$ (to nearest tenth)

44. $\frac{13.9}{8.4} = \frac{n}{9.2}$ (to nearest hundredth)

45. $\frac{2.7}{3.8} = \frac{5.9}{n}$ (to nearest tenth)

46. $\frac{12.2}{0.042} = \frac{x}{0.08}$ (to nearest hundredth)

ANSWERS

47. ______

48. ______

49. ______

50. ______

51. ______

52. ______

53. ______

54. ______

Solve the applications.

47. Salary. Bill earns \$248.40 for working 34.5 h. How much will he receive if he works at the same pay rate for 31.75 h?

48. Construction. Construction-grade lumber costs \$384.50 per 1,000 board-feet. What will be the cost of 686 board-feet?

49. Speed of sound. A speed of 88 ft/s is equal to a speed of 60 mi/h. If the speed of sound is 750 mi/h, what is the speed of sound in feet per second?

50. Manufacturing. A shipment of 75 parts is inspected, and 6 are found to be faulty. At the same rate, how many defective parts should be found in a shipment of 139? Round your result to the nearest whole number.

51. Taxes. The property tax on a \$67,250 home is \$2,315. At the same rate, what will be the tax on a home valued at \$87,625? Round your result to the nearest dollar.

52. Production. A machine produces 158 items in 12 min. At the same rate, how many items will it produce in 8 h?

53. Gas mileage. Sally travels 510 mi, using 11.6 gal of gas. How many gallons of gas will she need for a trip of 1,800 mi? Give your answer to the nearest tenth of a gallon.

54. Tom and Jerry operate a food concession stand at a local amusement park. They sell the most food when the attendance at the park is at maximum capacity. When this happens, they sell an average of 450 pork roll sandwiches and 550 cheese steak sandwiches. The company that owns the park is going to expand; they will increase the capacity of the park from 6,000 to 9,000 people next season. Tom and Jerry plan to expand their concession stand so they can sell more sandwiches.

(a) Using the same ratio of attendance to sandwiches, how many additional sandwiches of each kind would Tom and Jerry expect to sell?

(b) The following costs are associated with the anticipated expansion:

Item	Cost per Unit
Construction	$70/square foot
Supplies	$0.65/pork roll sandwich
	$1.25/steak sandwich
Employee costs	$8/hour

Currently, pork roll sandwiches sell for $1.50, and steak sandwiches sell for $2.75. Based on these prices and the information in part **(a)**, how would you plan the expansion, and what would you charge for a sandwich?

Find the best buy in each of the exercises.

55. Dishwashing liquid:

(a) 12 oz for 79¢
(b) 22 oz for $1.29

56. Canned corn:

(a) 10 oz for 21¢
(b) 17 oz for 39¢

57. Syrup:

(a) 12 oz for 99¢
(b) 24 oz for $1.59
(c) 36 oz for $2.19

58. Shampoo:

(a) 4 oz for $1.16
(b) 7 oz for $1.52
(c) 15 oz for $3.39

59. Salad oil (1 qt is 32 oz):

(a) 18 oz for 89¢
(b) 1 qt for $1.39
(c) 1 qt 16 oz for $2.19

60. Tomato juice (1 pt is 16 oz):

(a) 8 oz for 37¢
(b) 1 pt 10 oz for $1.19
(c) 1 qt 14 oz for $1.99

61. Peanut butter (1 lb is 16 oz):

(a) 12 oz for $1.25
(b) 18 oz for $1.72
(c) 1 lb 12 oz for $2.59
(d) 2 lb 8 oz for $3.76

62. Laundry detergent:

(a) 1 lb 2 oz for $1.99
(b) 1 lb 12 oz for $2.89
(c) 2 lb 8 oz for $4.19
(d) 5 lb for $7.99

ANSWERS

55. ____
56. ____
57. ____
58. ____
59. ____
60. ____
61. ____
62. ____

Answers

1. $60 **3.** $2.40 **5.** 100 players **7.** 1,920 no votes **9.** 18 in.
11. 264 mi **13.** 440 mi **15.** 35 teeth **17.** 110 mi

19. 215 mi **21.** 72 defective parts **23.** $\frac{212 \text{ yd}}{2 \text{ games}} = \frac{x \text{ yd}}{11 \text{ games}}$; $x = 1{,}166$ yd

25. 7 lb **27.** 600 mi **29.** $\left(\text{Write } 2\frac{1}{2} \text{ as } \frac{5}{2}\right) \frac{\frac{5}{2} \text{ qt}}{5000 \text{ mi}} = \frac{x \text{ qt}}{7200 \text{ mi}}$; $5{,}000x = 18{,}000$; $x = 3.6$ qt **31.** 50 g **33.** \$9.75

35. (Write 3 days as 72 h) $\frac{2 \text{ min}}{6 \text{ h}} = \frac{x \text{ min}}{72 \text{ h}}$; $x = 24$ min

37. 40 oz **39.** 80,500 cars with one person

41.  **43.** 3.3 **45.** 8.3 **47.** \$228.60 **49.** $1{,}100 \frac{\text{ft}}{\text{s}}$

51. \$3,016 **53.** 40.9 gal **55.** b **57.** c **59.** b **61.** c

6.6 Linear Measurement and Conversion

6.6 OBJECTIVES

1. Identify English system units of length
2. Use unit fractions to convert between English system units of length
3. Simplify and perform operations involving English system units of length
4. Identify metric system prefixes and units of length
5. Convert between metric system units of length
6. Simplify and perform operations involving metric system units of length

Many arithmetic problems involve **units of measure.** When we measure an object, we give it a number and some unit. For instance, we might say a board is 6 feet long, a container holds 4 quarts, or a package weighs 5 pounds. Feet, quarts, and pounds are the units of measure.

The system you are probably most familiar with is called the **English system of measurement.** This system is used in the United States and a few other countries. The table lists the units of measurement you should be familiar with.

NOTE Don't let the name mislead you. The English system is no longer used in England. England has converted to the metric system, which we will look at next.

English Units of Measure and Equivalents

Length	Weight
1 foot (ft) = 12 inches (in.)	1 pound (lb) = 16 ounces (oz)
1 yard (yd) = 3 ft	1 ton = 2,000 lb
1 mile (mi) = 5,280 ft	

Capacity
1 pint (pt) = 16 fluid ounces (fl oz)
1 quart (qt) = 2 pt
1 gallon (gal) = 4 qt

In this section, we will focus on English units of length. From the table, you can see that these units include inches, feet, yards, and miles.

You may want to use the equivalencies shown in the table to change from one unit to another. We will look at one approach.

Rules and Properties: Converting Units in the English System

To change from one unit to another, replace the unit of measure with the appropriate equivalent measure and multiply.

Example 1

Converting within the English System

NOTE We write 5 ft as 5(1 ft) and then change 1 ft to 12 in.

$5 \text{ ft} = 5(1 \text{ ft}) = 5(12 \text{ in.}) = 60 \text{ in.}$ Replace 1 ft with 12 in.

$48 \text{ in.} = 48(1 \text{ in.}) = 48\left(\frac{1}{12} \text{ ft}\right) = 4 \text{ ft}$ Because 12 in. = 1 ft, 1 in. = $\frac{1}{12}$ ft.

$4 \text{ yd} = 4(1 \text{ yd}) = 4(3 \text{ ft}) = 12 \text{ ft}$ Replace 1 yd with 3 ft.

CHECK YOURSELF 1

Complete each of the statements.

(a) 4 ft = _______ in. **(b)** 12 yd = _______ ft

(c) 144 in. = _______ ft **(d)** 3 ft = _______ in.

Here is another idea that may help you convert units. You can use a *unit rate* to convert from one unit to another. A *unit rate* is a fraction whose value is 1.

NOTE This is a variation on the method of units analysis discussed throughout this text.

Rules and Properties: Using Unit Rate

To decide which unit rate to use, just choose one with the unit you *want* in the numerator (inches in Example 2) and the unit you *want to remove* in the denominator (feet in Example 2).

Example 2

Using the Unit Rate to Convert

Convert 5 ft to inches.

To convert from feet to inches, you can multiply by the rate $\frac{12 \text{ in.}}{1 \text{ ft}}$. So, to convert 5 ft to inches, write

NOTE $\frac{12 \text{ in.}}{1 \text{ ft}}$ is a *unit rate*. It can be reduced to 1.

$5 \text{ ft} = 5 \cancel{\text{ft}}\left(\frac{12 \text{ in.}}{1 \cancel{\text{ft}}}\right)$ We are multiplying by 1, and so the value of the expression is not changed.

$= 60 \text{ in.}$ Note that we can divide out units just as we do numbers.

CHECK YOURSELF 2

Use a unit rate to complete each of the statements.

(a) 240 in. = _______ ft **(b)** 7 ft = _______ in.

NOTE Historically, units were associated with various things. A foot was the length of a foot, of course. The yard was the distance from the end of a nose to the fingertips of an outstretched arm. Objects were weighed by comparing them with grains of barley.

You have now had a chance to use two different methods for converting from one unit of measurement to another. Use whichever approach seems easier for you.

From our work so far, it should be clear that one big disadvantage of the English system is that the relationships between units are all different. One foot is 12 in., 1 lb is 16 oz, and so on. We will see that this problem does not exist in the metric system.

Example 3 shows the steps used to simplify English system units of length with multiple units.

Example 3

Simplifying English System Units of Length

Simplify 4 ft 18 in.

NOTE 18 in. is larger than 1 ft and can be simplified.

4 ft 18 in. = 4 ft + 1 ft + 6 in. — Write 18 in. as 1 ft 6 in. because 12 in. is 1 ft.

= 5 ft 6 in.

CHECK YOURSELF 3

(a) Simplify 2 yd 8 ft **(b)** Simplify 7 ft 20 in.

We can always add or subtract units of length according to this rule.

Step by Step: Adding Like Units of Length

Step 1 Arrange the numbers so that the like units are in the same vertical column.

Step 2 Add in each column.

Step 3 Simplify if necessary.

Example 4 illustrates this rule for adding English system units of length.

Example 4

Adding English System Units of Length

Add 5 ft 4 in., 6 ft 7 in., and 7 ft 9 in.

NOTE The columns here represent inches and feet.

		Notes
5 ft	4 in.	Arrange in a vertical column.
6 ft	7 in.	
+ 7 ft	9 in.	
18 ft	20 in.	Add in each column.
= 19 ft	8 in.	Simplify as before.

NOTE Be sure to simplify the results.

CHECK YOURSELF 4

Add 4 *ft* 6 *in.,* 2 *ft* 9 *in., and* 5 *ft* 11 *in.*

To subtract units of length, we have a similar rule.

Step by Step: Subtracting Like Units of Length

Step 1 Arrange the numbers so that the like units are in the same vertical column.

Step 2 Subtract in each column. You may have to borrow from the larger unit at this point.

Step 3 Simplify if necessary.

Consider Example 5 of subtracting English system units of length.

Example 5

Subtracting English System Units of Length

Subtract 3 ft 6 in. from 8 ft 10 in.

8 ft 10 in.	Arrange vertically.
− 3 ft 6 in.	
5 ft 4 in.	Subtract in each column.

CHECK YOURSELF 5

Subtract 5 ft 9 in. from 10 ft 11 in.

As step 2 points out, subtracting units of length may involve borrowing.

Example 6

Subtracting English System Units of Length

Subtract 5 ft 8 in. from 9 ft 3 in.

NOTE Borrowing with units of length is not the same as in the place-value system, in which we always borrowed 10.

9 ft 3 in.	
− 5 ft 8 in.	Do you see the problem? We cannot subtract in the inches column.

To complete the subtraction, we borrow 1 ft and rename. The "borrowed" number will depend on the units involved.

~~9~~ ft ~~3~~ in.	9 ft becomes 8 ft 12 in. Combine the 12 in. with the original 3 in.
8 ft 15 in.	
− 5 ft 8 in.	
3 ft 7 in.	We can now subtract.

CHECK YOURSELF 6

Subtract 4 ft 5 in. from 7 ft 2 in.

Certain types of problems involve multiplying or dividing units of length by abstract numbers, that is, numbers without a unit of measure attached. This rule is used.

Step by Step: Multiplying or Dividing by Abstract Numbers

Step 1 Multiply or divide each part of the unit of length by the abstract number.

Step 2 Simplify if necessary.

Examples 7 and 8 illustrate this procedure.

Example 7

Multiplying English System Units of Length by Abstract Numbers

(a) Multiply 4 · 5 in.

4 · 5 in. = 20 in. or 1 ft 8 in.

(b) Multiply 3(2 ft 7 in.).

NOTE Multiply each part of the unit of length by 3.

$$\begin{array}{r} 2\text{ ft} \quad 7\text{ in.} \\ \times \qquad\quad 3 \\ \hline 6\text{ ft} \quad 21\text{ in.} \end{array}$$

Simplify. The product is 7 ft 9 in.

CHECK YOURSELF 7

Multiply 2 *yd* 2 *ft* 2 *in. by* 6.

Division is illustrated in Example 8.

Example 8

Dividing English System Units of Length by Abstract Numbers

Divide 8 ft 4 in. by 4.

$$\frac{8\text{ ft }4\text{ in.}}{4} = 2\text{ ft }1\text{ in.}$$

CHECK YOURSELF 8

Divide 9 *ft* 6 *in. by* 3.

Thus far, we have studied the English system of measurement, which is used in the United States and a few other countries. Our work will now concentrate on the **metric system,** used throughout the rest of the world.

The metric system is based on one unit of length, the **meter (m).** In the eighteenth century the meter was defined to be one ten-millionth of the distance from the north pole to the equator. Today the meter is scientifically defined in terms of a wavelength in the spectrum of krypton-86 gas.

One big advantage of the metric system is that you can convert from one unit to another by simply multiplying or dividing by powers of 10. This advantage and the need for uniformity throughout the world have led to legislation that will promote the use of the metric system in the United States.

We will see how the metric system works by starting with measures of length and comparing a basic English unit, the yard, with the meter.

NOTE Even in the United States, the metric system is used in science, medicine, the automotive industry, the food industry, and many other areas.

NOTE The basic unit of length in the metric system is also spelled *metre* (the British spelling).

NOTE In the metric system, do not worry about things like 12 in. to 1 ft, or 5,280 ft to 1 mi.

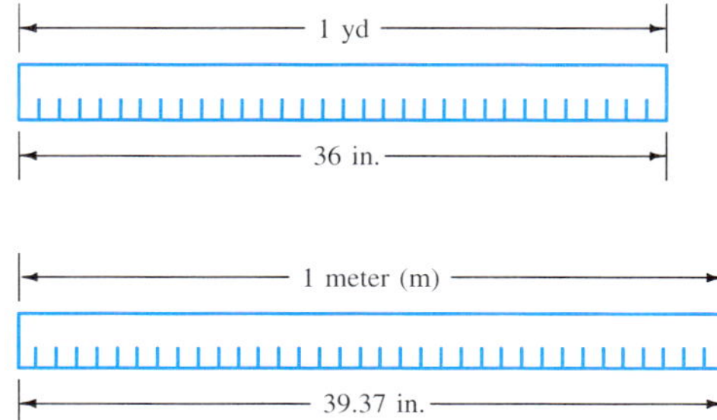

NOTE The meter is one of the basic units of the International System of Units (abbreviated SI). This is a standardization of the metric system agreed to by scientists in 1960.

NOTE There is a standard pattern of abbreviation in the metric system. We will introduce the abbreviation for each term as we go along. The abbreviation for meter is m (no period!).

As you can see, the meter is just slightly longer than the yard. It is used for measuring the same things you might measure in feet or yards. Look at Example 9 to get a feel for the size of the meter.

Example 9

Estimating Metric Length

A room might be 6 m long.

A building lot could be 30 m wide.

A fence is 2 m tall.

CHECK YOURSELF 9

Try to estimate each length in meters.

(a) A traffic lane is __________ m wide.
(b) A small car is __________ m long.
(c) You are __________ m tall.

For other units of length, the meter is multiplied or divided by powers of 10. One commonly used unit is the **centimeter (cm).**

NOTE The prefix *centi* means one hundredth. This should be no surprise. What is our cent? It is one hundredth of a dollar.

Definitions: Comparing Centimeters (cm) to Meters (m)

1 centimeter (cm) = $\frac{1}{100}$ meter (m)

The drawing relates the centimeter and the meter:

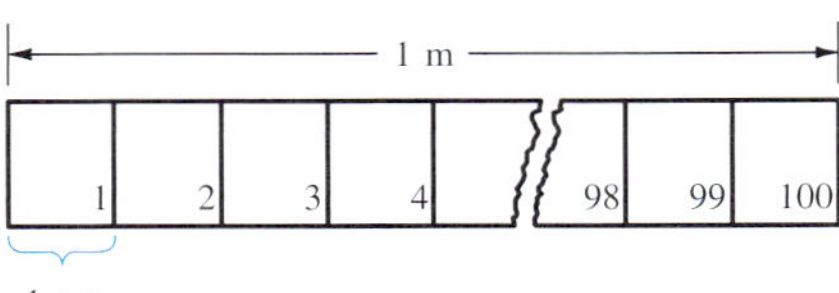

There are 100 cm in 1 m.

Just to give you an idea of the size of the centimeter, it is about the width of your little finger. There are about $2\frac{1}{2}$ cm to 1 in., and the unit is used to measure small objects. Look at Example 10 to get a feel for the length of a centimeter.

Example 10

Estimating Metric Length

A small paperback book is 10 cm wide.

A playing card is 8 cm long.

A ballpoint pen is 16 cm long.

CHECK YOURSELF 10

Try to estimate each value. Then use a metric ruler to check your guess.

(a) This page is __________ cm long.
(b) A dollar bill is __________ cm long.
(c) The seat of the chair you are on is __________ cm from the floor.

To measure *very* small things, the **millimeter (mm)** is used. To give you an idea of its size, the millimeter is about the thickness of a new dime.

NOTE The prefix *milli* means one thousandth.

Definitions: Comparing Millimeters (mm) to Meters (m)

1 millimeter (mm) = $\frac{1}{1000}$ m

The diagram will help you see the relationships of the three units we have looked at.

NOTE Notice that there are 10 mm to 1 cm.

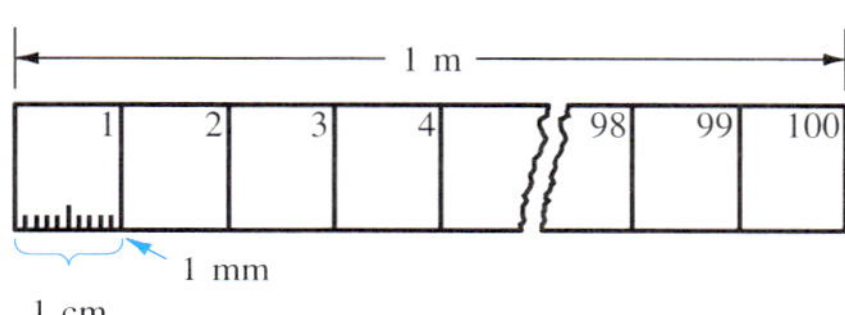

To get used to the millimeter, consider Example 11.

Example 11

Estimating Metric Length

Standard camera film is 35 mm wide.

A small paper clip is 5 mm wide.

A new pencil is about 200 mm long.

CHECK YOURSELF 11

Try to estimate each value. Then use a metric ruler to check your guess.

(a) Your pencil is __________ mm wide.
(b) The tabletop you are working on is __________ mm thick.

NOTE The prefix *kilo* means 1,000. You are already familiar with this. For instance, 1 kilowatt (kW) = 1,000 watts (W).

The **kilometer (km)** is used to measure long distances. The kilometer is about six-tenths of a mile.

Definitions: Comparing Kilometers (km) to Meters (m)

1 kilometer (km) = 1,000 m

Example 12 shows how to get used to the kilometer.

Example 12

Estimating Metric Length

The distance from New York to Boston is 325 km.

A popular distance for road races is 10 km.

Now that you have seen the four commonly used units of length in the metric system, you can review with the following Check Yourself exercise.

CHECK YOURSELF 12

Choose the most reasonable measure in each of the statements.

(a) The width of a doorway: 50 mm, 1 m, or 50 cm.
(b) The length of your pencil: 20 m, 20 mm, or 20 cm.
(c) The distance from your house to school: 500 km, 5 km, or 50 m.
(d) The height of a basketball center: 2.2 m, 22 m, or 22 cm.
(e) The width of a matchbook: 30 cm, 30 mm, or 3 mm.

NOTE Of course, this is easy. All we need to do is move the decimal point to the right or left the required number of places. Again, that's the big advantage of the metric system.

As we said earlier, to convert units of measure within the metric system, all we have to do is multiply or divide by the appropriate power of 10.

NOTE The *smaller* the unit, the *more* units it takes, so *multiply.*

Definitions: Converting Metric Measurements to Smaller Units

To convert to a *smaller* unit of measure, we *multiply* by a power of 10, moving the decimal point *to the right.*

Example 13

Converting Metric Length

For example, the first solution should be $5.2\ \cancel{m}\left(\frac{100\text{ cm}}{1\ \cancel{m}}\right) =$ 520 cm since 1 cm $= \frac{1}{100}$ m, 100 cm $=$ 1 m.

5.2 m = 520 cm — Multiply by 100 to convert from meters to centimeters.

8 km = 8,000 m — Multiply by 1,000.

6.5 m = 6,500 mm — Multiply by 1,000.

2.5 cm = 25 mm — Multiply by 10.

CHECK YOURSELF 13

Complete the statements. Remember, you don't need to do any calculation. Just move the decimal point the appropriate number of places, and write the answer.

(a) 3 km = __________ m
(b) 4.5 m = __________ cm
(c) 1.2 m = __________ mm
(d) 6.5 cm = __________ mm

Definitions: Converting Metric Measurements to Larger Units

To convert to a *larger* unit of measure, we *divide* by a power of 10, moving the decimal point *to the left.*

NOTE The *larger* the unit, the *fewer* units it takes, so *divide.*

Example 14

Converting Metric Length

43 mm = 4.3 cm   Divide by 10.

3,000 m = 3 km   Divide by 1,000.

450 cm = 4.5 m   Divide by 100.

CHECK YOURSELF 14

Complete the statements.

(a) 750 cm = __________ m **(b)** 5,000 m = __________ km

(c) 78 mm = __________ cm **(d)** 3,500 mm = __________ m

We have introduced all the commonly used units of linear measure in the metric system. There are other prefixes that can be used to form other linear measures. The prefix *deci* means $\frac{1}{10}$, *deka* means 10, and *hecto* means 100. Their use is illustrated in the chart.

Definitions: Using Metric Prefixes

1 *milli*meter (mm) = $\frac{1}{1000}$ m

1 *centi*meter (cm) = $\frac{1}{100}$ m

1 *deci*meter (dm) = $\frac{1}{10}$ m

1 meter (m)

1 *deka*meter (dam) = 10 m

1 *hecto*meter (hm) = 100 m

1 *kilo*meter (km) = 1,000 m

Example 15

Converting Between Metric Lengths

(a) 800 dm = ? m

To convert from decimeters to meters, you can see from the chart that you must move the decimal point *one place to the left.*

800 cm = 80.0 m = 80 m

(b) 500 m = ? km

To convert from meters to kilometers, move the decimal point *three places to the left.*

500 m = .500 km = 0.5 km

(c) 6 m = ? mm

To convert from meters to millimeters, move the decimal point *three places to the right.*

6 m = 6,000. mm

CHECK YOURSELF 15

Complete each statement.

(a) 300 cm = __________ m **(b)** 370 mm = __________ m

(c) 4,500 m = __________ km

CHECK YOURSELF ANSWERS

1. **(a)** 48 in.; **(b)** 36 ft; **(c)** 12 ft; **(d)** 36 in. **2.** **(a)** 20 ft; **(b)** 84 in.
3. **(a)** 4 yd 2 ft; **(b)** 8 ft 8 in. **4.** 13 ft 2 in. **5.** 5 ft 2 in.
6. 2 ft 9 in. **7.** 16 yd 1 ft **8.** 3 ft 2 in.
9. **(a)** About 3 m; **(b)** perhaps 5 m; **(c)** You are probably between 1.5 and 2 m tall.
10. **(a)** About 28 cm; **(b)** almost 16 cm; **(c)** about 45 cm
11. **(a)** About 8 mm; **(b)** probably between 25 and 30 mm
12. **(a)** 1 m; **(b)** 20 cm; **(c)** 5 km; **(d)** 2.2 m; **(e)** 30 mm
13. **(a)** 3,000 m; **(b)** 450 cm; **(c)** 1,200 mm; **(d)** 65 mm
14. **(a)** 7.5 m; **(b)** 5 km; **(c)** 7.8 cm; **(d)** 3.5 m **15.** **(a)** 3 m; **(b)** 0.37 m; **(c)** 4.5 km

Name ____________________

Section ________ Date ________

6.6 Exercises

Complete the statements.

1. 8 ft = ________ in.

2. 5 mi = ________ ft

3. 7 yd = ________ ft

4. 39 ft = ________ yd

5. 44 in. = ________ ft

6. 4.72 ft = ________ in.

Solve the application.

7. A unit of measurement used in surveying is the **chain.** There are 80 chains in a mile. If you measured the distance from your home to school, how many chains would you have traveled?

Simplify.

8. 4 ft 18 in.

9. 7 yd 50 in.

Add.

10. $\begin{array}{r} 9 \text{ ft } \ \ 7 \text{ in.} \\ + \ 3 \text{ ft } 10 \text{ in.} \\ \hline \end{array}$

11. $\begin{array}{r} 5 \text{ yd } 2 \text{ ft} \\ 4 \text{ yd} \phantom{\ 2 \text{ ft}} \\ + \ 6 \text{ yd } 1 \text{ ft} \\ \hline \end{array}$

12. 7 ft 8 in., 8 ft 5 in., and 9 ft 7 in.

Subtract.

13. $\begin{array}{r} 7 \text{ ft } 11 \text{ in.} \\ - \ 4 \text{ ft } \ \ 3 \text{ in.} \\ \hline \end{array}$

14. Subtract 2 yd 2 ft from 5 yd 1 ft.

Multiply.

15. $4 \cdot 10$ in.

16. 3(4 ft 5 in.)

ANSWERS

1. ____________________
2. ____________________
3. ____________________
4. ____________________
5. ____________________
6. ____________________
7. ____________________
8. ____________________
9. ____________________
10. ____________________
11. ____________________
12. ____________________
13. ____________________
14. ____________________
15. ____________________
16. ____________________

ANSWERS

17. ______________

18. ______________

19. ______________

20. ______________

21. ______________

22. ______________

23. ______________

24. ______________

25. ______________

26. ______________

27. ______________

28. ______________

29. ______________

Divide.

17. $\dfrac{4 \text{ ft } 6 \text{ in.}}{2}$

18. $\dfrac{16 \text{ mi } 28 \text{ yd}}{4}$

Solve each of the applications.

19. Construction. A railing for a deck requires pieces of cedar 4 ft 8 in., 11 ft 7 in., and 9 ft 3 in. long. What is the total length of material that is needed?

20. Sewing. A pattern requires a 2-ft 10-in. length of fabric. If a 2-yd length is used, what length remains?

21. Framing. A picture frame is to be 2 ft 6 in. long and 1 ft 8 in. wide. A 9-ft piece of molding is available for the frame. Will this be enough for the frame?

22. Plumbing. A plumber needs two pieces of plastic pipe that are 6 ft 9 in. long and 1 piece that is 2 ft 11 in. long. He has a 16-ft piece of pipe. Is this enough for the job?

23. Construction. A bookshelf requires four boards 3 ft 8 in. long and two boards 2 ft 10 in. long. How much lumber will be needed for the bookshelf?

Find the following.

24.
$$\begin{array}{r} 13 \text{ yd } 15 \text{ ft } 10 \text{ in.} \\ -\ \ 9 \text{ yd } 16 \text{ ft } 15 \text{ in.} \\ \hline \end{array}$$

25.
$$\begin{array}{r} 4 \text{ mi } 5 \text{ yd } 3 \text{ ft } 10 \text{ in.} \\ \times \qquad\qquad\qquad 2 \\ \hline \end{array}$$

Choose the most reasonable measure.

26. The height of a ceiling

(a) 25 m
(b) 2.5 m
(c) 25 cm

27. The height of a kitchen counter

(a) 9 m
(b) 9 cm
(c) 90 cm

28. The diagonal measure of a television screen

(a) 50 mm
(b) 50 cm
(c) 5 m

29. The height of a two-story building

(a) 7 m
(b) 70 m
(c) 70 cm

30. An hour's drive on a freeway

(a) 9 km
(b) 90 m
(c) 90 km

31. The width of a roll of cellophane tape

(a) 1.27 mm
(b) 12.7 mm
(c) 12.7 cm

32. The width of a sheet of typing paper

(a) 21.6 cm
(b) 21.6 mm
(c) 2.16 cm

33. The thickness of window glass

(a) 5 mm
(b) 5 cm
(c) 50 mm

34. The height of a refrigerator

(a) 16 m
(b) 16 cm
(c) 160 cm

35. The length of a ballpoint pen

(a) 16 mm
(b) 16 m
(c) 16 cm

36. The width of a handheld calculator key

(a) 1.2 mm
(b) 12 mm
(c) 12 cm

Complete each statement, using a metric unit of length.

37. A playing card is 6 __________ wide.

38. The diameter of a penny is 19 __________.

39. A doorway is 2 __________ high.

40. A table knife is 22 __________ long.

41. A basketball court is 28 __________ long.

42. A commercial jet flies 800 __________ per hour.

ANSWERS

30. __________
31. __________
32. __________
33. __________
34. __________
35. __________
36. __________
37. __________
38. __________
39. __________
40. __________
41. __________
42. __________

ANSWERS

43. ______

44. ______

45. ______

46. ______

47. ______

48. ______

49. ______

50. ______

51. ______

52. ______

53. ______

54. ______

55. ______

56. ______

57. ______

58. ______

59. ______

60. ______

61. ______

62. ______

63. ______

64. ______

43. The width of a nail file is 12 ______ .

44. The distance from New York to Washington, D.C., is 360 ______ .

45. A recreation room is 6 ______ long.

46. A ruler is 22 ______ wide.

47. A long-distance run is 35 ______ .

48. A paperback book is 11 ______ wide.

Complete each statement.

49. 3,000 mm = ______ m

50. 150 cm = ______ m

51. 8 m = ______ cm

52. 77 mm = ______ cm

53. 250 km = ______ cm

54. 500 cm = ______ m

55. 25 cm = ______ mm

56. 150 mm = ______ m

57. 7,000 m = ______ km

58. 9 m = ______ cm

59. 8 cm = ______ mm

60. 45 cm = ______ mm

61. 5 km = ______ m

62. 4,000 m = ______ km

63. 5 m = ______ mm

64. 7 km = ______ m

Use a metric ruler to measure the necessary dimensions, and complete the statements.

65. The perimeter of the parallelogram is __________ cm.

66. The perimeter of the triangle is __________ mm.

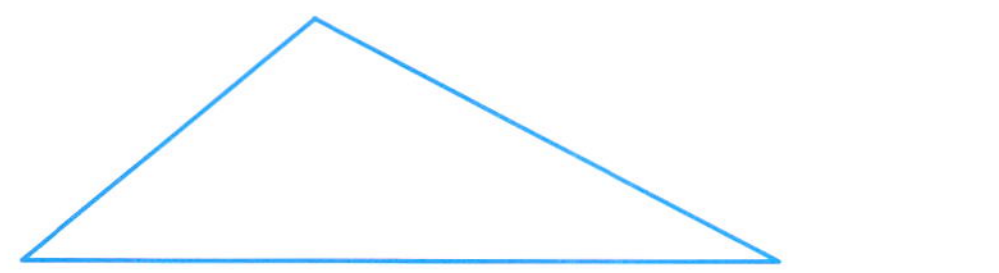

67. The perimeter of the rectangle is __________ cm.

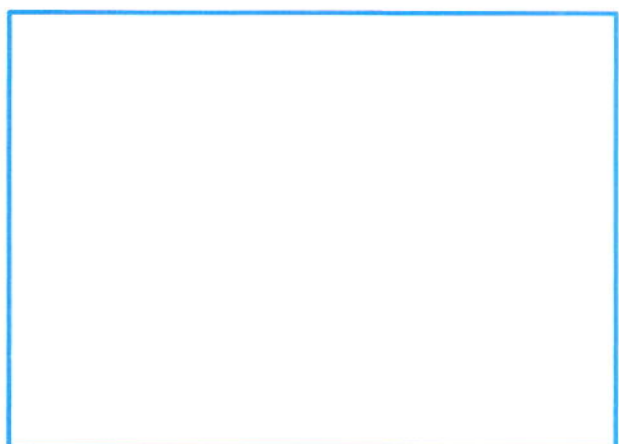

68. The area of the rectangle in exercise 67 is __________ cm^2.

69. The perimeter of the square is __________ mm.

70. The area of the square in exercise 69 is __________ mm^2.

71. What units in the metric system would you use to measure each of the quantities?

(a) Distance from Los Angeles to New York
(b) Your waist measurement
(c) Width of a hair
(d) Your height

ANSWERS

65. ______
66. ______
67. ______
68. ______
69. ______
70. ______

71. ______

Answers

1. 96 **3.** 21 **5.** $3\frac{2}{3}$ **7.**  **9.** 8 yd 14 in. or 8 yd 1 ft 2 in.

11. 16 yd **13.** 3 ft 8 in. **15.** 3 ft 4 in. **17.** 2 ft 3 in. **19.** 25 ft 6 in.
21. Yes, 8 in. will remain **23.** 20 ft 4 in. **25.** 8 mi 12 yd 1ft 8 in.
27. (c) **29.** (a) **31.** (b) **33.** (a) **35.** (c) **37.** cm **39.** m
41. m **43.** mm **45.** m **47.** km **49.** 3 **51.** 800
53. 25,000,000 **55.** 250 **57.** 7 **59.** 80 **61.** 5,000 **63.** 5,000
65. 12 **67.** 14 **69.** 100
71. **(a)** km; **(b)** cm; **(c)** mm; **(d)** m or cm

6 Summary

DEFINITION/PROCEDURE	EXAMPLE	REFERENCE
Ratios		**Section 6.1**
Ratio A means of comparing two numbers or quantities. A ratio can be written as a fraction	$\frac{4}{7}$ can be thought of as "the ratio of 4 to 7."	p. 463
Rates		**Section 6.2**
Rate A fraction involving two denominate numbers with different units.	$\frac{50 \text{ home runs}}{150 \text{ games}} = \frac{1}{3} \frac{\text{home run}}{\text{game}}$	p. 471
Unit price The cost per unit.	$\frac{\$2}{5 \text{ rolls}} = \$0.40$ per roll	p. 473
Proportions		**Section 6.3**
Proportion A statement that two ratios or rates are equal.	$\frac{3}{5} = \frac{6}{10}$ is a proportion read "three is to five as six is to ten."	p. 479
The Proportion Rule If $\frac{a}{b} = \frac{c}{d}$, then $a \cdot d = b \cdot c$	If $\frac{3}{5} = \frac{6}{10}$, then $5 \cdot 6 = 3 \cdot 10$	p. 480
To Solve a Proportion **1.** Use the proportion rule to write the equivalent equation $a \cdot d = b \cdot c$. **2.** Divide both terms of the equation by the coefficient of the variable. **3.** Use the value found to replace the unknown in the original proportion. Check that the ratios or rates are proportional.	To solve: $\frac{x}{5} = \frac{16}{20}$ $20x = 5 \cdot 16$ $20x = 80$ $\frac{\overset{1}{\cancel{20}}x}{\underset{1}{\cancel{20}}} = \frac{80}{20}$ $x = 4$	p. 483
Similar Triangles and Proportions		**Section 6.4**
A triangle in which two of the sides are perpendicular is called a *right triangle*.	is a right triangle.	p. 489
Two right triangles are similar if their corresponding sides are proportional.	3, 5; 6, 10 are proportional.	p. 489
If we know that two triangles are similar, we can use a proportion to find the length of a missing side.	2, 8; 3, x $\frac{2}{8} = \frac{3}{x}$ $2x = 3 \cdot 8$ $2x = 24$ $x = 12$	p. 490

Continued

DEFINITION/PROCEDURE	EXAMPLE	REFERENCE
Linear Measurement and Conversion		**Section 6.6**
The English system of measurement is in common use in the United States. *English Units of Measure and Equivalents* *Length* 1 foot (ft) = 12 inches (in.) 1 yard (yd) = 3 ft 1 mile (mi) = 5,280 ft *Weight* 1 pound (lb) = 16 ounces (oz) 1 ton = 2,000 lb *Capacity* 1 pint (pt) = 16 fluid ounces (fl oz) 1 quart (qt) = 2 pt 1 gallon (gal) = 4 qt		p. 511
Unit rates A fraction whose value is 1. Unit rates can be used to convert units.	$\frac{12 \text{ in.}}{1 \text{ ft}}$ and $\frac{1 \text{ ft}}{3 \text{ yd}}$ are unit rates.	p. 512
To Add Like Units of Length 1. Arrange the numbers so that the like units are in the same column. 2. Add in each column. 3. Simplify if necessary.	To add 4 ft 7 in. and 5 ft 10 in.: 4 ft 7 in. + 5 ft 10 in. 9 ft 17 in. = 10 ft 5 in.	p. 513
To Subtract Like Units of Length 1. Arrange the numbers so that the like units are in the same column. 2. Subtract in each column. You may have to borrow from the larger unit at this point. 3. Simplify if necessary.	To subtract: 4 ft 7 in. − 2 ft 9 in. Borrow and rename: 3 ft 19 in. − 2 ft 9 in. 1 ft 10 in.	p. 513
To Multiply or Divide Units by Abstract Numbers 1. Multiply or divide each part of the measurement by the abstract number. 2. Simplify if necessary.	2 · (3 yd 2 ft) = 6 yd 4 ft, or 7 yd 1 ft	p. 514
Metric Units of Length The metric system of measurement is used throughout most of the world. **Common metric units of length** are the meter (m), centimeter (cm), millimeter (mm), and kilometer (km).		p. 515
Basic Metric Prefixes *milli** means $\frac{1}{1000}$ *kilo** means 1,000 *centi** means $\frac{1}{100}$ *hecto* means 100 *deci* means $\frac{1}{10}$ *deka* means 10	*These are the most commonly used and should be memorized.	p. 519

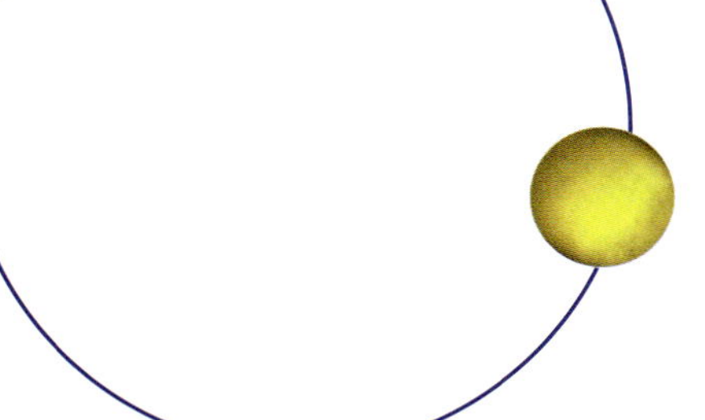

Summary and Review Exercises

You should now be reviewing the material in Chapter 6. These exercises will help in that process. Work all of the exercises carefully. References are provided to the section for each exercise. If you made an error, go back and review the related material.

[6.1] In exercises 1 to 8, write each ratio in simplest form.

1. The ratio of 4 to 17

2. The ratio of 28 to 42

3. For a football team that has won 10 of its 16 games, the ratio of wins to games played

4. For a rectangle of length 30 in. and width 18 in., the ratio of its length to its width

5. The ratio of $2\frac{1}{3}$ to $5\frac{1}{4}$

6. The ratio of 7.5 to 3.25

7. The ratio of 7 in. to 3 ft

8. The ratio of 72 h to 4 days

[6.2] In exercises 9 to 16, express each rate in simplest form.

9. $\frac{600 \text{ mi}}{6 \text{ h}}$

10. $\frac{270 \text{ mi}}{9 \text{ gal}}$

11. $\frac{350 \text{ calories}}{7 \text{ oz}}$

12. $\frac{36{,}000 \text{ dollars}}{9 \text{ yr}}$

13. $\frac{5000 \text{ ft}}{25 \text{ s}}$

14. $\frac{10{,}500 \text{ revolutions}}{3 \text{ min}}$

15. A baseball team has had 1[illegible]7 hits in 18 games. Find the team's hits per game rate.

16. A basketball team has scored 216 points in 8 quarters. Find the team's points per quarter rate.

17. Taniko scored 246 points in 20 games. Marisa scored 216 points in 16 games. Which player has the higher points per game rate?

18. One shop will charge $306 for a job that takes $4\frac{1}{2}$ h. A second shop can do the same job in 4 h, and will charge $290. Which shop has the higher cost per hour rate?

In exercises 19 to 24, find the unit price for each item.

19. A 32-oz bottle of dishwashing liquid costs \$2.88.

20. A 35-oz box of breakfast cereal costs \$5.60.

21. A 24-oz loaf of bread costs \$2.28.

22. Five large jars of fruit cost \$67.30.

23. Three CDs cost \$44.85.

24. Six tickets cost \$267.60.

[6.3] In exercises 25 to 28, write each proportion.

25. 4 is to 9 as 20 is to 45.

26. 7 is to 5 as 56 is to 40.

27. If Jorge can travel 110 mi in 2 h, he can travel 385 mi in 7 h.

28. If 4 gal of paint will cover 1,000 ft^2, it will take 10 gal of paint to cover 2,500 ft^2.

In exercises 29 to 34, determine whether the given fractions are proportional.

29. $\frac{4}{13} \stackrel{?}{=} \frac{7}{22}$

30. $\frac{8}{11} \stackrel{?}{=} \frac{24}{33}$

31. $\frac{9}{24} \stackrel{?}{=} \frac{12}{32}$

32. $\frac{7}{18} \stackrel{?}{=} \frac{35}{80}$

33. $\frac{5}{\frac{1}{6}} \stackrel{?}{=} \frac{120}{4}$

34. $\frac{0.8}{4} \stackrel{?}{=} \frac{12}{50}$

35. Is $\frac{156 \text{ francs}}{30 \text{ dollars}}$ equivalent to $\frac{442 \text{ francs}}{85 \text{ dollars}}$?

36. Is $\frac{188 \text{ words}}{8 \text{ min}}$ equivalent to $\frac{121 \text{ words}}{5 \text{ min}}$?

In exercises 37 to 45, solve for the unknown in each proportion.

37. $\frac{16}{24} = \frac{m}{3}$

38. $\frac{6}{a} = \frac{27}{18}$

39. $\frac{14}{35} = \frac{t}{10}$

40. $\frac{y}{22} = \frac{15}{55}$

41. $\frac{55}{88} = \frac{10}{p}$

42. $\frac{\frac{1}{2}}{18} = \frac{5}{w}$

43. $\frac{\frac{3}{2}}{9} = \frac{5}{a}$

44. $\frac{5}{x} = \frac{0.6}{12}$

45. $\frac{s}{2.5} = \frac{1.5}{7.5}$

[6.4] Find each missing length.

46.

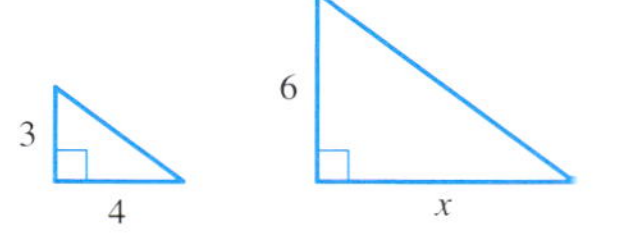

47.

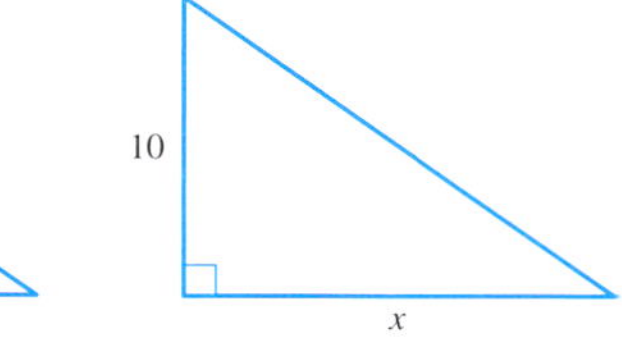

48.

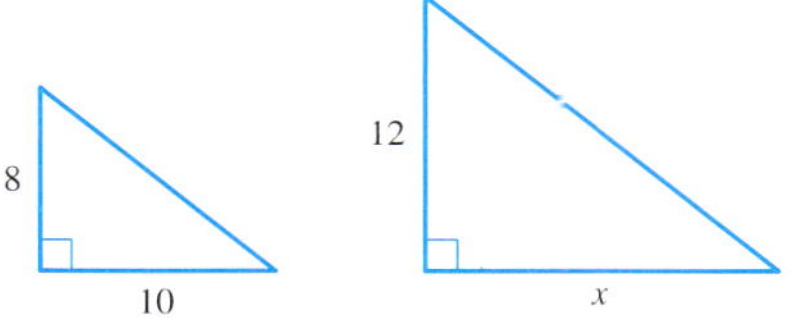

49.

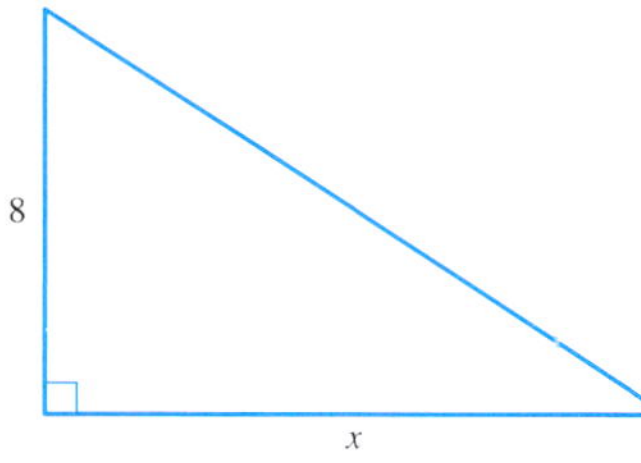

50.

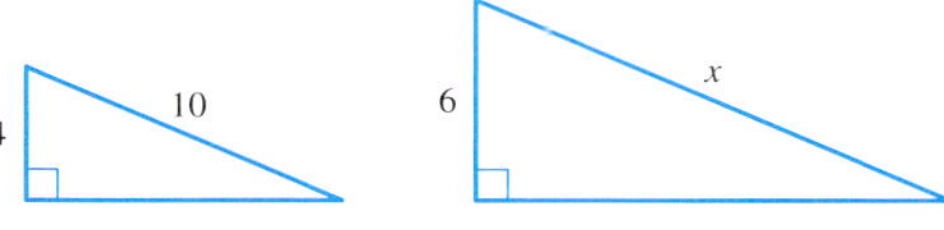

51.

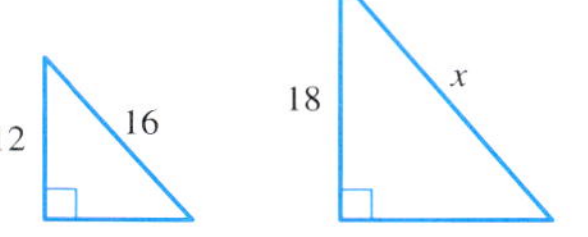

[6.5] In exercises 52 to 58, solve each application.

52. Ticket price. If 4 tickets to a civic theater performance cost $45, what will be the price for 6 tickets?

53. Enrollment. The ratio of first-year to second-year students at a school is 8 to 7. If there are 224 second-year students, how many first-year students are there?

54. Photo enlargement. A photograph that is 5 in. wide by 7 in. tall is to be enlarged so that the new height will be 21 in. What will be the width of the enlargement?

55. Worker output. Marcia assembles disk drives for a computer manufacturer. If she can assemble 11 drives in 2 h, how many can she assemble in a workweek (40 h)?

56. Defective parts. A firm finds 14 defective parts in a shipment of 400. How many defective parts can be expected in a shipment of 800 parts?

57. Distance. The scale on a map is $\frac{1}{4}$ in. = 10 mi. How many miles apart are two towns that are 3 in. apart on the map?

58. Weight. A piece of tubing that is 16.5 cm long weighs 55 g. What is the weight of a piece of the same tubing that is 42 cm long?

[6.6] Complete each of the statements.

59. 11 ft = _______ in.

60. 5 mi = _______ ft

Simplify.

61. 3 ft 23 in.

62. 1 mi 6,300 ft

Add.

63. $\begin{array}{r} 3\text{ ft }\ 9\text{ in.} \\ +\ 5\text{ ft }10\text{ in.} \\ \hline \end{array}$

64. $\begin{array}{r} 2\text{ ft }\ 8\text{ in.} \\ 5\text{ ft }10\text{ in.} \\ +\ 4\text{ ft }11\text{ in.} \\ \hline \end{array}$

Subtract.

65. $\begin{array}{r} 7\text{ ft }11\text{ in.} \\ -\ 2\text{ ft }\ 4\text{ in.} \\ \hline \end{array}$

66. $\begin{array}{r} 6\text{ yd }1\text{ ft} \\ -\ 3\text{ yd }2\text{ ft} \\ \hline \end{array}$

Multiply.

67. 3(1 ft 5 in.)

Divide.

68. $\dfrac{5\text{ ft }8\text{ in.}}{2}$

69. **Construction.** A room requires two pieces of floor molding 12 ft 8 in. long, one piece 6 ft 5 in. long, and one piece 10 ft long. Will 42 ft of molding be enough for the job?

Choose the most reasonable measure.

70. A marathon race

(a) 40 km
(b) 400 km
(c) 400 m

71. The distance around your wrist

(a) 15 mm
(b) 15 cm
(c) 1.5 m

72. The diameter of a penny

(a) 19 cm
(b) 1.9 mm
(c) 19 mm

73. The width of a portable television screen

(a) 28 mm
(b) 28 cm
(c) 2.8 m

Complete each statement, using a metric unit of length.

74. A matchbook is 39 ____________ wide.

75. The distance from San Francisco to Los Angeles is 618 ____________.

76. A 1-lb coffee can has a diameter of 10 ____________.

Complete each statement.

77. 2 km = ____________ m

78. 3 cm = ____________ mm

79. 3,000 mm = ____________ m

80. 8 m = ____________ mm

81. 6 cm = ____________ m

82. 8 m = ____________ km

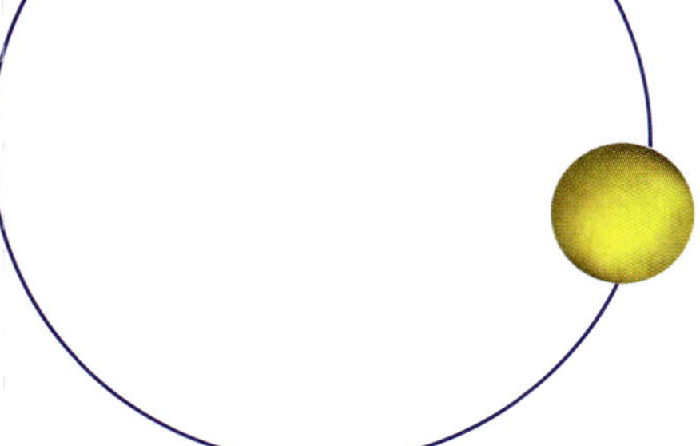

Chapter Test for Chapter 6

Name ______________________

Section __________ Date __________

ANSWERS

1. ______________________
2. ______________________
3. ______________________
4. ______________________
5. ______________________
6. ______________________
7. ______________________
8. ______________________
9. ______________________
10. ______________________
11. ______________________
12. ______________________
13. ______________________
14. ______________________
15. ______________________
16. ______________________
17. ______________________
18. ______________________
19. ______________________
20. ______________________

The purpose of this test is to help you check your progress and review for the in-class chapter test. When you are done with the test, check your answers in the back of the book. If you missed any answers, go back and review the appropriate sections of the chapter.

In exercises 1 to 5, write each ratio in simplest form.

1. The ratio of 7 to 19

2. The ratio of 75 to 45

3. The ratio of 8 ft to 4 yd

4. The ratio of 6 h to 3 days

5. A basketball team wins 26 of its 33 games during a season. What is the ratio of wins to games played? What is the ratio of wins to losses?

In exercises 6 to 8, express each rate in simplest form.

6. $\frac{840 \text{ mi}}{175 \text{ gal}}$

7. $\frac{132 \text{ dollars}}{16 \text{ h}}$

8. The unit price, if 11 gal of milk cost \$28.16.

In exercises 9 to 12, determine whether the given fractions are proportional.

9. $\frac{3}{9} \stackrel{?}{=} \frac{27}{81}$

10. $\frac{6}{7} \stackrel{?}{=} \frac{9}{11}$

11. $\frac{9}{10} \stackrel{?}{=} \frac{27}{30}$

12. $\frac{\frac{1}{2}}{5} \stackrel{?}{=} \frac{2}{18}$

In exercises 13 to 18, solve for the unknown in each proportion.

13. $\frac{45}{75} = \frac{12}{x}$

14. $\frac{a}{24} = \frac{45}{60}$

15. $\frac{\frac{1}{2}}{p} = \frac{5}{30}$

16. $\frac{\frac{5}{6}}{8} = \frac{5}{a}$

17. $\frac{x}{0.3} = \frac{60}{3}$

18. $\frac{3}{m} = \frac{0.9}{4.8}$

In exercises 19 to 26, solve each application, using a proportion.

19. Consumer affairs. If ballpoint pens are marked 5 for 95¢, how much will a dozen cost?

20. Basketball. A basketball player scores 207 points in her first 9 games. At the same rate, how many points will she score in the 28-game season?

ANSWERS

21. ______
22. ______
23. ______
24. ______
25. ______
26. ______
27. ______
28. ______
29. ______
30. ______
31. ______
32. ______

21. **Distance.** Your new compact car travels 324 mi on 9 gal of gas. If the tank holds 16 usable gallons, how far can you drive on a tankful of gas?

22. **Elections.** The ratio of yes to no votes in an election was 6 to 5. How many no votes were cast if 3,600 people voted yes?

23. **Mufflers installed.** An assembly line can install 5 car mufflers in 4 min. At this rate, how many mufflers can be installed in an 8-h shift?

24. **Mixing.** Instructions on a package of concentrated plant food call for 2 teaspoons (tsp) to 1 qt of water. We wish to use 3 gal of water. How much of the plant food concentrate should be added to the 3 gal of water?

25. A 10-ft fence casts a 16-ft shadow. How tall is a nearby tree that casts an 88-ft shadow?

26. A meterstick casts a shadow that is 0.6 m. How tall is a nearby pole that casts a shadow of 12 m?

In exercises 27 and 28, do the indicated operations.

27.
$$\begin{array}{r} 7 \text{ ft } 9 \text{ in.} \\ +\ 3 \text{ ft } 8 \text{ in.} \\ \hline \end{array}$$

28.
$$\begin{array}{r} 7 \text{ lb } \ 3 \text{ oz} \\ -\ 4 \text{ lb } 10 \text{ oz} \\ \hline \end{array}$$

29. **Total cost.** The Martins are fencing in a rectangular yard that is 110 ft long by 40 ft wide. If the fencing costs $3.50 per linear foot, what will be the total cost of the fencing?

In exercises 30 to 32, choose the reasonable measure.

30. The width of your hand

(a) 50 cm
(b) 10 cm
(c) 1 m

31. The speed limit on a freeway

(a) $9 \frac{\text{km}}{\text{h}}$
(b) $90 \frac{\text{km}}{\text{h}}$
(c) $90 \frac{\text{m}}{\text{h}}$

32. The height of a basketball player

(a) 21 cm
(b) 21 dm
(c) 21 m

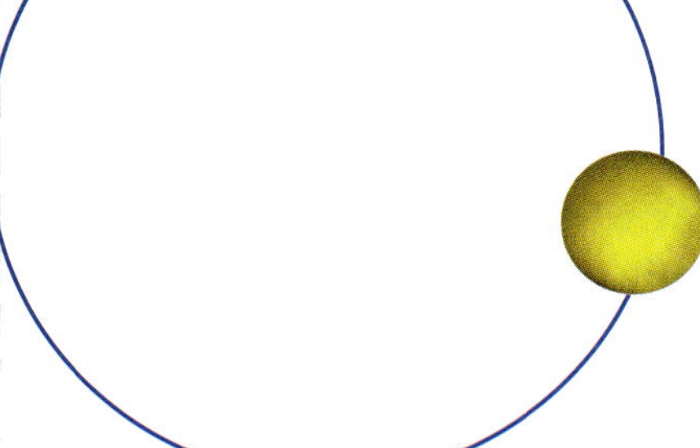

Cumulative Test for Chapters 1 to 6

Name ______________

Section ________ Date ________

ANSWERS

1. ______________
2. ______________
3. ______________
4. ______________
5. ______________
6. ______________
7. ______________
8. ______________
9. ______________
10. ______________
11. ______________
12. ______________
13. ______________
14. ______________
15. ______________
16. ______________
17. ______________

This test is provided to help you in the process of reviewing Chapters 1 through 6. Answers are provided in the back of the book. If you missed any answers, be sure to go back and review the appropriate chapter sections.

In exercises 1 to 3, name the property that is illustrated.

1. $(7 + 3) + 8 = 7 + (3 + 8)$

2. $6 \cdot 7 = 7 \cdot 6$

3. $5\,(2 + 4) = 5 \cdot 2 + 5 \cdot 4$

In exercises 4 and 5, round the numbers to the indicated place value.

4. 5,873 to the nearest hundred

5. 953,150 to the nearest ten thousand

6. Evaluate: $2 + 8 \cdot 3 \div 4$

7. Write the prime factorization of 264.

8. Find the least common multiple (LCM) of 6, 15, and 45.

9. Convert to a mixed number: $\frac{22}{7}$

10. Convert to an improper fraction: $6\frac{5}{8}$

In exercises 11 to 14, perform the indicated operations.

11. $\frac{2}{3} \cdot 1\frac{4}{5} \cdot \frac{5}{8}$

12. $2\frac{2}{7} \div 1\frac{11}{21}$

13. $4\frac{7}{8} + 3\frac{1}{6}$

14. $9 + \left(-5\frac{3}{8}\right)$

15. A $6\frac{1}{2}$-in. bolt is placed through a wall that is $5\frac{7}{8}$ in. thick. How far does the bolt extend beyond the wall?

16. You pay for purchases of \$13.99, \$18.75, \$9.20, and \$5 with a \$50 check. How much cash will you have left?

17. Find the area of a circle whose diameter is 3.2 ft. Use 3.14 for π, and round the result to the nearest hundredth.

ANSWERS

18. ______
19. ______
20. ______
21. ______
22. ______
23. ______
24. ______
25. ______
26. ______
27. ______
28. ______
29. ______
30. ______
31. ______
32. ______
33. ______
34. ______
35. ______

18. A 14-acre piece of land is being developed into home lots. If 2.8 acres of land will be used for roads, and each home site is to be 0.35 acre, how many lots can be formed?

19. Write the decimal equivalent of $\frac{8}{11}$. Use bar notation.

20. Solve for the unknown: $\frac{5}{m} = \frac{0.4}{9}$

21. You are using a photocopy machine to reduce an advertisement that is 14 in. wide by 21 in. long. If the new width is to be 8 in., what will the new length be?

22. The opposite of 8 is ________.

23. The absolute value of -20 is ________.

In exercises 24 to 29, evaluate.

24. $-(-12)$

25. $|-5|$

26. $-12 + (-6)$

27. $-8 - (-4)$

28. $(-6)(15)$

29. $48 \div (-12)$

Write using symbols.

30. 3 times the sum of x and y

31. The quotient when 5 less than n is divided by 3

Solve the equations, and check your results.

32. $9x + (-5) = 8x$

33. $-\frac{3}{4}x = 18$

34. $2x + 3 = 7x + 5$

35. $\frac{4}{3}x + (-6) = 4 + \left(-\frac{2}{3}x\right)$

7 PERCENT

INTRODUCTION

With a degree in literature and a love of books, Don set out looking for a career in publishing. He was surprised to find that he ended up selling books to bookstores. He was more surprised to find that he enjoyed the selling. He finds having to talk about all types of books (even math books!) with bookstore managers very challenging.

Don does admit that he wishes he had paid a little more attention in his math class. Thinking that math was not relevant to his field, he took it only because it was required. Now he has to do numerous calculations every day. Figuring discounts, quotas, bonuses, and commissions occupies much of Don's office time.

Name ____________________

Section ________ Date ________

Pre-Test Chapter 7

ANSWERS

1. ____________________
2. ____________________
3. ____________________
4. ____________________
5. ____________________
6. ____________________
7. ____________________
8. ____________________
9. ____________________
10. ____________________
11. ____________________
12. ____________________
13. ____________________

This pre-test will point out any difficulties you may be having with percents. Do all the problems, and then check your answers with those in the back of the book.

1. Write 7% as a fraction.

2. Write 23% as a decimal.

3. Write 0.035 as a percent.

4. Write $\frac{4}{5}$ as a percent.

5. What is 25% of 252?

6. What percent of 500 is 45?

7. 35% of a number is 210. What is the number?

8. **Interest.** How much simple interest will you pay on a $4,000 loan for 1 year if the interest rate is 14%?

9. **Commission.** A salesperson earns a $400 commission on sales of $8,000. What is the commission rate?

10. **Salary.** A salary increase of 5% amounts to a $60 monthly raise. What was the monthly salary before the increase?

11. A state sales tax is 6.5%. What is the sales tax on an item purchased for $174.00?

12. Adriana invests $5,000 in an account that pays 6% interest per year. How much will she have in the account after 3 years?

13. Julio gets a 15% discount at a local furniture store. How much will he pay for a sofa listed at $490?

7.1 Percents, Decimals, and Fractions

7.1 OBJECTIVES

1. Use percent notation
2. Convert between a percent and a fraction
3. Convert between a percent and a mixed number
4. Convert between a percent and a decimal

When we considered parts of a whole in earlier chapters, we used fractions and decimals. The idea of *percent* is another useful way of naming parts of a whole. We can think of percents as ratios whose denominators are 100. In fact, the word **percent** means "for each hundred." Look at the drawing:

Definitions: Percent

Percent means "for each hundred." The symbol for percent, %, can be read "out of each 100."

In the drawing, 25 of 100 squares are shaded. As a fraction, we write this as $\frac{25}{100}$. As a percent, we write 25%. So 25 percent of the squares are shaded.

Example 1

Using Percent Notation

(a) Four out of five geography students passed their midterm exams. Write this statement, using percent notation.

NOTE The ratio of students passing to students taking the class is $\frac{4}{5}$.

$$\frac{4}{5} = \frac{80}{100} = 80\%$$

Percent means "for each hundred." To obtain a denominator of 100, multiply the numerator and denominator of the original fraction by 20.

So we can say that 80% of the geography students passed.

(b) Of 50 automobiles sold by a dealer in 1 month, 35 were compact cars. Write this statement, using percent notation.

NOTE The ratio of compact cars to all cars is $\frac{35}{50}$.

$$\frac{35}{50} = \frac{70}{100} = 70\%$$

We can say that 70% of the cars sold were compact cars.

CHECK YOURSELF 1

Rewrite the statement, using percent notation: 4 *of the* 50 *parts in a shipment were defective.*

Because there are different ways of naming the parts of a whole, you need to know how to change from one of these ways to another. First we will look at changing a percent to a fraction. Because a percent is a fraction or a ratio with denominator 100, we can use this rule.

Rules and Properties: Changing a Percent to a Fraction

To change a percent to a common fraction, divide the number before the percent symbol by 100. Note that this is equivalent to multiplying by $\frac{1}{100}$.

The use of this rule is shown in Example 2.

Example 2

Changing a Percent to a Fraction

Change each percent to a fraction.

(a) $7\% = \frac{7}{100}$

NOTE Reduce $\frac{25}{100}$ to simplest form.

(b) $25\% = \frac{25}{100} = \frac{1}{4}$

CHECK YOURSELF 2

Write 12% as a fraction.

If a percent is *greater than 100,* the resulting fraction will be *greater than 1.* This is shown in Example 3.

Example 3

Changing a Percent to a Mixed Number

Change 150% to a mixed number.

$$150\% = \frac{150}{100} = 1\frac{50}{100} = 1\frac{1}{2}$$

CHECK YOURSELF 3

Write 125% as a mixed number.

The fractional equivalents of certain percents should be memorized.

$$33\frac{1}{3}\% = 33\frac{1}{3}\left(\frac{1}{100}\right) = \frac{100}{3} \cdot \frac{1}{100} = \frac{1}{3} \qquad 66\frac{2}{3}\% = 66\frac{2}{3}\left(\frac{1}{100}\right) = \frac{200}{3} \cdot \frac{1}{100} = \frac{2}{3}$$

It is best to try to remember these fractional equivalents.

In Example 2, we wrote percents as fractions by replacing the percent sign with $\frac{1}{100}$ and multiplying. How do we convert percents when we are working with decimals? Just move the decimal point two places to the left. This gives us a second rule for converting percents.

Rules and Properties: Changing a Percent to a Decimal

To change a percent to a decimal, multiply the number before the percent symbol by 0.01. This is equivalent to moving the decimal point two places to the left.

Example 4

Changing a Percent to a Decimal

Change each percent to a decimal equivalent.

(a) 25% = 0.25 The decimal point in 25% is understood to be after the 5.

(b) 8% = 0.08 We must add a zero to move the decimal point.

(c) 130% = 1.30

NOTE A percent greater than 100 gives a decimal greater than 1.

CHECK YOURSELF 4

Write as decimals.

(a) 5% **(b)** 32% **(c)** 115%

Look at Example 5, which involves fractions of a percent.

Example 5

Changing a Percent to a Decimal

Write as decimals.

(a) 4.5% = 0.045

(b) 0.5% = 0.005

CHECK YOURSELF 5

Write as decimals.

(a) 8.5% **(b)** 0.3%

Example 6

Changing a Percent to a Decimal

Write as decimals.

NOTE Write the common fractions as decimals. Then remove the percent symbol by using the Changing a Percent to a Decimal rule.

$$9\frac{1}{2}\% = 9.5\% = 0.095$$

$$\frac{3}{4}\% = 0.75\% = 0.0075$$

CHECK YOURSELF 6

Write as decimals.

(a) $7\frac{1}{2}\%$ **(b)** $\frac{1}{2}\%$

Changing a decimal to a percent is the opposite of changing from a percent to a decimal. We reverse the process we have just learned. Here is the rule:

Rules and Properties: Changing a Decimal to a Percent

To change a decimal to a percent, move the decimal point *two* places to the *right* and attach the percent symbol.

Example 7

Changing a Decimal to a Percent

Write 0.18 as a percent.

NOTE

$$0.18 = \frac{18}{100} = 18\left(\frac{1}{100}\right) = 18\%$$

$$0.18 = 18\%$$

CHECK YOURSELF 7

Write 0.27 *as a percent.*

Example 8

Changing a Decimal to a Percent

Write 0.03 as a percent.

NOTE

$$0.03 = \frac{3}{100} = 3\left(\frac{1}{100}\right) = 3\%$$

$$0.03 = 3\%$$

CHECK YOURSELF 8

Write 0.05 *as a percent.*

Example 9

Changing a Decimal to a Percent

NOTE

$1.25 = \frac{125}{100} = 125\left(\frac{1}{100}\right) = 125\%$

A decimal greater than 1 always gives a percent greater than 100.

Write 1.25 as a percent.

$1.25 = 125\%$

CHECK YOURSELF 9

Write 1.3 *as a percent.*

If the percent still includes a decimal after the decimal point is moved two places to the right, the fractional portion can be written as a decimal or as a fraction.

Example 10

Changing a Decimal to a Percent

NOTE

$0.045 = \frac{45}{1000} = \frac{45}{10}\left(\frac{1}{100}\right) = 4.5\%$

$0.003 = \frac{3}{1000} = \frac{3}{10}\left(\frac{1}{100}\right) = 0.3\%$

Write as a percent.

(a) $0.045 = 4.5\%$ or $4\frac{1}{2}\%$

(b) $0.003 = 0.3\%$ or $\frac{3}{10}\%$

CHECK YOURSELF 10

Write 0.075 *as a percent.*

This next rule allows us to change fractions to percents.

NOTE You may want to review Section 5.5 on writing decimal equivalents.

Rules and Properties: Changing a Fraction to a Percent

To change a fraction to a percent, write the decimal equivalent of the fraction. Then use the Changing a Decimal to a Percent rule to obtain the percent.

Example 11

Changing a Fraction to a Percent

Write $\frac{3}{5}$ as a percent.

First write the decimal equivalent.

$\frac{3}{5} = 0.6$ To find the decimal equivalent, just divide the denominator into the numerator.

Now write the percent.

NOTE Move the decimal point two places to the right and attach the percent symbol.

$\frac{3}{5} = 0.6 = 60\%$

CHECK YOURSELF 11

Write $\frac{3}{4}$ as a percent.

Again, you will find both decimals and fractions used in writing percents. Consider Example 12.

Example 12

Changing a Fraction to a Percent

Write $\frac{1}{8}$ as a percent.

$\frac{1}{8} = 0.125 = 12.5\%$ or $12\frac{1}{2}\%$

CHECK YOURSELF 12

Write $\frac{3}{8}$ as a percent.

Units Analysis

When computing a percentage, note that, in the result, the units are rarely expressed. Each time we say "percent" we are essentially saying, "numerator units per 100 denominator units."

Examples

Of 800 students, 200 were boys. What percent of the students were boys?

$\frac{\text{200 boys}}{\text{800 students}} = 0.25 = 25\%$

But what happened to our units? At the decimal, the units $\frac{\text{boys}}{\text{student}}$ (0.25 boys per student) wouldn't make much sense, but we can read the % as

25 "boys per 100 students"

and have a reasonable unit phrase.

Of 500 computers sold, 180 were equipped with a scanner. What percent were equipped with a scanner?

$\frac{180}{500} = 0.36 = 36\%$

36 computers were equipped with a scanner for each (per) 100 computers sold.

To write a mixed number as a percent, we use exactly the same steps.

Example 13

Changing a Mixed Number to a Percent

NOTE Notice that the resulting percent must be greater than 100 because the original mixed number was greater than 1.

Write $1\frac{1}{4}$ as a percent.

$1\frac{1}{4} = 1.25 = 125\%$

CHECK YOURSELF 13

Write $1\frac{2}{5}$ as a percent.

Some fractions have repeating-decimal equivalents. In writing these as percents, we will either round to some indicated place or use a fractional remainder form.

Example 14

Changing a Fraction to a Percent

Write $\frac{1}{3}$ as a percent.

NOTE We could say $\frac{1}{3}$ equals what part of 100?

$$\frac{1}{3} = \frac{x}{100}$$
$$3x = 100$$
$$x = 33\frac{1}{3}$$

$\frac{1}{3} = 0.33\overline{3} = 33\frac{1}{3}\%$

CHECK YOURSELF 14

Write $\frac{2}{3}$ as a percent.

Example 15

Changing a Fraction to a Percent

Write $\frac{5}{7}$ as a percent.

NOTE In this case, we round the decimal equivalent. Then we write the percent.

$\frac{5}{7} = 0.714$ (to the nearest thousandth)

$= 71.4\%$ (to the nearest tenth of a percent)

CHECK YOURSELF 15

Write $\frac{2}{9}$ to the nearest tenth of a percent.

CHECK YOURSELF ANSWERS

1. 8% were defective **2.** $12\% = \frac{12}{100} = \frac{3}{25}$ **3.** $1\frac{1}{4}$

4. **(a)** 0.05; **(b)** 0.32; **(c)** 1.15 **5.** **(a)** 0.085; **(b)** 0.003 **6.** **(a)** 0.075; **(b)** 0.005

7. 27% **8.** 5% **9.** 130% **10.** 7.5% or $7\frac{1}{2}\%$ **11.** $\frac{3}{4} = 0.75 = 75\%$

12. 37.5% or $37\frac{1}{2}\%$ **13.** $1\frac{2}{5} = 1.4 = 140\%$ **14.** $66\frac{2}{3}\%$ **15.** 22.2%

Name ____________

Section ______ Date ______

7.1 Exercises

Use percents to name the shaded portion of each drawing.

1. 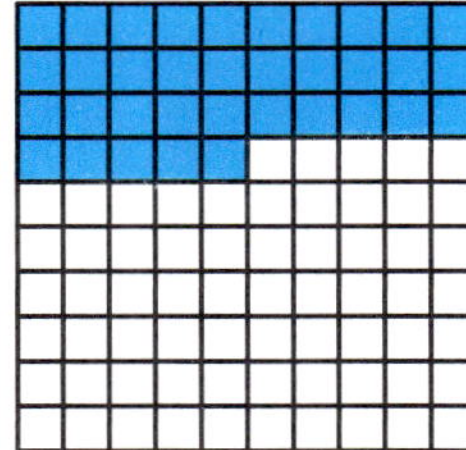

2. 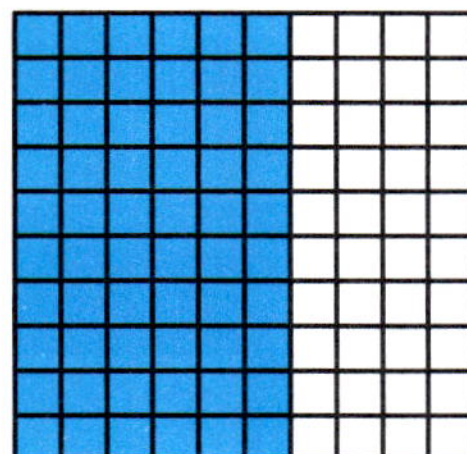

3. 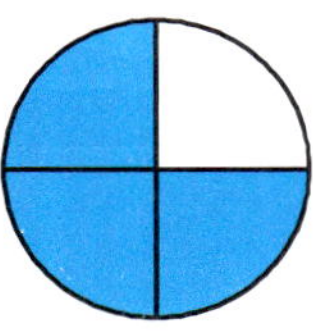

4. 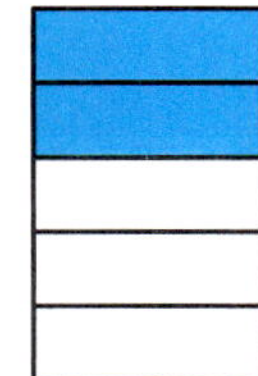

Rewrite each statement, using percent notation.

5. Out of every 100 eligible people, 53 voted in a recent election.

6. You receive \$5 in interest for every \$100 saved for 1 year.

7. Out of every 100 entering students, 74 register for English composition.

8. 17 out of 20 college students work at part-time jobs.

9. Of the 20 students in an algebra class, 5 receive a grade of A.

10. Of the 50 families in a neighborhood, 31 have children in public schools.

Write as fractions or mixed numbers.

11. 65% **12.** 48% **13.** 50%

14. 52% **15.** 46% **16.** 35%

17. 66% **18.** 4% **19.** 150%

20. 140% **21.** $166\frac{2}{3}\%$ **22.** $133\frac{1}{3}\%$

ANSWERS

1. ____
2. ____
3. ____
4. ____
5. ____
6. ____
7. ____
8. ____
9. ____
10. ____
11. ____
12. ____
13. ____
14. ____
15. ____
16. ____
17. ____
18. ____
19. ____
20. ____
21. ____
22. ____

ANSWERS

23. ______
24. ______
25. ______
26. ______
27. ______
28. ______
29. ______
30. ______
31. ______
32. ______
33. ______
34. ______
35. ______
36. ______
37. ______
38. ______
39. ______
40. ______

Write as decimals.

23. 20%

24. 70%

25. 5%

26. 7%

27. 135%

28. 250%

29. 23.6%

30. 10.5%

31. 6.4%

32. 3.5%

33. 0.2%

34. 0.5%

35. $7\frac{1}{2}\%$

36. $8\frac{1}{4}\%$

Solve the applications.

37. Travel. Automobiles account for 85% of the travel between cities in the United States. What fraction does this percent represent?

38. Travel. Automobiles and small trucks account for 84% of the travel to and from work in the United States. What fraction does this percent represent?

39. Explain the difference between $\frac{1}{4}$ of a quantity and $\frac{1}{4}\%$ of a quantity.

40. Match the percents in column A with their equivalent fractions in column B.

Column A	Column B
(a) $37\frac{1}{2}\%$	**(1)** $\frac{3}{5}$
(b) 5%	**(2)** $\frac{5}{8}$
(c) $33\frac{1}{3}\%$	**(3)** $\frac{1}{20}$
(d) $83\frac{1}{3}\%$	**(4)** $\frac{3}{8}$
(e) 60%	**(5)** $\frac{5}{6}$
(f) $62\frac{1}{2}\%$	**(6)** $\frac{1}{3}$

ANSWERS

41. ______

41. Complete the chart for the percentages given in the bar graph.

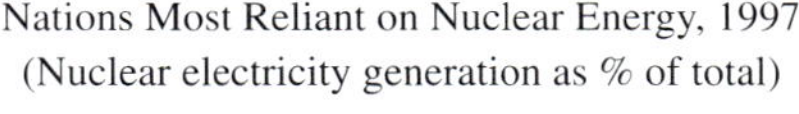

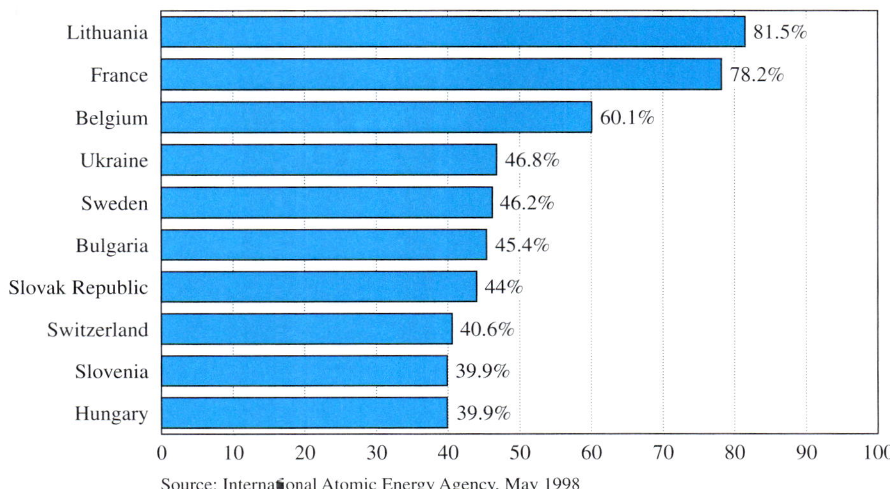

Source: International Atomic Energy Agency, May 1998

Country	Fraction Equivalent	Decimal Equivalent
Lithuania		
France		
Belgium		
Ukraine		
Sweden		
Bulgaria		
Slovak Republic		
Switzerland		
Slovenia		
Hungary		

42. (a) ____________

(b) ____________

(c) ____________

(d) ____________

(e) ____________

(f) ____________

42. The minimum daily values (MDV) for certain foods are given. They are based on a 2,000 calorie per day diet. Find decimal and fractional notation for the percent notation in each sentence.

(a) 1 ounce of Tostitos provides 9% of the MDV of fat.

(b) $\frac{1}{2}$ cup of B & M baked beans contains 15% of the MDV of sodium.

(c) $\frac{1}{2}$ cup of Campbells' New England clam chowder provides 6% of the MDV of iron.

(d) 2 ounces of Star Kist tuna provide 27% of the MDV of protein.

(e) Four 4-in. Aunt Jemima pancakes provide 33% of the MDV of sodium.

(f) 36 grams of Pop-Secret butter popcorn provide 2% of the MDV of sodium.

Write each decimal as a percent.

43. 0.05

44. 0.13

45. 0.18

46. 0.63

47. 0.7

48. 0.6

49. 1.10

50. 2.50

51. 4.40

52. 5

53. 0.004

54. 0.001

Write each fraction as a percent.

55. $\frac{1}{4}$

56. $\frac{4}{5}$

57. $\frac{2}{5}$

58. $\frac{1}{2}$

59. $3\frac{1}{2}$

60. $\frac{2}{3}$

61. $\frac{1}{6}$

62. $\frac{3}{16}$

63. $\frac{7}{9}$ (to nearest tenth of a percent)

64. $\frac{5}{11}$ (to nearest tenth of a percent)

In exercises 65 to 68, partially shaded decimal squares are given. Express the partially shaded region as (a) a decimal (b) a fraction, and (c) a percent.

65.

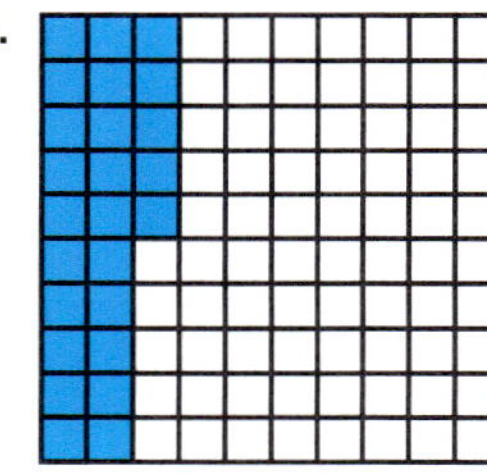

66.

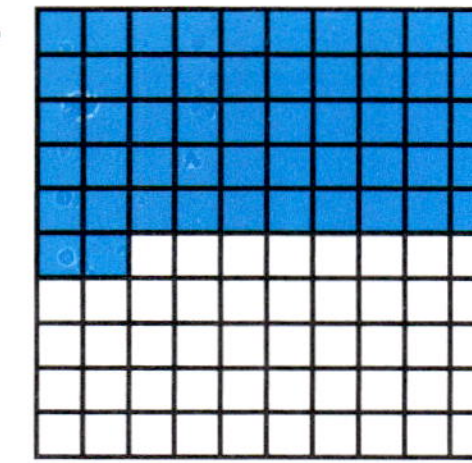

ANSWERS

43. ____
44. ____
45. ____
46. ____
47. ____
48. ____
49. ____
50. ____
51. ____
52. ____
53. ____
54. ____
55. ____
56. ____
57. ____
58. ____
59. ____
60. ____
61. ____
62. ____
63. ____
64. ____
65. ____
66. ____

67. ______

68. ______

69. ______

67. 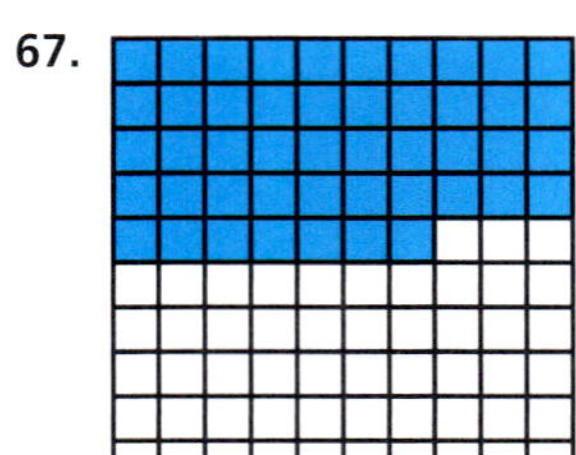

68. 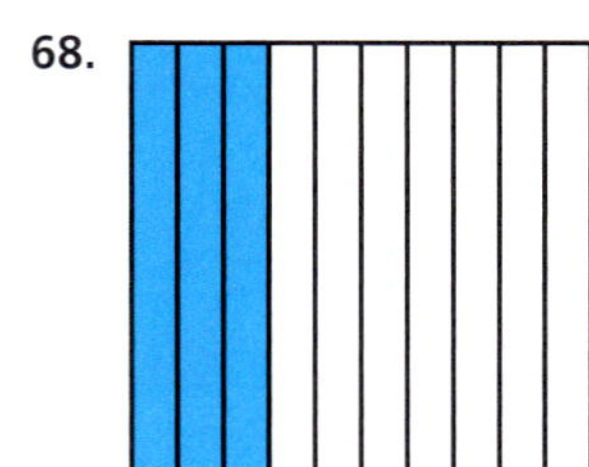

69. Complete the table of equivalents. Round decimals to the nearest ten thousandth. Round percents to the nearest hundredth of a percent.

Fraction	Decimal	Percent
$\frac{7}{12}$		
	0.08	
		35%
	0.265	
		$4\frac{3}{8}\%$
$\frac{11}{18}$		

Business travelers were asked how much they spent on different items during a business trip. The circle shows the results for every $1,000 spent. Use this information to answer exercises 70 to 73.

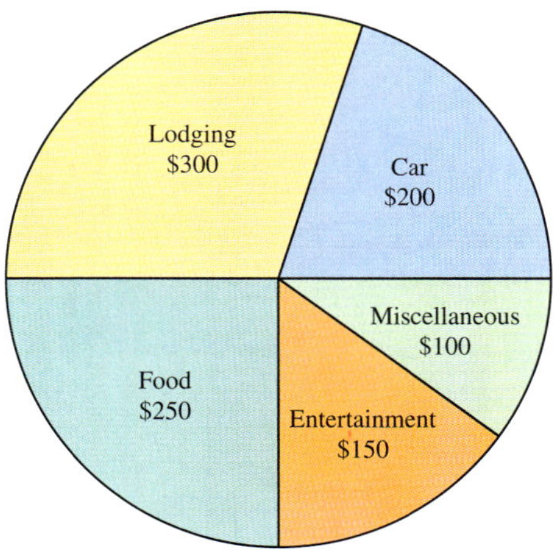

ANSWERS

70. ______

71. ______

72. ______

73. ______

70. What percent was spent on car expenses?

71. What percent was spent on food?

72. Where was the least amount of money spent? What percent was this?

73. What percent was spent on food and lodging together?

Answers

1. 35% **3.** 75% **5.** 53% of the eligible people voted.
7. 74% register for English composition.
9. $\frac{5}{20} = \frac{25}{100} = 25\%$; 25% of the students receive As. **11.** $\frac{13}{20}$ **13.** $\frac{1}{2}$
15. $\frac{23}{50}$ **17.** $\frac{33}{50}$ **19.** $1\frac{1}{2}$ **21.** $1\frac{2}{3}$ **23.** 0.2 **25.** 0.05 **27.** 1.35
29. 0.236 **31.** 0.064 **33.** 0.002 **35.** 0.075 **37.** $\frac{17}{20}$
39. 

41. See table.

Country	Fraction Equivalent	Decimal Equivalent
Lithuania	$\frac{163}{200}$	0.815
France	$\frac{391}{500}$	0.782
Belgium	$\frac{601}{1000}$	0.601
Ukraine	$\frac{117}{250}$	0.468
Sweden	$\frac{231}{500}$	0.462
Bulgaria	$\frac{227}{500}$	0.454
Slovak Republic	$\frac{11}{25}$	0.440
Switzerland	$\frac{203}{500}$	0.406
Slovenia	$\frac{399}{1000}$	0.399
Hungary	$\frac{399}{1000}$	0.399

43. 5% **45.** 18% **47.** 70% **49.** 110% **51.** 440% **53.** 0.4% or $\frac{2}{5}$%

55. $\frac{1}{4} = 0.25 = 25\%$ **57.** 40% **59.** 350% **61.** $16\frac{2}{3}\%$ **63.** 77.8%

65. 0.25, $\frac{1}{4}$, 25% **67.** 0.47, $\frac{47}{100}$, 47%

69. Fractions: $\frac{2}{25}, \frac{7}{20}, \frac{53}{200}, \frac{7}{160}$; decimals: 0.5833, 0.35, 0.0438, 0.6111; percents: 58.33%, 8%, 26.5%, 61.11%

71. 25% **73.** 55%

Solving Percent Problems Using Proportions

OBJECTIVES

1. Identify the parts of a percent problem
2. Solve percent problems using proportions

There are many practical applications of our work with percents. All these problems have three basic parts that need to be identified. We will look at some definitions that will help with that process.

Definitions: Base, Amount, and Rate

The **base** is the whole in a problem. It is the standard used for comparison.

The **amount** is the part of the whole being compared to the base.

The **rate** is the ratio of the amount to the base. It is written as a percent.

We will now look at an example of identifying the three parts in a percent problem.

Example 1

Identifying the Rate, Base, and Amount

Determine the rate, base, and amount in this problem:

12% of 800 is what number?

Finding the *rate* is not difficult. Just look for the percent symbol or the word *percent*. In this exercise, 12% is the rate.

The *base* is the whole. Here it follows the word *of*. 800 is the whole or the base.

The *amount* remains after the rate and the base have been found. Here the amount is the unknown. It follows the word *is*. "What number" asks for the unknown amount.

CHECK YOURSELF 1

Find the rate, base, and amount in the statements or questions.

(a) 75 is 25% of 300. **(b)** 20% of what number is 50?

We will use percents to solve a variety of applied problems. In all these situations, you will have to identify the three parts of the problem. We will work through some examples intended to help you build that skill.

Example 2

Identifying the Rate, Base, and Amount

Determine the rate, base, and amount in the application: In an algebra class of 35 students, 7 received a grade of A. What percent of the class received an A?

The *base* is the whole in the problem, or the number of students in the class. 35 is the base.

The *amount* is the portion of the base, here the number of students that receive the A grade. 7 is the amount.

The *rate* is the unknown in this example. "What percent" asks for the unknown rate.

CHECK YOURSELF 2

Determine the rate, base, and amount in the application: In a shipment of 150 parts, 9 of the parts were defective. What percent were defective?

We have now seen how to determine the amount, the rate, and the base. As you will see in the remainder of this section, our work in Chapter 6 with proportions will allow us to solve many types of percent problems in an identical fashion.

First, we will write what is called the **percent proportion.**

Rules and Properties: The Percent Proportion

$$\frac{\text{Amount}}{\text{Base}} = \frac{R}{100}$$

In symbols,

$$\frac{A}{B} = \frac{R}{100}$$

NOTE On the right, $\frac{R}{100}$ is the rate.

Because in any percent problem we know two of the three quantities (A, B, R), we can always solve for the unknown term. Consider the use of the percent proportion in Example 3.

Example 3

Solving a Problem Involving an Unknown Amount

NOTE This is an **unknown-amount problem.**

________ is 30% of 150.

↑ A ↑ R ↑ B

Substitute the values into the percent proportion.

$$\frac{A}{150} = \frac{30}{100}$$

(30 ← R; 150 ← B) The amount A is the unknown quantity of the proportion.

NOTE Recall that if $\frac{a}{b} = \frac{c}{d}$ then $ad = bc$

We solve the proportion with the methods of Section 6.3.

$$100A = 150 \cdot 30$$

$$100A = 4{,}500$$

Divide by the coefficient, 100.

$$\frac{\overset{1}{\cancel{100}}A}{\underset{1}{\cancel{100}}} = \frac{4500}{100} = 45$$

$$A = 45$$

The amount is 45. This means that 45 is 30% of 150.

CHECK YOURSELF 3

Use the percent proportion to answer this question: What is 24% of 300?

The same percent proportion will work if you want to find the rate.

Example 4

Solving a Problem Involving an Unknown Rate

NOTE This is an **unknown-rate problem.**

_______% of 400 is 72.

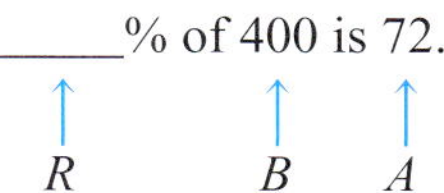

Substitute the known values into the percent proportion.

$$\frac{72}{400} = \frac{R}{100}$$

(A → 72, B → 400)

R, the rate, is the unknown quantity in this case.

Solving, we get

$$400R = 7{,}200$$

$$\frac{\overset{1}{\cancel{400}}R}{\underset{1}{\cancel{400}}} = \frac{7200}{400} = 18$$

$$R = 18$$

The rate is 18%. So 18% of 400 is 72.

CHECK YOURSELF 4

Use the percent proportion to answer this question: What percent of 50 is 12.5?

Finally, we use the same proportion to find an unknown base.

Example 5

Solving a Problem Involving an Unknown Base

NOTE This is an **unknown-base problem.**

40% of ______ is 200.

↑ R ↑ B ↑ A

Substitute the known values into the percent proportion.

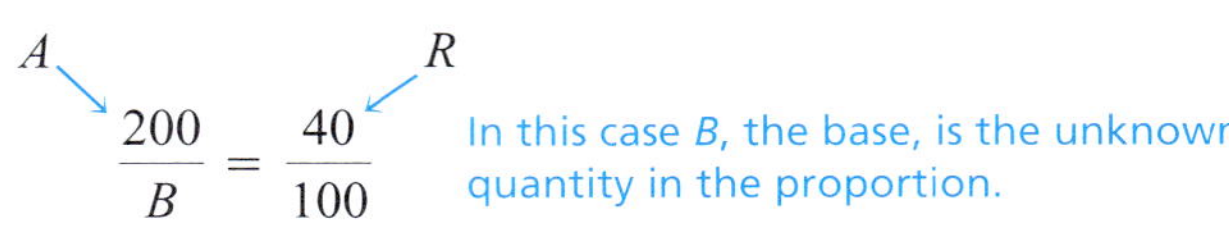

$$\frac{200}{B} = \frac{40}{100}$$

In this case B, the base, is the unknown quantity in the proportion.

Solving gives

$$40B = 200 \cdot 100$$

$$\frac{40B}{40} = \frac{20{,}000}{40} = 500$$

$$B = 500$$

The base is 500, and 40% of 500 is 200.

CHECK YOURSELF 5

288 is 60% of what number?

Remember that a percent (the rate) can be greater than 100.

Example 6

Solving a Percent Problem

NOTE The rate is 125%. The base is 300.

NOTE When the rate is greater than 100%, the amount will be *greater than* the base.

What is 125% of 300?

In the percent proportion, we have

$$\frac{A}{300} = \frac{125}{100}$$

So $100A = 300 \cdot 125$.

Dividing by 100 yields

$$A = \frac{37{,}500}{100} = 375$$

So 375 is 125% of 300.

CHECK YOURSELF 6

Find 150% of 500.

We next look at two examples of solving percent problems involving fractions of a percent.

Example 7

Solving a Percent Problem

34 is 8.5% of what number?

Using the percent proportion yields

NOTE The amount is 34; the rate is 8.5%. We want to find the base.

$$\frac{34}{B} = \frac{8.5}{100}$$

Solving, we have

$$8.5B = 34 \cdot 100$$

or

NOTE Divide by 8.5.

$$B = \frac{3400}{8.5} = 400$$

So 34 is 8.5% of 400.

CHECK YOURSELF 7

12.5% of what number is 75?

Estimating is an important and useful skill. When using a calculator, it is always a good idea to check to see whether the answer you get is reasonable. That is done by estimation. Sometimes operations with fractions can be used to estimate the result to a problem involving percents. Example 8 illustrates such a case.

Example 8

Estimating Percentages

Find 19.3% of 500.

Round the rate to 20% $\left(\text{as a fraction, } \frac{1}{5}\right)$. An estimate of the amount is then

$$\frac{1}{5} \cdot 500 = 100$$

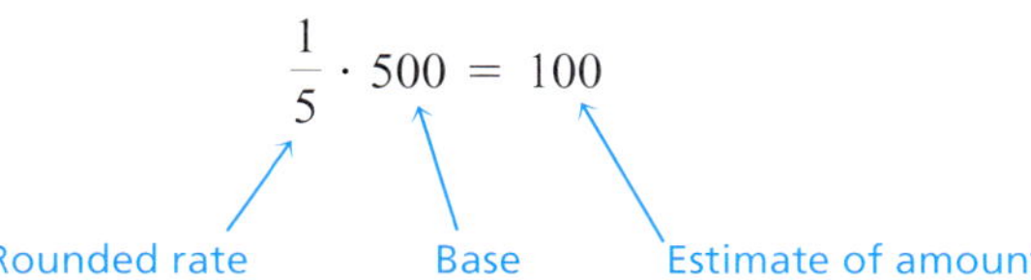

CHECK YOURSELF 8

Estimate the amount.

20.2% of 800

CHECK YOURSELF ANSWERS

1. **(a)** $R\% = 25\%$, $B = 300$, $A = 75$; **(b)** $R\% = 20\%$, $B =$ "what number," $A = 50$
2. $B = 150$, $A = 9$, $R\% =$ "what percent" (the unknown)
3. 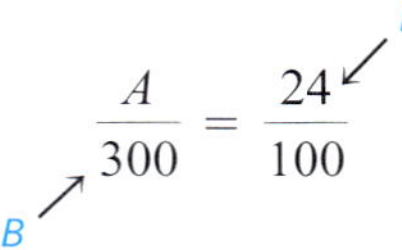

$100A = 7{,}200$; $A = 72$

4. A, B

$$\frac{12.5}{50} = \frac{R}{100}$$

$50R = 1{,}250$; $R\% = 25\%$

5. 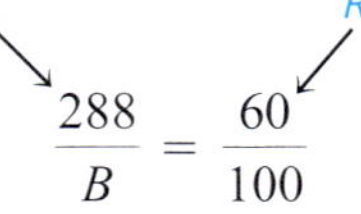

$60B = 28{,}800$; $B = 480$

6. 750
7. 600
8. 160

Name ______________________

7.2 Exercises

Section ________ Date ________

Solve each of the problems involving percent.

1. What is 35% of 600?

2. 20% of 400 is what number?

3. 45% of 200 is what number?

4. What is 40% of 1,200?

5. Find 40% of 2,500.

6. What is 75% of 120?

7. What percent of 50 is 4?

8. 51 is what percent of 850?

9. What percent of 500 is 45?

10. 14 is what percent of 200?

11. What percent of 200 is 340?

12. 392 is what percent of 2,800?

13. 46 is 8% of what number?

14. 7% of what number is 42?

15. Find the base if 11% of the base is 55.

16. 16% of what number is 192?

17. 58.5 is 13% of what number?

18. 21% of what number is 73.5?

19. Find 110% of 800.

20. What is 115% of 600?

21. What is 108% of 4,000?

22. Find 160% of 2,000.

23. 210 is what percent of 120?

24. What percent of 40 is 52?

25. 360 is what percent of 90?

26. What percent of 15,000 is 18,000?

27. 625 is 125% of what number?

28. 140% of what number is 350?

29. Find the base if 110% of the base is 935.

30. 130% of what number is 1,170?

31. Find 8.5% of 300.

32. $8\frac{1}{4}$% of 800 is what number?

33. Find $11\frac{3}{4}$% of 6,000.

34. What is 3.5% of 500?

35. What is 5.25% of 3,000?

36. What is 7.25% of 7,600?

37. 60 is what percent of 800?

38. 500 is what percent of 1,500?

39. What percent of 180 is 120?

40. What percent of 800 is 78?

ANSWERS

1. ______ 2. ______
3. ______ 4. ______
5. ______ 6. ______
7. ______ 8. ______
9. ______ 10. ______
11. ______ 12. ______
13. ______ 14. ______
15. ______ 16. ______
17. ______ 18. ______
19. ______ 20. ______
21. ______ 22. ______
23. ______ 24. ______
25. ______ 26. ______
27. ______ 28. ______
29. ______ 30. ______
31. ______ 32. ______
33. ______
34. ______
35. ______
36. ______
37. ______
38. ______
39. ______
40. ______

ANSWERS

41. ______
42. ______
43. ______
44. ______
45. ______
46. ______
47. ______
48. ______
49. ______
50. ______
51. ______
52. ______
53. ______
54. ______

55. ______
56. ______

41. What percent of 1,200 is 750?

42. 68 is what percent of 800?

43. 10.5% of what number is 420?

44. Find the base if $11\frac{1}{2}\%$ of the base is 46.

45. 58.5 is 13% of what number?

46. 6.5% of what number is 325?

47. 195 is 7.5% of what number?

48. 21% of what number is 73.5?

Estimate the amount in each of the exercises.

49. Find 25.8% of 4,000.

50. What is 48.3% of 1,500?

51. 74.7% of 600 is what number?

52. 9.8% of 1,200 is what number?

53. Find 152% of 400.

54. What is 118% of 5,000?

55. It is customary when eating in a restaurant to leave a 15% tip.

(a) Outline a method to do a quick approximation for the amount of tip to leave.

(b) Use this method to figure a 15% tip on a bill of \$47.76.

56. If the restaurant tax in a certain city is 9%, how might you use this to find a reasonable amount to leave as a tip?

Answers

1. 210 **3.** 90 **5.** 1,000 **7.** 8% **9.** 9% **11.** 170% **13.** 575 **15.** 500 **17.** 450 **19.** 880 **21.** 4,320 **23.** 175% **25.** 400% **27.** 500 **29.** 850 **31.** 25.5 **33.** 705 **35.** 157.5 **37.** 7.5% **39.** $66\frac{2}{3}\%$ **41.** 62.5% **43.** 4,000 **45.** 450 **47.** 2,600 **49.** 1,000 **51.** 450 **53.** 600 **55.**

7.3 Solving Percent Applications Using Equations

7.3 OBJECTIVES

1. Solve percent applications using equations
2. Solve percent applications involving sales tax
3. Solve percent applications involving commission
4. Solve percent applications involving discount

The concept of percent is perhaps the most frequently encountered mathematical idea that we will consider in this text. In this section and Sections 7.4 and 7.5, we will discuss a few of the most common applications. We will also discuss the special terms that are used in these applications. First, we will look at the equations generated from the percent proportion.

In Section 7.2, we found the answer to the question, "What is 30% of 150?" by setting up the proportion

$$\frac{A}{150} = \frac{30}{100}$$

Using the multiplication rule for equations from Chapter 4, we can rewrite this statement as

$$A = \frac{30}{100} \cdot 150$$

and solve for A. We could also obtain the same equation directly from the question: What is 30% of 150?

$$A = \frac{30}{100} \cdot 150$$

When we write an equation directly from a statement or question, we are using the **method of translation.**

Example 1

Translating an Application into an Equation

A salesman sells a used car for $9,500. His commission rate is 4%. What will be his commission for the sale?

First we must rephrase the problem as a single question. Here, we have, "What is 4% of $9,500?" Translate the question into an equation.

What is 4% of $9,500?

$$A = \frac{4}{100} \cdot 9{,}500$$

$$A = 4 \cdot 95$$

$$A = 380$$

His commission is $380.

CHECK YOURSELF 1

Jenny sells a $36,000 building lot. If her real estate commission rate is 5%, what commission will she receive for the sale?

If the question that is to be translated is not obvious, the equation can always be derived from the percent proportion.

Example 2

Solving a Percent Problem

A clerk sold \$3,500 in merchandise during 1 week. If he received a commission of \$140, what was the commission rate?

The base is \$3,500, and the amount is the commission of \$140. Using the percent proportion we have

$$\frac{140}{3500} = \frac{R}{100} \quad \text{or} \quad \frac{R}{100} = \frac{140}{3500}$$

Multiplying both sides by 100, we get

$$R = \frac{14{,}000}{3500}$$

$$R = 4$$

The commission rate is 4%.

CHECK YOURSELF 2

On a purchase of \$500 *you pay a sales tax of* \$21. *What is the tax rate?*

Example 3, involving a commission, shows how to find the total sold.

Example 3

Solving a Percent Problem

A saleswoman has a commission rate of 3.5%. To earn \$280, how much must she sell?

The rate is 3.5%. The amount is the commission, \$280. We want to find the base. In this case, this is the amount that the saleswoman needs to sell.

By the percent proportion

$$\frac{280}{B} = \frac{3.5}{100} \quad \text{or} \quad 3.5B = 280 \cdot 100$$

$$B = \frac{28{,}000}{3.5} = 8{,}000$$

The saleswoman must sell \$8,000 to earn \$280 in commissions.

CHECK YOURSELF 3

Kerri works with a commission rate of 5.5%. *If she wants to earn* \$825 *in commissions, find the total sales that she must make.*

Example 4

Solving a Percent Problem

A state taxes sales at 5.5%. How much sales tax will you pay on a purchase of $48?

The tax you pay is the amount (the part of the whole). Here the base is the purchase price, $48, and the rate is the tax rate, 5.5%. The question is,

NOTE In an application involving taxes, the tax paid is always the amount.

What is 5.5% of $48?

$$A = \frac{5.5}{100} \cdot 48$$

NOTE $48 \cdot 5.5 = 264$

$$A = \frac{264}{100} = 2.64$$

The sales tax paid is $2.64.

CHECK YOURSELF 4

Suppose that a state has a sales tax rate of $6\frac{1}{2}$%. If you buy a used car for $1,200, how much sales tax must you pay?

Percents are also used to deal with store markups or discounts. Consider Example 5.

Example 5

Solving a Percent Problem

A store marks up items to make a 30% profit. If an item costs $7.50 from the supplier, what will the selling price be?

The base is the cost of the item, $7.50, and the rate is 30%. In the percent proportion, the markup is the amount in this application.

The markup is 30% of $7.50.

$$A = \frac{30}{100} \cdot 7.5$$

$$A = 2.25$$

The markup is $2.25. Finally we have

Selling price = $7.50 + $2.25 = $9.75

Add the cost and the markup to find the selling price.

NOTE

Selling price = original cost + markup

CHECK YOURSELF 5

A store wants to discount (or mark down) an item by 25% for a sale. If the original price of the item was $45, find the sale price. [Hint: *Find the discount (the amount the item will be marked down), and subtract that from the original price.*]

One of the most common (and desirable!) applications of percents is the discount.

Example 6

Solving an Application Involving Discount

One of the benefits Vlade gets for working at worlddom.com is a 30% discount on all on-line purchases. Last month he ordered \$230 worth of software on-line. What was his total discount?

We rephrase the problem as a single question. Here, we have, "What is 30% of \$230?" Translate the question into an equation.

What is 30% of \$230?

$$A = \frac{30}{100} \cdot 230$$

$$A = 3 \cdot 23$$

$$A = 69$$

His discount was \$69.

CHECK YOURSELF 6

Indira receives a 26% discount for all of her tickets on AirGeorgia flights. What would be her discount on a \$580 flight?

In many everyday applications of percent, the computations required become quite lengthy, and so your calculator can be a great help. We will look at some examples.

Example 7

Solving a Problem Involving an Unknown Rate

In a test, 41 of 720 lightbulbs burn out before their advertised life of 700 h. What percent of the bulbs failed to last the advertised life?

We know the amount and base and want to find the percent (a rate). We will use the percent proportion for the solution.

$$\frac{41}{720} = \frac{R}{100}$$

(A → 41, B → 720)

$$720R = 4{,}100$$

$$R = \frac{4100}{720}$$

Now use your calculator to divide

4,100 [÷] 720 [=] 5.6944444

5.7% of the lightbulbs fail. We round the result to the nearest tenth of a percent.

CHECK YOURSELF 7

Last month, 35 of the 475 emergency calls received by the local police department were false alarms. What percent of the calls were false alarms?

Example 8

Solving a Problem Involving an Unknown Base

The price of a particular model of sofa has increased \$48.20. If this represents an increase of 9.65%, what was the price before the increase?

We want to find the base (the original price). Again, we will use the percent proportion for the solution.

A → \$48.20; R → 9.65

$$\frac{\$48.20}{B} = \frac{9.65}{100}$$

$$9.65B = 4{,}820$$

$$B = \frac{4820}{9.65}$$

Using the calculator gives

4,820 [÷] 9.65 [=] 499.48187

The original price was \$499.48.

Round to the nearest cent.

CHECK YOURSELF 8

The cost for medical insurance increased \$136.40 *last year. If this represents a* 12.35% *increase, what was the cost before the increase?*

Example 9 demonstrates an alternative method for solving percent problems.

Example 9

Solving a Problem Involving Markup

A store marks up items 22.5% to allow for profit. If an item costs a store \$36.40, what will be the selling price?

We can diagram the problem:

Cost	Markup
100% \$36.40	22.5% \$?

Selling price
122.5%

We could have done this example in two steps, finding the markup and then adding that amount to the cost.

This method allows you to do the problem in one step.

NOTE This approach may lead to time-consuming hand calculations, but using a calculator reduces the amount of work involved.

Now the base is $36.40 and the rate is 122.5%, and we want to find the amount (the selling price).

$$\frac{A}{36.40} = \frac{122.5}{100}$$

so

$$A = \frac{122.5 \cdot 36.40}{100} = \$44.59$$

The selling price should be $44.59.

CHECK YOURSELF 9

An item costs $75.40. If the markup is 36.2%, what is the selling price?

A similar approach will allow us to solve problems that involve a decrease in one step.

Example 10

Solving a Problem Involving an Unknown Amount

NOTE We could have done a problem like this by finding the decrease and then subtracting from the original value. Again, using this method requires just one step.

Paul invests $5,250 in a piece of property. At the end of a 6-month period, the value has decreased 7.5%. What is the property worth at the end of the period?

Again, we will diagram the problem.

Original value 100% or $5,250	
Decrease 7.5%	Ending value 100% − 7.5% = 92.5%

So the amount (ending value) is found as

$$\frac{A}{5250} = \frac{92.5}{100}$$

$$A = \frac{92.5 \cdot \$5250}{100} = \$4,856.25$$

The ending value is $4,856.25

CHECK YOURSELF 10

Tom buys a baseball card collection for $750. After 1 year, the value has decreased 8.2%. What is the value of the collection after 1 year?

CHECK YOURSELF ANSWERS

1. $1,800 **2.** 4.2% **3.** $15,000 **4.** $78 **5.** $33.75 **6.** $150.80
7. 7.4% **8.** $1,104.45 **9.** $102.69 **10.** $688.50

Name ______________________

Section __________ Date __________

Exercises

Solve each of the applications

1. **Commission.** If a salesman is paid a \$140 commission on the sale of a \$2,800 sailboat, what is his commission rate?

2. **Commission.** A real estate agent's commission rate is 6%. What will be the amount of the commission on the sale of an \$85,000 home?

3. **Sales tax.** A state sales tax is levied at a rate of 6.4%. How much tax would one pay on a purchase of \$260?

4. **Sales tax.** A state sales tax rate is 3.5%. If the tax on a purchase is \$7, what was the price of the purchase?

5. **Commission.** If a house sells for \$125,000 and the commission rate is $6\frac{1}{2}$%, how much will the salesperson make for the sale?

6. **Commission.** A saleswoman is working on a 5% commission basis. If she wants to make \$1,800 in 1 month, how much must she sell?

7. **Markup.** An appliance dealer marks up refrigerators 22% (based on cost). If the cost of one model was \$600, what will be its selling price?

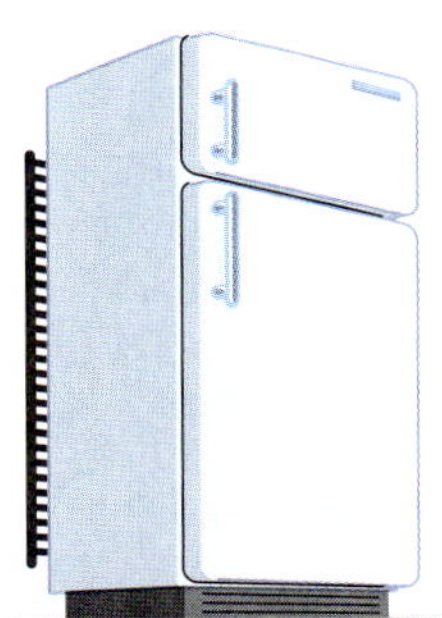

8. **Markdown.** A television set is marked down \$75, to be placed on sale. If this is a 12.5% decrease from the original price, what was the selling price before the sale?

9. **Markdown.** A stereo system is marked down from \$450 to \$382.50. What is the discount rate?

ANSWERS

1. ______________________
2. ______________________
3. ______________________
4. ______________________
5. ______________________
6. ______________________
7. ______________________
8. ______________________
9. ______________________

ANSWERS

10. ______

11. ______

12. ______

13. ______

14. ______

15. ______

16. ______

17. ______

18. ______

19. ______

20. ______

21. ______

22. ______

23. ______

24. ______

10. **Sales tax.** At True Grip hardware, you pay \$10 in tax for a barbecue grill, which is 6% of the purchase price. At Loose Fit hardware, you pay \$10 in tax for the same grill, but it is 8% of the purchase price. At which store do you get the better buy? Why?

11. **Markup.** A store marks up merchandise 25% to allow for profit. If an item costs the store \$11, what will be its selling price?

12. **Sales.** What were Jamal's total sales in a given month if he earned a commission of \$2,458 at a commission rate of 1.6%?

Calculator Exercises

Solve each of the percent problems.

13. What percent is 648 of 8,640?

14. 53.1875 is 9.25% of what number?

15. Find 7.65% of 375.

16. 17.4 is what percent (to the nearest tenth) of 81.5?

17. Find the base if 18.2% of the base is 101.01.

18. What is 3.52% of 2,450?

19. What percent (to the nearest tenth) of 1,625 is 182?

20. 22.5% of what number is 3,762?

21. **Markdown.** A dealer marks down the last year's model appliances 22.5% for a sale. If the regular price of an air conditioner was \$279.95, how much will it be discounted (to the nearest cent)?

22. **Population.** The population of a town increases 4.2% in 1 year. If the original population was 19,500, what is the population after the increase?

23. **Markup.** A store marks up items 42.5% to allow for profit. If an item costs a store \$24.40, what will be its selling price?

24. **Markdown.** A jacket that originally sold for \$98.50 is marked down by 12.5% for a sale. Find its sale price (to the nearest cent).

Answers

1. 5% **3.** \$16.64 **5.** \$8,125 **7.** \$732 **9.** 15% **11.** \$13.75
13. 7.5% **15.** 28.6875 **17.** 555 **19.** 11.2% **21.** \$62.99
23. \$34.77

Applications: Simple and Compound Interest

OBJECTIVES

1. Solve percent applications involving simple interest
2. Solve percent applications involving compound interest

NOTE The money borrowed or saved is called the **principal.**

As we said earlier, there are many applications of percents. One that almost all of us encounter involves **interest.** When you borrow money, you pay interest. When you place money in a savings account, you earn interest. Interest is a percent of the whole (in this case, the *principal*), and the percent is called the **interest rate.**

Example 1

Solving a Problem Involving an Unknown Amount

Find the interest you must pay if you borrow \$2,000 for 1 year with an interest rate of $9\frac{1}{2}\%$.

The base (the principal) is \$2,000, the rate is $9\frac{1}{2}\%$, and we want to find the interest (the amount). The question becomes,

REMEMBER:

$9\frac{1}{2}\% = 9.5\%$

What is 9.5% of \$2,000?

$$A = \frac{9.5}{100} \cdot 2{,}000 = 0.095 \cdot 2{,}000$$

$A = 190$ The interest (amount) is \$190.

CHECK YOURSELF 1

You invest \$5,000 *for* 1 *year at* $8\frac{1}{2}\%$. *How much interest will you earn?*

Example 2

Solving a Problem Involving an Unknown Base

Ms. Hobson agrees to pay 11% interest on a loan for her new automobile. She is charged \$2,200 interest on a loan for 1 year. How much did she borrow?

The rate is 11%. The amount, or interest, is \$2,200. We want to find the base, which is the principal, or the size of the loan. To solve the problem, we have

$$\frac{2200}{B} = \frac{11}{100}$$

$$11B = 2{,}200 \cdot 100$$

$$B = \frac{220{,}000}{11} = 20{,}000$$

She borrowed \$20,000.

CHECK YOURSELF 2

Sue paid \$210 *interest for a* 1-*year loan at* 10.5%. *What was the size of her loan?*

When looking at the total amount paid for a loan, it is helpful to be able to actually calculate the interest rate.

Example 3

Finding the Rate of Interest

Jaime borrowed \$1,200 from his brother to make a down payment on his car. At the end of a year, he paid back \$1,450. What was the rate of interest?

When finding rate, it is usually easiest to use the percent proportion. First, notice that the amount of interest (A) was \$250. He paid back \$250 more than he borrowed. With that in mind, we have the proportion

$$\frac{250}{1200} = \frac{R}{100} \qquad \text{so} \qquad R = \frac{250}{1200} \cdot 100$$

$$R = 20.83 \qquad \text{The rate was } 20.83\%.$$

CHECK YOURSELF 3

Dean borrowed \$500 from his sister to pay his fall tuition. He paid her back \$580 a year later. What was the rate of interest?

Many percent problems involve calculating what is known as **compound interest.**

Suppose that you invest \$1,000 at 5% in a savings account for 1 year. For year 1, the interest is 5% of \$1,000, or $0.05 \cdot \$1{,}000 = \$50$. At the end of year 1, you will have \$1,050 in the account.

Now if you leave that amount in the account for a second year, the interest will be calculated on the original principal, \$1,000, plus the first year's interest, \$50. This is called *compound interest.*

\$1,050 —At 5%→ \$1,102.50
Year 1 Year 2

For year 2, the interest is 5% of \$1,050, or $0.05 \cdot \$1{,}050 = \$52.50$. At the end of the second year, you will have \$1,102.50 in the account.

Example 4

Calculating Compound Interest

Brittney invested \$500 in a certificate that will pay her 7.5% interest per year. What will be the value of her investment in 2 years? To find the total, calculate the interest for each year and then add it to the previous year's total.

\$ 500.00	Start
+ 37.50	First-year interest = $500.00 \cdot 0.075 = 37.50$
537.50	Year 1
+ 40.31	Second-year interest = $537.50 \cdot 0.075 = 40.31$
577.81	Year 2

Her investment will have a value of \$577.81.

CHECK YOURSELF 4

Dominic invested \$1,000 at 8.5% per year. What will be the value of his investment in 3 years?

CHECK YOURSELF ANSWERS

1. \$425 **2.** \$2,000 **3.** 16% **4.** \$1,277.29

7.4 Exercises

Name ______________

Section ________ Date ________

Solve each of the applications.

1. Interest. What interest will you pay on a $3,400 loan for 1 year if the interest rate is 12%?

2. Interest. Ms. Jordan has been given a loan of $2,500 for 1 year. If the interest charged is $275, what is the interest rate on the loan?

3. Interest. Joan was charged $18 interest for 1 month on a $1,200 credit card balance. What was the monthly interest rate?

4. Loans. Patty pays $525 interest for a 1-year loan at 10.5%. How much was her loan?

5. Interest. A time-deposit savings plan gives an interest rate of 6.42% on deposits. If the interest on an account for 1 year was $545.70, how much was deposited?

Complete the table for exercises 6 to 14.

	Principal	Interest Rate	Interest
6.	$1,000	8.75%	
7.	$3,000	9.25%	
8.	$12,000	6.5%	
9.	$3,450		$258.75
10.	$36,500		$3,467.50
11.		6.75%	$238.95
12.		5.75%	$727.09
13.	$14,640		$1,207.80
14.	$9,850		$418.63

ANSWERS

1. ______________
2. ______________
3. ______________
4. ______________
5. ______________
6. ______________
7. ______________
8. ______________
9. ______________
10. ______________
11. ______________
12. ______________
13. ______________
14. ______________

ANSWERS

15. ______

16. ______

17. ______

18. ______

In exercises 15 to 18, assume the interest is compounded annually (at the end of each year), and find the amount in an account with the given interest rate and principal.

15. \$4,000, 6%, 2 years

16. \$3,000, 7%, 2 years

17. \$4,000, 5%, 3 years

18. \$5,000, 6%, 3 years

Answers

1. \$408 **3.** 1.5% **5.** \$8,500 **7.** \$277.50 **9.** 7.5% **11.** \$3,540
13. 8.25% **15.** \$4,494.40 **17.** \$4,630.50

7.5 More Applications of Percent

OBJECTIVE

1. Solve a variety of percent applications

Percents are used in too many ways for us to list. Look at the variety in the examples presented, which illustrate some additional situations in which you will find percents.

Example 1

Solving a Problem Involving an Unknown Amount

NOTE A *rate, base,* and *amount* will appear in *all* problems involving percents.

A student needs 70% to pass an examination containing 50 questions. How many questions must she answer correctly?

The *rate* is 70%. The *base* is the number of questions on the test, here 50. The *amount* is the number of questions that must be correct.

To find the amount, we will use the percent proportion from Section 7.2.

NOTE Substitute 50 for B and 70 for R.

$$B \longrightarrow \frac{A}{50} = \frac{70}{100} \longleftarrow R$$

so

$$100A = 50 \cdot 70$$

Dividing by 100 gives

$$A = \frac{3500}{100} = 35$$

She must answer 35 questions correctly to pass.

CHECK YOURSELF 1

Generally, 72% of the students in a chemistry class pass the course. If there are 150 students in the class, how many can be expected to pass?

We will look at an application that requires finding the rate.

Example 2

Solving a Problem Involving an Unknown Rate

Simon works at a restaurant called La Catalana. The \$45 tip he received from a family on Friday was the largest tip Simon has ever received. If the bill totaled \$250 before the tip was added, what percent of the total was the tip?

The base is the total of the bill, \$250. The amount is the \$45 tip. To find the percentage, we again use the percent proportion.

$$\frac{45}{250} = \frac{R}{100}$$

$$250 \cdot R = 45 \cdot 100$$

$$250R = 4{,}500$$

$$R = \frac{4500}{250} = 18$$

The tip was 18% of the bill.

CHECK YOURSELF 2

Last year, Xian reported an income of \$27,500 *on her tax return. Of that, she paid* \$6,600 *in taxes. What percent of her income went to taxes?*

Increases and decreases are often stated in terms of percents, as Examples 3 to 5 illustrate.

Example 3

Solving a Percent Problem

The population of a town increased 15% during a 3-year period. If the original population was 12,000, what was the population at the end of the 3 years?

First we find the increase in the population. That increase is the amount in the problem.

$$\frac{A}{12,000} = \frac{15}{100} \quad \text{so} \quad 100A = 15 \cdot 12,000$$

$$A = \frac{180,000}{100}$$

$$= 1,800$$

To find the population at the end of the period, we add

$$12,000 + 1,800 = 13,800$$

Original population (12,000) Increase (1,800) New population (13,800)

CHECK YOURSELF 3

A school's enrollment decreased by 8% *from a given year to the next. If the enrollment was* 550 *students the first year, how many students were enrolled the second year?*

Example 4

Solving a Percent Problem

Enrollment at a school increased from 800 to 888 students from a given year to the next. What was the rate of increase?

First we must subtract to find the amount of the increase.

Increase: $888 - 800 = 88$ students

Now to find the rate, we have

NOTE We use the *original* enrollment, 800, as our base.

$$\frac{88}{800} = \frac{R}{100} \quad \text{so} \quad 800R = 88 \cdot 100$$

$$R = \frac{8800}{800} = 11$$

The enrollment increased at a rate of 11%.

CHECK YOURSELF 4

Car sales at a dealership decreased from 350 units one year to 322 units the next. What was the rate of decrease?

Example 5

Solving a Percent Problem

A company hired 18 new employees in 1 year. If this was a 15% increase, how many employees did the company have before the increase?

The rate is 15%. The amount is 18, the number of new employees. The base in this problem is the number of employees *before the increase*. So

$$\frac{18}{B} = \frac{15}{100}$$

$$15B = 18 \cdot 100 \quad \text{or} \quad B = \frac{1800}{15} = 120$$

The company had 120 employees before the increase.

CHECK YOURSELF 5

A school had 54 new students in one term. If this was a 12% increase over the previous term, how many students were there before the increase?

There are many computer-related applications that include percentages as either part of the problem or as part of the solution.

Example 6

Solving a Computer Application

A computer loaded 60% of a new program in 120 seconds. How long should it take to load the entire program?

The rate is 60%, the amount is 120 seconds. The base is the total time taken to load the program.

$$\frac{120}{B} = \frac{60}{100}$$

$$60B = 12{,}000$$

$$B = 200$$

It will take 200 seconds to load the program.

CHECK YOURSELF 6

A virus scanning program is checking every computer file for viruses. It checked 30% of the files in 240 seconds. How long should it take to check all of the files?

CHECK YOURSELF ANSWERS

1. 108 **2.** 24% **3.** 506 **4.** 8% **5.** 450 **6.** 800 s (13 min 20 s)

Name ______________________

Section ________ Date ________

7.5 Exercises

Solve each of the applications.

ANSWERS

1. ______________
2. ______________
3. ______________
4. ______________
5. ______________

1. **Payroll deductions.** Roberto has 26% of his pay withheld for deductions. If he earns $550 per week, what amount is withheld?

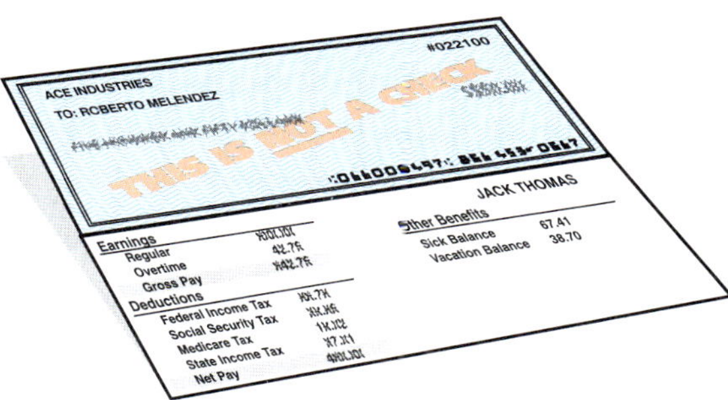

2. **Chemistry.** A chemist has 300 mL of solution that is 18% acid. How many milliliters of acid are in the solution?

3. **Test scores.** On a test, Alice had 80% of the problems right. If she had 20 problems correct, how many questions were on the test?

4. **Down payment.** Betty must make a $9\frac{1}{2}\%$ down payment on the purchase of a $2,000 motorcycle. How much must she pay down?

5. **Chemistry.** There are 117 mL of acid in 900 mL of a solution of acid and water. What percent of the solution is acid?

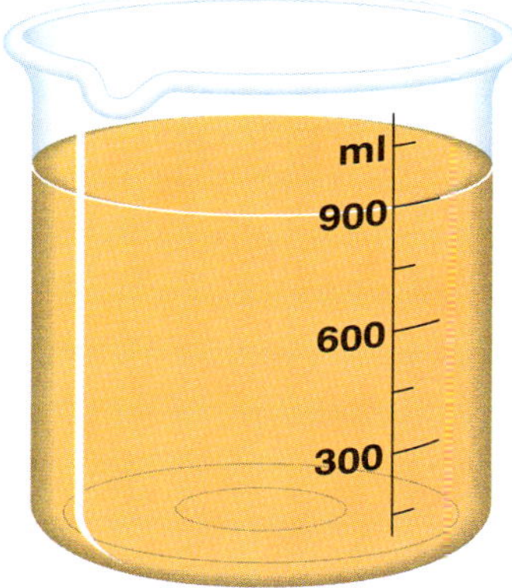

ANSWERS

6. ______
7. ______
8. ______
9. ______
10. ______
11. ______
12. ______
13. ______
14. ______
15. ______
16. ______
17. ______
18. ______

6. **Test scores.** Marla needs 70% on a final exam to pass the course. If the exam has 120 questions, how many questions must she answer correctly?

7. **Unemployment.** A study has shown that 102 of the 1,200 people in the workforce of a small town are unemployed. What is the town's unemployment rate?

8. **Surveys.** A survey of 400 people found that 66 were left-handed. What percent of those surveyed were left-handed?

9. **Dropout rate.** Of 60 people who started a training program, 45 completed the course. What was the dropout rate?

10. **Manufacturing.** In a shipment of 250 parts, 40 are found to be defective. What percent of the parts are faulty?

11. **Surveys.** In a recent survey, 65% of those responding were in favor of a freeway improvement project. If 780 people were in favor of the project, how many people responded to the survey?

12. **Enrollments.** A college finds that 42% of the students taking a foreign language are enrolled in Spanish. If 1,512 students are taking Spanish, how many foreign language students are there?

13. **Salary.** 22% of Samuel's monthly salary is deducted for withholding. If those deductions total $627, what is his salary?

14. **Budgets.** The Townsend's budget 36% of their monthly income for food. If they spend $864 on food, what is their monthly income?

15. **Land value.** A home lot purchased for $26,000 increased in value by 25% over 3 years. What was the lot's value at the end of the period?

16. **Depreciation.** New cars depreciate an average of 28% in their first year of use. What will a $9,000 car be worth after 1 year?

17. **Enrollment.** A school's enrollment was up from 950 students in 1 year to 1,064 students in the next. What was the rate of increase?

18. **Salary.** Under a new contract, the salary for a position increases from $22,000 to $23,780. What rate of increase does this represent?

ANSWERS

19. ____________
20. ____________
21. ____________
22. ____________
23. ____________
24. ____________
25. ____________
26. ____________
27. ____________
28. ____________

19. Price changes. The price of a new van has increased \$4,060, which amounts to a 14% increase. What was the price of the van before the increase?

20. Business. The electricity costs of a business decrease from \$12,000 one year to \$10,920 the next. What is the rate of decrease?

21. Workforce. A company had 66 fewer employees in July 2001 than in July 2000. If this represents a 5.5% decrease, how many employees did the company have in July 2000?

22. Salary. Carlotta received a monthly raise of \$162.50. If this represented a 6.5% increase, what was her monthly salary before the raise?

23. Stock. Mr. Hernandez buys stock for \$15,000. At the end of 6 months, the stock's value has decreased 7.5%. What is the stock worth at the end of the period?

24. Population. The population of a town increases 14% in 2 years. If the population was 6,000 originally, what is the population after the increase?

25. Computers. A computer loaded 80% of a new program in 80 seconds. How long should it take to load the entire program?

26. Payroll deductions. Frank's pay is \$450 per week. If deductions from his paycheck average 25%, what is the amount of his weekly paycheck (after deductions)?

27. Consumer affairs. The two ads pictured appeared last week and this week in the local paper. Is this week's ad accurate?

28. Computers. A virus scanning program is checking every file for viruses. It has completed checking 40% of the files in 300 seconds. How long should it take to check all the files?

ANSWERS

29. ______

30. ______

31. ______

32. ______

33. ______

34. ______

29. Retail sales. A pair of shorts is advertised for \$48.75 and as being 25% off the original price. What was the original price?

30. Tipping. If the total bill at a restaurant, including a 15% tip, is \$65.32, what was the cost of the meal alone?

The chart shows U.S. trade with Mexico from 1992 to 1997. Use this information for exercises 31 to 34.

U.S. Trade with Mexico, 1992–97

(millions of dollars) MEXICO			
Year	**Exports**	**Imports**	**Trade Balance**[1]
1992	\$40,592	\$35,211	\$5,381
1993	41,581	39,917	1,664
1994[2]	50,844	49,494	1,350
1995	46,292	61,685	−15,393
1996	56,792	74,297	−17,506
1997	71,388	85,938	−14,549

(1) Totals may not add due to rounding.
(2) NAFTA provisions began to take effect Jan. 1, 1994.
Source: Office of Trade and Economic Analysis. U.S. Dept. of Commerce.

31. What is the rate of increase (to the nearest whole percent) of exports from 1992 to 1997?

32. What is the rate of increase (to the nearest whole percent) of imports from 1992 to 1997?

33. By what percent did exports exceed imports in 1992?

34. By what percent did imports exceed exports in 1997?

Solve the applications.

35. **Automobiles.** In 1990, there were an estimated 145.0 million passenger cars registered in the United States. The total number of vehicles registered in the United States for 1990 was estimated at 194.5 million. What percent of the vehicles registered were passenger cars?

36. **Gasoline.** Gasoline accounts for 85% of the motor fuel consumed in the United States every day. If 8,882 thousand barrels (bbl) of motor fuel are consumed each day, how much gasoline is consumed each day in the United States?

37. **Petroleum.** In 1989, transportation accounted for 63% of U.S. petroleum consumption. If 10.85 million bbl of petroleum are used each day for transportation in the United States, what is the total daily petroleum consumption by all sources in the United States?

38. **Pollution.** Each year, 540 million metric tons (t) of carbon dioxide are added to the atmosphere by the United States. Burning gasoline and other transportation fuels is responsible for 35% of the carbon dioxide emissions in the United States. How much carbon dioxide is emitted each year by the burning of transportation fuels in the United States?

39. The progress of the local Lions club is shown. What percent of the goal has been achieved so far?

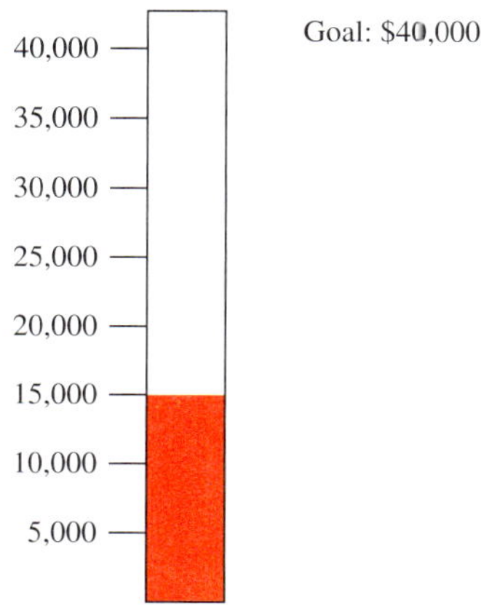

ANSWERS

35. ______

36. ______

37. ______

38. ______

39. ______

Answers

1. \$143 **3.** 25 questions **5.** 13% **7.** 8.5% **9.** 25%
11. 1,200 people **13.** \$2,850 **15.** \$32,500 **17.** 12% **19.** \$29,000
21. 1,200 employees **23.** \$13,875 **25.** 100 s **27.** **29.** \$65
31. 76% **33.** +15% **35.** 74.6% **37.** 17.22 million bbl **39.** 37.5%

7 Summary

DEFINITION/PROCEDURE	EXAMPLE	REFERENCE
Percents, Decimals, and Fractions		**Section 7.1**
Percent Another way of naming parts of a whole. Percent means per hundred.	Fractions and decimals are other ways of naming parts of a whole. $21\% = 21\left(\frac{1}{100}\right) = \frac{21}{100} = 0.21$	**p. 539**
1. *To convert a percent to a fraction,* replace the percent symbol with $\frac{1}{100}$ and multiply.	$37\% = 37\left(\frac{1}{100}\right) = \frac{37}{100}$	**p. 540**
2. *To convert a percent to a decimal,* remove the percent symbol, and move the decimal point two places to the left.	$37\% = 0.37$	**p. 541**
3. *To convert a decimal to a percent,* move the decimal point two places to the right and attach the percent symbol.	$0.58 = 58\%$	**p. 542**
4. *To convert a fraction to a percent,* write the decimal equivalent of the fraction, and then change that decimal to a percent.	$\frac{3}{5} = 0.60 = 60\%$	**p. 543**
Solving Percent Problems Using Proportions		**Section 7.2**
Every percent problem has the following three parts: **1.** *The base.* This is the whole in the problem. It is the standard used for comparison. Label the base B. **2.** *The amount.* This is the part of the whole being compared to the base. Label the amount A. **3.** *The rate.* This is the ratio of the amount to the base. The rate is written as a percent. Label the rate $\frac{R}{100}$ or $R\%$.	45 is 30% of 150. ↑ A ↑ R ↑ B	**p. 555**
Using the Percent Proportion The percent proportion is $\frac{A}{B} = \frac{R}{100}$ To solve a percent problem using this proportion: **1.** Substitute the two known values into the proportion. **2.** Solve the proportion to find the unknown value.	What is 24% of 300? $\frac{A}{300} = \frac{24}{100}$ $100A = 7{,}200$ $A = 72$	**p. 556**

Continued

DEFINITION/PROCEDURE	EXAMPLE	REFERENCE
Solving Percent Applications Using Equations		**Section 7.3**
Translating a question into an equation: What is R% of B? $A = \frac{R}{100} \cdot B$	What is 23% of 300? $A = \frac{23}{100} \cdot 300$ $A = 23 \cdot 3$ $= 69$	p. 563
Sales tax, commissions, and **discounts** are all rates. Applications can be solved by the percent proportion or by translating the question into an equation.	If a 12% commission is \$3,000, what is the total? $\frac{3000}{B} = \frac{12}{100}$ $12B = 300{,}000$ $B = \$25{,}000$	p. 563
Applications: Simple and Compound Interest		**Section 7.4**
Simple interest is a rate. **Compound interest** is interest on an amount that includes previous interest.		p. 571

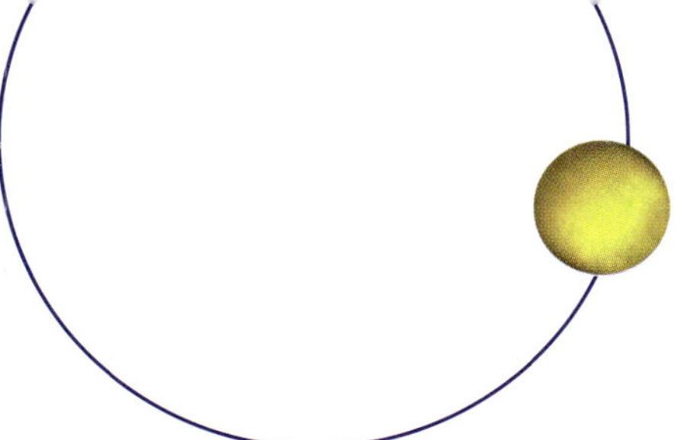

Summary and Review Exercises

You should now be reviewing the material in Chapter 7. The exercises will help in that process. Work all of the exercises carefully. References are provided to the section for each exercise. If you make an error, go back and review the related material.

[7.1]

1. Use a percent to name the shaded portion of the diagram.

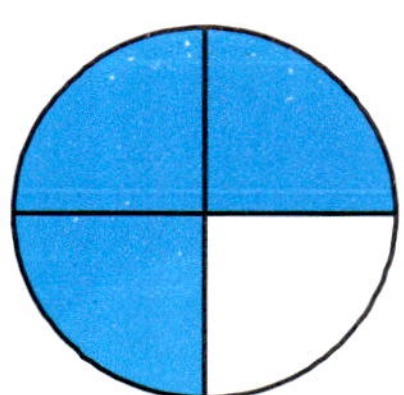

In exercises 2 to 7, write the percent as a common fraction or a mixed number.

2. 2%

3. 20%

4. 37.5%

5. 150%

6. $233\frac{1}{3}\%$

7. 300%

In exercises 8 to 13, write the percents as decimals.

8. 75%

9. 4%

10. 6.25%

11. 13.5%

12. 0.6%

13. 225%

In exercises 14 to 25, write as percents.

14. 0.06

15. 0.375

16. 2.4

17. 7

18. 0.035

19. 0.005

20. $\frac{43}{100}$

21. $\frac{7}{10}$

22. $\frac{2}{5}$

23. $1\frac{1}{4}$

24. $2\frac{2}{3}$

25. $\frac{3}{11}$ (to nearest tenth of a percent)

[7.2] In exercises 26 to 31, find the unknown using the percent proportion.

26. 80 is 4% of what number?

27. 70 is what percent of 50?

28. 11% of 3,000 is what number?

29. 24 is what percent of 192?

30. Find the base if 12.5% of the base is 625.

31. 90 is 120% of what number?

[7.3] In exercises 32 to 37, find the unknown using an equation.

32. What is 9.5% of 700?

33. Find 150% of 50.

34. Find the base if 130% of the base is 780.

35. 350 is what percent of 200?

36. What is 225% of 48?

37. 28.8 is what percent of 960?

Estimate the amount in the exercises.

38. 24.3% of 810

39. 109% of 592

[7.4] In exercises 40 to 43, solve the applications.

40. Commission. Joan works on a 4% commission basis. She sold $45,000 in merchandise during 1 month. What was the amount of her commission?

41. Discount rate. David buys a dishwasher that is marked down $77 from its original price of $350. What is the discount rate?

42. Markdown. A store advertises, "Buy the red-tagged items at 25% off their listed price." If you buy a coat marked $136, what will you pay for the coat during the sale?

43. Sales tax. A state sales tax rate is 7.5%. If the tax on a purchase is $9.75, what was the price of the purchase?

[7.5] In exercises 44 to 46, solve the applications.

44. Interest. A savings bank offers 5.25% on 1-year time deposits. If you place $3,000 in an account, how much will you have at the end of the year?

45. If you received $285 in interest on a CD that paid 9.5%, what was the original investment?

46. If $500 is compounded at 8% for 3 years, what will the final total be?

[7.6] In exercises 47 to 55, solve the applications.

47. Chemistry. A chemist prepares a 400-mL acid-water solution. If the solution contains 30 mL of acid, what percent of the solution is acid?

48. Price increase. The price of a new compact car has increased $819 over the previous year. If this amounts to a 4.5% increase, what was the price of the car before the increase?

49. Salary. Tom has 6% of his salary deducted for a retirement plan. If that deduction is $168, what is his monthly salary?

50. Enrollment. A college finds that 35% of its science students take biology. If there are 252 biology students, how many science students are there altogether?

51. Increase rate. A company finds that its advertising costs increased from $72,000 to $76,680 in 1 year. What was the rate of increase?

52. **Salary.** Maria's company offers her a 4% pay raise. This will amount to a $126 per month increase in her salary. What is her monthly salary before and after the raise?

53. **Computers.** A virus scanning program is checking every file for viruses. It has completed 30% of the files in 150 seconds. How long should it take to check all the files?

54. **Tipping.** If the total bill at a restaurant for 10 people is $572.89, including an 18% tip, what was the cost of the food itself?

55. **Sales.** A pair of running shoes is advertised as selling at 30% off the original price, for $80.15. What was the original price?

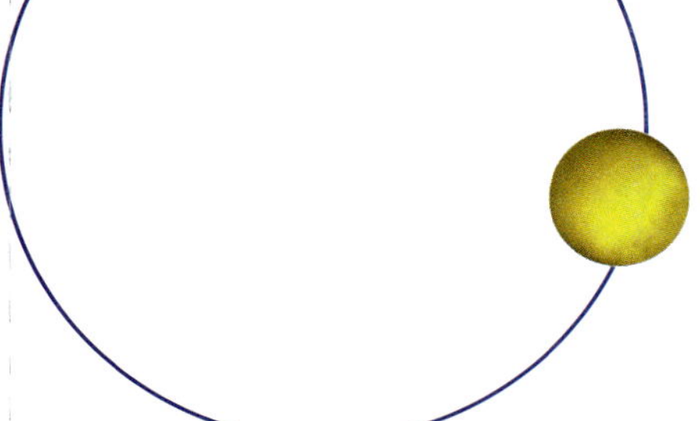

Chapter Test for Chapter 7

Name ______________

Section ______ Date ______

ANSWERS

1. ______
2. ______
3. ______
4. ______
5. ______
6. ______
7. ______
8. ______
9. ______
10. ______
11. ______
12. ______
13. ______
14. ______
15. ______
16. ______
17. ______
18. ______
19. ______
20. ______
21. ______

The purpose of this test is to help you check your progress and review for a chapter test in class. When you are done, check your answers in the back of the book. If you have missed any answers, be sure to go back and review the appropriate sections in the chapter.

1. Use a percent to name the shaded portion of the diagram.

In exercises 2 and 3, write as fractions.

2. 7%

3. 72%

In exercises 4 to 6, write as decimals.

4. 42%

5. 6%

6. 160%

In exercises 7 to 10, write as percents.

7. 0.03

8. 0.042

9. $\frac{2}{5}$

10. $\frac{5}{8}$

In exercises 11 to 13, identify the rate, base, and amount. *Do not solve* at this point.

11. 50 is 25% of 200.

12. What is 8% of 500?

13. **Purchase amount.** A state sales tax rate is 6%. If the tax on a purchase is $30, what is the amount of the purchase?

In exercises 14 to 21, solve the percent problems.

14. What is 4.5% of 250?

15. $33\frac{1}{3}$% of 1,500 is what number?

16. Find 125% of 600.

17. What percent of 300 is 60?

18. 4.5 is what percent of 60?

19. 875 is what percent of 500?

20. 96 is 12% of what number?

21. 8.5% of what number is 25.5?

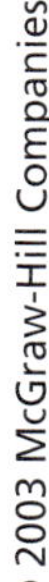

ANSWERS

22. ______________________

23. ______________________

24. ______________________

25. ______________________

26. ______________________

27. ______________________

28. ______________________

29. ______________________

30. ______________________

31. ______________________

32. ______________________

In exercises 22 to 32, solve the applications.

22. **Taxes.** A state taxes sales at 6.2%. What tax will you pay on an item that costs $80?

23. **Testing.** You receive a grade of 75% on a test of 80 questions. How many questions did you answer correctly?

24. **Markup.** An item that costs a store $54 is marked up 30% (based on cost). Find its selling price.

25. **Interests.** Mrs. Sanford pays $300 in interest on a $2,500 loan for 1 year. What is the interest rate for the loan?

26. **Salary.** Jovita's monthly salary is $2,200. If the deductions for taxes from her monthly paycheck are $528, what percent of her salary goes for these deductions?

27. **Markdown.** A car is marked down $1,552 from its original selling price of $19,400. What is the discount rate?

28. **Total sales.** Sarah earned $540 in commissions in 1 month. If her commission rate is 3%, what were her total sales?

29. **Enrollment.** A community college has 480 more students in fall 2000 than in fall 1999. If this is a 7.5% increase, what was the fall 1999 enrollment?

30. **Financing.** Shawn arranges financing for his new car. The interest rate for the financing plan is 12%, and he will pay $2,220 interest for 1 year. How much money did he borrow to finance the car?

31. Jalene earns a commission of 7% on property that she lists and sells. What is her commission on a $250,000 piece of property that she listed and sold?

32. If $2,650 is compounded at 7% for 3 years, what will the final total be?

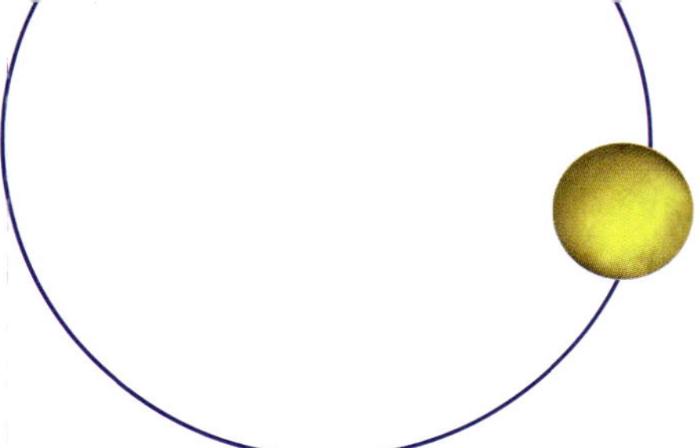

Cumulative Test for Chapters 1 to 7

Name ______________________

Section __________ Date __________

ANSWERS

1. ______________________
2. ______________________
3. ______________________
4. ______________________
5. ______________________
6. ______________________
7. ______________________
8. ______________________
9. ______________________
10. ______________________
11. ______________________
12. ______________________
13. ______________________
14. ______________________
15. ______________________
16. ______________________
17. ______________________
18. ______________________
19. ______________________

This test is provided to help you in the process of reviewing Chapters 1 through 7. Answers are provided in the back of the book. If you missed any answers, be sure to go back and review the appropriate chapter sections.

1. What is the place value of 4 in the number 234,768?

Perform the indicated operations.

2. $56 \cdot 203$

3. $3{,}026 \div 34$

Evaluate the expressions.

4. $-4 + 10$

5. $-5 - (-1)$

6. $\dfrac{-8 + 6}{-8 - (-10)}$

7. $5 + 3 \cdot (4 - 6)^2$

8. Write the prime factorization of 260.

9. Find the greatest common factor (GCF) of 84 and 140.

10. Find the least common multiple (LCM) of 18, 20, and 30.

Perform the indicated operations.

11. $3\frac{2}{5} \cdot 2\frac{1}{2}$

12. $5\frac{1}{3} \div 4$

13. $4\frac{3}{4} + 3\frac{5}{6}$

14. $7\frac{1}{6} - 2\frac{3}{8}$

15. A kitchen measures $5\frac{1}{2}$ yd by $3\frac{1}{4}$ yd. If vinyl flooring costs $16 per square yard, what will it cost to cover the floor?

16. If you drive 180 mi in $3\frac{1}{3}$ h, what is your average speed?

17. A bookshelf that is $54\frac{5}{8}$ in. long is cut from a board that is 8 ft long. If $\frac{1}{8}$ in. is wasted in the cut, what length board remains?

Find the indicated place values.

18. 8 in 4.2835

19. 4 in 6.09743

ANSWERS

20. ______
21. ______
22. ______
23. ______
24. ______
25. ______
26. ______
27. ______
28. ______
29. ______
30. ______
31. ______
32. ______
33. ______
34. ______
35. ______
36. ______
37. ______
38. ______
39. ______
40. ______

Complete each statement using the symbol <, =, or >.

20. 6.28 ______ 6.3

21. 3.75 ______ 3.750

Write as a common fraction or a mixed number. Simplify.

22. 0.36

23. 5.125

Perform the indicated operations.

24. $2.8 \cdot 4.03$

25. $54.528 \div 3.2$

26. A television set has an advertised price of \$599.95. You buy the set and agree to make payments of \$29.50 per month for 2 years. How much extra are you paying on this installment plan?

Solve each equation and check your result.

27. $9x + (-5) = 8x$

28. $\frac{2}{3}x = -24$

29. $2x + 3 = 7x + 5$

30. $\frac{4}{3}x + (-6) = 4 + \left(-\frac{2}{3}x\right)$

Solve for the unknown.

31. $\frac{3}{7} = \frac{8}{x}$

32. $\frac{1.9}{y} = \frac{5.7}{1.2}$

33. On a map the scale is $\frac{1}{4}$ in. = 25 mi. How many miles apart are two towns that are $3\frac{1}{2}$ in. apart on the map?

34. Diane worked 23.5 h on a part-time job and was paid \$131.60. She is asked to work 25 h the next week at the same pay rate. What salary will she receive?

35. Write 34% as a decimal and as a fraction.

36. Write $\frac{11}{20}$ as a decimal and as a percent.

37. Find 18% of 250.

38. 11% of what number is 55?

39. A company reduced the number of employees by 8% this year. There are now 115 employees. How many were there last year?

40. The sales tax on an item priced at \$72 is \$6.12. What percent is the tax rate?

GEOMETRY

8

INTRODUCTION

Every reporter must be prepared to deal with the interpretation of numerical information. Whether it is the details of a plane crash or the results of a new popularity poll, it is impossible to report on the findings if you cannot interpret them. But no reporter spends as much time working with numbers as one in the business or sports sections. Rebecca has worked in both areas.

From the time she worked on the school paper in high school, Rebecca wanted to become an editor for a city paper. When she got her first assignment as a reporter, the editor told her that she should try reporting for every section of the paper if she wanted to achieve her goal. So far she has reported for the life, business, and sports sections. Rebecca is enjoying covering sports, but she is ready to move on toward her coveted editorial position.

Name ____________________

Section ________ Date ________

ANSWERS

1. ____________________

2. ____________________

3. ____________________

4. ____________________

5. ____________________

6. ____________________

7. ____________________

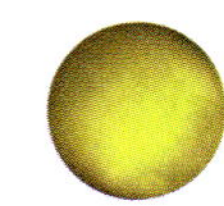

Pre-Test Chapter 8

This pre-test will point out any difficulties you may have with geometry and measurement. Do all the problems, and then check your answers with those in the back of the book.

1. Draw the line $\overleftrightarrow{AB}$.

A B

2. Find the measure of $\angle BCA$.

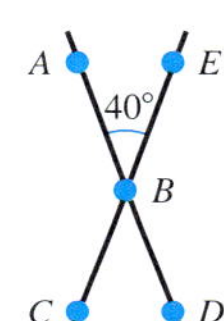

3. Find $m\angle A$.

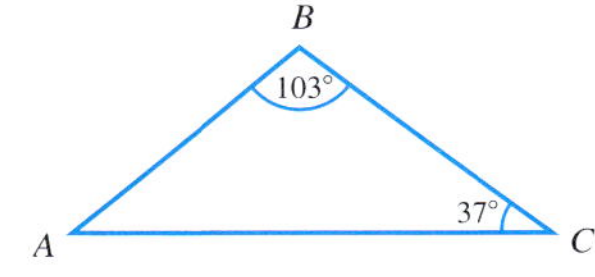

4. Find the measure of $\angle ABC$.

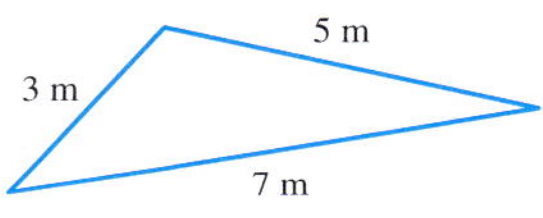

5. Find the perimeter of the triangle.

6. Find the circumference of the circle. Round to the nearest tenth.

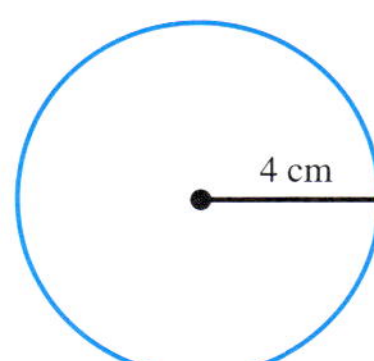

7. Determine the area of the triangle

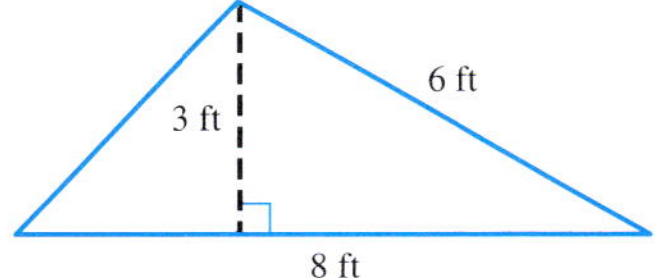

8.1 Lines and Angles

8.1 OBJECTIVES

1. Identify lines, line segments, and angles
2. Determine when lines are parallel or perpendicular
3. Determine whether an angle is right, acute, obtuse, or straight
4. Use a protractor to measure an angle
5. Determine when pairs of angles are complementary or supplementary
6. Identify vertical angles and adjacent angles
7. Find the measure of the third angle of a triangle

Once early communities had mastered the counting of their animals, they became interested in measuring their land. This is the foundation of geometry. Literally translated, **geometry** means earth measurement. Many of the topics we consider in geometry (topics such as angles, perimeter, and area) were first studied as part of surveying.

NOTE *Geo* means earth, just as it does in the words *geography* and *geology*.

As is usually the case, we start the study of a new topic by learning some vocabulary. Most of the terms we will discuss will be familiar to you. It is important that you understand what we mean when we use these words in the context of geometry.

We begin with the word **point.** A point is a location; it has no size and covers no area.

If we string points together forever, we create a **line.** In our studies we will consider only straight lines. We use arrowheads to indicate that a line goes on forever.

A piece of a line that has two endpoints is called a **line segment.**

Example 1

Recognizing Lines and Line Segments

NOTE The capital letters are labels for points.

Label each of the parts as a line or a line segment.

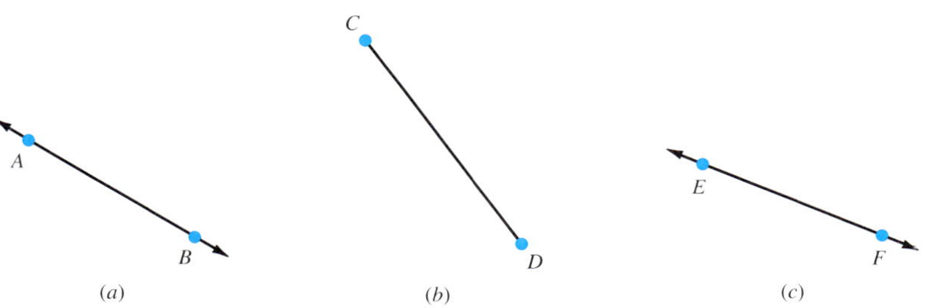

Both parts a and c continue forever in both directions. They are lines. We write $\overleftrightarrow{AB}$ and $\overleftrightarrow{EF}$, respectively, to indicate the lines through those points. Part b has two endpoints. It is a line segment. We designate that it is a segment by writing $\overline{CD}$.

CHECK YOURSELF 1

Label each as a line or a line segment.

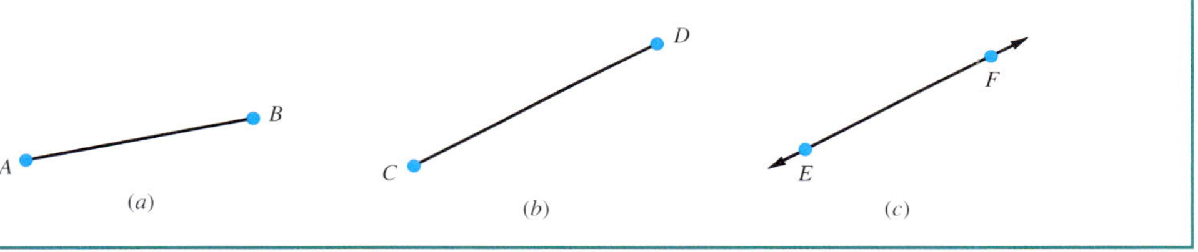

Definitions: Angle

An **angle** is a geometric figure consisting of two line segments that share a common endpoint.

The surveyor is using a transit to locate a property line.

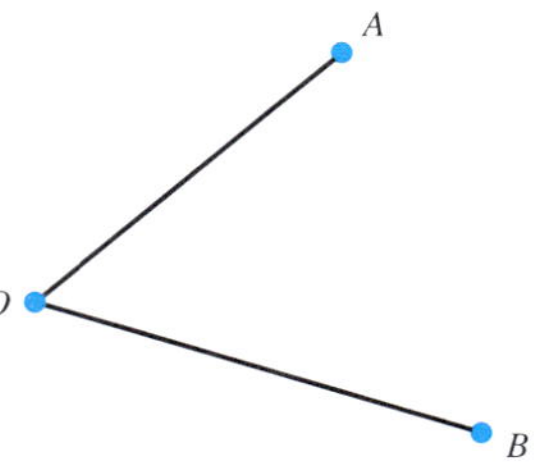

$\overline{OA}$ and $\overline{OB}$ are line segments. O is the **vertex** of the angle.

Surveyors use an instrument called a transit. A transit allows surveyors to measure angles so that, from a mathematical description, they can determine exactly where a property line is.

Definitions: Perpendicular Lines

When two lines cross (or intersect) they form four angles. If the lines intersect such that four equal angles are formed, we say that the two lines are **perpendicular.**

Perpendicular lines

At most street intersections, the two roads are perpendicular.

Definitions: Parallel Lines

If two lines are drawn so that they never intersect (even if we extend the lines forever), we say that the two lines are **parallel.**

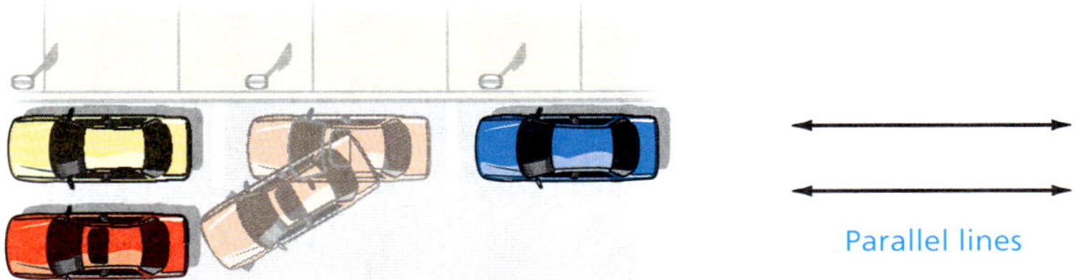

Parallel lines

Parallel parking gets its name from the fact that the parking spot is parallel to the traffic lane.

Example 2

Recognizing Parallel and Perpendicular Lines

Label each pair of lines as parallel, perpendicular, or neither.

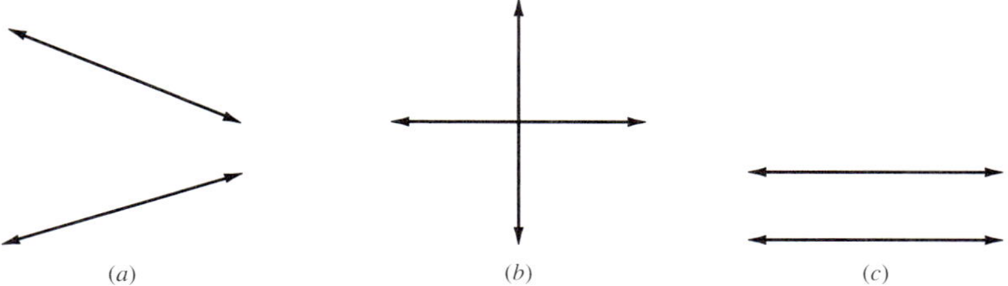

Although we don't see the lines in part *a* intersecting, if they were extended as the arrowheads indicate, they would. They are neither parallel nor perpendicular. The lines of part *b* are perpendicular because the four angles formed are equal. Only the lines in part *c* are parallel.

CHECK YOURSELF 2

Label each pair of lines as parallel, perpendicular, or neither.

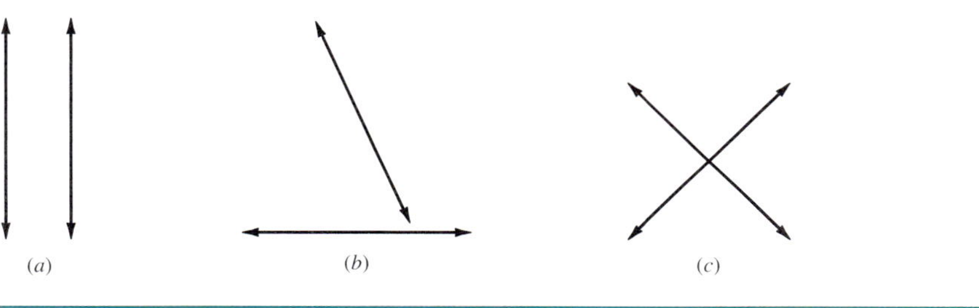

NOTE You may recall seeing this small square in Chapter 5.

We call the angle formed by two perpendicular lines or line segments a **right angle.** We designate a right angle by drawing a small square.

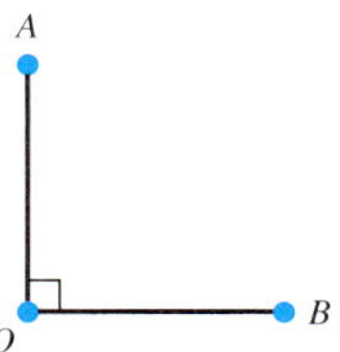

We can refer to a specific angle by naming three points. The middle point is the vertex of the angle. We can call this $\angle AOB$ or, if there is no confusion, we can call it $\angle O$.

Example 3

Naming an Angle

Name the highlighted angle.

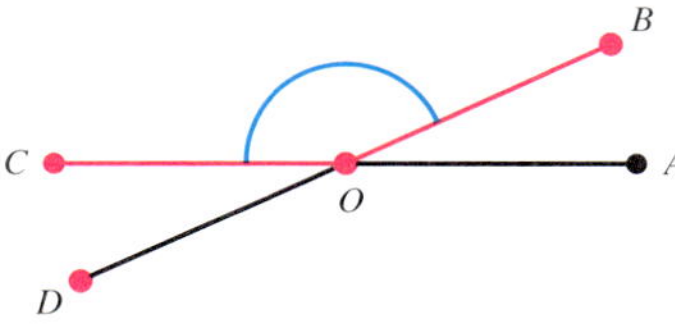

NOTE We could also call this angle $\angle BOC$.

The vertex of the angle is O, and the angle begins at C and ends at B, so we would name the angle $\angle COB$. We could not refer to this as $\angle O$, because it would not be clear to which angle we were referring.

CHECK YOURSELF 3

Name the highlighted angle.

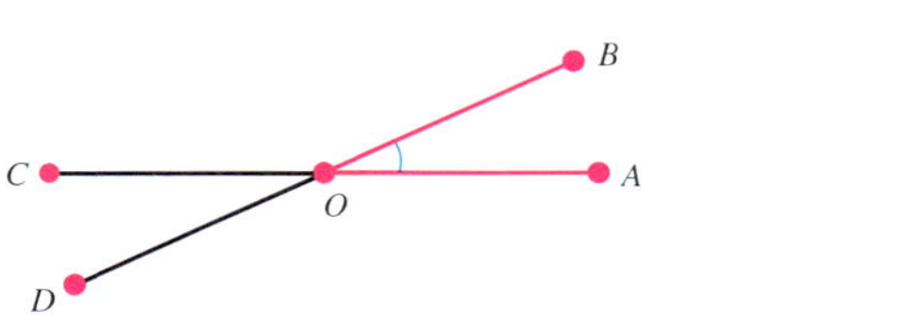

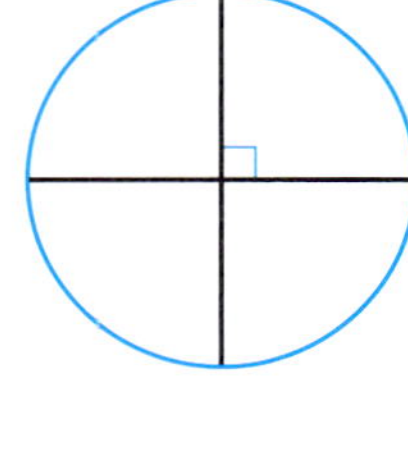

One way to measure an angle is to use a unit that we call a **degree.** There are 360 degrees (we write this as 360°) in a complete circle. Note in the picture on the left that there are four right angles in a circle. If we divide 360° by 4, we find that each right angle must measure 90°. Here are some other angles with their measurements.

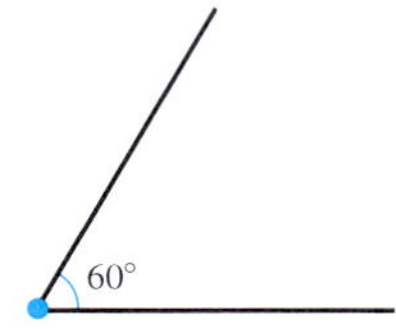

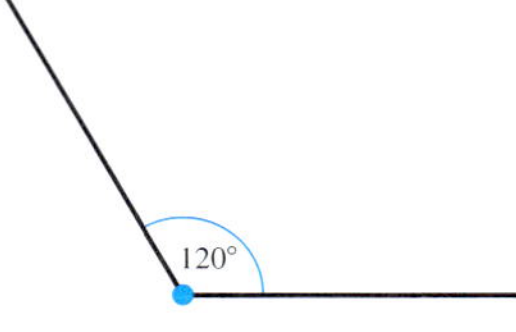

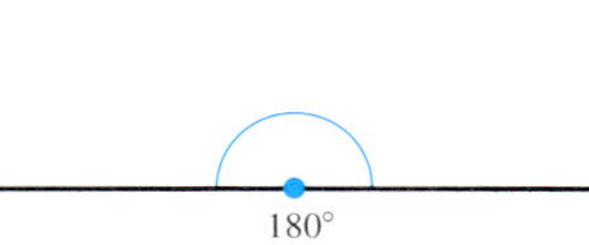

An **acute angle** measures between 0° and 90°. An **obtuse angle** measures between 90° and 180°. A **straight angle** measures 180°.

Example 4

Labeling Types of Angles

Label each of the angles as an acute, an obtuse, a right, or a straight angle.

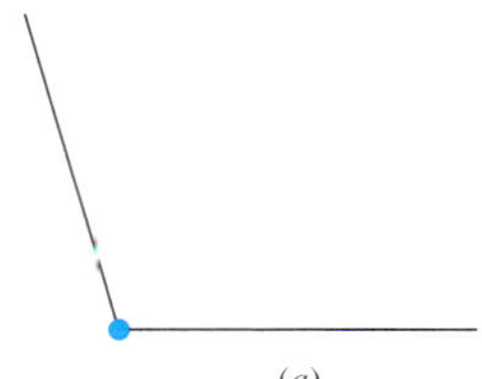
(*a*)

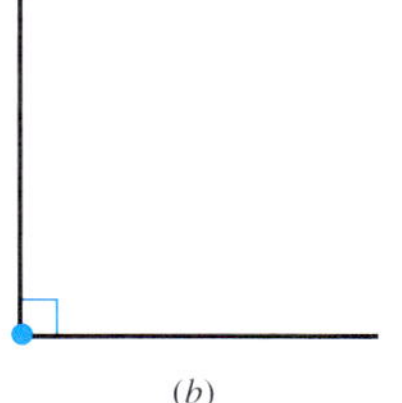
(*b*)

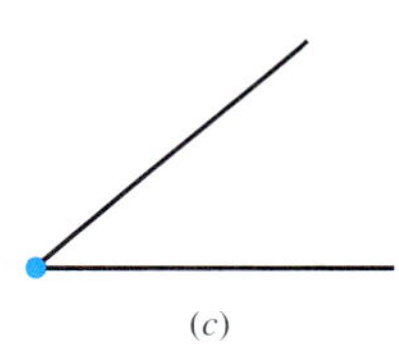
(*c*)

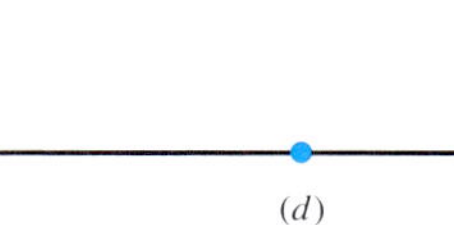
(*d*)

Part *a* is obtuse (the angle is more than 90°). Part *b* is a right angle (designated by the small square). Part *c* is an acute angle (it is less than 90°), and part *d* is a straight angle.

CHECK YOURSELF 4

Label each angle as an acute, an obtuse, a right, or a straight angle.

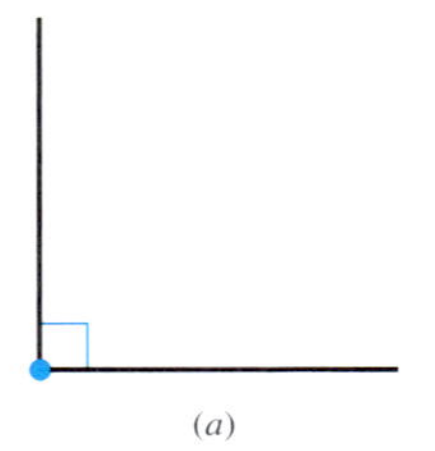
(*a*)

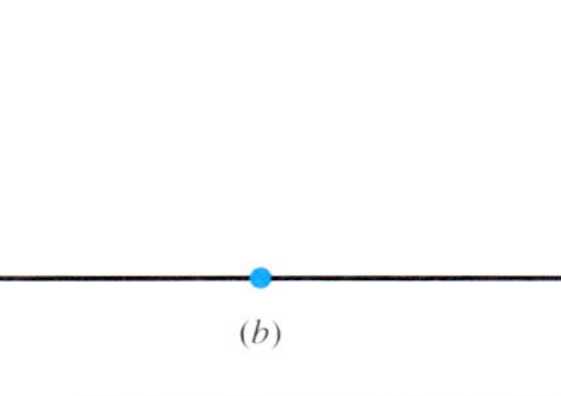
(*b*)

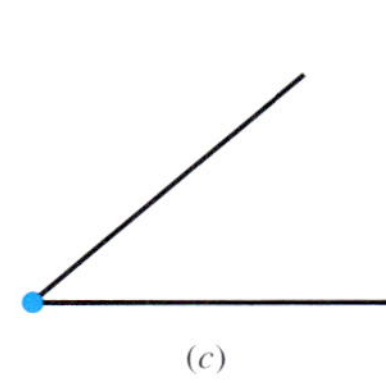
(*c*)

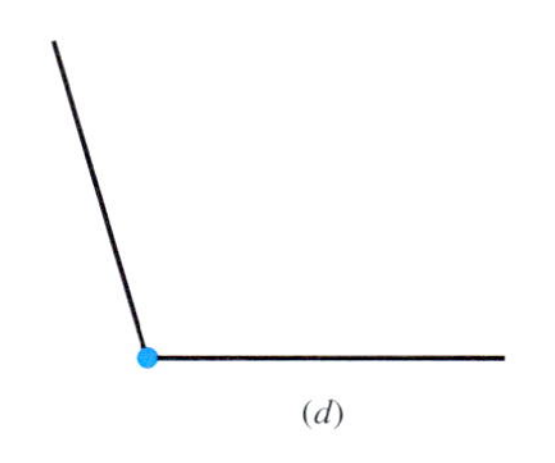
(*d*)

NOTE Your protractor may show the degree measures in both directions.

When assigning a measurement to an angle, we usually use a tool called a **protractor.**

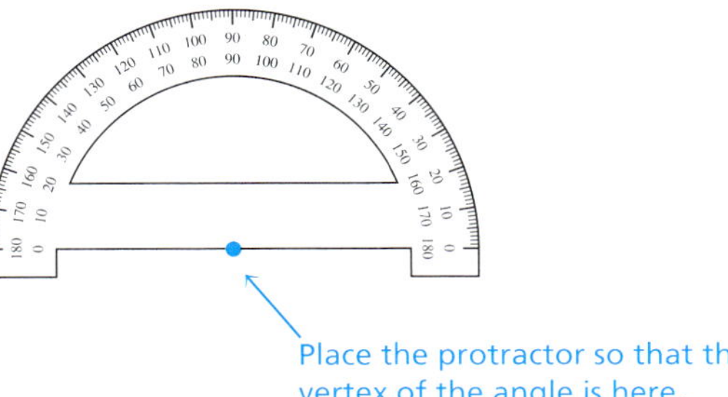

We read the protractor by placing one line segment of the angle at 0°. We then read the number that the other line segment passes through. This number represents the degree measurement of the angle. The point at the center of the protractor, the endpoint of the two line segments, is the vertex of the angle.

Example 5

Measuring an Angle

Use the protractor to estimate the measurement for each angle.

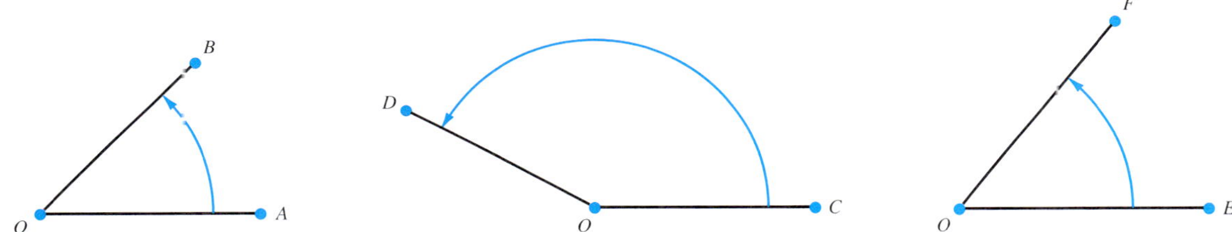

The measure of $\angle AOB$ is 45°. The measure of $\angle COD$ is between 150° and 155°. We could estimate it at 152°. The measure of $\angle EOF$ is between 50° and 55°. We could estimate that it is a 52° angle.

CHECK YOURSELF 5

Use a protractor to estimate the measurement for each angle.

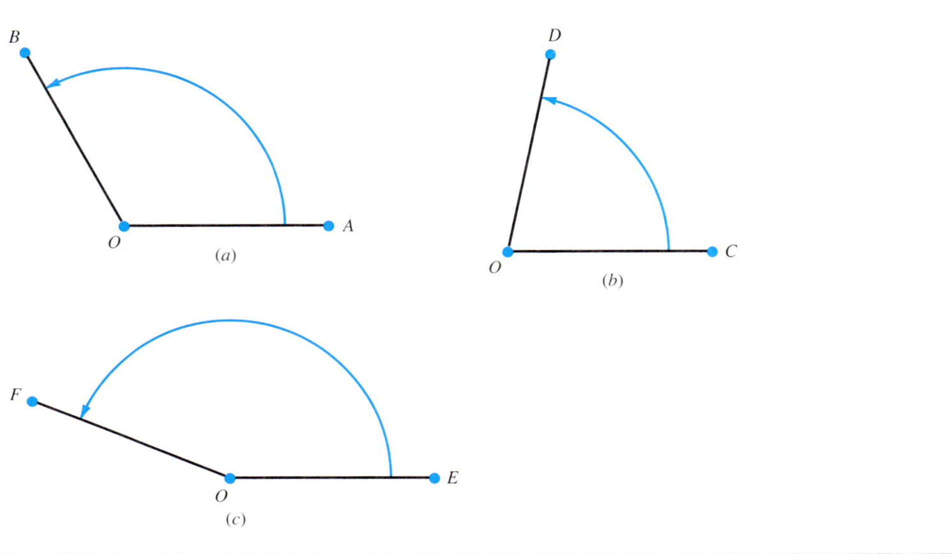

If we wish to refer to the degree measure of $\angle ABC$, we use $m\angle ABC$.

Example 6

Measuring an Angle

Find $m\angle AOB$.

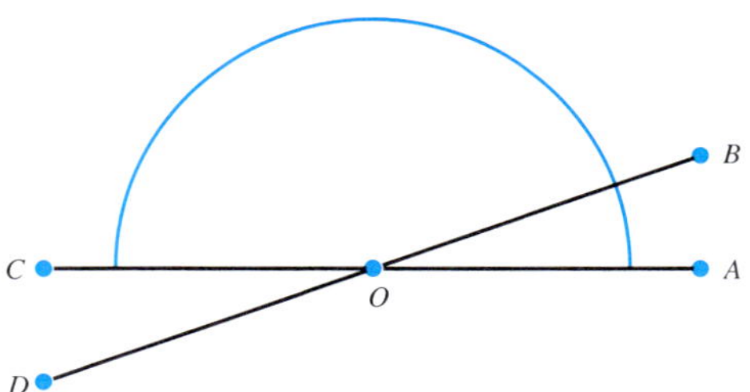

NOTE $m\angle AOB = 20°$ is read "the measure of angle *AOB* is 20 degrees".

Using the protractor, we find $m\angle AOB = 20°$.

CHECK YOURSELF 6

Find $m\angle AOC$.

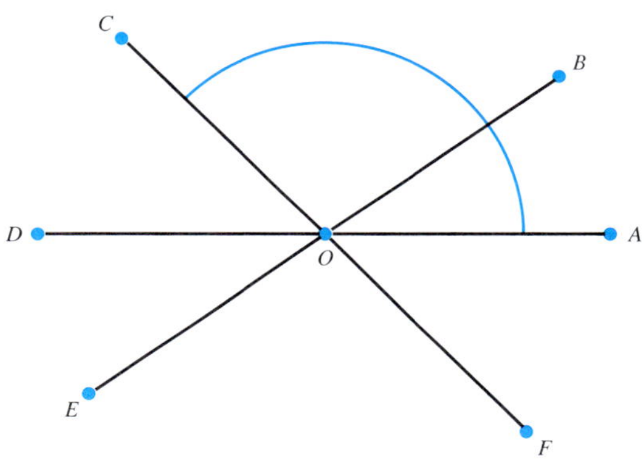

Definitions: Complementary Angles

Two angles are **complementary** if the sum of their angles is 90°.

Example 7

Using Complementary Angles

A and B are complementary angles. The measurement of $\angle A = 42°$; find the measure of $\angle B$.

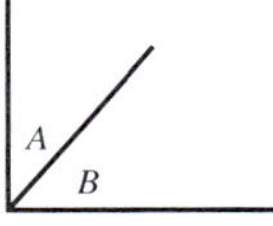

Because they are complementary, the sum of the angles is 90°.

$m\angle A + m\angle B = 90°$

$42° + m\angle B = 90°$

$m\angle B = 90° - 42°$

$m\angle B = 48°$

CHECK YOURSELF 7

A and B are complementary angles. The measure of $\angle A = 71°$; find the measure of $\angle B$.

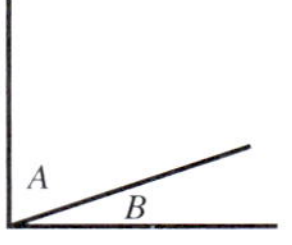

Definitions: Supplementary Angles

Two angles are **supplementary** if the sum of their measures is 180°.

Example 8

Using Supplementary Angles

A and *B* are supplementary angles. Find the measure of $\angle A$.

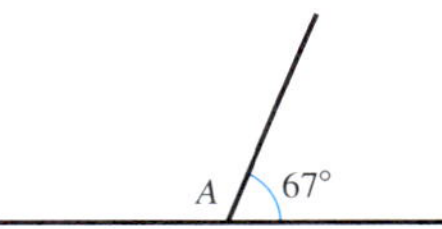

Because they are supplementary, the sum of the angles is 180°.

$m\angle A + m\angle B = 180°$

$m\angle A + 67° = 180°$

$m\angle A = 180° - 67°$

$m\angle A = 113°$

CHECK YOURSELF 8

A and B are supplementary angles. Find the measure of $\angle B$.

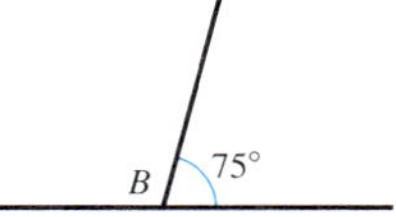

Rules and Properties

When two lines intersect, the adjacent angles are supplementary.

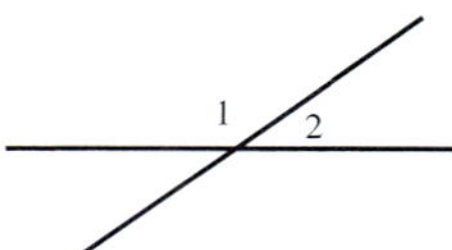

$\angle 1$ and $\angle 2$ are supplementary.

Example 9

Finding the Measure of an Adjacent Angle

Find the measure of $\angle A$.

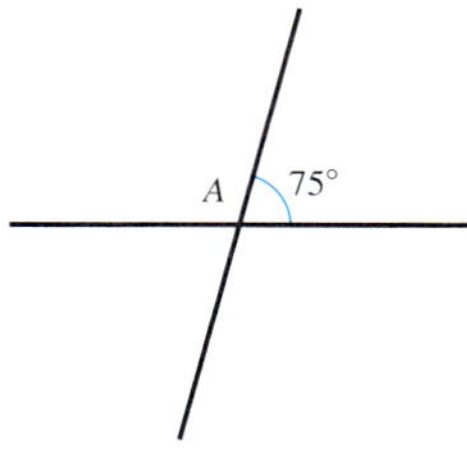

$m\angle A + 75° = 180°$

$m\angle A = 180° - 75°$

$m\angle A = 105°$

CHECK YOURSELF 9

Find the measure of $\angle A$.

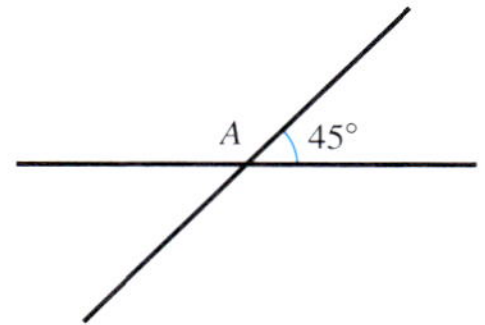

Definitions: Vertical Angles

When two lines intersect at a single point, the angles that are not adjacent are called **vertical angles.**

Rules and Properties

The measures of any two vertical angles are equal.

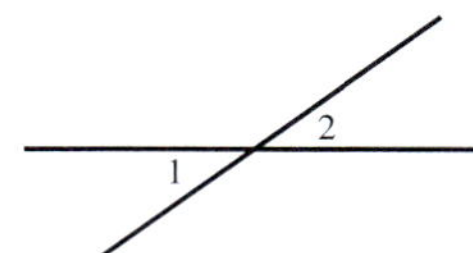

$\angle 1$ and $\angle 2$ are vertical angles.

Example 10

Finding the Measure of Vertical and Adjacent Angles

Find the measures of $\angle A$, $\angle B$, and $\angle C$.

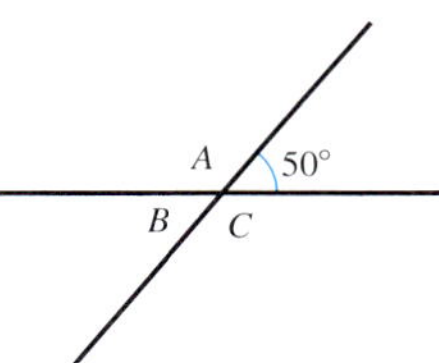

$\angle A$ is adjacent to a 50° angle, so the measure of $\angle A$ is 130°. $(180 - 50 = 130)$
$\angle B$ and the 50° angle are vertical angles, so $m\angle B = 50°$.
$\angle C$ is adjacent, so $m\angle C = 130°$.

CHECK YOURSELF 10

Find the measures of $\angle A$, $\angle B$, and $\angle C$.

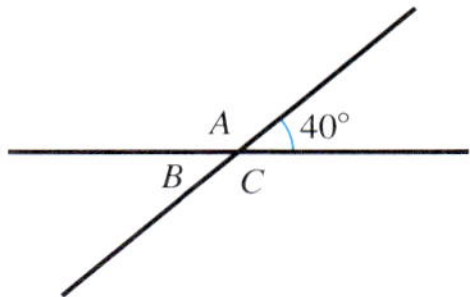

Now that you know something about angles, it is interesting to again look at triangles. Why is this shape called a triangle?

Literally, triangle means "three angles." Each of the triangles shown has three angles. For each triangle, the measure of each of its three angles is shown.

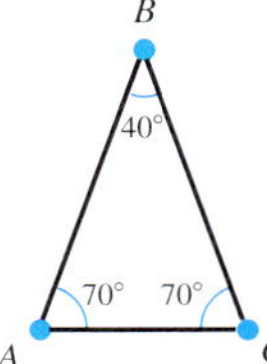

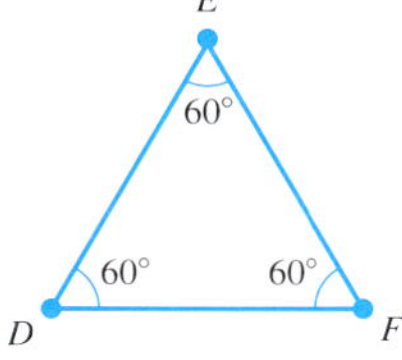

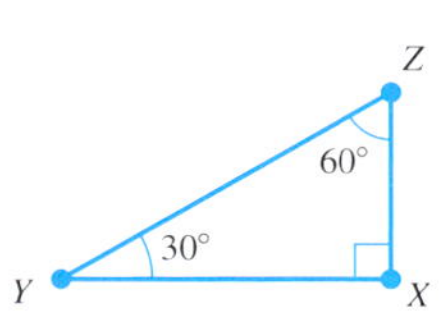

Now we will determine the sum of the angles inside each of the triangles. You will note that they always add up to 180°. No matter how we draw a triangle, the sum of the three angles inside the triangle will *always* be 180°.

Here is an experiment that might convince you that this is always the case.

1. Using a straight edge, draw any triangle you wish on a sheet of paper.

2. Use scissors to cut out the triangle.
3. Rip the three vertices off of the triangle.

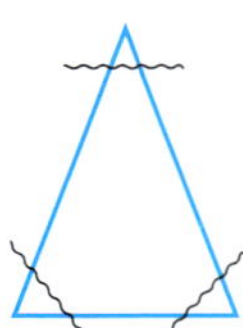

4. Lay the three vertices (with the points of the triangle touching) together. They will always form a straight angle, which we saw earlier in this section has a measure of 180°.

Rules and Properties: Angles of a Triangle

For any triangle *ABC*,

$m\angle A + m\angle B + m\angle C = 180°$

Example 11

Finding an Angle Measure

Find the measure of the third angle in this triangle.

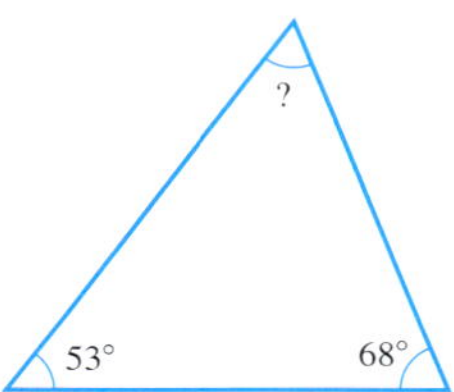

We need the three measurements to add to 180°, so we add the two given measurements (53° + 68° = 121°). Then we subtract that from 180° (180° − 121° = 59°). This gives us the measure of the third angle, 59°.

CHECK YOURSELF 11

Find the measure of $\angle ABC$.

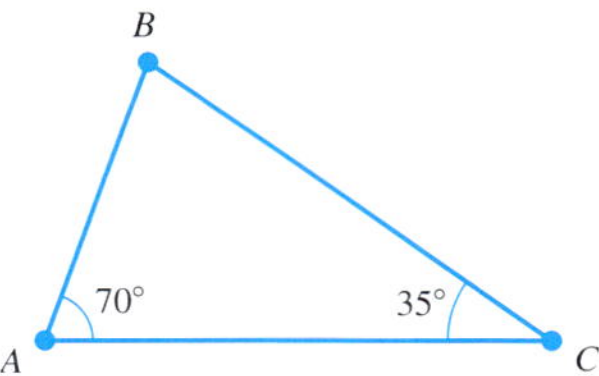

CHECK YOURSELF ANSWERS

1. **(a)** Line segment; **(b)** line segment; **(c)** line
2. **(a)** Parallel; **(b)** neither; **(c)** perpendicular **3.** $\angle BOA$ or $\angle AOB$
4. **(a)** Right; **(b)** straight; **(c)** acute; **(d)** obtuse **5.** **(a)** 120°; **(b)** 80°; **(c)** 160°
6. 135° **7.** 19° **8.** 105° **9.** 135° **10.** $m\angle A = 140°$, $m\angle B = 40°$, $m\angle C = 140°$ **11.** 75°

Name ____________

Section ________ Date ________

8.1 Exercises

1. Draw line segment $\overline{AB}$.

$\dot{A}$ $\dot{B}$

2. Draw line $\overleftrightarrow{EF}$.

$\dot{E}$ $\dot{F}$

3. Draw line $\overleftrightarrow{AC}$.

$\dot{A}$ $\dot{C}$

4. Draw line segment $\overline{BC}$.

$\dot{B}$ $\dot{C}$

Identify each object as a line or line segment.

5.

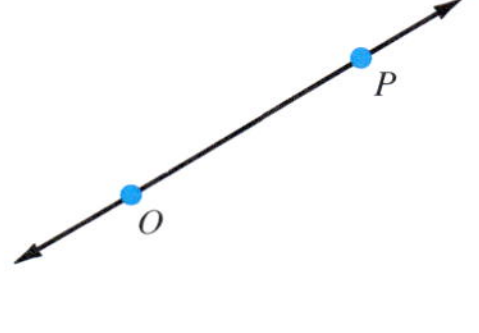

6.

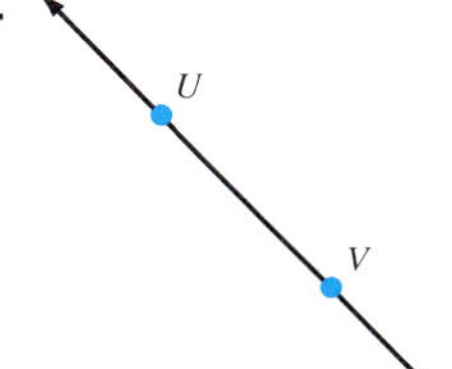

7.

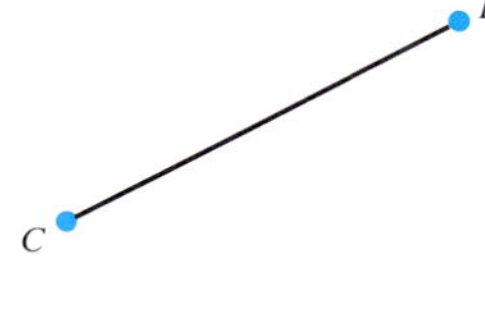

8.

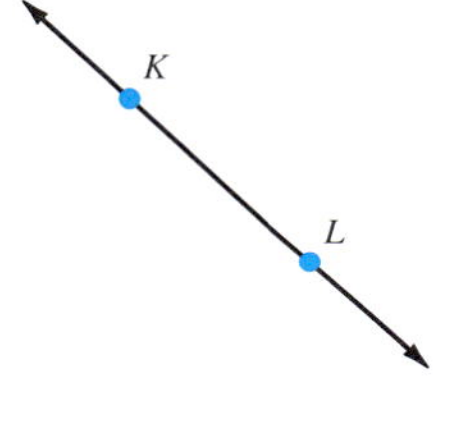

9.

10.

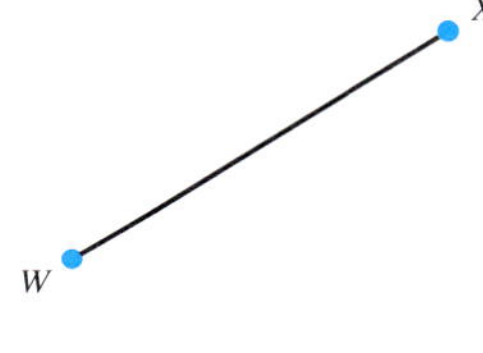

11.

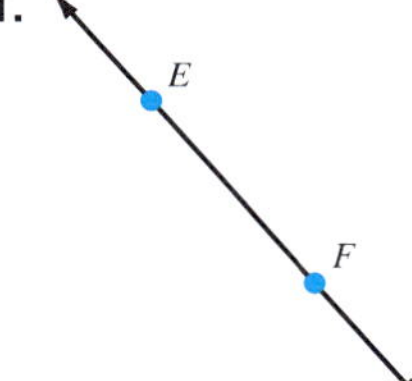

12.

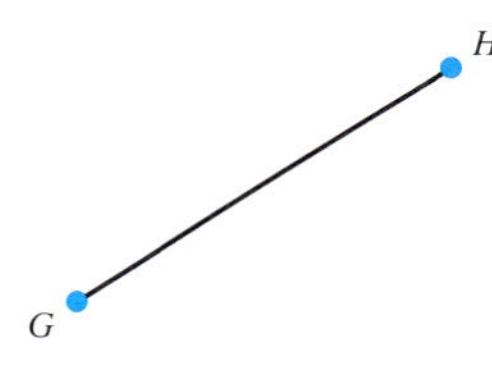

Label exercises 13 to 22 as true or false.

13. There are exactly two distinct line segments that can be drawn using two points.

14. There are exactly two distinct lines that can be drawn through two points.

15. Two opposite sides of a square are parallel line segments.

16. Two adjacent sides of a square are perpendicular line segments.

17. $\angle ABC$ will always have the same measure as $\angle CAB$.

18. Two acute angles have the same measure.

ANSWERS

1. ____________
2. ____________
3. ____________
4. ____________
5. ____________
6. ____________
7. ____________
8. ____________
9. ____________
10. ____________
11. ____________
12. ____________
13. ____________
14. ____________
15. ____________
16. ____________
17. ____________
18. ____________

ANSWERS

19. ______________

20. ______________

21. ______________

22. ______________

23. ______________

24. ______________

25. ______________

26. ______________

27. ______________

28. ______________

29. ______________

30. ______________

19. The sum of two vertical angles is 180°.

20. The sum of two complementary angles is 90°.

21. Two intersecting lines create two pairs of vertical angles.

22. Two intersecting lines create pairs of complementary angles.

23. Are the two lines parallel, perpendicular, or neither?

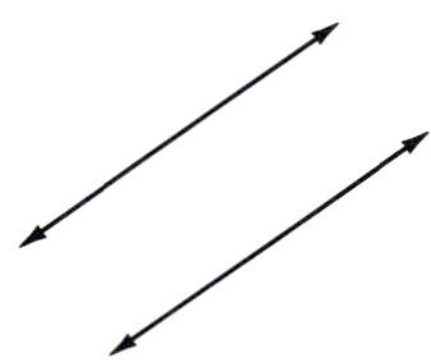

24. Are the two lines parallel, perpendicular, or neither?

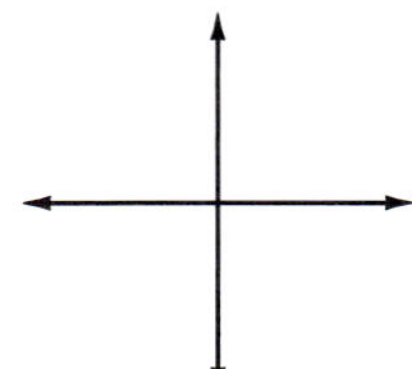

In exercises 25 to 32 label each indicated angle.

25.

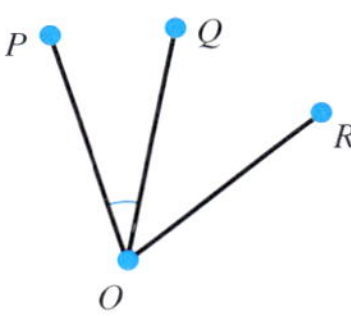

26.

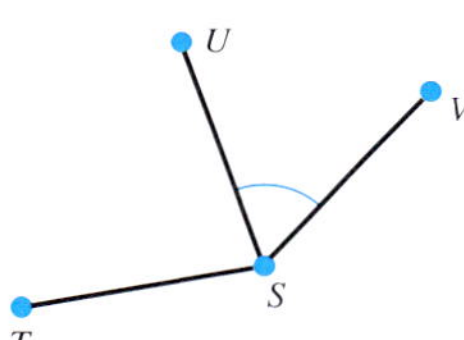

27.

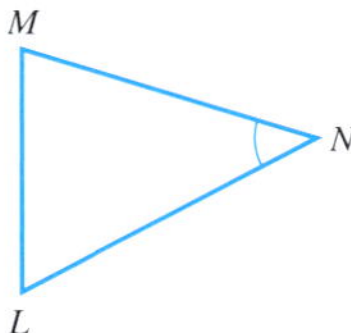

28.

29.

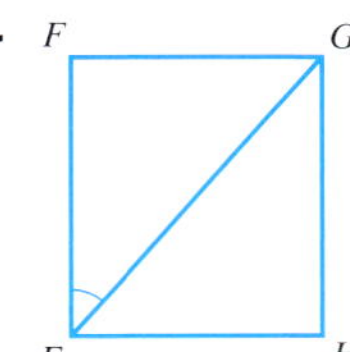

30.

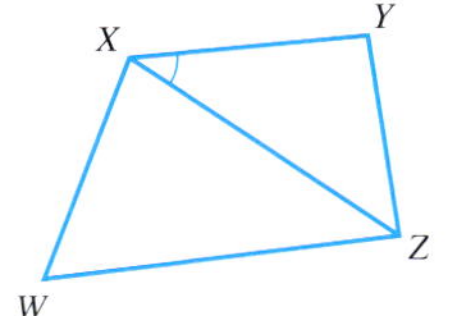

ANSWERS

31. ______

32. ______

33. ______

34. ______

35. ______

36. ______

37. ______

38. ______

39. ______

40. ______

41. ______

42. ______

31.

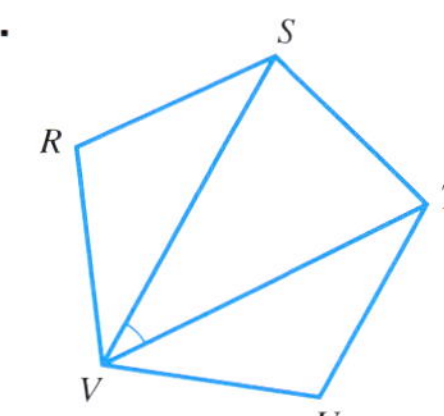

32.

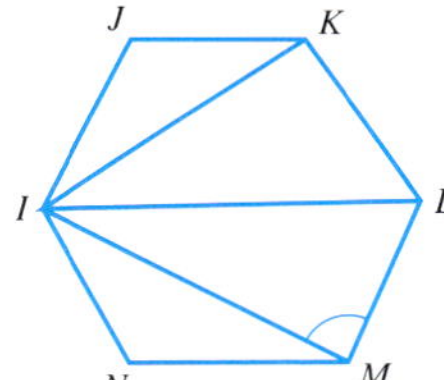

For each angle described, give its measure in degrees. One revolution is a full circle. Sketch the angle.

33. $\angle A$ represents $\frac{1}{6}$ of a revolution

34. $\angle B$ represents $\frac{1}{3}$ of a revolution

35. $\angle C$ represents $\frac{7}{12}$ of a revolution

36. $\angle D$ represents $\frac{11}{12}$ of a revolution

Measure each angle with a protractor. Identify the angle as acute, right, obtuse, or straight.

37.

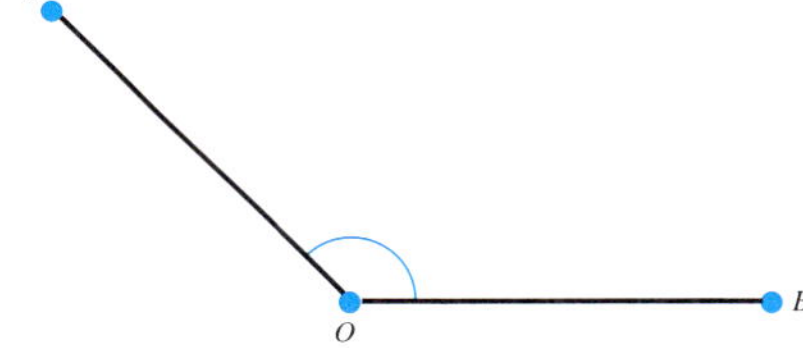

38.

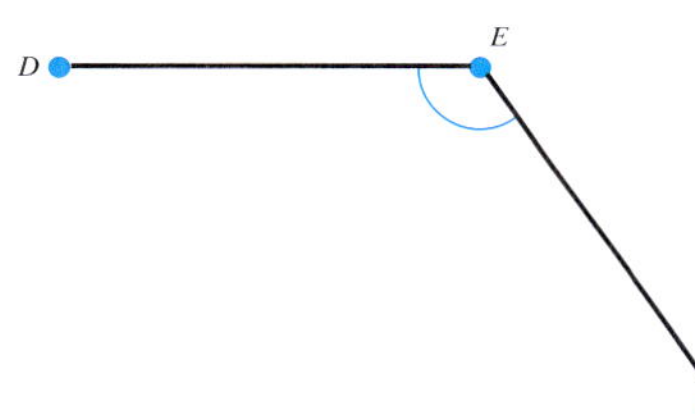

39.

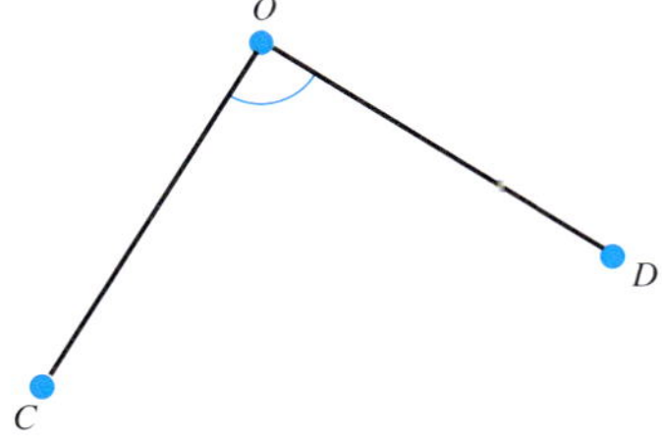

40.

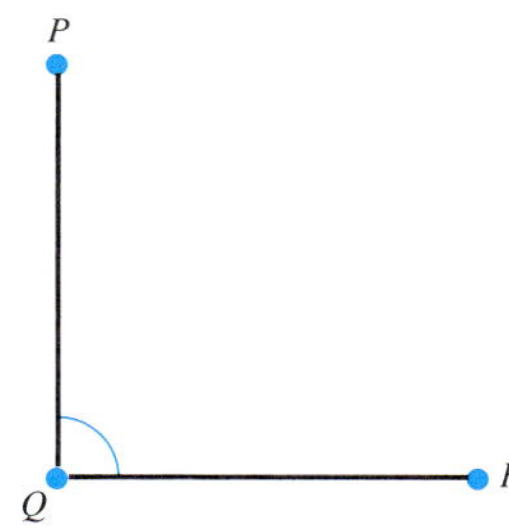

41.

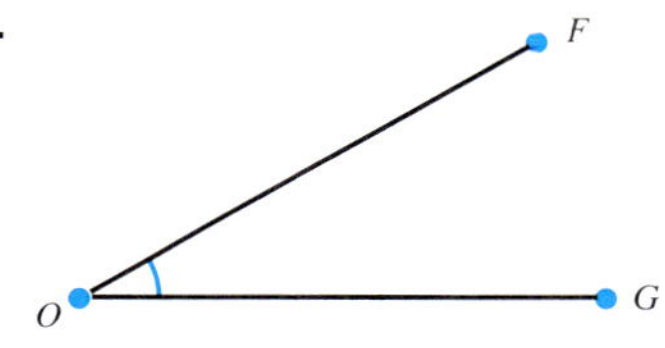

42.

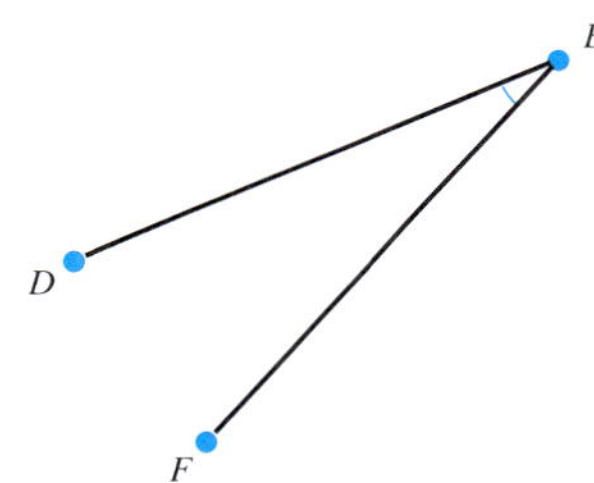

ANSWERS

43. ______

44. ______

45. ______

46. ______

47. ______

48. ______

49. ______

50. ______

In the figure, two parallel lines are intersected by a third line, forming eight angles. Draw lines like these on your paper.

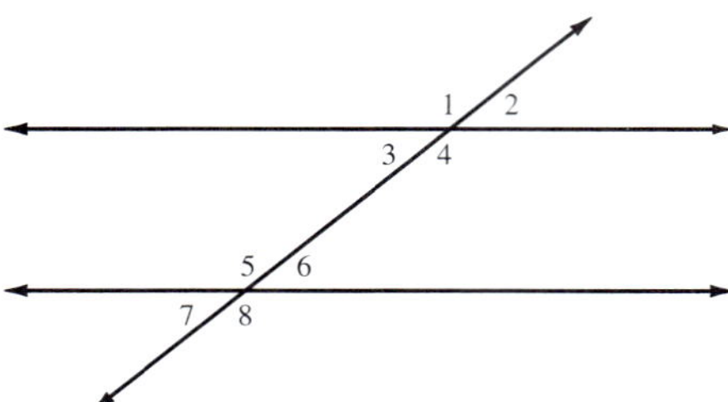

43. Use your protractor to measure $\angle 2$ and $\angle 6$. What do you notice?

44. Use your protractor to measure $\angle 3$ and $\angle 6$. What do you notice?

In exercises 45 to 50, find the missing angle.

45.

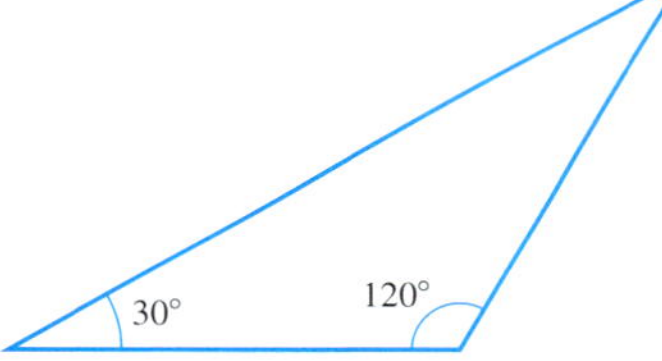

46.

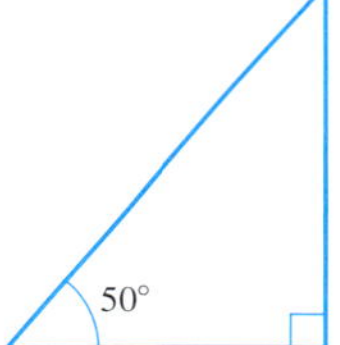

47.

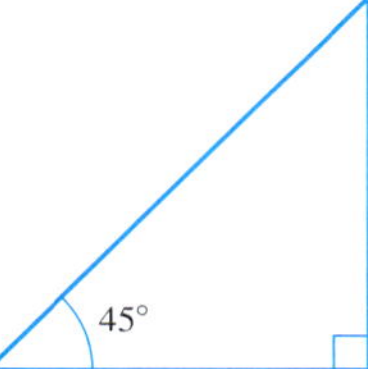

48.

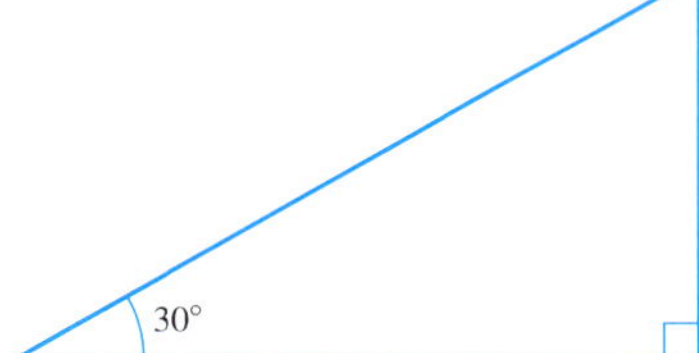

49.

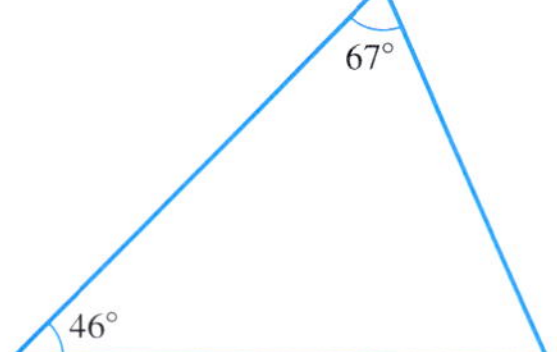

50.

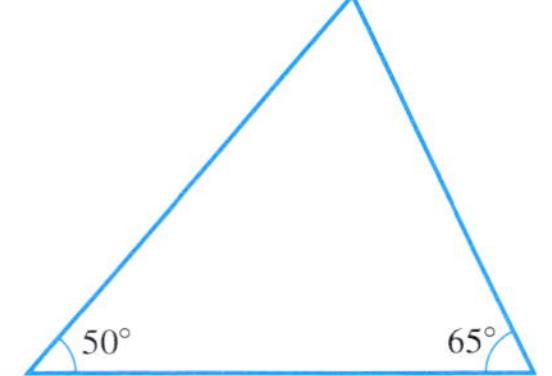

ANSWERS

51. __________

52. __________

53. __________

54. __________

55. __________

56. __________

57. __________

58. __________

59. __________

For each triangle shown, find the indicated angle.

51. Find $m\angle C$.

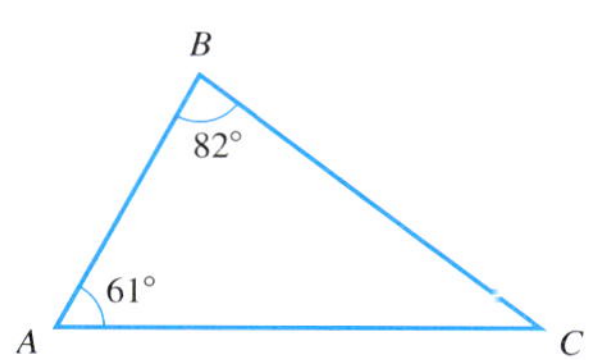

52. Find $m\angle B$.

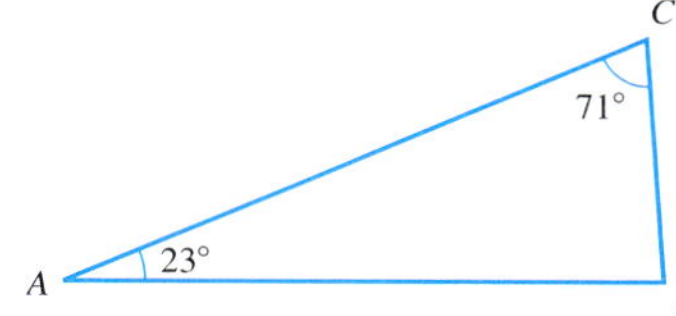

53. Find $m\angle A$.

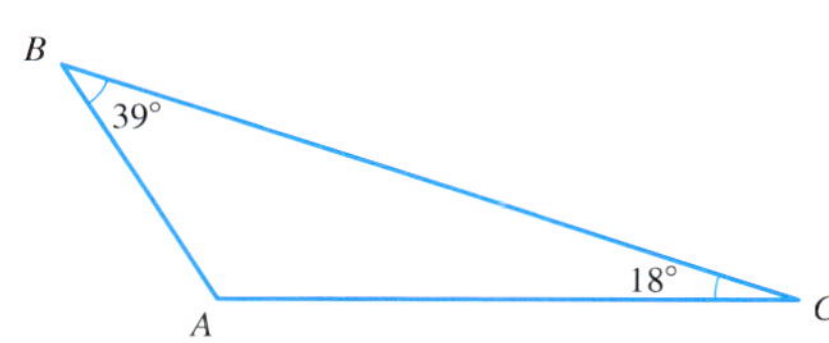

54. Find $m\angle B$.

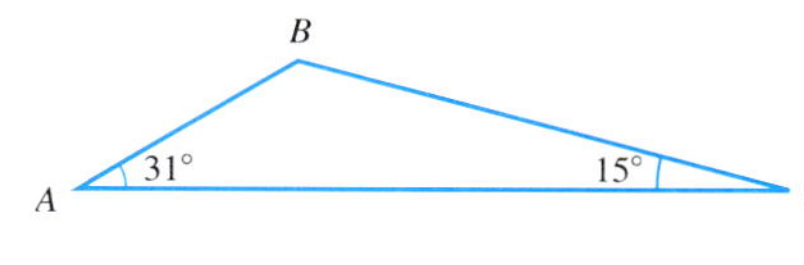

55. Find $m\angle B$.

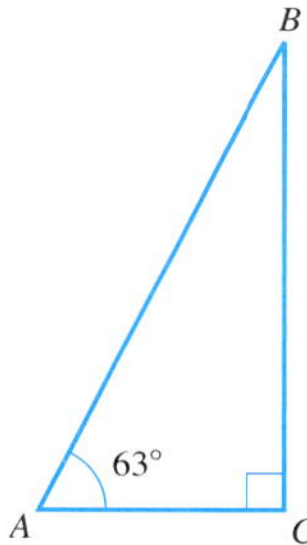

56. Find $m\angle A$.

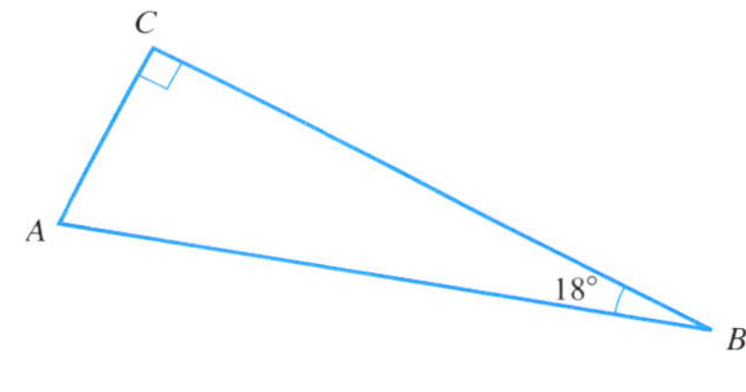

In exercises 57 to 59, one side of the triangle has been extended, forming what is called an *exterior angle*. In each case, find the measure of the indicated exterior angle.

57.

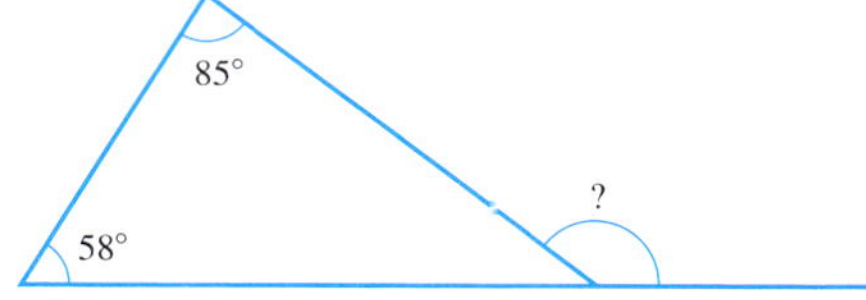

58.

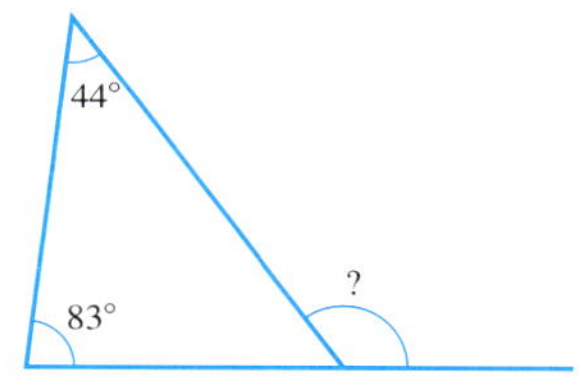

59.

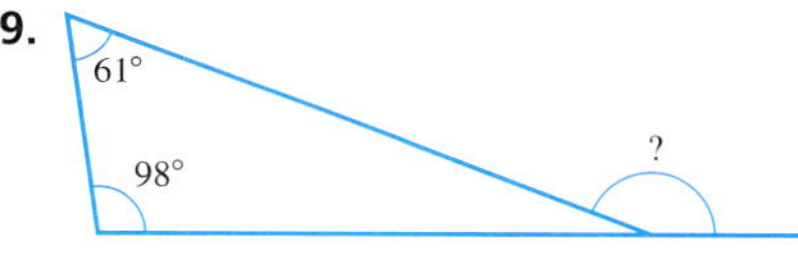

ANSWERS

60. ______________________

61. ______________________

62. ______________________

63. ______________________

64. ______________________

65. ______________________

66. ______________________

60. What do you observe from exercises 57 to 59? Write a general conjecture about the exterior angle of a triangle.

61. Draw any triangle using a ruler. With your protractor, carefully measure the three interior angles, and find their sum. Do this again with two more triangles of different shapes. What do you notice about the sums of the angles? Does it fit the rule of triangles?

62. A **quadrilateral** is a four-sided figure. Draw any quadrilateral, and measure the four interior angles with a protractor. Record these, and find their sum. Make a conjecture concerning the sum of the interior angles of *any* quadrilateral. Test your conjecture on another quadrilateral.

63. A **pentagon** is a five-sided figure. Draw any pentagon, and measure the five interior angles with a protractor. Record these, and find their sum. Make a conjecture concerning the sum of the interior angles of *any* pentagon. Test your conjecture on another pentagon.

64. A **hexagon** is a six-sided figure. Draw any hexagon, and measure the six interior angles with a protractor. Record these, and find their sum. Make a conjecture concerning the sum of the interior angles of *any* hexagon. Test your conjecture on another hexagon.

65. Argue that a triangle cannot have more than one obtuse angle.

66. Create an argument to support the following statement:

If $\triangle ABC$ is a right triangle, with $m\angle C = 90°$, then $\angle A$ and $\angle B$ must be acute and complementary.

Answers

1. 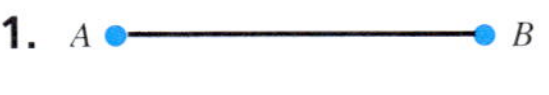

3. 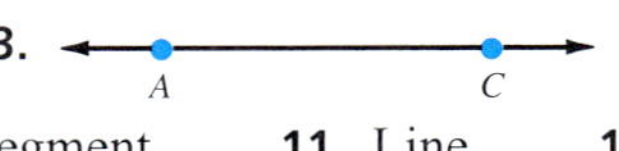

5. Line

7. Line segment **9.** Line segment **11.** Line **13.** False **15.** True **17.** False **19.** False **21.** True **23.** Parallel **25.** $\angle POQ$ **27.** $\angle MNL$ **29.** $\angle FEG$ **31.** $\angle SVT$ **33.** 60° **35.** 210° **37.** 135°; obtuse **39.** 90°; right **41.** 30°; acute **43.** 40°; 40° **45.** 30° **47.** 45° **49.** 67° **51.** 37° **53.** 123° **55.** 27° **57.** 143° **59.** 159° **61.** **63.** **65.** 

8.2 Perimeter and Circumference

OBJECTIVES

1. Determine perimeters of rectangles, squares, triangles, parallelograms, and other polygons
2. Determine the approximate circumference of a circle

We discussed squares and rectangles in Chapter 1 of this text. Each of these shapes is an example of a polygon.

Definitions: Polygon

A **polygon** is a closed geometric figure with three or more sides in which each side is a line segment.

In Section 3.2, we introduced the idea of a perimeter. In this section, we will expand our definition to include all polygons.

Definitions: Perimeter

The **perimeter** of any polygon is the sum of the lengths of its sides. We usually designate perimeter with the letter P.

The formulas given here can be used to find the perimeters of a triangle, a square, and a rectangle, respectively. Note that each is simply a way of finding the sum of the lengths of the sides.

Rules and Properties: Formula for Perimeter of a Triangle

$P = a + b + c$

where a, b, and c are the lengths of the three sides.

Rules and Properties: Formula for Perimeter of a Square

$P = 4s$

where s is the length of a side.

Rules and Properties: Formula for Perimeter of a Rectangle

$P = 2L + 2W$

where L is the length and W is the width.

Example 1

Finding the Perimeter of a Polygon

Find the perimeter for each of the polygons.

(a)

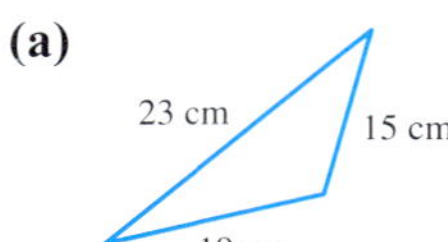

To find the perimeter of a triangle, we add the three lengths:

$P = 23 \text{ cm} + 15 \text{ cm} + 19 \text{ cm} = 57 \text{ cm}$

(b)

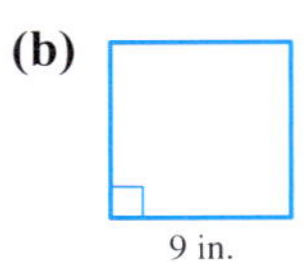

The perimeter of a square is four times the length of one side. Here

$P = 4(9 \text{ in.}) = 36 \text{ in.}$

(which could be written as 3 ft or 1 yd).

(c)

The perimeter of a rectangle is twice the length plus twice the width. Here,

$P = 2(3.6 \text{ ft}) + 2(1.7 \text{ ft}) = 7.2 \text{ ft} + 3.4 \text{ ft} = 10.6 \text{ ft}$

CHECK YOURSELF 1

Find the perimeter for each of the polygons.

(a)

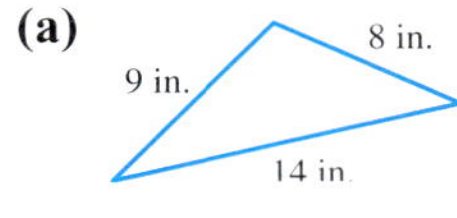

(b)

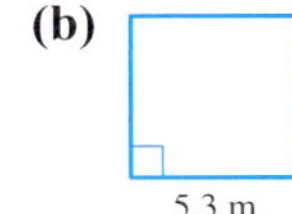

(c)

The distance around the outside of a circle is closely related to the concept of perimeter. We call the perimeter of a circle the **circumference.**

Definitions: Circumference of a Circle

The *circumference* of a circle is the distance around that circle.

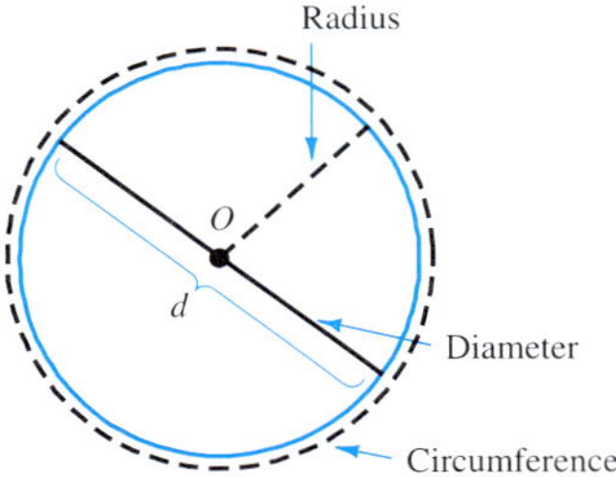

We will begin by defining some terms. In the given circle, d represents the **diameter.** This is the distance across the circle through its center (labeled with the letter O, for **origin**). The **radius** r is the distance from the center to a point on the circle. Note that, for a given circle, every radius is the same length. The diameter is always twice the radius.

It was discovered long ago that the ratio of the circumference of a circle to its diameter always stays the same. The ratio has a special name. It is named by the Greek letter π (pi). Pi is approximately 3.14 rounded to two decimal places. We can write this formula.

NOTE The formula comes from the ratio

$\frac{C}{d} = \pi$

Rules and Properties: Formula for the Circumference of a Circle

$C = \pi d$

Example 2

Finding the Circumference of a Circle

A circle has a diameter of 4.5 ft. Find its circumference, using 3.14 for π. If your calculator has a $\boxed{\pi}$ key, use that key instead of 3.14.

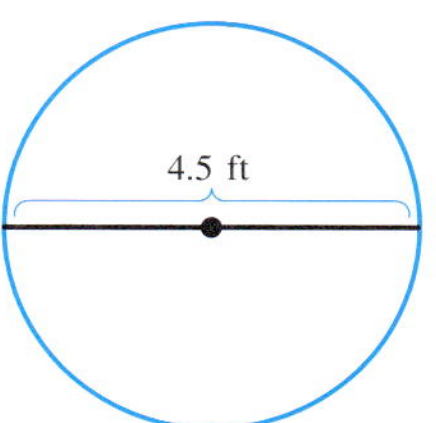

NOTE Because 3.14 is an approximation for pi, we can only say that the circumference is approximately 14.1 ft. The symbol $\approx$ means approximately.

$C = \pi d$

$\approx 3.14 \cdot 4.5$ ft

≈ 14.1 ft (rounded to one decimal place)

CHECK YOURSELF 2

A circle has a diameter of $3\frac{1}{2}$ *in. Find its circumference.*

NOTE If you want to approximate π, you needn't worry about running out of decimal places. The value for pi has been calculated to over 100,000,000 decimal places on a computer (the printout was some 20,000 pages long).

Rules and Properties: Formula for the Circumference of a Circle

$C = 2\pi r$

NOTE Because $d = 2r$ (the diameter is twice the radius) and $C = \pi d$, we have $C = \pi(2r)$, or $C = 2\pi r$.

Example 3

Finding the Circumference of a Circle

A circle has a radius of 8 cm. Find its circumference using 3.14 for π.

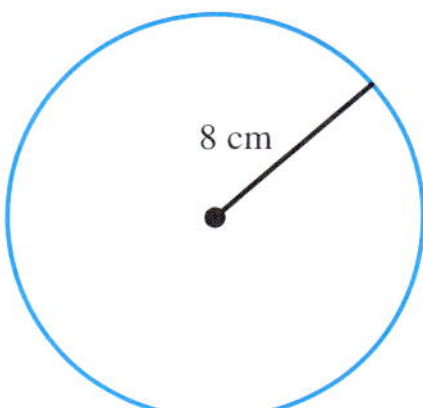

NOTE If we use the π button on a calculator and round to one decimal place, we get 50.3 cm.

$C = 2\pi r$

$\approx 2 \cdot 3.14 \cdot 8$ cm

≈ 50.2 cm (rounded to one decimal place)

CHECK YOURSELF 3

Find the circumference of a circle with a radius of 2.5 in.

Sometimes we will want to combine the ideas of perimeter and circumference to solve a problem.

Example 4

Finding Perimeter

We wish to build a wrought-iron frame gate according to the diagram. How many feet of material will be needed? (Round to the nearest tenth of a foot.)

NOTE The distance around the semicircle is $\frac{1}{2}\pi d$.

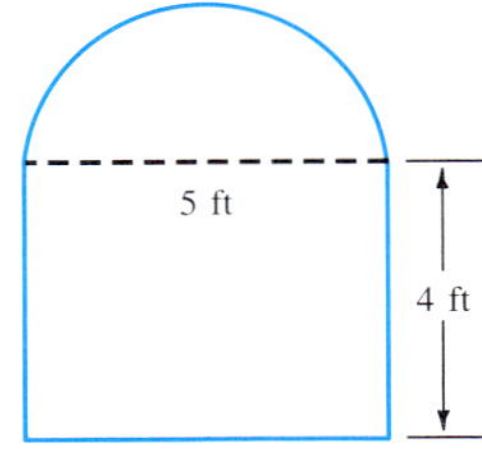

The problem can be broken into two parts. The upper part of the frame is a semicircle (half a circle). The remaining part of the frame is just three sides of a rectangle.

NOTE Using a calculator with a $\boxed{\pi}$ key,

1 $\boxed{\div}$ 2 $\boxed{\times}$ $\boxed{\pi}$ $\boxed{\times}$ 5

Circumference (upper part) $\approx \frac{1}{2} \cdot 3.14 \cdot 5 \text{ ft} \approx 7.9 \text{ ft}$

Perimeter (lower part) $= 4 + 5 + 4 = 13$ ft

Adding, we have

$7.9 + 13 = 20.9$ ft

We will need approximately 20.9 ft of material.

CHECK YOURSELF 4

Find the perimeter of the figure. Round to the nearest tenth of a meter.

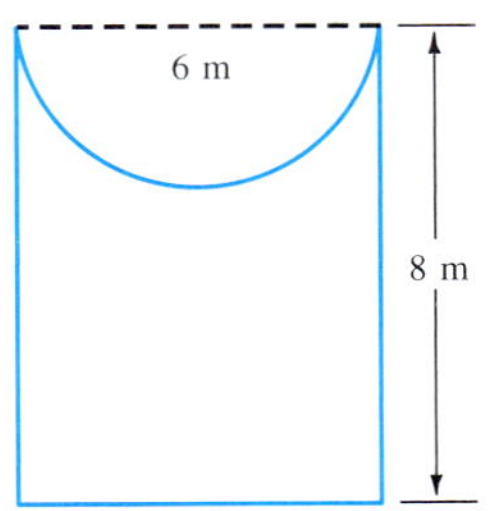

CHECK YOURSELF ANSWERS

1. **(a)** 31 in.; **(b)** 21.2 m; **(c)** 38 cm **2.** $C \approx 11$ in. **3.** $C \approx 15.7$ in.
4. $P \approx 31.4$ m

Name ____________

Section ______ Date ______

8.2 Exercises

Find the perimeter for each triangle.

1. 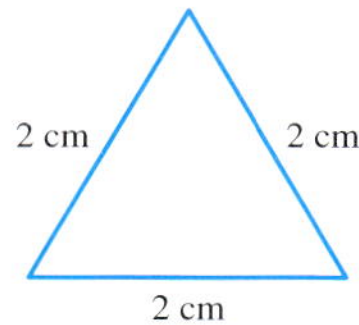

2. 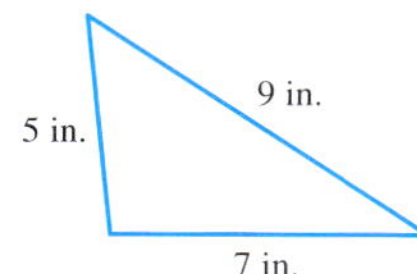

3. 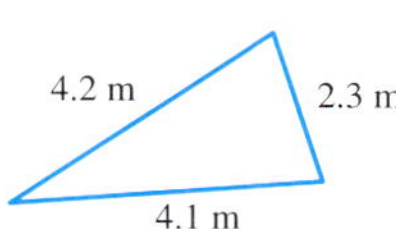

4. 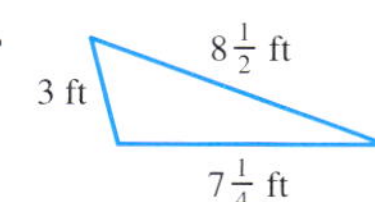

Find the perimeter for each polygon.

5. 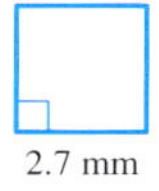

6. 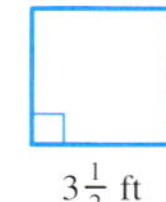

7. 

8. 

9. 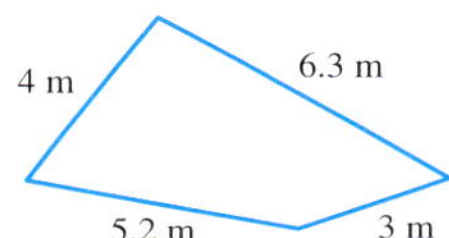

10. 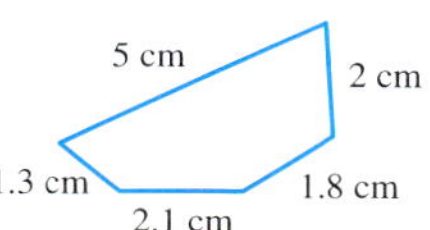

Find the circumference of each figure. Use 3.14 for π, and round your answer to one decimal place.

11. 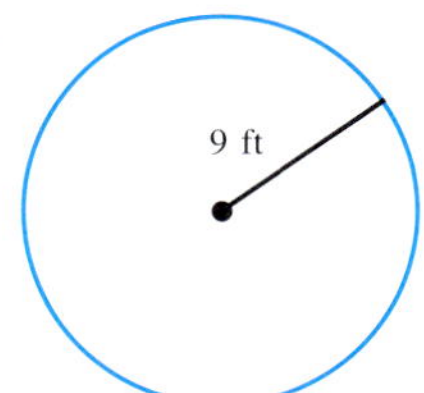

12. 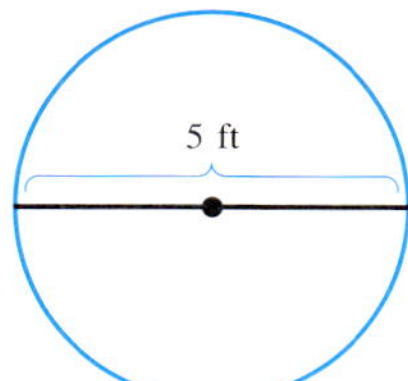

13. 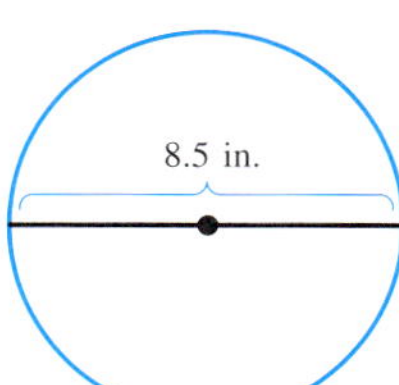

14. 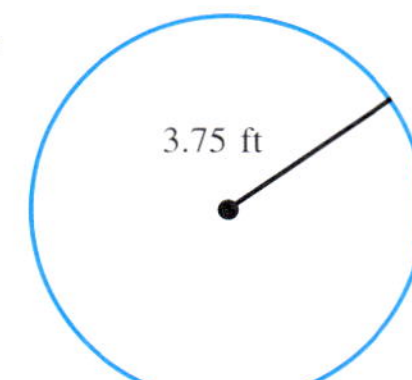

15. 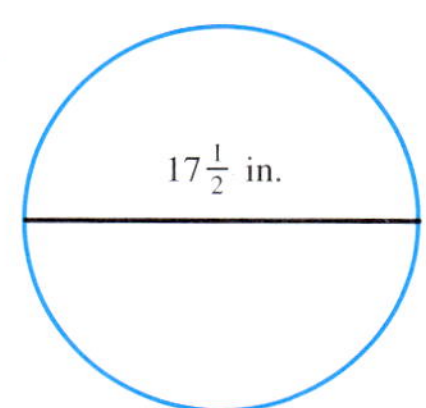

16. $3\frac{1}{2}$ ft

ANSWERS

1. ____________
2. ____________
3. ____________
4. ____________
5. ____________
6. ____________
7. ____________
8. ____________
9. ____________
10. ____________
11. ____________
12. ____________
13. ____________
14. ____________
15. ____________
16. ____________

ANSWERS

17. ______

18. ______

19. ______

20. ______

21. ______

22. ______

23. ______

24. ______

Find the perimeter of each figure. Use 3.14 to approximate π and round answers to one decimal place.

17.

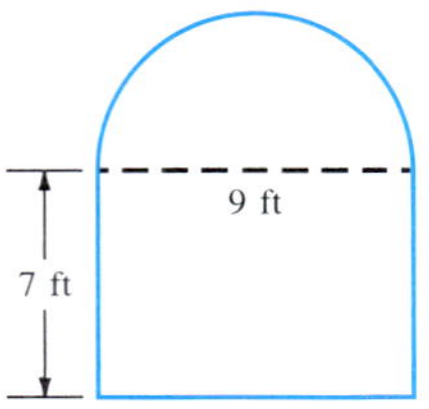

18.

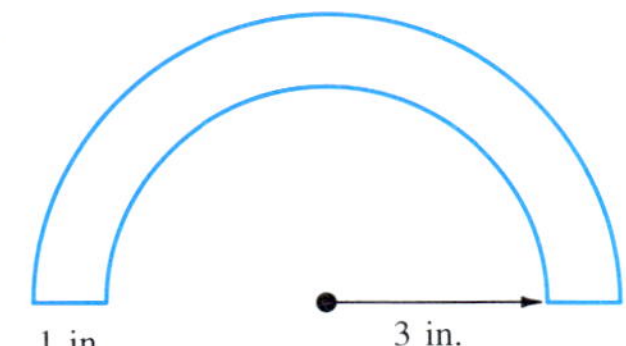

19.

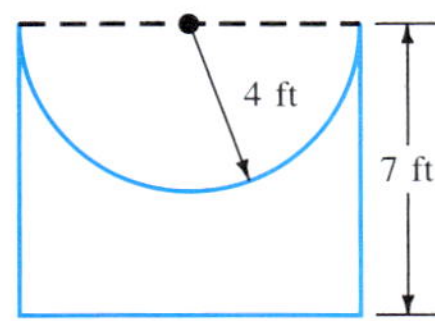

20.

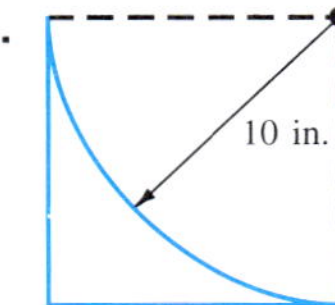

Solve the applications.

21. **Jogging.** A path runs around a circular lake with a diameter of 1,000 yd. Robert jogs around the lake three times for his morning run. How far has he run?

22. **Binding.** A circular rug is 6 ft in diameter. Binding for the edge costs $1.50 per yard. What will it cost to bind around the rug?

23. What happens to the circumference of a circle if you double the radius? If you double the diameter? If you triple the radius? Create some examples to demonstrate your answers.

24. The distance from Philadelphia to Sea Isle City is 100 mi. A car was driven this distance using tires with a radius of 14 in. How many revolutions of each tire occurred on the trip?

Answers

1. 6 cm **3.** 10.6 m **5.** 10.8 mm **7.** 47.8 cm **9.** 18.5 m
11. 56.5 ft **13.** 26.7 in. **15.** 55 in. **17.** 37.1 ft **19.** 34.6 ft
21. 9,420 yd **23.** Doubled; doubled; tripled

8.3 Area and Volume

8.3 OBJECTIVES

1. Determine areas of rectangles, squares, triangles, parallelograms, and other polygons
2. Convert between units of area
3. Determine the area of a circle
4. Determine volumes of rectangular solids, cubes, spheres, and cylinders

In Section 1.5, we examined the area of a rectangle. In this section, we will build on that idea. We begin with a definition.

Definitions: Area

The **area** of an object is the number of unit squares that cover its surface. We usually designate area with the letter A.

Here are the formulas for the areas of a square and a rectangle.

Rules and Properties: Formula for Area of a Square

$A = s^2$

where s is the length of a side.

Rules and Properties: Formula for Area of a Rectangle

$A = L \cdot W$

where L is the length and W is the width.

Example 1

Finding the Area of a Square

Find the area of the square.

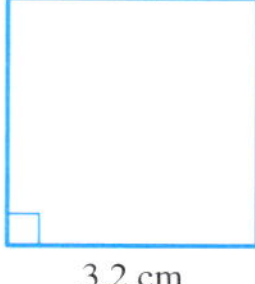

The area of a square is the square of the length of one side. Here we have

$$A = s^2 = (3.2 \text{ cm})^2 = 10.24 \text{ cm}^2$$

It is important to note that the units for area will *always* be square units.

CHECK YOURSELF 1

Find the area of the square.

In Example 2, we find the area of a rectangle.

Example 2

Finding the Area of a Rectangle

Find the area of the rectangle.

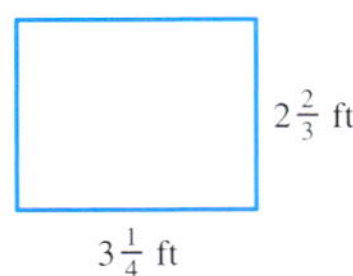

Using the formula $A = L \cdot W$, we get

$$A = 3\frac{1}{4}\text{ ft} \cdot 2\frac{2}{3}\text{ ft} = \frac{13}{4} \cdot \frac{8}{3}\text{ ft}^2 = \frac{104}{12}\text{ ft}^2 = \frac{26}{3}\text{ ft}^2$$

Note that the area is just under 9 ft^2.

CHECK YOURSELF 2

Find the area of the rectangle.

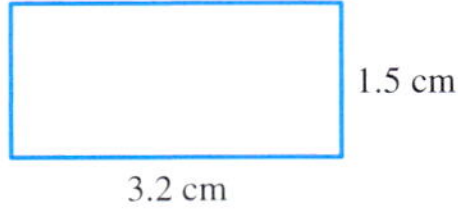

Two other figures that are frequently encountered are parallelograms and triangles.

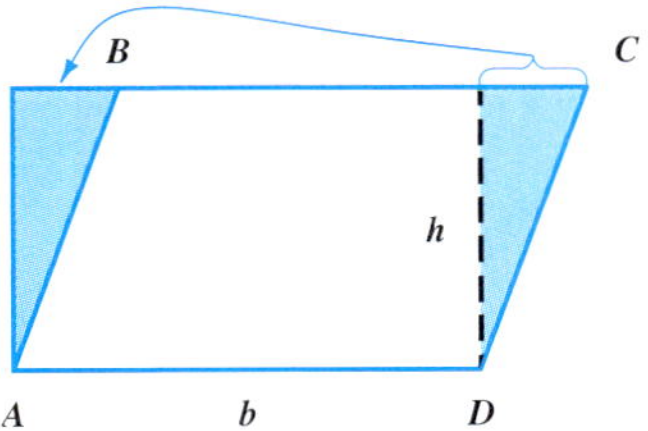

In the figure, $ABCD$ is a **parallelogram** (its opposite sides are parallel and of equal length). We can draw a line from D that forms a right angle with side BC. This cuts off one corner of the parallelogram. Now imagine that we move that corner over to the left side of the figure, as shown. This gives us a rectangle instead of a parallelogram. Because we haven't changed the area of the figure by moving the corner, the parallelogram has the same area as the rectangle, the product of the base and the height.

Rules and Properties: Formula for the Area of a Parallelogram

$A = b \cdot h$

where b is the base and h is the height.

Example 3

Finding the Area of a Parallelogram

A parallelogram has the dimensions shown in the figure. What is its area?

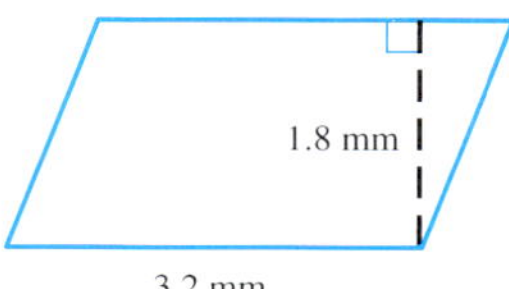

Use the formula for the area of a parallelogram, with $b = 3.2$ mm and $h = 1.8$ mm.

$$A = b \cdot h$$
$$= 3.2 \text{ mm} \cdot 1.8 \text{ mm} = 5.76 \text{ mm}^2$$

CHECK YOURSELF 3

If the base of a parallelogram is $3\frac{1}{2}$ in. and its height is $1\frac{1}{2}$ in., what is its area?

As we saw in Section 8.1, another common geometric figure is the **triangle.** It has three sides. An example is triangle ABC shown in the figure.

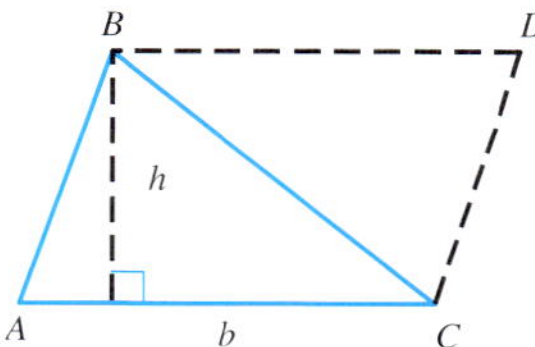

b is the base of the triangle.
h is the height, or the *altitude*, of the triangle.

Once we have a formula for the area of a parallelogram, it is not hard to find the area of a triangle. If we draw the dotted lines from B to D and from C to D parallel to two sides of the triangle, we form a parallelogram. The area of the triangle is then one-half the area of the parallelogram (which is $b \cdot h$).

Rules and Properties: Formula for the Area of a Triangle

$$A = \frac{1}{2} \cdot b \cdot h$$

Example 4

Finding the Area of a Triangle

A triangle has an altitude of 2.3 in., and its base is 3.4 in. What is its area?

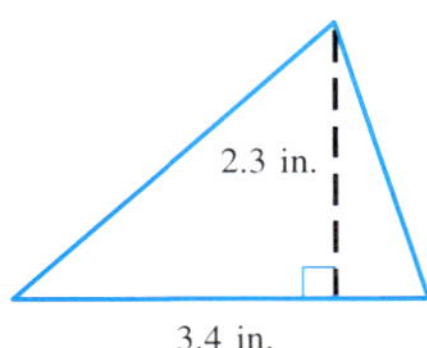

Using the formula for the area of a triangle, with $b = 3.4$ in. and $h = 2.3$ in.

$$A = \frac{1}{2} \cdot b \cdot h$$

$$= \frac{1}{2} \cdot 3.4 \text{ in.} \cdot 2.3 \text{ in.} = 3.91 \text{ in.}^2$$

CHECK YOURSELF 4

A triangle has a base of 10 kilometers (km) and an altitude of 6 km. Find its area.

Sometimes we will want to convert from one square unit to another. For instance, look at 1 yd^2 in the figure.

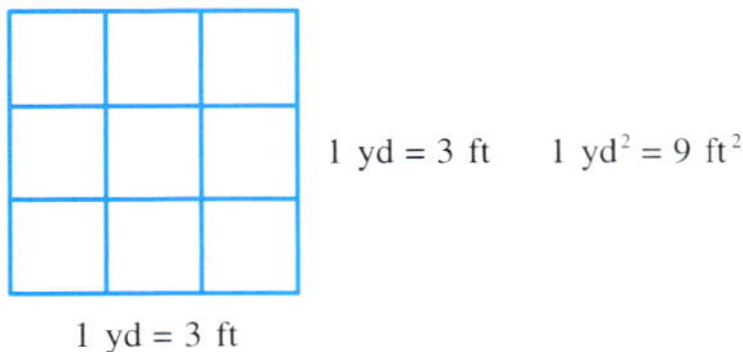

The table gives some useful relationships.

Square Units and Equivalents

1 square foot (ft^2)	= 144 square inches ($in.^2$)
1 square yard (yd^2)	= 9 ft^2
1 acre	= 4,840 yd^2 = 43,560 ft^2
1 square mile (mi^2)	= 640 acres

NOTE Originally the acre was the area that could be plowed by a team of oxen in a day!

Example 5

Converting Between Square Feet and Square Yards in Finding Area

A room has the dimensions 12 ft by 15 ft. How many square yards of linoleum will be needed to cover the floor?

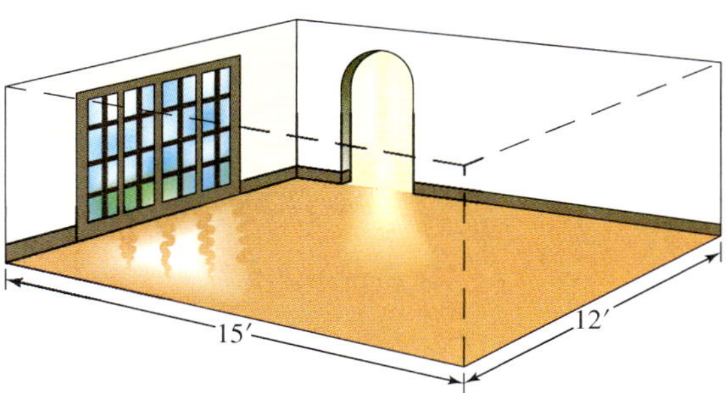

NOTE We first find the area in square feet, and then convert to square yards.

$$A = 12 \text{ ft} \cdot 15 \text{ ft} = 180 \text{ ft}^2$$

$$= \overset{20}{\cancel{180}} \text{ ft}^2 \cdot \frac{1}{\underset{1}{\cancel{9}}} \frac{\text{yd}^2}{\text{ft}^2}$$

$$= 20 \text{ yd}^2$$

Recall that $\frac{1 \text{ yd}^2}{9 \text{ ft}^2}$ is a unit fraction.

CHECK YOURSELF 5

A hallway is 27 ft long and 4 ft wide. How many square yards of carpeting will be needed to carpet the hallway?

Example 6 illustrates the use of a common unit of area, the acre.

Example 6

Converting Between Square Yards and Acres in Finding Area

A rectangular field is 220 yd long and 110 yd wide. Find its area in acres.

$$A = 220 \text{ yd} \cdot 110 \text{ yd} = 24{,}200 \text{ yd}^2$$

$$= \overset{5}{\cancel{24{,}200}} \text{ yd}^2 \cdot \frac{1}{\underset{1}{\cancel{4840}}} \frac{\text{acre}}{\text{yd}^2}$$

$$= 5 \text{ acres}$$

CHECK YOURSELF 6

A proposed site for an elementary school is 220 yd long and 198 yd wide. Find its area in acres.

The number pi (π), which we used to find circumference in Section 8.2, is also used in finding the area of a circle. If r is the radius of a circle, we have this formula.

Rules and Properties: Formula for the Area of a Circle

$A = \pi r^2$

NOTE This is read, "Area equals pi r squared." You can multiply the radius by itself and then by pi.

Example 7

Find the Area of a Circle

A circle has a radius of 7 in. What is its area?

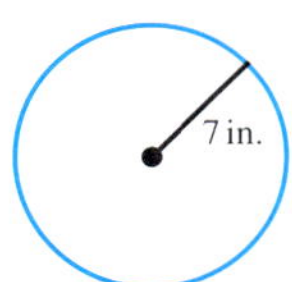

Use the formula for the area of a circle, 3.14 for π, and $r = 7$ in.

$$A \approx 3.14 \cdot (7 \text{ in.})^2$$
$$\approx 3.14 \cdot (49 \text{ in.}^2)$$
$$\approx 153.9 \text{ in.}^2$$

Again the area is an approximation because we use 3.14, an approximation for π.

CHECK YOURSELF 7

Find the area of a circle whose diameter is 4.8 cm. Remember that the formula refers to the radius. Use 3.14 for π, and round your result to the nearest tenth of a square centimeter.

Our next measurement deals with finding **volumes.** The volume of a **solid** is the measure of the space contained in the solid.

Definitions: Definition of a Solid

A *solid* is a three-dimensional figure. It has length, width, and height.

1 in.
1 in.
1 in.
1 cubic inch

Volume is measured in **cubic units.** Examples include cubic inches, cubic feet, and cubic centimeters. A cubic inch, for instance, is the measure of the space contained in a cube that is 1 in. on each edge.

In finding the volume of a figure, we want to know how many cubic units are contained in that figure. We will start with a simple example, a **rectangular solid.** A rectangular solid is a very familiar figure. A box, a crate, and most rooms are rectangular solids. Say that the dimensions of the solid are 5 in. by 3 in. by 2 in. as pictured. If we divide the solid into units of 1 in.^{3}, we have two layers, each containing 3 units by 5 units, or 15 in.^{3} Because there are two layers, the volume is 30 in.^{3}

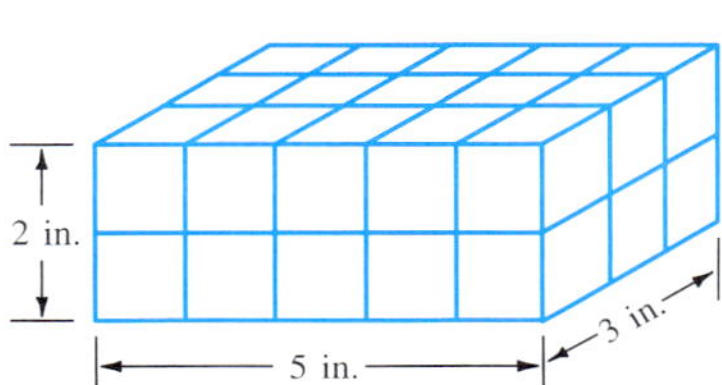

In general, we can see that the volume of a rectangular solid is the product of its length, width, and height.

Rules and Properties: Formula for the Volume of a Rectangular Solid

$V = L \cdot W \cdot H$

Example 8

Finding Volume

NOTE We are not particularly worried about which is the length, which is the width, and which is the height, because the order in which we multiply won't change the result.

A crate has dimensions 4 ft by 2 ft by 3 ft. Find its volume.

Use the formula for the volume of a rectangular solid, with $L = 4$ ft, $W = 2$ ft, and $H = 3$ ft.

$$
\begin{aligned}
V &= L \cdot W \cdot H \\
&= 4 \text{ ft} \cdot 2 \text{ ft} \cdot 3 \text{ ft} \\
&= 24 \text{ ft}^3
\end{aligned}
$$

CHECK YOURSELF 8

A room is 15 ft long, 10 ft wide, and 8 ft high. What is its volume?

The formulas for the volume of a sphere (the shape of a ball) and a right circular cylinder (the shape of a can) both involve the number π.

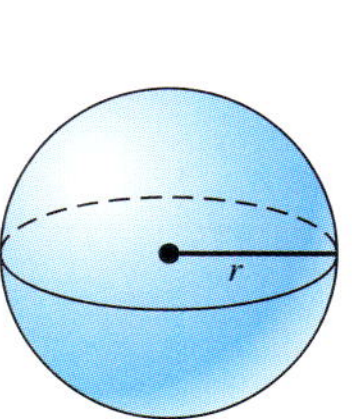

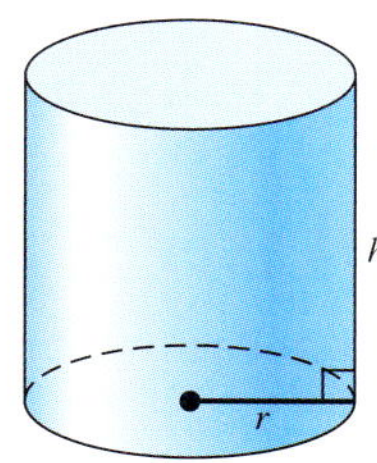

Rules and Properties: Volume of a Sphere

$V = \frac{4}{3}\pi r^3$

where r is the radius.

Rules and Properties: Volume of a Right Circular Cylinder

$V = \pi r^2 h$

where r is the radius and h is the height.

Example 9

Finding a Volume

Find the volume of each solid shape.

(a)

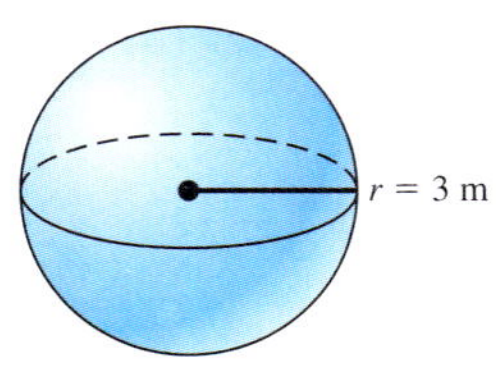

This is a sphere with radius 3 m. Using the formula for the volume of a sphere,

$$\begin{aligned} V &= \frac{4}{3}\pi r^3 \\ &= \frac{4}{3}\pi(3\text{ m})^3 \\ &= \frac{4}{3}\pi \cdot 27\text{ m}^3 \\ &= \frac{4}{3} \cdot \pi \cdot \frac{27}{1}\text{ m}^3 \\ &= 36\pi\text{ m}^3 \\ &\approx 36(3.14)\text{ m}^3 \\ &\approx 113.1\text{ m}^3 \end{aligned}$$

(b)

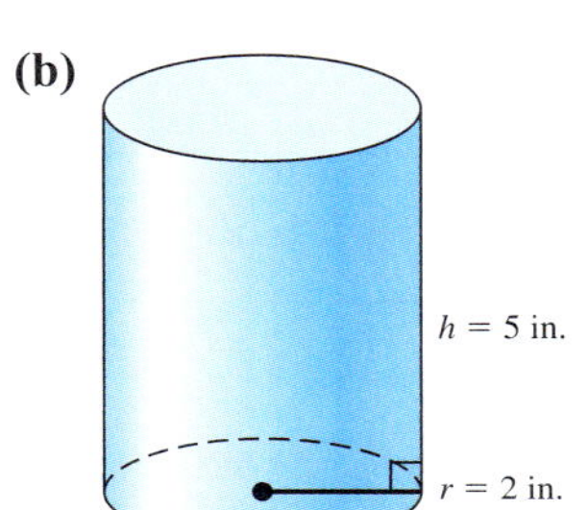

This is a right circular cylinder with radius 2 in. and height 5 in. Using the formula for the volume of a right circular cylinder,

$$\begin{aligned} V &= \pi r^2 h \\ &= \pi(2\text{ in.})^2\,(5\text{ in.}) \\ &= \pi 4\text{ in.}^2\,(5\text{ in.}) \\ &= 20\pi\text{ in.}^3 \\ &\approx 20(3.14)\text{ in.}^3 \\ &\approx 62.8\text{ in.}^3 \end{aligned}$$

CHECK YOURSELF 9

Find the volume of each solid shape.

(a)

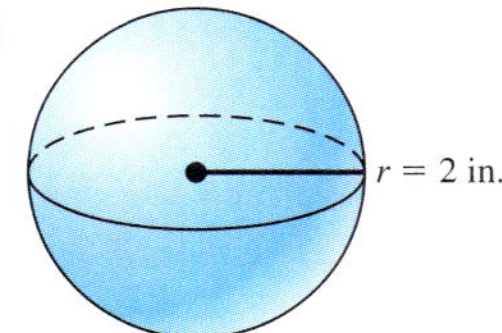

(b)

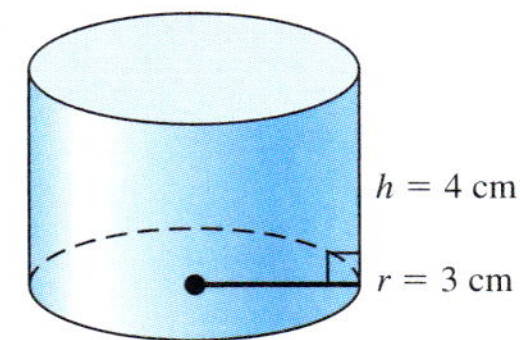

CHECK YOURSELF ANSWERS

1. 28.09 m^2 **2.** 4.8 cm^2 **3.** $A = 5\frac{1}{4}\text{ in.}^2$

4. $A = 30\text{ km}^2$ **5.** 12 yd^2 **6.** 9 acres **7.** $\approx 18.1\text{ cm}^2$

8. 1,200 ft^3 **9.** **(a)** $33.5\pi\text{ in.}^3$; **(b)** $113.1\pi\text{ cm}^3$

Name ______________________

Section ________ Date ________

8.3 Exercises

Find the area of each figure.

1. 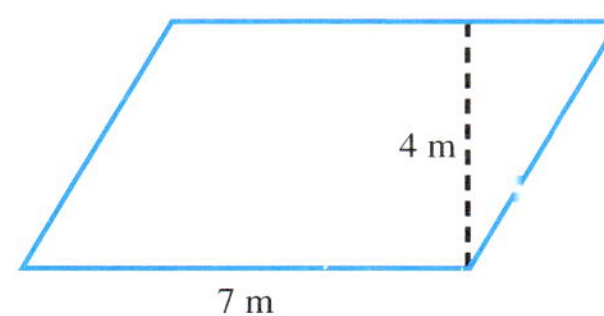

2. 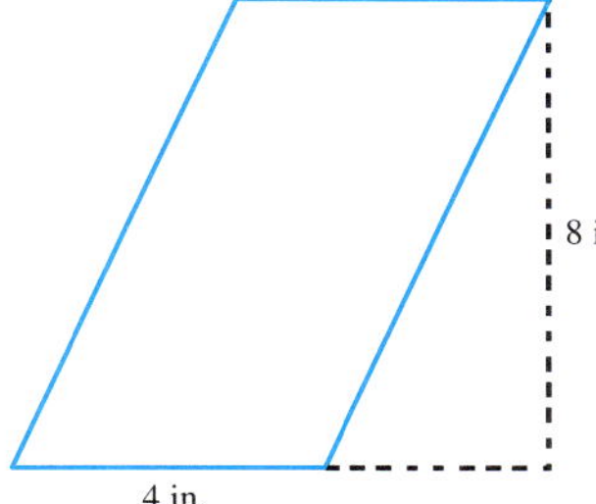

3. 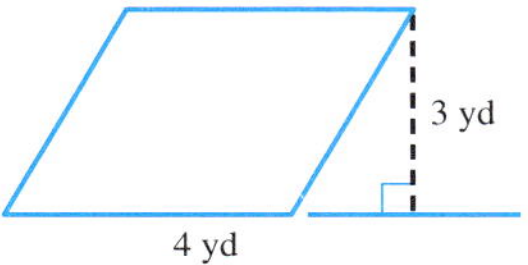

4. 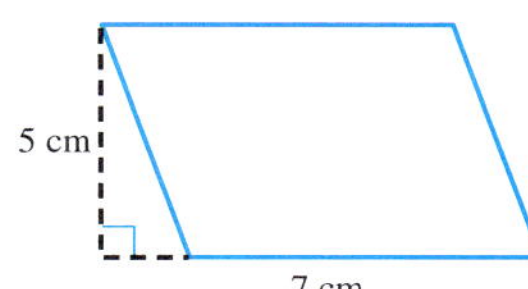

5. 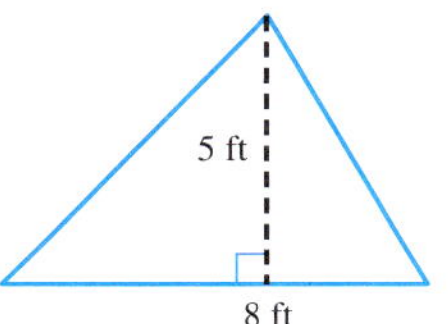

6. 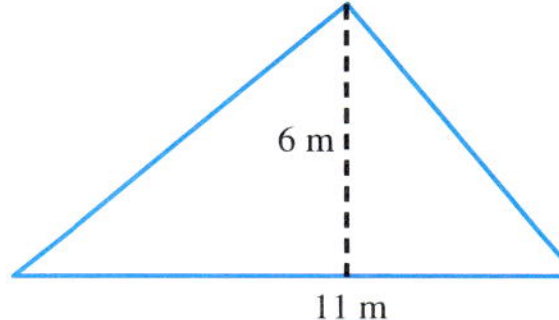

7. 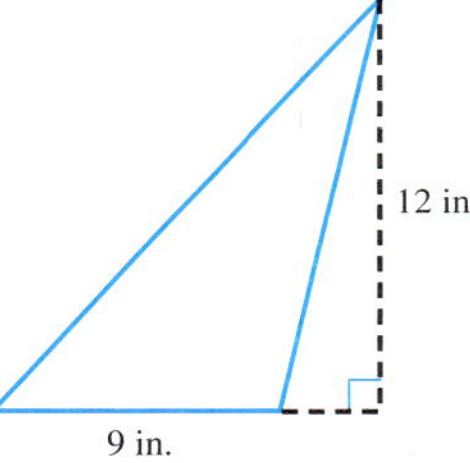

8. 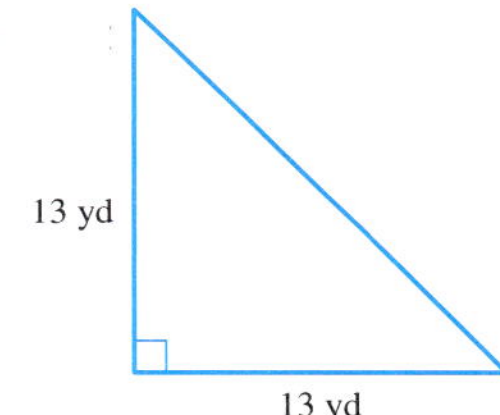

9. 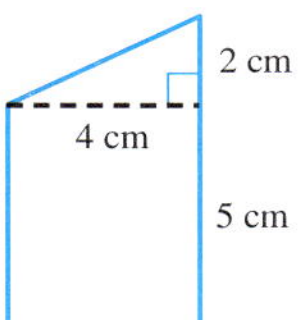

10. 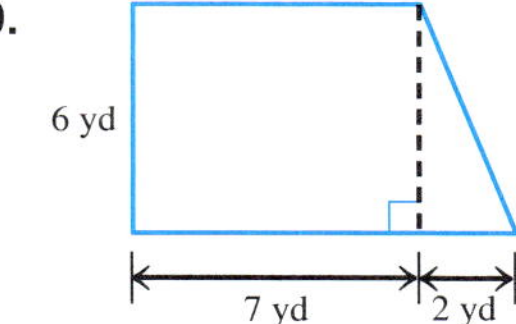

Find the area of each figure. Use 3.14 for π, and round your answer to one decimal place.

11. 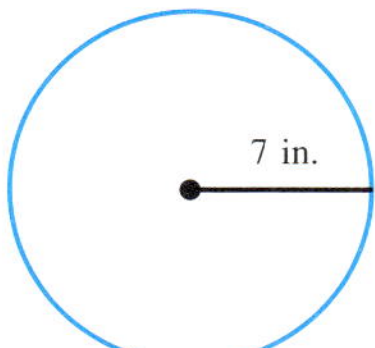

12. 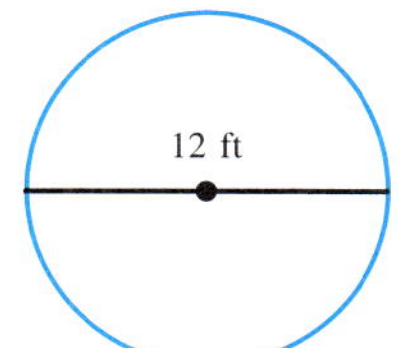

ANSWERS

1. ______________
2. ______________
3. ______________
4. ______________
5. ______________
6. ______________
7. ______________
8. ______________
9. ______________
10. ______________
11. ______________
12. ______________

ANSWERS

13. ______

14. ______

15. ______

16. ______

17. ______

18. ______

19. ______

20. ______

21. ______

13.

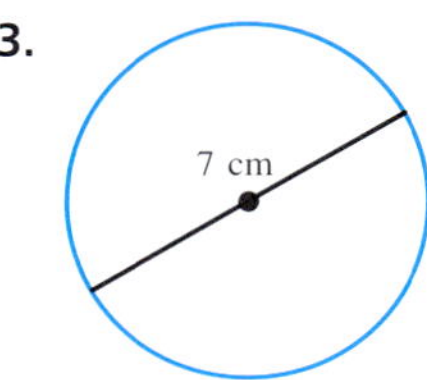

14.

8 m

15.

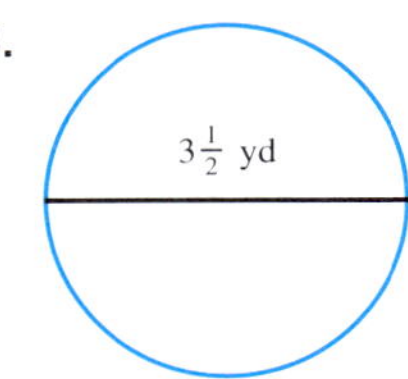

16.

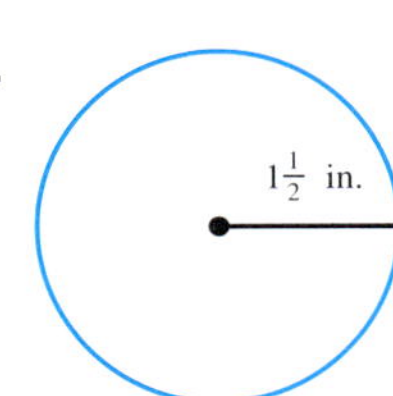

Solve the applications.

17. **Lawn care.** A circular piece of lawn has a radius of 28 ft. You have a bag of fertilizer that will cover 2,500 ft^2 of lawn. Do you have enough?

18. **Cost.** A circular coffee table has a diameter of 5 ft. What will it cost to have the top refinished if the company charges \$3 per square foot for the refinishing?

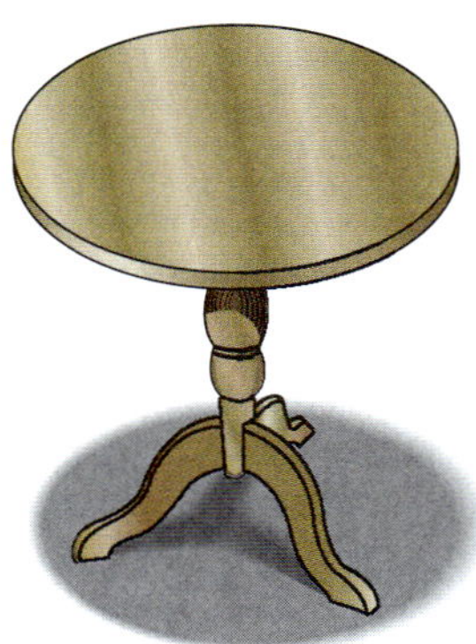

19. **Cost.** A circular terrace has a radius of 6 ft. If it costs \$1.50 per square foot to pave the terrace with brick, what will the total cost be?

20. **Area.** A house addition is in the shape of a semicircle (a half circle) with a radius of 9 ft. What is its area?

21. **Amount of material.** A Tetra-Kite uses 12 triangular pieces of plastic for its surface. Each triangle has a base of 12 in. and a height of 12 in. How much material is needed for the kite?

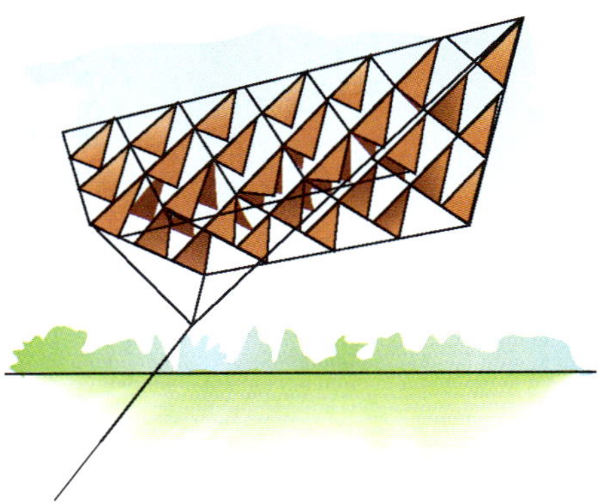

ANSWERS

22. ______
23. ______
24. ______
25. ______
26. ______
27. ______
28. ______
29. ______
30. ______
31. ______

22. **Acreage.** You buy a square lot that is 110 yd on each side. What is its size in acres?

23. **Area.** You are making rectangular posters 12 in. by 15 in. How many square feet of material will you need for four posters?

24. **Cost.** Andy is carpeting a recreation room 18 ft long and 12 ft wide. If the carpeting costs $15 per square yard, what will be the total cost of the carpet?

25. **Acreage.** A shopping center is rectangular, with dimensions of 550 yd by 440 yd. What is its size in acres?

26. **Cost.** An A-frame cabin has a triangular front with a base of 30 ft and a height of 20 ft. If the front is to be glass that costs $3 per square foot, what will the glass cost?

Find the area of the shaded part in each figure. Round your answers to one decimal place.

27. 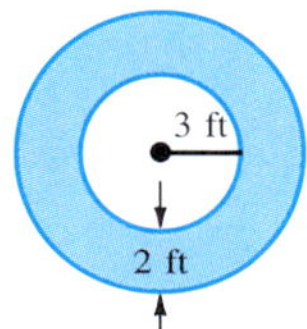

28. 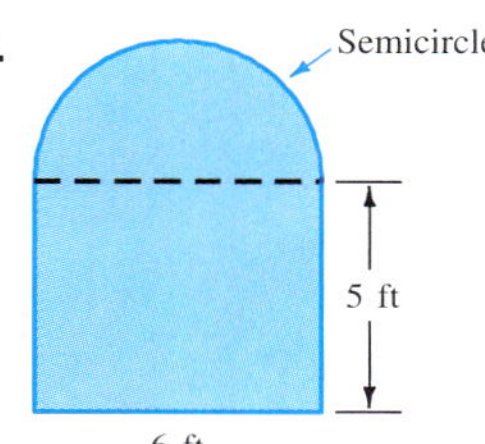

29. 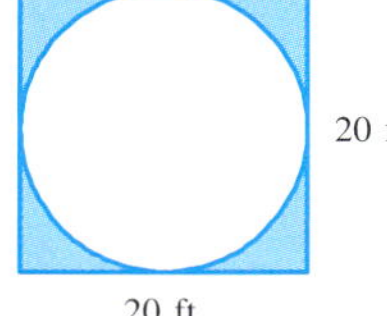

30. 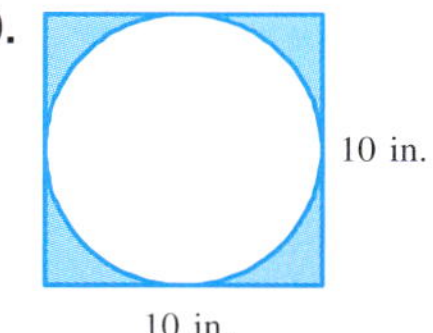

31. Papa Doc's delivers pizza. The price of an 8-in.-diameter pizza is $8.99, and the price of a 16-in.-diameter pizza is $17.98. Write a plan to determine which is the better buy.

ANSWERS

32. ______

33. ______

34. ______

35. ______

36. ______

32. An indoor track layout is shown.

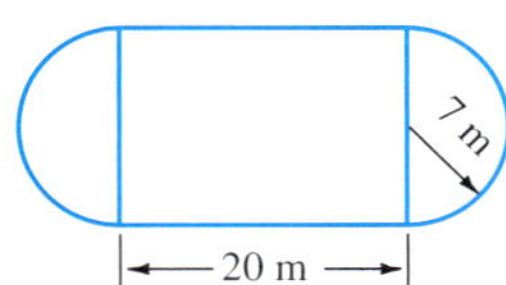

How much would it cost to lay down hardwood floor if the hardwood floor costs \$10.50 per square meter?

33. Find the area and the circumference (or perimeter) of each item:

(a) a penny **(b)** a nickel **(c)** a dime **(d)** a quarter **(e)** a half-dollar **(f)** a silver dollar **(g)** a Sacagawea dollar **(h)** a dollar bill **(i)** one face of the pyramid on the back of a \$1 bill.

34. How would you determine the cross-sectional area of a Douglas fir tree (at, say, 3 ft above the ground), without cutting it down? Use your method to solve this problem:

If the circumference of a Douglas fir is 6 ft 3 in., measured at a height of 3 ft above the ground, compute the cross-sectional area of the tree at that height.

35. What is the effect on the area of a triangle if the base is doubled and the altitude is cut in half? Create some examples to demonstrate your ideas.

36. What happens to the area of a circle if you double the radius? If you double the diameter? If you triple the radius? Create some examples to demonstrate your answers.

Find the volume of each solid shown. Where appropriate, use 3.14 as an approximation for π and round to the nearest tenth.

37.

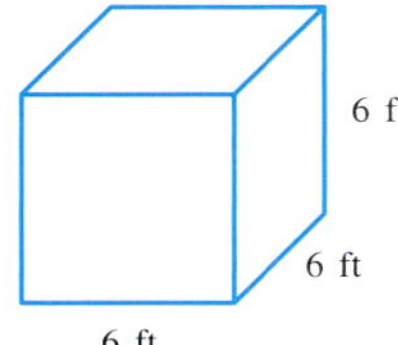

38.

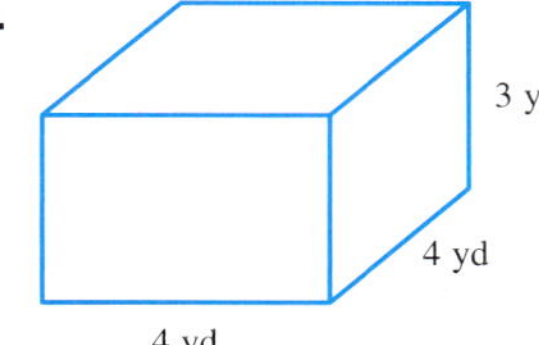

39.

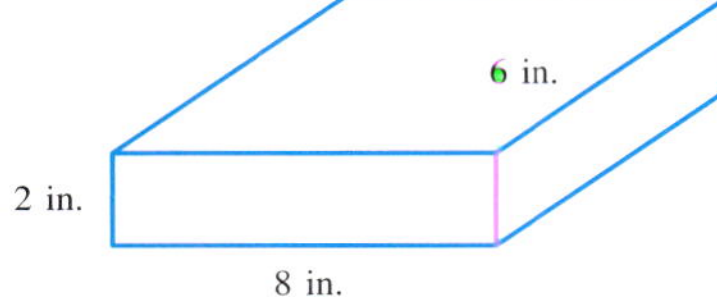

40.

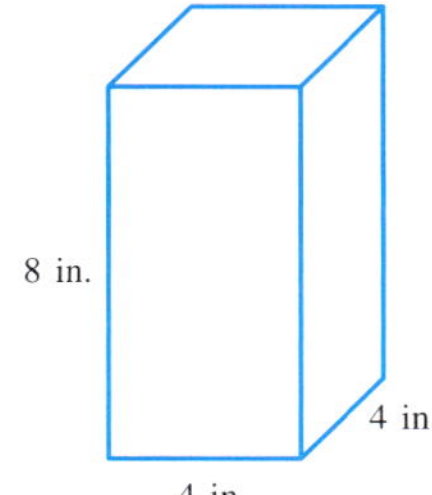

41.

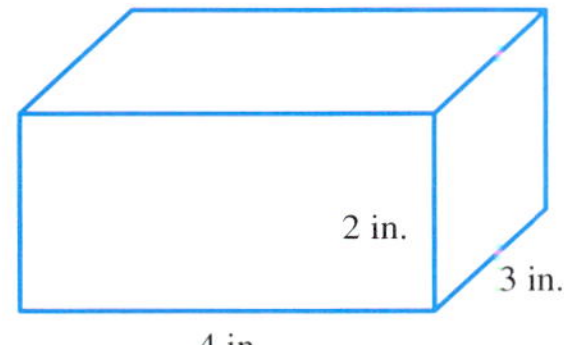

42.

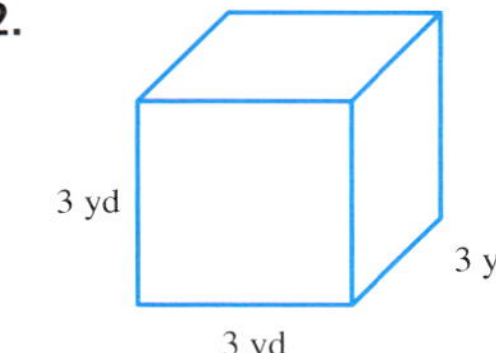

43.

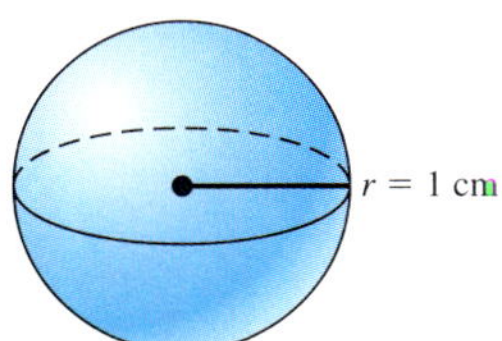

44.

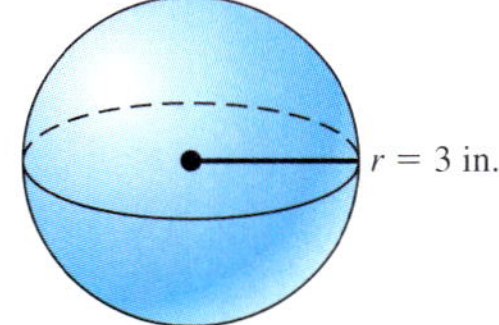

45.

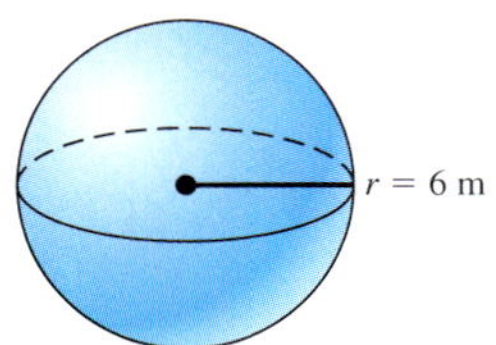

46.

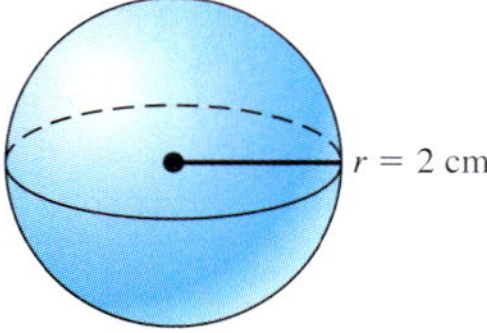

47.

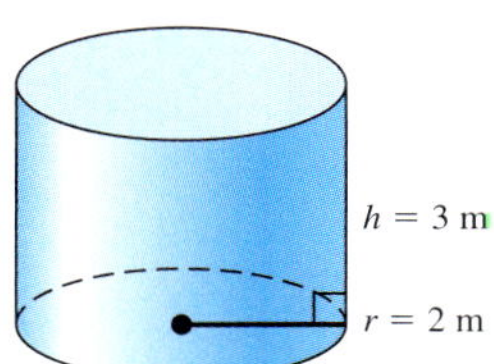

48.

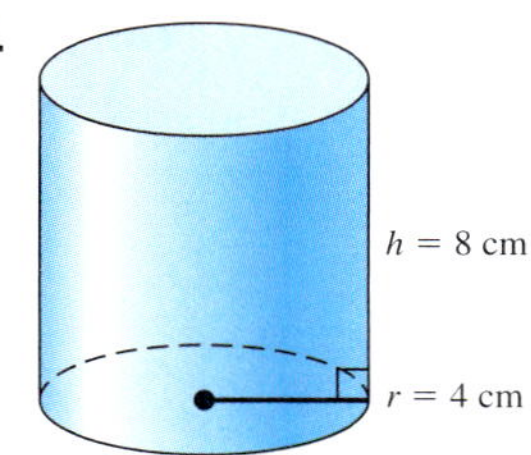

ANSWERS

37. ____________

38. ____________

39. ____________

40. ____________

41. ____________

42. ____________

43. ____________

44. ____________

45. ____________

46. ____________

47. ____________

48. ____________

ANSWERS

49. __________

50. __________

51. __________

52. __________

Solve the applications.

49. Shipping. A shipping container is 5 ft by 3 ft by 2 ft. What is its volume?

50. Size of a cord. A cord of wood is 4 ft by 4 ft by 8 ft. What is its volume?

51. Storage. The inside dimensions of a meat market's cooler are 3 m by 3 m by 2 m. What is the capacity of the cooler in cubic meters?

52. Storage. A storage bin is 6 m long, 2 m wide, and 2 m high. What is its volume in cubic meters?

Answers

1. 28 m^2 **3.** 12 yd^2 **5.** 20 ft^2 **7.** 54 $in.^2$ **9.** 24 cm^2 **11.** 153.9 $in.^2$ **13.** 38.5 cm^2 **15.** 9.6 yd^2 **17.** yes; ≈ 2,461.8 ft^2 **19.** ≈$169.56 **21.** 864 $in.^2$ **23.** 5 ft^2 **25.** 50 acres **27.** 50.2 ft^2 **29.** 86 ft^2 **31.** **33.** **35.** **37.** 216 ft^3 **39.** 96 $in.^3$ **41.** 24 $in.^3$ **43.** 4.2 cm^3 **45.** 803.8 m^3 **47.** 37.7 m^3 **49.** 30 ft^3 **51.** 18 m^3

8 Summary

Definition/Procedure	Example	Reference
Lines and Angles		**Section 8.1**
Line A series of points that goes on forever.	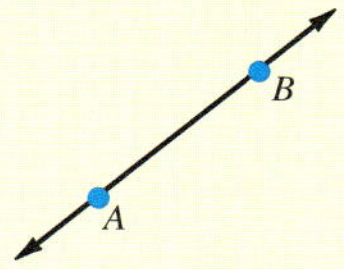 B, A	p. 597
Line segment A piece of a line that has two endpoints.	 C, D	p. 597
Angle A geometric figure consisting of two line segments that share a common endpoint.	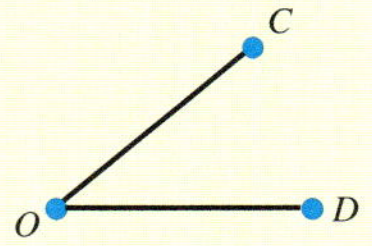 C, O, D	p. 598
Perpendicular lines Lines are *perpendicular* if they intersect to form four equal angles.	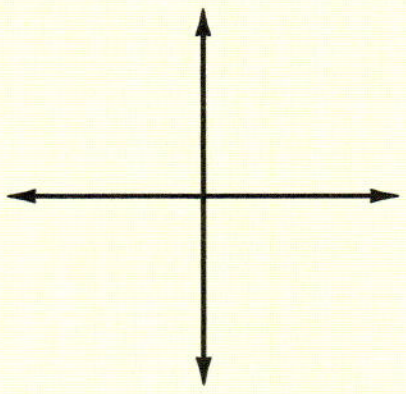	p. 598
Parallel lines Lines are *parallel* if they never intersect.	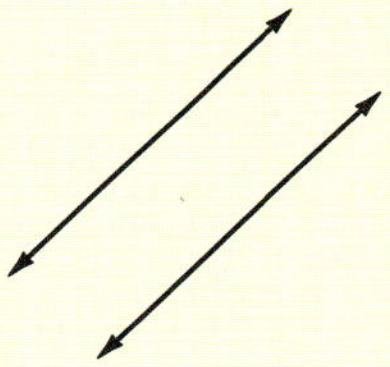	p. 598
Right angles have a measure of 90°.	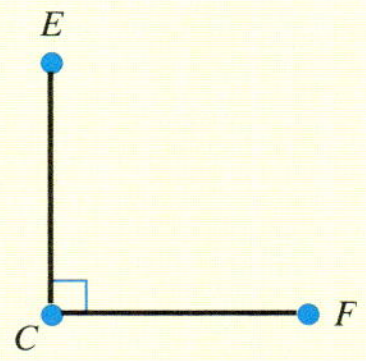 E, C, F	p. 599

Continued

DEFINITION/PROCEDURE	EXAMPLE	REFERENCE
Lines and Angles		**Section 8.1**
Acute angles have a measure less than 90°.	A, O, B	p. 600
Obtuse angles have a measure between 90° and 180°.	C, E, F	p. 600
Perimeter and Circumference		**Section 8.2**
A **polygon** is a figure with three or more sides. Each side is a line segment. The **perimeter** is the sum of the lengths of its sides: Triangle $P = a + b + c$ Square $P = 4s$ Rectangle $P = 2L + 2W$ Circle $C = 2\pi r$	Find the perimeter of the figure: 1.6 cm, 5 cm $P = 2L + 2W$ $= 2(1.6\text{ cm}) + 2(5\text{ cm})$ $= 3.2\text{ cm} + 10\text{ cm}$ $= 13.2\text{ cm}$	p. 613
Area and Volume		**Section 8.3**
The area of an object is the number of unit squares needed to cover its surface: Square $A = s^2$ Rectangle $A = L \cdot W$ Triangle $A = \frac{1}{2} \cdot b \cdot h$ Circle $A = \pi r^2$	Find the area of the figure: 2.4 in., 6 in. $A = \frac{1}{2}(6\text{ in.})(2.4\text{ in.})$ $= (3\text{ in.})(2.4\text{ in.})$ $= 7.2\text{ in.}^2$	p. 619
Some common volume formulas: Cube $V = s^3$ Rectanglar solid $V = L \cdot W \cdot H$ Sphere $V = \frac{4}{3}\pi r^2$ Cylinder $V = \pi r^2 h$	Find the volume: $h = 4$ cm, $r = 3$ cm $V = \pi r^2 h$ $= \pi(3\text{ cm})^2(4\text{ cm})$ $= 36\pi\text{ cm}^3$ $\approx 113.1\text{ cm}^3$	p. 625

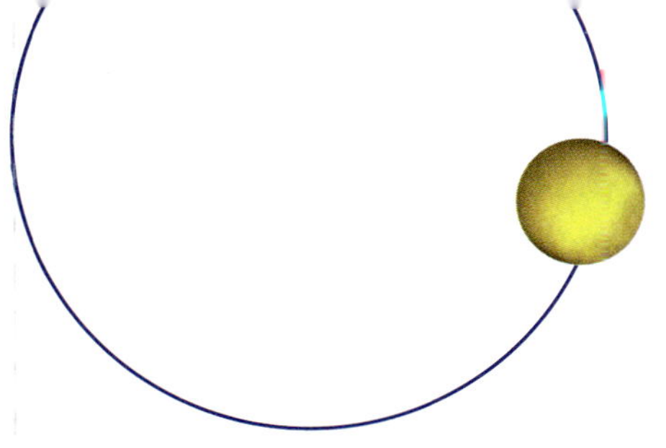

Summary and Review Exercises

This exercise set will give you practice with each of the objectives of the chapter. Each exercise is keyed to the appropriate chapter section. The answers are provided in the *Instructor's Manual*. Your instructor will give you guidelines on how to best use these exercises in your instructional setting.

[8.1] For exercises 1 to 6, name the angle; label it as acute, obtuse, right, or straight; then estimate its measure with a protractor to the nearest 10 degrees.

1.

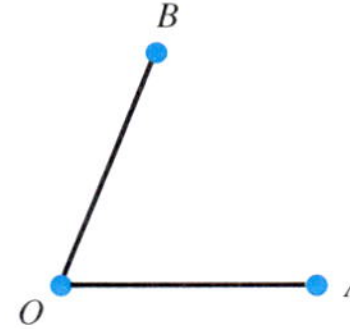

2.

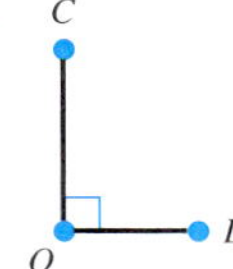

3.

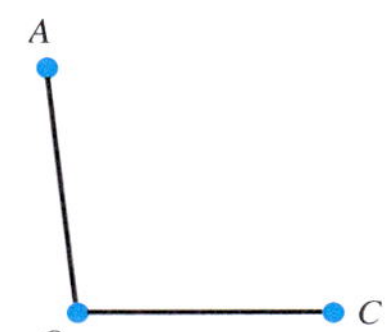

4.

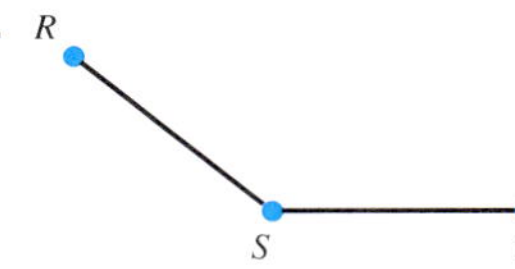

5.

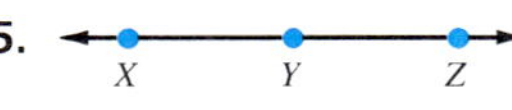

6.

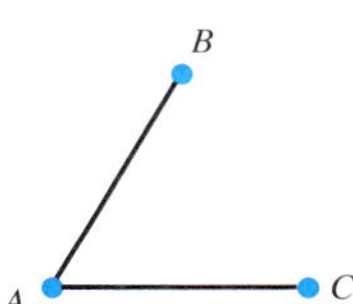

Give the measure of each angle in degrees. **Note:** A revolution is 360°.

7. $\angle A$ represents $\frac{3}{8}$ of a revolution

8. $\angle B$ represents $\frac{7}{10}$ of a revolution

[8.2] Find the perimeter or circumference of each figure.

9.

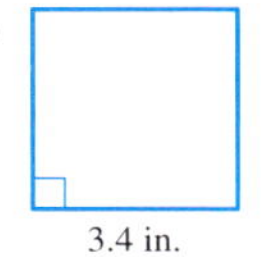

10.

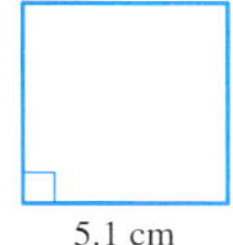

11.

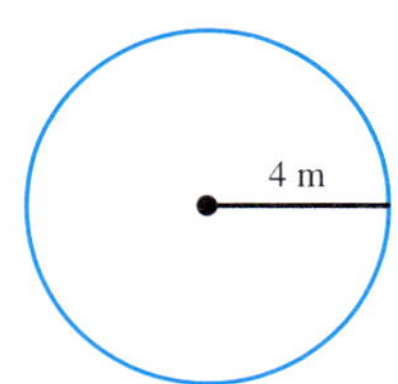

12.

[8.3] Find the area of each figure.

13.

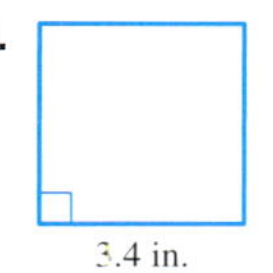

14.

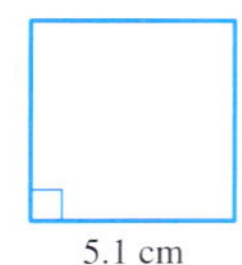

15.

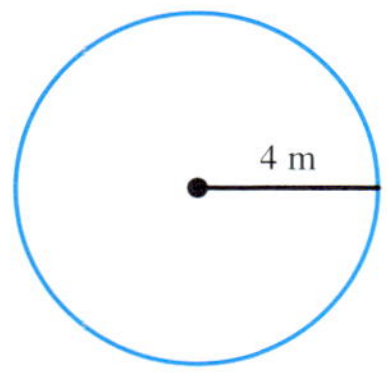

16.

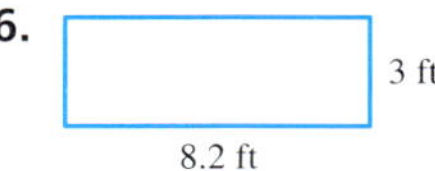

[8.2] Find the perimeter or circumference of each figure. Round answers to the nearest tenth.

17.

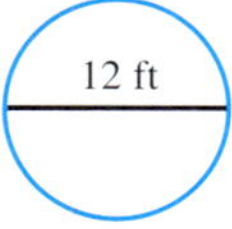

18.

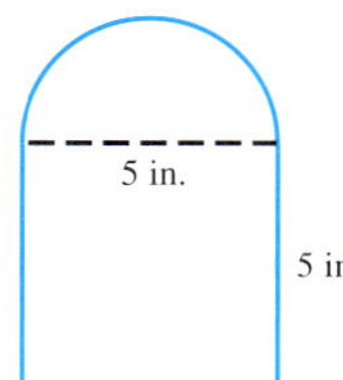

[8.3]

19. Find the area of a circle with radius 10 ft.

In exercises 20 and 21, find the area of each figure.

20.

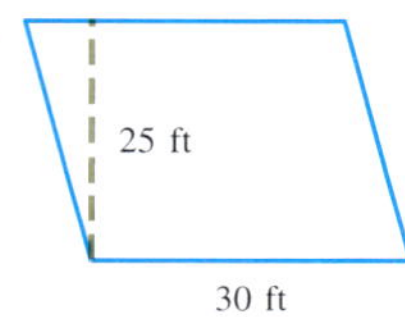

21.

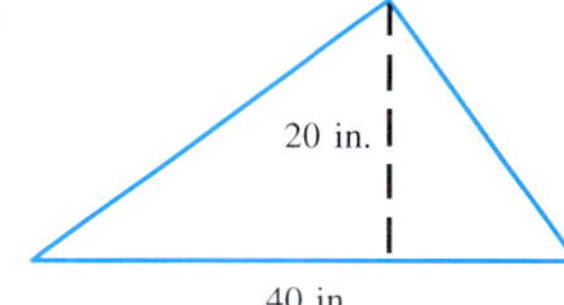

In exercises 22 and 23, solve the applications.

22. Area. How many square feet of vinyl floor covering will be needed to cover the floor of a room that is 10 ft by 18 ft? How many square yards will be needed? (*Hint:* How many square feet are in a square yard?)

23. Cost. A rectangular roof for a house addition measures 15 ft by 30 ft. A roofer will charge $175 per "square" (100 ft^2). Find the cost of the roofing for the addition.

In exercises 24 to 26, find the volume of each figure.

24.

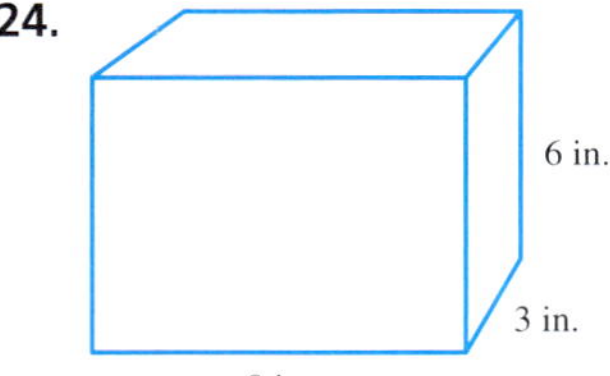

25.

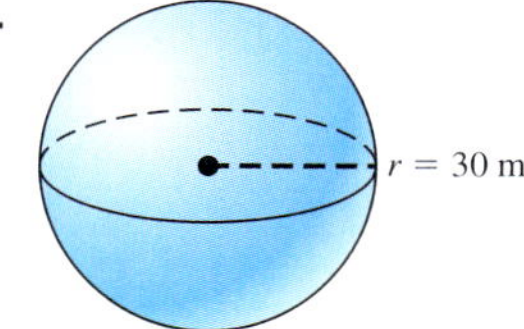

26.

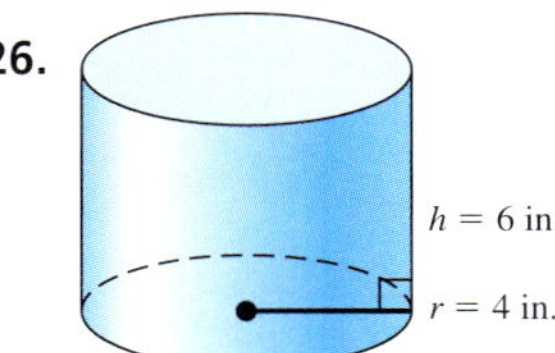

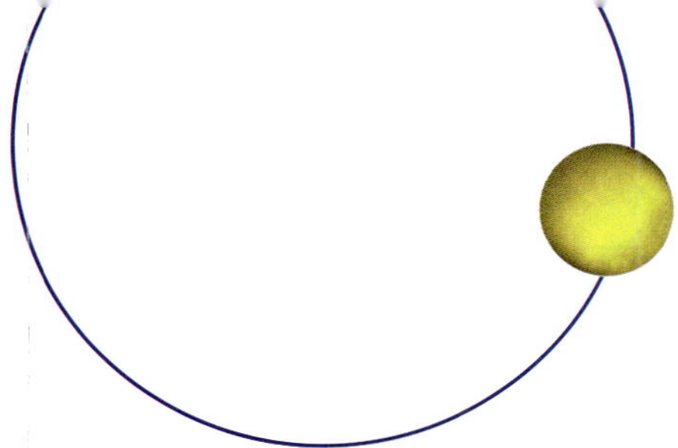

Chapter Test for Chapter 8

Name ______________

Section ________ Date ________

ANSWERS

1. ______________
2. ______________
3. ______________
4. ______________
5. ______________
6. ______________
7. ______________

The purpose of this test is to help you check your progress and review for a chapter test in class. When you are done, check your answers in the back of the book. If you missed any answers, be sure to go back and review the appropriate sections in the chapter.

1. Label each pair of lines as parallel, perpendicular, or neither.

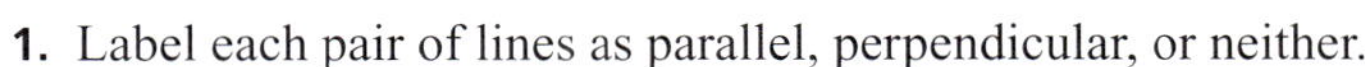

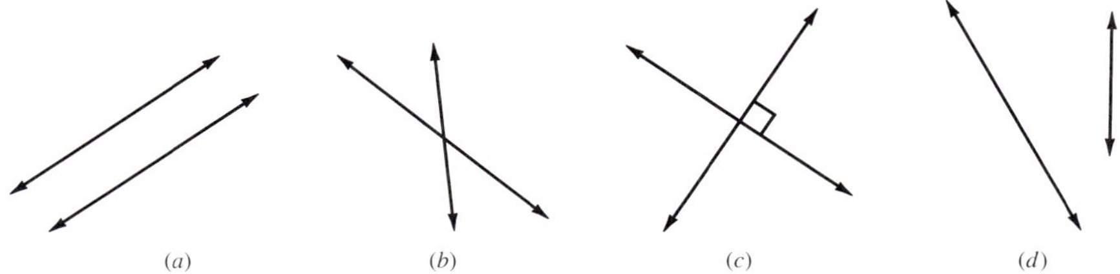

(*a*) (*b*) (*c*) (*d*)

2. Label each of the angles as acute, obtuse, right, or straight.

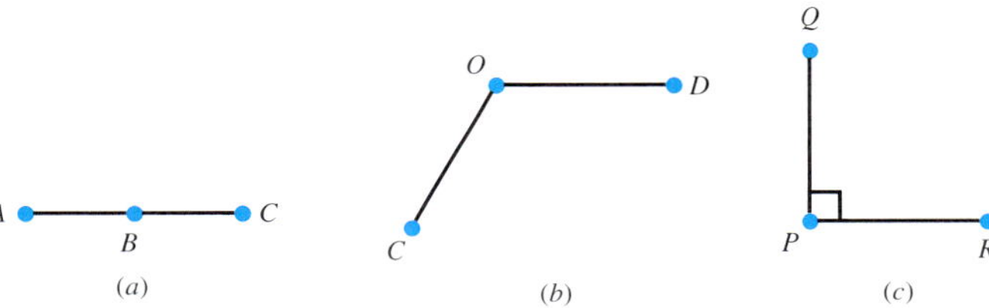

In exercises 3 and 4, use a protractor to estimate the measurement of the angles.

3.

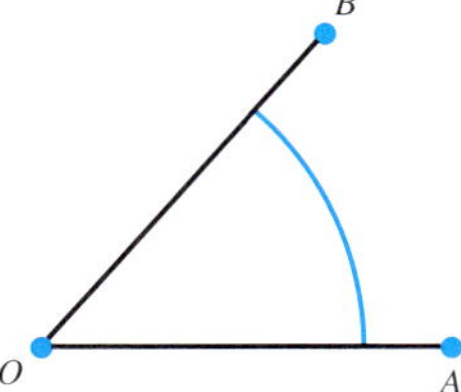

4.

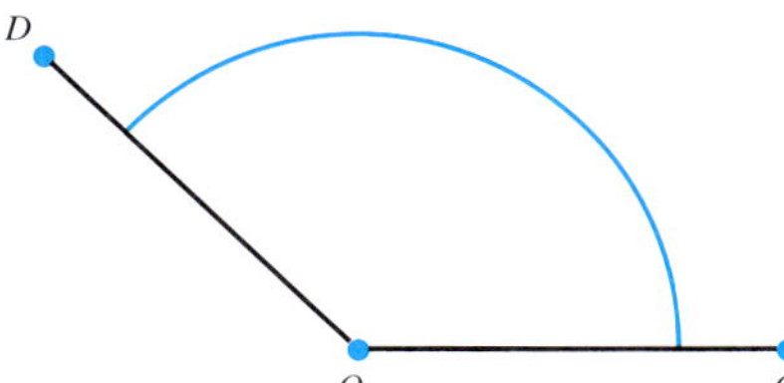

5. If $\angle A$ represents $\frac{5}{6}$ of a revolution, give the measure of $\angle A$ in degrees. (Note that one revolution is equal to 360°.)

Find the perimeter or circumference of each figure.

6.

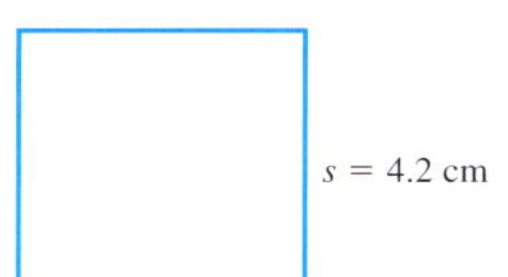

$s = 4.2$ cm

7.

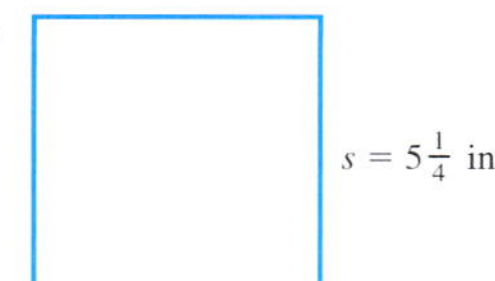

$s = 5\frac{1}{4}$ in.

ANSWERS

8. ______

9. ______

10. ______

11. ______

12. ______

13. ______

14. ______

15. ______

16. ______

17. ______

18. ______

19. ______

20. ______

21. ______

8.

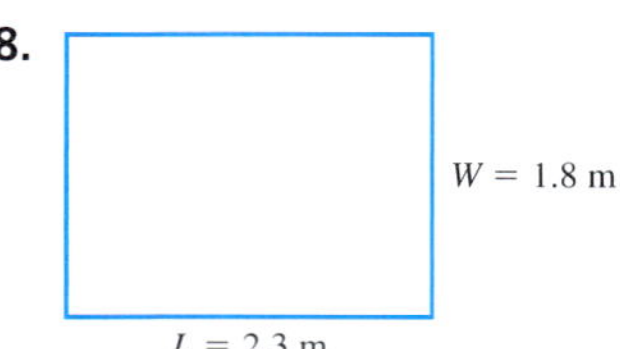

9.

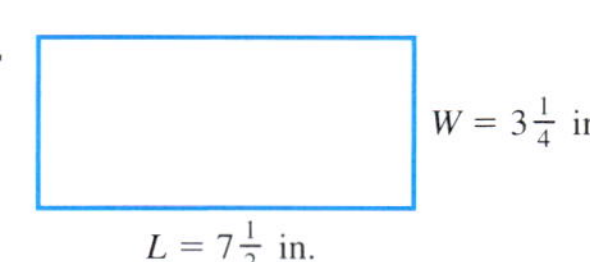

10.

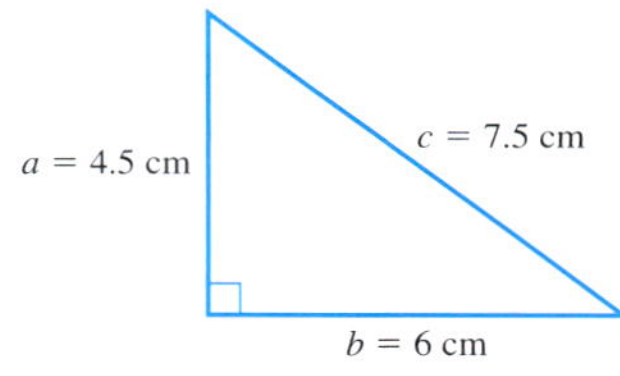

11.

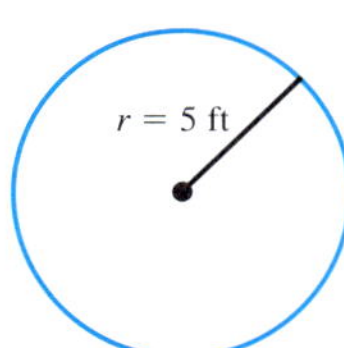

Find the area of each figure.

12.

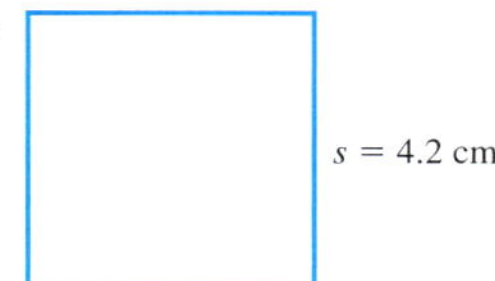

13.

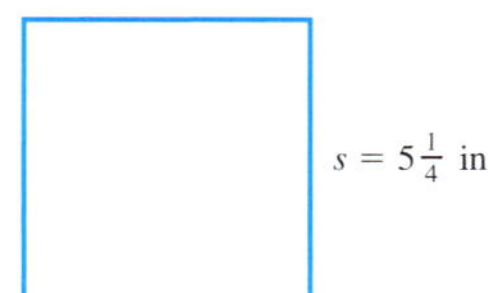

14.

15.

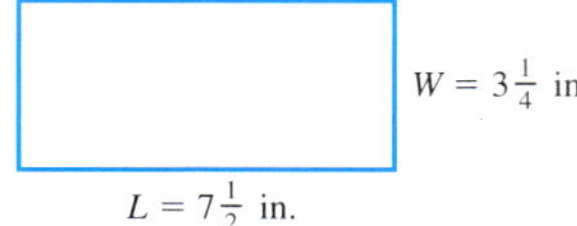

16.

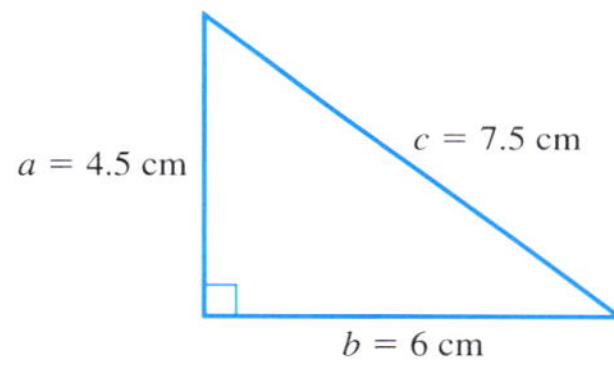

17.

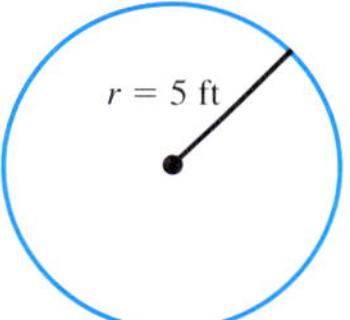

Find the volume of each figure.

18.

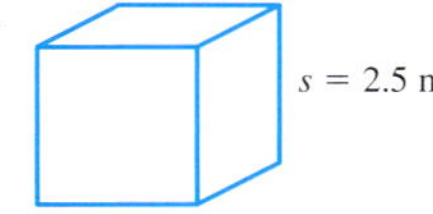

19.

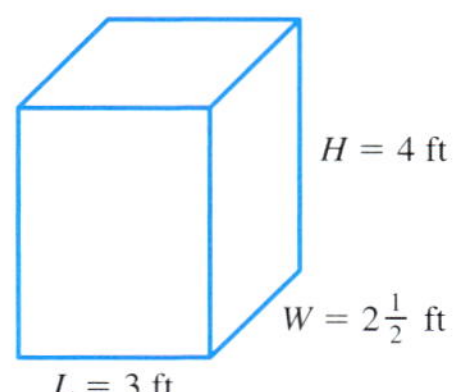

20.

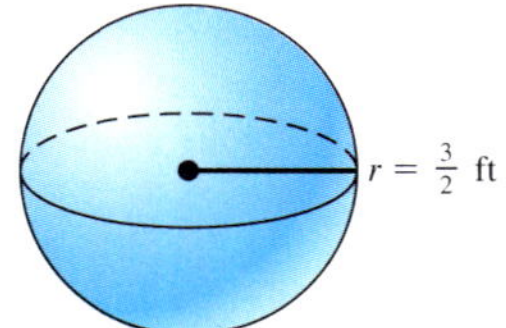

21.

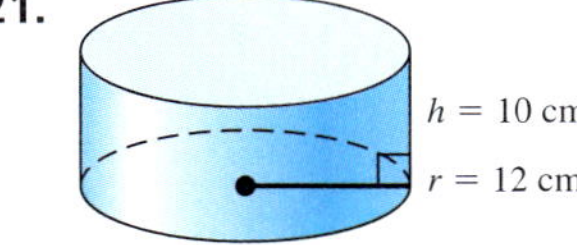

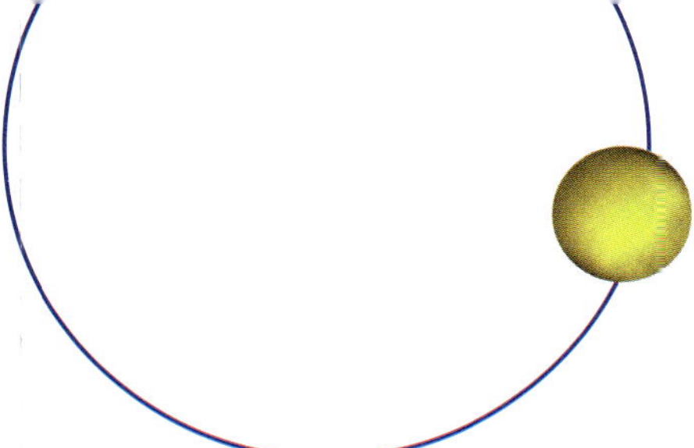

Cumulative Test for Chapters 1 to 8

Name ______________

Section ________ Date ________

ANSWERS

1. ______________
2. ______________
3. ______________
4. ______________
5. ______________
6. ______________
7. ______________
8. ______________
9. ______________
10. ______________
11. ______________
12. ______________
13. ______________
14. ______________
15. ______________
16. ______________
17. ______________
18. ______________
19. ______________
20. ______________
21. ______________
22. ______________

This test is provided to help you in the process of reviewing Chapters 1 through 8. Answers are provided in the back of the book. If you missed any answers, be sure to go back and review the appropriate chapter sections.

1. A classroom is 7 yd wide by 8 yd long. If the room is to be recarpeted with material costing $16 per square yard, find the cost of the carpeting.

2. Michael bought a washer-dryer combination that, with interest, cost $959. He paid $215 down and agreed to pay the balance in 12 monthly payments. Find the amount of each payment.

Evaluate the expressions.

3. $(-3) \cdot (-8)$

4. $-6 - 3 \cdot 5$

5. -4^2

6. $\dfrac{4 - (-3)^2}{-7 \cdot 4 + 3}$

7. Write the prime factorization for 168.

8. Find the greatest common factor of 12 and 20.

9. Arrange in order from smallest to largest: $\frac{5}{8}, \frac{3}{5}, \frac{2}{3}$

10. Multiply: $\frac{2}{3} \cdot 1\frac{4}{5} \cdot \frac{5}{8}$

11. Divide: $4\frac{1}{6} \div 10$

12. Find the least common multiple of 6, 15, and 20.

13. Add: $\frac{3}{5} + \frac{1}{6} + \frac{4}{15}$

14. Subtract: $7\frac{3}{8} - 3\frac{5}{6}$

Solve each equation and check your result.

15. $5 - 4x = 25$

16. $6x - 5 = 3x - 29$

17. $\frac{5}{7}x + 4 = 14$

18. $7x + 3(2x + 5) = 10x + 17$

19. You pay for purchases of $14.95, $18.50, $11.25, and $7 with a $70 check. How much cash will you have left?

20. Find the area of a rectangle with length 6.4 cm and width 4.35 cm.

21. Find the perimeter of a rectangle with length 6.4 cm and width 4.35 cm.

22. Find the circumference of a circle whose radius is 3.2 ft. Use 3.14 for π, and round the result to the nearest tenth of a foot.

ANSWERS

23. ______
24. ______
25. ______
26. ______
27. ______
28. ______
29. ______
30. ______
31. ______
32. ______
33. ______
34. ______
35. ______
36. ______
37. ______
38. ______
39. ______
40. ______

23. Find the decimal equivalent of $\frac{9}{16}$.

24. Write the decimal form of $\frac{7}{13}$. Round to the nearest thousandth.

25. If two legs of a right triangle have lengths 8 ft and 15 ft, find the length of the hypotenuse.

26. Solve for the unknown in the following proportion: $\frac{15}{x} = \frac{10}{16}$

27. If the scale on a map is $\frac{1}{4}$ in. equals 20 mi, how far apart are two towns that are 5 in. apart on the map?

28. Felipe traveled 342 mi using 19 gal of gas. At this rate, how far can he travel on 25 gal?

29. Write as a simplified fraction: 12.5%

30. What is 43% of 8,200?

31. 315 is what percent of 140?

32. 120% of what number is 180?

33. A home that was purchased for $125,000 increased in value by 14% over a 3-year period. What was its value at the end of that period?

34. Find the sum of 8 lb 14 oz and 12 lb 13 oz.

35. Find the difference between 7 ft 2 in. and 4 ft 5 in.

Complete each statement.

36. 62 kg = ______ g

37. 740 mm = ______ cm

38. Use a protractor to find the measure of the given angle.

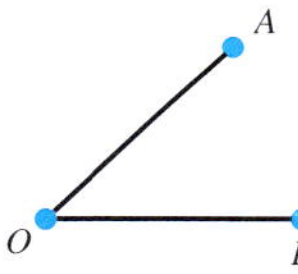

39. Find the missing angle.

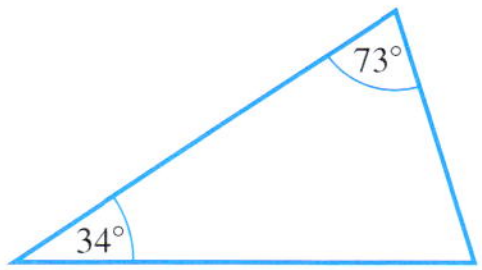

40. The given triangles are similar. Find x.

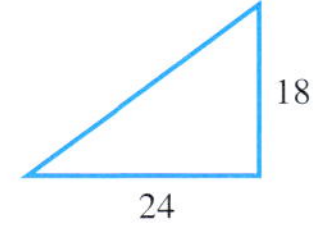

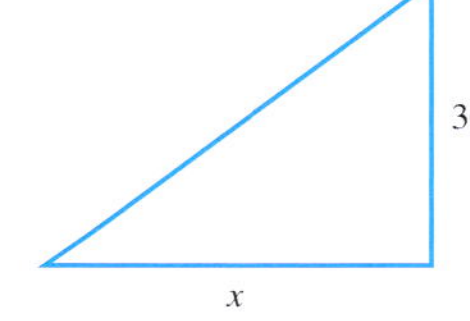

Graphing and Introduction to Statistics

9

Introduction

A graphic artist might work for a sign shop, an advertising firm, or a magazine publisher. Since getting his degree, Donovan has worked in these three places, although many other jobs also require graphic arts.

Currently, Donovan does the graphic art for a "house organ," which is a magazine published by an employer for its employees only. Donovan says that the most difficult part of the job is producing graphs. Almost every issue contains several graphs, each with information the employer wants the employees to see. Donovan must create a design that is pleasing to look at, uncluttered, and full of information. To do this, he must first make certain that he thoroughly understands the numbers to be graphed.

Name ____________________

Section ________ Date ________

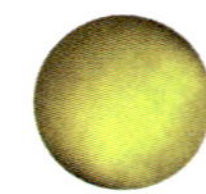

Pre-Test Chapter 9

ANSWERS

1. ____________________
2. ____________________
3. ____________________
4. ____________________
5. ____________________
6. ____________________
7. ____________________

This pre-test will point out any difficulties you may have with reading graphs, plotting points, solving two-variable equations, and finding measures of central tendency. Do all the problems, and then check your answers with those in the back of the book.

The following circle graph represents the portion of the $40,000,000,000 tourist industry that each country accounts for.

1. How many dollars do U.S. travelers account for?

2. How many dollars are accounted for by non-U.S. travelers?

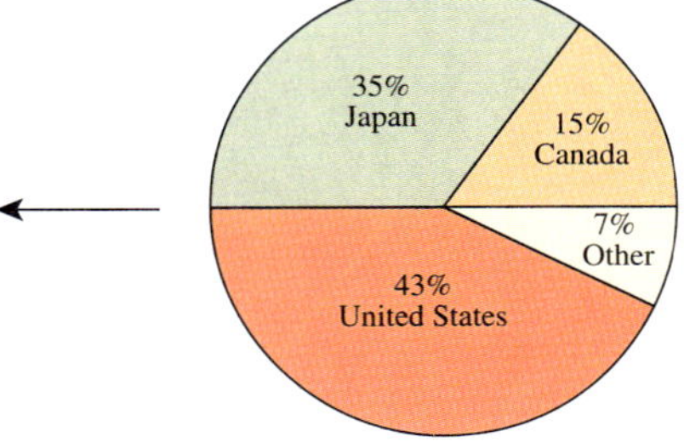

Use the bar graph to answer exercises 3 and 4.

3. What is the difference in tuition between GCC and BSU?

4. What is the average (mean) tuition for these five schools?

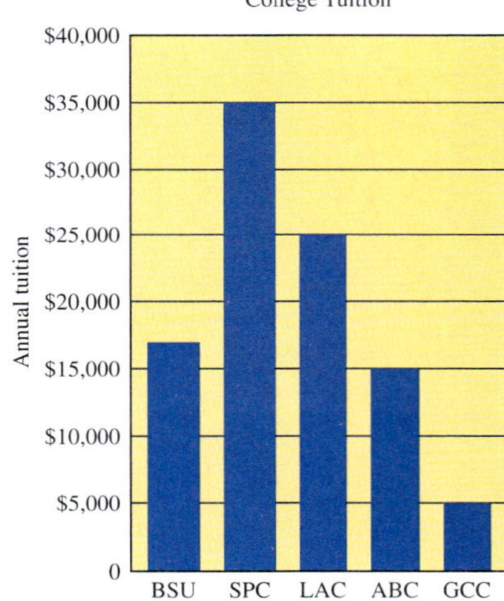

Use the line graph to answer exercises 5 to 7.

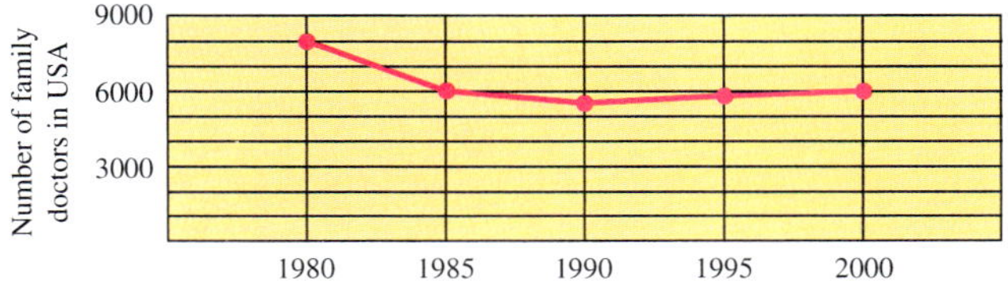

5. How many fewer family doctors were there in the United States in 1990 than in 1980?

6. What was the total change in family doctors between 2000 and 1980?

7. In what 5-year period was the decrease in family doctors the greatest? What was the decrease?

ANSWERS

8. ______

9. ______

10. ______

11. ______

12. ______

13. ______

14. ______

15. ______

Use the pictograph to answer Exercises 8 and 9.

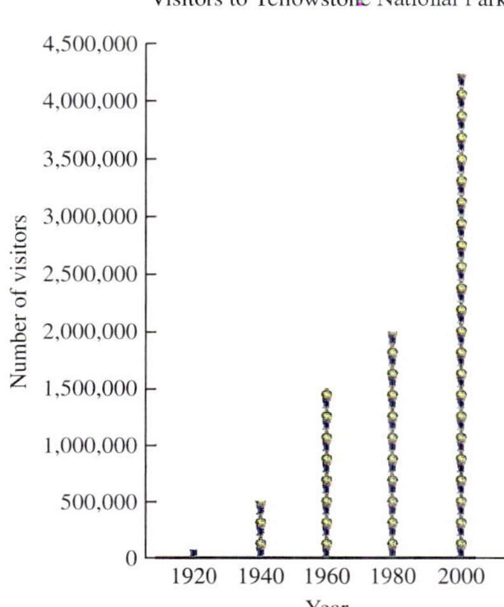

8. How many visitors did Yellowstone have in 1960?

9. What was the percent increase in visitors between 1940 and 2000?

Determine which of the ordered pairs are solutions for the given equations.

10. $x - y = 12$ (15, 3), (9, 6), (18, 6)

11. $3x + 2y = 6$ (1, 2), (0, 3), (2, 0)

12. Complete the ordered pairs so that each is a solution for the given equation.

$2x + y = 5$ (1,), (0,), (, 11)

13. Find three solutions for each of the equations.

$2x - 3y = 8$ $6x + y = 11$

14. Give the coordinates of the graphed points.

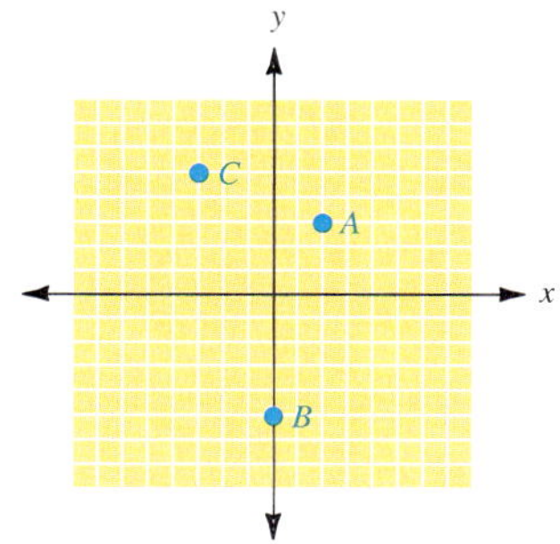

15. Plot the points with the given coordinates. $S(-1, 2)$, $T(3, 0)$

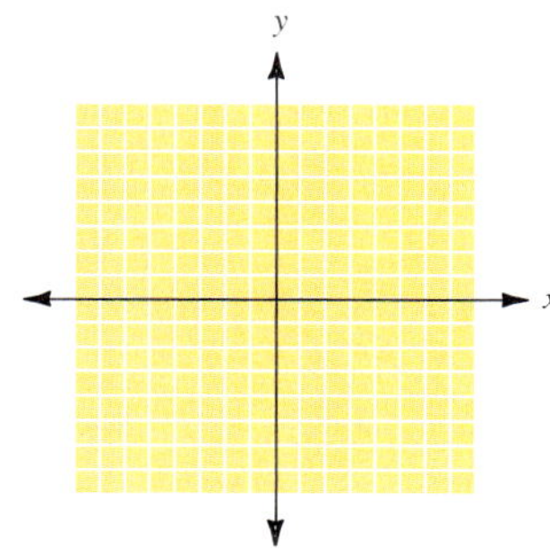

ANSWERS

16. ______

17. ______

18. ______

19. ______

Graph each of the equations.

16. $x + y = 5$

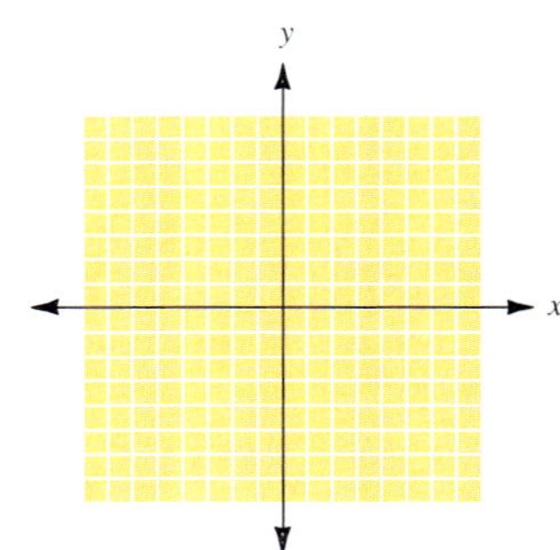

17. $y = \frac{1}{2}x - 1$

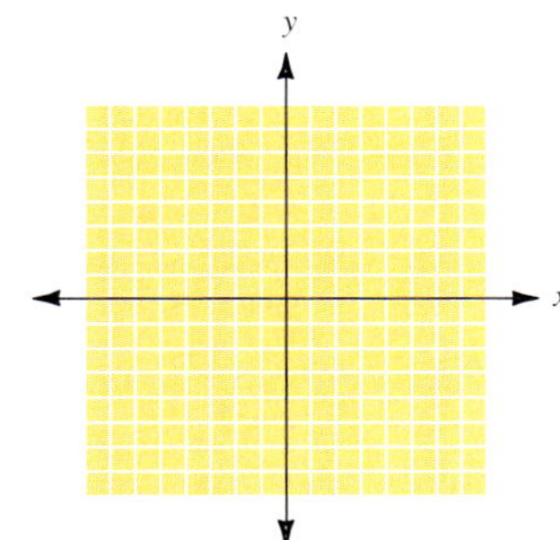

18. $y = -2$

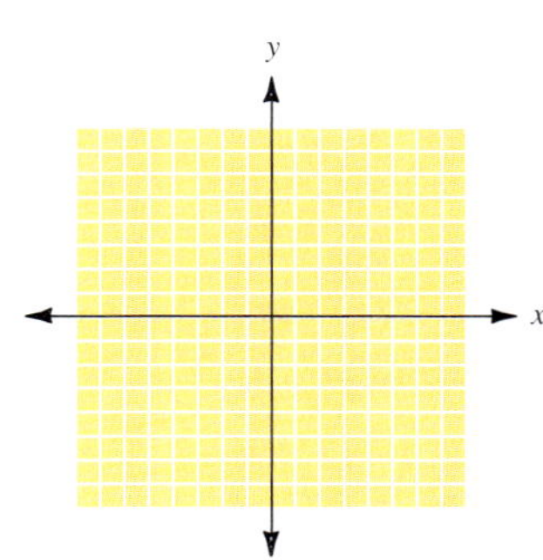

19. Find the mean, median, and mode for the set of numbers.

12, 16, 17, 18, 24, 24, 29, 33, 42, 42, 42

9.1 Circle Graphs

9.1 OBJECTIVES

1. Interpret a circle graph
2. Create a circle graph

A **graph** is a diagram that represents the connection between two or more things. When a graph is needed to represent how some unit is divided, a *circle graph* is frequently used. As you might expect, a **circle graph** is created using a circle. Wedges, called **sectors,** are drawn in the circle to show how much of the whole each part makes up.

Example 1

Reading a Circle Graph

This circle graph represents the results of a survey that asked students how they get to school most often.

NOTE The percents in a circle graph should add up to 100%.

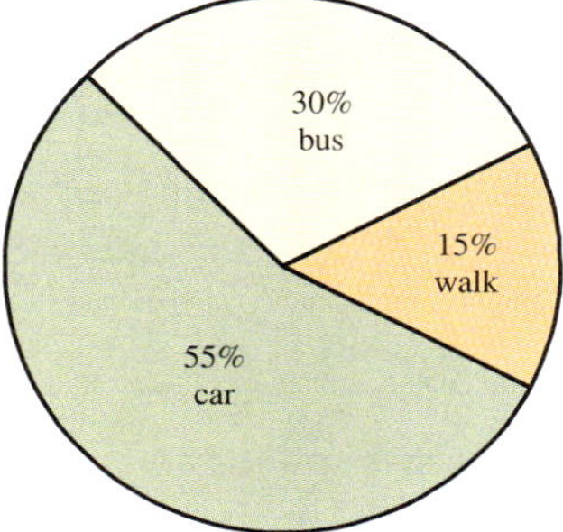

(a) What percent of the students walk to school?

We see that 15% walk to school.

(b) What percent of the students do not arrive by car?

Because 55% arrive by car, there are 100% − 55%, or 45%, who do not.

CHECK YOURSELF 1

This circle graph represents the results of a survey that asked students whether they bought lunch, brought it, or skipped lunch altogether.

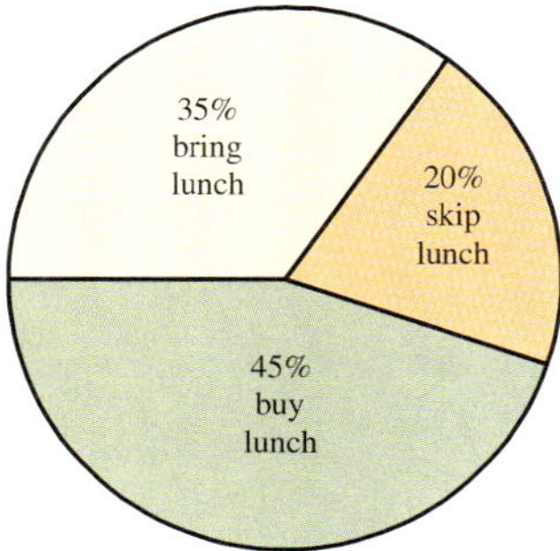

(a) What percent of the students skipped lunch?
(b) What percent of the students did not buy lunch?

If we know what the whole circle represents, we can also find out more about what each sector represents. Example 2 illustrates this point.

Example 2

Interpreting a Circle Graph

This circle graph shows how Sarah spent her $12,000 college scholarship.

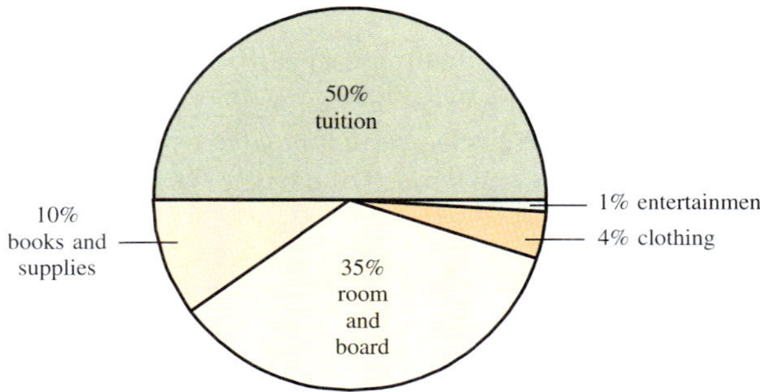

(a) How much money did she spend on tuition?

50% of her $12,000 scholarship, or $6,000.

(b) How much money did she spend on clothing and entertainment?

Together, 5% of the money was spent on clothing and entertainment, and $0.05 \cdot 12{,}000 = 600$. Therefore, $600 was spent on clothing and entertainment.

CHECK YOURSELF 2

This circle graph shows how Rebecca spends an average 24-h school day.

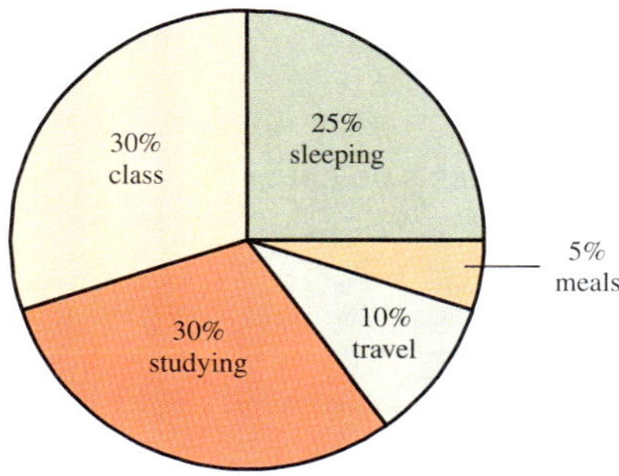

(a) How many hours does she spend sleeping each day?

(b) How many hours does she spend altogether studying and in class?

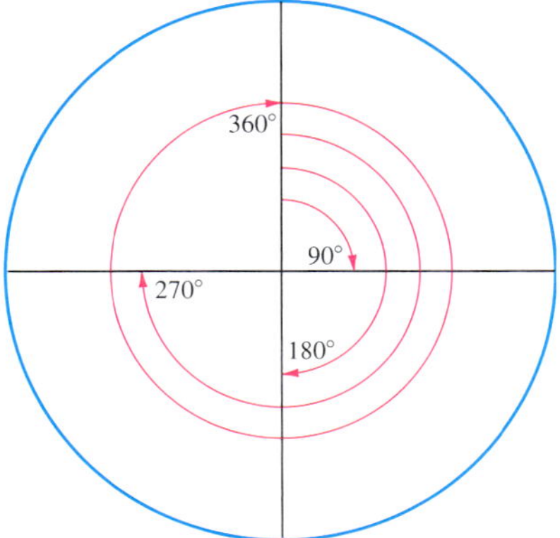

If we are creating a circle graph, how do we know how much of the circle to use for each piece? For us to make this decision, the circle must be scaled. A standard scale has been established for all circles. As we saw in Chapter 8, each circle has 360°. That means that $\frac{1}{4}$ of the circle has $\frac{1}{4}$ of 360°, which is 90°.

With a protractor, we can now create our own circle graph.

Example 3

Creating a Circle Graph

The table represents the source of automobiles purchased in the United States in 1997. Create a circle graph that represents the same data.

Source of Automobiles Purchased in 1997

Country of Origin	Number	% of total
U.S.	6,500,000	80
Japan	800,000	10
Germany	400,000	5
All Others	400,000	5

(*Source:* Amer. Auto. Manuf. Assn.)

To find the size of the sector for each country, we take the given percent of 360°. We will create another table column to represent the degrees needed.

NOTE 80% of 360° is $0.80 \cdot 360° = 288°$

Source of Automobiles Purchased in 1997

Country of Origin	Number	% of total	Degrees
U.S.	6,500,000	80	288
Japan	800,000	10	36
Germany	400,000	5	18
All Others	400,000	5	18

(*Source:* Amer. Auto. Manuf. Assn.)

Using a protractor, we start with Japan, and mark a sector that is 36°.

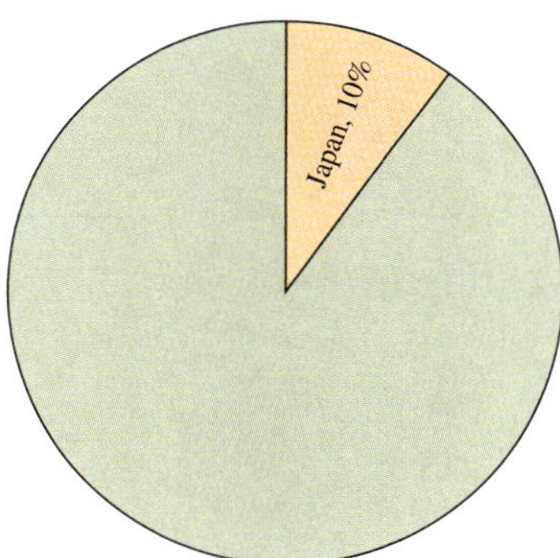

Again, using the protractor we mark the 18° sector for Germany and the 18° sector for the other countries.

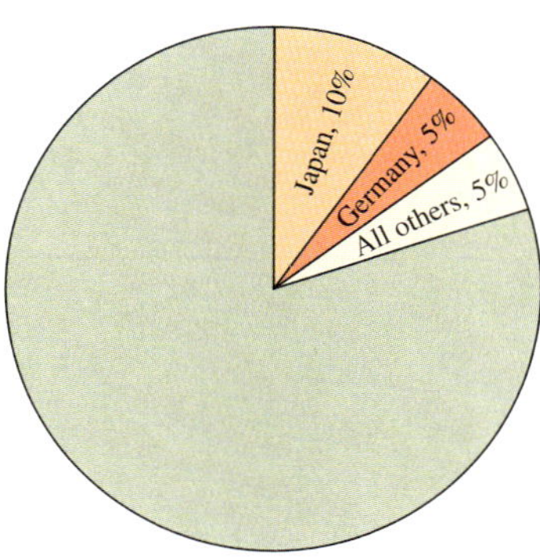

There is no need to measure the remainder of the circle. What is left is the 288° sector for U.S.-made cars. Notice that we saved the largest sector for last. It is much easier to mark the smaller sectors and leave the largest for last.

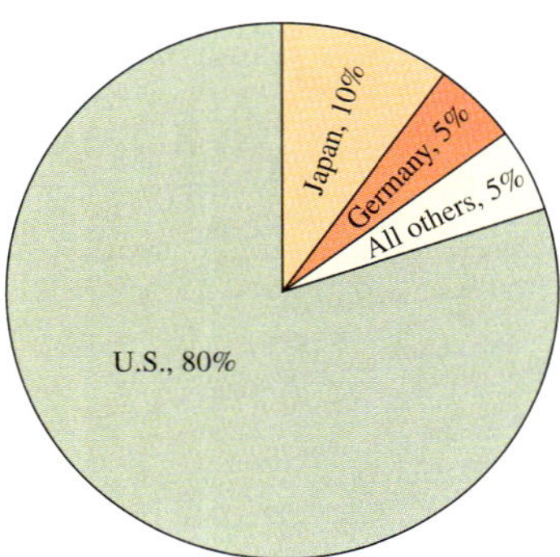

CHECK YOURSELF 3

Create a circle graph for the table, which shows TV ownership for all United States homes.

TV Ownership (Nielsen Media Research)

Number of TVs	% of U.S. Homes
0	2%
1	22%
2	34%
3 or more	42%

CHECK YOURSELF ANSWERS

1. **(a)** 20%; **(b)** 55% 2. **(a)** 6 h; **(b)** 14.4 h 3.

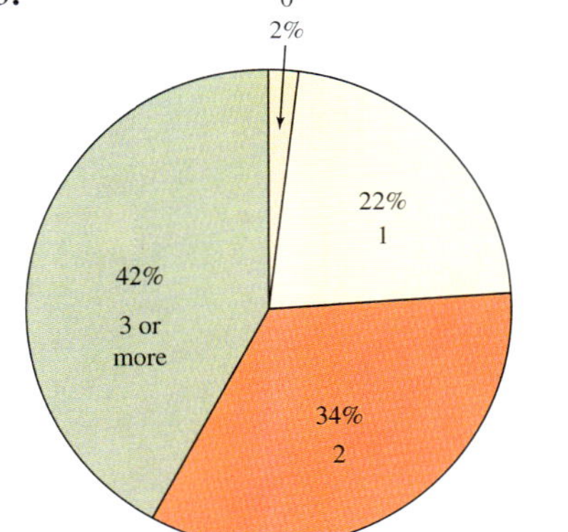

9.1 Exercises

Name ______

Section ______ Date ______

The circle graph shows the budget for a local company. The total budget is $600,000. Find the dollar amount budgeted in each of the categories.

1. Production
2. Taxes
3. Research
4. Operating expenses
5. Miscellaneous

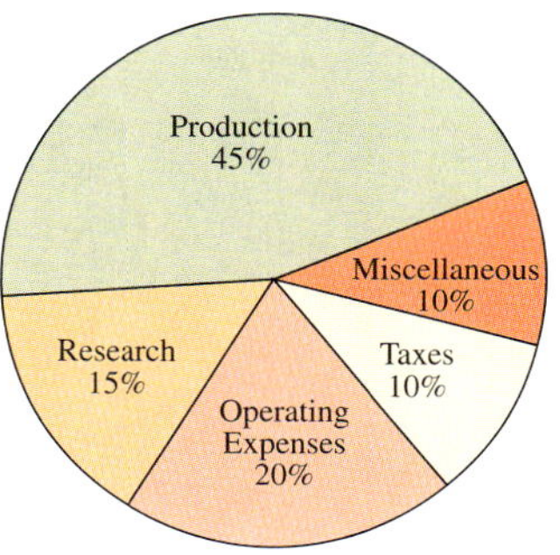

The circle graph shows the distribution of a person's total yearly income of $24,000. Find the dollar amount budgeted for each category.

6. Food
7. Rent
8. Utilities
9. Transportation
10. Clothing
11. Entertainment

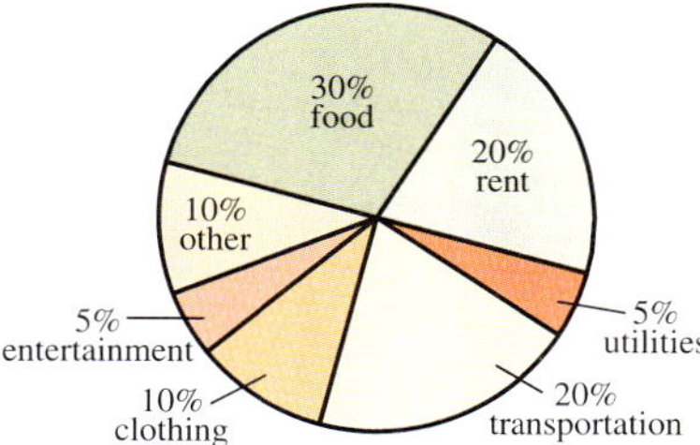

12. The table represents women on active duty in 1998.

Service	Number of Women
Army	56,800
Navy	45,000
Marines	18,600
Air Force	65,700
Coast Guard	36,000

Create a circle graph for the information.

13. The table represents the number of Nobel Prize laureates during the years 1901 to 1993.

Country	Number
United States	170
United Kingdom	69
Germany	59
France	24
USSR	10
Others	88

Create a circle graph for the information.

ANSWERS

1. ______
2. ______
3. ______
4. ______
5. ______
6. ______
7. ______
8. ______
9. ______
10. ______
11. ______
12. ______
13. ______

Answers

1. $270,000 **3.** $90,000 **5.** $60,000 **7.** $4,800 **9.** $4,800

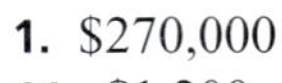

11. $1,200 **13.**

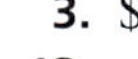
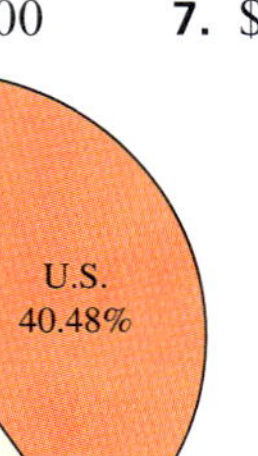

Pictographs, Bar Graphs, and Line Graphs

OBJECTIVES

1. Interpret a pictograph
2. Create a pictograph
3. Interpret a bar graph
4. Create a bar graph
5. Interpret a line graph
6. Create a line graph

A **table** is a display of information in parallel rows or columns. Tables can be used anywhere that information is to be summarized.

The given table describes land area and world population. Each entry in the table is sometimes called a **cell.**

Continent or Region	Land Area (1,000 mi^2)	% of Earth	Population 1900	Population 1950	Population 2000
North America	9,400	16.2	106,000,000	221,000,000	477,000,000
South America	6,900	11.9	38,000,000	111,000,000	339,000,000
Europe	3,800	6.6	400,000,000	392,000,000	729,000,000
Asia (including Russia)	17,400	30.1	932,000,000	1,591,000,000	3,788,000,000
Africa	11,700	20.2	118,000,000	229,000,000	771,000,000
Oceana (including Australia)	3,300	5.7	6,000,000	12,000,000	31,000,000
Antarctica	5,400	9.3	Uninhabited		
World Total	57,900		1,600,000,000	2,556,000,000	6,135,000,000

(*Source:* Bureau of the Census, U.S. Dept. of Commerce.)

Example 1

Reading a Table

Using the land area and world population table, answer each question.

(a) What was the population of Africa in 1950?

Looking at the cell that is in the row labeled "Africa" and the column labeled "Population 1950," we find a population of 229,000,000.

(b) What is the land area of Asia in square miles?

The cell in the row Asia and column Land Area says 17,400. But note that the column is labeled "1,000 mi^2." The land area is 17,400 thousand square miles, or 17,400,000 mi^2.

CHECK YOURSELF 1

Use the table above to answer each question.

(a) What was the population of South America in 1900?
(b) What is the land area of Africa as a percent of Earth's land area?

Example 2

Creating a Pictograph

Create a pictograph that displays the information in the table. (The numbers in the table have been rounded.)

Cars Registered in the USA (Federal Highway Admin., U.S. Dept. of Transportation)

Year	Cars Registered
1970	90,000,000
1975	106,700,000
1980	121,600,000
1985	131,700,000
1990	143,500,000
1995	136,000,000
2000	130,000,000

First, we must decide what to use as our picture unit. A car is the obvious choice in this case.

Second, we must decide the "value" of each car. Given the units in the table, we will let each car represent 10,000,000 registrations.

(Federal Highway Admin., U.S. Dept. of Transportation)

Year	Cars Pictured
1970	9 cars
1975	$10\frac{2}{3}$ cars
1980	$12\frac{1}{6}$ cars
1985	$13\frac{1}{6}$ cars
1990	$14\frac{1}{3}$ cars
1995	$13\frac{2}{3}$ cars
2000	13 cars

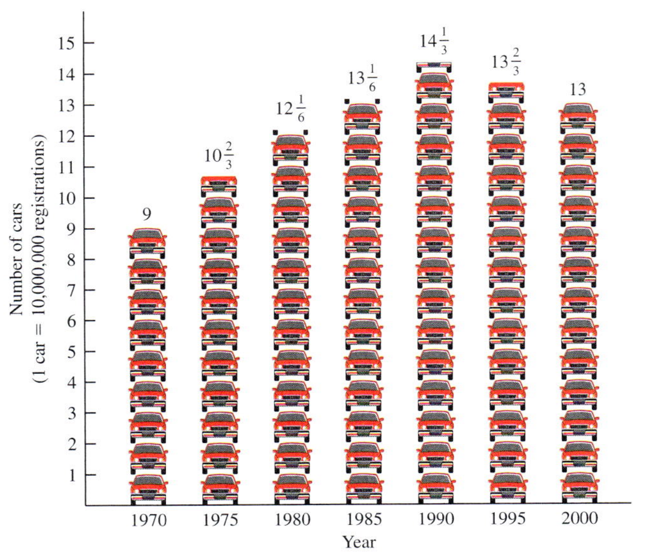

CHECK YOURSELF 2

Create a pictograph to represent the table, which gives the percent of U.S. workers employed as farm workers.

(Economic Research Service, U.S. Dept. of Agriculture)

Year	Percent of Workers in Farm Occupations
1800	73.2
1840	68.6
1880	57.1
1920	27.0
1960	6.1
2000	2.3

Other kinds of graphs also show the relationship of two sets of data. Perhaps the most common is the **bar graph.**

A bar graph is read in much the same manner as a pictograph.

Example 3

Reading a Bar Graph

The given bar graph represents the response to a 2001 Gallup poll that asked people what their favorite spectator sport was. In the graph, the information at the bottom describes the sport, and the information along the side describes the percent of people surveyed. The height of the bar indicates the percent of people who favor that particular sport.

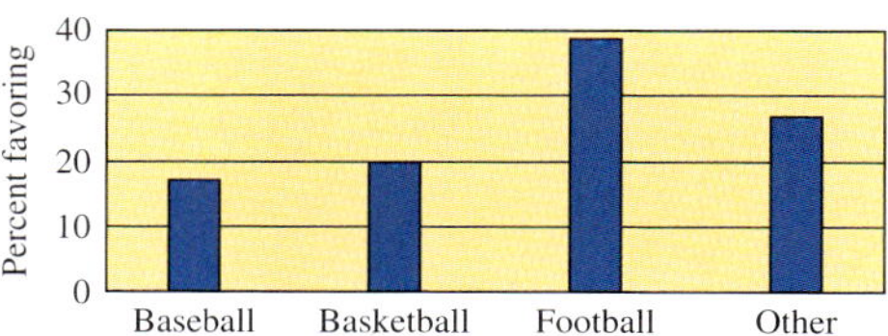

(a) Find the percent of people for whom football is their favorite spectator sport.

As is the case with pictographs, we frequently have to estimate our answer when reading a bar graph. In this case, 38% would be a good estimate.

(b) Find the percent of people for whom baseball is their favorite spectator sport.

Again, we can only estimate our answer. It appears to be approximately 17% of the people responding who favored baseball.

CHECK YOURSELF 3

This bar graph represents the number of students who majored in each of five areas at Experimental Community College.

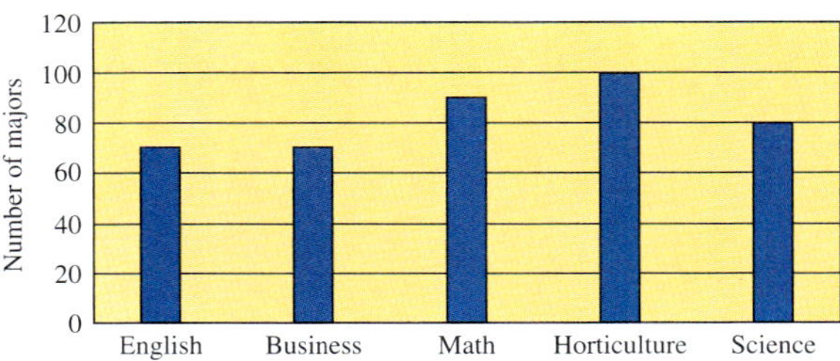

(a) How many mathematics majors were there?

(b) How many English majors were there?

Some bar graphs display additional information by using different colors or shading for different bars. With such graphs it is important to read the legend. The **legend** is the key that describes what each color or shade of the bar represents.

Example 4

Reading the Legend of a Graph

The bar graph represents the average student age at ECC.

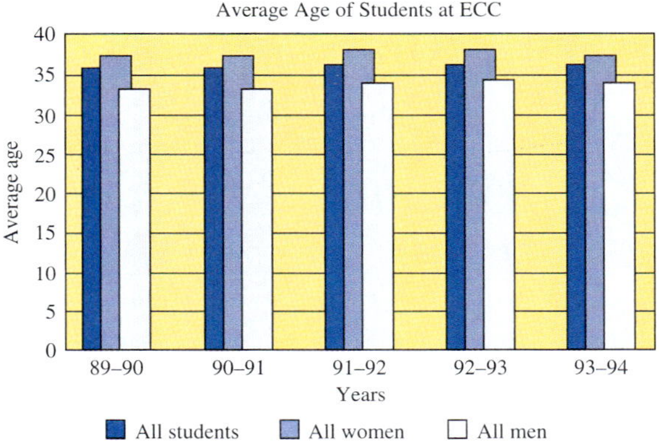

(a) What was the average age of female students in 93–94?

The legend tells us that the ages of all women are represented as the medium blue color. Looking at the height of the medium blue column for the year 93–94, we see the average age was about 38.

(b) Who tends to be older, male students or female students?

The medium blue bar is higher than the light blue bar in every year. Female students tend to be older than male students at ECC.

CHECK YOURSELF 4

Use the graph in Example 4 to answer the questions.

(a) Did the average age of female students increase or decrease between 92–93 and 93–94?

(b) What was the average age of male students in 91–92?

We have just learned to read a bar graph. In Example 5, we will create one.

Example 5

Creating a Bar Graph

The table represents the 1995 population of the six most populated urban areas in the world. Each population is the population of the city plus the population of all of its suburbs. Create a bar graph from the information in the table.

Population of the World's Largest Urban Areas (U.N. Dept. for Economic and Social Info.)

City	1995 Population
Tokyo, Japan	27,500,000
Mexico City, Mexico	17,500,000
São Paulo, Brazil	16,500,000
New York City, USA	16,000,000
Bombay, India	15,000,000
Shanghai, China	14,000,000

We will let the vertical axis, the vertical line to the left of the graph, represent population. The six urban areas will be placed along the horizontal axis. To create a graph, we must decide on the scale for the vertical axis. The given steps will accomplish that.

NOTE Each $\frac{1}{2}$ in. therefore represents 5,000,000 people.

1. Pick a number that is slightly larger than the biggest number we are to graph. 30,000,000 is slightly larger than 27,500,000.
2. Decide how long the axis will be. It is best if this length easily divides into the number of step 1. To accomplish this division, we will choose 3 in.
3. Scale the axis by dividing it with tick marks. Label each tick mark with the appropriate number. In this graph, each inch will represent 10,000,000 people (the 30,000,000 divided by the 3 in. results in 10,000,000 people per inch).

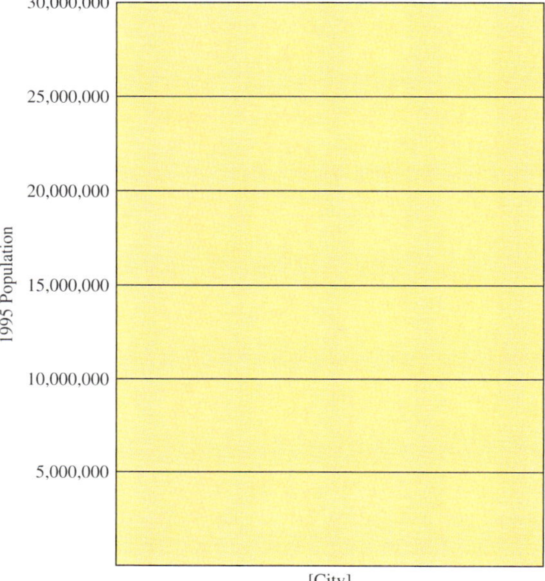

Now, the height of each bar is determined by using the scale created for the axis. Remembering that we have 10,000,000 people per inch, we divide each population by 10,000,000. The result is the height of each bar. The height for Mexico City is 1.75 in. That would be $1\frac{3}{4}$ in. Remember, all we can get from a bar graph is a rough approximation of the actual number.

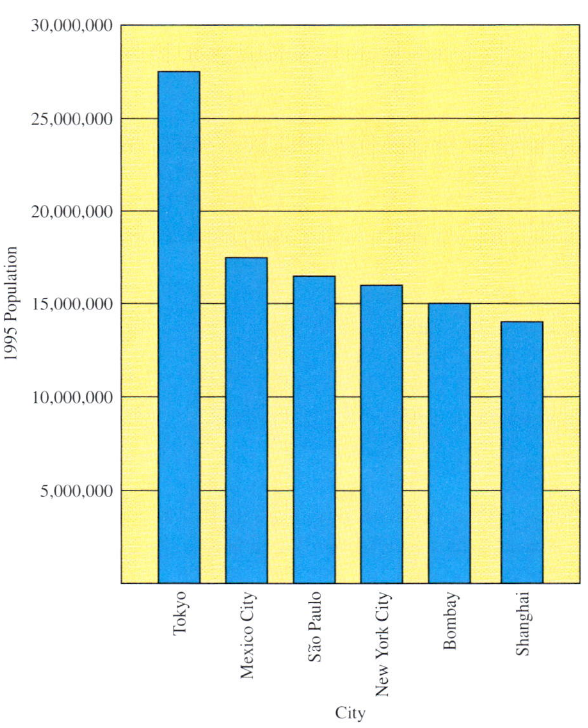

CHECK YOURSELF 5

The table represents the 1995 *population of the six most populated cities in the United States. Each population is the population within the city limits, which is why the New York population is so different from that in the table in Example* 5. *Create a bar graph from the information in the table.*

Population of the Largest Cities in the United States (Bureau of the Census, U.S. Dept. of Commerce)

City	1995 Population
New York, NY	7,500,000
Los Angeles, CA	3,500,000
Chicago, IL	2,750,000
Houston, TX	2,750,000
Philadelphia, PA	1,500,000
San Diego, CA	1,250,000

We have seen that data can be represented graphically with a circle graph, pictograph, or bar graph. Another useful type of graph is called a **line graph.** In a line graph, one of the pieces of information is almost always related to time (clock time, day, month, or year).

Example 6

Reading a Line Graph

This graph represents the number of regular season games won by the Dallas Cowboys each year of the 1990s. Note that the information across the bottom indicates the year and the information along the side indicates the number of victories.

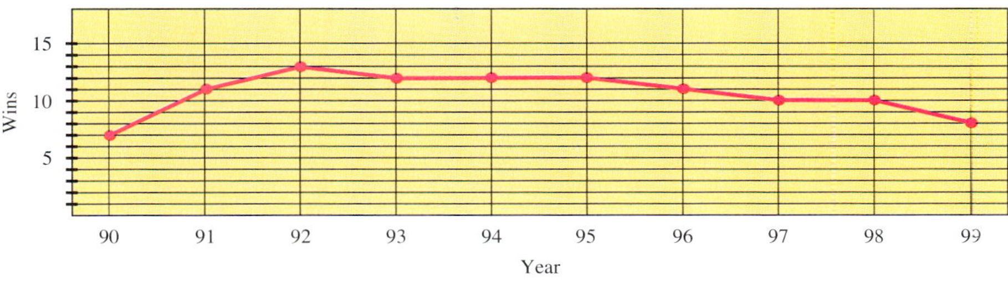

(a) How many games did they win in 1994?

We look across the bottom until we find 1994. We then look straight up until we see the line of the graph. Following across to the left, we see that they won 12 games in 1994.

(b) Find the average (mean) number of games won by the Cowboys in the 1990s by adding the games won in each of the 10 years, and then dividing by 10.

For each x (or dot) on the line, we look to the left to find how many victories it represents. We then add, so

$$\bar{x} = \frac{7 + 11 + 13 + 12 + 12 + 12 + 11 + 10 + 10 + 8}{10}$$

$$= \frac{106}{10} = 10\frac{6}{10} = 10\frac{3}{5}$$

NOTE As a decimal, we could write this as 10.6.

The mean number of games won is $10\frac{3}{5}$.

CHECK YOURSELF 6

The graph indicates the high temperatures in Baltimore, Maryland, for a week in September.

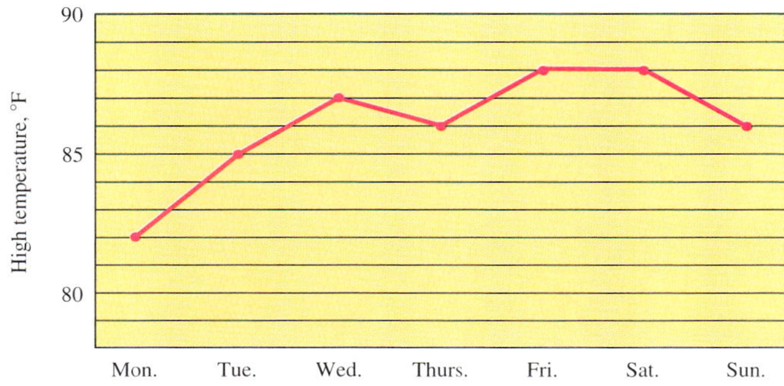

(a) What was the high temperature on Friday?
(b) Find the average (mean) high temperature for that week.

It is often tempting, and sometimes useful, to use a line graph to predict a future value. Using an earlier trend to predict a future value is called **extrapolation.** This is something that statisticians warn us not to rely on, but most of us do it anyway.

Example 7

Making a Prediction Using a Line Graph

Use the given line graph and table to predict the number of Social Security Beneficiaries in the year 2005.

(Social Security Admin.)

Year	Beneficiaries
1955	6,000,000
1965	18,000,000
1975	28,000,000
1985	36,000,000
1995	41,000,000
2005	?

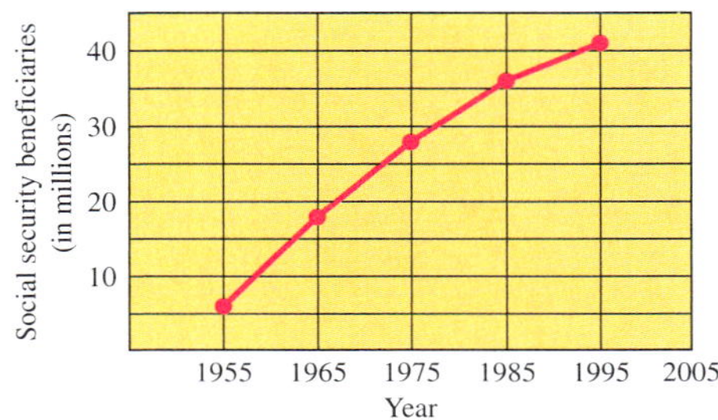

From the shape of the line graph, it would be reasonable to guess that the next point on the graph would continue on the same "curve."

(Social Security Admin.)

Year	Beneficiaries
2005	44,000,000

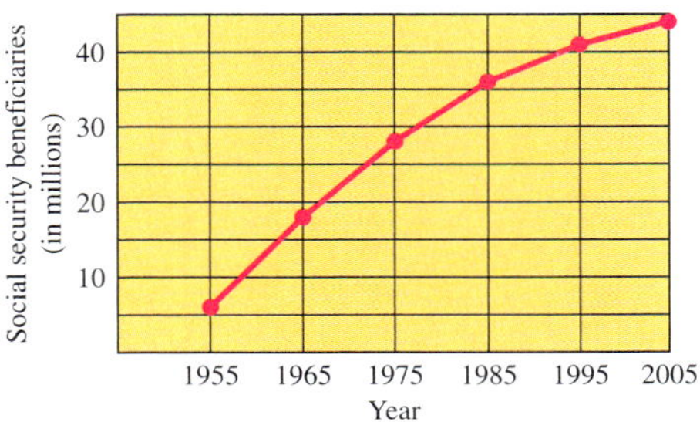

This point indicates that, by 2005, we should expect about 44,000,000 Social Security beneficiaries. This number closely matches more sophisticated predictions for the number of beneficiaries in the year 2005.

CHECK YOURSELF 7

The graph and table show the amount spent on health care (in billions of dollars) in the United States every 5 *years from* 1965 *to* 1995. *Use that information to predict the amount spent in the year* 2000.

(National Center for Health Stats)

Year	Health Care Expenditures (in billions of $)
1965	41
1970	73
1975	130
1980	247
1985	428
1990	700
1995	991
2000	?

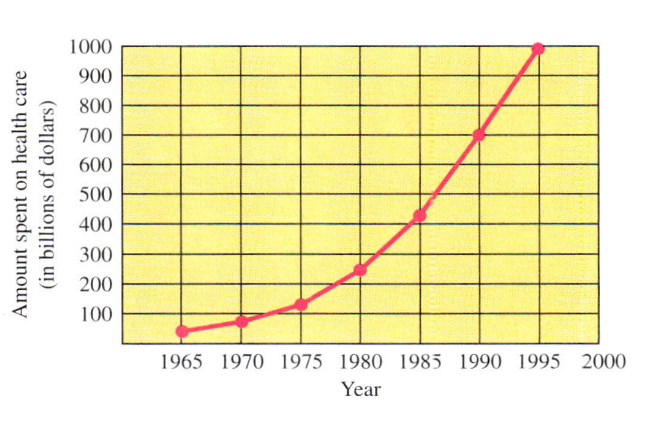

Now, we will look at an example that yields a result that is not quite as useful.

Example 8

Making a Prediction Using a Line Graph

Use the line graph and table to "predict" the cost of a first-class stamp on January 1, 2000.

Year	Cost, cents
1960	4
1965	5
1970	6
1975	10
1980	15
1985	22
1990	25
1995	32
2000	?

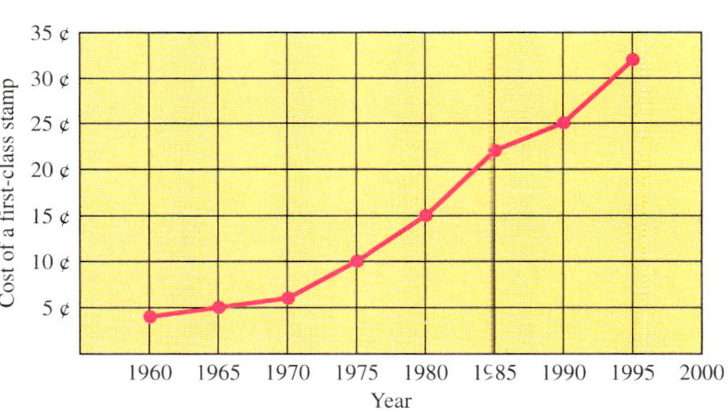

Place a straightedge so that it comes close to going through all of the points.

Year	Cost, cents
1960	4
1965	5
1970	6
1975	10
1980	15
1985	22
1990	25
1995	32
2000	38

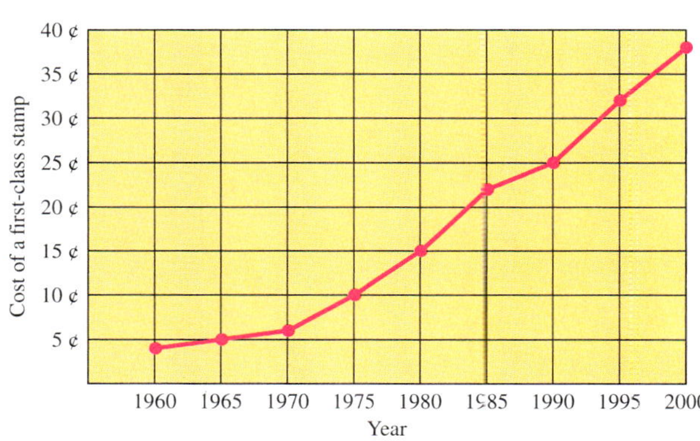

From the line graph, it would be reasonable to guess that the cost would be 38 cents. In fact, on January 1, 2000, the cost was 33 cents. This is evidence of the danger of extrapolation, a danger we mentioned before Example 7.

CHECK YOURSELF 8

The graph represents the number of larceny-theft cases in the United States each 5 years from 1975 *to* 1990. *Use the graph to predict the number of cases in* 1995.

(FBI Uniform Crime Report)

Year	Larceny-Theft Cases (in hundred thousands)
1975	56
1980	66
1985	73
1990	79
1995	?

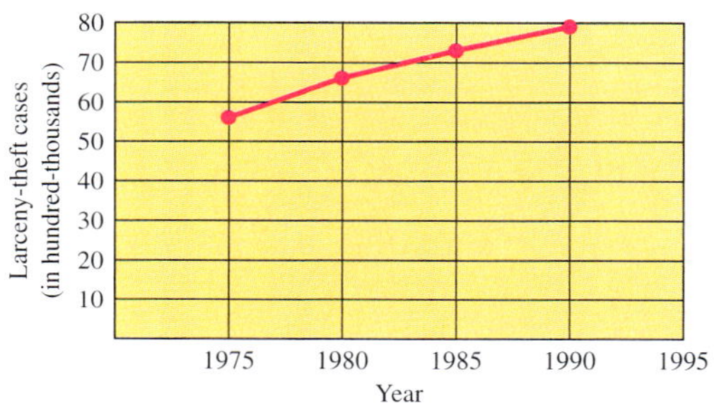

Now that we have seen how to read a line graph and use it to make predictions, let us now create one from a table.

Example 9

Creating a Line Graph

Create a line graph from the table, which gives the average low temperature in Fairbanks, Alaska, for each calendar month.

Month	Jan	Feb	Mar	Apr	May	Jun	Jul	Aug	Sep	Oct	Nov	Dec
Temp. in °F	−8°	−6°	19°	28°	37°	42°	45°	48°	34°	27°	15°	−1°

The first step in creating a line graph is to set up and label the two axes. We must be certain that all of our values will fit on the graph. The vertical axis must accommodate all of the numbers from −8 to 48. We will sketch axes so that all 12 months will be labeled, and the vertical axis marks every 10° from −10 to 50.

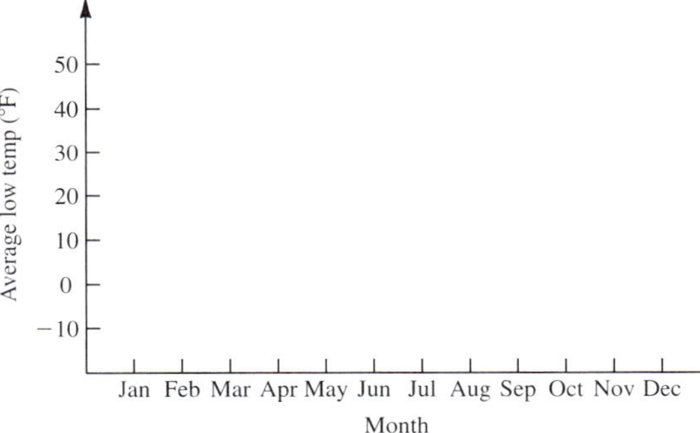

Now we mark each of the 12 points indicated on the table.

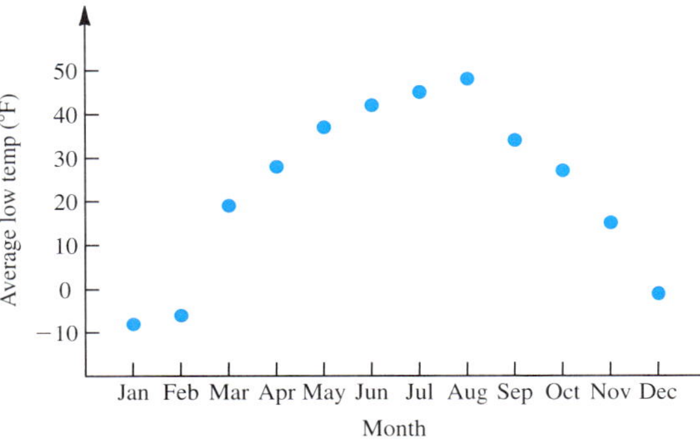

Finally, we connect each adjacent pair of points.

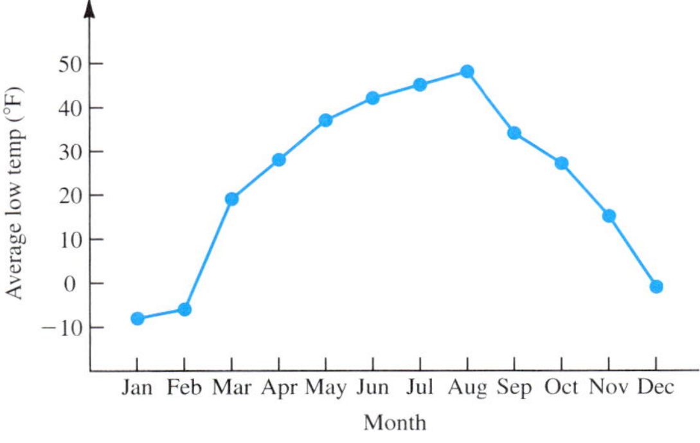

CHECK YOURSELF 9

Create a line graph from the table, which gives the average high temperature in Johannesburg, South Africa, for each calendar month.

Month	Jan	Feb	Mar	Apr	May	Jun	Jul	Aug	Sep	Oct	Nov	Dec
Temp. in °F	83°	87°	78°	65°	55°	48°	45°	48°	55°	62°	71°	78°

CHECK YOURSELF ANSWERS

1. **(a)** 38,000,000; **(b)** 20.2%

2.

1 Worker = 10% (Economic Research Service U.S. Dept. of Agriculture)

Year	Workers Pictured
1800	$7\frac{1}{3}$ workers
1840	$6\frac{2}{3}$ workers
1880	$5\frac{2}{3}$ workers
1920	$2\frac{2}{3}$ workers
1960	$\frac{2}{3}$ worker
2000	$\frac{1}{3}$ worker

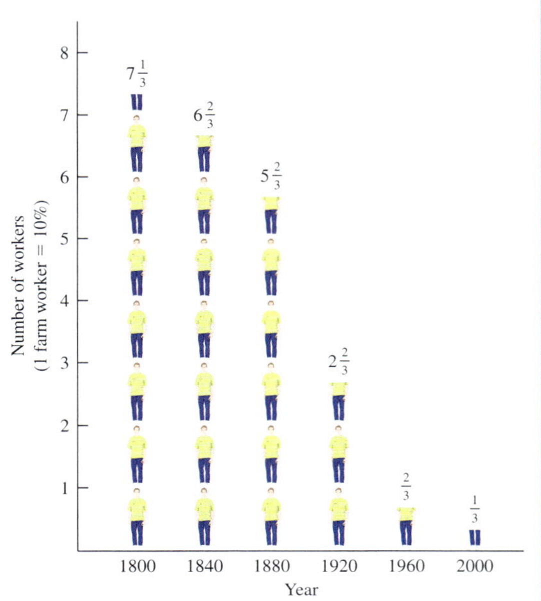

3. **(a)** 90; **(b)** 70

4. **(a)** It decreased; **(b)** 34

5.

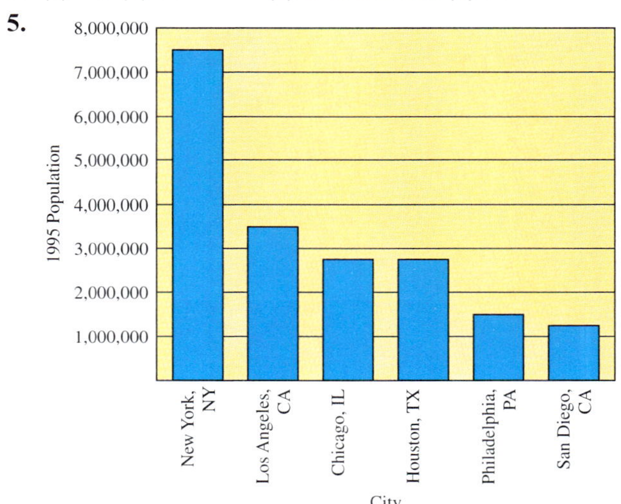

6. **(a)** 88°; **(b)** 86°

7. A reasonable prediction from the data would be about 1,300 billion dollars. That is very close to the actual costs.

8. A reasonable prediction from the data would be about 85 hundred thousand cases. There were actually 80 hundred thousand.

9.

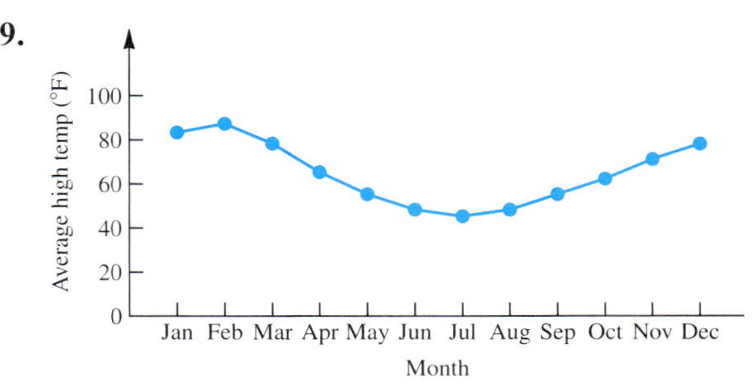

Name ______________

9.2 Exercises

Section ________ Date ________

ANSWERS

(a)
(b)

1. (c) ______________

2. ______________

1. The pictograph represents the world population, by continent, in 2000.

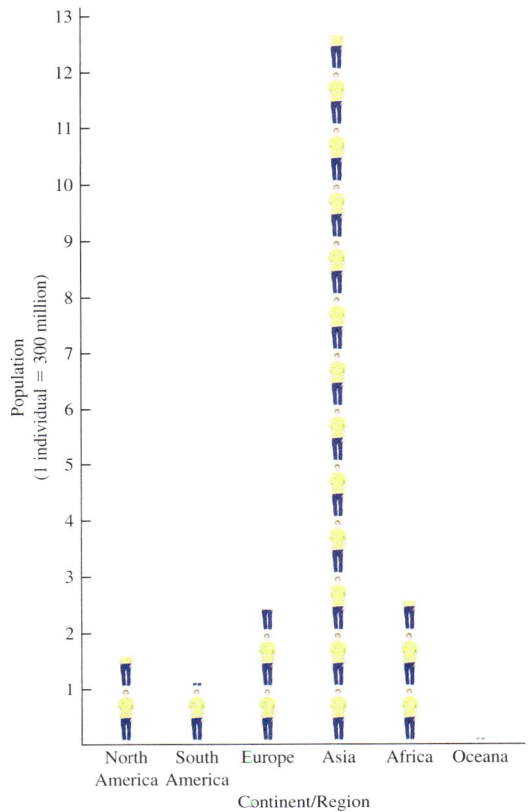

(a) What was the population of Africa?

(b) What is the total population of North and South America?

(c) What other information would help you understand the significance of this data?

2. Create a pictograph for the total world population in 1950 using the information given in the world population and land area table in the text (see page 651).

1 Individual ≈ 200 Million	
Continent/ Region	**Population Individuals**
North America	1.1
South America	0.5
Europe	1.96
Asia	7.9
Africa	1.14
Oceana	0.06

3. Use the information in the table to create a pictograph.

U.S. Car Sales

Year	Car Sales
1970	8,403,000
1975	8,538,000
1980	8,979,000
1985	11,039,000
1990	9,484,000
1995	8,686,000

3. __________

4. Use the information in the table to create a pictograph.

World Motor Vehicle Production

Year	Vehicles
1970	29,419,000
1980	38,565,000
1985	44,909,000
1990	48,554,000
1995	49,983,000

Use the bar graph, showing automobile production in millions of cars from 1989 to 1995, to solve exercises 5 to 8.

4. __________

5. __________

6. __________

7. __________

8. __________

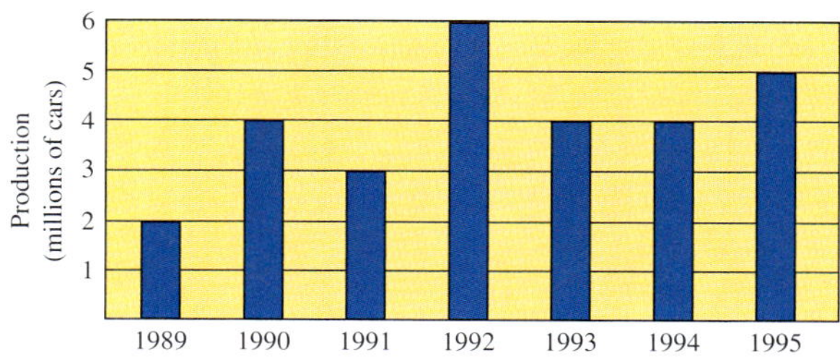

5. What was the production in 1991?

6. In what year did the greatest production occur?

7. How much did production increase from 1991 to 1992?

8. In what year was the production decline the greatest compared to the previous year?

9. ____

10. ____

11. ____

12. ____

13. ____

Use the bar graph, showing the attendance at a circus for 7 days in August, to solve exercises 9 to 12.

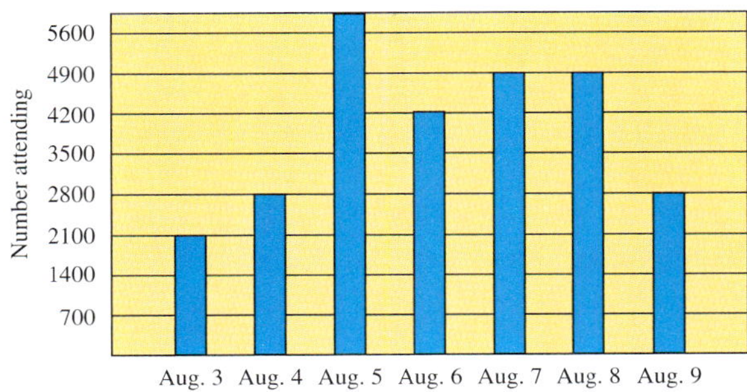

9. Find the attendance on August 4.

10. Which day had the greatest attendance?

11. Which day had the lowest attendance?

12. How much did attendance drop from August 8 to August 9?

For exercises 13 to 16, use the given bar graph.

Sport Utility Vehicle Sales in the U.S., 1988–97

In 1988, 960,852 sport utility vehicles (SUVs) were sold in the United States, accounting for 6.3% of all sales of light vehicles (SUVs, minivans, vans, pickup trucks, and trucks under 14,000 lb). By 1997, sales of SUVs in the United States increased to 2,435,301, accounting for 16.1% of total light vehicle sales.

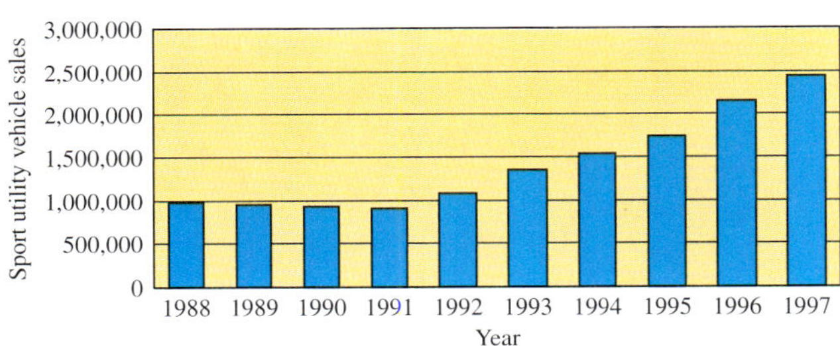

Source: American Automobile Manufacturers Assn.

13. What were the sales of SUVs in 1997?

ANSWERS

14. ______

15. ______

16. ______

17. ______

18. ______

19. ______

20. ______

14. In what year did the greatest sales occur?

15. What was the percent increase in sales from 1988 to 1997?

16. In what year did the greatest increase in sales occur?

17. The table represents the top six metropolitan areas where immigrants were admitted to the United States in 1996.

Area	Number
New York, NY	133,168
Los Angeles, CA	64,285
Miami, FL	41,527
Chicago, IL	39,989
Washington, DC	34,327
Houston, TX	21,387

Create a bar graph from this information.

18. The information given in the table represents the 1997 population of the six largest counties in the United States.

County	Population
Los Angeles, CA	9,145,219
Cook, IL	5,076,786
Harris, TX	3,158,095
San Diego, CA	2,722,650
Maricopa, AZ	2,696,198
Orange, CA	2,674,091

Create a bar graph from this information.

Use the line graph, showing the yearly utility costs of a family, for exercises 19 to 22.

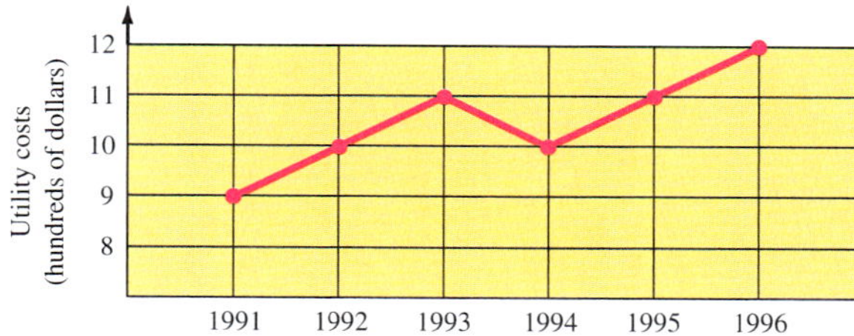

19. What was the cost in 1994?

20. Identify the years for which utility costs were the same (two answers).

21.
22.
23.
24.
25.
26.
27.

21. What was the decrease in the cost of utilities from 1993 to 1994?

22. In what year was the cost of utilities the smallest?

Use the line graph, showing the number of robberies in a town during the last 6 months of a year, for exercises 23 to 26.

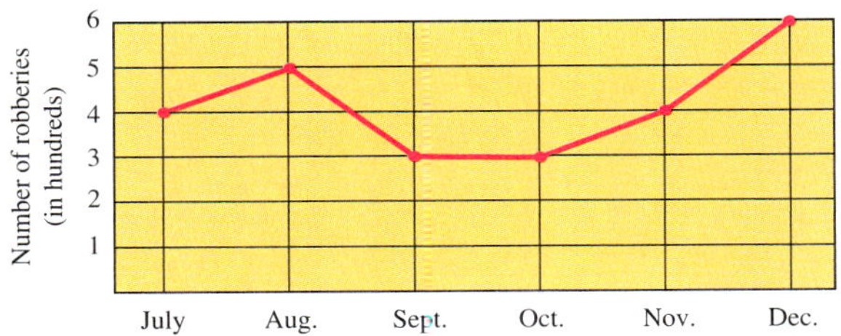

23. In which month did the greatest number of robberies occur?

24. How many robberies occurred in November?

25. Find the decrease in the number of robberies between August and September.

26. What was the total number of robberies for the 6 months?

27. The line graph and table show the income to the Hospital Insurance Trust Fund. Use this information to predict the income in the year 2000.

Year	Total Income (in billions)
1975	$12,568
1980	$25,415
1990	$79,563
1995	$114,847
2000	

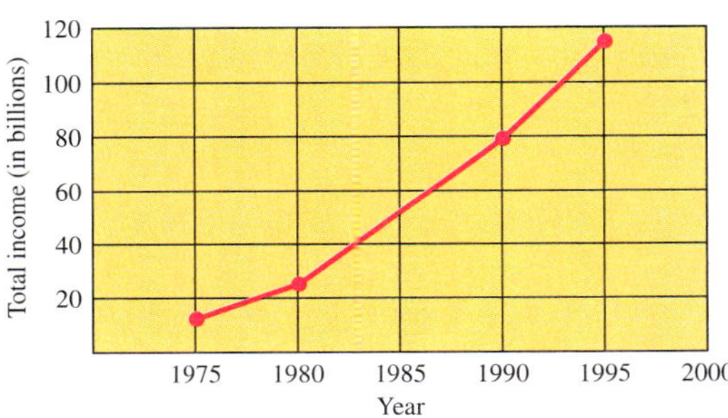

28. ______________

29. ______________

28. The line graph and table give the monthly principal and interest payments for a mortgage from 1993 to 1998. Use this information to predict the payment for 1999.

Year	Payment
1993	$578
1994	$613
1995	$654
1996	$675
1997	$706
1998	$730
1999	

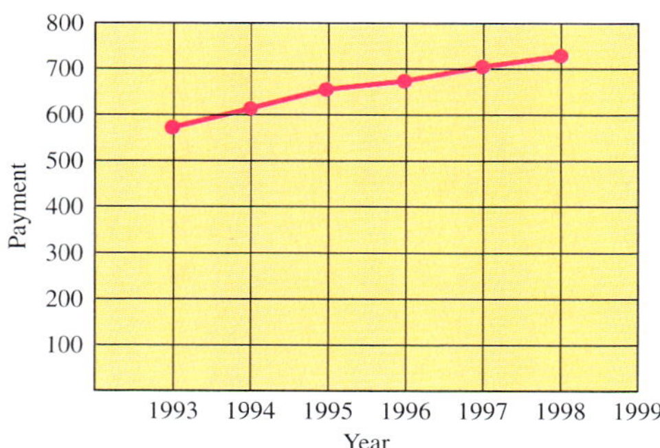

29. The given information shows a relationship between years of formal education and typical income (in thousands of dollars) at age 30:

Years	8	10	12	14	16
Income (×1000)	16	21	23	28	31

Use this information to make a line graph, and then predict the yearly income associated with 18 years of formal education.

ANSWERS

30. ____________

31. ____________

30. The given information shows a relationship between the amount spent per week on advertising by a small fast-food shop and the total sales per week:

Amount spent	10	20	30	40
Sales	200	380	625	790

Use this information to make a line graph, and then predict the weekly sales associated with the expenditure of \$50 (per week) on advertising.

31. The given information shows a relationship between the number of weeks on a special diet and the number of pounds lost during that time:

Number of weeks	2	4	6	8
Number of pounds	2	5	9	11

Use this information to make a line graph, and then predict the number of pounds lost associated with 10 weeks on the special diet.

32. ______________

32. The given information shows a relationship between the speed of a certain car over a 100-mi test trip and the gas mileage (in miles per gallon) obtained:

Speed	40	45	50	55	60
Gas mileage	28	25	21	18	16

Use this information to make a line graph, and then predict the gas mileage associated with a speed of 65 $\frac{\text{mi}}{\text{h}}$.

Answers

1. **(a)** ≈ 750,000,000; **(b)** ≈ 800,000,000; **(c)** 

3.

Year	# of Cars
1970	4.2
1975	4.3
1980	4.5
1985	5.5
1990	4.7
1995	4.3

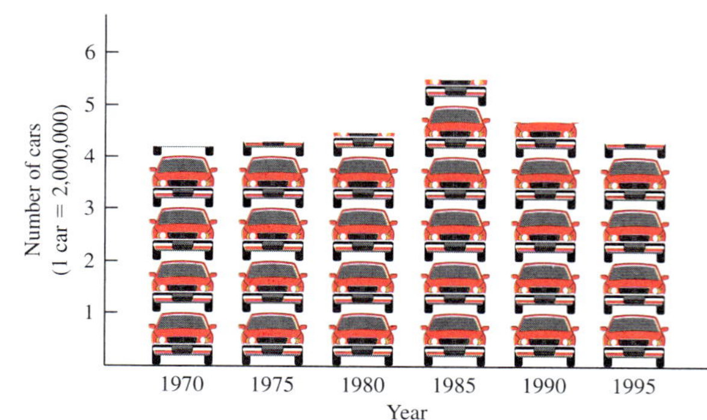

5. 3,000,000 **7.** 3,000,000 **9.** 2,800 **11.** August 3 **13.** 2,400,000
15. 140% **17.**

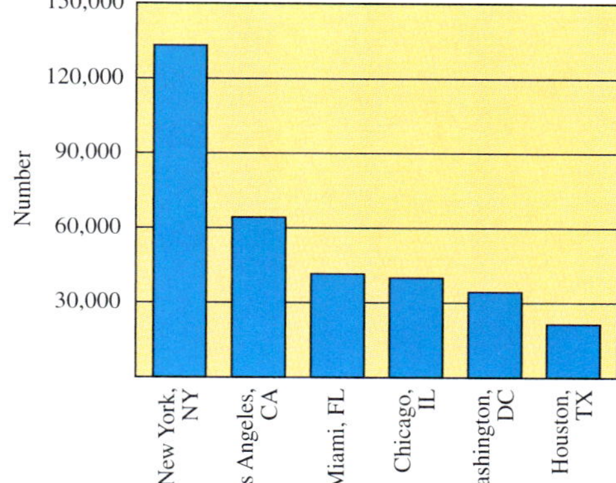

19. $1,000 **21.** $100 **23.** December **25.** 200 **27.** $148,000,000,000
29. $34,000 **31.** 14 lb

9.3 The Rectangular Coordinate System

9.3 OBJECTIVES

1. Plot a point corresponding to an ordered pair
2. Determine the coordinates of a point plotted on the coordinate plane

In Chapter 3, we looked at equations such as

$$2x + 3 = 7$$

For such an equation, we found a solution (here $x = 2$), which was a value that made the equation a true statement.

In algebra, we also encounter equations with two variables, such as

$$y = 2x + 1$$

What would the solution to such an equation be? If $x = 0$ and $y = 1$, the equation would be a true statement. However, if $x = 1$ and $y = 3$, the equation would also be a true statement. Because solutions with two variables require two numbers, we need a way to indicate that two numbers together make a solution. To do this, we write the numbers as an ordered pair.

Instead of $x = 0$ and $y = 1$, we write (0, 1).

Instead of $x = 1$ and $y = 3$, we write (1, 3).

Each is called an ordered pair, because the order is important (the x always comes before the y). The pair (1, 0) is different from the pair (0, 1). For the pair (1, 0), $x = 1$ and $y = 0$, and we can check that (1, 0) is *not* a solution for the equation $y = 2x + 1$. In this section, we will learn to graph information that comes in ordered pairs.

NOTE We call the flat surface on which we are working *the plane.*

Because there are two numbers (one for x and one for y), we will need two number lines. One line is drawn horizontally, and the other is drawn vertically; their point of intersection (at their respective zero points) is called the **origin.** The horizontal line is called the ***x*-axis,** and the vertical line is called the ***y*-axis.** Together the lines form the **rectangular coordinate system.**

The axes divide the plane into four regions called **quadrants,** which are numbered (usually by Roman numerals) counterclockwise from the upper right.

NOTE This system is also called the **cartesian coordinate system,** named in honor of its inventor, René Descartes (1596–1650), a French mathematician and philosopher.

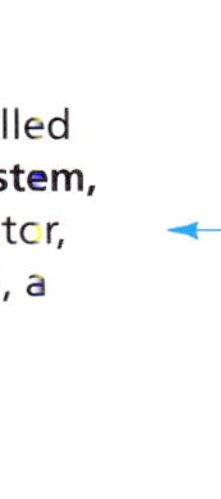

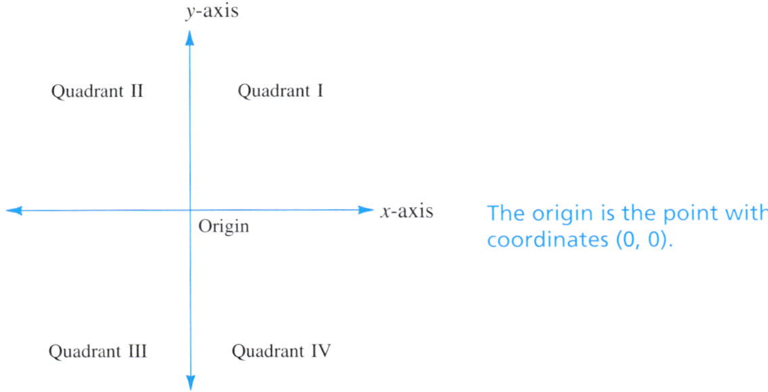

We now want to establish correspondences between ordered pairs of numbers (x, y) and points in the plane.

For any ordered pair

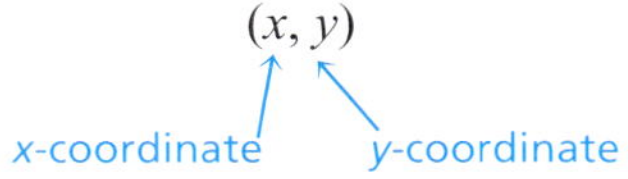

the following are true:

1. If the x-coordinate is

Positive, the point corresponding to that pair is located x units to the *right* of the y-axis.
Negative, the point is x units to the *left* of the y-axis.
Zero, the point is on the y-axis.

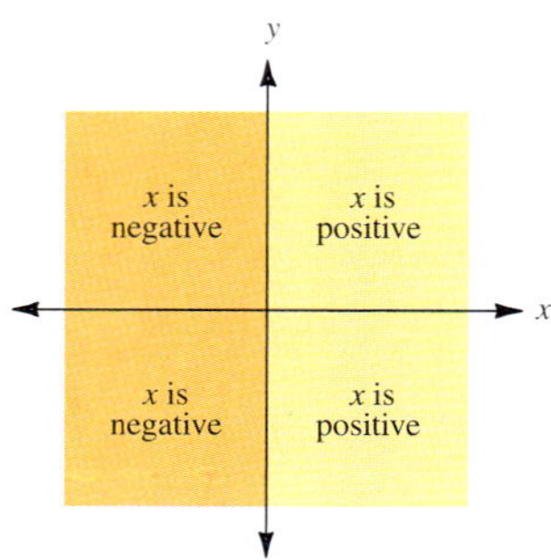

2. If the y-coordinate is

Positive, the point is y units *above* the x-axis.
Negative, the point is y units *below* the x-axis.
Zero, the point is on the x-axis.

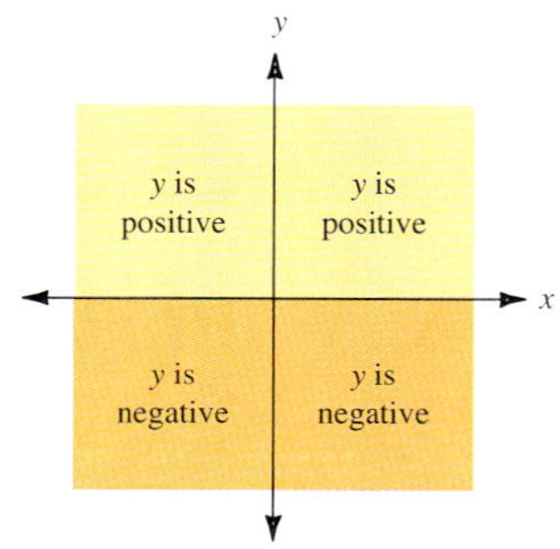

Example 1 illustrates how to use these guidelines to give coordinates to points in the plane.

REMEMBER: The x-coordinate gives the *horizontal* distance from the y-axis. The y coordinate gives the *vertical* distance from the x-axis.

Example 1

Identifying the Coordinates for a Given Point

Give the coordinates for the given point.

(a)

Point A is 3 units to the *right* of the y-axis and 2 units *above* the x-axis. Point A has coordinates (3, 2).

(b)

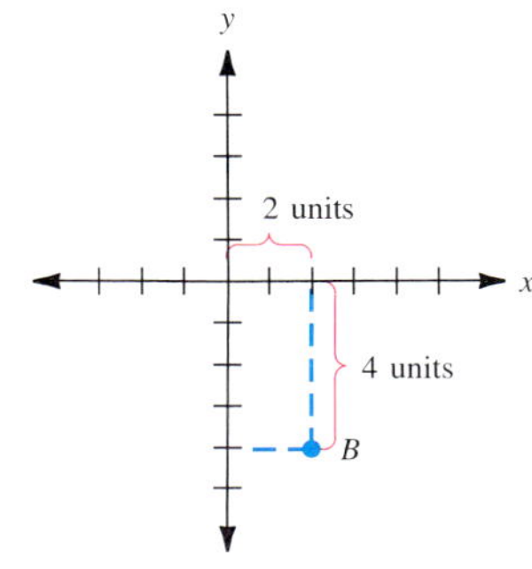

Point B is 2 units to the *right* of the y-axis and 4 units *below* the x-axis. Point B has coordinates $(2, -4)$.

(c)

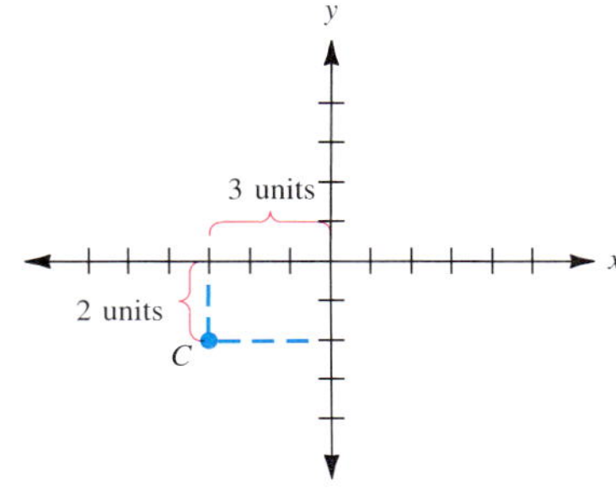

Point C is 3 units to the *left* of the y-axis and 2 units *below* the x-axis. C has coordinates $(-3, -2)$.

(d)

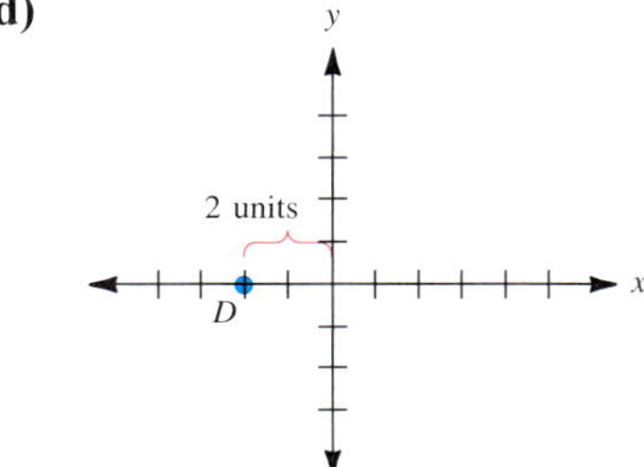

Point D is 2 units to the *left* of the y-axis and *on* the x-axis. Point D has coordinates $(-2, 0)$.

CHECK YOURSELF 1

Give the coordinates of points P, Q, R, and S.

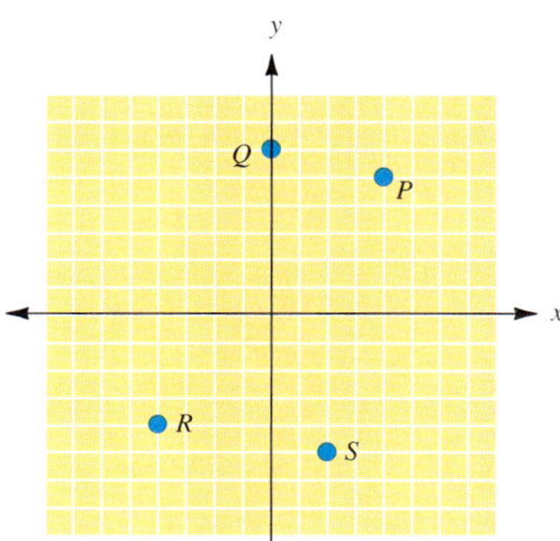

P __________

Q __________

R __________

S __________

Reversing this process will allow us to graph (or plot) a point in the plane given the coordinates of the point. You can use these steps.

NOTE The graphing of individual points is sometimes called **point plotting.**

Step by Step: To Graph a Point in the Plane

Step 1 Start at the origin.
Step 2 Move right or left according to the value of the *x*-coordinate.
Step 3 Move up or down according to the value of the *y*-coordinate.

Example 2

Graphing Points

(a) Graph the point corresponding to the ordered pair (4, 3).

Move 4 units to the right of the origin on the x-axis. Then move 3 units up from the point you stopped at on the x-axis. This locates the point corresponding to (4, 3).

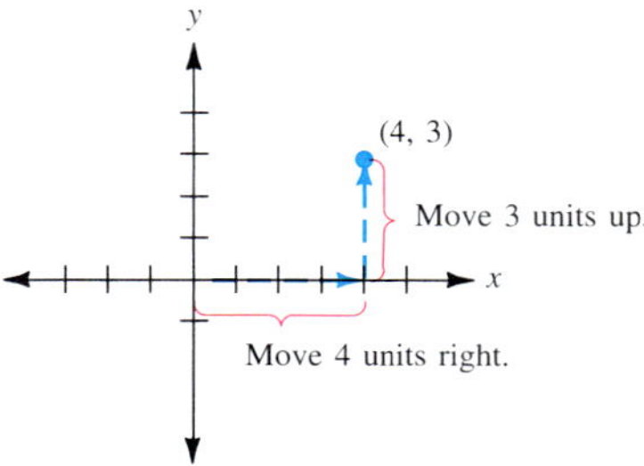

(b) Graph the point corresponding to the ordered pair (−5, 2).

In this case, move 5 units *left* of the origin (because the x-coordinate is negative) and then 2 units *up*.

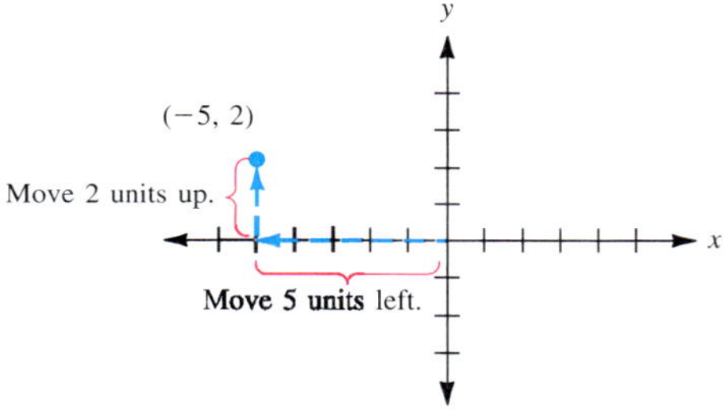

(c) Graph the point corresponding to (−4, −2).

Here move 4 units *left* of the origin and then 2 units *down* (the y-coordinate is negative).

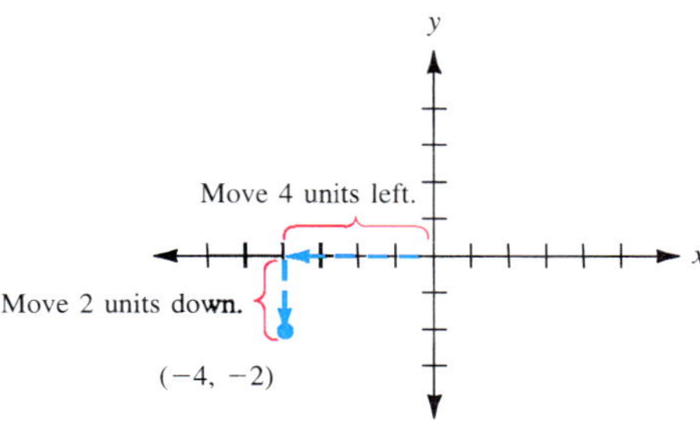

NOTE Any point on an axis will have 0 for one of its coordinates.

(d) Graph the point corresponding to $(0, -3)$.

There is *no* horizontal movement because the x-coordinate is 0. Move 3 units *down* from the origin.

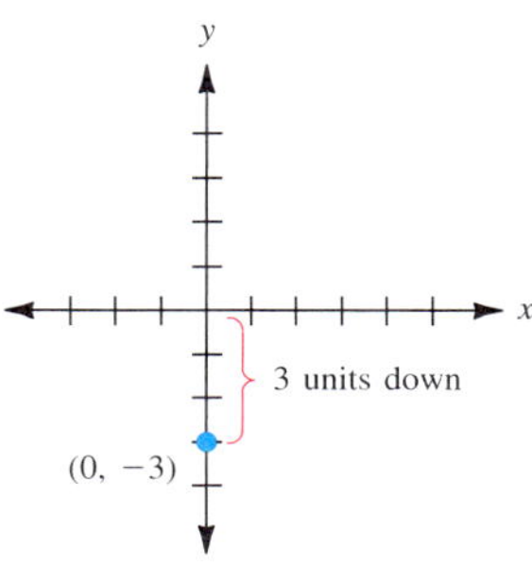

(e) Graph the point corresponding to $(5, 0)$.

Move 5 units *right* of the origin. The desired point is on the x-axis because the y-coordinate is 0.

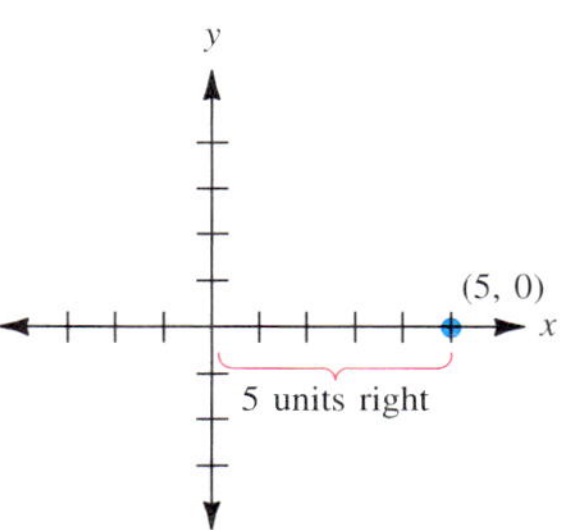

CHECK YOURSELF 2

Graph the points corresponding to $M(4, 3)$, $N(-2, 4)$, $P(-5, -3)$, and $Q(0, -3)$.

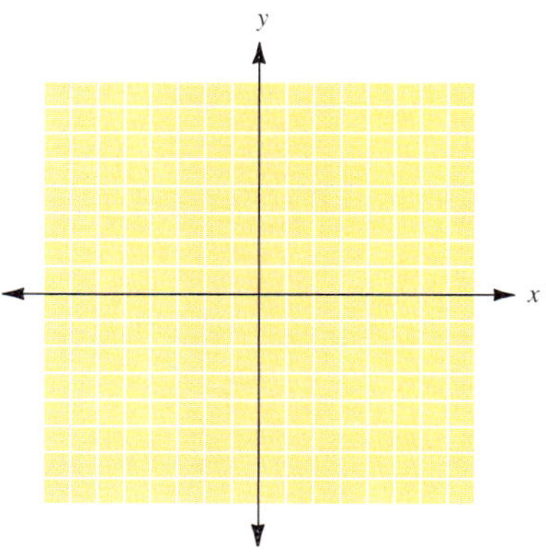

CHECK YOURSELF ANSWERS

1. $P(4, 5)$, $Q(0, 6)$, $R(-4, -4)$, and $S(2, -5)$
2. 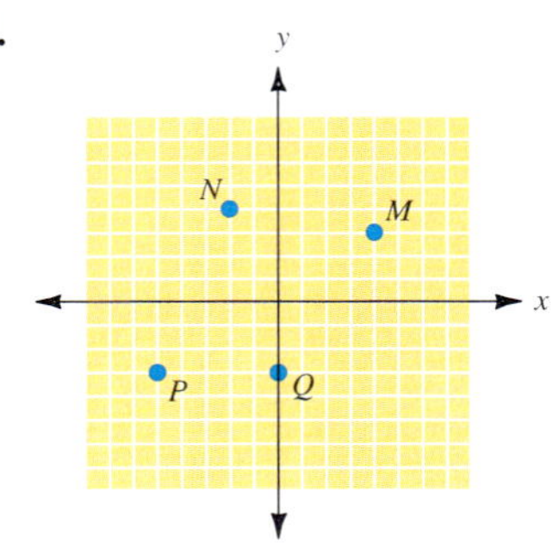

Name ________________

Section ________ Date ________

9.3 Exercises

Give the coordinates of the graphed points.

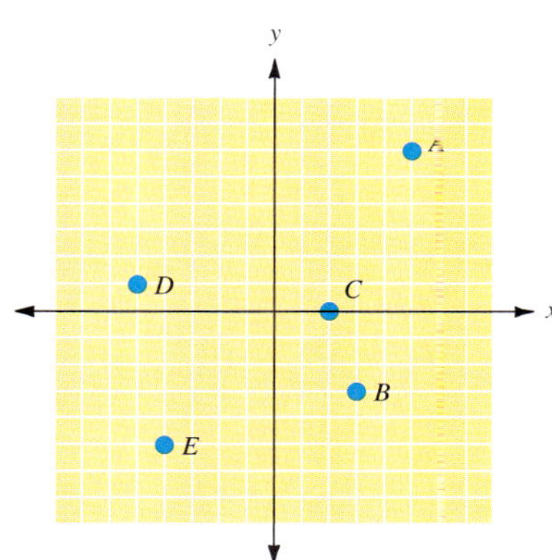

1. A

2. B

3. C

4. D

5. E

Give the coordinates of the graphed points.

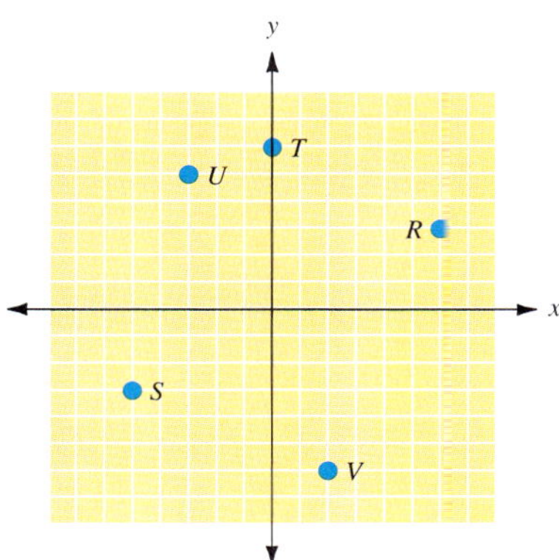

6. R

7. S

8. T

9. U

10. V

Plot points with the given coordinates on the graph provided.

11. $M(5, 3)$ **12.** $N(0, -3)$

13. $P(-2, 6)$ **14.** $Q(5, 0)$

15. $R(-4, -6)$ **16.** $S(-3, -4)$

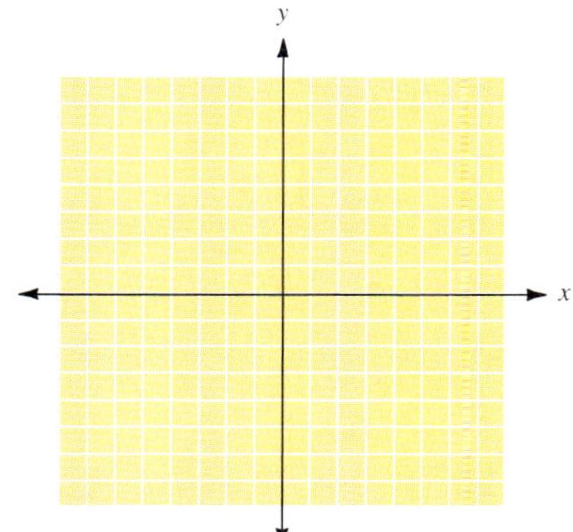

ANSWERS

1. ________
2. ________
3. ________
4. ________
5. ________
6. ________
7. ________
8. ________
9. ________
10. ________
11. ________
12. ________
13. ________
14. ________
15. ________
16. ________

ANSWERS

17. ______________

18. ______________

19. ______________

20. ______________

21. ______________

22. ______________

23. ______________

24. ______________

Plot points with the given coordinates on the graph provided.

17. $F(-3, -1)$

18. $G(4, 3)$

19. $H(5, -2)$

20. $I(-3, 0)$

21. $J(-5, 3)$

22. $K(0, 6)$

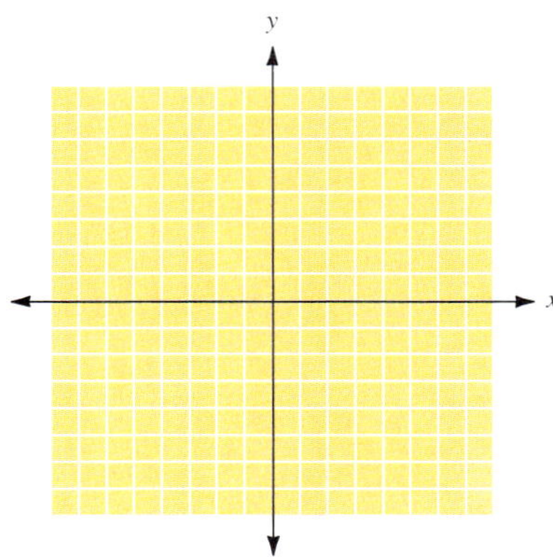

23. Graph points with coordinates (2, 3), (3, 4), and (4, 5) on the graph provided. What do you observe? Can you give the coordinates of another point with the same property?

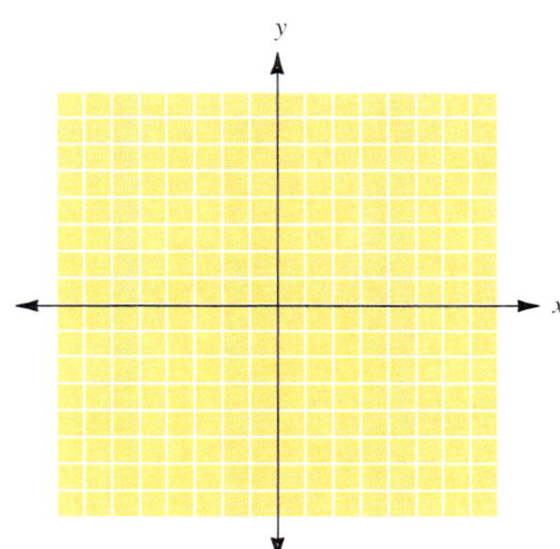

24. Graph points with coordinates (−1, 4), (0, 3), and (1, 2) on the graph provided. What do you observe? Can you give the coordinates of another point with the same property?

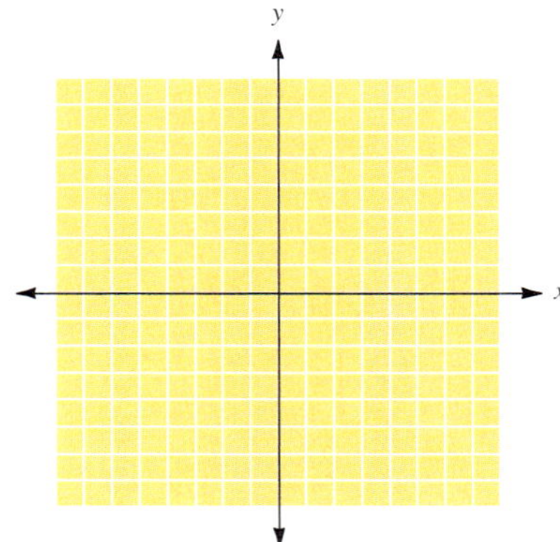

25. ______

26. ______

27. ______

25. Graph points with coordinates $(-1, 3)$, $(0, 0)$, and $(1, -3)$ on the graph provided. What do you observe? Can you give the coordinates of another point with the same property?

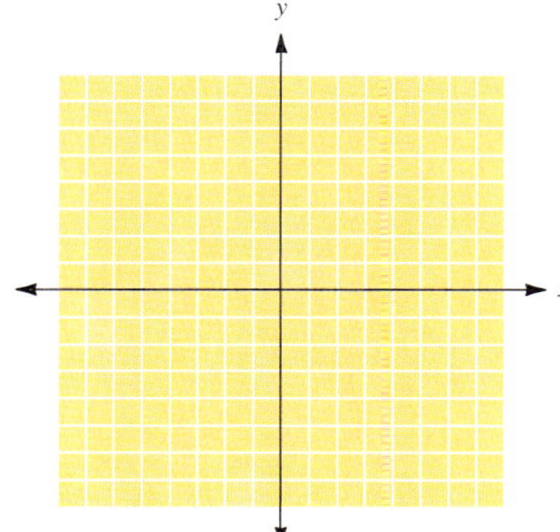

26. Graph points with coordinates $(1, 5)$, $(-1, 3)$, and $(-3, 1)$ on the graph provided. What do you observe? Can you give the coordinates of another point with the same property?

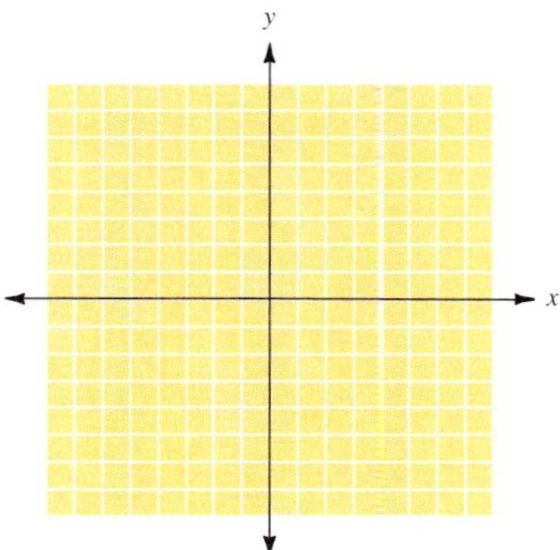

27. **Environment.** A local plastics company is sponsoring a plastics recycling contest for the local community. The focus of the contest is collecting plastic milk, juice, and water jugs. The company will award \$200 plus the current market price of the jugs collected to the group that collects the most jugs in a single month. The number of jugs collected and the amount of money won can be represented as an ordered pair.

(a) In April, group A collected 1,500 lb of jugs to win first place. The prize for the month was \$350. On the graph on the next page, x represents the pounds of jugs and y represents the amount of money that the group won. Graph the point that represents the winner for April.

ANSWERS

28. ______________________

29. ______________________

(b) In May, group *B* collected 2,300 lb of jugs to win first place. The prize for the month was \$430. Graph the point that represents the May winner on the same axis you used in part (a).

(c) In June, group *C* collected 1,200 lb of jugs to win the contest. The prize for the month was \$320. Graph the point that represents the June winner on the same axis as used before.

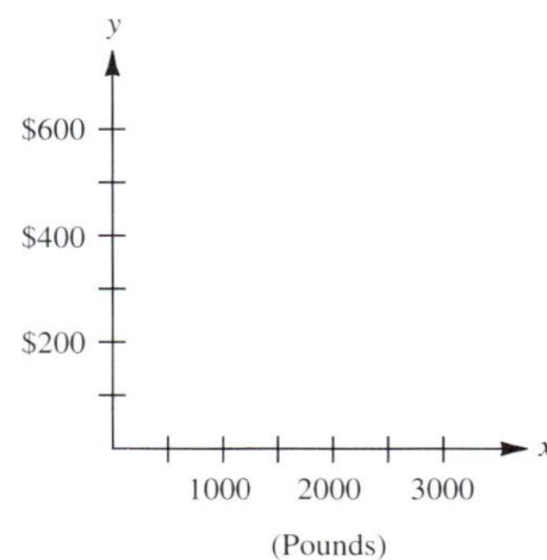

28. Education. The table gives the hours, x, that Damien studied for five different math exams and the resulting grades, y. Plot the data given in the table.

x	4	5	5	2	6
y	83	89	93	75	95

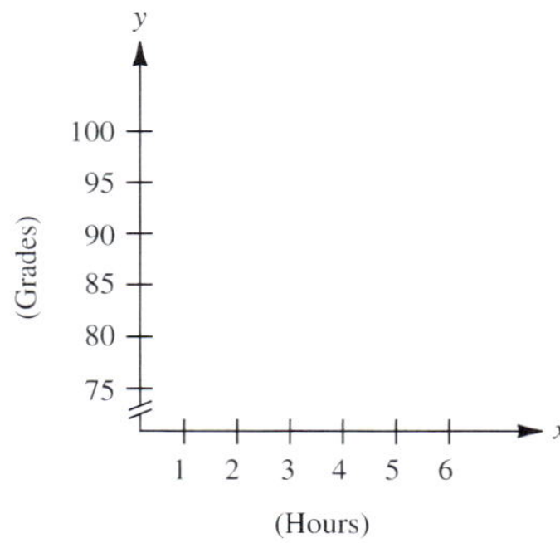

29. Science. The table gives the average temperature y (in degrees Fahrenheit) for the first 6 months of the year, x. The months are numbered 1 through 6, with 1 corresponding to January. Plot the data given in the table.

x	1	2	3	4	5	6
y	4	14	26	33	42	51

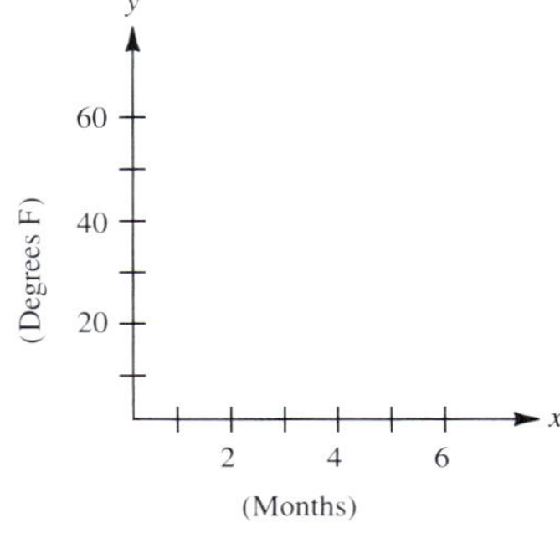

30. ______

31. ______

30. Business. The table gives the total salary of a salesperson, y, for each of the four quarters of the year, x. Plot the data given in the table.

x	1	2	3	4
y	\$6,000	\$5,000	\$8,000	\$9,000

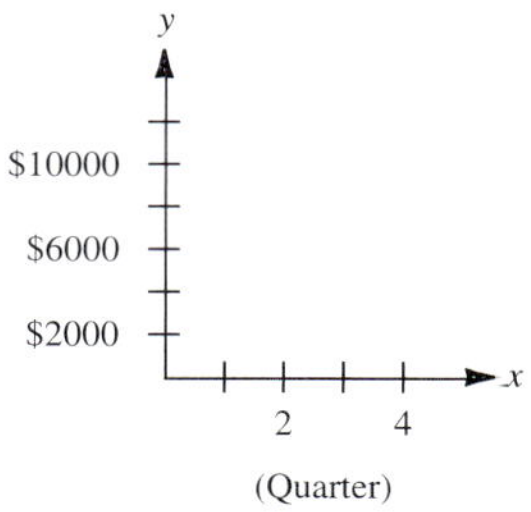

31. Sports. The table shows the number of runs scored by the New York Yankees in the 1999 World Series.

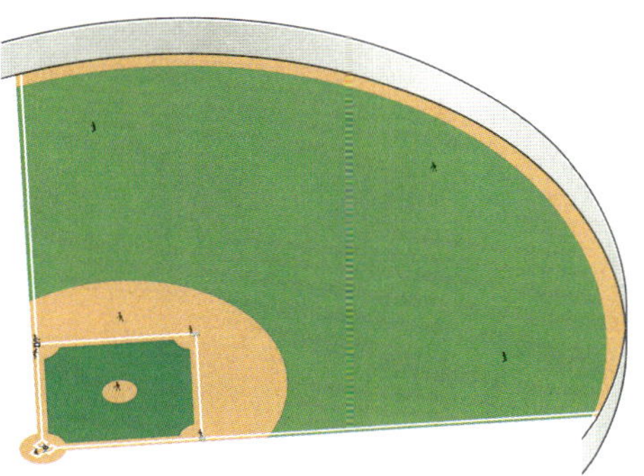

Game	1	2	3	4
Runs	4	7	6	4

Plot the data given in the table.

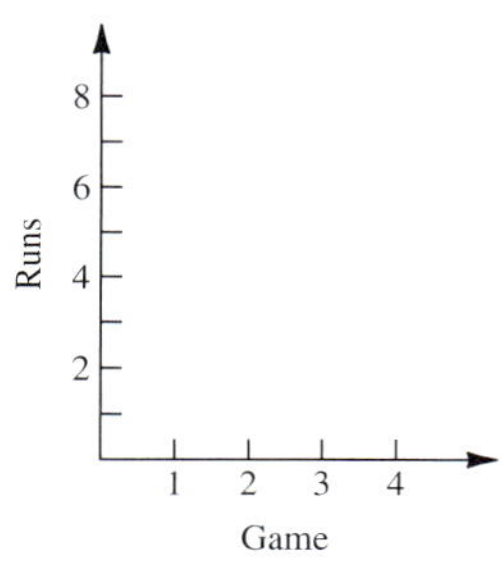

ANSWERS

32.

33.

34.

32. Sports. The table shows the number of wins and total points for the five teams in the Atlantic Division of the National Hockey League in the early part of the 1999–2000 season.

Team	Wins	Points
New Jersey Devils	5	12
Philadelphia Flyers	4	10
New York Rangers	4	9
Pittsburgh Penguins	2	6
New York Islanders	2	5

Plot the data given in the table.

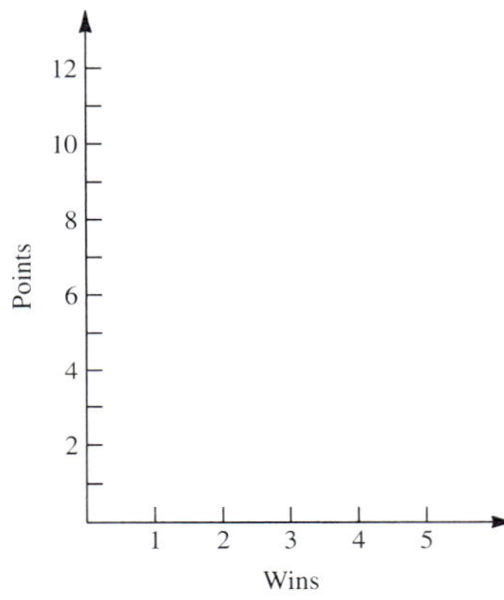

33. How would you describe a rectangular coordinate system? Explain what information is needed to locate a point in a coordinate system.

34. Some newspapers have a special day that they devote to automobile ads. Use this special section or the Sunday classified ads from your local newspaper to find all the want ads for a particular automobile model. Make a list of the model year and asking price for 10 ads, being sure to get a variety of ages for this model. After collecting the information, make a graph of the age and the asking price for the car.

Describe your graph, including an explanation of how you decided which variable to put on the vertical axis and which on the horizontal axis. What trends or other information are given by the graph?

ANSWERS

(a)
(b)
35. ______________________

35. The map shown below uses letters and numbers to label a grid that helps to locate a city. For instance, Salem is located at E-4.

(a) Find the coordinates for the following: White Swan, Newport, and Wheeler.

(b) What cities correspond to the following coordinates: A2, F4, and A5?

Answers

1. (5, 6) **3.** (2, 0) **5.** (−4, −5) **7.** (−5, −3) **9.** (−3, 5)

11–21.

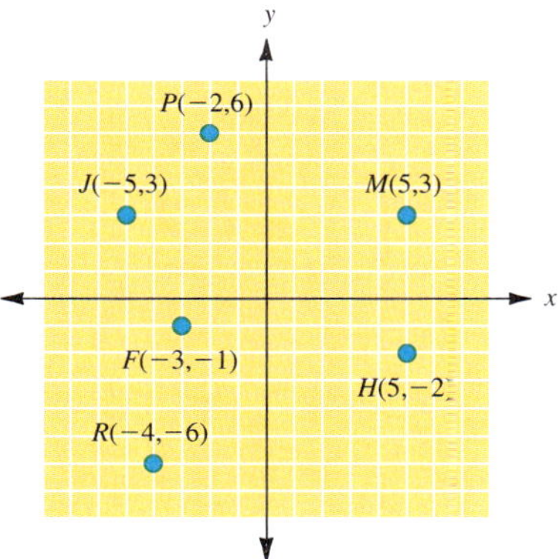

23. The points lie on a line; sample answer is (1, 2)

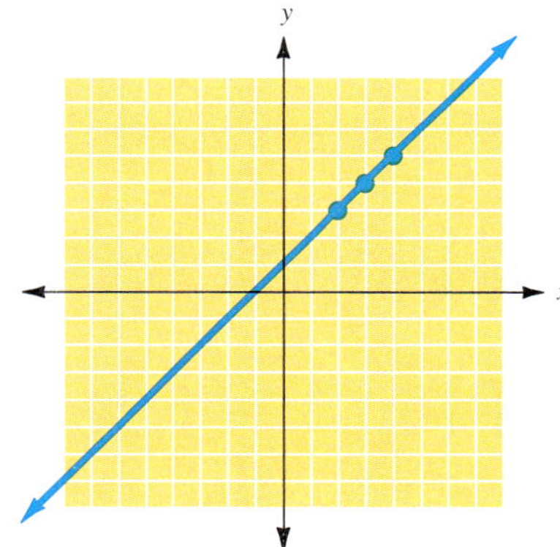

25. The points lie on a line; sample answer is (2, −6)

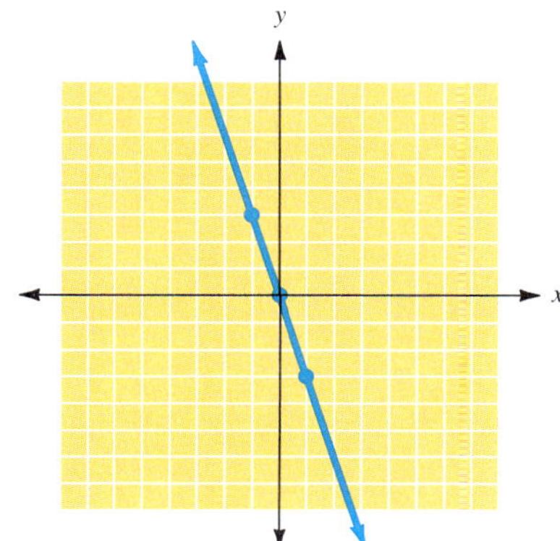

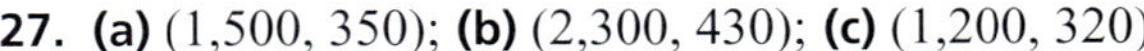
27. **(a)** (1,500, 350); **(b)** (2,300, 430); **(c)** (1,200, 320)

29.

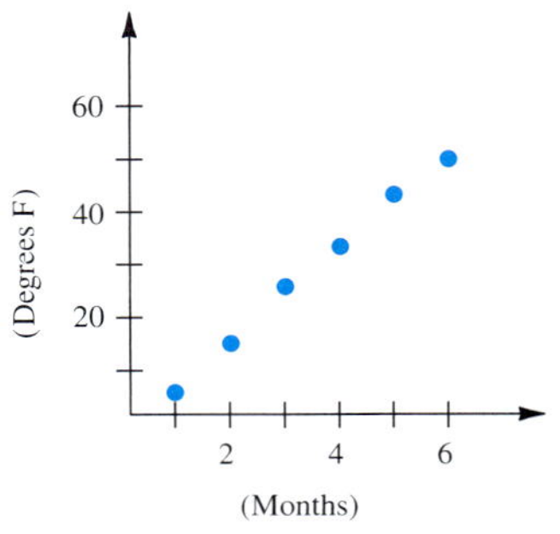

31.

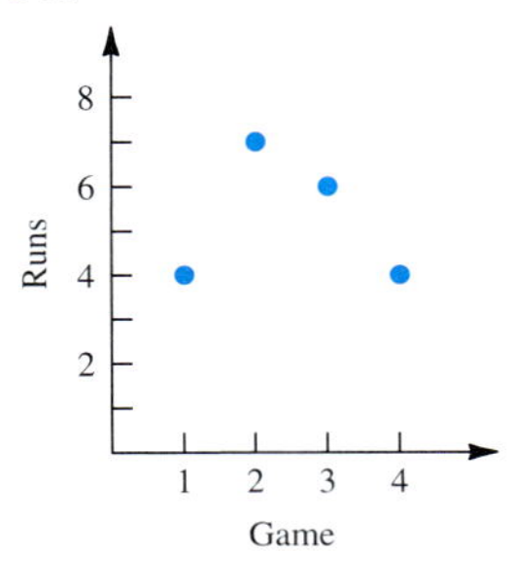

33.

35. **(a)** A7, F2, C2; **(b)** Oysterville, Sweet Home, Mineral

9.4 Linear Equations in Two Variables

OBJECTIVES

1. Find solutions for an equation in two variables
2. Use ordered pair notation to write solutions for an equation in two variables
3. Graph a linear equation by plotting points

We discussed finding solutions for equations in Section 2.9. Recall that a solution is a value for the variable that "satisfies" the equation, or makes the equation a true statement. For instance, we know that 4 is a solution of the equation

$$2x + 5 = 13$$

We know this is true because, when we replace x with 4, we have

$$2 \cdot 4 + 5 \stackrel{?}{=} 13$$

$$8 + 5 \stackrel{?}{=} 13$$

$$13 = 13 \quad \text{A true statement}$$

We now want to study **equations in two variables.** An example is

$$x + y = 5$$

As we saw in Section 9.3, a solution is not going to be a single number, because there are two variables. Here a solution will be a pair of numbers—one value for each of the variables, x and y. Suppose that x has the value 3. In the equation $x + y = 5$, you can substitute 3 for x.

$$3 + y = 5$$

Solving for y gives

$$y = 2$$

NOTE An equation in two variables "pairs" two numbers, one for x and one for y.

So the pair of values $x = 3$ and $y = 2$ satisfies the equation because

$$3 + 2 = 5$$

That pair of numbers is then a *solution* for the equation in two variables.

How many such pairs are there? Choose any value for x (or for y). You can always find the other *paired* or *corresponding* value in an equation of this form. We say that there are an *infinite* number of pairs that will satisfy the equation. Each of these pairs is a solution. We will find some other solutions for the equation $x + y = 5$ in Example 1.

Example 1

Solving for Corresponding Values

For the equation $x + y = 5$, find (a) y if $x = 5$ and (b) x if $y = 4$.

(a) If $x = 5$

$$5 + y = 5 \qquad \text{or} \qquad y = 0$$

(b) If $y = 4$,

$x + 4 = 5$ or $x = 1$

So the pairs $x = 5, y = 0$ and $x = 1, y = 4$ are both solutions.

CHECK YOURSELF 1

For the equation $2x + 3y = 26$,

(a) If $x = 4, y = ?$ **(b)** If $y = 0, x = ?$

NOTE We introduced ordered pairs in Section 9.3.

To simplify writing the pairs that satisfy an equation, we use the **ordered-pair notation.** Recall that the numbers are written in parentheses and are separated by a comma. For example, we know that the values $x = 3$ and $y = 2$ satisfy the equation $x + y = 5$. So we write the pair as

CAUTION

(3, 2) means $x = 3$ and $y = 2$. (2, 3) means $x = 2$ and $y = 3$. (3, 2) and (2, 3) are entirely different. That's why we call them *ordered pairs.*

(3, 2)

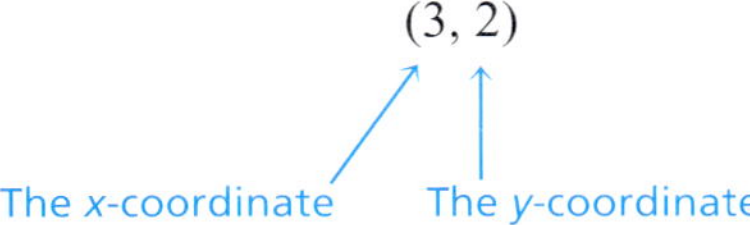

The first number of the pair is *always* the value for x and is called the **x-coordinate.** The second number of the pair is *always* the value for y and is the **y-coordinate.**

Using this ordered-pair notation, we can say that (3, 2), (5, 0), and (1, 4) are all *solutions* for the equation $x + y = 5$. Each pair gives values for x and y that will satisfy the equation.

Example 2

Identifying Solutions of Two-Variable Equations

Which of the ordered pairs (a) (2, 5), (b) (5, −1), and (c) (3, 4) are solutions for the equation $2x + y = 9$?

(a) To check whether (2, 5) is a solution, let $x = 2$ and $y = 5$ and see if the equation is satisfied.

$2x + y = 9$ The original equation.

$2 \cdot 2 + 5 \stackrel{?}{=} 9$ Substitute 2 for x and 5 for y.

$4 + 5 \stackrel{?}{=} 9$

$9 = 9$ A true statement

NOTE (2, 5) is a solution because a *true statement* results.

(2, 5) is a solution for the equation.

(b) For $(5, -1)$, let $x = 5$ and $y = -1$.

$$2 \cdot 5 + (-1) \stackrel{?}{=} 9$$

$$10 - 1 \stackrel{?}{=} 9$$

$$9 = 9 \quad \text{A true statement}$$

So $(5, -1)$ is a solution.

(c) For $(3, 4)$, let $x = 3$ and $y = 4$. Then

$$2 \cdot 3 + 4 \stackrel{?}{=} 9$$

$$6 + 4 \stackrel{?}{=} 9$$

$$10 = 9 \quad \textit{Not} \text{ a true statement}$$

So $(3, 4)$ is *not* a solution for the equation.

CHECK YOURSELF 2

Which of the ordered pairs (3, 4), (4, 3), (1, −2), *and* (0, −5) *are solutions for the equation?*

$$3x - y = 5$$

If the equation contains only one variable, then the missing variable can take on any value.

Example 3

Identifying Solutions of One-Variable Equations

Which of the ordered pairs $(2, 0)$, $(0, 2)$, $(5, 2)$, $(2, 5)$, and $(2, -1)$ are solutions for the equation $x = 2$?

A solution is any ordered pair in which the x-coordinate is 2. That makes $(2, 0)$, $(2, 5)$, and $(2, -1)$ solutions for the given equation.

CHECK YOURSELF 3

Which of the ordered pairs (3, 0), (0, 3), (3, 3), (−1, 3), *and* (3, −1) *are solutions for the equation* $y = 3$*?*

Remember that, when an ordered pair is presented, the first number is always the x-coordinate and the second number is always the y-coordinate.

Example 4

Completing Ordered Pair Solutions

Complete the ordered pairs (a) (9,), (b) (, −1), (c) (0,), and (d) (, 0) for the equation $x - 3y = 6$.

NOTE The x-coordinate is sometimes called the **abscissa** and the y-coordinate the **ordinate.**

(a) The first number, 9, appearing in (9,) represents the x value. To complete the pair (9,), substitute 9 for x and then solve for y.

$$9 - 3y = 6$$
$$-3y = -3$$
$$y = 1$$

(9, 1) is a solution.

(b) To complete the pair (, −1), let y be −1 and solve for x.

$$x - 3(-1) = 6$$
$$x + 3 = 6$$
$$x = 3$$

(3, −1) is a solution.

(c) To complete the pair (0,), let x be 0.

$$0 - 3y = 6$$
$$-3y = 6$$
$$y = -2$$

(0, −2) is a solution.

(d) To complete the pair (, 0), let y be 0.

$$x - 3 \cdot 0 = 6$$
$$x - 0 = 6$$
$$x = 6$$

(6, 0) is a solution.

CHECK YOURSELF 4

Complete the ordered pairs so that each is a solution for the equation $2x + 5y = 10$.

(10,), (, 4), (0,), and (, 0)

Example 5

Finding Some Solutions of a Two-Variable Equation

Find four solutions for the equation

$$2x + y = 8$$

NOTE Generally, you'll want to pick values for *x* (or for *y*) so that the resulting equation in one variable is easy to solve.

In this case the values used to form the solutions are *up to you.* You can assign any value for x (or for y). We'll demonstrate with some possible choices.

(a) Solution with $x = 2$:

$$2 \cdot 2 + y = 8$$
$$4 + y = 8$$
$$y = 4$$

(2, 4) is a solution.

(b) Solution with $y = 6$:

$$2x + 6 = 8$$
$$2x = 2$$
$$x = 1$$

(1, 6) is a solution.

(c) Solution with $x = 0$:

$$2 \cdot 0 + y = 8$$
$$y = 8$$

NOTE The solutions (0, 8) and (4, 0) will have special significance later in graphing. They are also easy to find!

(0, 8) is a solution.

(d) Solution with $y = 0$:

$$2x + 0 = 8$$
$$2x = 8$$
$$x = 4$$

(4, 0) is a solution.

CHECK YOURSELF 5

Find four solutions for $x - 3y = 12$.

We are now ready to combine our work of Section 9.3 with our knowledge of two-variable equations. You just learned to write the solutions of equations in two variables as ordered pairs. In Section 9.3, these ordered pairs were graphed in the plane. Putting these ideas together will let us graph certain equations. Example 6 illustrates this approach.

Example 6

Graphing a Linear Equation

Graph $x + 2y = 4$.

NOTE We are going to find *three* solutions for the equation. We'll point out why shortly.

Step 1 Find some solutions for $x + 2y = 4$. To find solutions, we choose any convenient values for x, say $x = 0$, $x = 2$, and $x = 4$. Given these values for x, we can substitute and then solve for the corresponding value for y. So

If $x = 0$, then $y = 2$, so (0, 2) is a solution.

If $x = 2$, then $y = 1$, so (2, 1) is a solution.

If $x = 4$, then $y = 0$, so (4, 0) is a solution.

A handy way to show this information is in a table such as this:

NOTE The table is just a convenient way to display the information. It is the same as writing (0, 2), (2, 1), and (4, 0).

x	y
0	2
2	1
4	0

Step 2 We now graph the solutions found in step 1.

$x + 2y = 4$

x	y
0	2
2	1
4	0

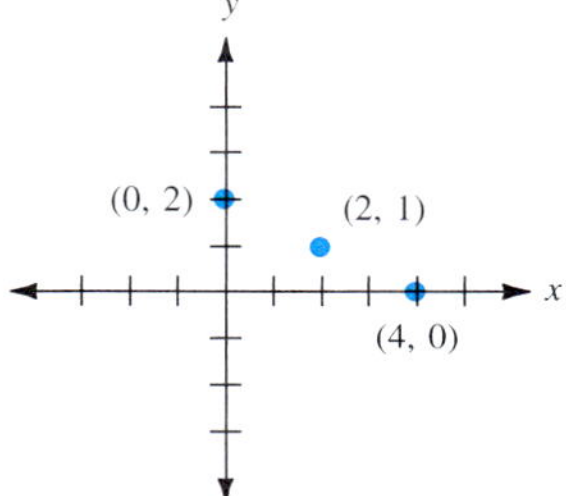

What pattern do you see? It appears that the three points lie on a straight line, and that is in fact the case.

Step 3 Draw a straight line through the three points graphed in step 2.

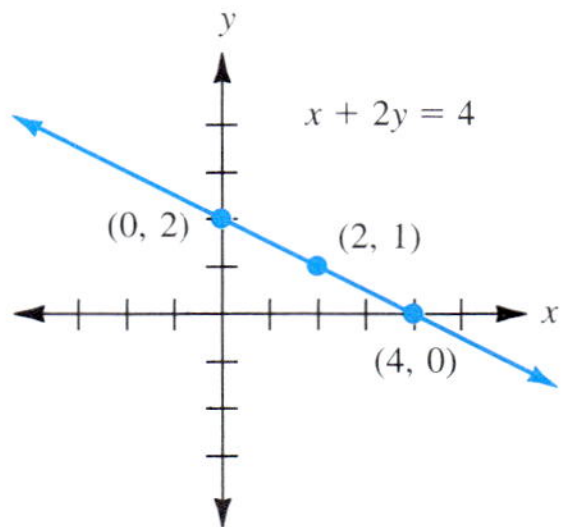

NOTE The arrows on the end of the line mean that the line extends indefinitely in either direction.

NOTE The graph is a "picture" of the solutions for the given equation.

The line shown is the **graph** of the equation $x + 2y = 4$. It represents *all* of the ordered pairs that are solutions (an infinite number) for that equation.

Every ordered pair that is a solution will have its graph on this line. Any point on the line will have coordinates that are a solution for the equation.

Note: Why did we suggest finding *three* solutions in step 1? Two points determine a line, so technically you need only two. The third point that we find is a check to catch any possible errors.

CHECK YOURSELF 6

Graph $2x - y = 6$, *using the steps shown in Example* 6.

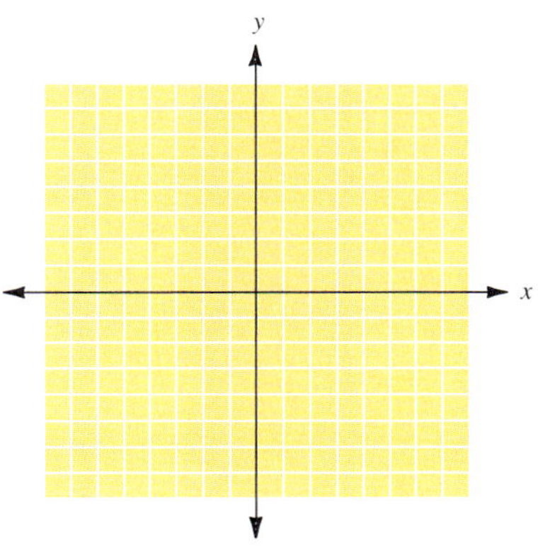

An equation that can be written in the form

$$Ax + By = C$$

in which A, B, and C are real numbers and A and B cannot both be 0 is called a **linear equation in two variables.** The graph of this equation is a *straight line.*

Here are the steps of graphing.

Step by Step: To Graph a Linear Equation

Step 1 Find at least three solutions for the equation, and put your results in tabular form.

Step 2 Graph the solutions found in step 1.

Step 3 Draw a straight line through the points determined in step 2 to form the graph of the equation.

Example 7

Graphing a Linear Equation

Graph $y = 3x$.

Step 1 Some solutions are

x	y
0	0
1	3
2	6

NOTE Let $x = 0$, 1, and 2, and substitute to determine the corresponding y values. Again the choices for x are simply convenient. Other values for x would serve the same purpose.

Step 2 Graph the points.

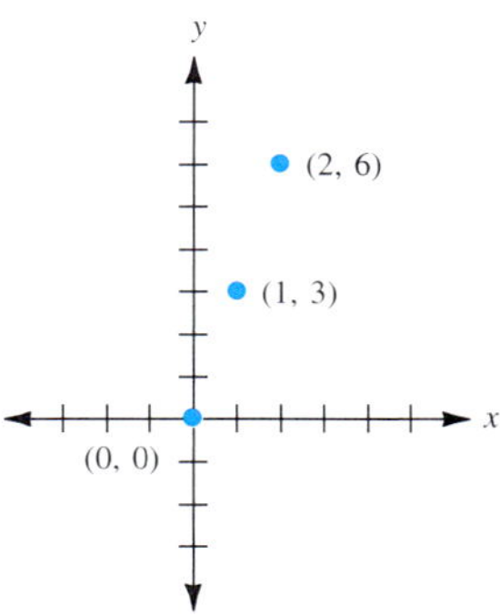

NOTE Notice that connecting any two of these points produces the same line.

Step 3 Draw a line through the points.

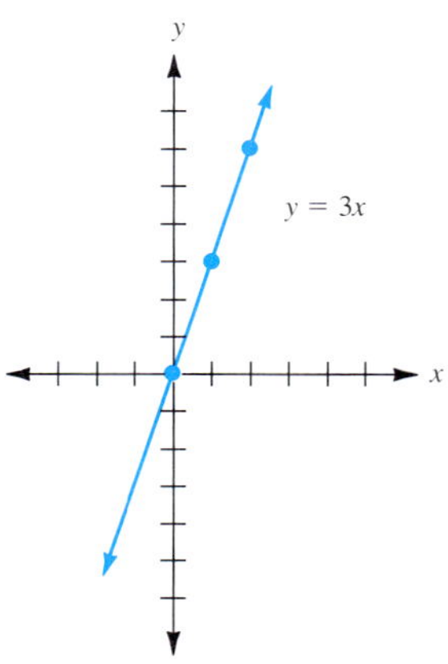

CHECK YOURSELF 7

Graph the equation $y = -2x$ after completing the table of values.

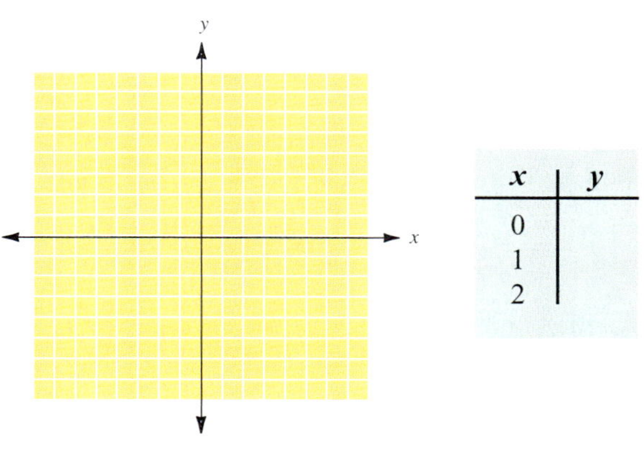

x	y
0	
1	
2	

We will work through another example of graphing a line from its equation.

Example 8

Graphing a Linear Equation

Graph $y = 2x + 3$.

Step 1 Some solutions are

x	y
0	3
1	5
2	7

Step 2 Graph the points corresponding to these values.

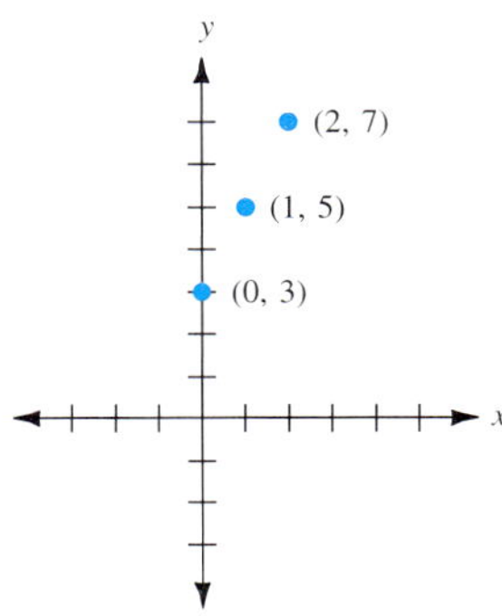

Step 3 Draw a line through the points.

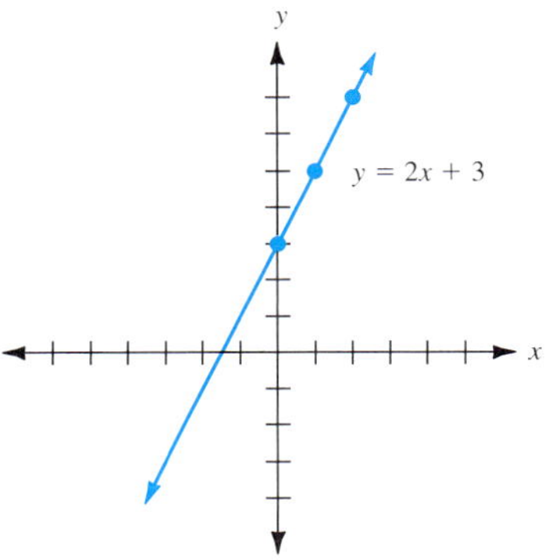

CHECK YOURSELF 8

Graph the equation $y = 3x - 2$ after completing the table of values.

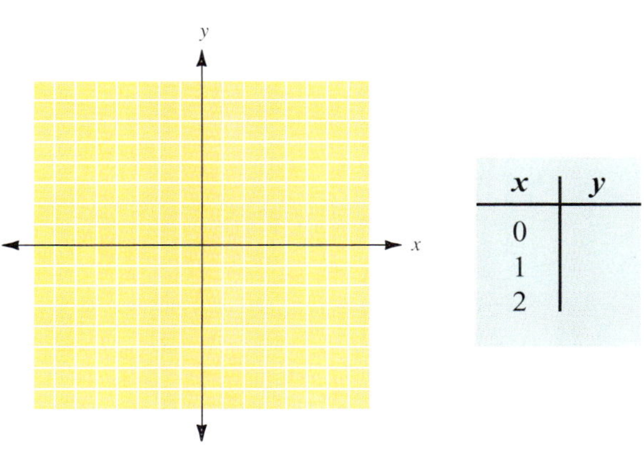

x	y
0	
1	
2	

In graphing equations, particularly when fractions are involved, a careful choice of values for x can simplify the process. Consider Example 9.

Example 9

Graphing a Linear Equation

Graph

$$y = \frac{3}{2}x - 2$$

As before, we want to find solutions for the given equation by picking convenient values for x. Note that in this case, choosing *multiples of 2* will avoid fractional values for y and make the plotting of those solutions much easier. For instance, here we might choose values of -2, 0, and 2 for x.

Step 1 If $x = -2$:

$$y = \frac{3}{2}(-2) - 2$$

$$= -3 - 2 = -5$$

If $x = 0$:

$$y = \frac{3}{2}(0) - 2$$

$$= 0 - 2 = -2$$

If $x = 2$:

$$y = \frac{3}{2}(2) - 2$$

$$= 3 - 2 = 1$$

In tabular form, the solutions are

x	y
-2	-5
0	-2
2	1

NOTE Suppose we do *not* choose a multiple of 2, say, $x = 3$. Then

$$y = \frac{3}{2}(3) - 2$$

$$= \frac{9}{2} - 2$$

$$= \frac{5}{2}$$

$\left(3, \frac{5}{2}\right)$ is still a valid solution, but we must graph a point with fractional coordinates.

Step 2 Graph the points determined in Step 1.

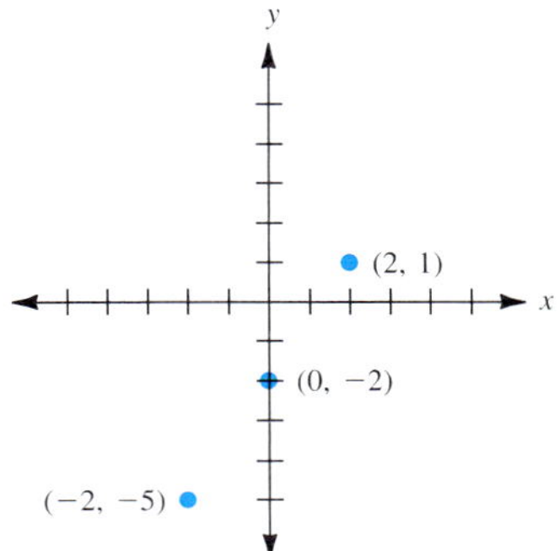

Step 3 Draw a line through the points.

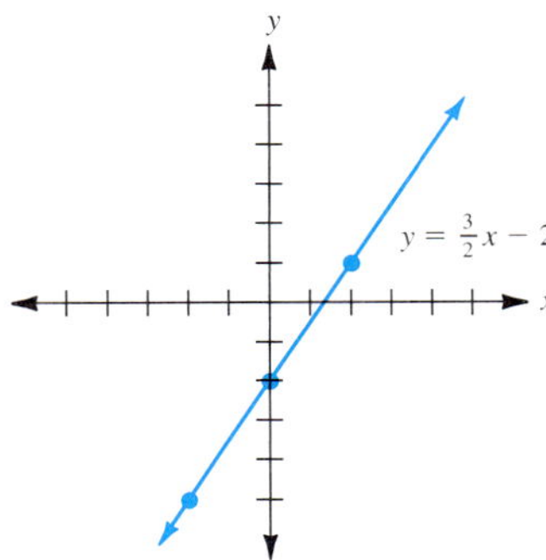

CHECK YOURSELF 9

Graph the equation $y = -\frac{1}{3}x + 3$ after completing the table of values.

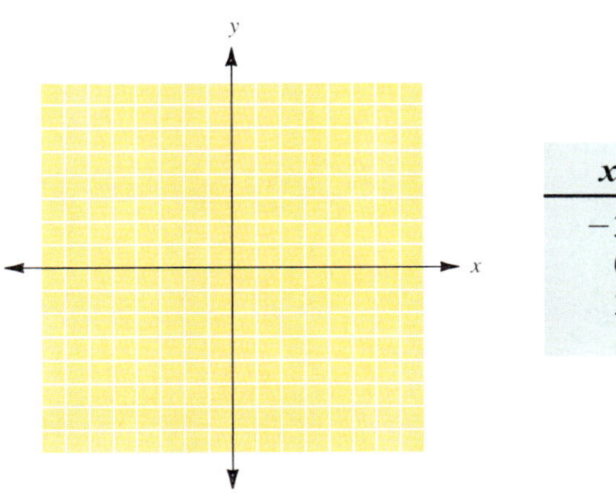

x	y
−3	
0	
3	

Some special cases of linear equations are illustrated in Examples 10 and 11.

Example 10

Graphing an Equation That Results in a Vertical Line

Graph $x = 3$.

The equation $x = 3$ is equivalent to $x + 0 \cdot y = 3$. We will look at scme solutions.

If $y = 1$:	If $y = 4$:	If $y = -2$:
$x + 0 \cdot 1 = 3$	$x + 0 \cdot 4 = 3$	$x + 0(-2) = 3$
$x = 3$	$x = 3$	$x = 3$

In tabular form,

x	y
3	1
3	4
3	−2

What do you observe? The variable x has the value 3, regardless of the value of y. Look at the given graph.

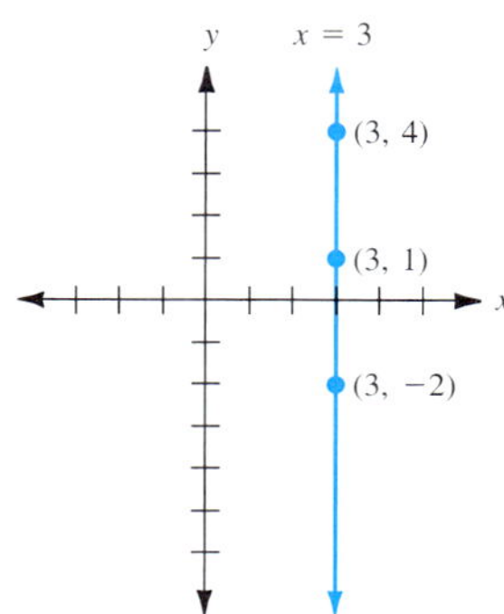

The graph of $x = 3$ is a vertical line crossing the x-axis at (3, 0).

Note that graphing (or plotting) points in this case is not really necessary. Simply recognize that the graph of $x = 3$ *must* be a vertical line (parallel to the y-axis) that intercepts the x-axis at (3, 0).

CHECK YOURSELF 10

Graph the equation $x = -2$.

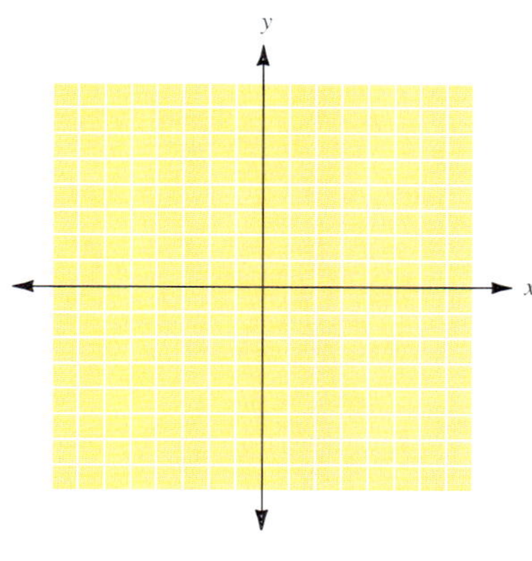

Example 11 is a related example involving a horizontal line.

Example 11

Graphing an Equation That Results in a Horizontal Line

Graph $y = 4$.

Because $y = 4$ is equivalent to $0 \cdot x + y = 4$, any value for x paired with 4 for y will form a solution. A table of values might be

x	y
−2	4
0	4
2	4

Here is the graph.

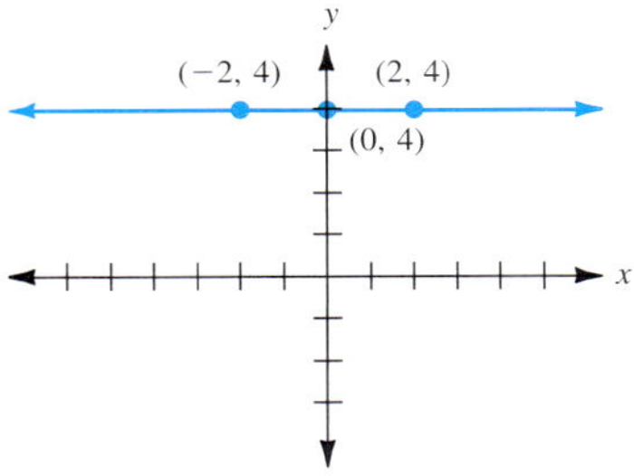

This time the graph is a horizontal line that crosses the y-axis at (0, 4). Again the graphing of points is not required. The graph of $y = 4$ *must* be horizontal (parallel to the x-axis) and intercepts the y-axis at (0, 4).

CHECK YOURSELF 11

Graph the equation $y = -3$.

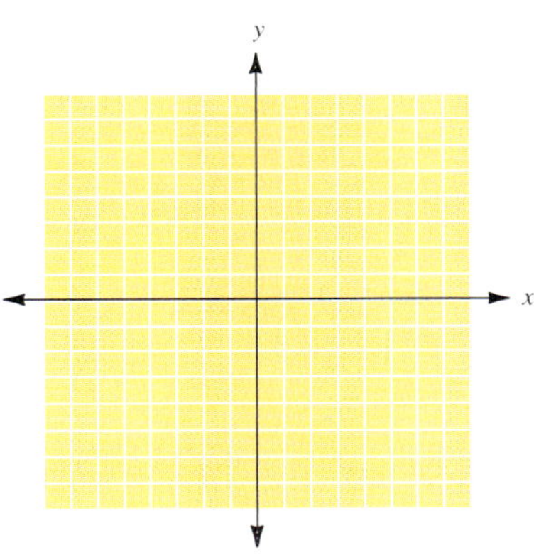

The Definitions box summarizes our work in Examples 10 and 11.

Definitions: Vertical and Horizontal Lines

1. The graph of $x = a$ is a *vertical line* crossing the x-axis at $(a, 0)$.
2. The graph of $y = b$ is a *horizontal line* crossing the y-axis at $(0, b)$.

CHECK YOURSELF ANSWERS

1. **(a)** $y = 6$; **(b)** $x = 13$ **2.** (3, 4), (1, −2), and (0, −5) are solutions
3. (0, 3), (3, 3), and (−1, 3) are solutions **4.** (10, −2), (−5, 4), (0, 2), and (5, 0)
5. (6, −2), (3, −3), (0, −4), and (12, 0) are four possibilities

6.

x	y
1	−4
2	−2
3	0

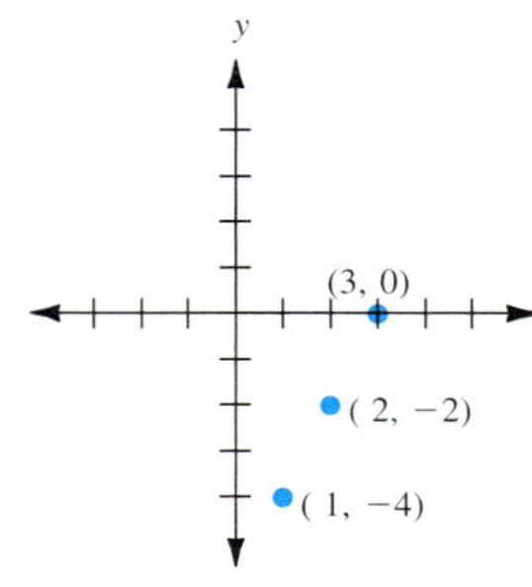

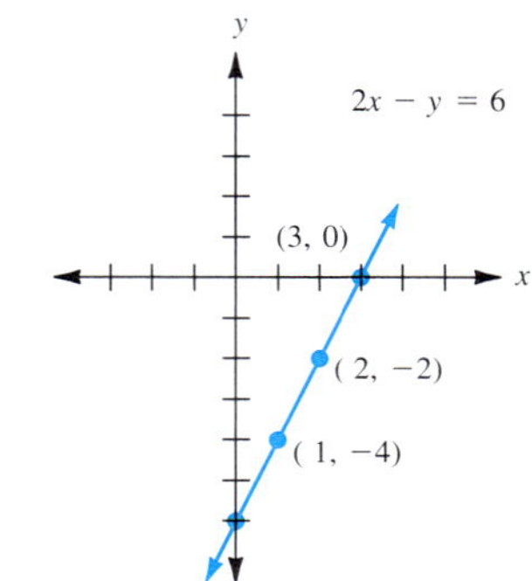

7.

x	y
0	0
1	−2
2	−4

8.

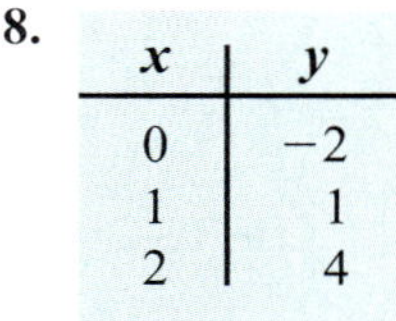

x	y
0	−2
1	1
2	4

9.

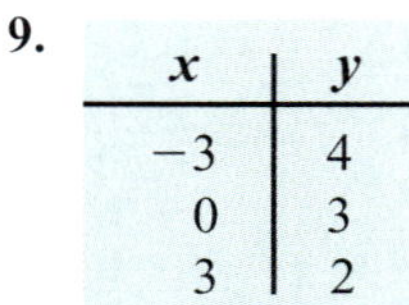

x	y
−3	4
0	3
3	2

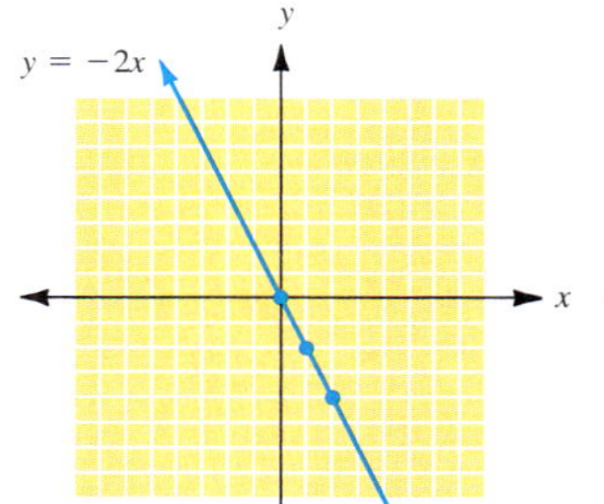

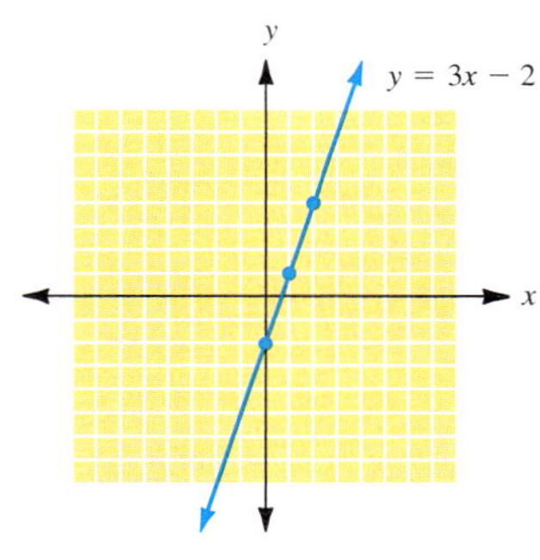

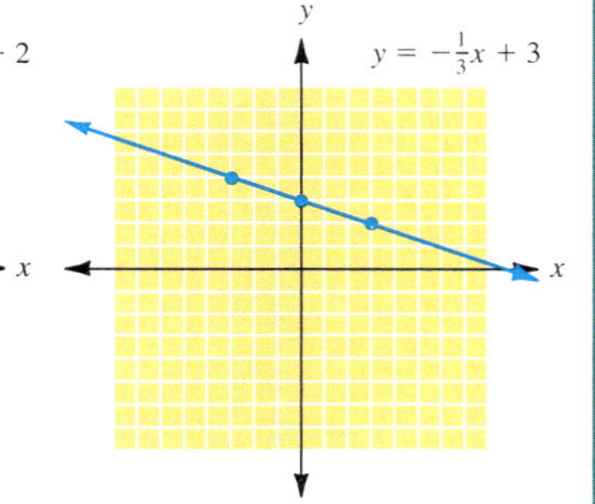

10.

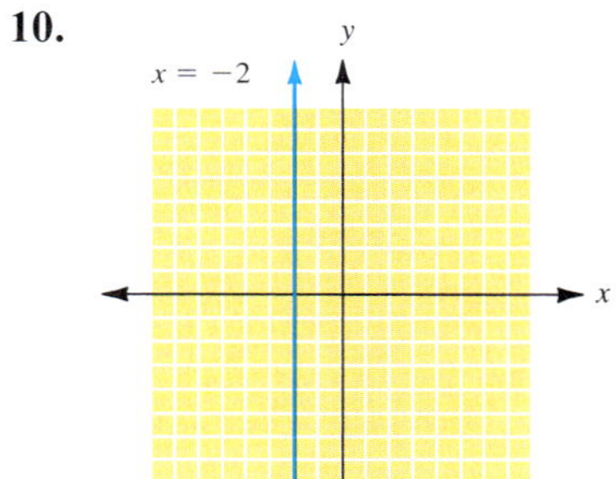

11.

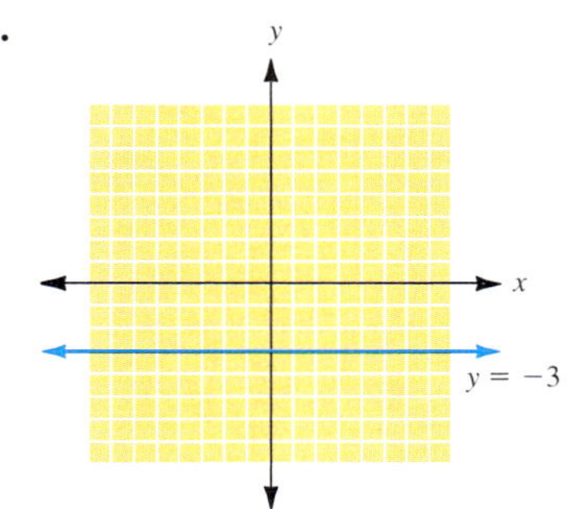

Name ______________

Section ________ Date ________

9.4 Exercises

Determine which of the ordered pairs are solutions for the given equation.

1. $x + y = 6$ (4, 2), (−2, 4), (0, 6), (−3, 9)

2. $x - y = 12$ (13, 1), (13, −1), (12, 0), (6, 6)

3. $2x - y = 8$ (5, 2), (4, 0), (0, 8), (6, 4)

4. $x + 5y = 20$ (10, −2), (10, 2), (20, 0), (25, −1)

5. $3x - 2y = 12$ $(4, 0), \left(\frac{2}{3}, -5\right), (0, 6), \left(5, \frac{3}{2}\right)$

6. $3x + 4y = 12$ $(-4, 0), \left(\frac{2}{3}, \frac{5}{2}\right), (0, 3), \left(\frac{2}{3}, 2\right)$

7. $y = 4x$ (0, 0) (1, 3), (2, 8), (8, 2)

8. $y = 2x - 1$ $(0, -2), (0, -1), \left(\frac{1}{2}, 0\right), (3, -5)$

9. $x = 3$ (3, 5), (0, 3), (3, 0), (3, 7)

10. $y = 5$ (0, 5), (3, 5), (−2, −5), (5, 5)

Complete the ordered pairs so that each is a solution for the given equation.

11. $x + y = 12$ (4,), (, 5), (0,), (, 0)

12. $x - y = 7$ (, 4), (15,), (0,), (, 0)

13. $3x + y = 9$ (3,), (, 9), (, −3), (0,)

14. $x + 5y = 20$ (0,), (, 2), (10,), (, 0)

15. $3x - 2y = 12$ (, 0), (, −6), (2,), (, 3)

16. $2x + 5y = 20$ (0,), (5,), (, 0), (, 6)

17. $y = 3x - 4$ $(0,\ ), (\ , 5), (\ , 0), \left(\frac{5}{3},\ \right)$

18. $y = -2x + 5$ $(0,\ ), (\ , 5), \left(\frac{3}{2},\ \right), (\ , 1)$

ANSWERS

1. ______
2. ______
3. ______
4. ______
5. ______
6. ______
7. ______
8. ______
9. ______
10. ______
11. ______
12. ______
13. ______
14. ______
15. ______
16. ______
17. ______
18. ______

19. ______
20. ______
21. ______
22. ______
23. ______
24. ______
25. ______
26. ______
27. ______
28. ______
29. ______
30. ______
31. ______
32. ______
33. ______
34. (a) ______
(b) ______

Find four solutions for each of the equations. **Note:** Your answers may vary from those shown in the answer section.

19. $x - y = 7$

20. $x + y = 18$

21. $x + 4y = 8$

22. $x + 3y = 12$

23. $2x - 5y = 10$

24. $2x + 7y = 14$

25. $y = 2x + 3$

26. $y = 8x - 5$

27. $x = -5$

28. $y = 8$

An equation in three variables has an ordered triple as a solution. For example, (1, 2, 2) is a solution to the equation $x + 2y - z = 3$. Complete the ordered-triple solutions for each equation.

29. $x + y + z = 0$ $(2, -3, \ )$

30. $2x + y + z = 2$ $(\ , -1, 3)$

31. $x + y + z = 0$ $(1, \ , 5)$

32. $x + y - z = 1$ $(-2, 1, \ )$

33. The number of programs for the disabled in the United States from 1993 to 1997 is approximated by the equation $y = 162x + 4365$ in which x is the number of years after 1993. Complete the table.

x	1	2	3	4	6
y					

34. Your monthly pay as a car salesman is determined using the equation $S = 200x + 1{,}500$ in which x is the number of cars you can sell each month.

(a) Complete the table.

x	12	15	17	18
S				

(b) You are offered a job at a salary of \$56,400 per year. How many cars would you have to sell per month to equal this salary?

ANSWERS

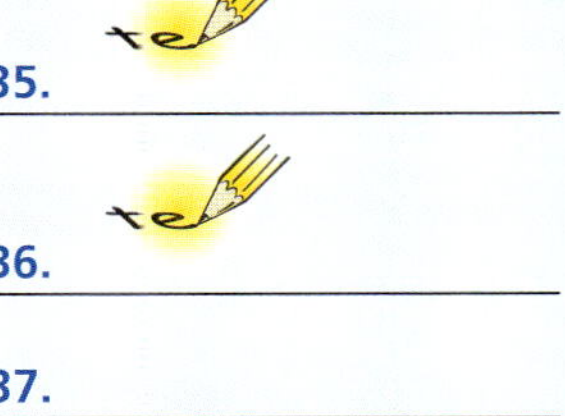

35. ______

36. ______

37. ______

38. ______

39. ______

40. ______

35. You now have had practice solving equations with one variable and equations with two variables. Compare equations with one variable to equations with two variables. How are they alike? How are they different?

36. Each of the sentences describes pairs of numbers that are related. After completing the sentences in parts (a) to (g), write two of your own sentences in (h) and (i).

(a) The *number of hours you work* determines the *amount you are* ______________.

(b) The *number of gallons of gasoline* you put in your car determines *the amount you* ______________.

(c) The *amount of the* ______________ in a restaurant is related to *the amount of the tip.*

(d) The *sales amount of a purchase in a store* determines ______________.

(e) The *age of an automobile* is related to ______________.

(f) The *amount of electricity you use in a month* determines ______________.

(g) The *cost of food for a family of four* and ______________.

Think of two more:

(h) ____.

(i) ____.

Graph each of the equations by plotting points and finding the line.

37. $x + y = 6$

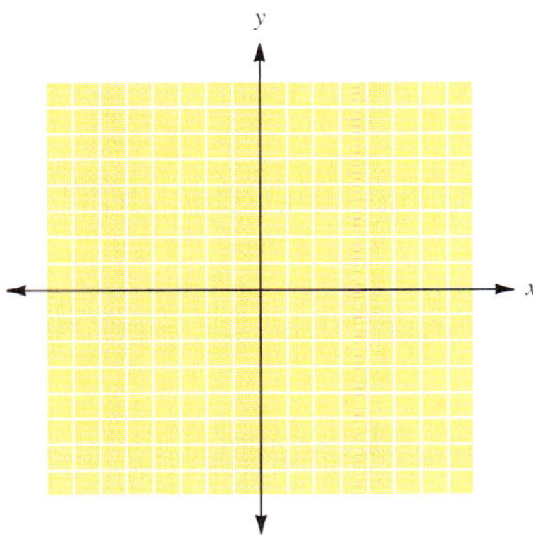

38. $x - y = 5$

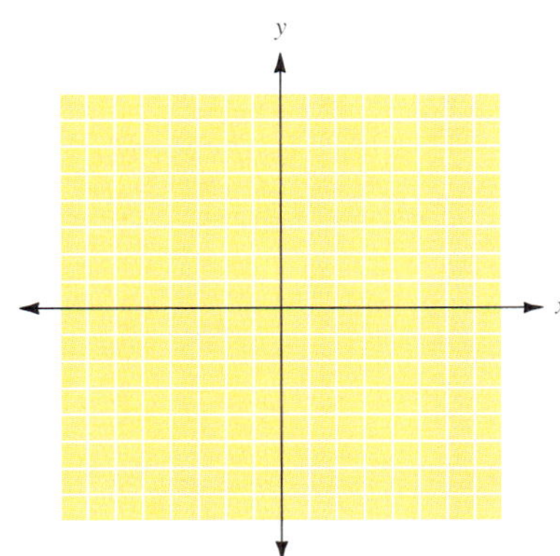

39. $x - y = -3$

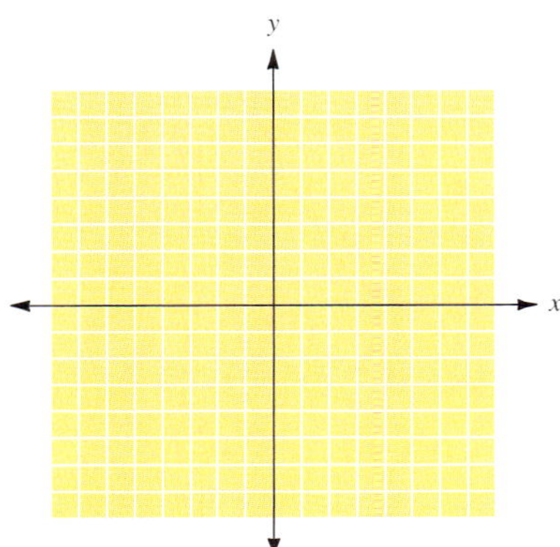

40. $x + y = -3$

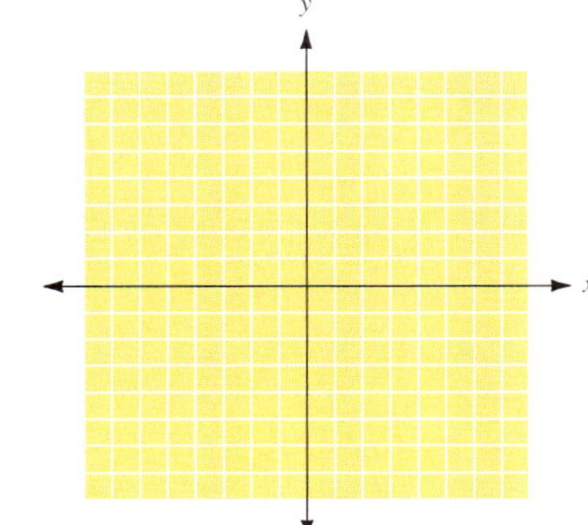

ANSWERS

41. ______

42. ______

43. ______

44. ______

45. ______

46. ______

41. $2x + y = 2$

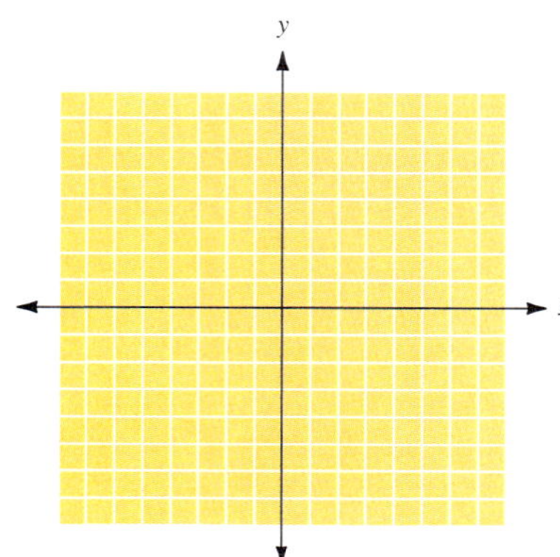

42. $x - 2y = 6$

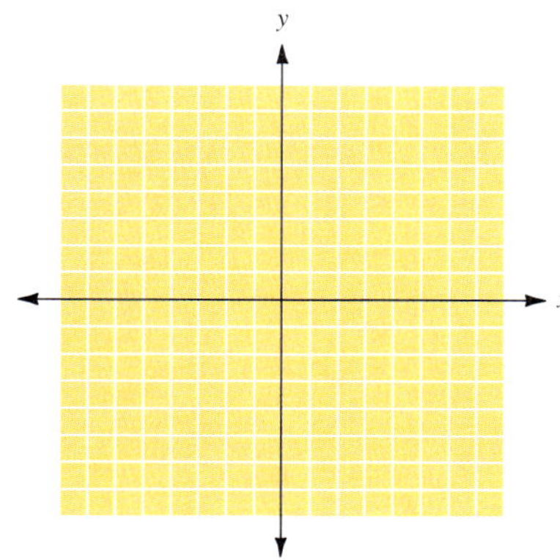

43. $3x + y = 0$

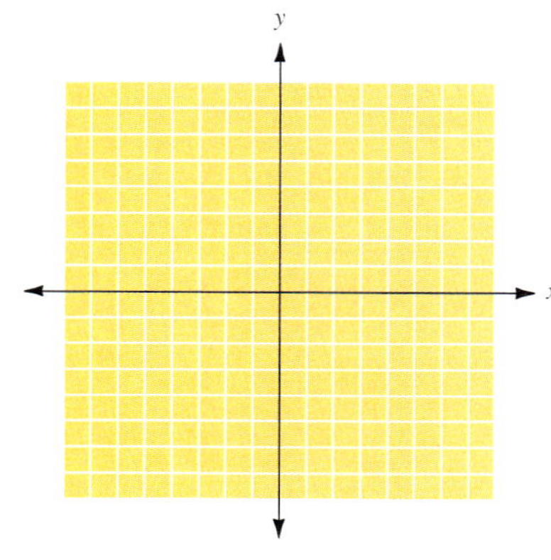

44. $2x - 3y = 6$

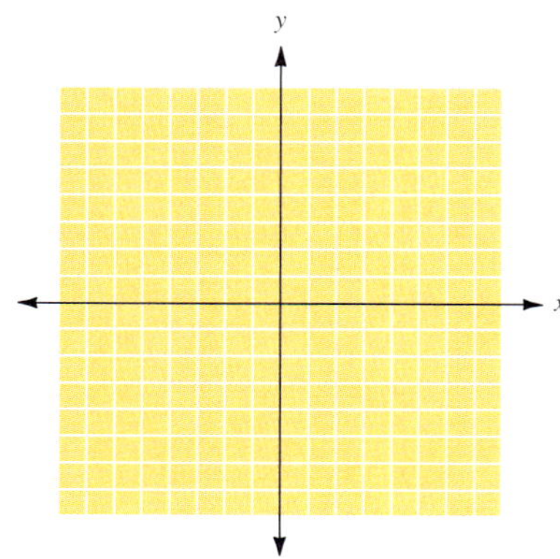

45. $y = 5x$

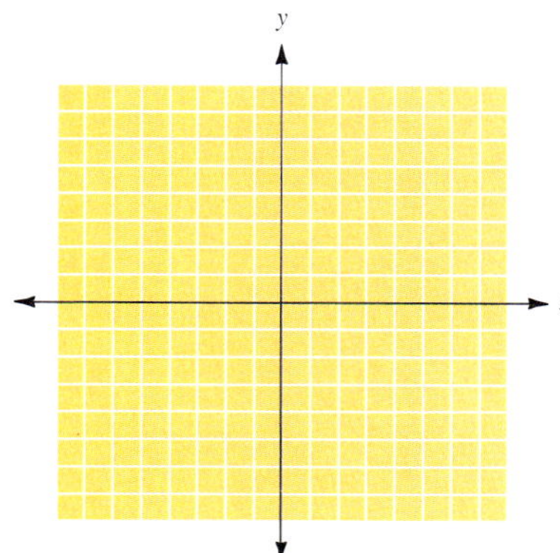

46. $y = -4x$

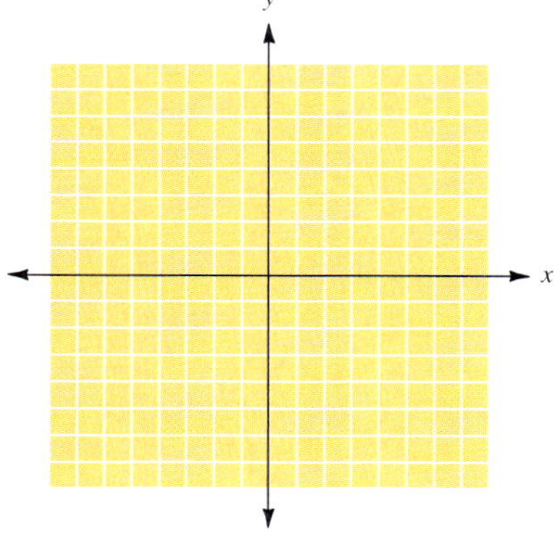

47. $y = 2x - 1$

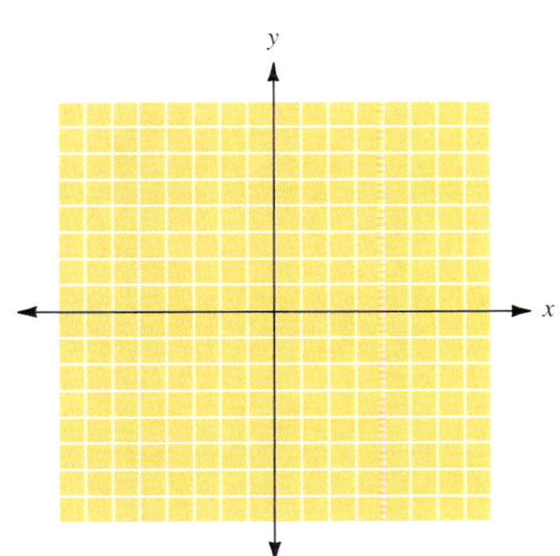

48. $y = -3x - 3$

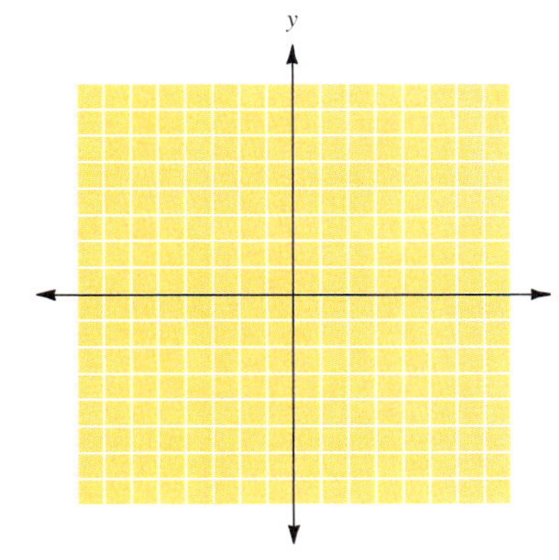

49. $y = \frac{1}{3}x$

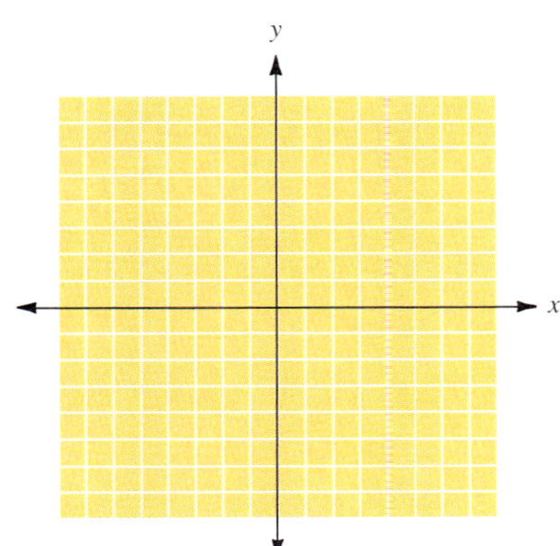

50. $y = -\frac{1}{4}x$

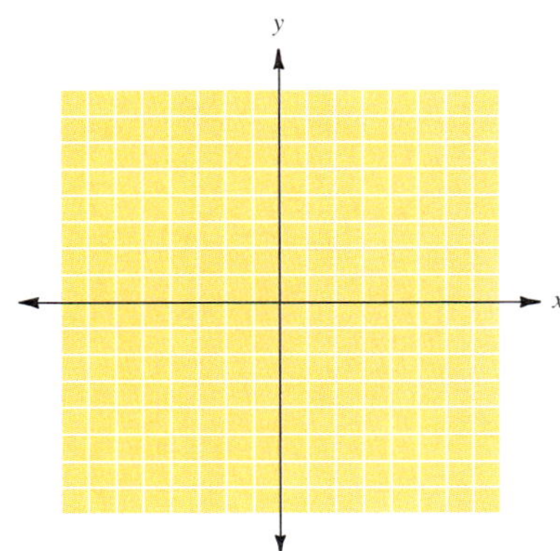

51. $y = \frac{2}{3}x - 3$

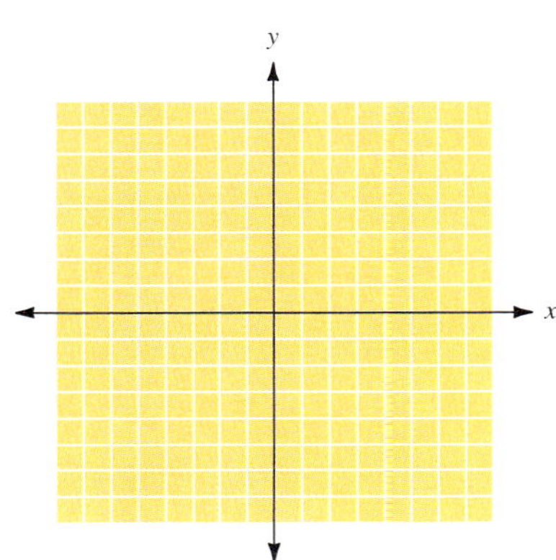

52. $y = \frac{3}{4}x + 2$

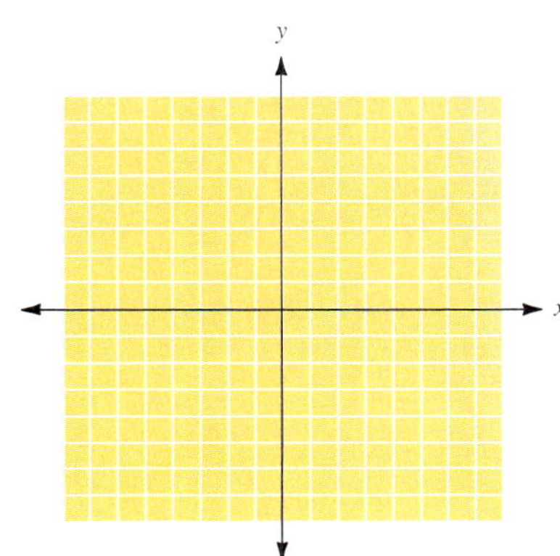

ANSWERS

47. ______________________

48. ______________________

49. ______________________

50. ______________________

51. ______________________

52. ______________________

ANSWERS

53. ______

54. ______

55. ______

56. ______

57. ______

58. ______

53. $x = 5$

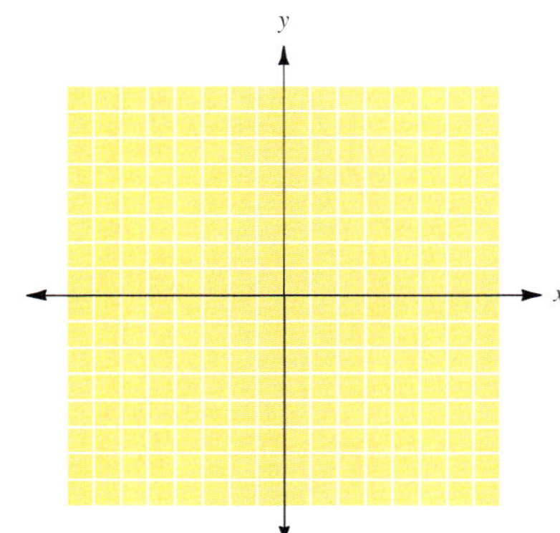

54. $y = -3$

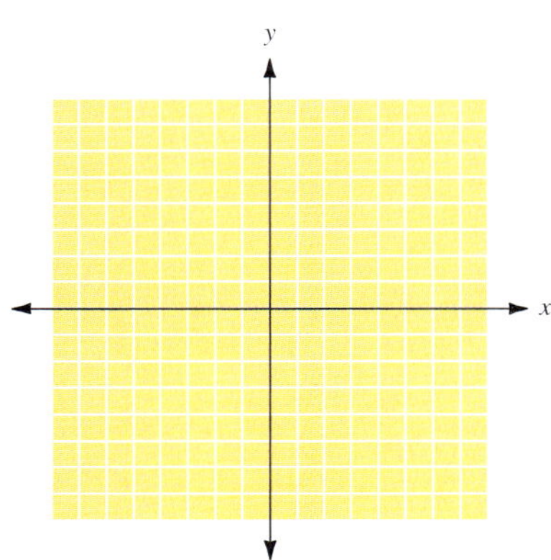

55. $y = 1$

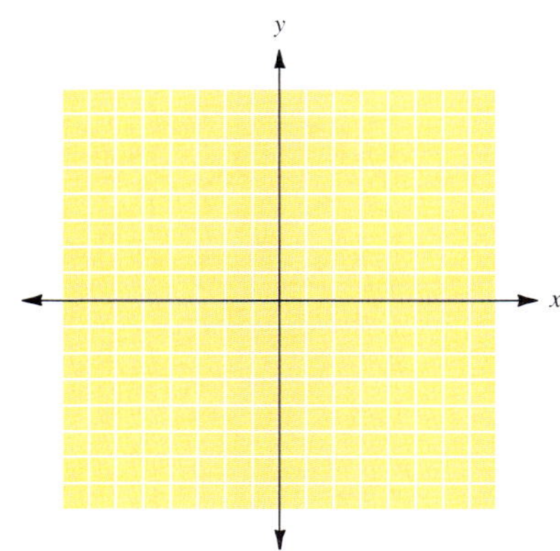

56. $x = -2$

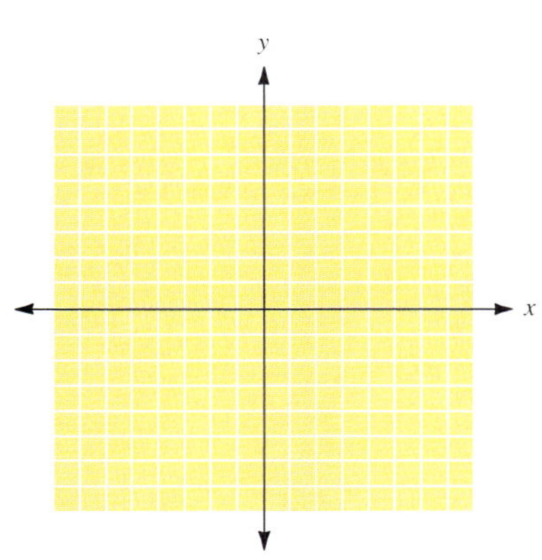

Write an equation that describes the given relationships between x and y. Then graph each relationship.

57. y is twice x.

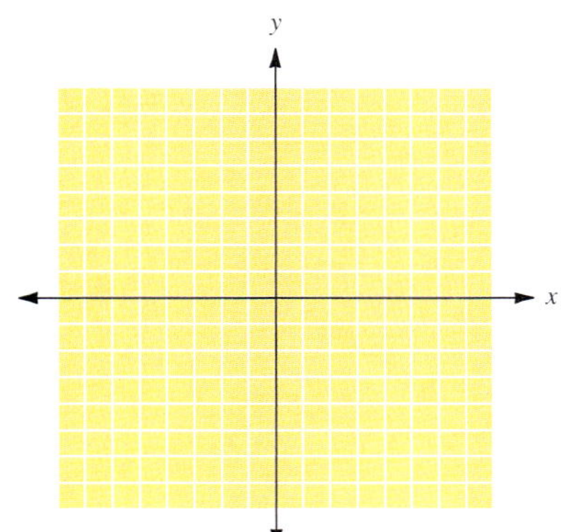

58. y is 3 times x.

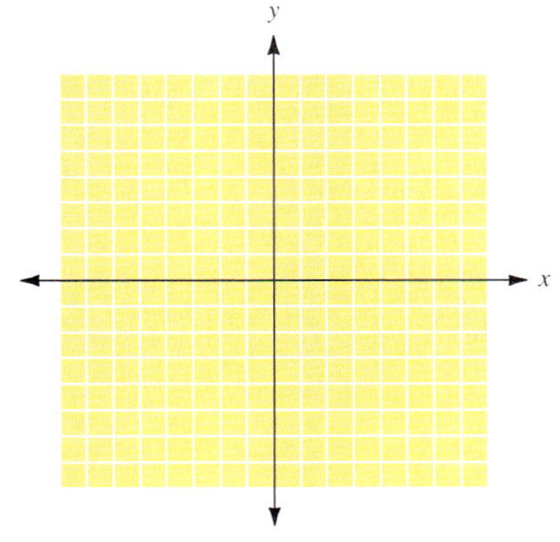

ANSWERS

59. ____________

60. ____________

61. ____________

62. ____________

63. ____________

64. ____________

59. y is 3 more than x.

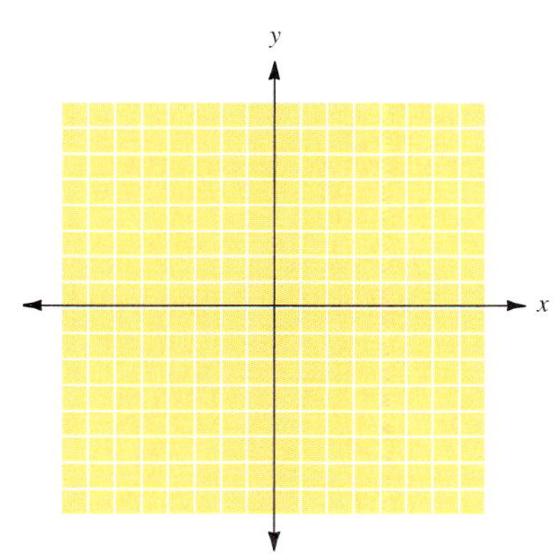

60. y is 2 less than x.

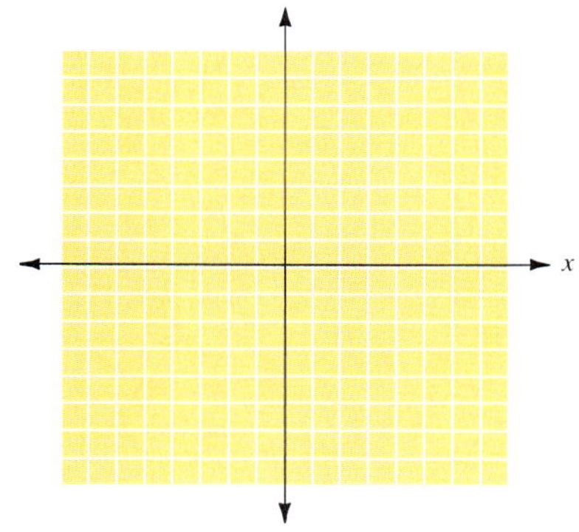

61. y is 3 less than 3 times x.

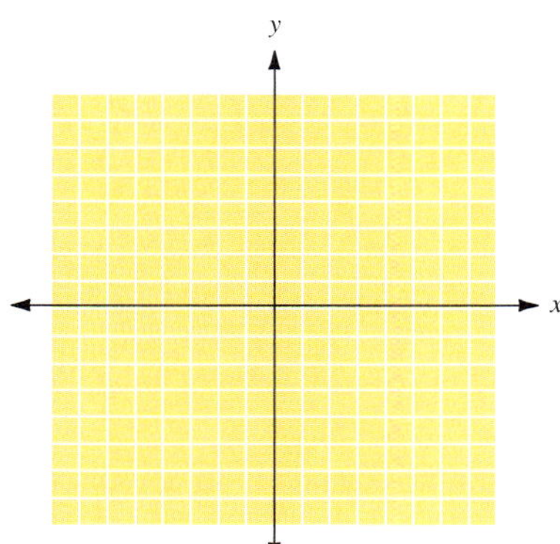

62. y is 4 more than twice x.

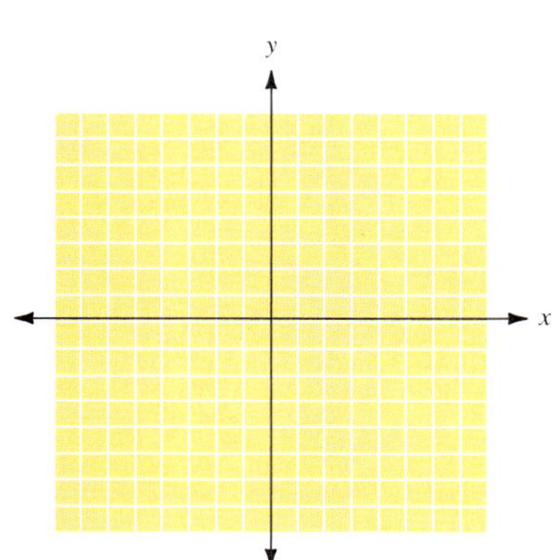

Graph each pair of equations on the same axes. Estimate the coordinates of the point where the lines intersect.

63. $x + y = 4$
$x - y = 2$

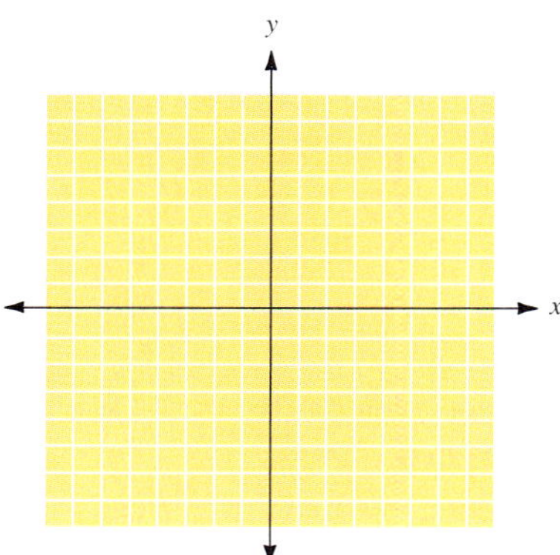

64. $x - y = 3$
$x + y = 5$

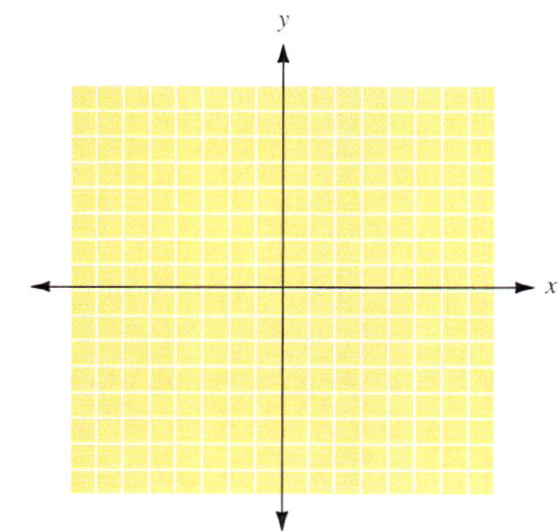

ANSWERS

65. (a) ______

(b) ______

(c) ______

(d) ______

66. (a) ______

(b) ______

(c) ______

(d) ______

65. Fundraising. A high school class wants to raise some money by recycling newspapers. They decide to rent a truck for a weekend and to collect the newspapers from homes in the neighborhood. The market price for recycled newsprint is currently \$11 per ton. The equation $y = 11x - 100$ describes the amount of money the class will make, in which y is the amount of money made in dollars, x is the number of tons of newsprint collected, and 100 is the cost in dollars to rent the truck.

(a) Using the axes below, draw a graph that represents the relationship between newsprint collected and money earned.

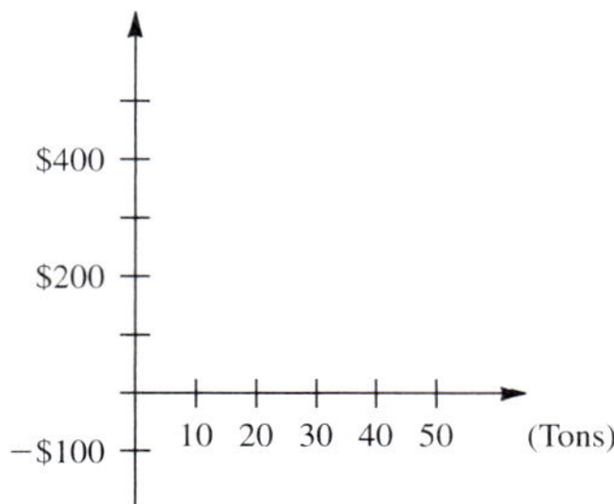

(b) The truck is costing the class \$100. How many tons of newspapers must the class collect to break even on this project?

(c) If the class members collect 16 tons of newsprint, how much money will they earn?

(d) Six months later the price of newsprint is \$17 a ton, and the cost to rent the truck has risen to \$125. Write the equation that describes the amount of money the class might make at that time.

66. Production costs. The cost of producing a number of items x is given by $C = mx + b$, in which b is the fixed cost and m is the variable cost (the cost of producing one more item).

(a) If the fixed cost is \$40 and the variable cost is \$10, write the cost equation.

(b) Graph the cost equation.

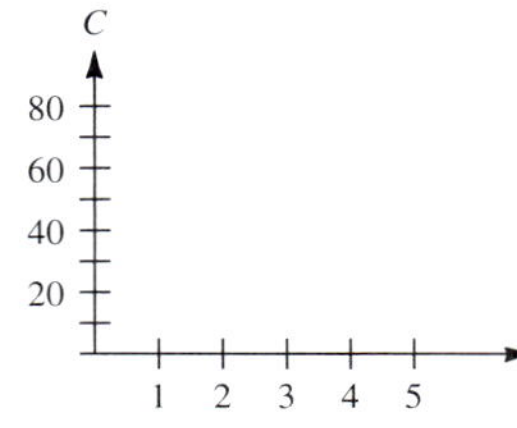

(c) The revenue generated from the sale of x items is given by $R = 50x$. Graph the revenue equation on the same set of axes as the cost equation.

(d) How many items must be produced for the revenue to equal the cost (the break-even point)?

Answers

1. (4, 2), (0, 6), (−3, 9) **3.** (5, 2), (4, 0), (6, 4) **5.** (4, 0), $\left(\frac{2}{3}, -5\right)$, $\left(5, \frac{3}{2}\right)$

7. (0, 0), (2, 8) **9.** (3, 5), (3, 0), (3, 7) **11.** 8, 7, 12, 12 **13.** 0, 0, 4, 9

15. 4, 0, −3, 6 **17.** $-4, 3, \frac{4}{3}, 1$ **19.** (0, −7), (2, −5), (4, −3), (6, −1)

21. (8, 0), (−4, 3), (0, 2), (4, 1) **23.** (−5, −4), (0, −2), (5, 0), (10, 2)

25. (0, 3), (1, 5), (2, 7), (3, 9) **27.** (−5, 0), (−5, 1), (−5, 2), (−5, 3)

29. (2, −3, 1) **31.** (1, −6, 5) **33.** 4527, 4689, 4851, 5013, 5337

35. 

37. $x + y = 6$

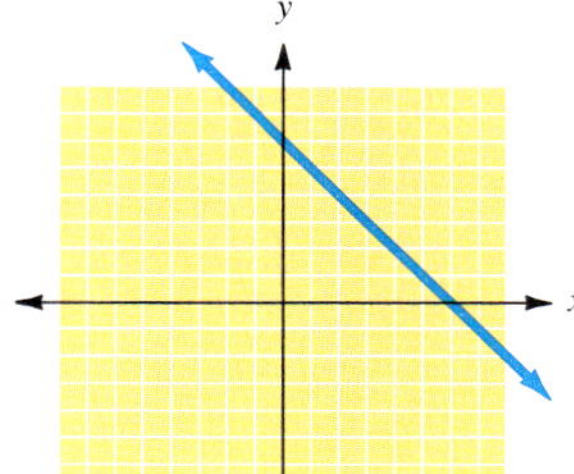

39. $x - y = -3$

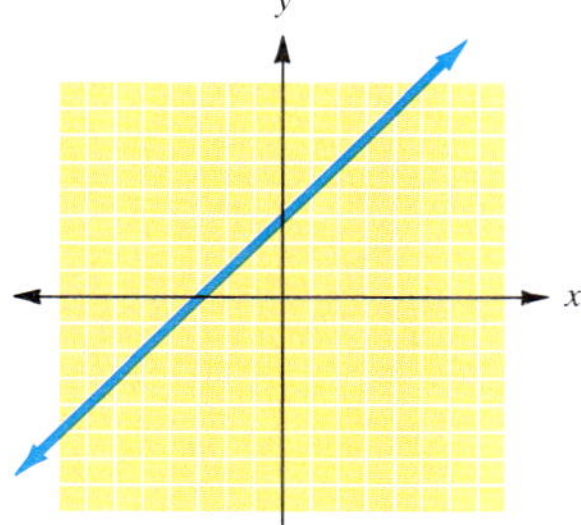

41. $2x + y = 2$

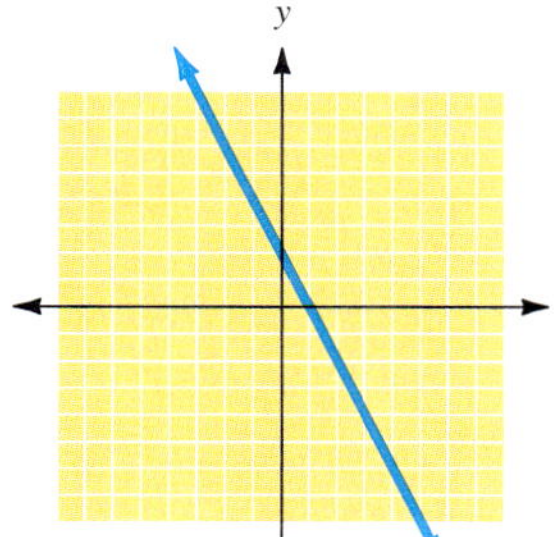

43. $3x + y = 0$

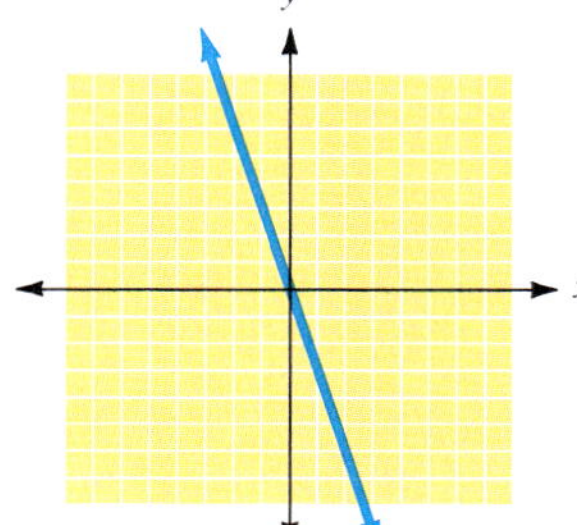

45. $y = 5x$

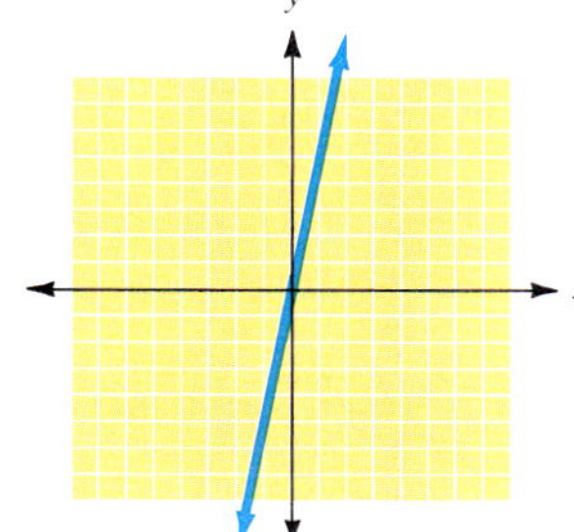

47. $y = 2x - 1$

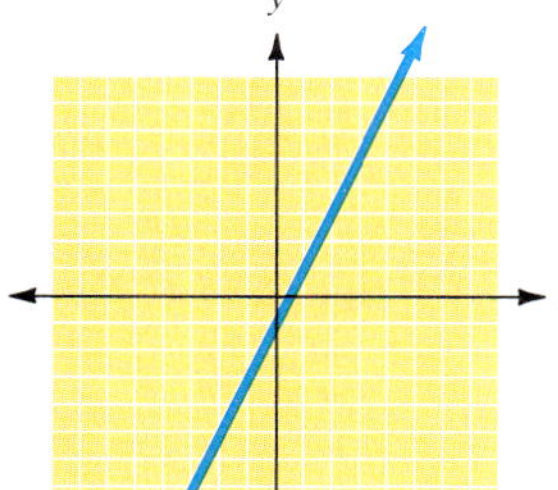

49. $y = \frac{1}{3}x$

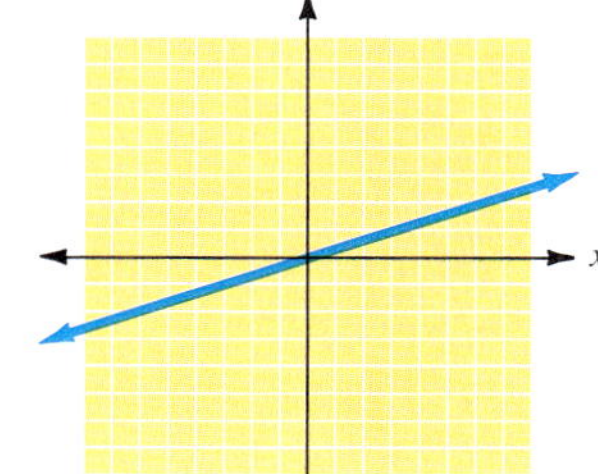

51. $y = \frac{2}{3}x - 3$

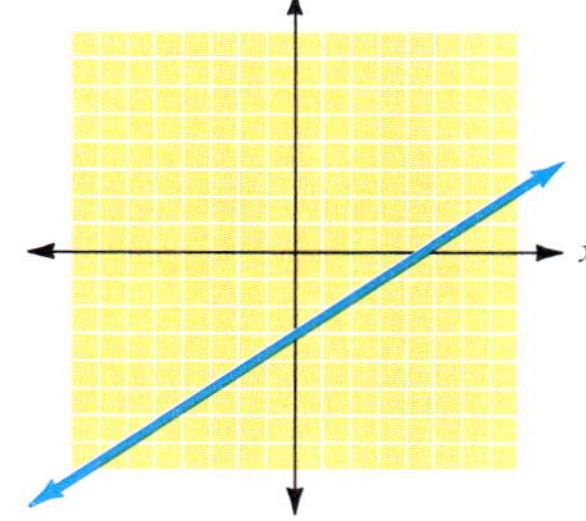

53. $x = 5$

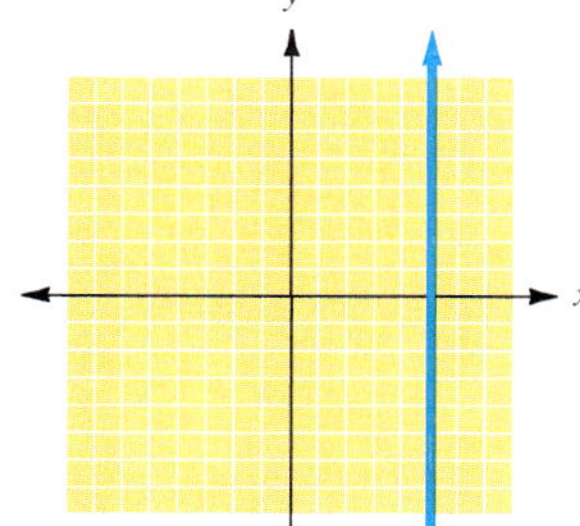

55. $y = 1$

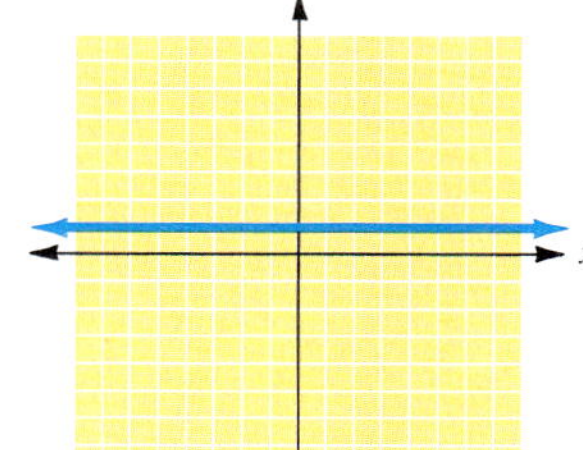

57. $y = 2x$

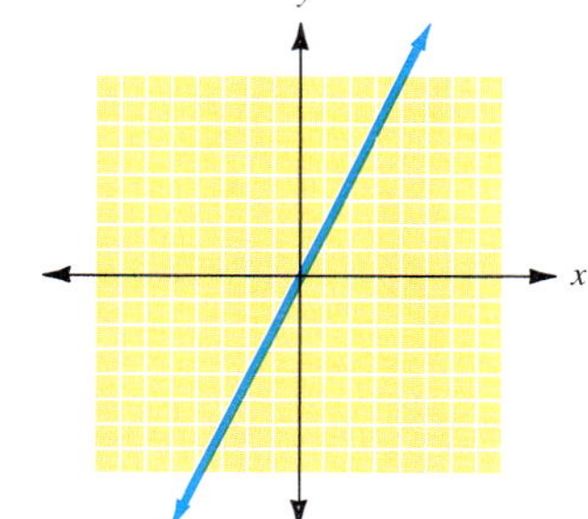

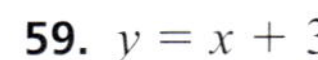

59. $y = x + 3$

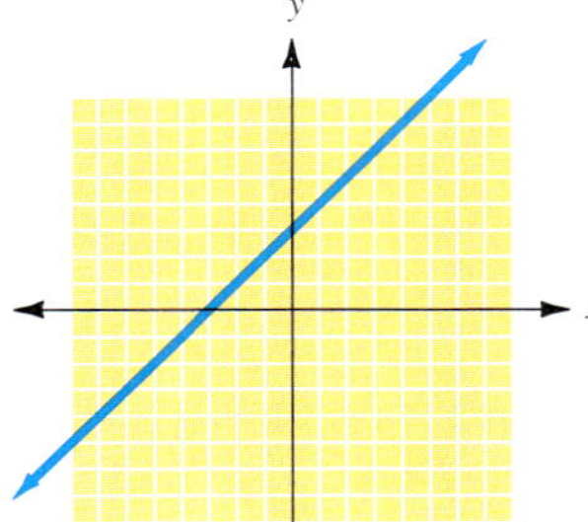

61. $y = 3x - 3$

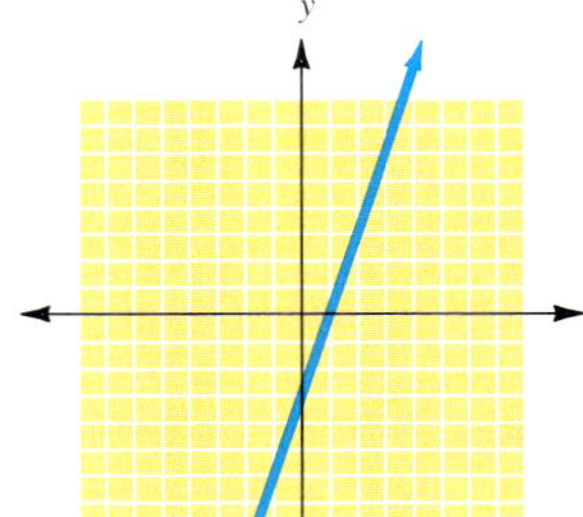

63. $(3, 1)$

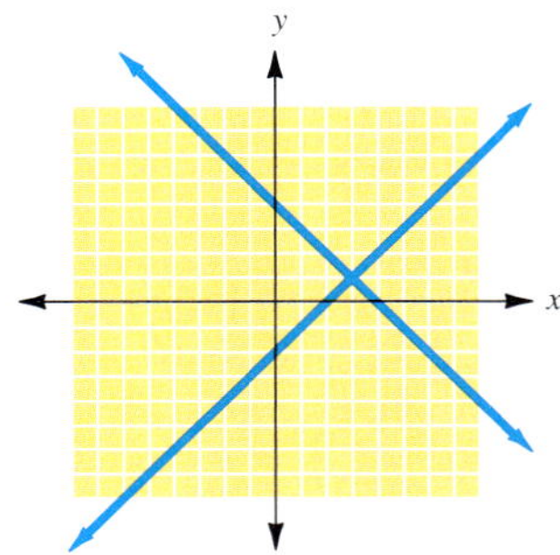

65. **(a)** Graph; **(b)** $\frac{100}{11}$ or ≈ 9 tons; **(c)** \$76; **(d)** $y = 17x - 125$

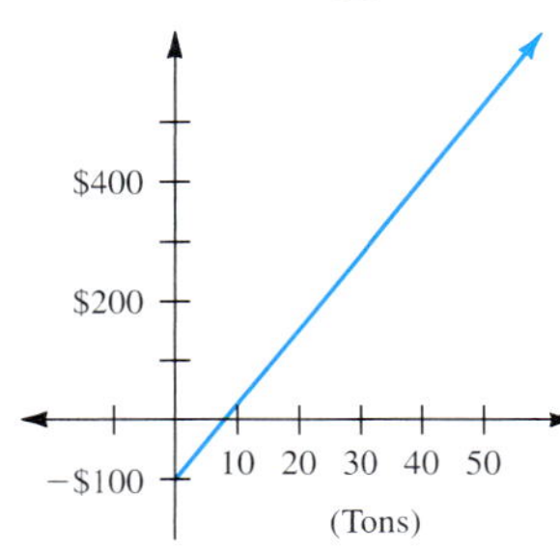

Mean, Median, and Mode

OBJECTIVES

1. Calculate the mean
2. Interpret the mean
3. Find the median
4. Interpret the median
5. Find a mode

A very useful concept is the **average** of a group of numbers. An average is a number that is typical of a larger group of numbers. In mathematics, we have several different kinds of averages that we can use to represent a larger group of numbers. The first of these is the **mean.**

Step by Step: Finding the Mean

To find the mean for a group of numbers, follow these two steps:

Step 1 Add all of the numbers in the group.

Step 2 Divide that sum by the number of items in the group.

Example 1

Finding the Mean

Find the mean of the group of numbers 12, 19, 15, and 14.

Step 1 Add all of the numbers.

$12 + 19 + 15 + 14 = 60$

Step 2 Divide that sum by the number of items.

$60 \div 4 = 15$ There are four items in this group.

The mean of this group of numbers is 15.

CHECK YOURSELF 1

Find the mean of the group of numbers 17, 24, 19, *and* 20.

We will apply the mean to a word problem.

Example 2

Finding the Mean

The ticket prices (in dollars) for the nine concerts held at the Civic Arena this school year were

33, 31, 30, 59, 30, 35, 32, 36, 56

What was the mean price for these tickets?

Step 1 Add all the numbers.

$33 + 31 + 30 + 59 + 30 + 35 + 32 + 36 + 56 = 342$

NOTE Divide by 9 because there are 9 ticket prices.

Step 2 Divide by 9.

$342 \div 9 = 38$

The mean ticket price was $38.

CHECK YOURSELF 2

The costs (in dollars) of the six textbooks that Aaron needs for the fall quarter are

75, 69, 57, 87, 76, 80

Find the mean cost of these books.

Although the mean is probably the most common way to find an average for a group of numbers, it is not always the most representative. Another kind of average is called the **median.**

Step by Step: Finding the Median

The *median* is the number for which there are as many instances that are above that number as there are instances below it. To find the median, follow these steps:

Step 1 Rewrite the numbers in order from smallest to largest.

Step 2 Count from both ends to find the number in the middle.

Step 3 If there are two numbers in the middle, add them together and find their mean.

Example 3

Finding the Median

Find the median for the following groups of numbers.

(a) 35, 18, 27, 38, 19, 63, 22

Step 1 Rewrite the numbers in order from smallest to largest.

18, 19, 22, 27, 35, 38, 63

Step 2 Count from both ends to find the number in the middle.

Counting from both ends, we find that 27 is the median. There are three numbers above 27 and three numbers below it.

(b) 29, 88, 74, 81, 62, 37

Step 1 Rewrite the numbers in order from smallest to largest.

29, 37, 62, 74, 81, 88

Step 2 Count from both ends to find the number in the middle.

Counting from both ends, we find that there are two numbers in the middle, 62 and 74. We go on to step 3.

Step 3 If there are two numbers in the middle, find their mean.

$(62 + 74) \div 2 = 136 \div 2 = 68$

CHECK YOURSELF 3

Find the median for each group of numbers.

(a) 8, 6, 19, 4, 21, 5, 27 **(b)** 43, 29, 13, 37, 29, 53

There are times in which the median is a better representative of a group of numbers than the mean is. Example 4 illustrates such a case.

Example 4

Comparing the Mean and the Median

The given numbers represent the hourly wage of seven employees of a local chip manufacturing plant.

12, 11, 14, 16, 32, 13, 14

(a) Find the mean hourly wage.

Step 1 Add all of the numbers in the group.

$12 + 11 + 14 + 16 + 32 + 13 + 14 = 112$

Step 2 Divide that sum by the number of items in the group.

$112 \div 7 = 16$

The mean wage is $16 an hour.

(b) Find the median wage for the seven workers.

Step 1 Rewrite the numbers in order from smallest to largest.

11, 12, 13, 14, 14, 16, 32

Step 2 Count from both ends to find the number in the middle.

The middle number is 14. There are three numbers above it and three numbers below it. The median salary is $14 per hour. Which salary do you think is more typical of the workers? Why?

CHECK YOURSELF 4

The following are Jessica's phone bills for each month of 2000.

26, 67, 31, 24, 15, 17, 41, 27, 17, 22, 26, 47

(a) Find the mean amount of her phone bills.
(b) Find the median amount of her phone bills.

Another measure used as an "average" is called the **mode.**

Definitions: Mode

The *mode* of a set of data is the item or number that appears most frequently.

NOTE A set with two different modes is called **bimodal.**

Example 5

Finding a Mode

Find the mode for the set of numbers given.

22, 24, 24, 24, 24, 27, 28, 32, 32

The mode, 24, is the number that appears most frequently.

CHECK YOURSELF 5

Find the mode for the set of numbers given.

7, 7, 7, 9, 11, 13, 13, 15, 15, 15, 15, 21

One advantage of the mode is that it can be used with data that are not a set of numbers.

Example 6

Finding a Mode

NOTE If each eye color appeared three times, there would be no mode! Not every data set has a mode.

Following are the eye colors from a class of 12 students. Which color is the mode?

blue, brown, hazel, blue, brown, brown, brown, brown, blue, brown, hazel, green

Because brown occurs most frequently, it is the mode.

CHECK YOURSELF 6

The given types of computers were available in the lab. Which type was the mode?

Apple, IBM, Compaq, Dell, Apple, IBM, Apple, Compaq, Dell, Apple, IBM, Apple, Dell, Apple, Compaq

Using Your Calculator to Find an Average

Most electronic calculators have statistical functions that allow you to calculate means, medians, and other statistical values. Because these features vary so much from one calculator to another, you will need to consult your owner's manual to learn how to access these features.

In this section, we will focus on using a calculator to compute the mean.

Example 7

Calculating a Mean

Find the mean for the set of numbers.

45, 48, 53, 59, 67, 76

When entering these numbers in your calculator, keep in mind the order of operations. There are two different techniques you may use.

Method 1

Add the numbers

$45 + 48 + 53 + 59 + 67 + 76 = 348$

then divide the sum by 6 (there are six numbers)

$348 \div 6 = 58$

Method 2

Use parentheses to find the mean.

$(45 + 48 + 53 + 59 + 67 + 76) \div 6 = 58$

CHECK YOURSELF 7

Find the mean for the set of numbers.

132, 144, 156, 158, 279, 337

Unfortunately, not all calculations result in answers that are whole numbers.

Example 8

Calculating a Mean

Find the mean of the set of numbers below.

13, 15, 16, 19, 26, 28, 38

Entering this as a single expression, we have

$(13 + 15 + 16 + 19 + 26 + 28 + 38) \div 7 \approx 22$

Your calculator is probably displaying something like 22.14285714. Remember that this is just another (although closer) approximation.

We could approximate the answer as 22, as we've indicated. We could also subtract 22 from the calculator display, leaving only the calculator's internal decimal approximation, which is approximately

0.142857142

If we multiply this by the divisor, 7, we will get the remainder, which is 1. The exact answer is

$22\frac{1}{7}$

CHECK YOURSELF 8

Find an approximate and exact mean for the set of numbers.

17, 23, 33

CHECK YOURSELF ANSWERS

1. 20 **2.** \$74 **3.** **(a)** 8; **(b)** 33 **4.** **(a)** \$30; **(b)** \$26 **5.** 15 **6.** Apple **7.** 201 **8.** Approximately 24, exactly $24\frac{1}{3}$

Name ______________________

Section ________ Date ________

9.5 Exercises

Find the mean for each set of numbers.

1. 6, 9, 10, 8, 12

2. 13, 15, 17, 17, 18

3. 13, 15, 17, 19, 24, 26

4. 41, 43, 56, 67, 69, 72

5. 12, 14, 15, 16, 16, 16, 17, 22, 25, 27

6. 21, 25, 27, 32, 36, 37, 43, 44, 44, 51

7. 5, 8, 9, 11, 12

8. 7, 18, 11, 7, 12

9. 9, 8, 11, 14, 8

10. 21, 23, 25, 27, 22, 20

Find the median for each set of numbers.

11. 2, 3, 5, 6, 10

12. 12, 13, 15, 17, 18

13. 23, 24, 27, 31, 36, 38, 41

14. 1, 4, 9, 16, 25, 36, 49

15. 46, 13, 47, 25, 68, 51, 71

16. 26, 71, 33, 69, 71, 25, 75

Find the mode for each set of numbers.

17. 17, 13, 16, 18, 17

18. 41, 43, 56, 67, 69, 72

19. 21, 44, 25, 27, 32, 36, 37, 44

20. 9, 8, 10, 9, 9, 10, 8

21. 12, 13, 7, 14, 4, 11, 9

22. 8, 2, 3, 3, 4, 9, 9, 3

Solve the applications.

23. Temperature. High temperatures (in °F) of 86°, 91°, 92°, 103°, and 98° were recorded for the first 5 days of July. What was the mean high temperature?

ANSWERS

1. ______
2. ______
3. ______
4. ______
5. ______
6. ______
7. ______
8. ______
9. ______
10. ______
11. ______
12. ______
13. ______
14. ______
15. ______
16. ______
17. ______
18. ______
19. ______
20. ______
21. ______
22. ______
23. ______

ANSWERS

24. ______

25. ______

26. ______

27. ______

28. ______

29. ______

30. ______

24. Travel. A salesperson drove 238, 159, 87, 163, and 198 mi on a 5-day trip. What was the mean number of miles driven per day?

25. Mileage rating. Highway mileage ratings for seven new diesel cars were 43, 29, 51, 36, 33, 42, and 32 $\frac{\text{mi}}{\text{gal}}$. What was the mean rating?

26. Enrollments. The enrollments in the four elementary schools of a district are 278, 153, 215, and 198 students. What is the mean enrollment?

27. Test scores. To get an A in history, you must have a mean of 90 on five tests. Your scores thus far are 83, 93, 88, and 91. How many points must you have on the final test to receive an A? (*Hint:* First find the total number of points you need to get an A.)

28. Test scores. To pass biology, you must have a mean of 70 on six quizzes. So far your scores have been 65, 78, 72, 66, and 71. How many points must you have on the final quiz to pass biology?

29. Test scores. Louis had scores of 87, 82, 93, 89, and 84 on five tests. Tamika had scores of 92, 83, 89, 94, and 87 on the same five tests. Who had the higher mean score? By how much?

30. Heating bills. The Wong family had heating bills of $105, $110, $90, and $67 in the first 4 months of 1999. The bills for the same months of 2000 were $110, $95, $75, and $76. In which year was the mean monthly bill higher? By how much?

ANSWERS

31. ____

32. ____

33. ____

34. ____

35. ____

36. ____

37. ____

38. ____

Monthly energy use, in kilowatt-hours (kWh), by appliance type for four typical U.S. families is shown in the table.

	Wong Family	McCarthy Family	Abramowitz Family	Gregg Family
Electric range	97	115	80	96
Electric heat	1,200	1,086	1,103	975
Water heater	407	386	368	423
Refrigerator	127	154	98	121
Lights	75	99	108	94
Air conditioner	123	117	96	120
Color TV	39	45	21	47

31. **Heating.** What is the mean number of kilowatt-hours used each month by the four families for heating their homes?

32. **Heating.** What is the mean number of kilowatt-hours used each month by the four families for hot water?

33. **Heating.** What is the mean number of kilowatt-hours used per appliance by the McCarthy family?

34. **Heating.** What is the mean number of kilowatt-hours used per appliance by the Gregg family?

Use your calculator for exercises 35 and 36.

35. **Utility bills.** Fred kept the following records of his utility bills for 12 months: $53, $51, $43, $37, $32, $29, $34, $41, $58, $55, $49, and $58. What was the mean monthly bill?

36. **Test scores.** The following scores were recorded on a 200-point final examination: 193, 185, 163, 186, 192, 135, 158, 174, 188, 172, 168, 183, 195, 165, 183. What was the mean of the scores?

37. The following are eye colors from a class of eight students. Which color is the mode?

    Hazel, green, brown, brown, blue, green, hazel, green

38. The weather in Philadelphia over the last 7 days was as follows:

    Rain, sunny, cloudy, rain, sunny, rain, rain. What type of weather was the mode?

ANSWERS

39. ____________

40. ____________

41. ____________

42. ____________

43. ____________

44. ____________

45. ____________

46. ____________

47. ____________

48. ____________

49. ____________

50. ____________

51. ____________

52. ____________

39. List the advantages and disadvantages of the mean, median, and mode.

40. In a certain math class, you take four tests and the final, which counts as two tests. Your grade is the average of the six test scores. At the end of the course, you compute both the mean and the median.

(a) You want to convince the teacher to use the mean to compute your average. Write a note to your teacher explaining why this is a better choice. Choose numbers that make a convincing argument.

(b) You want to convince the teacher to use the median to compute your average. Write a note to your teacher explaining why this is a better choice. Choose numbers that make a convincing argument.

41. Create a set of five numbers such that the mean is equal to the median.

42. Create a set of five numbers such that the mean is greater than the median.

43. Create a set of five numbers such that the mean is less than the median.

Calculator Exercises

In exercises 44 to 52, find the mean for each set of numbers.

44. 20, 18, 17, 24, 22, 19

45. 108, 113, 109, 113, 110, 101, 112, 114

46. 211, 213, 215, 208, 209, 220, 215, 221

47. 1,560, 1,540, 1,570, 1,555, 1,565, 1,545, 1,557

48. 346, 351, 353, 347, 341, 382, 373, 363

49. 2,357, 2,361, 2,372, 2,371, 2,357, 2,375, 2,364, 2,371

50. 16,430, 16,211, 16,149, 16,232, 16,317, 16,113

51. 24,637, 24,251, 24,454, 24,580, 24,324, 24,478

52. 311,431, 286,356, 356,090, 292,007, 301,857, 299,005

In exercises 53 to 58, find the approximate and exact mean for each set of numbers.

53. 18, 21, 20, 22

54. 36, 41, 43, 39, 40, 37, 39

55. 125, 121, 129, 126, 128, 123

56. 356, 371, 366, 373, 359, 363

57. 1,898, 1,913, 1,875, 1,937

58. 15,865, 16,270, 16,090, 15,904

53. __________
54. __________
55. __________
56. __________
57. __________
58. __________
59. __________
60. __________

Solve the applications.

59. The revenue for the leading apparel companies in the United States in 1997 is given in the table.

Company	Revenue (in millions)
Nike	\$9,187
Vanity Fair	\$5,222
Liz Claiborne	\$2,413
Reebok	\$3,637
Fruit of the Loom	\$2,140
Nine West	\$1,865
Kellwood	\$1,521
Warmaio	\$1,437
Jones Apparel	\$1,387

What is the mean revenue taken in by these companies?

60.

Unemployment in the United States (in thousands)

Year	Employed	Unemployed
1989	117,342	6,528
1990	118,793	7,047
1991	117,718	8,628
1992	118,482	9,613
1993	120,259	8,940
1994	123,060	7,996
1995	124,900	7,404
1996	126,708	7,236
1997	129,558	6,739

Find the mean number of employed and unemployed per year from 1989 to 1997.

ANSWERS

61. ____________

62. ____________

The work stoppages (strikes and lockouts) in the United States from 1990 to 1997 are given in the table.

Year	No. of Stoppages	Work Days Idle
1990	185	5,926
1991	392	4,584
1992	364	3,989
1993	182	3,981
1994	322	5,020
1995	192	5,771
1996	273	4,889
1997	339	4,493

61. Find the mean number of work stoppages per year from 1990 to 1997.

62. Find the mean number of work days idle from 1990 to 1997.

Answers

1. 9 **3.** 19 **5.** 18 **7.** 9 **9.** 10 **11.** 5 **13.** 31 **15.** 47 **17.** 17 **19.** 44 **21.** No mode **23.** 94° **25.** 38 $\frac{\text{mi}}{\text{gal}}$ **27.** 95 points **29.** Louis's mean score was 87, Tamika's was 89. Tamika's average score was 2 points higher than Louis's **31.** 1,091 kWh **33.** 286 kWh **35.** \$45 **37.** Green **39.** **41.** **43.**

45. 110 **47.** 1,556 **49.** 2,366 **51.** 24,454 **53.** 20; $20\frac{1}{4}$ **55.** 125; $125\frac{1}{3}$ **57.** 1,906; $1,905\frac{3}{4}$ **59.** \$3,201,000,000 **61.** 281

Summary

Definition/Procedure	**Example**	**Reference**
Circle Graphs		**Section 9.1**
Circle graph Circle graphs are graphs that show the component parts of a whole.	30% of 360° = $0.30 \cdot 360 = 108°$ 20% of 360° = $0.20 \cdot 360 = 72°$	p. 645
Pictographs, Bar Graphs, and Line Graphs		**Section 9.2**
A **table** is a display of information in parallel columns or rows.		p. 651
A **graph** is a diagram that relates two different pieces of information. One of the most common graphs is the **bar graph.**		p. 653
Line graph In line graphs, one of the axes is usually related to time.		p. 656
The Rectangular Coordinate System		**Section 9.3**
The Rectangular Coordinate System The rectangular coordinate system is a system formed by two perpendicular axes that intersect at a point called the **origin.** The horizontal line is called the ***x*-axis.** The vertical line is called the ***y*-axis.**		p. 671
Graphing Points from Ordered Pairs The coordinates of an ordered pair allow you to associate a point in the plane with every ordered pair. To graph a point in the plane, **Step 1** Start at the origin. **Step 2** Move right or left according to the value of the x-coordinate: to the right if x is positive or to the left if x is negative. **Step 3** Then move up or down according to the value of the y-coordinate: up if y is positive and down if y is negative.	To graph the point corresponding to (2, 3):	p. 674

Continued

DEFINITION/PROCEDURE	EXAMPLE	REFERENCE
Linear Equations in Two Variables		**Section 9.4**
Solutions of Linear Equations A pair of values that satisfies the equation. Solutions for linear equations in two variables are written as *ordered pairs.* An ordered pair has the form (x, y) x-coordinate, y-coordinate	If $2x - y = 10$, $(6, 2)$ is a solution for the equation, because substituting 6 for x and 2 for y gives a true statement.	p. 685
Linear Equation An equation that can be written in the form $Ax + By = C$ in which A and B are not both 0.	$2x - 3y = 4$ is a linear equation.	p. 691
Graphing Linear Equations **Step 1** Find at least three solutions for the equation, and put your results in tabular form. **Step 2** Graph the solutions found in step 1. **Step 3** Draw a straight line through the points determined in step 2 to form the graph of the equation.		p. 691
Mean, Median, and Mode		**Section 9.5**
Finding the Mean To find the **mean** for a group of numbers follow these two steps: **Step 1** Add all the numbers in the group. **Step 2** Divide that sum by the number of items in the group.	Given the numbers 4, 8, 17, 23 $4 + 8 + 17 + 23 = 52$ $\text{Mean} = \frac{52}{4} = 13$	p. 709
Finding the Median The **median** is the number for which there are as many instances that are above that number as there are instances below it. To find the median follow these steps: **Step 1** Rewrite the numbers in order from smallest to largest. **Step 2** Count from both ends to find the number in the middle. **Step 3** If there are two numbers in the middle, add them together and find their mean.	Given the numbers 9, 2, 5, 13, 7, 3 Rewrite them 2, 3, 5, 7, 9, 13 The middle numbers are 5 and 7 $\frac{5 + 7}{2} = \frac{12}{2} = 6$ Median = 6	p. 710
Finding the Mode The **mode** is the number that occurs most frequently in a set of numbers.	Given the numbers 2, 3, 3, 3, 5, 5, 7, 7, 9, 11. 3 is the mode.	p. 712

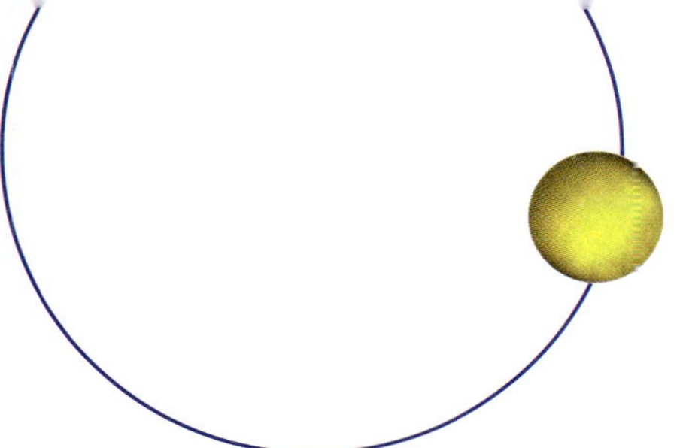

Summary and Review Exercises

This exercise set will give you practice with each of the objectives of the chapter. Each exercise is keyed to the appropriate chapter section. The answers are provided in the *Instructor's Manual*. Your instructor will give you guidelines on how to best use these exercises in your instructional setting.

[9.1] The table gives the production of different models for the Ford Motor Company in 1997.

Model	Number	Percent
Cougar	24,019	6
Mystique	43,501	12
Sable	123,873	33
Lincoln Town Car	94,594	25
Mark	17,467	5
Continental	34,421	9
Tracer	38,473	10

1. Create a circle graph from the table.

The circle graphs show the sales of U.S. cars by size in 1983 and 1993.

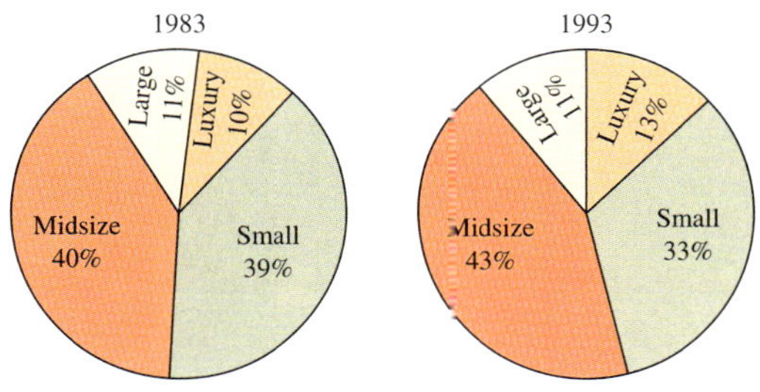

2. What was the percent of midsize and small cars sold in 1993?

3. What was the percent of large and luxury cars sold in 1993?

4. Sales of which type of car increased the most between 1983 and 1993?

[9.2] In exercises 5 and 6, create a pictograph to represent the data.

5. The table represents the 10 most expensive areas for home prices in the United States in 1999.

City	Avg. Price of a home
Palo Alto, Calif.	$843,500
Beverly Hills North	743,825
Beverly Hills South	743,200
San Francisco	720,250
Greenwich, Conn.	674,625
San Mateo, Calif.	671,972
La Jolla, Calif.	652,500
Wellesley, Mass.	586,625
Darien, Conn.	561,333
Palos Verdes, Calif.	537,750

6. The table represents the 10 most affordable areas for home prices in the United States in 1999.

City	Avg. Price of a home
Oklahoma City	$107,850
Sioux City, Iowa	109,000
Arlington, Texas	111,500
Killeen, Texas	111,875
Houston	115,000
Fort Wayne, Ind.	118,950
Lubbock, Texas	120,972
Tulsa, Okla.	121,250
Kearney, Neb.	122,200

Use the bar graph for exercises 7 and 8.

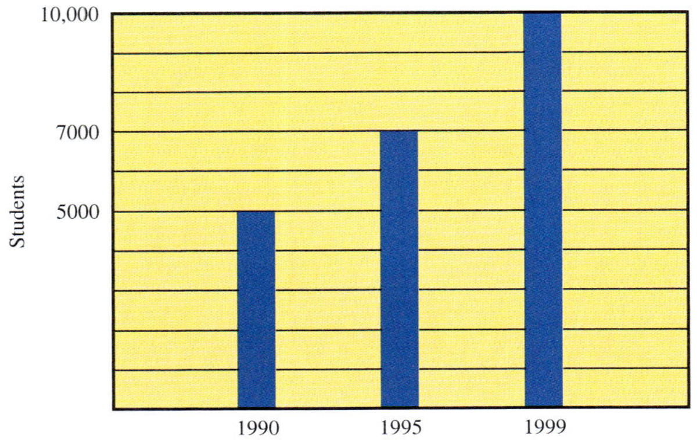

7. How many more students were enrolled in 1999 than in 1990?

8. What was the percent increase from 1990 to 1995?

9. Create a bar graph from the table in Exercise 1 on page 723.

Use the line graph for exercises 10 to 12.

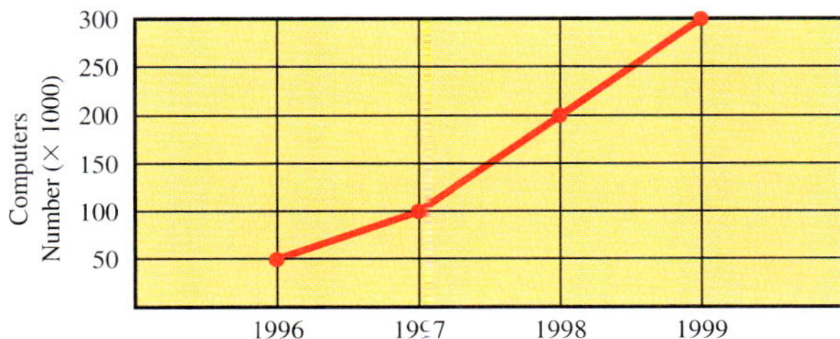

10. How many more personal computers were sold in 1999 than in 1996?

11. What was the percent increase in sales from 1996 to 1999?

12. Predict the sales of personal computers in the year 2000.

13. *Create* a line graph from the given data showing attendance at a county fair.

Day	Attendance
Monday	4,600
Tuesday	5,100
Wednesday	4,800
Thursday	4,900
Friday	5,700
Saturday	6,500
Sunday	6,200

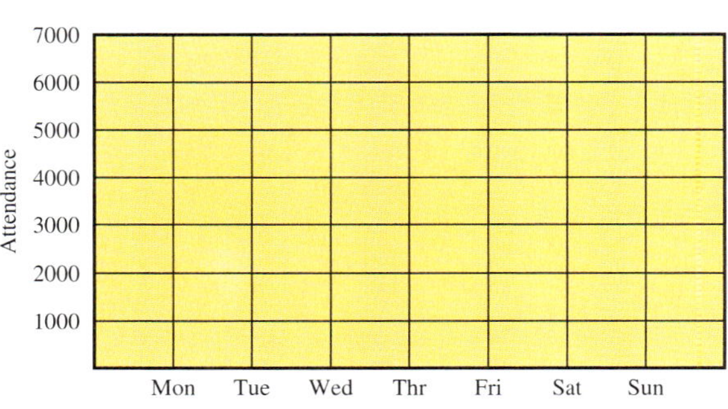

[9.3] Give the coordinates of the graphed points.

14. A

15. B

16. E

17. F

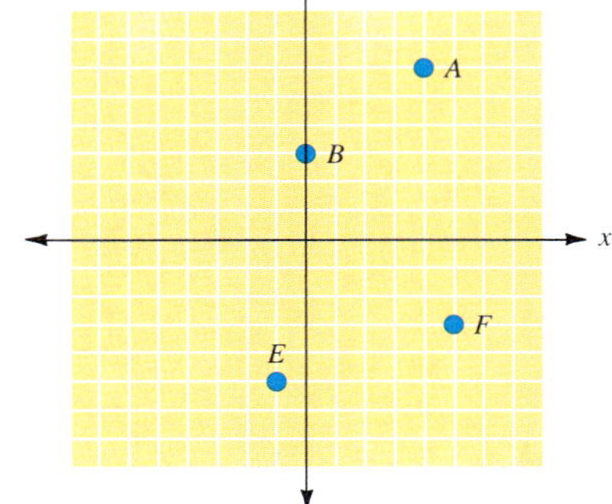

Plot points with the coordinates shown.

18. $P(6, 0)$

19. $Q(5, 4)$

20. $T(-2, 4)$

21. $U(4, -2)$

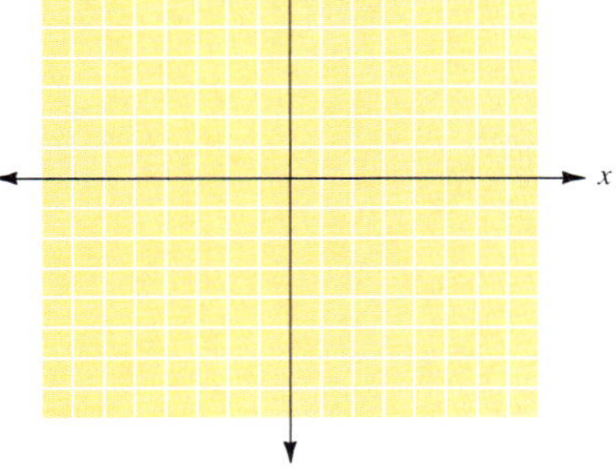

[9.4] Tell whether the number shown in parentheses is a solution for the given equation.

22. $7x + 2 = 16$ (2)

23. $5x - 8 = 3x + 2$ (4)

24. $7x - 2 = 2x + 8$ (2)

25. $\frac{2}{3}x - 2 = 10$ (21)

Determine which of the ordered pairs are solutions for the given equations.

26. $2x + y = 8$ $(4, 0), (2, 2), (2, 4), (4, 2)$

27. $2x + 3y = 6$ $(3, 0), (6, 2), (-3, 4), (0, 2)$

28. $2x - 5y = 10$ $(5, 0), \left(\frac{5}{2}, -1\right), \left(2, \frac{2}{5}\right), (0, -2)$

Complete the ordered pairs so that each is a solution for the given equation.

29. $x - 2y = 10$ $(0, \ ), (12, \ ), (\ , -2), (8, \ )$

30. $2x + 3y = 6$ $(3, \ ), (6, \ ), (\ , -4), (-3, \ )$

31. $y = 3x + 4$ $(2, \ ), (\ , 7), \left(\frac{1}{3}, \ \right), \left(\frac{4}{3}, \ \right)$

Find four solutions for each of the equations.

32. $2x + y = 8$

33. $2x - 3y = 6$

34. $y = -\frac{3}{2}x + 2$

Graph each of the equations.

35. $x + y = 5$

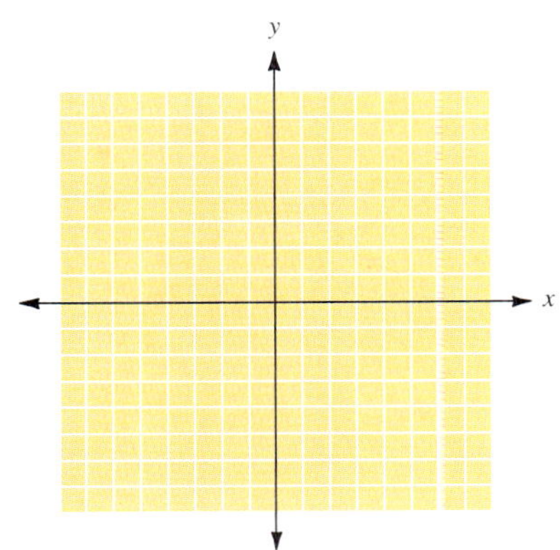

36. $x - y = 6$

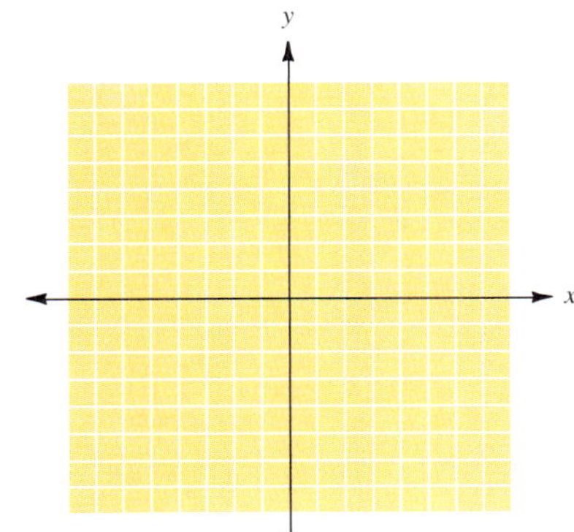

37. $y = 2x$

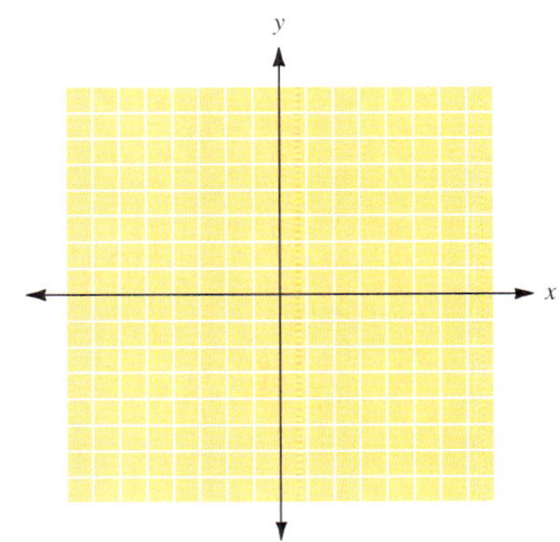

38. $y = -3x$

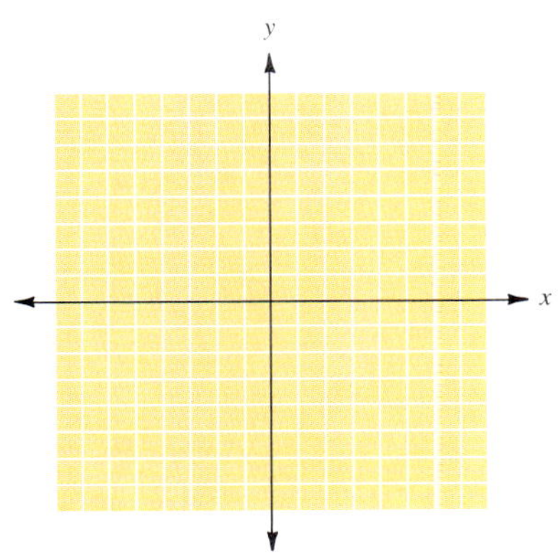

39. $y = \frac{3}{2}x$

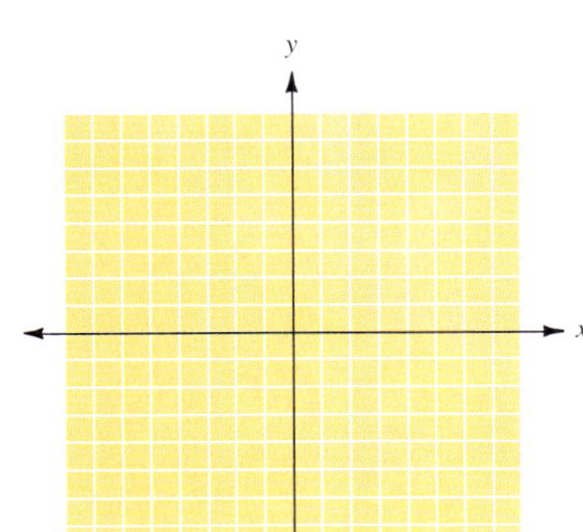

40. $y = 3x + 2$

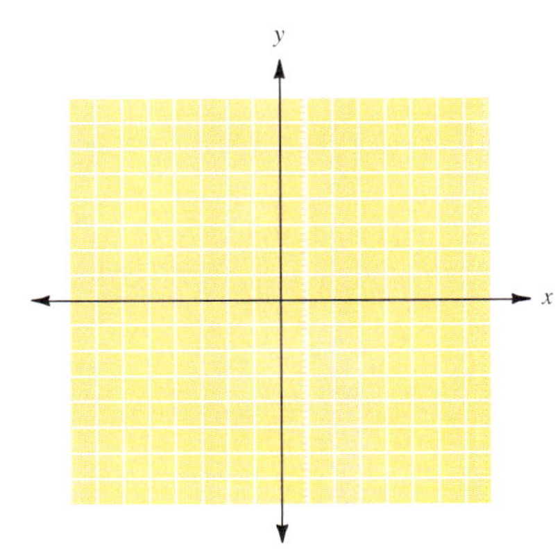

41. $y = -3x + 4$

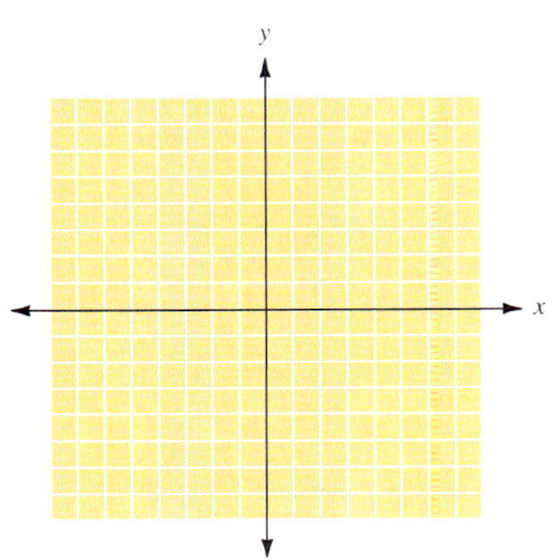

42. $y = \frac{2}{3}x + 2$

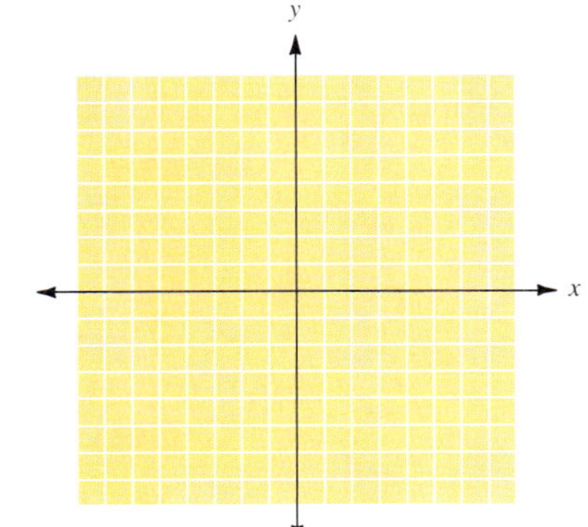

43. $3x - y = 3$

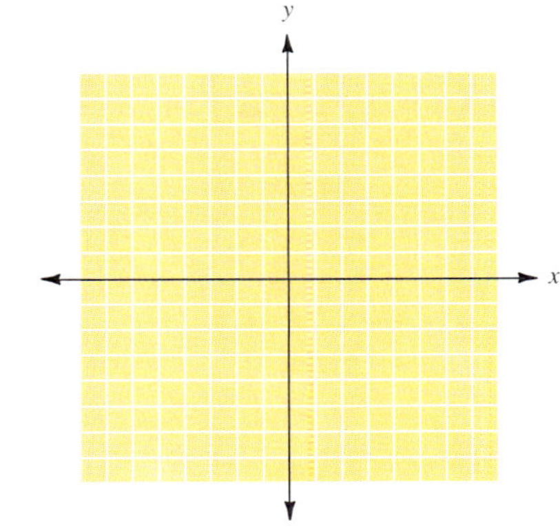

44. $3x + 2y = 12$

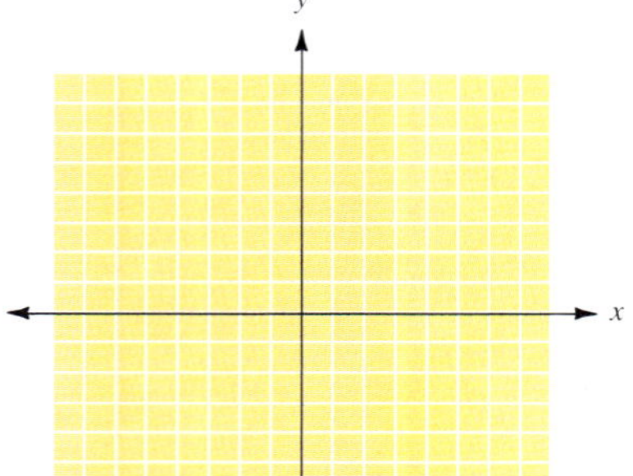

45. $x = 3$

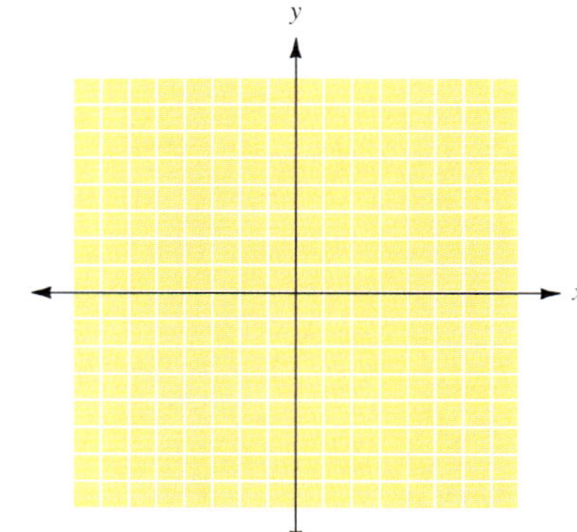

46. $y = -2$

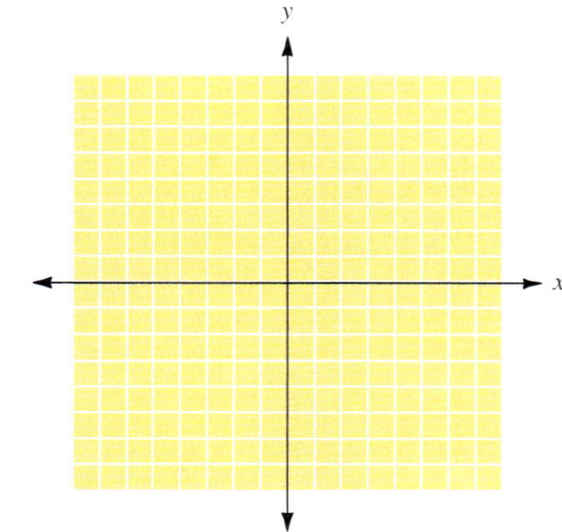

[9.5] In exercises 47 to 50, find the mean for each set of numbers.

47. 8, 6, 7, 4, 5

48. 12, 14, 17, 19, 13

49. 117, 121, 122, 118, 115, 125, 123, 119

50. 134, 126, 128, 129, 133, 125, 122, 127

51. Elmer had test scores of 89, 71, 93, and 87 on his four math tests. What was his mean score?

52. The costs (in dollars) of the seven textbooks that Jacob needs for the spring semester are 77, 66, 55, 49, 85, 80, and 78. Find the mean cost of these books.

In exercises 53 to 56, find the median and the mode for each set of data.

53. 16, 20, 20, 19, 18

54. 8, 9, 9, 11, 11, 8, 7, 11, 12, 14, 10

55. 26, 31, 28, 35, 27, 28, 31, 30, 28, 30

56. 15, 18, 21, 23, 17, 19, 30, 35, 15, 32

57. Anita's first four test scores in her mathematics class were 88, 91, 86, and 93. What score must she get on her next test to have a mean of 90?

58. The weekly sales of a small company for 3 weeks were \$2,400, \$2,800, and \$3,300. How much do sales need to be in the fourth week to achieve a mean of \$3,000?

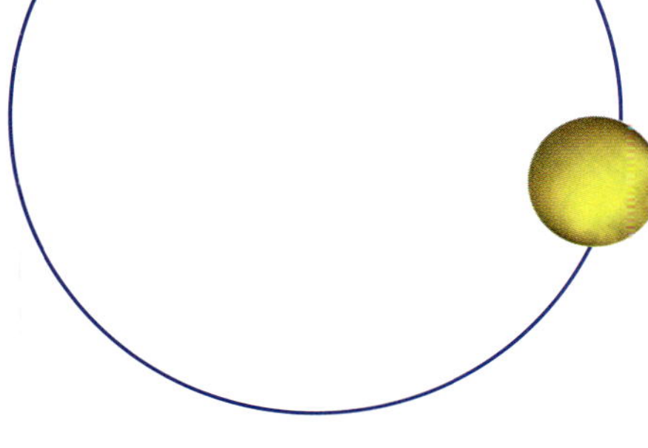

Chapter Test for Chapter 9

Name ____________

Section ______ Date ______

ANSWERS

1. ____________
2. ____________
3. ____________
4. ____________
5. ____________

The purpose of this test is to help you check your progress and review for a chapter test in class. When you are done, check your answers in the back of the book. If you missed any answers, be sure to go back and review the appropriate sections in the chapter.

The circle graph represents the way a new company ships its goods.

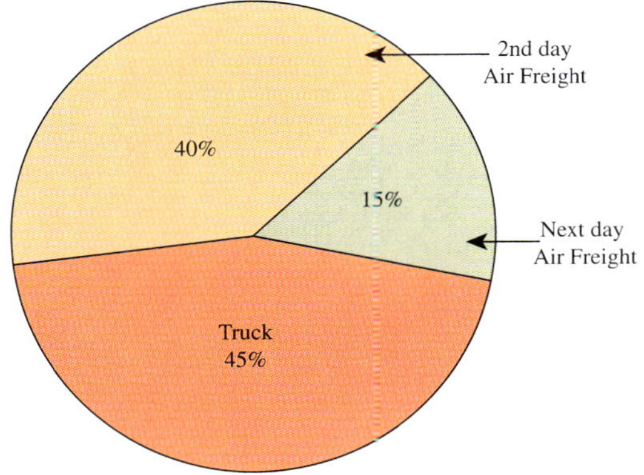

1. What percent was shipped by air freight?
2. What percent was shipped by truck?
3. What percent was shipped by truck or second-day air freight?
4. The table represents the number of injuries suffered by students in classes involving participation.

Type of Injury	Number
Knee	24
Ankle	14
Elbow	4
Wrist	10
Others	8

Create a circle graph to show the distribution of the injury types.

5. Use the information in the table to create a pictograph.

Year	Population of United States (in millions)
1950	151
1960	179
1970	203
1980	227
1990	248

Let each figure represent 10,000,000 people.

ANSWERS

6. ____________________

7. ____________________

8. ____________________

9. ____________________

10. ____________________

11. ____________________

The bar graph represents the number of bankruptcy filings during a recent 5 years.

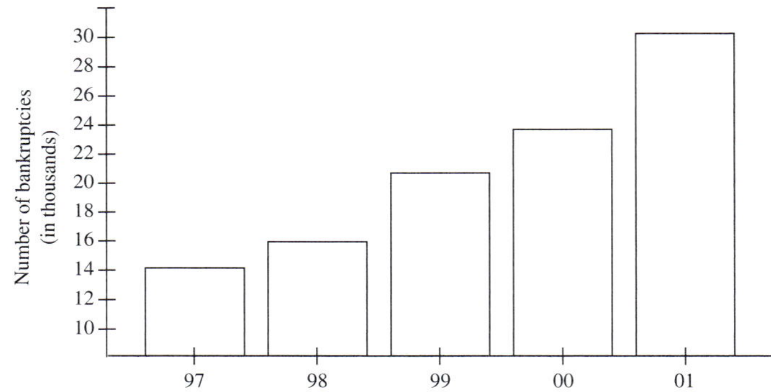

6. How many people filed for bankruptcy in 1998?

7. What was the increase in filings from 1997 to 2001?

8. Between which 2 years did the greatest increase in filings occur?

9. The given information represents a relationship between age and college education. Create a bar graph from this information.

Age Group	Percent with 4 Years of College
25–34	24
35–44	27
45–54	21
55–64	15
65+	11

The line graph shows ticket sales for the last 6 months of the year.

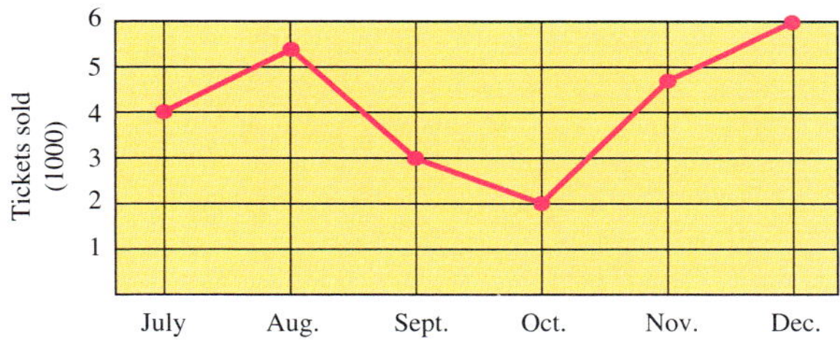

10. What month had the greatest number of ticket sales?

11. Between what 2 months did the greatest decrease in ticket sales occur?

12. The information in the table shows a relationship between the number of workers absent from the assembly line and the number of defects coming off the line. Use this information to create a line graph and then predict the number of defects coming off the line if five workers are absent.

Number of Workers Absent	Number of Defects
0	9
1	10
2	12
3	16
4	18

Give the coordinates of the graphed points.

13. A

14. B

15. C

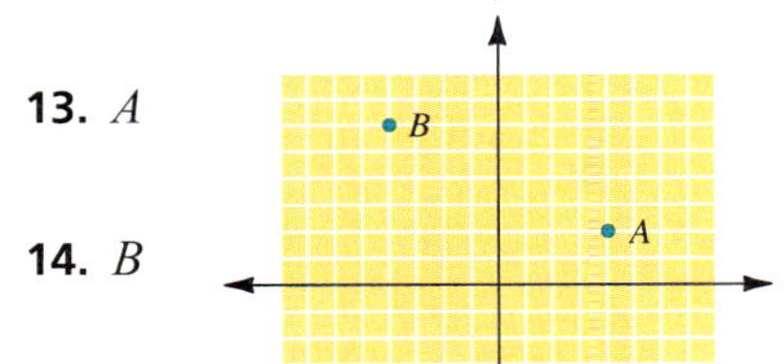

Plot points with the coordinates shown.

16. $S(1, -2)$

17. $T(0, 3)$

18. $U(-2, -3)$

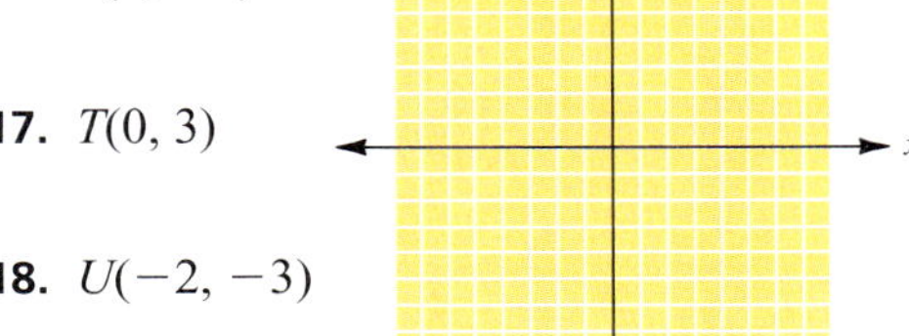

Determine which of the ordered pairs are solutions for the given equation.

19. $4x - y = 16$ $\quad (4, 0), (3, -1), (5, 4)$

Complete the ordered pairs so that each is a solution for the given equation.

20. $x + 3y = 12$ $\quad (3, \ \ ), (\ \ , 2), (9, \ \ )$

Find four solutions for the given equation.

21. $x - y = 7$

Graph each of the given equations.

22. $x + y = 4$

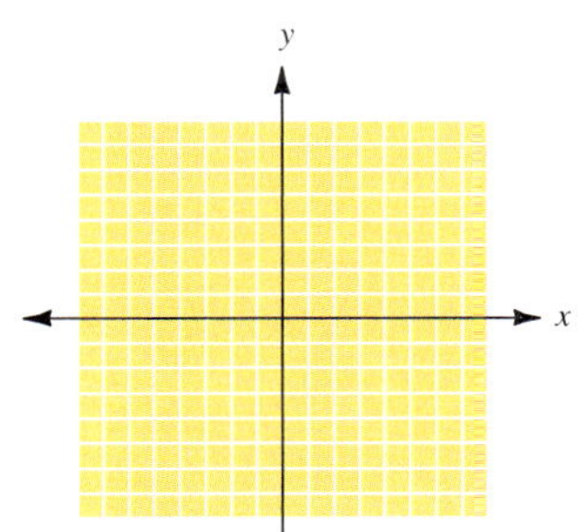

23. $y = 3x$

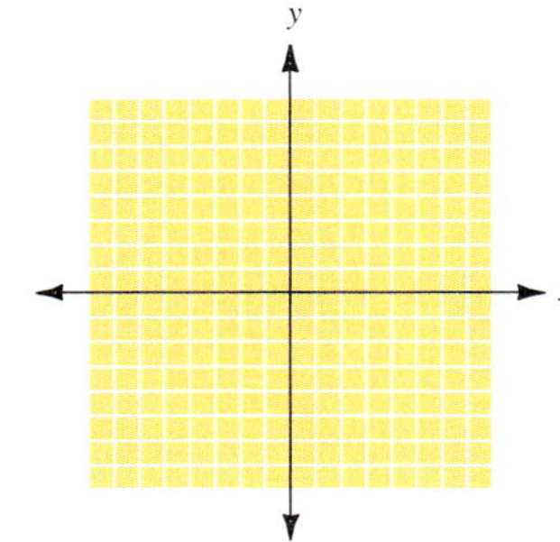

12. ____________

13. ____________

14. ____________

15. ____________

16. ____________

17. ____________

18. ____________

19. ____________

20. ____________

21. ____________

22. ____________

23. ____________

ANSWERS

24. ____________

25. ____________

26. ____________

27. ____________

28. ____________

29. ____________

30. ____________

24. $2x + 5y = 10$

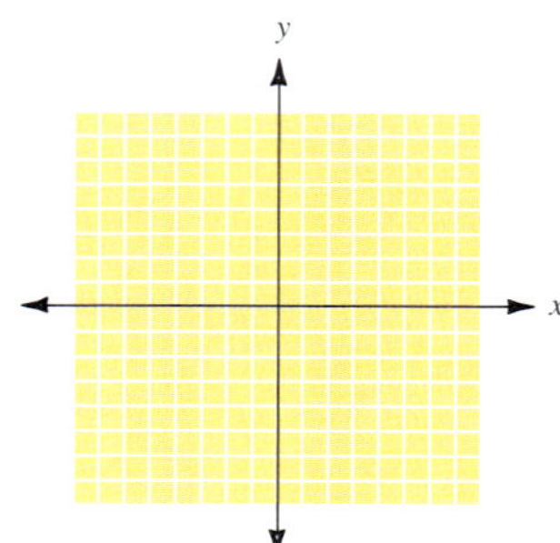

25. $y = -4$

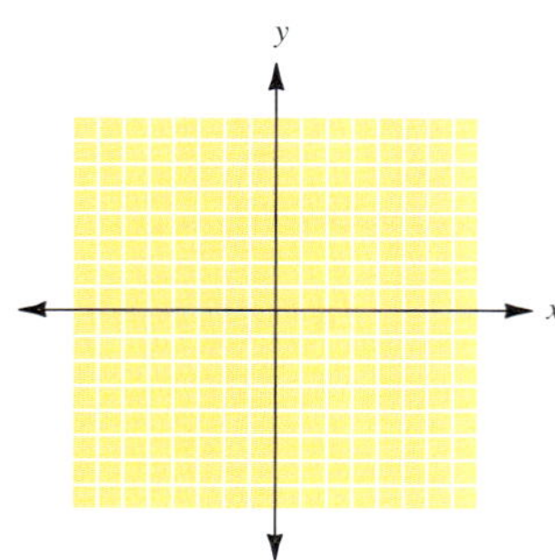

26. Find the mean of the numbers 12, 19, 15, 20, 11, and 13.

27. Find the median of the numbers 12, 18, 9, 10, 16, 6.

28. Find the mode of the numbers 6, 2, 3, 6, 2, 9, 2, 6, 6.

29. Bus riders. A bus carried 234 passengers on the first day of a newly scheduled route. The next 4 days there were 197, 172, 203, and 214 passengers. What was the mean number of riders per day?

30. The following hair colors are from a class of eight students. What color is the mode?

Brown, black, red, blonde, brown, brown, blue, gray

Cumulative Test for Chapters 1 to 9

Name ______________________

Section __________ Date __________

ANSWERS

1. ______________________
2. ______________________
3. ______________________
4. ______________________
5. ______________________
6. ______________________
7. ______________________
8. ______________________
9. ______________________
10. ______________________
11. ______________________
12. ______________________
13. ______________________
14. ______________________
15. ______________________
16. ______________________
17. ______________________
18. ______________________
19. ______________________
20. ______________________
21. ______________________
22. ______________________

This test is provided to help you in the process of reviewing Chapters 1 through 9. Answers are provided in the back of the book. If you missed any answers, be sure to go back and review the appropriate chapter sections.

1. What is the place value of 6 in the numeral 126,489?

In exercises 2 to 7, perform the indicated operation.

2. $20 \div (-4)$ **3.** $15 \div 0$ **4.** $2.45 \cdot 30.7$

5. $\frac{4}{7} \cdot \frac{28}{24}$ **6.** $\frac{11}{15} \div \frac{121}{90}$ **7.** $3\frac{2}{3} + 5\frac{5}{6} - 2\frac{5}{12}$

Evaluate the following expressions if $x = 5$, $y = 2$, and $w = -4$.

8. $2xy$ **9.** $4(x + 3w)$

Combine like terms.

10. $7x - x$ **11.** $6x + 5y - 2x - 8y$

Solve each equation and check your result.

12. $-\frac{3}{4}x = 18$ **13.** $6x - 8 = 2x - 3$

In exercises 14 and 15, solve for the unknown.

14. $\frac{4}{7} = \frac{8}{x}$ **15.** $\frac{3}{5} = \frac{x}{15}$

16. Write 18% as a decimal and fraction.

17. Write $\frac{17}{40}$ as a decimal and percent.

In exercise 18, do the indicated operation.

18.
```
   7 lb  9 oz
 + 3 lb 12 oz
```

In exercises 19 to 22, complete each statement.

19. 8 km = __________ m **20.** 3,000 mg = __________ g

21. 500 cm = __________ m **22.** 25 cL = __________ mL

ANSWERS

23. ______

24. ______

25. ______

26. ______

27. ______

28. ______

29. ______

30. ______

31. ______

32. ______

33. ______

34. ______

35. ______

23. According to the line graph, between what 2 years was the increase in benefits the greatest?

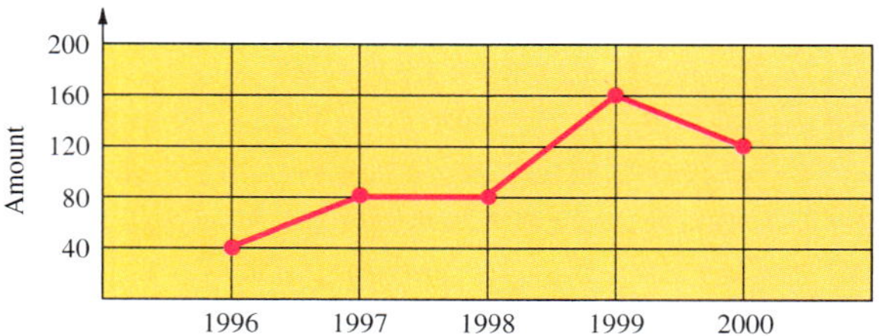

24. Construct a bar graph to represent the following data.

Type of Stock	Number of Stocks
Industrial capital gains	110
Industrial consumer's goods	184
Public utilities	60
Railroads	15
Banks	25
Property liability insurance	16

25. Calculate the mean, median, and mode for the given data.

11, 9, 3, 6, 7, 9, 8, 11, 12, 13, 11, 11, 4, 8, 12

26. If a boat uses 14 gal of gas to go 102 mi, how many gallons would be needed to go 510 mi?

27. The floor of a room that is 12 ft by 18 ft is to be carpeted. If the price of the carpet is \$17 per square yd, what will the carpet cost?

28. If you drive 152 mi in $3\frac{1}{6}$ h, what is your average speed?

29. A rectangle has length $8\frac{3}{5}$ cm and width $5\frac{7}{10}$ cm. Find the perimeter.

30. The sides of a square each measure $13\frac{5}{6}$ ft. Find the perimeter of the square.

31. What is $9\frac{1}{2}\%$ of 1,400?

32. 15 is what percent of 7,500?

33. 111 is 60% of what number?

34. Find $\frac{2}{3}$ of $6\frac{1}{2}$.

35. The number of students attending a small college increased 6% since last year. This year there are 2,968 students. How many students attended last year?

POLYNOMIALS

10

INTRODUCTION

The U.S. Post Office limits the size of rectangular boxes it will accept for mailing. The regulations state that "length plus girth cannot exceed 108 inches." *Girth* means the distance around a cross section; in this case, this measurement is $2h + 2w$. Using the polynomial $l + 2w + 2h$ to describe the measurement required by the Post Office, the regulations say that $l + 2w + 2h \leq 108$ in.

The volume of a rectangular box is expressed by another polynomial: $V = lwh$

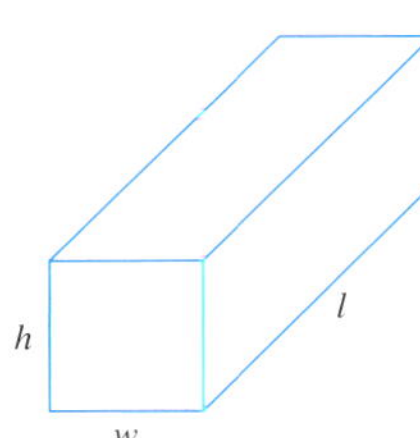

A company that wishes to produce boxes for use by postal patrons must use these formulas and do a statistical survey about the shapes that are useful to the most customers. The surface area, expressed by another polynomial expression, $2lw + 2wh + 2lh$, is also used so each box can be manufactured with the least amount of material, to help lower costs.

Name ______________________

Section ________ Date ________

Pre-Test Chapter 10

ANSWERS

1. ______________
2. ______________
3. ______________
4. ______________
5. ______________
6. ______________
7. ______________
8. ______________
9. ______________
10. ______________
11. ______________
12. ______________
13. ______________
14. ______________
15. ______________
16. ______________

This pre-test will point out any difficulties you may be having with polynomials. Do all the problems, and then check your answers with those in the back of the book.

Classify each of the polynomials as monomial, binomial, or trinomial.

1. $6x^2 - 7x$

2. $-4x^3 + 5x - 9$

3. Evaluate $3x^2 + 2x - 7$ when $x = -1$.

Add.

4. $4x^2 - 7x + 5$ and $-2x^2 + 5x - 7$

Subtract.

5. $-2x^2 + 3x - 1$ from $7x^2 - 8x + 5$

Simplify each of the expressions.

6. x^5x^7

7. $(2x^3y^2)(4x^2y^5)$

Multiply.

8. $3xy(4x^2y^2 - 2xy + 7xy^3)$

9. $(3x + 2)(2x - 5)$

10. $(x + 2y)(x - 2y)$

11. $(4m + 5)(4m + 5)$

12. $(3x - 2y)(x^2 - 4xy + 3y^2)$

Completely factor each of the polynomials.

13. $15c + 35$

14. $8q^4 - 20q^3$

15. $6x^2 - 12x + 24$

16. $7c^3d^2 - 21cd + 14cd^3$

10.1 Introduction to Polynomials

10.1 OBJECTIVES

1. Identify polynomials, terms, and coefficients
2. Identify types of polynomials
3. Evaluate a polynomial

Our work in this chapter deals with the most common kind of algebraic expression, a *polynomial*. To define a polynomial, we should first define the word *term.*

Definition: Term

A **term** is a number or the product of a number and one or more variables.

For example, x^5, $3x$, $-4xy^2$, and 8 are terms.

NOTE The whole numbers are 0, 1, 2, 3, . . . , and so on.

Definition: Polynomial

A **polynomial** is a single term, or the sum of two or more terms, in which the only allowable exponents of any variables are whole numbers.

NOTE Now consider a nonpolynomial term such as $\frac{4}{x}$. As you will see in a later algebra class, when x appears in the denominator, we cannot write this as a product and still have whole-number exponents. Thus an expression containing a term such as $\frac{4}{x}$ is not a polynomial.

Consider the expression $x^3 - 2x^2 - 7$. Recall that subtraction may always be rewritten as addition of an opposite. So we can rewrite the expression as $x^3 + (-2x^2) + (-7)$. This expression is then a sum of three terms. The exponents are whole numbers, so the expression is a polynomial. In general, we can say that the terms of a polynomial are separated by addition signs.

Definition: Numerical Coefficient

In each term of a polynomial, the number in front of the variable is called the **numerical coefficient,** or more simply the **coefficient,** of that term.

Example 1

Identifying Polynomials, Terms, and Coefficients

NOTE We can rewrite x as $1 \cdot x$. Thus the coefficient of x is 1.

(a) $x + 3$ is a polynomial. The terms are x and 3. The coefficients are 1 and 3.

(b) $3x^2 - 2x + 5$, or $3x^2 + (-2x) + 5$, is also a polynomial. Its terms are $3x^2$, $-2x$, and 5. The coefficients are 3, -2, and 5.

(c) $5x^3 + 2 - \frac{3}{x}$ is *not* a polynomial because of the presence of x in the denominator.

CHECK YOURSELF 1

Which of the given expressions are polynomials?

(a) $5x^2$ **(b)** $3y^3 - 2y + \frac{5}{y}$ **(c)** $4x^2 - 2x + 3$

NOTE The prefix *mono-* means 1. The prefix *bi-* means 2. The prefix *tri-* means 3. There are no special names for polynomials with four or more terms.

Definitions: Monomial, Binomial, and Trinomial

A polynomial with one term is called a **monomial.**
A polynomial with two terms is called a **binomial.**
A polynomial with three terms is called a **trinomial.**

Example 2

Identifying Types of Polynomials

(a) $3x^2y$ is a monomial. It has one term.

(b) $2x^3 + 5x$ is a binomial. It has two terms, $2x^3$ and $5x$.

(c) $5x^2 - 4x + 3$, or $5x^2 + (-4x) + 3$, is a trinomial. Its three terms are $5x^2$, $-4x$, and 3.

CHECK YOURSELF 2

Classify each of these as monomial, binomial, or trinomial.

(a) $5x^4 - 2x^3$ **(b)** $4x^7$ **(c)** $2x^2 + 5x - 3$

The value of a polynomial depends on the value given to the variable.

Example 3

Evaluating Polynomials

Given the polynomial

$3x^3 - 2x^2 - 4x + 1$

Find the value of the polynomial when $x = 2$.

Substituting 2 for x, we have

$$3(2)^3 - 2(2)^2 - 4(2) + 1$$
$$= 3(8) - 2(4) - 4(2) + 1$$
$$= 24 - 8 - 8 + 1$$
$$= 9$$

NOTE Again note how the rules for the order of operations are applied. See Section 1.7 for a review.

CHECK YOURSELF 3

Find the value of the polynomial

$4x^3 - 3x^2 + 2x - 1$ *when* $x = 3$

CHECK YOURSELF ANSWERS

1. **(a)** and **(c)** are polynomials. **2.** **(a)** Binomial; **(b)** monomial; **(c)** trinomial
3. 86

Name ____________

Section ______ Date ______

10.1 Exercises

Which of the expressions are polynomials?

1. $7x^3$
2. $5x^3 - \frac{3}{x}$
3. $4x^4y^2 - 3x^3y$
4. 7
5. -7
6. $4x^3 + x$
7. $\frac{3 + x}{x^2}$
8. $5a^2 - 2a + 7$

For each of the polynomials, list the terms and the coefficients.

9. $2x^2 - 3x$
10. $5x^3 + x$
11. $4x^3 - 3x + 2$
12. $7x^2$

Classify each as monomial, binomial, or trinomial where possible.

13. $7x^3 - 3x^2$
14. $4x^7$
15. $7y^2 + 4y + 5$
16. $2x^2 + 3xy + y^2$
17. $2x^4 - 3x^2 + 5x - 2$
18. $x^4 + \frac{5}{x} + 7$
19. $6y^8$
20. $4x^4 - 2x^2 + 5x - 7$
21. $x^5 - \frac{3}{x^2}$
22. $4x^2 - 9$

Find the values of each of the polynomials for the given values of the variable.

23. $6x + 1$, $x = 1$ and $x = -1$
24. $5x - 5$, $x = 2$ and $x = -2$
25. $x^3 - 2x$, $x = 2$ and $x = -2$
26. $3x^2 + 7$, $x = 3$ and $x = -3$

ANSWERS

1. ____
2. ____
3. ____
4. ____
5. ____
6. ____
7. ____
8. ____
9. ____
10. ____
11. ____
12. ____
13. ____
14. ____
15. ____
16. ____
17. ____
18. ____
19. ____
20. ____
21. ____
22. ____
23. ____ 24. ____
25. ____ 26. ____

27. ____________

28. ____________

29. ____________

30. ____________

31. ____________

32. ____________

33. ____________

34. ____________

35. ____________

36. ____________

37. ____________

38. ____________

39. ____________

40. ____________

41. ____________

42. ____________

27. $3x^2 + 4x - 2, x = 4$ and $x = -4$

28. $2x^2 - 5x + 1, x = 2$ and $x = -2$

29. $-x^2 - 2x + 3, x = 1$ and $x = -3$

30. $-x^2 - 5x - 6, x = -3$ and $x = -2$

Indicate whether each of the statements is always true, sometimes true, or never true.

31. A monomial is a polynomial.

32. A binomial is a trinomial.

33. A polynomial has four or more terms.

34. A binomial must have two coefficients.

Capital italic letters such as P or Q are often used to name polynomials. For example, we might write $P(x) = 3x^3 - 5x^2 + 2$ in which $P(x)$ is read "P of x." The notation permits a convenient shorthand. We write $P(2)$, read "P of 2," to indicate the value of the polynomial when $x = 2$. Here

$$
\begin{aligned}
P(2) &= 3(2)^3 - 5(2)^2 + 2 \\
&= 3 \cdot 8 - 5 \cdot 4 + 2 \\
&= 6
\end{aligned}
$$

Use this information in exercises 35 to 48. If $P(x) = x^3 - 2x^2 + 5$ and $Q(x) = 2x^2 + 3$, find:

35. $P(1)$

36. $P(-1)$

37. $Q(2)$

38. $Q(-2)$

39. $P(3)$

40. $Q(-3)$

41. $P(0)$

42. $Q(0)$

43. $P(2) + Q(-1)$

44. $P(-2) + Q(3)$

45. $P(3) - Q(-3) \div Q(0)$

46. $Q(-2) \div Q(2) \cdot P(0)$

47. $|Q(4)| - |P(4)|$

48. $\dfrac{P(-1) + Q(0)}{P(0)}$

Solve the applications.

49. **Cost of typing.** The cost, in dollars, of typing a term paper is given as 3 times the number of pages plus 20. Use y as the number of pages to be typed and write a polynomial to describe this cost. Find the cost of typing a 50-page paper.

50. **Manufacturing.** The cost, in dollars, of making suits is described as 20 times the number of suits plus 150. Use s as the number of suits and write a polynomial to describe this cost. Find the cost of making seven suits.

51. **Revenue.** The revenue, in dollars, when x pairs of shoes are sold is given by $3x^2 - 95$. Find the revenue when 12 pairs of shoes are sold. What is the average (mean) revenue per pair of shoes?

52. **Manufacturing.** The cost in dollars of manufacturing w wing nuts is given by the expression $0.07w + 13.3$. Find the cost when 375 wing nuts are made. What is the average (mean) cost to manufacture one wing nut?

ANSWERS

43. ________

44. ________

45. ________

46. ________

47. ________

48. ________

49. ________

50. ________

51. ________

52. ________

Answers

1. Polynomial **3.** Polynomial **5.** Polynomial **7.** Not a polynomial
9. $2x^2, -3x; 2, -3$ **11.** $4x^3, -3x, 2; 4, -3, 2$ **13.** Binomial
15. Trinomial **17.** Polynomial **19.** Monomial **21.** Not a polynomial
23. $7, -5$ **25.** $4, -4$ **27.** 62, 30 **29.** 0, 0 **31.** Always
33. Sometimes **35.** 4 **37.** 11 **39.** 14 **41.** 5
43. 10 **45.** 7 **47.** -2 **49.** $3y + 20$, \$170 **51.** \$337, \$28.08

Addition and Subtraction of Polynomials

OBJECTIVES

1. Add two polynomials
2. Subtract two polynomials

NOTE To review combining like terms, see Section 2.8.

Addition is always a matter of combining like quantities (two apples plus three apples, four books plus five books, and so on). If you keep that basic idea in mind, adding polynomials will be easy. It is just a matter of combining like terms. Suppose that you want to add

$5x^2 + 3x + 4$ and $4x^2 + 5x - 6$

Parentheses are sometimes used in adding, so for the sum of these polynomials, we can write

NOTE The plus sign between the parentheses indicates the addition.

$(5x^2 + 3x + 4) + (4x^2 + 5x - 6)$

Now what about the parentheses? You can use this rule.

Rules and Properties: Removing Signs of Grouping Case 1

If a plus sign (+) or nothing at all appears in front of parentheses, just remove the parentheses. No other changes are necessary.

Now we will return to the addition.

NOTE Just remove the parentheses. No other changes are necessary.

$(5x^2 + 3x + 4) + (4x^2 + 5x - 6)$

$= 5x^2 + 3x + 4 + 4x^2 + 5x - 6$

Note that the addition sign in front of the parentheses remains.

Like terms Like terms Like terms

NOTE Note the use of the associative and commutative properties in reordering and regrouping.

Collect like terms. (*Remember:* Like terms have the same variables raised to the same power.)

$= (5x^2 + 4x^2) + (3x + 5x) + (4 - 6)$

Combine like terms for the result:

NOTE Here we use the distributive property. For example,

$5x^2 + 4x^2 = (5 + 4)x^2$
$= 9x^2$

$= 9x^2 + 8x - 2$

As should be clear, much of this work can be done mentally. You can then write the sum directly by locating like terms and combining. Example 1 illustrates this approach.

Example 1

Combining Like Terms

NOTE We call this the **horizontal method** because the entire problem is written on one line.
$3 + 4 = 7$ is the horizontal method.

$$\begin{array}{r} 3 \\ +\,4 \\ \hline 7 \end{array}$$

is the vertical method.

Add $3x - 5$ and $2x + 3$.

Write the sum.

$(3x - 5) + (2x + 3)$

$= 3x - 5 + 2x + 3 = 5x - 2$

Like terms Like terms

CHECK YOURSELF 1

Add $6x^2 + 2x$ and $4x^2 - 7x$.

The same technique is used to find the sum of two trinomials.

Example 2

Adding Polynomials Using the Horizontal Method

Add $4a^2 - 7a + 5$ and $3a^2 + 3a - 4$.

Write the sum.

$$(4a^2 - 7a + 5) + (3a^2 + 3a - 4)$$

$$= 4a^2 - 7a + 5 + 3a^2 + 3a - 4 = 7a^2 - 4a + 1$$

Like terms

Like terms

Like terms

REMEMBER Only the like terms are combined in the sum.

CHECK YOURSELF 2

Add $5y^2 - 3y + 7$ and $3y^2 - 5y - 7$.

Example 3

Adding Polynomials Using the Horizontal Method

Add $2x^2 + 7x$ and $4x - 6$.

Write the sum.

$$(2x^2 + 7x) + (4x - 6)$$

$$= 2x^2 + 7x + 4x - 6$$

These are the only like terms; $2x^2$ and -6 cannot be combined.

$$= 2x^2 + 11x - 6$$

CHECK YOURSELF 3

Add $5m^2 + 8$ and $8m^2 - 3m$.

It is sometimes helpful to rewrite the polynomials in descending exponent form before adding. This is illustrated in Example 4.

Example 4

Adding Polynomials Using the Horizontal Method

Add $3x - 2x^2 + 7$ and $5 + 4x^2 - 3x$.

Write the polynomials in descending exponent form, and then add.

$$(-2x^2 + 3x + 7) + (4x^2 - 3x + 5)$$

$$= 2x^2 + 12$$

CHECK YOURSELF 4

Add $8 - 5x^2 + 4x$ *and* $7x - 8 + 8x^2$.

Subtracting polynomials requires another rule for removing signs of grouping.

Rules and Properties: Removing Signs of Grouping Case 2

If a minus sign (−) appears in front of a set of parentheses, the parentheses can be removed by changing the sign of each term inside the parentheses.

The use of this rule is illustrated in Example 5.

Example 5

Removing Parentheses

Remove the parentheses.

NOTE This uses the distributive property, because

$$-(2x + 3y) = (-1)(2x + 3y)$$
$$= -2x - 3y$$

(a) $-(2x + 3y) = -2x - 3y$ Change each sign to remove the parentheses.

(b) $m - (5n - 3p) = m \underbrace{- 5n + 3p}_{\text{Signs change.}}$ Note that the subtraction sign in front of the parentheses is dropped.

(c) $2x - (-3y + z) = 2x \underbrace{+ 3y - z}_{\text{Signs change.}}$

CHECK YOURSELF 5

Remove the parentheses.

(a) $-(3m + 5n)$ **(b)** $-(5w - 7z)$

(c) $3r - (2s - 5t)$ **(d)** $5a - (-3b - 2c)$

Subtracting polynomials is now a matter of using case 2 of the removing signs of grouping rule to remove the parentheses and then combining the like terms. Consider Example 6.

Example 6

Subtracting Polynomials Using the Horizontal Method

(a) Subtract $5x - 3$ from $8x + 2$.

NOTE The expression following "from" is written first in the problem.

Write

$(8x + 2) - (5x - 3)$

$= 8x + 2 \underbrace{- 5x + 3}_{\text{Signs change.}}$ Recall that subtracting $5x$ is the same as adding $-5x$.

$= 3x + 5$

(b) Subtract $4x^2 - 8x + 3$ from $8x^2 + 5x - 3$.

Write

$(8x^2 + 5x - 3) - (4x^2 - 8x + 3)$

$= 8x^2 + 5x - 3 \underbrace{- 4x^2 + 8x - 3}$

Signs change.

$= 4x^2 + 13x - 6$

CHECK YOURSELF 6

(a) Subtract $7x + 3$ from $10x - 7$.

(b) Subtract $5x^2 - 3x + 2$ from $8x^2 - 3x - 6$.

Again, writing all polynomials in descending exponent form will make locating and combining like terms much easier. Look at Example 7.

Example 7

Subtracting Polynomials Using the Horizontal Method

(a) Subtract $4x^2 - 3x^3 + 5x$ from $8x^3 - 7x + 2x^2$.

Write

$(8x^3 + 2x^2 - 7x) - (-3x^3 + 4x^2 + 5x)$

$= 8x^3 + 2x^2 - 7x \underbrace{+ 3x^3 - 4x^2 - 5x}$

Signs change.

$= 11x^3 - 2x^2 - 12x$

(b) Subtract $8x - 5$ from $-5x + 3x^2$.

Write

$(3x^2 - 5x) - (8x - 5)$

$= 3x^2 \underbrace{- 5x - 8x} + 5$

Only the like terms can be combined.

$= 3x^2 - 13x + 5$

CHECK YOURSELF 7

(a) Subtract $7x - 3x^2 + 5$ from $5 - 3x + 4x^2$.

(b) Subtract $3a - 2$ from $5a + 4a^2$.

If you think back to addition and subtraction in arithmetic, you'll remember that the work was arranged vertically. That is, the numbers being added or subtracted were placed under one another so that each column represented the same place value. This meant that in adding or subtracting columns you were always dealing with "like quantities."

It is also possible to use a vertical method for adding or subtracting polynomials. First rewrite the polynomials in descending exponent form, and then arrange them one under another, so that each column contains like terms. Then add or subtract in each column.

Example 8

Adding Using the Vertical Method

Add $2x^2 - 5x$, $3x^2 + 2$, and $6x - 3$.

Like terms

$$\begin{array}{rrr} 2x^2 & - 5x & \\ 3x^2 & & + 2 \\ & 6x & - 3 \\ \hline 5x^2 & + \ x & - 1 \end{array}$$

CHECK YOURSELF 8

Add $3x^2 + 5$, $x^2 - 4x$, *and* $6x + 7$.

Example 9 illustrates subtraction by the vertical method.

Example 9

Subtracting Using the Vertical Method

(a) Subtract $5x - 3$ from $8x - 7$.

Write

$$\begin{array}{rr} & 8x - 7 \\ (-) & 5x - 3 \\ \hline & 3x - 4 \end{array}$$

To subtract, change each sign of $5x - 3$ to get $-5x + 3$, and then add.

$$\begin{array}{rr} & 8x - 7 \\ (+) & -5x + 3 \\ \hline & 3x - 4 \end{array}$$

(b) Subtract $5x^2 - 3x + 4$ from $8x^2 + 5x - 3$.

Write

$$\begin{array}{rr} & 8x^2 + 5x - 3 \\ (-) & 5x^2 - 3x + 4 \\ \hline & 3x^2 + 8x - 7 \end{array}$$

To subtract, change each sign of $5x^2 - 3x + 4$ to get $-5x^2 + 3x - 4$, and then add.

$$\begin{array}{rr} & 8x^2 + 5x - 3 \\ (+) & -5x^2 + 3x - 4 \\ \hline & 3x^2 + 8x - 7 \end{array}$$

Subtracting using the vertical method takes some practice. Take time to study the method carefully.

CHECK YOURSELF 9

Subtract, using the vertical method.

(a) $4x^2 - 3x$ from $8x^2 + 2x$ **(b)** $8x^2 + 4x - 3$ from $9x^2 - 5x + 7$

CHECK YOURSELF ANSWERS

1. $10x^2 - 5x$ **2.** $8y^2 - 8y$ **3.** $13m^2 - 3m + 8$ **4.** $3x^2 + 11x$
5. **(a)** $-3m - 5n$; **(b)** $-5w + 7z$; **(c)** $3r - 2s + 5t$; **(d)** $5a + 3b + 2c$
6. **(a)** $3x - 10$; **(b)** $3x^2 - 8$ **7.** **(a)** $7x^2 - 10x$; **(b)** $4a^2 + 2a + 2$
8. $4x^2 + 2x + 12$ **9.** **(a)** $4x^2 + 5x$; **(b)** $x^2 - 9x + 10$

Name ______________

Section ________ Date ________

10.2 Exercises

Add.

1. $6a - 5$ and $3a + 9$

2. $9x + 3$ and $3x - 4$

3. $8b^2 - 11b$ and $5b^2 - 7b$

4. $2m^2 + 3m$ and $6m^2 - 8m$

5. $3x^2 - 2x$ and $-5x^2 + 2x$

6. $3p^2 + 5p$ and $-7p^2 - 5p$

7. $2x^2 + 5x - 3$ and $3x^2 - 7x + 4$

8. $4d^2 - 8d + 7$ and $5d^2 - 6d - 9$

9. $2b^2 + 8$ and $5b + 8$

10. $4x - 3$ and $3x^2 - 9x$

11. $8y^3 - 5y^2$ and $5y^2 - 2y$

12. $9x^4 - 2x^2$ and $2x^2 + 3$

13. $2a^2 - 4a^3$ and $3a^3 + 2a^2$

14. $9m^3 - 2m$ and $-6m - 4m^3$

15. $4x^2 - 2 + 7x$ and $5 - 8x - 6x^2$

16. $5b^3 - 8b + 2b^2$ and $3b^2 - 7b^3 + 5b$

Remove the parentheses in each of the expressions, and simplify when possible.

17. $-(2a + 3b)$

18. $-(7x - 4y)$

19. $5a - (2b - 3c)$

20. $7x - (4y + 3z)$

21. $9r - (3r + 5s)$

22. $10m - (3m - 2n)$

23. $5p - (-3p + 2q)$

24. $8d - (-7c - 2d)$

ANSWERS

1. ______________
2. ______________
3. ______________
4. ______________
5. ______________
6. ______________
7. ______________
8. ______________
9. ______________
10. ______________
11. ______________
12. ______________
13. ______________
14. ______________
15. ______________
16. ______________
17. ______________
18. ______________
19. ______________
20. ______________
21. ______________
22. ______________
23. ______________
24. ______________

ANSWERS

25. ________

26. ________

27. ________

28. ________

29. ________

30. ________

31. ________

32. ________

33. ________

34. ________

35. ________

36. ________

37. ________

38. ________

39. ________

40. ________

41. ________

42. ________

43. ________

44. ________

45. ________

46. ________

47. ________

48. ________

49. ________

50. ________

Subtract.

25. $x + 4$ from $2x - 3$

26. $x - 2$ from $3x + 5$

27. $3m^2 - 2m$ from $4m^2 - 5m$

28. $9a^2 - 5a$ from $11a^2 - 10a$

29. $6y^2 + 5y$ from $4y^2 + 5y$

30. $9n^2 - 4n$ from $7n^2 - 4n$

31. $x^2 - 4x - 3$ from $3x^2 - 5x - 2$

32. $3x^2 - 2x + 4$ from $5x^2 - 8x - 3$

33. $3a + 7$ from $8a^2 - 9a$

34. $3x^3 + x^2$ from $4x^3 - 5x$

35. $4b^2 - 3b$ from $5b - 2b^2$

36. $7y - 3y^2$ from $3y^2 - 2y$

37. $x^2 - 5 - 8x$ from $3x^2 - 8x + 7$

38. $4x - 2x^2 + 4x^3$ from $4x^3 + x - 3x^2$

Perform the indicated operations.

39. Subtract $3b + 2$ from the sum of $4b - 2$ and $5b + 3$.

40. Subtract $5m - 7$ from the sum of $2m - 8$ and $9m - 2$.

41. Subtract $3x^2 + 2x - 1$ from the sum of $x^2 + 5x - 2$ and $2x^2 + 7x - 8$.

42. Subtract $4x^2 - 5x - 3$ from the sum of $x^2 - 3x - 7$ and $2x^2 - 2x + 9$.

43. Subtract $2x^2 - 3x$ from the sum of $4x^2 - 5$ and $2x - 7$.

44. Subtract $5a^2 - 3a$ from the sum of $3a - 3$ and $5a^2 + 5$.

45. Subtract the sum of $3y^2 - 3y$ and $5y^2 + 3y$ from $2y^2 - 8y$.

46. Subtract the sum of $7r^3 - 4r^2$ and $-3r^3 + 4r^2$ from $2r^3 + 3r^2$.

Add, using the vertical method.

47. $2w^2 + 7$, $3w - 5$, and $4w^2 - 5w$

48. $3x^2 - 4x - 2$, $6x - 3$, and $2x^2 + 8$

49. $3x^2 + 3x - 4$, $4x^2 - 3x - 3$, and $2x^2 - x + 7$

50. $5x^2 + 2x - 4$, $x^2 - 2x - 3$, and $2x^2 - 4x - 3$

ANSWERS

51. ______
52. ______
53. ______
54. ______
55. ______
56. ______
57. ______
58. ______
59. ______
60. ______
61. ______
62. ______
63. ______
64. ______

Subtract, using the vertical method.

51. $3a^2 - 2a$ from $5a^2 + 3a$

52. $6r^3 + 4r^2$ from $4r^3 - 2r^2$

53. $5x^2 - 6x + 7$ from $8x^2 - 5x + 7$

54. $8x^2 - 4x + 2$ from $9x^2 - 8x + 6$

55. $5x^2 - 3x$ from $8x^2 - 9$

56. $7x^2 + 6x$ from $9x^2 - 3$

Perform the indicated operations.

57. $[(9x^2 - 3x + 5) - (3x^2 + 2x - 1)] - (x^2 - 2x - 3)$

58. $[(5x^2 + 2x - 3) - (-2x^2 + x - 2)] - (2x^2 + 3x - 5)$

Find values for a, b, c, and d so that the equations are true.

59. $3ax^4 - 5x^3 + x^2 - cx + 2 = 9x^4 - bx^3 + x^2 - 2d$

60. $(4ax^3 - 3bx^2 - 10) - 3(x^3 + 4x^2 - cx - d) = x^2 - 6x + 8$

61. **Geometry.** A rectangle has sides of $8x + 9$ and $6x - 7$. Find the polynomial that represents its perimeter.

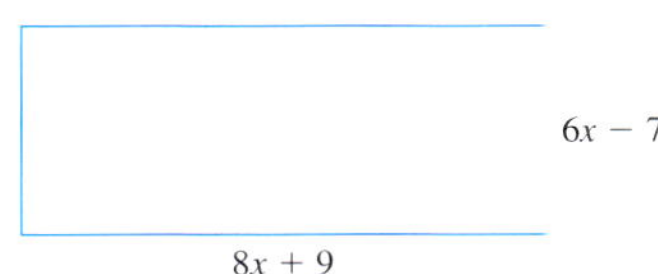

62. **Geometry.** A triangle has sides $3x + 7$, $4x - 9$, and $5x + 6$. Find the polynomial that represents its perimeter.

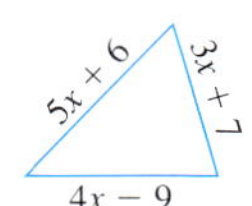

63. **Business.** The cost of producing x units of an item is $C = 150 + 25x$. The revenue for selling x units is $R = 90x - x^2$. The profit is given by the revenue minus the cost. Find the polynomial that represents profit.

64. **Business.** The revenue for selling y units is $R = 3y^2 - 2y + 5$ and the cost of producing y units is $C = y^2 + y - 3$. Find the polynomial that represents profit.

Answers

1. $9a + 4$ **3.** $13b^2 - 18b$ **5.** $-2x^2$ **7.** $5x^2 - 2x + 1$
9. $2b^2 + 5b + 16$ **11.** $8y^3 - 2y$ **13.** $-a^3 + 4a^2$ **15.** $-2x^2 - x + 3$
17. $-2a - 3b$ **19.** $5a - 2b + 3c$ **21.** $6r - 5s$ **23.** $8p - 2q$
25. $x - 7$ **27.** $m^2 - 3m$ **29.** $-2y^2$ **31.** $2x^2 - x + 1$
33. $8a^2 - 12a - 7$ **35.** $-6b^2 + 8b$ **37.** $2x^2 + 12$ **39.** $6b - 1$
41. $10x - 9$ **43.** $2x^2 + 5x - 12$ **45.** $-6y^2 - 8y$ **47.** $6w^2 - 2w + 2$
49. $9x^2 - x$ **51.** $2a^2 + 5a$ **53.** $3x^2 + x$ **55.** $3x^2 + 3x - 9$
57. $5x^2 - 3x + 9$ **59.** $a = 3, b = 5, c = 0, d = -1$ **61.** $28x + 4$
63. $-x^2 + 65x - 150$

10.3 Multiplying Polynomials

10.3 OBJECTIVES

1. Use properties of exponents
2. Find the product of a monomial and a polynomial
3. Find the product of two polynomials

In Chapter 1, we introduced the idea of exponents. Recall that the exponent notation indicates repeated multiplication and that the exponent tells us how many times the base is to be used as a factor.

Exponent

$$3^5 = \underbrace{3 \cdot 3 \cdot 3 \cdot 3 \cdot 3}_{\text{5 factors}} = 243$$

Base

Now, we will look at the properties of exponents.

Consider the expression $4^2 \cdot 4^3$. Since this represents $\underbrace{(4 \cdot 4)}_{\text{2 factors}} \cdot \underbrace{(4 \cdot 4 \cdot 4)}_{\text{3 factors}}$, we see that

$$4^2 \cdot 4^3 = \underbrace{4 \cdot 4 \cdot 4 \cdot 4 \cdot 4}_{\text{5 factors}} = 4^5$$

This suggests the first property of exponents. It is used when multiplying two values with the same base.

Rules and Properties: Product Property of Exponents

For any real number a and positive integers m and n,

$$a^m \cdot a^n = a^{m+n}$$

For example,

$$2^5 \cdot 2^7 = 2^{12}$$

Example 1

Using the Product Property of Exponents

Simplify each expression.

(a) $8^5 \cdot 8^9 = 8^{5+9} = 8^{14}$

(b) $x^6 \cdot x^2 = x^{6+2} = x^8$

(c) $5y^3 \cdot 2y^4 = 5 \cdot 2 \cdot y^3 \cdot y^4 = 10 \cdot y^{3+4} = 10y^7$

CHECK YOURSELF 1

Simplify each expression.

(a) $7^3 \cdot 7^6$ **(b)** $z^5 \cdot z^3$ **(c)** $4x^6 \cdot 7x^5$

Now consider the following:

$(x^2)^4 = x^2 \cdot x^2 \cdot x^2 \cdot x^2 = x^8$

NOTE Notice that this means that the base, x^2, is used as a factor *4* times.

This leads us to our second property for exponents.

Rules and Properties: Power Property of Exponents

For any real number *a* and positive integers *m* and *n*,

$(a^m)^n = a^{m \cdot n}$

For example,

$(2^3)^2 = 2^{3 \cdot 2} = 2^6$

The use of this new property is illustrated in Example 2.

Example 2

Using the Power Property of Exponents

Simplify each expression.

(a) $(x^4)^5 = x^{4 \cdot 5} = x^{20}$ Multiply the exponents.

(b) $(2^3)^4 = 2^{3 \cdot 4} = 2^{12}$

CAUTION

Be careful! Be sure to distinguish between the correct use of the product property and the power property.

$(x^4)^5 = x^{4 \cdot 5} = x^{20}$

but

$x^4 \cdot x^5 = x^{4+5} = x^9$

CHECK YOURSELF 2

Simplify each expression.

(a) $(m^5)^6$ **(b)** $(m^5)(m^6)$

(c) $(3^2)^4$ **(d)** $(3^2)(3^4)$

NOTE Here the base is $3x$.

Suppose we now have a product raised to a power. Consider an expression such as $(3x)^4$. We know that

NOTE Here we have applied the commutative and associative properties.

$$(3x)^4 = (3x)(3x)(3x)(3x)$$
$$= (3 \cdot 3 \cdot 3 \cdot 3)(x \cdot x \cdot x \cdot x)$$
$$= 3^4 \cdot x^4 = 81x^4$$

Note that the power, here 4, has been applied to each factor, 3 and x. In general, we have

Rules and Properties: Power of a Product Property of Exponents

For any real numbers *a* and *b* and positive integer *m*,

$(ab)^m = a^m b^m$

For example,

$(3x)^3 = 3^3 \cdot x^3 = 27x^3$

The use of this property is shown in Example 3.

Example 3

Using the Power of a Product Property of Exponents

Simplify each expression.

NOTE Notice that $(2x)^5$ and $2x^5$ are entirely different expressions. For $(2x)^5$, the base is $2x$, so we raise each factor to the fifth power. For $2x^5$, the base is x, and so the exponent applies only to x.

(a) $(2x)^5 = 2^5 \cdot x^5 = 32x^5$

(b) $(3ab)^4 = 3^4 \cdot a^4 \cdot b^4 = 81a^4b^4$

(c) $5(2r)^3 = 5 \cdot 2^3 \cdot r^3 = 40r^3$

CHECK YOURSELF 3

Simplify each expression.

(a) $(3y)^4$ **(b)** $(2mn)^6$

(c) $3(4x)^2$ **(d)** $5x^3$

We may have to use more than one of our properties in simplifying an expression involving exponents. Consider Example 4.

Example 4

Using the Properties of Exponents

NOTE To help you understand each step of the simplification, we refer to the property being applied. Make a list of the properties now to help you as you work through the remainder of this section.

Simplify each expression.

(a) $(r^4s^3)^3 = (r^4)^3 \cdot (s^3)^3$ Power of a product property of exponents

$= r^{12}s^9$ Power property of exponents

(b) $(3x^2)^2 \cdot (2x^3)^3$

$= 3^2(x^2)^2 \cdot 2^3 \cdot (x^3)^3$ Power of a product property of exponents

$= 9x^4 \cdot 8x^9$ Power property of exponents

$= 72x^{13}$ Multiply the coefficients and apply the product property of exponents.

CHECK YOURSELF 4

Simplify each expression.

(a) $(m^5n^2)^3$ **(b)** $(2p)^4(4p^2)^2$

The table summarizes the three properties of exponents that have been discussed in this section:

General Form	Example
1. $a^m a^n = a^{m+n}$	$x^2 \cdot x^3 = x^5$
2. $(a^m)^n = a^{mn}$	$(z^5)^4 = z^{20}$
3. $(ab)^m = a^m b^m$	$(4x)^3 = 4^3x^3 = 64x^3$

We are now ready to combine what we learned about polynomials in Sections 10.1 and 10.2 with what we have just learned about properties of exponents. We will begin by finding the product of two monomials.

Step by Step: To Find the Product of Monomials

Step 1 Multiply the coefficients.

Step 2 Use the product property of exponents to combine the variables.

NOTE The product property of exponents:

$x^m \cdot x^n = x^{m+n}$

Example 5

Multiplying Monomials

Multiply $3x^2y$ and $2x^3y^5$.

Write

NOTE Once again we have used the commutative and associative properties to rewrite the problem.

$(3x^2y)(2x^3y^5)$

$= (3 \cdot 2)(x^2 \cdot x^3)(y \cdot y^5)$

Multiply the coefficients. Add the exponents.

$= 6x^5y^6$

CHECK YOURSELF 5

Multiply.

(a) $(5a^2b)(3a^2b^4)$ **(b)** $(-3xy)(4x^3y^5)$

NOTE You might want to review Section 1.5 before going on.

Our next task is to find the product of a monomial and a polynomial. Here we use the distributive property, which we introduced in Section 1.5. That property leads us to this rule for multiplication.

Rules and Properties: To Multiply a Polynomial by a Monomial

Use the distributive property to multiply each term of the polynomial by the monomial.

NOTE Distributive property:

$a(b + c) = ab + ac$

Example 6

Multiplying a Monomial and a Binomial

(a) Multiply $2x + 3$ by x.

Write

NOTE With practice you will do this step mentally.

$x(2x + 3)$

$= x \cdot 2x + x \cdot 3$

$= 2x^2 + 3x$

Multiply x by $2x$ and then by 3, the terms of the polynomial. That is, "distribute" the multiplication over the sum.

(b) Multiply $2a^3 + 4a$ by $3a^2$.

Write

$3a^2(2a^3 + 4a)$

$= 3a^2 \cdot 2a^3 + 3a^2 \cdot 4a = 6a^5 + 12a^3$

CHECK YOURSELF 6

Multiply.

(a) $2y(y^2 + 3y)$ **(b)** $3w^2(2w^3 + 5w)$

The patterns of Example 6 extend to *any* number of terms.

Example 7

Multiplying a Monomial and a Polynomial

Multiply the following.

(a) $3x(4x^3 + 5x^2 + 2)$

$= 3x \cdot 4x^3 + 3x \cdot 5x^2 + 3x \cdot 2 = 12x^4 + 15x^3 + 6x$

NOTE Again we have shown all the steps of the process. With practice you can write the product directly, and you should try to do so.

(b) $5y^2(2y^3 - 4)$

$= 5y^2 \cdot 2y^3 - 5y^2 \cdot 4 = 10y^5 - 20y^2$

(c) $-5c(4c^2 - 8c)$

$= (-5c)(4c^2) - (-5c)(8c) = -20c^3 + 40c^2$

(d) $3c^2d^2(7cd^2 - 5c^2d^3)$

$= 3c^2d^2 \cdot 7cd^2 - 3c^2d^2 \cdot 5c^2d^3 = 21c^3d^4 - 15c^4d^5$

CHECK YOURSELF 7

Multiply.

(a) $3(5a^2 + 2a + 7)$ **(b)** $4x^2(8x^3 - 6)$

(c) $-5m(8m^2 - 5m)$ **(d)** $9a^2b(3a^3b - 6a^2b^4)$

Example 8

Multiplying Binomials

(a) Multiply $x + 2$ by $x + 3$.

NOTE Note that this ensures that each term, x and 2, of the first binomial is multiplied by each term, x and 3, of the second binomial.

We can think of $x + 2$ as a single quantity and apply the distributive property.

$(x + 2)(x + 3)$ — Multiply $x + 2$ by x and then by 3.

$= (x + 2)x + (x + 2)3$ — Apply the distributive property.

$= x \cdot x + 2 \cdot x + x \cdot 3 + 2 \cdot 3$

$= x^2 + 2x + 3x + 6$ — Combine like terms.

$= x^2 + 5x + 6$

(b) Multiply $a - 3$ by $a - 4$. (Think of $a - 3$ as a single quantity and distribute.)

$(a - 3)(a - 4)$

$= (a - 3)a - (a - 3)(4)$

$= a \cdot a - 3 \cdot a - [(a \cdot 4) - (3 \cdot 4)]$

$= a^2 - 3a - (4a - 12)$ — Note that the parentheses are needed here because a *minus sign* precedes the binomial.

$= a^2 - 3a - 4a + 12$

$= a^2 - 7a + 12$

CHECK YOURSELF 8

Multiply.

(a) $(x + 4)(x + 5)$ **(b)** $(y + 5)(y - 6)$

Fortunately, there is a pattern to this kind of multiplication that allows you to write the product of the two binomials directly without going through all these steps. We call it the **FOIL method** of multiplying. The reason for this name will be clear as we look at the process in more detail.

To multiply $(x + 2)(x + 3)$:

1. $(x + 2)(x + 3)$ → $x \cdot x$ — Find the product of the *first* terms of the factors.

NOTE Remember this by F!

2. $(x + 2)(x + 3)$ → $x \cdot 3$ — Find the product of the *outer* terms.

NOTE Remember this by O!

3. $(x + 2)(x + 3)$ → $2 \cdot x$ — Find the product of the *inner* terms.

NOTE Remember this by I!

4. $(x + 2)(x + 3)$ → $2 \cdot 3$ — Find the product of the *last* terms.

NOTE Remember this by L!

Combining the four steps, we have

NOTE Of course, these are the same four terms found in Example 8*a*.

$(x + 2)(x + 3)$

$= x^2 + 3x + 2x + 6$

$= x^2 + 5x + 6$

NOTE It's called FOIL to give you an easy way of remembering the steps: *F*irst, *O*uter, *I*nner, and *L*ast.

With practice, the FOIL method will let you write the products quickly and easily. Consider Example 9 which illustrates this approach.

Example 9

Using the FOIL Method

Find the following products, using the FOIL method.

F: $x \cdot x$ L: $4 \cdot 5$

(a) $(x + 4)(x + 5)$

I: $4x$

O: $5x$

NOTE When possible, you should combine the outer and inner products mentally and write just the final product.

$= x^2 + 5x + 4x + 20$

F O I L

$= x^2 + 9x + 20$

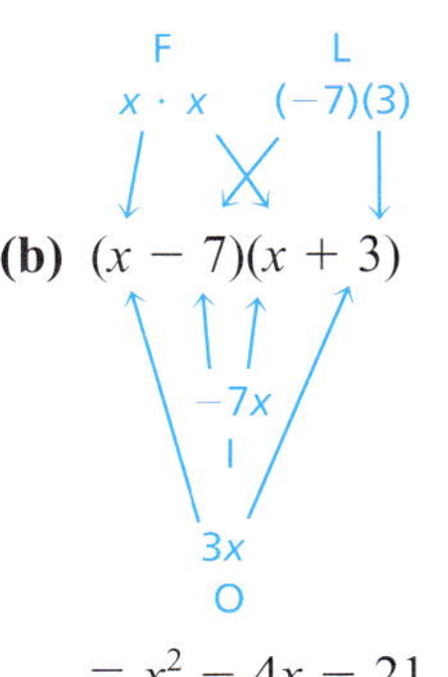

Combine the outer and inner products as $-4x$.

$= x^2 - 4x - 21$

CHECK YOURSELF 9

Multiply.

(a) $(x + 6)(x + 7)$ **(b)** $(x + 3)(x - 5)$ **(c)** $(x - 2)(x - 8)$

Using the FOIL method, you can also find the product of binomials with coefficients other than 1 or with more than one variable.

Example 10

Using the FOIL Method

Find the following products, using the FOIL method.

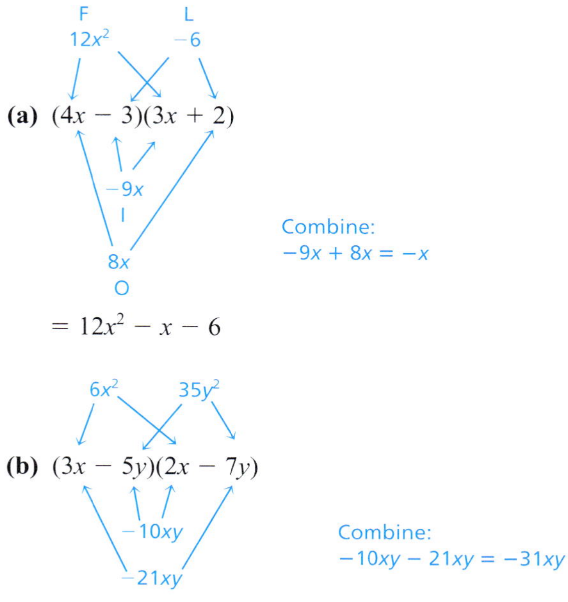

$= 12x^2 - x - 6$

(b) $(3x - 5y)(2x - 7y)$

$6x^2$, $35y^2$, $-10xy$, $-21xy$

Combine:
$-10xy - 21xy = -31xy$

$= 6x^2 - 31xy + 35y^2$

CHECK YOURSELF 10

Multiply.

(a) $(5x + 2)(3x - 7)$ **(b)** $(4a - 3b)(5a - 4b)$
(c) $(3m + 5n)(2m + 3n)$

This rule summarizes our work in multiplying binomials.

Step by Step: To Multiply Two Binomials

Step 1 Find the first term of the product of the binomials by multiplying the first terms of the binomials (F).

Step 2 Find the outer and inner terms of the product. If they are like terms, add them to create the middle term (O + I).

Step 3 Find the last term of the product by multiplying the last terms of the binomials (L).

Sometimes, especially with larger polynomials, it is easier to use the vertical method to find their product. This is the same method you originally learned when multiplying two large integers.

Example 11

Multiplying Using the Vertical Method

Use the vertical method to find the product of $(3x + 2)(4x - 1)$.

First, we rewrite the multiplication in vertical form.

NOTE We have chosen to write $4x - 1$ as $4x + (-1)$.

$$\begin{array}{l} 3x + 2 \\ \underline{4x + (-1)} \end{array}$$

Multiplying the quantity $3x + 2$ by -1 yields

$$\begin{array}{r} 3x + 2 \\ \underline{4x + (-1)} \\ -3x + (-2) \end{array}$$

Note that we maintained the columns of the original binomial when we found the product. We will continue with those columns as we multiply by the $4x$ term.

$$\begin{array}{r} 3x + 2 \\ \underline{4x + (-1)} \\ -3x + (-2) \\ \underline{12x^2 + 8x \qquad\quad} \\ 12x^2 + 5x + (-2) \end{array}$$

We could write the product as $(3x + 2)(4x - 1) = 12x^2 + 5x - 2$.

CHECK YOURSELF 11

Use the vertical method to find the product of $(5x - 3)(2x + 1)$.

We'll use the vertical method again in Example 12. This time, we will multiply a binomial and a trinomial. Note that the FOIL method can never work for anything but the product of two binomials.

Example 12

Using the Vertical Method

Multiply $x^2 - 5x + 8$ by $x + 3$.

Step 1

$$\begin{array}{rrr} x^2 - & 5x + & 8 \\ & x + & 3 \\ \hline 3x^2 - & 15x + & 24 \end{array}$$

Multiply each term of $x^2 - 5x + 8$ by 3.

Step 2

$$\begin{array}{rrrr} & x^2 - & 5x + & 8 \\ & & x + & 3 \\ \hline & 3x^2 - & 15x + & 24 \\ x^3 - & 5x^2 + & 8x & \\ \hline \end{array}$$

Now multiply each term by x.

Note that this line is shifted over so that like terms are in the same columns.

NOTE Using this vertical method ensures that each term of one factor multiplies each term of the other. That's why it works!

Step 3

$$\begin{array}{rrrr} & x^2 - & 5x + & 8 \\ & & x + & 3 \\ \hline & 3x^2 - & 15x + & 24 \\ x^3 - & 5x^2 + & 8x & \\ \hline x^3 - & 2x^2 - & 7x + & 24 \end{array}$$

Now add like terms to write the product.

CHECK YOURSELF 12

Multiply $2x^2 - 5x + 3$ by $3x + 4$.

CHECK YOURSELF ANSWERS

1. **(a)** 7^9; **(b)** z^8; **(c)** $28x^{11}$ **2.** **(a)** m^{30}; **(b)** m^{11}; **(c)** 3^8; **(d)** 3^6
3. **(a)** $81y^4$; **(b)** $64m^6n^6$; **(c)** $48x^2$; **(d)** $5x^3$ **4.** **(a)** $m^{15}n^6$; **(b)** $256p^8$
5. **(a)** $15a^4b^5$; **(b)** $-12x^4y^6$ **6.** **(a)** $2y^3 + 6y^2$; **(b)** $6w^5 + 15w^3$
7. **(a)** $15a^2 + 6a + 21$; **(b)** $32x^5 - 24x^2$; **(c)** $-40m^3 + 25m^2$; **(d)** $27a^5b^2 - 54a^4b^5$
8. **(a)** $x^2 + 9x + 20$; **(b)** $y^2 - y - 30$
9. **(a)** $x^2 + 13x + 42$; **(b)** $x^2 - 2x - 15$; **(c)** $x^2 - 10x + 16$
10. **(a)** $15x^2 - 29x - 14$; **(b)** $20a^2 - 31ab + 12b^2$; **(c)** $6m^2 + 19mn + 15n^2$
11. $10x^2 - x - 3$ **12.** $6x^3 - 7x^2 - 11x + 12$

Name ______________

Section ______ Date ______

10.3 Exercises

Use the product property of exponents to simplify each of the expressions.

1. $x^2 \cdot x^3$

2. $a^5 \cdot a^3$

3. $5^7 \cdot 5^8$

4. $8^4 \cdot 8^6$

5. $2y^4 \cdot 6y^3$

6. $5m^9 \cdot 3m^2$

Use the power property of exponents to simplify each of the expressions.

7. $(x^2)^3$

8. $(a^5)^3$

9. $(m^4)^4$

10. $(p^7)^2$

11. $(2^4)^2$

12. $(3^3)^2$

13. $(5^3)^5$

14. $(7^2)^4$

Use the properties of exponents to simplify each of the expressions.

15. $(3x)^3$

16. $(4m)^2$

17. $(2xy)^4$

18. $(5pq)^3$

19. $5(3ab)^3$

20. $4(2rs)^4$

21. $(2x^2)^4$

22. $(3y^2)^5$

23. $(a^8b^6)^2$

24. $(p^3q^4)^2$

25. $(4x^2y)^3$

26. $(4m^4n^4)^2$

Solve the problems.

27. Write x^{12} as a power of x^2.

28. Write y^{15} as a power of y^3.

29. Write a^{16} as a power of a^2.

30. Write m^{20} as a power of m^5.

ANSWERS

1. ______ 2. ______
3. ______ 4. ______
5. ______ 6. ______
7. ______ 8. ______
9. ______ 10. ______
11. ______ 12. ______
13. ______
14. ______
15. ______
16. ______
17. ______
18. ______
19. ______
20. ______
21. ______
22. ______
23. ______
24. ______
25. ______
26. ______
27. ______
28. ______
29. ______
30. ______

ANSWERS

31. __________

32. __________

33. __________

34. __________

35. __________

36. __________

37. __________

38. __________

39. __________

40. __________

41. __________

42. __________

43. __________

44. __________

45. __________

46. __________

47. __________

48. __________

49. __________

50. __________

51. __________

52. __________

53. __________

54. __________

55. __________

56. __________

31. Write each as a power of 8. (Remember that $8 = 2^3$.)

$2^{12}, 2^{18}, (2^5)^3, (2^7)^6$

32. Write each as a power of 9.

$3^8, 3^{14}, (3^5)^8, (3^4)^7$

33. Write an explanation of why $(x^3)(x^4)$ is *not* x^{12}.

34. Your algebra study partners are confused. "Why isn't $x^2 \cdot x^3 = 2x^5$?", they ask you. Write an explanation that will convince them.

Multiply.

35. $(-2b^2)(14b^8)$

36. $(14y^4)(-4y^6)$

37. $(-10p^6)(-4p^7)$

38. $(-6m^8)(9m^7)$

39. $(-3m^5n^2)(2m^4n)$

40. $(7a^3b^5)(-6a^4b)$

41. $5(2x + 6)$

42. $4(7b - 5)$

43. $3a(4a + 5)$

44. $5x(2x - 7)$

45. $3s^2(4s^2 - 7s)$

46. $9a^2(3a^3 + 5a)$

47. $2x(4x^2 - 2x + 1)$

48. $5m(4m^3 - 3m^2 + 2)$

49. $3xy(2x^2y + xy^2 + 5xy)$

50. $5ab^2(ab - 3a + 5b)$

Multiply.

51. $(x + 3)(x + 2)$

52. $(a - 3)(a - 7)$

53. $(m - 5)(m - 9)$

54. $(b + 7)(b + 5)$

55. $(p - 8)(p + 7)$

56. $(x - 10)(x + 9)$

57. $(3x - 5)(x - 8)$

58. $(w + 5)(4w - 7)$

59. $(2x - 3)(3x + 4)$

60. $(5a + 1)(3a + 7)$

61. $(3p - 4q)(7p + 5q)$

62. $(5x - 4y)(2x - y)$

63. $(x + 5)(x + 5)$

64. $(y + 8)(y + 8)$

65. $(y - 9)(y - 9)$

66. $(2a + 3)(2a + 3)$

67. $(6m + n)(6m + n)$

68. $(7b - c)(7b - c)$

69. $(a - 5)(a + 5)$

70. $(x - 7)(x + 7)$

71. $(x - 2y)(x + 2y)$

72. $(7x + y)(7x - y)$

Multiply, using the vertical method.

73. $(x + 2)(3x + 5)$

74. $(a - 3)(2a + 7)$

75. $(2m - 5)(3m + 7)$

76. $(5p + 3)(4p + 1)$

77. $(a^2 + 3ab - b^2)(a^2 - 5ab + b^2)$

78. $(m^2 - 5mn + 3n^2)(m^2 + 4mn - 2n^2)$

79. $(x - 2y)(x^2 + 2xy + 4y^2)$

80. $(m + 3n)(m^2 - 3mn + 9n^2)$

Label each statement as true or false.

81. $(x + y)^2 = x^2 + y^2$

82. $(x - y)^2 = x^2 - y^2$

83. $(x + y)^2 = x^2 + 2xy + y^2$

84. $(x - y)^2 = x^2 - 2xy + y^2$

85. **Area.** The length of a rectangle is given by $3x + 5$ cm and the width is given by $2x - 7$ cm. Express the area of the rectangle in terms of x.

ANSWERS

57. ______
58. ______
59. ______
60. ______
61. ______
62. ______
63. ______
64. ______
65. ______
66. ______
67. ______
68. ______
69. ______
70. ______
71. ______
72. ______
73. ______
74. ______
75. ______
76. ______
77. ______
78. ______
79. ______
80. ______
81. ______ 82. ______
83. ______ 84. ______
85. ______

86. ______

87. ______

88. ______

86. Area. The base of a triangle measures $3y + 7$ in. and the height is $2y - 3$ in. Express the area of the triangle in terms of y.

87. Revenue. The price of an item is given by $p = 2x - 10$. If the revenue generated is found by multiplying the number of items (x) sold by the price of an item, find the polynomial which represents the revenue.

88. Revenue. The price of an item is given by $p = 2x^2 - 100$. Find the polynomial that represents the revenue generated from the sale of x items.

Answers

1. x^5 **3.** 5^{15} **5.** $12y^7$ **7.** x^6 **9.** m^{16} **11.** 2^8 **13.** 5^{15}
15. $27x^3$ **17.** $16x^4y^4$ **19.** $135a^3b^3$ **21.** $16x^8$ **23.** $a^{16}b^{12}$
25. $64x^6y^3$ **27.** $(x^2)^6$ **29.** $(a^2)^8$ **31.** $8^4, 8^6, 8^5, 8^{14}$ **33.**
35. $-28b^{10}$ **37.** $40p^{13}$ **39.** $-6m^9n^3$ **41.** $10x + 30$
43. $12a^2 + 15a$ **45.** $12s^4 - 21s^3$ **47.** $8x^3 - 4x^2 + 2x$
49. $6x^3y^2 + 3x^2y^3 + 15x^2y^2$ **51.** $x^2 + 5x + 6$ **53.** $m^2 - 14m + 45$
55. $p^2 - p - 56$ **57.** $3x^2 - 29x + 40$ **59.** $6x^2 - x - 12$
61. $21p^2 - 13pq - 20q^2$ **63.** $x^2 + 10x + 25$ **65.** $y^2 - 18y + 81$
67. $36m^2 + 12mn + n^2$ **69.** $a^2 - 25$ **71.** $x^2 - 4y^2$ **73.** $3x^2 + 11x + 10$
75. $6m^2 - m - 35$ **77.** $a^4 - 2a^3b - 15a^2b^2 + 8ab^3 - b^4$ **79.** $x^3 - 8y^3$
81. False **83.** True **85.** $6x^2 - 11x - 35$ cm^2 **87.** $2x^2 - 10x$

10.4 Introduction to Factoring Polynomials

OBJECTIVES

1. Determine the greatest common factor (GCF) of a list of terms
2. Factor the GCF from the terms of a polynomial

In Section 10.3 you were given factors and asked to find a product. We are now going to reverse the process. You will be given a polynomial and asked to find its factors. This is called **factoring.**

We will start with an example from arithmetic. To *multiply* $5 \cdot 7$, you write

$$5 \cdot 7 = 35$$

To *factor* 35, you would write

$$35 = 5 \cdot 7$$

Factoring is the *reverse* of multiplication.

Now we will look at factoring in algebra. You have used the distributive property as

$$a(b + c) = ab + ac$$

For instance,

NOTE 3 and $x + 5$ are the factors of $3x + 15$.

$$3(x + 5) = 3x + 15$$

To use the distributive property in factoring, we apply that property in the opposite fashion, as

$$ab + ac = a(b + c)$$

The property lets us remove the common monomial factor a from the terms of $ab - ac$. To use this in factoring, the first step is to see whether each term of the polynomial has a common monomial factor. We can apply this to our equation $3(x + 5) = 3x + 15$.

$$3x + 15 = 3 \cdot x + 3 \cdot 5$$

Common factor (the 3 in each term)

So, by the distributive property,

$$3x + 15 = 3(x + 5)$$

The original terms are each divided by the greatest common factor to determine the terms in parentheses.

NOTE Again, factoring is the reverse of multiplication.

To check this, multiply $3(x + 5)$.

NOTE This diagram relates the idea of multiplication and factoring.

Multiplying →

$$3(x + 5) = 3x + 15$$

← Factoring

The first step in factoring is to identify the *greatest common factor* (GCF) of a set of terms. This is the monomial with the largest common numerical coefficient and the largest power common to any variables.

Definition: Greatest Common Factor

The **greatest common factor (GCF)** of a polynomial is the monomial with the largest powers and the largest numerical coefficient that is a factor of each term of the polynomial.

Example 1

Finding the GCF

Find the GCF for each set of terms.

(a) 9 and 12

NOTE This technique for finding the GCF was studied earlier in Chapter 3.

The prime factorizations of 9 and 12 are

$9 = 3 \cdot \textcircled{3}$

$12 = 2 \cdot 2 \cdot \textcircled{3}$

We have circled the common prime factors of 9 and 12, and we see that the GCF of 9 and 12 is 3.

(b) 10, 25, and 150

The prime factorizations of 10, 25, and 150 are

$10 = 2 \cdot \textcircled{5}$

$25 = \textcircled{5} \cdot 5$

$150 = 2 \cdot 3 \cdot \textcircled{5} \cdot 5$

We see from this that the GCF is 5.

(c) x^4 and x^7

The prime factorizations of x^4 and x^7 are

$x^4 = \textcircled{x} \cdot \textcircled{x} \cdot \textcircled{x} \cdot \textcircled{x}$

$x^7 = \textcircled{x} \cdot \textcircled{x} \cdot \textcircled{x} \cdot \textcircled{x} \cdot x \cdot x \cdot x$

We see from this that the GCF is $x \cdot x \cdot x \cdot x$, or x^4.

(d) $12a^3$ and $18a^2$

NOTE With practice, you will be able to find the GCF entirely in your head.

The prime factorizations of $12a^3$ and $18a^2$ are

$12a^3 = \textcircled{2} \cdot 2 \cdot \textcircled{3} \cdot \textcircled{a} \cdot \textcircled{a} \cdot a$

$18a^2 = \textcircled{2} \cdot \textcircled{3} \cdot 3 \cdot \textcircled{a} \cdot \textcircled{a}$

We see from this that the GCF is $2 \cdot 3 \cdot a \cdot a$, or $6a^2$.

CHECK YOURSELF 1

Find the GCF for each set of terms.

(a) 14, 24 **(b)** 9, 27, 81

(c) a^9, a^5 **(d)** $10x^5, 35x^4$

Step by Step: To Factor a Monomial from a Polynomial

Step 1 Find the GCF for all the terms.

Step 2 Factor the GCF from each term, then apply the distributive property.

Step 3 Mentally check your factoring by multiplication.

NOTE Checking your answer is always important and perhaps is never easier than after you have factored.

Example 2

Finding the GCF of a Binomial

(a) Factor $8x^2 + 12x$.

The largest common numerical factor of 8 and 12 is 4, and the greatest common variable factor is x. So $4x$ is the GCF. Write

NOTE It is always a good idea to check your answer by multiplying to make sure that you get the original polynomial. Try it here. Multiply $4x$ by $2x + 3$.

$$8x^2 + 12x = 4x \cdot 2x + 4x \cdot 3$$

GCF

Now, by the distributive property, we have

$$8x^2 + 12x = 4x(2x + 3)$$

(b) Factor $6a^4 - 18a^2$.

The GCF in this case is $6a^2$. Write

NOTE It is also true that $6a^4 + (-18a^2) = 3a(2a^3 + (-6a))$. However, this is *not completely factored.* Do you see why? You want to find the common monomial factor with the *largest possible* coefficient and the *largest* exponent, in this case $6a^2$.

$$6a^4 + (-18a^2) = 6a^2 \cdot a^2 + 6a^2 \cdot (-3)$$

GCF

Again, using the distributive property yields

$$6a^4 - 18a^2 = 6a^2(a^2 - 3)$$

You should check this by multiplying.

CHECK YOURSELF 2

Factor each of the polynomials.

(a) $5x + 20$ **(b)** $6x^2 - 24x$ **(c)** $10a^3 - 15a^2$

The process is exactly the same for polynomials with more than two terms. Consider Example 3.

Example 3

Finding the GCF of a Polynomial

(a) Factor $5x^2 - 10x + 15$.

NOTE The GCF is 5.

$$5x^2 - 10x + 15 = 5 \cdot x^2 - 5 \cdot 2x + 5 \cdot 3$$

GCF

$$= 5(x^2 - 2x + 3)$$

(b) Factor $6ab + 9ab^2 - 15a^2$.

NOTE The GCF is $3a$.

$$6ab + 9ab^2 - 15a^2 = 3a \cdot 2b + 3a \cdot 3b^2 - 3a \cdot 5a$$

GCF

$$= 3a(2b + 3b^2 - 5a)$$

(c) Factor $4a^4 + 12a^3 - 20a^2$.

NOTE The GCF is $4a^2$.

$$4a^4 + 12a^3 - 20a^2 = 4a^2 \cdot a^2 + 4a^2 \cdot 3a - 4a^2 \cdot 5$$

GCF

$$= 4a^2(a^2 + 3a - 5)$$

NOTE In each of these examples, you will want to check the result by multiplying the factors.

(d) Factor $6a^2b + 9ab^2 + 3ab$.

Mentally note that 3, a, and b are factors of each term, so

$$6a^2b + 9ab^2 + 3ab = 3ab(2a + 3b + 1)$$

CHECK YOURSELF 3

Factor each of the polynomials.

(a) $8b^2 + 16b - 32$

(b) $4xy - 8x^2y + 12x^3$

(c) $7x^4 - 14x^3 + 21x^2$

(d) $5x^2y^2 - 10xy^2 + 15x^2y$

CHECK YOURSELF ANSWERS

1. **(a)** 2; **(b)** 9; **(c)** a^5; **(d)** $5x^4$ **2.** **(a)** $5(x + 4)$; **(b)** $6x(x - 4)$; **(c)** $5a^2(2a - 3)$

3. **(a)** $8(b^2 + 2b - 4)$; **(b)** $4x(y - 2xy + 3x^2)$; **(c)** $7x^2(x^2 - 2x + 3)$; **(d)** $5xy(xy - 2y + 3x)$

Name ______________

Section ________ Date ________

10.4 Exercises

Find the greatest common factor for each of the sets of terms.

1. 10, 12

2. 15, 35

3. 16, 32, 88

4. 55, 33, 132

5. x^2, x^5

6. y^7, y^9

7. a^3, a^6, a^9

8. b^4, b^6, b^8

9. $5x^4, 10x^5$

10. $8y^9, 24y^3$

11. $8a^4, 6a^6, 10a^{10}$

12. $9b^3, 6b^5, 12b^4$

13. $9x^2y, 12xy^2, 15x^2y^2$

14. $12a^3b^2, 18a^2b^3, 6a^4b^4$

15. $15ab^3, 10a^2bc, 25b^2c^3$

16. $9x^2, 3xy^3, 6y^3$

17. $15a^2bc^2, 9ab^2c^2, 6a^2b^2c^2$

18. $18x^3y^2z^3, 27x^4y^2z^3, 81xy^2z$

19. xy^2z^3, x^3y^2z

20. $12ab^9, 9ab^{12}$

Factor each of the polynomials.

21. $8a + 4$

22. $5x - 15$

23. $24m - 32n$

24. $7p - 21q$

25. $12m^2 + 8m$

26. $24n^2 - 32n$

27. $10s^2 + 5s$

28. $12y^2 - 6y$

29. $12x^2 + 24x$

30. $14b^2 - 28b$

ANSWERS

1. ______ 2. ______
3. ______ 4. ______
5. ______ 6. ______
7. ______ 8. ______
9. ______ 10. ______
11. ______ 12. ______
13. ______
14. ______
15. ______
16. ______
17. ______
18. ______
19. ______
20. ______
21. ______
22. ______
23. ______
24. ______
25. ______
26. ______
27. ______
28. ______
29. ______
30. ______

ANSWERS

31. ________
32. ________
33. ________
34. ________
35. ________
36. ________
37. ________
38. ________
39. ________
40. ________
41. ________
42. ________
43. ________
44. ________
45. ________
46. ________
47. ________
48. ________
49. ________
50. ________
51. ________
52. ________

31. $15a^3 - 25a^2$

32. $36b^4 + 24b^2$

33. $6pq + 18p^2q$

34. $8ab - 24ab^2$

35. $7m^3n - 21mn^3$

36. $36p^2q^2 - 9pq$

37. $6x^2 - 18x + 30$

38. $7a^2 + 21a - 42$

39. $3a^3 + 6a^2 - 12a$

40. $5x^3 - 15x^2 + 25x$

41. $6m + 9mn - 15mn^2$

42. $4s + 6st - 14st^2$

43. $10x^2y + 15xy - 5xy^2$

44. $3ab^2 + 6ab - 15a^2b$

45. $10r^3s^2 + 25r^2s^2 - 15r^2s^3$

46. $28x^2y^3 - 35x^2y^2 + 42x^3y$

47. $9a^5 - 15a^4 + 21a^3 - 27a$

48. $8p^6 - 40p^4 + 24p^3 - 16p^2$

49. $15m^3n^2 - 20m^2n + 35mn^3 - 10mn$

50. $14ab^4 + 21a^2b^3 - 35a^3b^2 + 28ab^2$

51. **Dimensions of a rectangle.** The area of a rectangle with width t is given by $33t - t^2$. Factor the expression and determine the length of the rectangle in terms of t.

52. **Dimensions of a rectangle.** The area of a rectangle of length x is given by $3x^2 + 5x$. Find the width of the rectangle.

Answers

1. 2 **3.** 8 **5.** x^2 **7.** a^3 **9.** $5x^4$ **11.** $2a^4$ **13.** $3xy$
15. $5b$ **17.** $3abc^2$ **19.** xy^2z **21.** $4(2a + 1)$ **23.** $8(3m - 4n)$
25. $4m(3m + 2)$ **27.** $5s(2s + 1)$ **29.** $12x(x + 2)$ **31.** $5a^2(3a - 5)$
33. $6pq(1 + 3p)$ **35.** $7mn(m^2 - 3n^2)$ **37.** $6(x^2 - 3x + 5)$
39. $3a(a^2 + 2a - 4)$ **41.** $3m(2 + 3n - 5n^2)$ **43.** $5xy(2x + 3 - y)$
45. $5r^2s^2(2r + 5 - 3s)$ **47.** $3a(3a^4 - 5a^3 + 7a^2 - 9)$
49. $5mn(3m^2n - 4m + 7n^2 - 2)$ **51.** $t(33 - t)$; $33 - t$

10 Summary

Definition/Procedure	Example	Reference
Introduction to Polynomials		**Section 10.1**
Term A term is a number, or the product of a number and variables.		
Polynomial A polynomial is an algebraic expression made up of terms in which the exponents are whole numbers. These terms are connected by plus or minus signs. Each sign (+ or −) is attached to the term following that sign.	$4x^3 - 3x^2 + 5x$ is a polynomial. The terms of $4x^3 - 3x^2 + 5x$ are $4x^3$, $-3x^2$, and $5x$.	p. 737
Coefficient In each term of a polynomial, the number in front of the variable is called the *numerical coefficient* or, more simply, the *coefficient* of that term.	The coefficients of $4x^3 - 3x^2$ are 4 and −3.	p. 737
Types of Polynomials A polynomial can be classified according to the number of terms it has. A *mono*mial has one term. A *bi*nomial has two terms. A *tri*nomial has three terms.	$2x^3$ is a monomial. $3x^2 - 7x$ is a binomial. $5x^5 - 5x^3 + 2$ is a trinomial.	p. 738
Addition and Subtraction of Polynomials		**Section 10.2**
Removing Signs of Grouping 1. If a plus sign (+) or no sign at all appears in front of parentheses, just remove the parentheses. No other changes are necessary. 2. If a minus sign (−) appears in front of parentheses, the parentheses can be removed by changing the sign of each term inside the parentheses.	$3x + (2x - 3)$ $= 3x + 2x - 3$ $2x - (x - 4)$ $= 2x - x + 4$	p. 743 p. 745
Adding Polynomials Remove the signs of grouping. Then collect and combine any like terms.	$(2x + 3) + (3x - 5)$ $= 2x + 3 + 3x - 5 = 5x - 2$	p. 743
Subtracting Polynomials Remove the signs of grouping by changing the sign of each term in the polynomial being subtracted. Then combine any like terms.	$(3x^2 + 2x) - (2x^2 + 3x - 1)$ $= 3x^2 + 2x - 2x^2 - 3x + 1$ Signs change $3x^2 - 2x^2 + 2x - 3x + 1$ $x^2 - x + 1$	p. 745

Continued

DEFINITION/PROCEDURE	EXAMPLE	REFERENCE
Multiplying Polynomials		**Section 10.3**
Properties of Exponents **1.** $a^m \cdot a^n = a^{m+n}$ **2.** $(a^m)^n = a^{mn}$ **3.** $(ab)^m = a^m b^m$	$3^3 \cdot 3^4 = 3^7$ $(2^3)^5 = 2^{15}$ $(3x)^2 = 9x^2$	**p. 753** **p. 754** **p. 754**
To Multiply a Polynomial by a Monomial Multiply each term of the polynomial by the monomial, and simplify the results.	$3x(2x + 3)$ $= 3x \cdot 2x + 3x \cdot 3$ $= 6x^2 + 9x$	**p. 756**
To Multiply a Binomial by a Binomial Use the FOIL method: F O I L $(a + b)(c + d) = a \cdot c + a \cdot d + b \cdot c + b \cdot d$	$(2x - 3)(3x + 5)$ $= 6x^2 + 10x - 9x - 15$ F O I L $= 6x^2 + x - 15$	**p. 758**
To Multiply a Polynomial by a Polynomial Arrange the polynomials vertically. Multiply each term of the upper polynomial by each term of the lower polynomial, and add the results.	$x^2 - 3x + 5$ $2x - 3$ $-3x^2 + 9x - 15$ $2x^3 - 6x^2 + 10x$ $2x^3 - 9x^2 + 19x - 15$	**p. 761**
Introduction to Factoring Polynomials		**Section 10.4**
Common Monomial Factor A common monomial factor is a single term that is a factor of every term of the polynomial. The greatest common factor (GCF) of a polynomial is the common monomial factor that has the largest possible numerical coefficient and the largest possible exponents.	$4x^2$ is the greatest common factor of $8x^4 - 12x^3 + 16x^2$.	**p. 767**
Factoring a Monomial from a Polynomial **1.** Determine the GCF for all terms. **2.** Factor the GCF from each term, and then apply the distributive property in the form $ab + ac = a(b + c)$ ↑ The greatest common factor **3.** Mentally check by multiplication.	$8x^4 - 12x^3 + 16x^2$ $= 4x^2(2x^2 - 3x + 4)$	**p. 768**

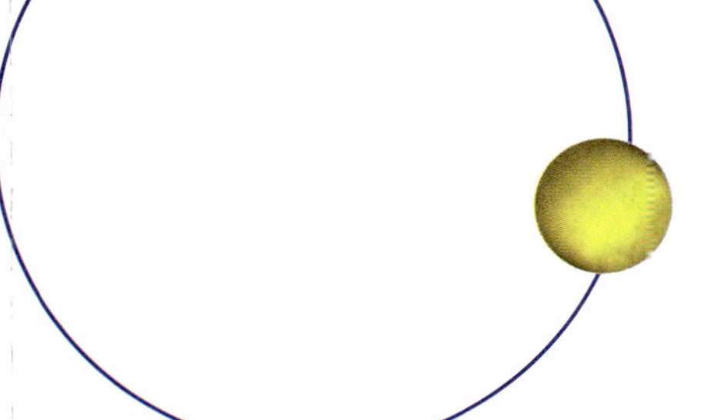

Summary and Review Exercises

This exercise set will give you practice with each of the objectives of the chapter. Each exercise is keyed to the appropriate chapter section. The answers are provided in the *Instructor's Manual.* Your instructor will give you guidelines on how to best use these exercises in your instructional setting.

[10.1] Classify each of the polynomials as monomial, binomial, or trinomial, where possible.

1. $5x^3 - 2x^2$

2. $7x^5$

3. $4x^5 - 8x^3 + 5$

4. $x^3 + 2x^2 - 5x$

5. $9a^2 - 18a^2$

Find the value of each of the polynomials for the given value of the variable.

6. $5x + 1; x = -1$

7. $2x^2 + 7x - 5; x = 2$

8. $-x^2 + 3x - 1; x = 6$

9. $4x^2 + 5x + 7; x = -4$

[10.2] Add.

10. $9a^2 - 5a$ and $12a^2 + 3a$

11. $5x^2 + 3x - 5$ and $4x^2 - 6x - 2$

12. $5y^3 - 3y^2$ and $4y + 3y^2$

Subtract.

13. $4x^2 - 3x$ from $8x^2 + 5x$

14. $2x^2 - 5x - 7$ from $7x^2 - 2x + 3$

15. $5x^2 + 3$ from $9x^2 - 4x$

Perform the indicated operations.

16. Subtract $5x - 3$ from the sum of $9x + 2$ and $-3x - 7$.

17. Subtract $5a^2 - 3a$ from the sum of $5a^2 + 2$ and $7a - 7$.

18. Subtract the sum of $16w^2 - 3w$ and $8w + 2$ from $7w^2 - 5w + 2$.

Add, using the vertical method.

19. $x^2 + 5x - 3$ and $2x^2 + 4x - 3$

20. $9b^2 - 7$ and $8b + 5$

21. $x^2 + 7$, $3x - 2$, and $4x^2 - 8x$

Subtract, using the vertical method.

22. $5x^2 - 3x + 2$ from $7x^2 - 5x - 7$

23. $8m - 7$ from $9m^2 - 7$

[10.3] Simplify each of the expressions.

24. $(2ab)^2$

25. $(p^2q^3)^3$

26. $(2x^2y^2)^3(3x^3y)^2$

27. $(4w^2t)^2\,(3wt^2)^3$

28. $(y^3)^2(3y^2)^3$

Multiply.

29. $(5a^3)(a^2)$

30. $(2x^2)(3x^5)$

31. $(-9p^3)(-6p^2)$

32. $(3a^2b^3)(-7a^3b^4)$

33. $5(3x - 8)$

34. $4a(3a + 7)$

35. $(-5rs)(2r^2s - 5rs)$

36. $7mn(3m^2n - 2mn^2 + 5mn)$

37. $(x + 5)(x + 4)$

38. $(w - 9)(w - 10)$

39. $(a - 7b)(a + 7b)$

40. $(p - 3q)(p - 3q)$

41. $(a + 4b)(a + 3b)$

42. $(b - 8)(2b + 3)$

43. $(3x - 5y)(2x - 3y)$

44. $(5r + 7s)(3r - 9s)$

45. $(y + 2)(y^2 - 2y + 3)$

46. $(b + 3)(b^2 - 5b - 7)$

47. $(x - 2)(x^2 + 2x + 4)$

48. $(m^2 - 3)(m^2 + 7)$

49. $2x(x + 5)(x - 6)$

50. $a(2a - 5b)(2a - 7b)$

51. $(x + 7)(x + 7)$

52. $(a - 8)(a - 8)$

53. $(2w - 5)(2w - 5)$

54. $(3p + 4)(3p + 4)$

55. $(a + 7b)(a + 7b)$

56. $(8x - 3y)(8x - 3y)$

57. $(x - 5)(x + 5)$

58. $(y + 9)(y - 9)$

59. $(2m + 3)(2m - 3)$

60. $(3r - 7)(3r + 7)$

61. $(5r - 2s)(5r + 2s)$

62. $(7a + 3b)(7a - 3b)$

[10.4] Factor each of the polynomials.

63. $18a + 24$

64. $9m^2 - 21m$

65. $24s^2t - 16s^2$

66. $18a^2b + 36ab^2$

67. $35s^3 - 28s^2$

68. $3x^3 - 6x^2 + 15x$

69. $18m^2n^2 - 27m^2n + 45m^2n^3$

70. $121x^8y^3 + 77x^6y^3$

71. $8a^2b + 24ab - 16ab^2$

72. $3x^2y - 6xy^3 + 9x^3y - 12xy^2$

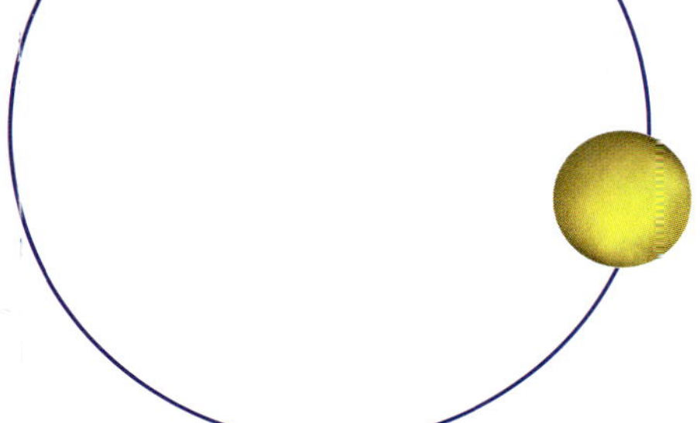

Chapter Test for Chapter 10

Name ______________________

Section ________ Date ________

ANSWERS

1. ______________________
2. ______________________
3. ______________________
4. ______________________
5. ______________________
6. ______________________
7. ______________________
8. ______________________
9. ______________________
10. ______________________
11. ______________________
12. ______________________
13. ______________________
14. ______________________

The purpose of the Chapter Test is to help you check your progress and review for a chapter test in class. When you are done, check your answers in the back of the book. If you missed any answers, be sure to go back and review the appropriate sections in the chapter.

Classify each of the polynomials as monomial, binomial, or trinomial.

1. $6x^2 + 7x$

2. $5x^2 + 8x - 8$

3. Find the value of the polynomial $y = -3x^2 - 5x + 8$ if $x = -2$.

Add.

4. $3x^2 - 7x + 2$ and $7x^2 - 5x - 9$

5. $7a^2 - 3a$ and $7a^3 + 4a^2$

Subtract.

6. $5x^2 - 2x + 5$ from $8x^2 + 9x - 7$

7. $2b^2 + 5$ from $3b^2 - 7b$

8. $5a^2 + a$ from the sum of $3a^2 - 5a$ and $9a^2 - 4a$

Add, using the vertical method.

9. $x^2 + 3$, $5x - 7$, and $3x^2 - 2$

Subtract, using the vertical method.

10. $3x^2 - 5$ from $5x^2 - 7x$

Simplify each of the expressions.

11. $a^5 \cdot a^9$

12. $3x^2y^3 \cdot 5xy^4$

13. $(3x^2y)^3$

14. $(2x^3y^2)^4(x^2y^3)^3$

15. ____
16. ____
17. ____
18. ____
19. ____
20. ____
21. ____
22. ____
23. ____
24. ____
25. ____

Multiply.

15. $5ab(3a^2b - 2ab + 4ab^2)$

16. $(x - 2)(3x + 7)$

17. $(a - 7b)(a + 7b)$

18. $(4x + 3y)(2x - 5y)$

19. $x(3x - y)(4x + 5y)$

20. $(3m + 2n)(3m + 2n)$

21. $(2x + y)(x^2 + 3xy - 2y^2)$

Factor each of the polynomials.

22. $12b + 18$

23. $9p^3 - 12p^2$

24. $5x^2 - 10x + 20$

25. $6a^2b - 18ab + 12ab^2$

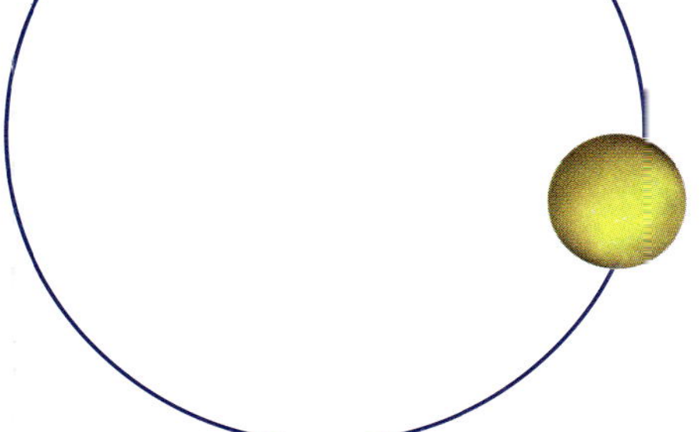

Practice Final Exam

Name ______________________

Section ________ Date ________

ANSWERS

1. ______________________
2. ______________________
3. ______________________
4. ______________________
5. ______________________
6. ______________________
7. ______________________
8. ______________________
9. ______________________
10. ______________________
11. ______________________
12. ______________________
13. ______________________
14. ______________________
15. ______________________
16. ______________________
17. ______________________
18. ______________________

This test is provided to help you in the process of reviewing Chapters 1 through 10. It should help you prepare for a final exam in your course. Answers are provided in the back of the book. If you missed any answers, be sure to go back and review the appropriate chapter sections.

Write the word name for each number.

1. 806,015

2. 17,608

In exercises 3 to 5, perform the indicated operations.

3. $425 - 67$

4. $48 \cdot 65$

5. $247 \div 13$

Solve the exercises.

6. Carlotta made car payments of $355 each month for 48 months. What is the total amount of money paid?

7. Minh's annual salary of $30,240 is paid in 12 equal installments. How much does he receive each month?

Evaluate.

8. $3 \cdot 5^2$

9. $9 - 2^3$

10. $18 \div 2 \cdot 3 - 1 + 2$

In exercises 11 to 21, evaluate each expression.

11. $-7 - (-4)$

12. $8 - 15 + 3$

13. $(-6) \cdot (-9)$

14. $\frac{4}{5} \div \left(-\frac{7}{15}\right)$

15. $-24 \div (8 - 12)$

16. $-18 - 11$

17. $23 - (-8)$

18. $|-16| - |-7|$

ANSWERS

19. ______
20. ______
21. ______
22. ______
23. ______
24. ______
25. ______
26. ______
27. ______
28. ______
29. ______
30. ______
31. ______
32. ______
33. ______
34. ______
35. ______
36. ______
37. ______

19. $(-2)(-4)(-7)$

20. $\dfrac{8 - (4)(-5)}{(5)(-2) + 6}$

21. $9 - 5 \cdot 3^2$

In exercises 22 to 24, evaluate the expressions for the given values of the variables.

22. $b^2 - 4ac$, for $a = 2$, $b = -5$, and $c = -3$

23. $-2x^2 - 3x + 5$, for $x = -3$

24. $x^2 + 2z^2$, for $x = -5$ and $z = 3$

25. Write the prime factorization of 1,260.

26. Find the least common multiple (LCM) of 4, 18, and 45.

27. Reduce this fraction to lowest terms: $\dfrac{60}{132}$.

In exercises 28 to 36, perform the indicated operations.

28. $\dfrac{4}{15} + \dfrac{8}{15}$

29. $\dfrac{3}{7} \cdot \dfrac{5}{12}$

30. $4\dfrac{1}{3} \cdot 2\dfrac{3}{5}$

31. $\dfrac{2}{3} \div \dfrac{4}{9}$

32. $\dfrac{1}{5} \div 5$

33. $\dfrac{3}{4} + \dfrac{5}{8}$

34. $\dfrac{11}{12} - \dfrac{2}{3}$

35. $8\dfrac{1}{4} - 2\dfrac{7}{8}$

36. $\dfrac{5}{9} \cdot \dfrac{36}{15}$

Solve the exercise.

37. Marissa has $6\dfrac{1}{2}$ yd of material. Her new skirt will take $1\dfrac{4}{5}$ yd. How much material will she have left after the skirt is made?

ANSWERS

38. ____
39. ____
40. ____
41. ____
42. ____
43. ____
44. ____
45. ____
46. ____
47. ____
48. ____
49. ____
50. ____
51. ____
52. ____
53. ____
54. ____
55. ____
56. ____
57. ____
58. ____
59. ____

In exercises 38 to 43, solve each equation for x.

38. $x + 3 = -5$

39. $-6x = 42$

40. $5 - 4x = -7$

41. $\frac{x}{4} - 9 = -3$

42. $4x - 5 = -6x + 3$

43. $4x - 2(x - 5) = 16$

44. Write 0.125 as a reduced fraction.

In exercises 45 to 50, perform the indicated operations.

45. $17.289 + 4.93$

46. $0.62 \cdot 0.095$

47. $2.4 \div 0.08$

48. $239.95 - 35.99$

49. $2,537.74 \div 113.8$

50. $9.35 \cdot 34.06$

Solve the exercises.

51. Toni had \$189.75 in her checking account. She then wrote a check for \$39.95. How much money did she have left in her account?

52. A car rental agency charges \$45 per day, plus \$0.29 per mile for a certain type of car. How much is the rental charge for a 6-day trip of 462 mi?

53. Round 0.426839 to the nearest ten thousandth.

54. Write $\frac{3}{8}$ in decimal form.

55. Simplify: $\sqrt{81}$.

56. The length of one of the legs in a right triangle is 8 in. If the hypotenuse is 17 in. long, find the length of the other leg.

57. Solve the proportion for n: $\frac{6}{n} = \frac{9}{57}$.

58. Express $\frac{17}{20}$ as a percent.

59. Write 185% as a decimal.

ANSWERS

60. ______
61. ______
62. ______
63. ______
64. ______
65. ______
66. ______
67. ______
68. ______
69. ______
70. ______
71. ______
72. ______

60. Write 0.08 as a percent.

61. Write $1\frac{1}{5}$ as a percent.

62. Marcia worked on her homework from 7 P.M. until 11 P.M. What percent of the total day did she work on her homework?

63. What is 16% of 80?

64. 64 is 4% of what number?

65. Johanna bought a pair of slacks priced at $42. If the sales tax rate in Johanna's state is 7%, what is the sales tax on the purchase?

66. What is a salesperson's commission on a $1,200 sale if the commission rate is 15%?

67. An article regularly selling for $90 is advertised at 25% off. Find the sales price.

68. A store marks up items to make a 20% profit. If an item sells for $22.20, what was the cost before the markup?

69. Convert 630 cm to m.

70. Find the perimeter of a rectangle 84.5 cm long and 52.5 cm wide.

71. Find the area of the triangle shown.

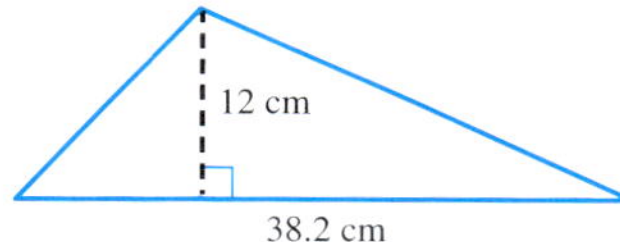

72. The given circle graph represents a family's monthly budget. If the total monthly income is $3,500, how much money is spent on food?

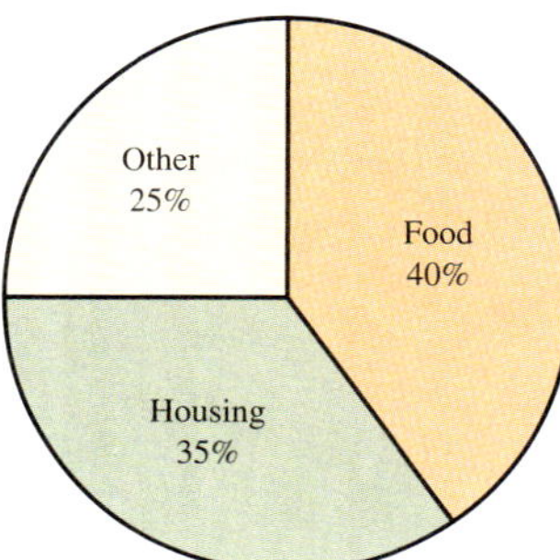

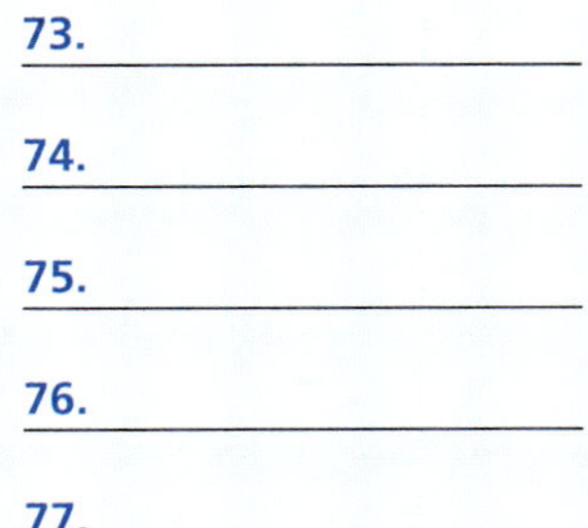

73. The bar graph shown indicates how many spelling words Joseph learned in a 5-week period. How many words did Joseph learn after the second week?

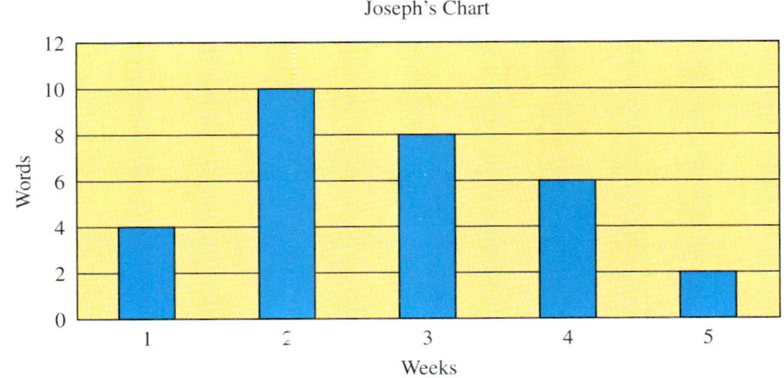

74. The line graph shows sales (in millions of dollars) for XYZ Corporation. How much greater were the sales in 2002 than in 2000?

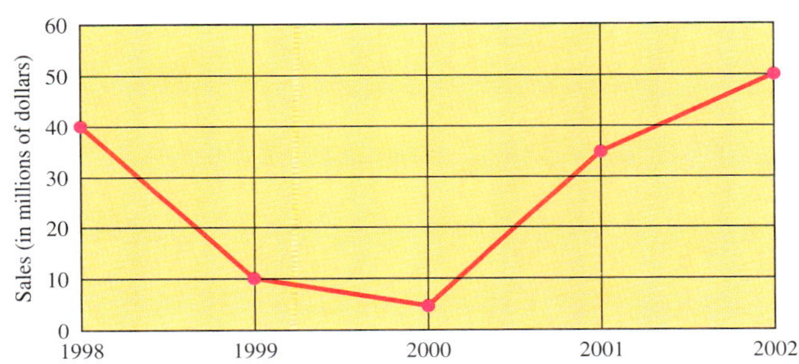

75. Hiro's scores on his first four math tests were 82, 95, 77, and 86. What is the mean of his scores?

76. Hiro's scores on his first four math tests were 82, 95, 77, and 86. What is the median of his scores?

77. Graph the line whose equation is $2x - 3y = 6$.

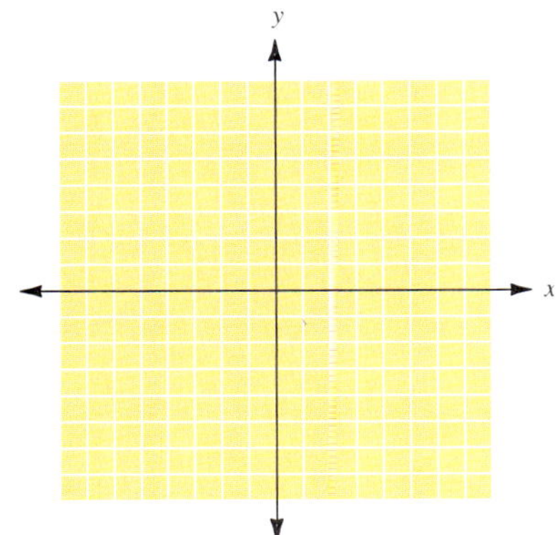

ANSWERS

78. ______

79. ______

80. ______

81. ______

82. ______

83. ______

In exercises 78 to 83, perform the indicated operations. Write each answer in simplified form.

78. $3x(2x^2 - x + 4)$

79. $(3x^2 - 5x - 9) - (4x^2 + 2x - 13)$

80. $(5x - 2)(6x + 5)$

81. $(7x - 2y)(7x + 2y)$

82. $(3x - 4y)(3x - 4y)$

83. $2x(x + 3y) - 4y(2x - y)$

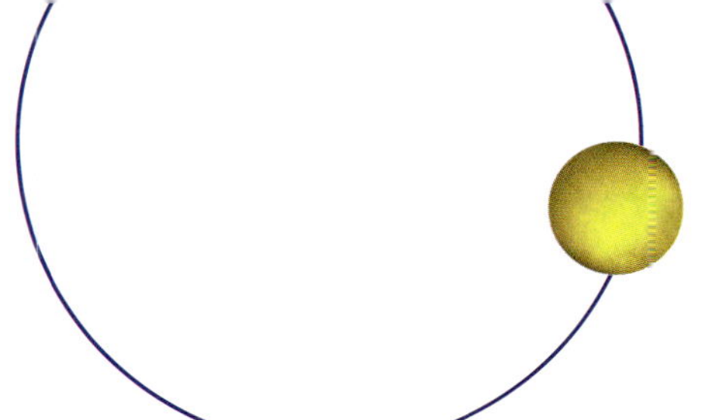

Practice Final Exam Answers

1. Eight hundred six thousand, fifteen
2. Seventeen thousand, six hundred eight **3.** 358 **4.** 3,120
5. 19 **6.** \$17,040 **7.** \$2,520 **8.** 75 **9.** 1 **10.** 28 **11.** -3
12. -4 **13.** 54 **14.** $-\frac{12}{7}$ **15.** 6 **16.** -29 **17.** 31 **18.** 9
19. -56 **20.** -7 **21.** -36 **22.** 49 **23.** -4 **24.** 43
25. $2 \cdot 2 \cdot 3 \cdot 3 \cdot 5 \cdot 7$ **26.** 180 **27.** $\frac{5}{11}$ **28.** $\frac{4}{5}$ **29.** $\frac{5}{28}$ **30.** $11\frac{4}{15}$
31. $1\frac{1}{2}$ **32.** $\frac{1}{25}$ **33.** $1\frac{3}{8}$ **34.** $\frac{1}{4}$ **35.** $5\frac{3}{8}$ **36.** $1\frac{1}{3}$ **37.** $4\frac{7}{10}$ yd
38. $x = -8$ **39.** $x = -7$ **40.** $x = 3$ **41.** $x = 24$ **42.** $x = \frac{4}{5}$
43. $x = 3$ **44.** $\frac{1}{8}$ **45.** 22.219 **46.** 0.0589 **47.** 30 **48.** 203.96
49. 22.3 **50.** 318.461 **51.** \$149.80 **52.** \$403.98 **53.** 0.4268
54. 0.375 **55.** 9 **56.** 15 in. **57.** $n = 38$ **58.** 85% **59.** 1.85
60. 8% **61.** 120% **62.** $16\frac{2}{3}\%$ **63.** 12.8 **64.** 1,600 **65.** \$2.94
66. \$180 **67.** \$67.50 **68.** \$18.50 **69.** 6.3 m **70.** 274 cm

71. 229.2 cm^2 **72.** \$1,400 **73.** 16 words **74.** 45 million dollars
75. 85 **76.** 84
77.

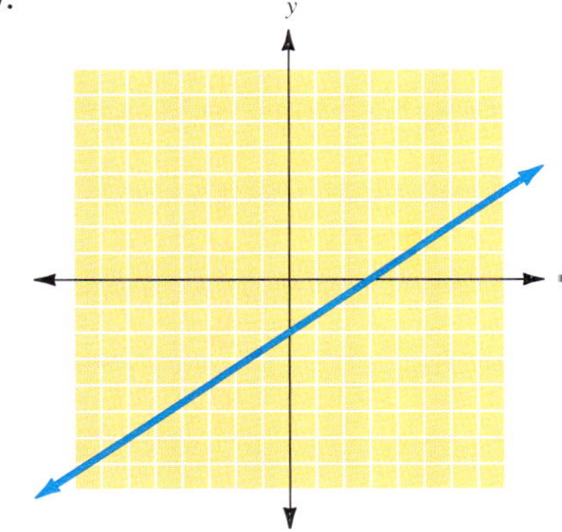

78. $6x^3 - 3x^2 + 12x$ **79.** $-x^2 - 7x + 4$ **80.** $30x^2 + 13x - 10$
81. $49x^2 - 4y^2$ **82.** $9x^2 - 24xy + 16y^2$ **83.** $2x^2 - 2xy + 4y^2$

Answers to Pre-Tests, Chapter Tests and Cumulative Tests

Pre-Test Chapter 1

1. One hundred seven thousand, nine hundred forty-five **2.** Associative property of addition **3.** Commutative property of multiplication **4.** 27,614 **5.** 19,992 **6.** 56,100 **7.** 462 **8.** **(a)** 26; **(b)** 56 **9.** 1204 r3 **10.** 81 r131 **11.** 93 **12.** \$151 **13.** 6 **14.** $P = 14$ yd; $A = 10$ yd^2 **15.** Estimate: 2800; Actual: 2802

Chapter Test for Chapter 1

1. Hundred thousands **2.** Three hundred two thousand, five hundred twenty-five **3.** 2,430,000 **4.** Commutative property of addition **5.** Associative property of addition **6.** 1918 **7.** 13,103 **8.** 21,696 **9.** 55,978 **10.** 235 **11.** 12,220 **12.** 30,770 **13.** 40,555 **14.** 72 lb **15.** 7700 **16.** $>$ **17.** $<$ **18.** Associative property of multiplication **19.** Distributive property of multiplication over addition **20.** Multiplicative identity **21.** 4984 **22.** 55,414 **23.** \$308,750 **24.** 76 r7 **25.** \$223 **26.** 15 **27.** 12 in. **28.** 12 in.^{2} **29.** True **30.** $3^2 - 2 = 7$

Pre-Test Chapter 2

1.

2. $-4, -2, -1, 0, 1, 5$ **3.** Max: 7; Min: -5 **4.** 5 **5.** 6 **6.** 6 **7.** 6 **8.** 6 **9.** 16 **10.** -23 **11.** $x - 8$ **12.** $\frac{w}{17x}$ **13.** No **14.** Yes **15.** -10 **16.** -1 **17.** -5 **18.** -3 **19.** -19 **20.** 12 **21.** 0 **22.** 21 **23.** 7 **24.** -3 **25.** 55 **26.** -7 **27.** $8w^2t$ **28.** $-a^2 + 4a + 3$

Chapter Test for Chapter 2

1.

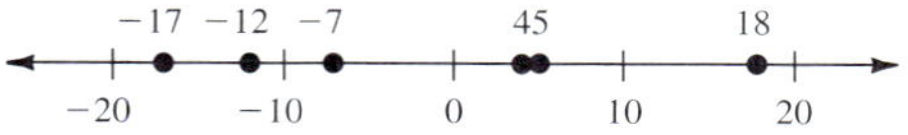

2. $-6, -3, -2, 0, 2, 4, 5$ **3.** 7 **4.** 7 **5.** 11 **6.** 11 **7.** -19 **8.** -40 **9.** 19 **10.** -13 **11.** -3 **12.** -21 **13.** -6 **14.** -24 **15.** 9 **16.** 0 **17.** -40 **18.** 63 **19.** -24 **20.** -25 **21.** 3 **22.** -5 **23.** Undefined **24.** 3 **25.** 65 **26.** 144 **27.** -9 **28.** -4 **29.** $a - 5$ **30.** $6m$ **31.** $4(m + n)$ **32.** $\frac{a + b}{3}$ **33.** Not an expression **34.** Expression **35.** $15a$ **36.** $19x + 5y$ **37.** $8a^2$ **38.** $2x - 8$ **39.** $2w + 4$ **40.** No **41.** Yes **42.** 11 **43.** 12 **44.** 7

Cumulative Test for Chapters 1 and 2

1. Hundred thousands **2.** Three hundred two thousand, five hundred twenty-five **3.** 2,430,000 **4.** Commutative property of addition **5.** Additive identity **6.** Associative property of addition **7.** 966 **8.** 23,351 **9.** 5900 **10.** 950,000 **11.** 7700 **12.** 9 **13.** -9 **14.** 3 **15.** -18 **16.** 0 **17.** -2 **18.** -2 **19.** 0 **20.** 10 **21.** 96 **22.** -90 **23.** -2 **24.** Undefined **25.** -60 **26.** 27 **27.** 24 **28.** 3 **29.** $3x + 19$ **30.** $2x + y$ **31.** 22 **32.** -8

Pre-Test Chapter 3

1. 1, 2, 3, 6, 7, 14, 21, 42 **2.** $2 \cdot 5 \cdot 5 \cdot 7$ **3.** 4 **4.** Yes **5.** $\frac{-3}{5}$ **6.** $\frac{-4}{7}$ **7.** $\frac{4}{5}$ **8.** $\frac{2}{21}$ **9.** -1 **10.** 7

Chapter Test for Chapter 3

1. $\frac{5}{6}$ 5 is the numerator; 6 is the denominator.

2. $\frac{5}{8}$ 5 is the numerator; 8 is the denominator.

3. $\frac{3}{5}$ 3 is the numerator; 5 is the denominator.

4. Proper: $\frac{10}{11}, \frac{1}{8}$; improper: $\frac{9}{5}, \frac{7}{7}, \frac{8}{1}$; mixed number: $2\frac{3}{5}$ **5.** Prime: 5, 13, 17, 31; composite: 9, 22, 27, 45 **6.** 2 and 3 **7.** $2 \cdot 2 \cdot 2 \cdot 3 \cdot 11$ **8.** 12 **9.** 8 **10.** Yes **11.** Yes **12.** No **13.** $\frac{7}{9}$ **14.** $\frac{3}{7}$ **15.** $\frac{8}{23}$ **16.** 4 **17.** 3 **18.** 7 **19.** $\frac{9}{16}$ **20.** $\frac{-4}{15}$ **21.** $\frac{5}{8}$ **22.** $\frac{-3}{20}$ **23.** $\frac{2}{5}$ **24.** 1 **25.** $\frac{49}{9}$ **26.** 7 **27.** -12 **28.** 25 **29.** 3 **30.** 4 **31.** -5 **32.** -5 **33.** 21 **34.** $\frac{1}{2}$ **35.** $\frac{-57}{8}$

Cumulative Test for Chapters 1 to 3

1. 7173 **2.** 1918 **3.** 2731 **4.** 13,103 **5.** -235 **6.** 12,220 **7.** 429 **8.** 3239 **9.** -174 **10.** 1911 **11.** 4984 **12.** 55,414 **13.** 24 r191 **14.** 22 r21 **15.** 209 r145 **16.** 5 **17.** 7 **18.** 3 **19.** 16 **20.** 20 **21.** 3 **22.** Proper: $\frac{5}{7}, \frac{2}{5}$; Improper: $\frac{15}{9}, \frac{8}{8}, \frac{11}{1}$ **23.** Yes **24.** No **25.** $\frac{1}{9}$ **26.** $\frac{1}{6}$ **27.** $\frac{-3}{2}$ **28.** $\frac{-15}{8}$ **29.** $\frac{5}{2}$ **30.** $\frac{-3}{13}$ **31.** $5x$ **32.** $-x + 23$ **33.** $-2x + 1$ **34.** $-2x + 3y$ **35.** $13x + 12y$ **36.** 5 **37.** -24 **38.** $\frac{5}{4}$ **39.** $\frac{-2}{5}$ **40.** 5

Pre-Test Chapter 4

1. **(a)** $\frac{5}{7}$; **(b)** $\frac{5}{9}$ **2.** 48 **3.** **(a)** 200; **(b)** 120 **4.** $\frac{17}{8}$ **5.** $\frac{5}{72}$ **6.** $9\frac{1}{4}$ **7.** $\frac{46}{7}$ **8.** $\frac{21}{5} = 4\frac{1}{5}$ **9.** $\frac{2}{21}$ **10.** $6\frac{17}{24}$ **11.** $1\frac{1}{36}$ **12.** $\frac{9}{20}$ **13.** 14 blocks **14.** $18\frac{1}{4}$ yd^2 **15.** 40 **16.** 30

Chapter Test for Chapter 4

1. $\frac{9}{10}$ **2.** $\frac{2}{3}$ **3.** 72 **4.** 60 **5.** 36 **6.** $\frac{4}{5}$ **7.** $\frac{-25}{42}$ **8.** $\frac{63}{40}$ **9.** $\frac{11}{9}$ **10.** $\frac{-2}{3}$ **11.** $\frac{23}{30}$ **12.** $4\frac{1}{4}$ **13.** $\frac{74}{9}$ **14.** $3\frac{3}{7}$ **15.** 4

16. $2\frac{2}{3}$ **17.** $7\frac{11}{12}$ **18.** $1\frac{3}{4}$ **19.** $1\frac{8}{15}$ **20.** $\frac{4}{5}$ **21.** $\frac{18}{43}$ **22.** 47 homes
23. 48 books **24.** Around 39,000 cups **25.** -32 **26.** $\frac{-4}{3}$ **27.** 50
28. $\frac{28}{3}$ **29.** 18 and 19 **30.** $35

Cumulative Test for Chapters 1 to 4

1. 4 **2.** -12 **3.** 6 **4.** $\frac{2}{15}$ **5.** -18 **6.** $\frac{1}{5}$ **7.** $\frac{-2}{7}$ **8.** 10
9. $\frac{-4}{3}$ **10.** Undefined **11.** 0 **12.** $\frac{37}{60}$ **13.** 27 **14.** $\frac{1}{7}$
15. $1\frac{7}{9}$ **16.** $7\frac{1}{5}$ **17.** $\frac{23}{4}$ **18.** $\frac{55}{9}$ **19.** Expression **20.** Equation
21. Expression **22.** Equation **23.** $-2\frac{1}{8}$ **24.** $11\frac{1}{3}$ **25.** $6\frac{2}{7}$
26. $-4\frac{5}{9}$ **27.** $3\frac{5}{8}$ **28.** $3\frac{13}{24}$ **29.** $\frac{-3}{4}$ **30.** $-1\frac{1}{2}$ **31.** $\frac{5}{8}$ in.
32. $1\frac{11}{12}$ h **33.** -3 **34.** 8 **35.** 13 **36.** 42, 43 **37.** 7 **38.** $420
39. 5 cm, 17 cm **40.** 8 in., 13 in., 16 in.

Pre-Test Chapter 5

1. Thousandths **2.** 2.371; two and three hundred seventy-one thousandths **3. (a)** 52.31; **(b)** 2.375 **4.** $2.36 **5. (a)** -0.86037; **(b)** 536.2 **6.** 2.36 **7.** $11.95 **8.** $x = -5$ **9.** $x = -3.536$
10. 4.25 **11.** 2.435 **12.** $7.60 **13.** 0.0534 **14. (a)** 0.375; **(b)** 0.29
15. 10

Chapter Test for Chapter 5

1. Ten thousandths **2.** 0.049 **3.** Two and fifty-three hundredths
4. 12.017 **5.** $<$ **6.** $>$ **7.** 16.64 **8.** 27.547 **9.** 12.803
10. 0.598 **11.** 23.57 **12.** 10.54 **13.** 24.375 **14.** -3.888
15. 17.437 **16.** 0.02793 **17.** -1.4575 **18.** 735 **19.** 12,570
20. 0.465 **21.** 2.35 **22.** 0.051 **23.** 2.55 **24.** 2.385 **25.** -7.35
26. 0.067 **27.** 0.004983 **28.** 0.00523 **29.** 0.5625 **30.** 0.571
31. $0.\overline{63}$ **32.** $\frac{9}{125}$ **33.** $4\frac{11}{25}$ **34.** $>$ **35.** $\frac{229}{500}$ **36.** 1.45
37. -3.6 **38.** -1.07 **39.** 2.15 **40.** -5.3 **41.** 21 **42.** 65 m
43. 112 mm **44.** 50.2 gal **45.** $3.06 **46.** 7.525 in.^{2} **47.** $543
48. 32 lots **49.** $573.40

Cumulative Test for Chapters 1 to 5

1. 12 **2.** 20 **3.** 0 **4.** 15 **5.** -16 **6.** $\frac{3}{4}$ **7.** $\frac{-13}{12}$ **8.** $\frac{-4}{3}$
9. 20.82 **10.** 10.83 **11.** 12.75 **12.** 4.2 **13.** 0.5326 **14.** 6.821
15. 6.0 **16.** $<$ **17.** $=$ **18.** $\frac{3}{20}$ **19.** $\frac{2}{25}$ **20.** 0.625 **21.** 0.86
22. $2 \cdot 3 \cdot 5 \cdot 7$ **23.** 6 **24.** 90 **25.** Equation **26.** Expression
27. 27 **28.** -3 **29.** $3x - y$ **30.** $2x + 7$ **31.** $x = -4$
32. $x = -5$ **33.** $x = 48$ **34.** $x = -25$ **35.** $x = -2.5$
36. $x = -8$ **37.** 26.4 ft **38.** 41 ft^2 **39.** 15 **40.** $550
41. $18.2\frac{\text{mi}}{\text{gal}}$ **42.** $W = 12$ cm; $L = 32$ cm **43.** 17 **44.** 22.2 in.

Pre-Test Chapter 6

1. $\frac{7}{10}$ **2.** $\frac{4}{3}$ **3.** $29\frac{\text{mi}}{\text{gal}}$ **4.** $\frac{\$0.79}{\text{can}}$ **5.** Yes **6.** No **7.** 10 **8.** 15
9. 12 **10.** 12 **11.** 132 ft **12.** $6.30 **13.** 32 **14.** 11 gal

Chapter Test for Chapter 6

1. $\frac{7}{19}$ **2.** $\frac{5}{3}$ **3.** $\frac{2}{3}$ **4.** $\frac{1}{12}$ **5.** $\frac{26}{33}, \frac{26}{7}$ **6.** $4.8\frac{\text{mi}}{\text{gal}}$ **7.** $8.25\frac{\text{dollars}}{\text{h}}$
8. $\frac{\$2.56}{\text{gal}}$ **9.** yes **10.** no **11.** yes **12.** no **13.** $x = 20$
14. $a = 18$ **15.** $p = 3$ **16.** $a = 48$ **17.** $x = 6$ **18.** $m = 16$
19. $2.28 **20.** 644 points **21.** 576 mi **22.** 3,000 no votes
23. 600 mufflers **24.** 24 tsp **25.** 55 feet **26.** 20 meters
27. 11 ft 5 in. **28.** 2 lb 9 oz **29.** $1,050 **30.** (b) **31.** (b)
32. (b)

Cumulative Test for Chapters 1 to 6

1. Associative property of addition **2.** Commutative property of multiplication **3.** Distributive property **4.** 5,900 **5.** 950,000
6. 8 **7.** $2 \cdot 2 \cdot 2 \cdot 3 \cdot 11$ **8.** 90 **9.** $3\frac{1}{7}$ **10.** $\frac{53}{8}$ **11.** $\frac{3}{4}$
12. $1\frac{1}{2}$ **13.** $8\frac{1}{24}$ **14.** $3\frac{5}{8}$ **15.** $\frac{5}{8}$ in. **16.** $3.06 **17.** 8.04 ft^2
18. 32 **19.** $0.\overline{72}$ **20.** $m = 112.5$ **21.** 12 in. **22.** -8
23. 20 **24.** 12 **25.** 5 **26.** -18 **27.** -4 **28.** -90 **29.** -4
30. $3(x + y)$ **31.** $\frac{n - 5}{3}$ **32.** 5 **33.** -24 **34.** $-\frac{2}{5}$ **35.** 5

Pre-Test Chapter 7

1. $\frac{7}{100}$ **2.** 0.23 **3.** 3.5% or $3\frac{1}{2}$% **4.** 80% **5.** 63 **6.** 9%
7. 600 **8.** $560 **9.** 5% **10.** $1,200 **11.** $11.31 **12.** $5,955.08
13. $416.50

Chapter Test for Chapter 7

1. 80% **2.** $\frac{7}{100}$ **3.** $\frac{72}{100}$ or $\frac{18}{25}$ **4.** 0.42 **5.** 0.06 **6.** 1.6
7. 3% **8.** 4.2% **9.** 40% **10.** 62.5% **11.** Rate: 25%; base: 200; amount: 50 **12.** Rate: 8%; base: 500; amount: unknown
13. Rate: 6%; base: amount of purchase; amount: $30 **14.** 11.25
15. 500 **16.** 750 **17.** 20% **18.** 7.5% **19.** 175% **20.** 800
21. 300 **22.** $4.96 **23.** 60 questions **24.** $70.20 **25.** 12%
26. 24% **27.** 8% **28.** $18,000 **29.** 6,400 students **30.** $18,500
31. $17,500 **32.** $3,246.37

Cumulative Test for Chapters 1 to 7

1. Thousands **2.** 11,368 **3.** 89 **4.** 6 **5.** -4 **6.** -1 **7.** 17
8. $2 \cdot 2 \cdot 5 \cdot 13$ **9.** 28 **10.** 180 **11.** $8\frac{1}{2}$ **12.** $1\frac{1}{3}$ **13.** $8\frac{7}{12}$
14. $4\frac{19}{24}$ **15.** $286 **16.** $54\frac{\text{mi}}{\text{h}}$ **17.** $41\frac{1}{4}$ in **18.** Hundredths
19. Ten thousandths **20.** $<$ **21.** $=$ **22.** $\frac{9}{25}$ **23.** $5\frac{1}{8}$ **24.** 11.284
25. 17.04 **26.** $108.05 **27.** 5 **28.** -36 **29.** $-\frac{2}{5}$ **30.** 5
31. $18\frac{2}{3}$ **32.** 0.4 **33.** 350 mi **34.** $140 **35.** 0.34; $\frac{17}{50}$
36. 0.55; 55% **37.** 45 **38.** 500 **39.** 125 **40.** 8.5%

Pre-Test Chapter 8

1. 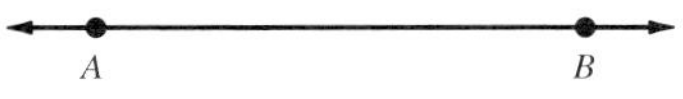 A B **2.** 100° **3.** 40° **4.** 140°
5. 15 m **6.** $8\pi \approx 25.1$ cm **7.** 12 ft^2

Chapter Test for Chapter 8

1. **(a)** Parallel; **(b)** neither; **(c)** perpendicular; **(d)** neither
2. **(a)** Straight; **(b)** obtuse; **(c)** right **3.** 50° **4.** 135° **5.** 300°
6. 16.8 cm **7.** 21 in. **8.** 8.2 m **9.** $21\frac{1}{2}$ in. **10.** 18 cm
11. $10\pi \approx 31.4$ ft **12.** 17.64 cm^2 **13.** $27\frac{9}{16}$ in.^{2} **14.** 4.14 m^2
15. $24\frac{3}{8}$ in.^{2} **16.** 13.5 cm^2 **17.** $25\pi \approx 78.5$ ft^2 **18.** 15.625 m^3
19. 30 ft^3 **20.** $\approx$14.1 ft^3 **21.** $\approx$4,524.6 cm^3

Cumulative Test for Chapters 1 to 8

1. \$896 **2.** \$62 **3.** 24 **4.** -21 **5.** -16
6. $\frac{1}{5}$ **7.** $2 \cdot 2 \cdot 2 \cdot 3 \cdot 7$ **8.** 4 **9.** $\frac{3}{5}, \frac{5}{8}, \frac{2}{3}$ **10.** $\frac{3}{4}$ **11.** $\frac{5}{12}$
12. 60 **13.** $\frac{31}{30}$ **14.** $\frac{85}{24}$ **15.** -5 **16.** -8 **17.** 14 **18.** $\frac{2}{3}$
19. \$18.30 **20.** 27.84 cm^2 **21.** 21.5 cm **22.** 20.1 ft **23.** 0.5625
24. 0.538 **25.** 17 ft **26.** $x = 24$ **27.** 400 mi **28.** 450 mi
29. $\frac{1}{8}$ **30.** 3,526 **31.** 225% **32.** 150 **33.** \$142,500
34. 21 lb 11 oz **35.** 2 ft 9 in. **36.** 62,000 **37.** 74 **38.** 43°
39. 73° **40.** $x = 40$

Pre-Test Chapter 9

1. \$17,200,000,000 **2.** \$22,800,000,000 **3.** \$12,000
4. \$19,400 **5.** $\approx$2,500 **6.** $-2{,}000$ **7.** 1980–1985; 2000
8. 1,500,000 **9.** 750% **10.** (15, 3); (18, 6) **11.** (0, 3); (2, 0)
12. (1, 3); (0, 5); $(-3, 11)$ **13.** Answers will vary. Sample answer: $(-2, -4), (1, -2), (4, 0); (0, 11), (1, 5), (2, -1)$
14. $A(2, 3)$, $B(0, -5)$, $C(-3, 5)$

15.

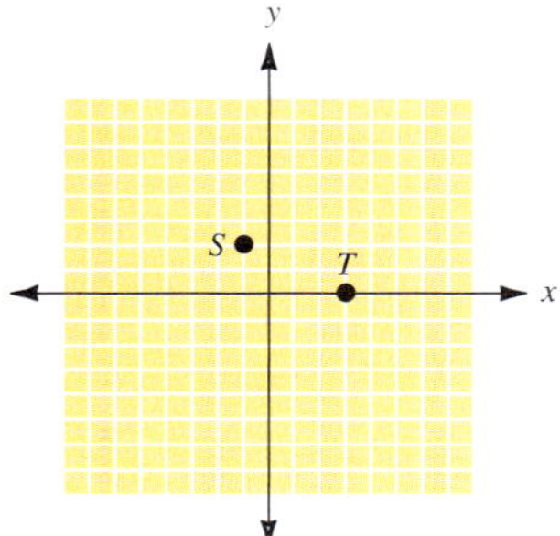

16.

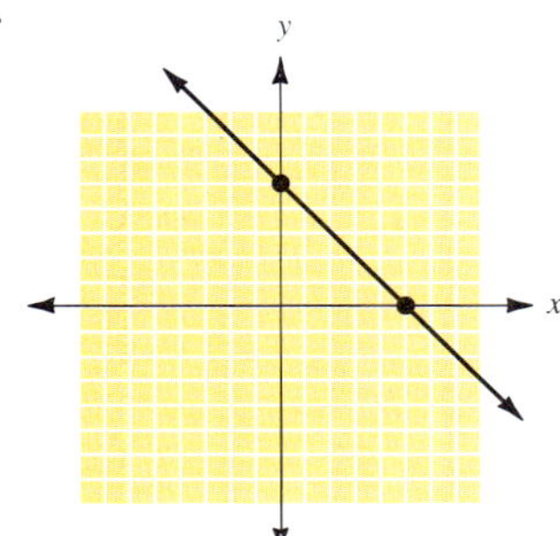

17.

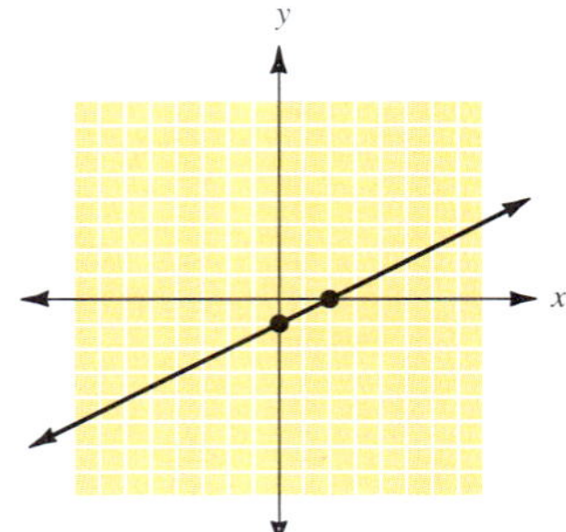

18.

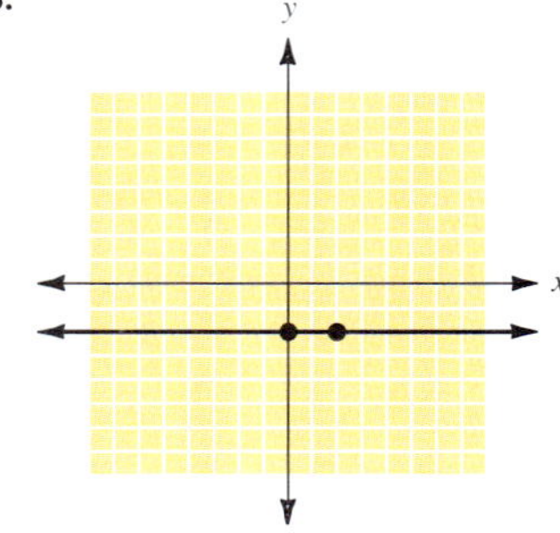

19. Mean: 27; median: 24; mode: 42

Chapter Test for Chapter 9

1. 55% **2.** 45% **3.** 85%

4.

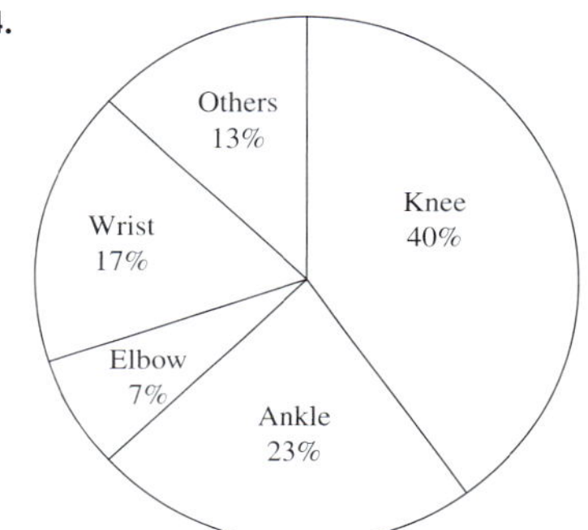

5.

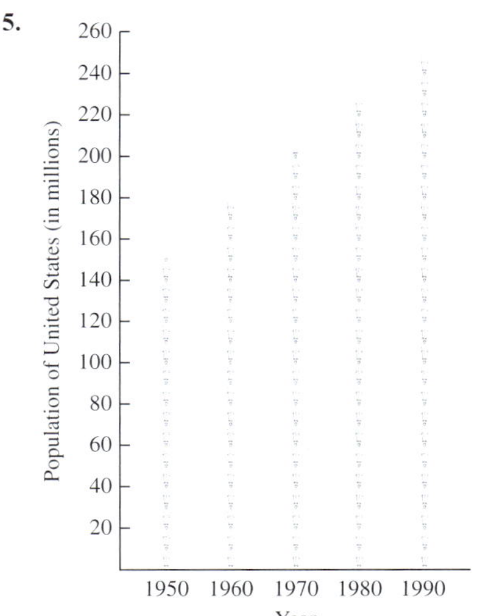

6. 16,000 **7.** 16,000 **8.** 2000 to 2001

9.

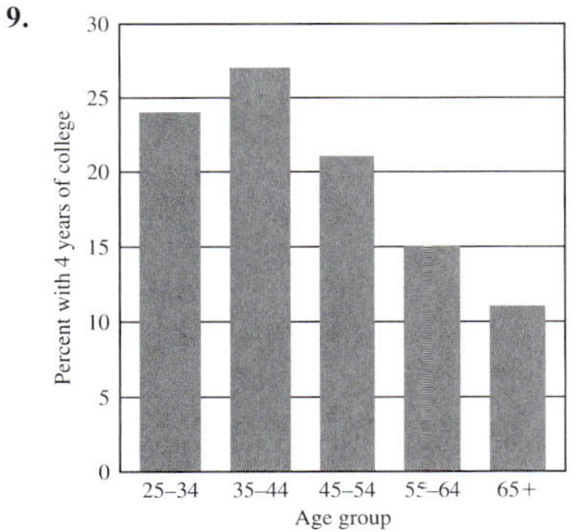

10. December **11.** August and September

12.

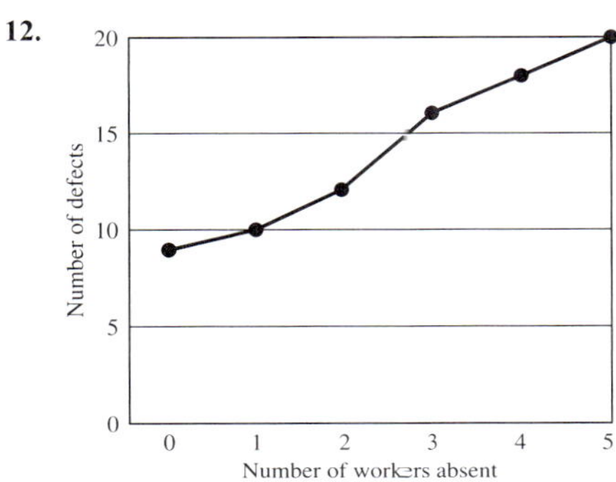

13. (4, 2) **14.** (−4, 6) **15.** (0, −7)

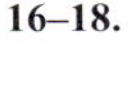

16–18.

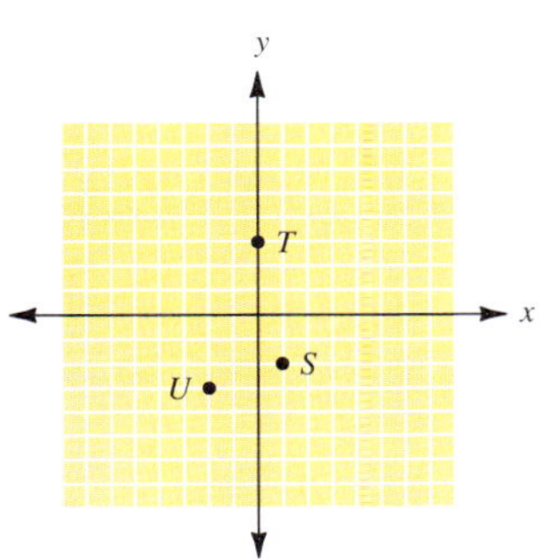

19. (4, 0), (5, 4) **20.** (3, 3), (6, 2), (9, 1)
21. Different answers are possible. Sample answer: (−1, −8), (0, −7), (1, −6), (2, −5)

22.

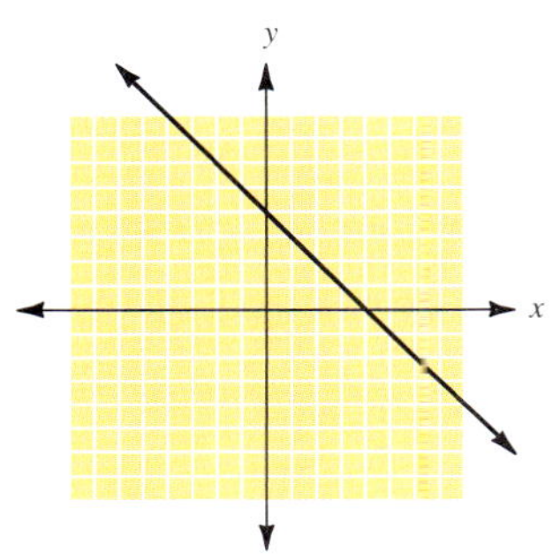

23.

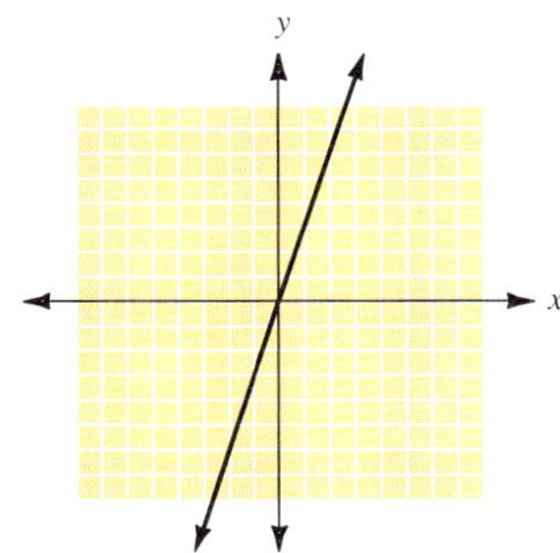

24.

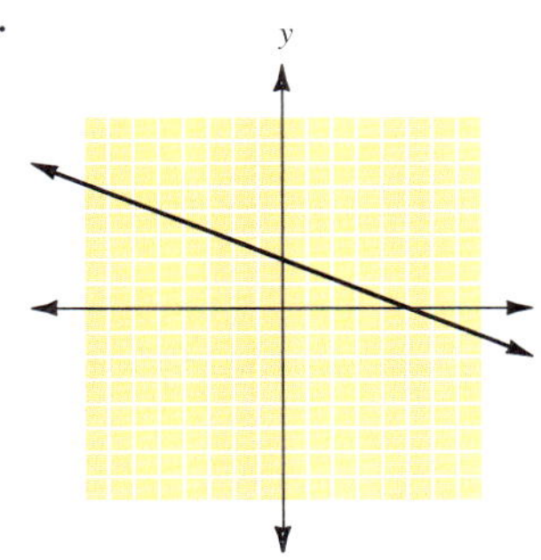

25.

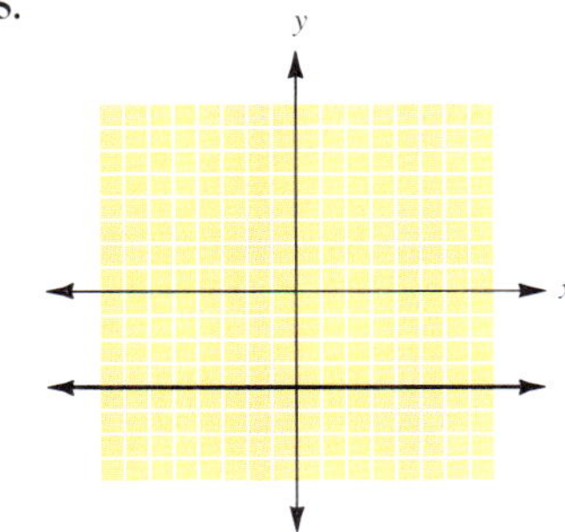

26. 15 **27.** 11 **28.** 6 **29.** 204 **30.** Brown

Cumulative Test for Chapters 1 to 9

1. Thousands **2.** −5 **3.** Undefined **4.** 75.215 **5.** $\frac{2}{3}$ **6.** $\frac{6}{11}$ **7.** $7\frac{1}{12}$ **8.** 20 **9.** −28 **10.** $6x$ **11.** $4x - 3y$ **12.** −24 **13.** $\frac{5}{4}$ **14.** 14 **15.** 9 **16.** 0.18; $\frac{9}{50}$ **17.** 0.425; 42.5% **18.** 11 lb 5 oz **19.** 8,000 **20.** 3 **21.** 5 **22.** 250 **23.** 1998 and 1999

24.

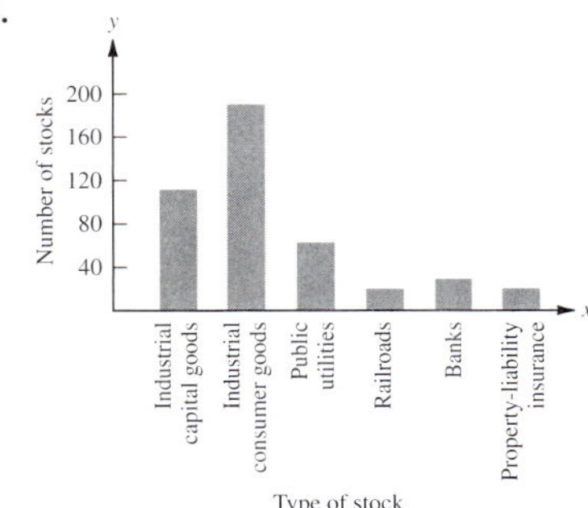

25. Mean: 9; Median: 9; Mode: 11 **26.** 70 **27.** \$408 **28.** $48 \frac{\text{mi}}{\text{hr}}$
29. $28\frac{3}{5}$ cm **30.** $55\frac{1}{3}$ ft **31.** 133 **32.** 0.2% **33.** 185 **34.** $4\frac{1}{3}$
35. 2,800 students

Pre-Test Chapter 10

1. Binomial **2.** Trinomial **3.** -6 **4.** $2x^2 - 2x - 2$
5. $9x^2 - 11x + 6$ **6.** x^{12} **7.** $8x^5y^7$ **8.** $12x^3y^3 - 6x^2y^2 + 21x^2y^4$
9. $6x^2 - 11x - 10$ **10.** $x^2 - 4y^2$ **11.** $16m^2 + 40m + 25$
12. $3x^3 - 14x^2y + 17xy^2 - 6y^3$ **13.** $5(3c + 7)$ **14.** $4q^3(2q - 5)$
15. $6(x^2 - 2x + 4)$ **16.** $7cd(c^2d - 3 + 2d^2)$

Chapter Test for Chapter 10

1. Binomial **2.** Trinomial **3.** 6 **4.** $10x^2 - 12x - 7$
5. $7a^3 + 11a^2 - 3a$ **6.** $3x^2 + 11x - 12$ **7.** $b^2 - 7b - 5$
8. $7a^2 - 10a$ **9.** $4x^2 + 5x - 6$ **10.** $2x^2 - 7x + 5$ **11.** a^{14}
12. $15x^3y^7$ **13.** $27x^6y^3$ **14.** $16x^{18}y^{17}$ **15.** $15a^3b^2 - 10a^2b^2 + 20a^2b^3$
16. $3x^2 + x - 14$ **17.** $a^2 - 49b^2$ **18.** $8x^2 - 14xy - 15y^2$
19. $12x^3 + 11x^2y - 5xy^2$ **20.** $9m^2 + 12mn + 4n^2$
21. $2x^3 + 7x^2y - xy^2 - 2y^3$ **22.** $6(2b + 3)$ **23.** $3p^2(3p - 4)$
24. $5(x^2 - 2x + 4)$ **25.** $6ab(a - 3 + 2b)$

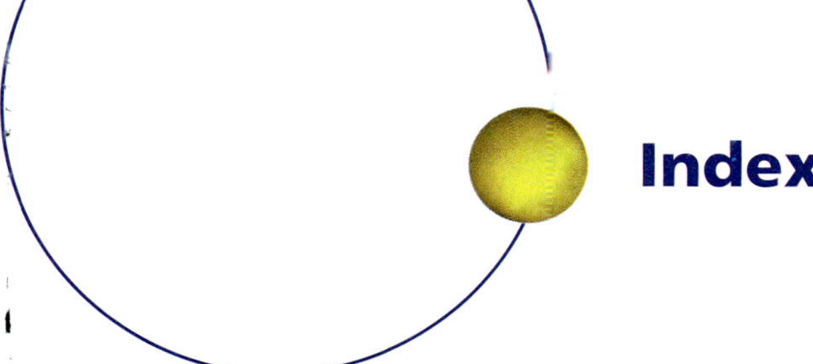

Index